PROBABILITY THEORY

THIRD EDITION

MICHEL LOÈVE

DOVER PUBLICATIONS
GARDEN CITY, NEW YORK

Bibliographical Note

This Dover edition, first published in 2017, is an unabridged republication of the work originally published in 1963 by the D. Van Nostrand Company, Inc., Princeton, New Jersey, as part of "The University Series in Higher Mathematics." The book was first published by Van Nostrand in 1955 under the title *Probability Theory: Foundations, Random Sequences*.

Library of Congress Cataloging-in-Publication Data

Names: Loève, M. (Michel), 1907–1979.
Title: Probability theory / Michel Loève.
Description: Third edition, Dover edition. | Garden City, New York : Dover Publications, 2017. | Originally published: Princeton, New Jersey : D. Van Nostrand Company, Inc., 1963. | Includes bibliographical references and index.
Identifiers: LCCN 2017003066| ISBN 9780486814889 | ISBN 0486814882
Subjects: LCSH: Probabilities.
Classification: LCC QA273 .L63 2017 | DDC 519.2—dc23 LC record available at https://lccn.loc.gov/2017003066

Manufactured in the United States of America
81488204 2022
www.doverpublications.com

To Line

and

To the students and teachers
of the School in
the Camp de Drancy

PREFACE TO THE THIRD EDITION

This book is intended as a text for graduate students and as a reference for workers in Probability and Statistics. The prerequisite is honest calculus. The material covered in Parts Two to Five inclusive requires about three to four semesters of graduate study. The introductory part may serve as a text for an undergraduate course in elementary probability theory.

The Foundations are presented in:

the Introductory Part on the background of the concepts and problems, treated without advanced mathematical tools;
Part One on the Notions of Measure Theory that every probabilist and statistician requires;
Part Two on General Concepts and Tools of Probability Theory.

Random sequences whose general properties are given in the Foundations are studied in:

Part Three on Independence devoted essentially to sums of independent random variables and their limit properties;
Part Four on Dependence devoted to the operation of conditioning and limit properties of sums of dependent random variables. The last section introduces random functions of second order.

Random functions and processes are discussed in:

Part Five on Elements of random analysis devoted to the basic concepts of random analysis and to the martingale, decomposable, and Markov types of random functions.

Since the primary purpose of the book is didactic, methods are emphasized and the book is subdivided into:

unstarred portions, independent of the remainder; starred portions, which are more involved or more abstract;
complements and details, including illustrations and applications of the material in the text, which consist of propositions with fre-

quent hints; most of these propositions can be found in the articles and books referred to in the Bibliography.

Also, for teaching and reference purposes, it has proved useful to name most of the results.

Numerous historical remarks about results, methods, and the evolution of various fields are an intrinsic part of the text. The purpose is purely didactic: to attract attention to the basic contributions while introducing the ideas explored. Books and memoirs of authors whose contributions are referred to and discussed are cited in the Bibliography, which parallels the text in that it is organized by parts and, within parts, by chapters. Thus the interested student can pursue his study in the original literature.

This work owes much to the reactions of the students on whom it has been tried year after year. However, the book is definitely more concise than the lectures, and the reader will have to be armed permanently with patience, pen, and calculus. Besides, in mathematics, as in any form of poetry, the reader has to be a poet *in posse*.

This third edition differs from the second (1960) in a number of places. Modifications vary all the way from a prefix ("sub"martingale in lieu of "semi"-martingale) to an entire subsection (§36.2). To preserve pagination, some additions to the text proper (especially 9, p. 656) had to be put in the Complements and Details. It is hoped that moreover most of the errors have been eliminated and that readers will be kind enough to inform the author of those which remain.

I take this opportunity to thank those whose comments and criticisms led to corrections and improvements: for the first edition, E. Barankin, S. Bochner, E. Parzen, and H. Robbins; for the second edition, Y. S. Chow, R. Cogburn, J. L. Doob, J. Feldman, B. Jamison, J. Karush, P. A. Meyer, J. W. Pratt, B. A. Sevastianov, J. W. Woll; for the third edition, S. Dharmadhikari, J. Fabius, D. Freedman, A. Maitra, U. V. Prokhorov. My warm thanks go to Cogburn, whose constant help throughout the preparation of the second edition has been invaluable. This edition has been prepared with the partial support of the Office of Naval Research and of the National Science Foundation.

M. L.

April, 1962
Berkeley, California

CONTENTS

PART TWO: GENERAL CONCEPTS AND TOOLS OF PROBABILITY THEORY

CHAPTER III: PROBABILITY CONCEPTS

CHAPTER IV: DISTRIBUTION FUNCTIONS AND CHARACTERISTIC FUNCTIONS

PART THREE: INDEPENDENCE

CHAPTER V: SUMS OF INDEPENDENT RANDOM VARIABLES

CHAPTER VI: CENTRAL LIMIT PROBLEM

PART FOUR: DEPENDENCE

CHAPTER VII: CONDITIONING

CHAPTER VIII: FROM INDEPENDENCE TO DEPENDENCE

CHAPTER IX: ERGODIC THEOREMS

CHAPTER X: SECOND ORDER PROPERTIES

PART FIVE: ELEMENTS OF RANDOM ANALYSIS

CHAPTER XI: FOUNDATIONS; MARTINGALES AND DECOMPOSABILITY

CHAPTER XII: MARKOV PROCESSES

some ideas derived from everyday experience—especially from games of chance—we shall arrive at an elementary axiomatic setup; we leave the illustrations with coins, dice, cards, darts, etc., to the reader. Then, we shall apply this axiomatic setup to describe in a precise manner and to investigate in a rigorous fashion a few of the "intuitive notions" relative to randomness. No special tools will be needed, whereas in the nonelementary setup measure-theoretic concepts and Fourier-Stieltjes transforms play a prominent role.

I. INTUITIVE BACKGROUND

1. Events. The primary notion in the understanding of nature is that of *event*—the occurrence or nonoccurrence of a phenomenon. The abstract concept of event pertains only to its occurrence or nonoccurrence and not to its nature. This is the concept we intend to analyze. We shall denote events by $A, B, C, \cdots$ with or without affixes.

To every event A there corresponds a contrary event "not A," to be denoted by A^c; A^c occurs if, and only if, A does not occur. An event may imply another event: *A implies B* if, when A occurs, then B necessarily occurs; we write $A \subset B$. If A implies B and also B implies A, then we say that A and B are *equivalent;* we write $A = B$. The nature of two equivalent events may be different, but as long as we are concerned only with occurrence or nonoccurrence, they can and will be identified. Events are combined into new events by means of operations expressed by the terms "and," "or" and "not."

A "and" B is an event which occurs if, and only if, both the event A and the event B occur; we denote it by $A \cap B$ or, simply, by AB. If AB cannot occur (that is, if A occurs, then B does not occur, and if B occurs, then A does not occur), we say that the event A and the event B are *disjoint* (exclude one another, are mutually exclusive, are incompatible).

A "or" B is an event which occurs if, and only if, at least one of the events A, B occurs; we denote it by $A \cup B$. If, and only if, A and B are disjoint, we replace "or" by $+$. Similarly, more than two events can be combined by means of "and," "or"; we write

$$A_1 \cap A_2 \cap \cdots \cap A_n \quad \text{or} \quad A_1A_2 \cdots A_n \quad \text{or} \quad \bigcap_{k=1}^{n} A_k,$$

$$A_1 \cup A_2 \cup \cdots \cup A_n \quad \text{or} \quad \bigcup_{k=1}^{n} A_k, \quad A_1 + A_2 + \cdots A_n \quad \text{or} \quad \sum_{k=1}^{n} A_k.$$

There are two combinations of events which can be considered as "boundary events"; they are the first and the last events—in terms of implication. Events of the form $A + A^c$ can be said to represent an "always occurrence," for they can only occur. Since, whatever be the event A, the events $A + A^c$ and the events they imply are equivalent, all such events are to be identified and will be called the *sure event*, to be denoted by Ω. Similarly, events of the form AA^c and the events which imply them, which can be said to represent a "never occurrence" for they cannot occur, are to be identified, and will be called the *impossible event*, to be denoted by $\emptyset$; thus, the definition of disjoint events A and B can be written $AB = \emptyset$. The impossible and the sure events are "first" and "last" events, for, whatever be the event A, we have $\emptyset \subset A \subset \Omega$.

The interpretation of symbols $\subset$, $=$, $\cap$, $\cup$, in terms of occurrence and nonoccurrence, shows at once that

$$\text{if} \quad A \subset B, \quad \text{then} \quad B^c \subset A^c, \quad \text{and conversely;}$$

$$AB = BA, \quad A \cup B = B \cup A;$$

$$(AB)C = A(BC), \quad (A \cup B) \cup C = A \cup (B \cup C);$$

$$A(B \cup C) = AB \cup AC, \quad A \cup BC = (A \cup B)(A \cup C);$$

$$(AB)^c = A^c \cup B^c, \quad (A \cup B)^c = A^cB^c, \quad A \cup B = A + A^cB;$$

more generally

$$(\bigcap_{k=1}^{n} A_k)^c = \bigcup_{k=1}^{n} A_k^c, \quad (\bigcup_{k=1}^{n} A_k)^c = \bigcap_{k=1}^{n} A_k^c,$$

and so on.

We recognize here the rules of operations on sets. In terms of sets, Ω is the space in which lie the sets $A, B, C, \cdots, \emptyset$ is the *empty* set, A^c is the set *complementary* to the set A; AB is the *intersection*, $A \cup B$ is the *union* of the sets A and B, and $A \subset B$ means that A is contained in B.

In science, or, more precisely, in the investigation of "laws of nature," events are classified into conditions and outcomes of an experiment. *Conditions* of an experiment are events which are known or are made to occur. *Outcomes* of an experiment are events which *may* occur when the experiment is performed, that is, when its conditions occur. All (finite) combinations of outcomes by means of "not," "and," "or," are outcomes; in the terminology of sets, the outcomes of an experiment form a *field* (or an "algebra" of sets). The conditions of an experiment, together with its field of outcomes, constitute a *trial*. Any (finite) number of trials can be combined by "conditioning," as follows:

The collective outcomes are combinations by means of "not," "and," "or," of the outcomes of the constituent trials. The conditions are conditions of the first constituent trial together with conditions of the second to which are added the observed outcomes of the first, and so on. Thus, given the observed outcomes of the preceding trials, every constituent trial is performed under supplementary conditions: it is conditioned by the observed outcomes. When, for every constituent trial, any outcome occurs if, and only if, it occurs without such conditioning, we say that the trials are *completely independent*. If, moreover, the trials are identical, that is, have the same conditions and the same field of outcomes, we speak of *repeated trials* or, equivalently, *identical and completely independent trials*. The possibility of repeated trials is a basic assumption in science, and in games of chance: *every trial can be performed again and again, the knowledge of past and present outcomes having no influence upon future ones*.

2. Random events and trials. Science is essentially concerned with permanencies in repeated trials. For a long time *Homo sapiens* investigated *deterministic trials* only, where the conditions (causes) determine completely the outcomes (effects). Although another type of permanency has been observed in games of chance, it is only recently that *Homo sapiens* was led to think of a rational interpretation of nature in terms of these permanencies: nature plays the greatest of all games of chance with the observer. This type of permanency can be described as follows:

Let the *frequency* of an outcome A in n repeated trials be the ratio n_A/n of the number n_A of occurrences of A to the total number n of trials. If, in repeating a trial a large number of times, the observed frequencies of any one of its outcomes A cluster about some number, the trial is then said to be *random*. For example, in a game of dice (two homogeneous ones) "double-six" occurs about once in 36 times, that is, its observed frequencies cluster about 1/36. The number 1/36 is a permanent numerical property of "double-six" under the conditions of the game, and the observed frequencies are to be thought of as measurements of the property. This is analogous to stating that, say, a bar at a fixed temperature has a permanent numerical property called its "length" about which the measurements cluster.

The outcomes of a random trial are called *random* (chance) *events*. The number measured by the observed frequencies of a random event A is called the *probability* of A and is denoted by PA. Clearly, $P\emptyset = 0$,

$P\Omega = 1$ and, for every A, $0 \leqq PA \leqq 1$. Since the frequency of a sum $A_1 + A_2 + \cdots + A_n$ of disjoint random events is the sum of their frequencies, we are led to assume that

$$P(A_1 + A_2 + \cdots + A_n) = PA_1 + PA_2 + \cdots + PA_n.$$

Furthermore, let n_A, n_B, n_{AB} be the respective numbers of occurrences of outcomes A, B, AB in n repeated random trials. The frequency of outcome B in the n_A trials in which A occurs is

$$\frac{n_{AB}}{n_A} = \frac{n_{AB}}{n} : \frac{n_A}{n}$$

and measures the ratio PAB/PA, to be called probability of B *given* A (given that A occurs); we denote it by P_AB and have

$$PAB = PA \cdot P_AB.$$

Thus, when to the original conditions of the trial is added the fact that A occurs, then the probability PB of B is transformed into the probability P_AB of B given A. This leads to defining B as being *stochastically independent* of A if $P_AB = PB$ or

$$PAB = PA \cdot PB.$$

Then it follows that A is stochastically independent of B, for

$$P_BA = \frac{PAB}{PB} = PA,$$

and it suffices to say that *A and B are stochastically independent.* (We assumed in the foregoing ratios that the denominators were not null.)

Similarly, if a collective trial is such that the probability of any outcome of any constituent random trial is independent of the observed outcomes of preceding constituents, we say that the constituent random trials are *stochastically independent.* Clearly, complete independence defined in terms of occurrences implies stochastic independence defined in terms of probability alone. Thus, as long as we are concerned with stochastic independence only, the concept of repeated trials reduces to that of *identical and stochastically independent trials.*

3. Random variables. For a physicist, the outcomes are, in general, values of an observable. From the gambler's point of view, what counts is not the observed outcome of a random trial but the corresponding gain or loss. In either case, when there is only a finite number of possible outcomes, the sure outcome Ω is partitioned into a number of disjoint outcomes $A_1, A_2, \cdots, A_m$. The *random variable* X, say, the chance gain of the gambler, is stated by assigning to these outcomes numbers $x_{A_1}, x_{A_2}, \cdots, x_{A_m}$, which may be positive, null, or negative. The "average gain" in n repeated random trials is

$$x_{A_1}\frac{n_{A_1}}{n} + x_{A_2}\frac{n_{A_2}}{n} + \cdots + x_{A_m}\frac{n_{A_m}}{m}.$$

Since the trial is random, this average clusters about $x_{A_1}PA_1 + x_{A_2}PA_2 + \cdots + x_{A_m}PA_m$ which is defined as the *expectation* EX of the random variable X. It is easily seen that the averages of a sum of two random variables X and Y cluster about the sum of their averages, that is,

$$E(X+Y) = EX + EY.$$

The concept of random variable is more general than that of a random event. In fact, we can assign to every random event A a random variable—its indicator $I_A = 1$ or 0 according as A occurs or does not occur. Then, the observed value of I_A tells us whether or not A occurred, and conversely. Furthermore, we have $EI_A = 1 \cdot PA + 0 \cdot PA^c = PA$.

A physical observable may have an infinite number of possible values, and then the foregoing simple definitions do not apply. The evolution of probability theory is due precisely to the consideration of more and more complicated observables.

II. AXIOMS; INDEPENDENCE AND THE BERNOULLI CASE

We give now a consistent model for the intuitive concepts which appeared in the foregoing brief analysis; we shall later see that this model has to be extended.

1. Axioms of the finite case. Let Ω or the *sure event* be a space of points ω; the empty set (set containing no points ω) or the *impossible event* will be denoted by $\emptyset$. Let $\mathfrak{A}$ be a nonempty class of sets in Ω, to be called *random events* or, simply, *events*, since no other type of events will be considered. Events will be denoted by capitals $A, B, \cdots$ with or without affixes. Let P or *probability* be a numerical function defined on $\mathfrak{A}$; the value of P for an event A will be called the *probability of A* and will be denoted by PA. The pair $(\mathfrak{A}, P)$ is called a *probability field* and the triplet $(\Omega, \mathfrak{A}, P)$ is called a *probability space.*

Axiom I. *$\mathfrak{A}$ is a field:* complements A^c, finite intersections $\bigcap_{k=1}^{n} A_k$, and finite unions $\bigcup_{k=1}^{n} A_k$ of events are events.

Axiom II. *P on $\mathfrak{A}$ is normed, nonnegative, and finitely additive:*

$$P\Omega = 1, \quad PA \geqq 0, \quad P\sum_{k=1}^{n} A_k = \sum_{k=1}^{n} PA_k.$$

It suffices to assume additivity for two arbitrary disjoint events, since the general case follows by induction.

Since $\emptyset$ is disjoint from any event A and $A + \emptyset = A$, we have

$$PA = P(A + \emptyset) = PA + P\emptyset,$$

so that $P\emptyset = 0$. Furthermore, it is immediate that, if $A \subset B$, then $PA \leqq PB$, and also that

$$P\bigcup_{k=1}^{n} A_k = PA_1 + PA_1^cA_2 + \cdots + PA_1^cA_2^c \cdots A_{n-1}^cA_n \leqq \sum_{k=1}^{n} PA_k.$$

The axioms are consistent.

To see this, it suffices to construct an example in which the axioms are both verified: take as the field $\mathcal{A}$ of events Ω and $\emptyset$ only, and set $P\Omega = 1$, $P\emptyset = 0$. A less trivial example is that of a *simple probability field:* 1° The events, except $\emptyset$, are formed by all sums of disjoint events $A_1, A_2, \cdots, A_n$ which form a finite partition of the sure event: $A_1 + A_2 + \cdots + A_n = \Omega$; 2° to every event A_k of the partition is assigned a probability $p_k = PA_k$ such that every $p_k \geqq 0$ and $\sum_{k=1}^{n} p_k = 1$—this is always possible. Then P is defined on $\mathcal{A}$, consistently with axiom II, by assigning to every event A as its probability the sum of probabilities of those A_k whose sum is A.

2. Simple random variables. Let the probability field $(\mathcal{A}, P)$ be fixed. In order to introduce the concept of random variables, it will be convenient to begin with very special ones, which permit operations on events to be transformed into ordinary algebraic operations.

To every event A we assign a function I_A on Ω with values $I_A(\omega)$, such that $I_A(\omega) = 1$ or 0 according as ω belongs or does not belong to A; I_A will be called the *indicator* of A (in terms of occurrences, $I_A = 1$ or 0, according as A occurs or does not occur). Thus, $I_A{}^2 = I_A$ and the boundary cases are those of $I_\emptyset = 0$ and $I_\Omega = 1$ (if, in a relation containing functions of an argument, the argument does not figure, then the relation holds for all values of the argument unless otherwise stated).

The following properties are immediate:

$$\text{if} \quad A \subset B, \quad \text{then} \quad I_A \leqq I_B, \quad \text{and conversely;}$$

$$\text{if} \quad A = B, \quad \text{then} \quad I_A = I_B, \quad \text{and conversely;}$$

$$I_{A^c} = 1 - I_A, \quad I_{AB} = I_A I_B, \quad I_{A+B} = I_A + I_B,$$

$$I_{A \cup B} = I_{A+A^cB} = I_A + I_B - I_{AB}$$

and, more generally,

$$I_{\bigcap\limits_{k=1}^{n} A_k} = \prod_{k=1}^{n} I_{A_k}, \quad I_{\sum\limits_{k=1}^{n} A_k} = \sum_{k=1}^{n} I_{A_k}$$

$$I_{\bigcup\limits_{k=1}^{n} A_k} = I_{A_1} + (1 - I_{A_1})I_{A_2} + \cdots + (1 - I_{A_1}) \cdots (1 - I_{A_{n-1}})I_{A_n}.$$

Linear combinations $X = \sum_{j=1}^{m} x_j I_{A_j}$ of indicators of events A_j of a finite partition of Ω, where the x_j are (finite) numbers, are called *simple random variables*, to be denoted by capitals $X, Y, \cdots$, with or without affixes. By convention, every written linear combination of indicators will be that of indicators of disjoint events whose sum is the sure event; however, when $x_j = 0$, we may drop the corresponding null term $x_j I_{A_j} = 0$ from the linear combination. The set of values PA_j which correspond to the values x_j of X, assumed all distinct, is called the *probability distribution* of X. The *expectation* EX of a simple random variable $X = \sum_{j=1}^{m} x_j I_{A_j}$ is defined by

$$EX = \sum_{j=1}^{m} x_j PA_j.$$

Clearly, any constant c is a simple random variable, and the sum or the product of two simple random variables is a simple random variable; $E(c) = c$, $EcX = cEX$; if $X \geqq 0$, that is, all its values $x_j \geqq 0$, then $EX \geqq 0$; if $X \leqq Y$, then $EX \leqq EY$. Furthermore, expectations possess the following basic property.

ADDITION PROPERTY. *The expectation of a sum of (a finite number of) simple random variables is the sum of their expectations.*

It suffices to prove the assertion for a sum of two simple random variables

$$X = \sum_{j=1}^{m} x_j I_{A_j}, \quad Y = \sum_{k=1}^{n} y_k I_{B_k},$$

since the general case follows by induction. Because of the properties of probabilities and indicators given above,

$$EX + EY = \sum_{j=1}^{m} x_j PA_j + \sum_{k=1}^{n} y_k PB_k = \sum_{j=1}^{m} \sum_{k=1}^{n} (x_j + y_k) PA_j B_k$$

while

$$E(X + Y) = E \sum_{j=1}^{m} \sum_{k=1}^{n} (x_j + y_k) I_{A_j B_k}$$

$$= \sum_{j=1}^{m} \sum_{k=1}^{n} (x_j + y_k) PA_j B_k.$$

and the conclusion is reached.

Application to probabilities of combinations of events. To begin with, we observe that

$$EI_A = 1 \cdot PA + 0 \cdot PA^c = PA.$$

Therefore, from

$$I_{A \cup B} = I_A + I_B - I_{AB}$$

it follows, upon taking expectations of both sides, that

$$P(A \cup B) = PA + PB - PAB.$$

Similarly, from

$$I_{A \cup B \cup C} = I_A + (1 - I_A)I_B + (1 - I_A)(1 - I_B)I_C$$

it follows, upon expanding the right-hand side and taking expectations, that

$$P(A \cup B \cup C) = PA + PB + PC - PAB - PBC - PCA + PABC,$$

and so on.

The foregoing properties of expectations lead to the celebrated

TCHEBICHEV INEQUALITY. *If X is a simple random variable, then, for every $\epsilon > 0$,*

$$P[|X| \geqq \epsilon] \leqq \frac{1}{\epsilon^2} EX^2.$$

$[|X| \geqq \epsilon]$ is to be read: the union of all those events for which the values of $|X|$ are $\geqq \epsilon$.

The inequality follows from

$$EX^2 = E(X^2 I_{[|X| \geqq \epsilon]}) + E(X^2 I_{[|X| < \epsilon]}) \geqq E(X^2 I_{[|X| \geqq \epsilon]}) \geqq \epsilon^2 E I_{[|X| \geqq \epsilon]}$$
$$= \epsilon^2 P[|X| \geqq \epsilon].$$

3. Independence. Two events A_1, A_2 are said to be *stochastically independent* or, simply, *independent* (no other type of independence of events will be considered) if

$$PA_1A_2 = PA_1PA_2.$$

More generally, events A_k, $k = 1, 2, \cdots, n$ are *independent*, if, for every $m \leqq n$ and for arbitrary distinct integers $k_1, k_2, \cdots, k_m \leqq n$,

$$PA_{k_1}A_{k_2} \cdots A_{k_m} = PA_{k_1}PA_{k_2} \cdots PA_{k_m}.$$

If this property holds for all events A_k selected arbitrarily each within a different class $\mathfrak{A}_k$, we say that these classes are *independent.* Simple random variables X_k, $k = 1, 2, \cdots, n$, are said to be *independent* if the partitions on which they are defined are independent. A basic property of independent simple random variables is the following

MULTIPLICATION PROPERTY. *The expectation of a product (of a finite number) of independent simple random variables is the product of their expectations.*

It suffices to give the proof for two independent simple random variables,

$$X = \sum_{j=1}^{m} x_j I_{A_j}, \quad Y = \sum_{k=1}^{n} y_k I_{B_k}, \text{ all } x_j(y_k) \text{ distinct},$$

since the general case follows by induction. Because of independence,

$$EXY = E\sum_{j=1}^{m}\sum_{k=1}^{n} x_j y_k I_{A_j B_k} = \sum_{j=1}^{m}\sum_{k=1}^{n} x_j y_k PA_j PB_k$$

$$= (\sum_{j=1}^{m} x_j PA_j)(\sum_{k=1}^{n} y_k PB_k) = EXEY,$$

and the conclusion is reached.

The expectation $E(X - EX)^2$, called the *variance* of X, is denoted by $\sigma^2 X$. By the additive property,

$$\sigma^2 X = E(X^2 - 2XEX + E^2 X) = EX^2 - E^2 X.$$

The celebrated Bienaymé equality follows from the additive and multiplicative properties.

Bienaymé equality. *If* X_k, $k = 1, 2, \cdots, n$, *are independent, then*

$$\sigma^2 \sum_{k=1}^{n} X_k = \sum_{k=1}^{n} \sigma^2 X_k.$$

Since

$$E(X_k - EX_k) = EX_k - EX_k = 0$$

and independence of the X_k implies independence of the $X_k - EX_k$, it follows that

$$\sigma^2 \sum_{k=1}^{n} X_k = E(\sum_{k=1}^{n} X_k - \sum_{k=1}^{n} EX_k)^2 = E\{\sum_{k=1}^{n}(X_k - EX_k)\}^2$$

$$= \sum_{k=1}^{n} E(X_k - EX_k)^2 + \sum_{j\neq k=1}^{n} E(X_j - EX_j)(X_k - EX_k)$$

$$= \sum_{k=1}^{n} \sigma^2 X_k + \sum_{j\neq k=1}^{n} E(X_j - EX_j)E(X_k - EX_k) = \sum_{k=1}^{n} \sigma^2 X_k.$$

Observe that we used independence of the X_k considered *two by two* only.

4. Bernoulli case. A simple case of independence has played a central role in the evolution of probability theory. This is the *Bernoulli* *case* of events A_k, $k = 1, 2, \cdots$, which are independent whatever be their total number n under consideration and such that their probabilities PA_k have the same value p.

We observe that independence of the A_k, $k = 1, 2, \cdots, n$ implies independence of the $\overline{A}_k = A_k$ or A_k^c, and, more generally, of the n fields $\mathcal{A}_k = \{\emptyset, A_k, A_k^c, \Omega\}$. For example,

$$\begin{aligned} PA_{k_1}^c A_{k_2} \cdots A_{k_m} &= PA_{k_2}A_{k_3} \cdots A_{k_m} - PA_{k_1}A_{k_2} \cdots A_{k_m} \\ &= PA_{k_2}PA_{k_3} \cdots PA_{k_m} - PA_{k_1}PA_{k_2} \cdots PA_{k_m} \\ &= (1 - PA_{k_1})PA_{k_2} \cdots PA_{k_m} = PA_{k_1}^c PA_{k_2} \cdots PA_{k_m}, \end{aligned}$$

where the subscripts are all distinct and $\leqq n$. These fields correspond to repeated random trials where an outcome A at the kth trial is represented by A_k.

The number of occurrences of outcome A in n repeated trials is represented by a simple random variable $S_n = \sum_{k=1}^{n} I_{A_k}$. To write S_n in the usual form, that is, with values assigned to events of a partition of the sure event, we observe that

$$I_{A_k} = I_{A_k} \prod_{\substack{j=1 \\ j \neq k}}^{n} (I_{A_j} + I_{A_j^c}).$$

It follows, upon substituting in S_n and expanding, that

$$S_n = \sum_{j=0}^{n} j I_{B_j},$$

where

$$I_{B_j} = \sum I_{A_{k_1}} \cdots I_{A_{k_j}} I_{A_{k_{j+1}}^c} \cdots I_{A_{k_n}^c}.$$

The summation is over all permutations of subscripts $k = 1, 2, \cdots, n$, classified into two groups, one having j terms and the other having $n - j$ terms.

On account of the independence, the expectations of the terms under the summation sign are

$$PA_{k_1}PA_{k_2} \cdots PA_{k_j}PA_{k_{j+1}}^c \cdots PA_{k_n}^c = p^j q^{n-j}, \quad q = 1 - p,$$

and, therefore, the probability of j occurrences in n trials is given by

$$P[S_n = j] = PB_j = \frac{n!}{j!(n-j)!} p^j q^{n-j}, \quad j = 0, 1, \cdots, n.$$

With this result we can compute directly the expectation and variance of S_n, but we prefer to use the additive property which gives

$$ES_n = E\sum_{k=1}^{n} I_{A_k} = \sum_{k=1}^{n} PA_k = np,$$

and the Bienaymé equality for independent random variables I_{A_k} which gives

$$\sigma^2 S_n = \sum_{k=1}^{n} \sigma^2 I_{A_k} = npq,$$

since

$$\sigma^2 I_{A_k} = E(I_{A_k} - EI_{A_k})^2 = EI_{A_k}{}^2 - E^2 I_{A_k}$$
$$= EI_{A_k} - E^2 I_{A_k} = p - p^2 = pq.$$

In order to justify the model investigated so far, we ought to give a precise and acceptable "meaning" to the notion of "clustering of frequencies" which, as we have seen, is at the very root of the interpretation of randomness. The most celebrated interpretation, and rightly so, is the following

Bernoulli law of large numbers (1713). *In the Bernoulli case, for every* $\epsilon > 0$, *as* $n \to \infty$,

$$P\left[\left|\frac{S_n}{n} - p\right| \geqq \epsilon\right] \to 0.$$

In other words, the probability distribution of values of the frequency S_n/n of an outcome in n repeated trials concentrates at the value p of the probability of the outcome, as the number of trials increases indefinitely.

The proof is immediate for, upon applying the Tchebichev inequality, we have, as $n \to \infty$,

$$P\left[\left|\frac{S_n}{n} - p\right| \geqq \epsilon\right] = P[|S_n - ES_n| \geqq \epsilon n] \leqq \frac{1}{\epsilon^2 n^2}\sigma^2 S_n = \frac{pq}{\epsilon^2 n} \to 0.$$

Observe that only independence two by two has been required.

A particular sequence of Bernoulli cases, introduced by Poisson, shows that the finite setup considered so far is not satisfactory, at least from the sophisticated mathematician's point of view.

Consider a sequence of Bernoulli cases of independent events A_{nk}, $k = 1, 2, \cdots, n$; $n = 1, 2, \cdots$, of the same probability p_n which varies with the number n of trials in such a manner that the expectation of the number of occurrences $S_n = \sum_{k=1}^{n} I_{A_{nk}}$ remains constant: $ES_n = np_n = \lambda$. Then, as $n \to \infty$ while j remains fixed,

$$
\begin{aligned}
P[S_n = j]& \\
&= \frac{n!}{j!(n-j)!} p_n^j q_n^{n-j} \\
&= \frac{n(n-1)\cdots(n-j+1)}{j!}\left(\frac{\lambda}{n}\right)^j\left(1-\frac{\lambda}{n}\right)^{n-j} \\
&= \frac{\lambda^j}{j!}\left(1-\frac{\lambda}{n}\right)^n\cdot\left(1-\frac{1}{n}\right)\left(1-\frac{2}{n}\right)\cdots\left(1-\frac{j-1}{n}\right)\left(1-\frac{\lambda}{n}\right)^{-j} \\
&\to \frac{\lambda^j}{j!}e^{-\lambda},
\end{aligned}
$$

and we have the following

Poisson theorem (1832). *If* $S_n = \sum_{k=1}^{n} I_{A_{nk}}$ *is the sum of indicators of independent equiprobable events, such that the expectation* $ES_n = \lambda > 0$ *remains constant as* n *varies, then, as* $n \to \infty$,

$$P[S_n = j] \to \frac{\lambda^j}{j!}e^{-\lambda}, \quad j = 0, 1, 2, \cdots.$$

Since

$$\sum_{j=0}^{\infty} \frac{\lambda^j}{j!}e^{-\lambda} = e^{-\lambda}\sum_{j=0}^{\infty}\frac{\lambda^j}{j!} = 1,$$

we can say that, in the foregoing passage to the limit, no positive probability escapes to infinity. The total probability is now distributed among a *denumerable* number of values $j = 0, 1, 2, \cdots$, provided we assume that the probability of the sum of a denumerable number of disjoint events $[S_n = j]$ is the sum of their probabilities. However, in the setup of § 1 neither a denumerable sum of events nor the property just stated has content. Thus, if we want to give an interpretation to Poisson's result, we have to expand the model so as to include the preceding possibilities.

5. Axioms for the countable case. As soon as the concept of infinity appears, intuition fails and the vague everyday idea of randomness yields nothing. A first and obvious way to pass from the finite to the infinite is to extrapolate, that is, to postulate that properties of the finite case continue to hold in the infinite case. Yet these extrapolations have to be meaningful and consistent.

In set theory, intersections $\bigcap_{n=1}^{\infty} A_n$ and unions $\bigcup_{n=1}^{\infty} A_n$ of sets A_n, where n runs over the denumerable set of integers, continue to be defined as the sets of points which belong to every A_n and to at least one A_n, respectively. We still have that

$$(\bigcap_{n=1}^{\infty} A_n)^c = \bigcup_{n=1}^{\infty} A_n^c, \quad (\bigcup_{n=1}^{\infty} A_n)^c = \bigcap_{n=1}^{\infty} A_n^c,$$

$$\bigcup_{n=1}^{\infty} A_n = A_1 + A_1^c A_2 + A_1^c A_2^c A_3 + \cdots \quad \text{ad infinitum}$$

and, correspondingly,

$$I_{\bigcap_{n=1}^{\infty} A_n} = \prod_{n=1}^{\infty} I_{A_n}, \quad I_{\sum_{n=1}^{\infty} A_n} = \sum_{n=1}^{\infty} I_{A_n}$$

$$I_{\bigcup_{n=1}^{\infty} A_n} = I_{A_1} + I_{A_1^c} I_{A_2} + I_{A_1^c} I_{A_2^c} I_{A_3} + \cdots.$$

If we want all *countable* (finite or denumerable) combinations of events by means of "not," "and," "or," to be events and their probabilities to be defined, then axioms I and II become

AXIOM I′. *Events form a σ-field* $\mathcal{A}$: Complements A^c, countable intersections $\bigcap_j A_j$, and countable unions $\bigcup_j A_j$ of events are events.

AXIOM II′. *Probability P on* $\mathcal{A}$ *is normed, nonnegative, and σ-additive:*

$$P\Omega = 1, \quad PA \geqq 0, \quad P\sum_j A_j = \sum_j PA_j.$$

It follows that

COVERING RULE: $P\bigcup_j A_j = PA_1 + PA_1^c A_2 + PA_1^c A_2^c A_3 + \cdots$ $\leqq \sum_j PA_j.$

These axioms are consistent, since the examples constructed for the finite case continue to apply trivially. A nontrivial example in the infinite case is that of nonsimple *elementary probability fields*: 1° The events, except $\emptyset$, are formed by all countable sums of events A_n which form a denumerable partition of the sure event: $\sum_{n=1}^{\infty} A_n = \Omega$; 2° to every event A_n of the partition is assigned probability $p_n = PA_n$ such that every $p_n \geqq 0$ and $\sum_{n=1}^{\infty} p_n = 1$—this is always possible. Then P is defined on $\mathcal{A}$, consistently with axiom II′, by assigning to every event A as its probability the sum (finite sum or convergent series) of probabilities of those A_n whose sum is A.

6. Elementary random variables. A linear combination $X = \sum_j x_j I_{A_j}$ of a countable number of indicators of disjoint events A_j is an *elementary random variable* X; if j varies over a finite set, then X reduces to a simple random variable. Clearly a sum or a product of two elementary random variables is an elementary random variable. We may still try to define the expectation EX by

$$EX = \sum_j x_j PA_j.$$

But, if the sum is a divergent series, it has no content or is infinite. Furthermore, even if it is a convergent series, it may not be absolutely convergent, so that by changing the order of terms we can change its value, and the expectation is no longer well defined if no ordering is specified; this is undesirable according to the very meaning of an expectation. We are therefore led to define EX by the foregoing expression *only when the right-hand side is absolutely convergent*, so that

if EX exists and is finite, then $E|X|$ exists and is finite; and conversely.

(We recognize here an integrable elementary function in the sense of Lebesgue with respect to the measure P.)

The argument used to prove the addition property of simple random variables continues to apply to finite sums of elementary random variables whose expectations exist and are finite, provided σ-additivity of P is used. We obtain:

If the expectations of a finite number of elementary random variables exist and are finite, then the expectation of their sum exists and is finite and is the sum of their expectations.

Also, Tchebichev's inequality remains valid, provided its right-hand side exists and is finite.

Independence of a countable number of events A_j, or σ-fields $\mathcal{A}_j$ contained in $\mathcal{A}$, is defined to be independence of every finite number of these events, or σ-fields. Independence of a countable number of elementary random variables $X_k = \sum_j x_{j_k} I_{A_{jk}}$ is defined to be independence of every finite number of events $A_{j_k,k}$ as k varies. The argument used to prove the multiplication property yields:

If the expectations of a finite number of independent elementary random variables exist and are finite, then the expectation of their product exists and is the finite product of their expectations.

Also, Bienaymé's equality remains valid, provided its right-hand side exists and is finite.

In the Bernoulli law of large numbers only simple random variables figure and only finite additivity of the probability P is used, so that nothing is to be changed. However, now we can introduce probabilities of denumerable combinations of events and use the supplementary requirement that the additive property of P remains valid for denumerable sums. Therefore, in the present setup we can expect a more precise interpretation of the "clustering of frequencies." This is the celebrated Borel strong law of large numbers derived below.

Let $X_1, X_2, \cdots$ be a sequence of elementary random variables. We investigate the convergence to 0 of the sequence; the limits are taken as $n \to \infty$. It will be more convenient to consider the contrary case—X_n does not converge to 0 or, equivalently, there exists at least one integer m such that to every integer n there corresponds at least one integer ν for which $|X_{n+\nu}| \geqq \frac{1}{m}$. Since "at least one" corresponds to "$\bigcup$" while "every" corresponds to "$\bigcap$," we can write

$$[X_n \nrightarrow 0] = \bigcup_{m=1}^{\infty} \bigcap_{n=1}^{\infty} \bigcup_{\nu=1}^{\infty} \left[|X_{n+\nu}| \geqq \frac{1}{m} \right];$$

the right-hand side is an event. Thus, the condition $X_n \nrightarrow 0$ determines the event $[X_n \nrightarrow 0]$, the contrary condition $X_n \to 0$ determines the complementary event $[X_n \to 0]$, and the probabilities of these two events add up to 1.

We are interested in $X_n \to 0$ with probability 1 or, equivalently, $X_n \nrightarrow 0$ with probability 0, and require the following proposition.

If, for every integer m, $\sum_{n=1}^{\infty} P\left[|X_n| \geqq \frac{1}{m} \right] < \infty$, *then* $P[X_n \nrightarrow 0] = 0$.

We set $A_{nm} = \bigcup_{\nu=1}^{\infty} \left[|X_{n+\nu}| \geqq \frac{1}{m} \right]$ and $A_m = \bigcap_{n=1}^{\infty} A_{nm}$ and observe that, by the covering rule and the hypothesis, for every m,

$$PA_{nm} = P \bigcup_{k=n+1}^{\infty} \left[| X_k | \geqq \frac{1}{m} \right]$$

$$\leqq \sum_{k=n+1}^{\infty} P\left[| X_k | \geqq \frac{1}{m} \right] \to 0 \quad \text{as} \quad n \to \infty.$$

Since whatever be n',

$$PA_m = P \bigcap_{n=1}^{\infty} A_{nm} \leqq PA_{n'm}.$$

it follows upon letting $n' \to \infty$ that $PA_m = 0$. Therefore, by the covering rule

$$P[X_n \nrightarrow 0] = P \bigcup_{m=1}^{\infty} A_m \leqq \sum_{m=1}^{\infty} PA_m = 0$$

and the proposition is proved.

We can now pass to

Borel's strong law of large numbers (1909). *In the Bernoulli case*

$$P\left[\frac{S_n}{n} \to p\right] = 1.$$

We recall that in the Bernoulli case

$$X_n = \frac{S_n}{n} = \frac{1}{n} \sum_{j=1}^{n} I_{A_j}$$

where the A_j are independent events of common probability p whatever be n, and $EX_n = p$, $\sigma^2 X_n = pq/n$ (observe that only independence two by two is used). Since for every m

$$\sum_{k=1}^{\infty} P\left[| X_{k^2} - p | \geqq \frac{1}{m} \right] \leqq m^2 pq \sum_{k=1}^{\infty} \frac{1}{k^2} < \infty,$$

it follows by the foregoing proposition that $X_{k^2} \to p$ with probability 1 as $k \to \infty$. But to every n there corresponds an integer $k = k(n)$ with $k^2 \leqq n < (k+1)^2$; hence $0 \leqq n - k^2 \leqq 2k$ and $n \to \infty$ implies $k \to \infty$.

Since

$$| X_n - X_{k^2} | = \left| \left(\frac{1}{n} - \frac{1}{k^2} \right) \sum_{j=1}^{k^2} I_{A_j} + \frac{1}{n} \sum_{j=k^2+1}^{n} I_{A_j} \right|$$

$$\leqq \frac{(n-k^2)k^2}{nk^2} + \frac{n-k^2}{n} \leqq \frac{4}{k}$$

so that

$$| X_n - p | \leqq | X_n - X_{k^2} | + | X_{k^2} - p | \leqq \frac{4}{k} + | X_{k^2} - p |,$$

it follows that $X_n \to p$ with probability 1 as $n \to \infty$, and Borel's result is proved.

**Application.* Let X be an elementary random variable. We set $F(x - 0) = F(x) = P[X < x]$, $F(x + 0) = P[X \leqq x]$ so that $P[X = x] = F(x + 0) - F(x)$. The function F so defined determines the probability distribution of X, that is, the probabilities of all values of X; it is called the *distribution function* of X. We organize repeated independent trials where we observe the values of X; in other words, we consider independent random variables $X_1, X_2, \cdots$ with the same probability distribution as X.

If k is the number of values observed in n of those trials and which are less than x or, equivalently, if k is the number of independent events $[X_1 < x], [X_2 < x], \cdots [X_n < x]$ (with common probability $p = F(x)$) which occur, we set $F_n(x - 0) = F_n(x) = k/n$. Thus, $F_n(x)$ is a random variable with

$$P\left[F_n(x) = \frac{k}{n} \right] = \frac{n!}{k!(n-k)!} \{F(x)\}^k \{1 - F(x)\}^{n-k}.$$

The function F_n is called *empirical distribution function* of X in n trials. According to Borel's strong law of large numbers, this frequency $F_n(x)$ of occurrences of the outcome $[X < x]$ converges to $F(x)$ with probability 1. In other words, the observations permit us to find with probability 1 every value $F(x)$ of the distribution function of X. In fact, Borel's result yields more (Glivenko-Cantelli):

Central statistical theorem. *If F is the distribution function of a random variable X and F_n is the empirical distribution function of X in n independent and identical trials, then*

$$P[\sup_{-\infty < x < +\infty} | F_n(x) - F(x) | \to 0] = 1.$$

In other words, with probability 1, $F_n(x) \to F(x)$ uniformly in x.

Let x_{jk} be the smallest value x such that

$$F(x) \leqq \frac{j}{k} \leqq F(x+0).$$

Since the frequency of the event $[X < x_{jk}]$ is $F_n(x_{jk})$ and its probability is $F(x_{jk})$, it follows by Borel's result that $PA'_{jk} = 1$ where $A'_{jk} = [F_n(x_{jk}) \to F(x_{jk})]$. Similarly, $PA''_{jk} = 1$ where $A''_{jk} = [F_n(x_{jk}+0) \to F(x_{jk}+0]$. Let $A_{jk} = A'_{jk}A''_{jk}$ and let $\theta = \pm 0$

$$A_k = \bigcap_{j=1}^{k} A_{jk} = [\sup_{1 \leqq j \leqq k} | F_n(x_{jk}+\theta) - F(x_{jk}+\theta) | \to 0].$$

By the covering rule and by what precedes

$$PA_k^c = P\bigcup_{i=1}^{k} A_{jk}^c \leqq \sum_{j=1}^{k} PA_{jk}^c = 0$$

and, hence, $PA_k = 1$. Upon setting $A = \bigcap_{k=1}^{\infty} A_k$, it follows similarly that $PA = 1$.

On the other hand, for every x between x_{jk} and $x_{j+1,k}$

$$F(x_{jk}+0) \leqq F(x) \leqq F(x_{j+1,k}), \quad F_n(x_{jk}+0) \leqq F_n(x) \leqq F_n(x_{j+1,k})$$

while for every x_{jk}

$$0 \leqq F(x_{j+1,k}) - F(x_{jk}+0) \leqq \frac{1}{k}.$$

Therefore,

$$F_n(x) - F(x) \leqq F_n(x_{j+1,k}) - F(x_{jk}+0) \leqq F_n(x_{j+1,k}) - F(x_{j+1,k}) + \frac{1}{k}$$

and

$$F_n(x) - F(x) \geqq F_n(x_{jk}+0) - F(x_{j+1,k})$$
$$\geqq F_n(x_{jk}+0) - F(x_{jk}+0) - \frac{1}{k}.$$

It follows that, whatever be x and k,

$$| F_n(x) - F(x) | \leqq \sup_{1 \leqq j \leqq k} | F_n(x_{jk}+\theta) - F(x_{jk}+\theta) | + \frac{1}{k}$$

or

$$\Delta_n = \sup_{-\infty < x < +\infty} | F_n(x) - F(x) | \leqq \sup_{1 \leqq j \leqq k} | F_n(x_{jk}+\theta) - F(x_{jk}+\theta) | + \frac{1}{k}.$$

Hence $P[\Delta_n \to 0] \geqq PA = 1$, and the theorem is proved.

*REMARK. The foregoing proof and hence the theorem remain valid when the random variable X is not elementary.

7. Need for nonelementary random variables. The sophisticated mathematician prefers to work with "closed" models—such that the operations defined for the entities within the model yield only entities within the model. While elementary random variables can be obtained as limits of sequences of simple random variables, *all* limits of sequences of simple and, more generally, of elementary random variables are not necessarily elementary—families of elementary random variables are not necessarily closed under passages to the limit. If this closure is required, then the concept of a random variable has to be extended so as to include "measurable functions." This will be done in the following parts. In fact, the need for further expansion of the model in order to include random variables with a noncountable set of values appeared quite early in the development of probability theory, once more in connection with the Bernoulli case. This is the celebrated (as the reader observes, all results obtained in or used for the Bernoulli case are "celebrated")

De Moivre-Laplace theorem. *In the Bernoulli case with* $p > 0$, $q = 1 - p > 0$, *as* $n \to \infty$,

de Moivre (1732):

$$P_n(x) = P[S_n = j] \sim \frac{1}{\sqrt{2\pi npq}} e^{-x^2/2}, \quad x = \frac{j - np}{\sqrt{npq}},$$

uniformly on every finite interval $[a, b]$ *of values of* x;

Laplace (1801):

$$P\left[a \leqq \frac{S_n - np}{\sqrt{npq}} \leqq b\right] \to \frac{1}{\sqrt{2\pi}} \int_a^b e^{-x^2/2}\, dx.$$

The relation $a_n \sim b_n$ means that $a_n/b_n \to 1$. The integer j varies with n, so that $x = x(n)$ remains within a fixed finite interval $[a, b]$ and

$$j = np + x\sqrt{npq} \to \infty, \quad k = n - j = nq - x\sqrt{npq} \to \infty.$$

We apply Stirling's formula

$$m! = \sqrt{2\pi m} \cdot m^m e^{-m} e^{\theta_m}, \quad 0 < \theta_m < \frac{1}{12m}$$

to the binomial probabilities $P_n(x) = \dfrac{n!}{j!k!} p^j q^k$. Thus

$$P_n(x) = \frac{\sqrt{2\pi n}\cdot n^n e^{-n}}{\sqrt{2\pi j}\cdot j^j e^{-j}\sqrt{2\pi k}\cdot k^k e^{-k}} p^j q^k e^{\theta_n - \theta_j - \theta_k}$$

$$= \frac{1}{\sqrt{2\pi}}\sqrt{\frac{n}{jk}}\left(\frac{np}{j}\right)^j\left(\frac{nq}{k}\right)^k e^{\theta}$$

where, uniformly on $[a, b]$,

$$|\theta| < \frac{1}{12}\left(\frac{1}{n}+\frac{1}{j}+\frac{1}{k}\right)$$

and

$$\frac{jk}{n} = n\left(p + x\sqrt{\frac{pq}{n}}\right)\left(q - x\sqrt{\frac{pq}{n}}\right) \sim npq.$$

Therefore, uniformly on $[a, b]$,

$$P_n(x) \sim \frac{1}{\sqrt{2\pi npq}}\left(\frac{np}{j}\right)^j\left(\frac{nq}{k}\right)^k$$

and

$$\log\left(\frac{np}{j}\right)^j\left(\frac{nq}{k}\right)^k = -(np + x\sqrt{npq})\left[x\sqrt{\frac{q}{np}} - \frac{1}{2}\frac{qx^2}{np} + O\left(\frac{1}{n^{3/2}}\right)\right]$$

$$- (nq - x\sqrt{npq})\left[-x\sqrt{\frac{p}{nq}} - \frac{1}{2}\frac{px^2}{nq} + O\left(\frac{1}{n^{3/2}}\right)\right]$$

$$= -\frac{x^2}{2} + O\left(\frac{1}{\sqrt{n}}\right).$$

The first assertion follows.

Let x_{nj} be those numbers of the form $\frac{j - np}{\sqrt{npq}}$ which belong to the interval $[a, b]$; consecutive x_{nj}'s differ by $1/\sqrt{npq}$. On account of the first assertion, uniformly in j,

$$P_n(x_{nj}) \sim \frac{1}{\sqrt{2\pi npq}} e^{-x_{nj}^2/2}$$

and

$$P\left[a \leq \frac{S_n - np}{\sqrt{npq}} \leq b\right] = \sum_j P_n(x_{nj}) \sim \frac{1}{\sqrt{2\pi}}\cdot\frac{1}{\sqrt{npq}}\sum_j e^{-x^2_{nj}/2}.$$

Since the last expression is a Riemann sum approximating the integral $\frac{1}{\sqrt{2\pi}}\int_a^b e^{-x^2/2}\,dx$, the second assertion follows.

III. DEPENDENCE AND CHAINS

1. Conditional probabilities. Let A be an event with $PA > 0$. The ratio PAB/PA is called the *conditional probability of B given A* or, simply, *probability of B given A* and is denoted by P_AB, so that

$$PAB = PAP_AB.$$

By induction we obtain the *multiplication rule:*

$$P(AB \cdots KL) = PAP_AB \cdots P_{AB\cdots K}L.$$

Furthermore, if $\sum_j A_j = \Omega$, then, from

$$PB = P\Omega B = \sum_j PA_jB,$$

follows the *total probability rule:*

$$PB = \sum_j PA_jP_{A_j}B.$$

Bayes' theorem,

$$P_BA_k = \frac{PA_kP_{A_k}B}{\sum_j PA_jP_{A_j}B},$$

follows upon replacing PB by the foregoing expression in the relation

$$PA_kB = PA_kP_{A_k}B = PBP_BA_k.$$

All events which figure as subscripts are supposed to be of positive probability. However, *if, say, P_AB is given, then every given PA, whether zero or not, determines correctly PAB by $PAB = PAP_AB$, since $PA = 0$ implies $PAB = 0$.*

The set of all probabilities of events given a fixed A with $PA > 0$ defines a function P_A on $\mathcal{A}$, to be called the *conditional probability given A* or, simply, the *probability given A*. It follows at once from the definition that P_A obeys axiom II′: it is normed, nonnegative, and σ-additive on $\mathcal{A}$. Therefore, the pair $(\mathcal{A}, P_A)$ is a probability field *given A* for which all definitions and general properties of probability fields remain valid. In particular, if $X = \sum_j x_j I_{A_j}$ is an elementary random variable, the expectation of X with respect to P_A or the *conditional expectation of X given A* or, simply, the *expectation of X given A* is defined by

$$E_A X = \sum_j x_j P_A A_j = \frac{1}{PA} \sum_j x_j PAA_j;$$

clearly, if EX exists and is finite, then $E_A X$ exists and is finite. In terms of trials, the probability field given A represents the original trial with the occurrence of outcome A added to the original conditions.

It is easily verified that the events A_j of a countable set are independent if, and only if, for every finite subset $j_1, j_2, \cdots, j_k$ of indices

$$P_{A_{j_1} A_{j_2} \cdots A_{j_{k-1}}}(A_{j_k}) = PA_{j_k},$$

provided the "given" events have positive probability.

2. Asymptotically Bernoullian case. Let A_n, $n = 1, 2, \cdots$, be an arbitrary sequence of events, and let $X_n = \frac{1}{n} \sum_{k=1}^{n} I_{A_k}$ be the random frequency of occurrence of the n first ones. We set

$$p_1(n) = \frac{1}{n} \sum_{k=1}^{n} PA_k, \quad p_2(n) = \frac{2}{n(n-1)} \sum_{1 \leqq j < k \leqq n} PA_j A_k$$

so that $p_1(n)$ and $p_2(n)$ are bounded by 0 and 1. It follows, by elementary computations, that

$$EX_n = p_1(n), \quad \sigma^2 X_n = p_2(n) - p_1^2(n) + \frac{p_1(n) - p_2(n)}{n}.$$

In the Bernoulli case

$$d_n = p_2(n) - p_1^2(n) = p^2 - p^2 = 0,$$

and we can consider the quantity d_n as some sort of measure of "deviation" from the Bernoulli case. To make this precise, let us first prove a

Kolmogorov inequality. *If X is an elementary random variable bounded by* 1 *(in absolute value), then, for every* $\epsilon > 0$,

$$P[|X| \geqq \epsilon] \geqq EX^2 - \epsilon^2.$$

We proceed as for the proof of Tchebichev's inequality: the inequality follows from

$$\begin{aligned} EX^2 &= E(X^2 I_{[|X| \geqq \epsilon]}) + E(X^2 I_{[|X| < \epsilon]}) \leqq EI_{[|X| \geqq \epsilon]} + \epsilon^2 \\ &= P[|X| \geqq \epsilon] + \epsilon^2. \end{aligned}$$

EXTENDED BERNOULLI LAW OF LARGE NUMBERS. *Bernoulli's result, that for every* $\epsilon > 0$

$$P[|X_n - EX_n| \geqq \epsilon] \to 0,$$

remains valid for the sequence of events A_n, *independent or not, if, and only if,*

$$d_n = p_2(n) - p_1^2(n) \to 0.$$

Since $|X_n| \leqq 1$, we can apply Kolmogorov's inequality as well as Tchebichev's, so that

$$\sigma^2 X_n - \epsilon^2 \leqq P[|X_n - EX_n| \geqq \epsilon] \leqq \sigma^2 X_n/\epsilon^2.$$

Therefore, the asserted property holds if, and only if, $\sigma^2 X_n \to 0$. But

$$|\sigma^2 X_n - d_n| = \frac{|p_1(n) - p_2(n)|}{n} \leqq \frac{1}{n} \to 0,$$

and the extension follows.

If $d_n \to 0$ at least as fast as $1/n$, then (asymptotically) we are even "closer" to the Bernoulli case. In fact,

EXTENDED BOREL STRONG LAW OF LARGE NUMBERS. *If* $d_n = O(1/n)$, *then Borel's result remains valid:*

$$P[X_n - EX_n \to 0] = 1.$$

The hypothesis means that there exists a fixed finite number c such that $|nd_n| \leqq c$. Upon referring to the proof of Borel's result, we observe that it suffices to show that $\sum_{k=1}^{\infty} \sigma^2 X_{k^2} < \infty$. Since

$$n\sigma^2 X_n \leqq |nd_n| + |p_1(n) - p_2(n)| \leqq c + 1,$$

it follows by setting $n = k^2$ that

$$\sum_{k=1}^{\infty} \sigma^2 X_{k^2} \leqq (c+1) \sum_{k=1}^{\infty} \frac{1}{k^2} < \infty,$$

and the extension follows.

It is easily shown that both extensions apply to the events A_n which are independent but otherwise arbitrary.

3. Recurrence. The decomposition

$$\sigma^2 X_n = p_2(n) - p_1^2(n) + \frac{p_1(n) - p_2(n)}{n}$$

yields at once a proposition which leads very simply to the celebrated

Poincaré's recurrence theorem and its known refinements. Since $\sigma^2 X_n \geqq 0$ and $p_1(n)$, $p_2(n)$ are bounded by 0 and 1, it follows that, for any fixed $\epsilon > 0$, if $n \geqq 1/\epsilon$, then

$$p_2(n) = p_1^2(n) + \frac{p_2(n) - p_1(n)}{n} + \sigma^2 X_n \geqq p_1^2(n) - \frac{1}{n} \geqq p_1^2(n) - \epsilon.$$

But $p_2(n)$ is the arithmetic mean of PA_jA_k for $1 \leqq j < k \leqq n$. Therefore,

Whatever be the events A_n, if $n \geqq 1/\epsilon$, then there exist at least two events A_j, A_k, $1 \leqq j < k \leqq n$, such that $PA_jA_k \geqq p_1^2(n) - \epsilon$.

In particular, if $PA_n \geqq p > 0$ whatever be n, then every subsequence of these events contains at least two events A_j, A_k such that $PA_jA_k \geqq p^2 - \epsilon$; if this inequality holds, we say that A_j "ϵ-intersects" A_k. In fact, there exists then a subsequence whose first term ϵ-intersects every other term. For, if there is no such subsequence, then there exist integers m_n such that no event A_n ϵ-intersects events $A_{n'}$ with $n' \geqq n + m_n$, no two events of the subsequence $A_{n_1}, A_{n_2}, A_{n_3}, \cdots$ with $n_1 = 1$, $n_2 = n_1 + m_{n_1}$, $n_3 = n_2 + m_{n_2}$, $\cdots$, ϵ-intersect, and this contradicts the particular case of the foregoing proposition. Thus, let $A_{11}, A_{21}, A_{31}, \cdots$, be a subsequence such that the first term ϵ-intersects every other term. Let $A_{12}, A_{22}, A_{32}, \cdots$, be a subsequence of $A_{21}, A_{31}, \cdots$, with same property, and so on indefinitely. The sequence $A_{11}, A_{12}, \cdots$, is such that every one of its terms ϵ-intersects every other term. Hence

RECURRENCE THEOREM. *If $PA_n \geqq p > 0$ whatever be n, then for every $\epsilon > 0$ there exists a subsequence of events A_n such that $PA_jA_k \geqq p^2 - \epsilon$ whatever be the terms A_j, A_k of this subsequence.*

We observe that, if $PA_n = p$, then $PA_jA_k \geqq p^2 - \epsilon$ while, if the A_n are two by two independent, then $PA_jA_k = p^2$. Thus, however small be $\epsilon > 0$, for every sequence A_n of events, independent or not, there exists a subsequence which behaves as if its terms were two by two semi-independent *up to* ϵ ("semi" only since we do not have necessarily $PA_jA_k \leqq p^2 + \epsilon$).

A phenomenological interpretation of the foregoing theorem is as follows. Consider integer values of time and an incompressible fluid in motion filling a container of unit volume. Any portion of the fluid which at time 0 occupies a position A of volume $PA = p > 0$ occupies at time m a position A_m of same volume $PA_m = p$. The theorem says that, for every $\epsilon > 0$, the portion occupies in its motion an infinity of positions such that the volume of the intersection of any two of these positions is $\geqq p^2 - \epsilon$. In particular, if the motion is "second order stationary," that is, $PA_jA_{j+k} = PAA_k$, then it intersects infinitely often its initial position—this is Poincaré's recurrence theorem (he assumes "stationarity")—and the intersections may be selected to be of volume $\geqq p^2 - \epsilon$—this is Khintchine's refinement.

4. Chain dependence. There is a type of dependence, studied by Markov and frequently called Markov dependence, which is of considerable phenomenological interest. It represents the chance (random, stochastic) analogue of nonhereditary systems, mechanical, optical, $\cdots$, whose known properties constitute the bulk of the present knowledge of laws of nature.

A system is subject to laws which govern its evolution. For example, a particle in a given field of forces is subject to Newton's laws of motion, and its positions and velocities at times 1, 2, $\cdots$, describe the "states" (events) that we observe; crudely described, a very small particle in a given liquid is subject to Brownian laws of motion, and its positions (or positions and velocities) at times $t = 1, 2, \cdots$, are the "states" (events) that we observe. While Newton's laws of motion are deterministic in the sense that, given the present state of the particle, the future states are uniquely determined (are sure outcomes), Brownian laws of motion are stochastic in the sense that only the *probabilities* of future states are determined. Yet both systems are "nonhereditary" in the sense that the future (described by the sure outcomes or probabilities of outcomes, respectively) is determined by the last observed state only—the "present." It is sometimes said that nonhereditary systems obey the "Huygens principle." The mathematical concept of nonheredity in a stochastic context is that of Markov or chain dependence, and appears as a "natural" generalization of that of independence.

Events A_j, where j runs over an ordered countable set, are said to be *chained* if the probability of every A_j given any finite set of the preceding ones depends only upon the last given one; in symbols, for every finite subset of indices $j_1 < j_2 < \cdots j_k$, we have

$$P_{A_{j_1}A_{j_2}\cdots A_{j_{k-1}}}(A_{j_k}) = P_{A_{j_{k-1}}}(A_{j_k}).$$

Classes $\mathcal{C}_j = \{A_{j1}, A_{j2}, \cdots\}$ of events are said to be *chained* if events A_{jk} selected arbitrarily—one in each $\mathcal{C}_j$—are chained.

An elementary chain is a sequence of chained elementary partitions $\sum_k A_{nk} = \Omega$, $n = 1, 2, \cdots$; in particular, if $X_n = \sum_k x_{nk} I_{A_{nk}}$ with distinct $x_{n1}, x_{n2}, \cdots$ are elementary random variables, then the X_n are said to be *chained*, or to form a *chain*, when the corresponding partitions are chained.

It will be convenient to use a phenomenological terminology. Events of the nth partition will be called *states at time n*, or *at the nth* step, of the system described by the chain. The totality of all states of the system is countable; we shall denote them by the letters $j, k, h, \cdots$, and summations over, say, states k will be over the set of all states, unless otherwise stated.

The *evolution* of the system is described by the probabilities of its states given the last known one. The probabilities $P_{jk}^{m,n}$ of passage from a state j at time m to a state k at time $m + n$ (in n steps) form a matrix $P^{m,n}$. Since "probability given j" is a probability, and the probability given j at time m to pass to some state in n steps is one, we have

$$P_{jk}^{m,n} \geqq 0, \quad \sum_k P_{jk}^{m,n} = 1.$$

Furthermore, by the definition of chain dependence, the probability given j at time m to pass to state k in $n + n'$ steps equals the probability given j at time m to pass to some state in n steps and then to pass to k in n' steps, we have

$$P_{jk}^{m,n+n'} = \sum_h P_{jh}^{m,n} P_{hk}^{m+n,n'}$$

or, in matrix notation,

$$P^{m,n+n'} = P^{m,n} P^{m+n,n'}$$

An elementary chain is said to be *constant* if $P_{jk}^{m,n}$ is independent of m whatever be j, k, and n. Then we denote this probability by P_{jk}^n, and call it *transition probability from j to k in n steps*. The corresponding matrix P^n is called *transition matrix in n steps*; if $n = 1$ we drop it. The foregoing relations become the *basic constant chain relations:*

$$P_{jk}^n \geqq 0, \quad \sum_k P_{jk}^n = 1, \quad P_{jk}^{n+n'} = \sum_h P_{jh}^n P_{hk}^{n'}.$$

The last one can also be written as a matrix product $P^{n+n'} = P^n P^{n'}$. Hence P^n is the nth power of the transition matrix $P = P^1$, so that P determines all transition probabilities. In fact, for an elementary chain to be constant it suffices that the matrix $P^{m,1}$ be independent of m: $P^{m,1} = P$, since then

$$P^{m,2} = P^{m,1}P^{m+1,1} = P^2, \quad P^{m,3} = P^{m,2}P^{m+2,1} = P^3, \quad \cdots.$$

We observe that in P_{jk}^n and in every symbol to be introduced below, superscripts are not power indices, unless so stated.

We investigate the evolution of a system subject to constant chain laws described by a transition matrix P. In particular, we want to find its asymptotic behavior according to the state from which it starts. In phenomenological terms the system is a nonhereditary one subject to constant laws (independent of the time) and we ask what happens to the system in the long run. The "direct" method we use—requiring no special tools and which has a definite appeal to the intuition—has been developed by Kolmogorov (1936) and by Döblin (1936, 1937) after Hadamard (1928) introduced it. But the concept of chain and the basic pioneering work are due to Markov (1907).

***5. Types of states and asymptotic behavior.** According to the total probability rule and the definition of chain dependence, the probability Q_{jk}^n of passage from j to k *in exactly* n steps, that is, without passing through k before the nth step, is given by

$$Q_{jk}^n = \sum_{h_1 \neq k, h_2 \neq k, \cdots, h_{n-1} \neq k} P_{jh_1} P_{h_1 h_2} \cdots P_{h_{n-1}k}.$$

The *central relation* in our investigation is

$$(1) \qquad P_{jk}^n = \sum_{m=1}^{n} Q_{jk}^m P_{kk}^{n-m}, \quad n = 1, 2, \cdots,$$

the expressions $P_{kk}^0 = 1$ (obtained for $m = n$) are the diagonal elements of the unit matrix P^0.

The proof is immediate upon applying the total probability rule. The system passes from j to k in n steps if, and only if, it passes from j to k for the first time in exactly m steps, $m = 1, 2, \cdots, n$, and then passes from k to k in the remaining $n - m$ steps. These "paths" are disjoint events, and their probabilities are given by $Q_{jk}^m P_{kk}^{n-m}$.

Summing over $n = 1, 2, \cdots, N$, the central relation yields

$$\sum_{n=1}^{N} P_{jk}^n = \sum_{n=1}^{N} \sum_{m=1}^{n} Q_{jk}^m P_{kk}^{n-m} = \sum_{m=1}^{N} (Q_{jk}^m \sum_{n=m}^{N} P_{kk}^{n-m})$$

and, therefore,

$$(1 + \sum_{n=1}^{N} P_{kk}^n) \sum_{m=1}^{N} Q_{jk}^m \geq \sum_{n=1}^{N} P_{jk}^n \geq (1 + \sum_{n=1}^{N-N'} P_{kk}^n) \sum_{m=1}^{N'} Q_{jk}^m, \quad N' < N$$

It follows. upon dividing by $1 + \sum_{n=1}^{N} P_{kk}^n$ and letting first $N \to \infty$ and then $N' \to \infty$, that

$$(2) \qquad \sum_{m=1}^{\infty} Q_{jk}^m = \lim_{N \to \infty} \frac{\sum_{n=1}^{N} P_{jk}^n}{1 + \sum_{n=1}^{N} P_{kk}^n};$$

in particular,

$$(3) \qquad 1 - \sum_{m=1}^{\infty} Q_{jj}^m = \lim_{N \to \infty} \frac{1}{1 + \sum_{n=1}^{N} P_{jj}^n}.$$

The sum

$$q_{jk} = \sum_{m=1}^{\infty} Q_{jk}^m$$

is the probability, starting at j, of passing through k *at least once;* for $k = j$ it is the probability of *returning to j at least once.* More generally, the probability q_{jk}^n, starting at j, of passing through k *at least n times* is given by

$$q_{jk}^n = \left(\sum_{m=1}^{\infty} Q_{jk}^m\right) q_{kk}^{n-1} = q_{jk} q_{kk}^{n-1}.$$

In particular, the probability q_{jj}^n of *returning to j at least n times* is given by

$$q_{jj}^n = q_{jj} q_{jj}^{n-1} = (q_{jj})^2 q_{jj}^{n-2} = \cdots = (q_{jj})^n.$$

Its limit,

$$(4) \quad r_{jj} = \lim_{n \to \infty} (q_{jj})^n = 0 \quad \text{or} \quad 1, \quad \text{according as} \quad q_{jj} < 1 \quad \text{or} \quad q_{jj} = 1,$$

is the probability of *returning to j infinitely often.* It follows that the probability, starting at j, of passing through k *infinitely often* is

$$r_{jk} = \lim_{n \to \infty} q_{jk}^n = q_{jk} \lim_{n \to \infty} q_{kk}^{n-1} = q_{jk} r_{kk},$$

so that

$$(5) \qquad r_{jk} = 0 \quad \text{or} \quad q_{jk}, \quad \text{according as} \quad q_{kk} < 1 \quad \text{or} \quad q_{kk} = 1.$$

Upon singling out the states j such that $q_{jj} = 0$ (noreturn) and $q_{jj} = 1$ (return with probability 1), we are led to two dichotomies of states:

j is a *return state* or a *noreturn state* according as $q_{jj} > 0$ or $q_{jj} = 0$; j is a *recurrent state* or a *nonrecurrent state* according as $q_{jj} = 1$ or $q_{jj} < 1$ or, on account of (4), according as $r_{jj} = 1$ or $r_{jj} = 0$.

Clearly, noreturn states are boundary cases of nonrecurrent states and recurrent states are boundary cases of return states. In terms of transition probabilities, we have the following criteria.

RETURN CRITERION. *A state j is a return or a noreturn state according as $P_{jj}^n > 0$ for at least one n or $P_{jj}^n = 0$ for all n.*

This follows at once from the fact that

$$\sup_{n \geqq 1} P_{jk}^n \leqq q_{jk} \leqq \sum_{n=1}^{\infty} P_{jk}^n.$$

RECURRENCE CRITERION. *A state j is a recurrent or a nonrecurrent state according as the series $\sum_{n=1}^{\infty} P_{jj}^n$ is divergent or convergent.*
This follows from (3).

Less obvious *types* of states are described in terms of "mean frequency of returns," as follows:

Let ν_{jk} be the *passage time*, from j to k, taking values $m = 1, 2, \cdots$, with probabilities Q_{jk}^m. If $q_{jk} = 1$, then ν_{jk} are elementary random variables. If $q_{jk} < 1$, then, to avoid exceptions, we say that $\nu_{jk} = \infty$ with probability $1 - q_{jk}$. The symbol ∞ is subject to the rules

$$\frac{1}{\infty} = 0,\ \infty + c = \infty, \text{ and } \infty \times c = \infty \text{ or } 0 \text{ according as } c > 0 \text{ or } c = 0.$$

We define the expected passage time τ_{jk} from j to k by

$$\tau_{jk} = \sum_{m=1}^{\infty} mQ_{jk}^m + \infty(1 - q_{jk});$$

we call τ_{jj} the *expected return time to j and the mean frequency of returns to j* is $\dfrac{1}{\tau_{jj}}$.

We can now define the following dichotomy of states. A state j is *null* or *positive* according as $\dfrac{1}{\tau_{jj}} = 0$ or $\dfrac{1}{\tau_{jj}} > 0$. Clearly, a noreturn and, more generally, a nonrecurrent state is null while a positive state is recurrent.

We shall now establish a criterion for this new dichotomy of states in terms of transition probabilities. To make it precise, we have to introduce the concept of period of a state.

Let j be a return state; then let d_j be the period of the Q_{jj}^m, that is, the greatest integer such that a return to j can occur with positive probability only after multiples of d_j steps: $Q_{jj}^n = 0$ for all $n \neq 0$ (modulo d_j), and $Q_{jj}^{nd_j} > 0$ for some n. Let d_j' be the period of the P_{jj}^n defined similarly. We prove that $d_j = d_j'$ and call it the *period* (of return) *of j*.

The proof is immediate. If $Q_{jj}^{nd_j} > 0$, then $P_{jj}^{nd_j} \geqq Q_{jj}^{nd_j} > 0$ so that $d_j' \leqq d_j$. Thus, if $d_j = 1$, then $d_j' = 1$. If $d_j > 1$ and $r = 1, \cdots, d_j - 1$, then the central relation yields

$$P_{jj}^r = 0, \quad P_{jj}^{d_j+r} = Q_{jj}^{d_j} P_{jj}^r = 0,$$

$$P_{jj}^{2d_j+r} = Q_{jj}^{d_j} P_{jj}^{d_j+r} + Q_{jj}^{2d_j} P_{jj}^r = 0, \text{ etc. } \cdots,$$

so that $d_j \leqq d_j'$ and, hence, $d_j = d_j'$.

If j is a noreturn state, then we say that its period is infinite.

POSITIVITY CRITERION. *A state j is null or positive according as* $\limsup_{n \to \infty} P_{jj}^n = 0$ *or* > 0.

More precisely, if j is a null state, then $P_{jj}^n \to 0$, *and if j is a positive state, then* $P_{jj}^{nd_j} \to \dfrac{d_j}{\tau_{jj}} > 0$, *while* $P_{jj}^n = 0$ *for all* $n \neq 0$ (*modulo* d_j).

Since the proof is involved, we give it in several steps.

1° If j is nonrecurrent, then it is null and, by the recurrence criterion, the series $\sum_{n=1}^{\infty} P_{jj}^n$ converges so that $P_{jj}^n \to 0$.

If j is recurrent, then, by definition of its period d_j, $P_{jj}^n = 0$ for all $n \neq 0$ (modulo d_j). Therefore, it suffices to prove that, if j is recurrent, then $P_{jj}^{nd_j} \to \dfrac{d_j}{\tau_{jj}}$; for, if j is null, then $\dfrac{1}{\tau_{jj}} = 0$ implies $\dfrac{d_j}{\tau_{jj}} = 0$, and if j is positive, then $\dfrac{d_j}{\tau_{jj}} > 0$.

Assume, for the moment, that, if the period d_j of the positive recurrent state j is 1, then $P_{jj}^n \to \dfrac{1}{\tau_{jj}}$. In the general case, take d_j for the unit step and set $P' = P^{d_j}$, so that $P_{jj}'^n = P_{jj}^{nd_j}$; hence $Q_{jj}'^n = Q_{jj}^{nd_j}$. Then, since $\tau_{jj}' = \sum_{n=1}^{\infty} n Q_{jj}'^n = \dfrac{\tau_{jj}}{d_j}$, the assertion follows by

$$P_{jj}^{nd_j} = P_{jj}'^n \to \frac{1}{\tau_{jj}'} = \frac{d_j}{\tau_{jj}}.$$

Thus, it suffices to prove that, if j is recurrent with period $d_j = 1$, then $P_{jj}^n \to \dfrac{1}{\tau_{jj}}$.

2° Let j be recurrent with $d_j = 1$. To simplify the writing, we drop the subscripts j and, to avoid confusion with matrices, we write superscripts as subscripts. We follow now Erdös, Feller, and Pollard.

Let $\alpha = \limsup P_n$ so that there is a subsequence n' of integers such that $P_{n'} \to \alpha$ as $n' \to \infty$. Since $q = \sum_{m=1}^{\infty} Q_m = 1$, it follows that, given $\epsilon > 0$, there exists n_ϵ such that, for $n \geqq n_\epsilon$, $\sum_{m=n+1}^{\infty} Q_m < \epsilon$. Therefore, for $n' \geqq n \geqq n_\epsilon$ and every p for which $Q_p > 0$, the central relation yields

$$P_{n'} \leqq Q_p P_{n'-p} + \sum_{m \leqq n, m \neq p} Q_m P_{n'-m} + \epsilon.$$

Since for n' sufficiently large, $P_{n'} > \alpha - \epsilon$ and $P_{n'-m} < \alpha + \epsilon$ for $m \leqq n$, it follows that

$$\alpha - \epsilon \leqq Q_p P_{n'-p} + (1 - Q_p)(\alpha + \epsilon) + \epsilon$$

hence

$$\alpha + \epsilon - \frac{3\epsilon}{Q_p} < P_{n'-p} < \alpha + \epsilon.$$

Therefore, letting $n' \to \infty$ and then $\epsilon \to 0$, we obtain $P_{n'-p} \to \alpha$, and, repeating the argument, we have, for every fixed integer m,

$$P_{n'-mp} \to \alpha \quad \text{as} \quad n' \to \infty.$$

3° Let us assume, for the moment, that $Q_1 > 0$ so that $P_{n'-m} \to \alpha$ for every fixed m. We introduce the expected return time τ and use the fact that j is recurrent, so that, setting $q_n = \sum_{m=n+1}^{\infty} Q_m$, we have $q_0 = 1$. The expected return time τ can be written

$$\tau = \sum_{m=1}^{\infty} mQ_m = \sum_{m=1}^{\infty} m(q_{m-1} - q_m) = \sum_{m=0}^{\infty} q_m,$$

and the central relation can be written

$$P_n = \sum_{m=1}^{n} Q_m P_{n-m} = \sum_{m=1}^{n} (q_{m-1} - q_m) P_{n-m},$$

so that

$$\sum_{m=0}^{n} q_m P_{n-m} = \sum_{m=0}^{n-1} q_m P_{n-1-m} = \cdots = q_0 P_0 = 1.$$

Therefore, for $n < n'$,
$$\sum_{m=0}^{n} q_m P_{n'-m} \leqq 1$$
and, letting $n' \to \infty$ and then $n \to \infty$, we obtain $\alpha \leqq 1/\tau$. If $\tau = \infty$, then $\alpha = 0$; hence $P_n \to 1/\tau = 0$. Thus let $\tau < \infty$. The same argument for $\beta = \liminf_{n\to\infty} P_n$ shows that, for a subsequence n'' such that $P_{n''} \to \beta$ as $n'' \to \infty$, we have $P_{n''-m} \to \beta$ for every fixed m and, from
$$\sum_{m=0}^{n} q_m P_{n''-m} + \epsilon \geqq 1 \quad \text{for} \quad n_\epsilon \leqq n < n'',$$
it follows as above that $\beta \geqq \dfrac{1}{\tau}$. Therefore, $P_n \to \dfrac{1}{\tau}$, and the assertion is proved under the assumption that $Q_1 > 0$.

4° To get rid of the last assumption, we appeal to elementary number theory. Consider the set of all those p for which $Q_p > 0$. It contains a finite subset $\{p_i\}$ whose greatest common divisor is the period $d(=1)$. As above, if $P_{n'} \to \alpha$, then $P_{n'-m_ip_i} \to \alpha$ for every fixed m_i and p_i, and it follows that $P_{n'-m} \to \alpha$ for every fixed linear combination $m = \sum_i m_ip_i$. But every multiple of the period $md = m \geqq \prod_i p_i$ can be written in this form, so that, starting with n' sufficiently large, $P_{n'-m} \to \alpha$ for every fixed m, and the assertion follows as above. This concludes the proof.

Since, for a state j with period d_j there exists a finite number of integers p_i such that $P_{jj}^{p_i} > 0$ and, for m sufficiently large $md_j = \sum m_ip_i$, it follows, by $P_{jj}^{md_j} \geqq \prod_i P_{jj}^{m_ip_i} > 0$, that

If d_j is the period of j, then $P_{jj}^{md_j} > 0$ for all sufficiently large values of m.

In other words, after some time elapses the system returns to j with positive probability after *every* interval of time d_j.

We can now describe the asymptotic behavior of the system. If k is a return state of period d_k, set
$$q_{jk}(r) = \sum_{m=0}^{\infty} Q_{jk}^{md_k+r}, \quad r = 1, 2, \cdots, d_k,$$
so that $q_{jk}(r)$ is the probability of passage from j to k in $n = r$ (modulo d_k) steps and
$$\sum_{r=1}^{d_k} q_{jk}(r) = q_{jk}.$$

ASYMPTOTIC PASSAGE THEOREM. *For every state j*

if k is a null state, then $P^n_{jk} \to 0$;

if k is a positive state, then $P^{nd_k+r}_{jk} \to q_{jk}(r)\dfrac{d_k}{\tau_{kk}}$;

and, whatever be the state k,

$$\frac{1}{n}\sum_{m=1}^{n} P^m_{jk} \to \bar{P}_{jk} = \frac{q_{jk}}{\tau_{kk}}.$$

The theorem results from the positivity criterion and the central relation, as follows:

If k is a null state, then $P^n_{kk} \to 0$. Therefore,

$$P^n_{jk} \leqq \sum_{m=1}^{n'} Q^m_{jk} P^{n-m}_{kk} + \sum_{m=n'+1}^{n} Q^m_{jk},$$

and it follows, upon letting $n \to \infty$ and then $n' \to \infty$, that $P^n_{jk} \to 0$.

If k is a positive state, then $P^{nd_k+r}_{kk} = 0$ for $r < d_k$ and $P^{nd_k}_{kk} \to d_k/\tau_{kk}$. Therefore, from

$$0 \leqq P^{nd_k+r}_{jk} - \sum_{m=1}^{n'} Q^{md_k+r}_{jk} P^{(n-m)d_k}_{kk} \leqq \sum_{m=n'+1}^{n} Q^{md_k+r}_{jk}$$

it follows, upon letting $n \to \infty$ and then $n' \to \infty$, that $P^{nd_k+r}_{jk} \to q_{jk}(r)d_k/\tau_{kk}$.

The last assertion follows from the first two assertions.

***6. Motion of the system.** To investigate the motion of the system we have to consider the probabilities of passage from one state to another. But, first, let us introduce a convenient terminology.

A state j is an *everreturn* state if, for every state k such that $q_{jk} > 0$, we have $q_{kj} > 0$. Two states j and k are *equivalent* and we write $j \sim k$ if $q_{jk} > 0$ and $q_{kj} > 0$; they are *similar* if they have the same period and are of the same type. A class of similar states will be qualified according to the common type of its states.

A class of states is *indecomposable* if any two of its states are equivalent, and it is *closed* if the probability of staying within the class is one. For example, the class of all states is closed but not necessarily indecomposable.

The motion of the system is described by the foregoing asymptotic behavior of the probabilities of passage from a given state to another given state, and also by the following theorem.

Decomposition theorem. *The class of all return states splits into equivalence classes which are indecomposable classes of similar states.*

A not everreturn equivalence class is not closed. An everreturn equivalence class is closed; if its period $d > 1$, then it splits into d cyclic subclasses $C(1), C(2), \cdots, C(d)$ such that the system passes from a state in $C(r)$ to a state in $C(r+1)$ $(C(d+1) = C(1))$ with probability 1.

The proof is simple but somewhat long. To begin with, we observe that, if j and k are two equivalent states, distinct or not, then there exist two integers, say m and p, such that $P_{jk}^m > 0$, $P_{kj}^p > 0$.

1° The set of all states which are equivalent to some state coincides with the set of all return states. For, on the one hand, every return state is equivalent to itself and, on the other hand, if $j \sim k$, then $q_{jj} \geqq P_{jj}^{m+p} \geqq P_{jk}^m P_{kj}^p > 0$. Thus, the relation $j \sim k$, symmetric by definition, is reflexive: $j \sim j$. It is also transitive, for $j \sim k$ implies $P_{jk}^m > 0$, $k \sim h$ implies $P_{kh}^n > 0$ for some integer n and, hence, $q_{jh} \geqq P_{jh}^{m+n} \geqq P_{jk}^m P_{kh}^n > 0$; similarly for q_{hj}. Therefore, the relation $j \sim k$ has the usual properties of an equivalence relation and the set of all return states splits into indecomposable equivalence classes.

We prove now that, if $j \sim k$, then they are similar. We know already that they are both return states; let d_j and d_k be their respective periods. There exists an integer n such that $P_{kk}^n > 0$; hence $P_{kk}^{2n} \geqq P_{kk}^n P_{kk}^n > 0$ and $P_{jj}^{m+n+p} \geqq P_{jk}^m P_{kk}^n P_{kj}^p > 0$; similarly, $P_{jj}^{m+2n+p} > 0$. Therefore, d_j, being a divisor of $m + n + p$ and of $m + 2n + p$, is a divisor of every such n and hence of d_k. By interchanging j and k, it follows that j and k have the same period.

If j is an everreturn state and $P_{kh}^q > 0$, then, from $P_{jh}^{m+q} \geqq P_{jk}^m P_{kh}^q > 0$, it follows that there exists an integer r such that $P_{hj}^r > 0$; hence $P_{hk}^{r+p} \geqq P_{hj}^r P_{jk}^p > 0$, and k is an everreturn state. By interchanging j and k, it follows that they are both either everreturn or not everreturn states.

If k is recurrent, then, by the recurrence criterion,

$$\sum_{n=1}^{\infty} P_{jj}^n \geqq \sum_{n=1}^{\infty} P_{jj}^{m+n+p} \geqq P_{jk}^m \left(\sum_{n=1}^{\infty} P_{kk}^n\right) P_{kj}^p = \infty$$

and j is recurrent. By interchanging j and k, it follows that they are both either recurrent or nonrecurrent.

If d is the common period of the two equivalent states j and k, then, from

$$P_{jj}^{m+nd+p} \geqq P_{jk}^m P_{kk}^{nd} P_{kj}^p,$$

it follows that d is a divisor of $m + p$ and $\lim_{n \to \infty} P_{kk}^{nd} > 0$ implies $\lim_{n \to \infty} P_{jj}^{nd}$

> 0. Hence, upon applying the positivity criterion and then interchanging j and k, both are either positive or null. This completes the proof of the first assertion.

2° If j is a return but not an everreturn state, then there exists a state h such that $q_{jh} > 0$ while $q_{hj} = 0$, so that h is not equivalent to j and there is a positive probability of leaving the equivalence class of j.

If j is an everreturn state, then $q_{jk} > 0$ entails $q_{kj} > 0$ so that k belongs to the equivalence class of j. Therefore, the probability of passage from j to a state which does not belong to the equivalence class of j is zero and, the class of all states being countable, the probability of leaving this class is zero.

Finally, we split an everreturn equivalence class C of period $d > 1$ as follows: Let j and k belong to C. Since $P_{jj}^{m+p} \geqq P_{jk}^m P_{kj}^p > 0$, d is a divisor of $m + p$ and, if m_1 and m_2 are two values of m, then $m_1 = m_2$ (modulo d). Thus, fixing j, to every k belonging to C there corresponds a unique integer $r = 1$ or $2, \cdots,$ or d such that, if $P_{jk}^m > 0$, then $m = r$ (modulo d). The states belonging to C with the same value of r form a subclass $C(r)$ and C splits into subclasses $C(1), C(2), \cdots C(d)$. It follows that, if k and k' belong respectively to $C(r)$ and $C(r')$, then $P_{kk'}^n$ can be positive only for $n = |r - r'|$ (modulo d). Moreover, according to the proposition which follows the positivity criterion, $P_{kk'}^n > 0$ for all such n sufficiently large. Thus no subclass $C(r)$ is empty and the system moves cyclically from $C(r)$ to $C(r + 1) \cdots$ with $C(d + 1) = C(1)$. This proves the second assertion.

Corollary 1. *The states of an everreturn equivalence class C are linked in a constant chain whose transition matrix is obtained from the initial transition matrix P by deleting all those P_{jk} for which j or k or both do not belong to C.*

Corollary 2. *The states of a cyclic subclass $C(r)$ of an everreturn equivalence class with period d are linked in a constant chain whose transition matrix P' is obtained from P^d by deleting all those P_{jk}^d for which j or k or both do not belong to $C(r)$.*

Corollary 3. *An everreturn null equivalence class C is either empty or infinite. In particular, a finite chain has no everreturn null states.*

Let C be finite nonempty. By the asymptotic passage theorem, $P_{jk}^n \to 0$ for $k \in C$. But C is closed, so that $1 = \sum_{k \in C} P_{jk}^n \to 0$ for $j \in C$, and we reach a contradiction.

Corollary 4. *If j and k are nonequivalent everreturn states, then* $P^n_{jk} = 0$.

If j and k are equivalent positive states, with period d, then

$$P^{nd+r}_{jk} \to d/\tau_{kk} \quad \textit{for some} \quad r = r(j, k)$$

$$P^{nd+r'}_{jk} = 0 \quad \textit{for} \quad r' \neq r \ (\textit{modulo } d).$$

This follows by the asymptotic passage theorem.

***7. Stationary chains.** The evolution of a system is determined by the laws which govern the system. In the case of constant elementary chains these laws are represented by the transition matrix P with elements P_{jk}. While P determines probabilities of *passage* from one state to another, it does not determine the probability that at a given time the system *be* in a given state. To obtain such probabilities we have to know the initial conditions. In the deterministic case this is the state at time 0. In our case it is the probability distribution at time 0, that is, the set of probabilities P_j for the system to be in the state j at time 0. Then, according to the total probability rule, the probability P^n_k that the system be in the state k at time $n = 1, 2, \cdots$, is

$$P^n_k = \sum_j P_j P^n_{jk}.$$

The notion of statistical equilibrium corresponds to the concept of stationarity in time. In our case of a constant elementary chain with transition matrix P, it is *stationary* if $P^n_k = P_k$ for every state k and every $n = 1, 2, \cdots$.

Given the laws of evolution represented by a transition matrix, the problem arises whether or not there exist initial conditions represented by the initial probability distribution such that the chain is stationary; in other words, whether or not there exists a probability distribution $\{\bar{P}_j\}$ which remains *invariant* under transitions. In general, one expects that if, under given laws of evolution, an equilibrium is possible, then it is attained in the long run. To this somewhat vague idea corresponds the following

Invariance theorem. *For states j belonging to a cyclic subclass of a positive equivalence class with period d, the set of values $\bar{P}_j = \dfrac{d}{\tau_{jj}}$ is an invariant and the only invariant distribution under the transition matrix of the subclass.*

According to Corollary 2 of the decomposition theorem, it suffices to consider the chain formed by the subclass, that is, by one cyclic positive class with some transition matrix P of period one. According to the asymptotic passage theorem,

$$P_{jk}^n \to \frac{1}{\tau_{kk}} = \bar{P}_k > 0.$$

Since

$$\sum_k P_{jk}^n = 1 \quad \text{and} \quad P_{jk}^{n+m} = \sum_h P_{jh}^n P_{hk}^m,$$

it follows, upon taking arbitrary but finite sets of states and letting $n \to \infty$, that

$$\sum_k \bar{P}_k \leqq 1, \quad \bar{P}_k \geqq \sum_h \bar{P}_h P_{hk}^m.$$

But if, for some k, the second inequality is strict, then summing over all states k, we obtain

$$1 \geqq \sum_k \bar{P}_k > \sum_h \bar{P}_h$$

so that, *ab contrario*,

$$\bar{P}_k = \sum_h \bar{P}_h P_{hk}^m.$$

Since $\sum_h \bar{P}_h$ is finite, we can pass to the limit under the summation sign, so that, by letting $m \to \infty$, we obtain

$$\bar{P}_k = (\sum_h \bar{P}_h)\bar{P}_k$$

and, $\bar{P}_k$ being positive, it follows that $\sum_h \bar{P}_h = 1$. Thus, the set of values $\bar{P}_k$ is a probability distribution invariant under P.

It remains for us to prove that, if a set of values P_k has the same properties, then $P_k = \bar{P}_k$. But from

$$P_k = \sum_h P_h P_{hk}^m$$

it follows, as before, that $P_k = (\sum_h P_h)\bar{P}_k = \bar{P}_k$, and the conclusion is reached.

Corollary. *If C is a positive equivalence class, then*

$$\sum_{j \in C} \frac{1}{\tau_{jj}} = 1.$$

This follows from

$$d \sum_{j \in C} \frac{1}{\tau_{jj}} = \sum_{r=1}^{d} \sum_{j \in C(r)} \frac{d}{\tau_{jj}} = d.$$

STATIONARITY THEOREM. *A constant elementary chain with transition matrix P is stationary with initial probability distribution $\{\bar{P}_k\}$ if, and only if, $\bar{P}_k = 0$ for all null states k, and $\bar{P}_k = \dfrac{p_t}{\tau_{kk}}$ for all states belonging to positive equivalence classes C_t, with $\sum_t p_t = 1$.*

Let the probability distribution $\{\bar{P}_k\}$ be invariant under the transition matrix P so that

$$\bar{P}_k = \sum_j \bar{P}_j P_{jk}^n.$$

If k is a null state, then, by the asymptotic passage theorem, $P_{jk}^n \to 0$. $\sum_j \bar{P}_j$ being finite, we can pass to the limit under the summation sign. It follows, upon letting $n \to \infty$, that $\bar{P}_k = 0$. Hence, by summing over positive states only, $\sum' \bar{P}_k = 1$.

If k belongs to a positive equivalence class C_t, then, by the asymptotic passage theorem, we have that $P_{jk}^n = 0$ for every j which does not belong to C_t and $\dfrac{1}{n}\sum_{m=1}^{n} P_{jk}^m \to \dfrac{1}{\tau_{kk}}$ for every j belonging to C_t. It follows that

$$\bar{P}_k = \sum_{j \in C_t} \bar{P}_j P_{jk}^n = \sum_{j \in C_t} \bar{P}_j \left(\frac{1}{n}\sum_{m=1}^{n} P_{jk}^m\right) \to \frac{p_t}{\tau_{kk}}$$

where

$$p_t = \sum_{j \in C_t} \bar{P}_j \quad \text{and} \quad \sum_t p_t = \sum{}' \bar{P}_k = 1.$$

This proves the "only if" assertion.

Conversely, let the conditions on the $\bar{P}_k$ hold and use

$$P_k^m = \sum_j{}' \bar{P}_j P_{jk}^m$$

where the summation is over positive states j only, since $\bar{P}_j = 0$ for j null.

Therefore, if k is null, then $P_{jk}^m = 0$ and $P_k^m = 0$ for every m. If k belongs to a positive equivalence class C_t, then, since C_t is closed, $P_{jk}^m = 0$ for all states j which do not belong to C_t, and, C'_t being a finite subclass of C_t such that $\sum \bar{P}_j < \epsilon$ with sum over $j \in C_t - C'_t$, we have

$$P_k^m = \sum_{j \in C_t} \bar{P}_j P_{jk}^m \leqq p_t \sum_{j \in C'_t} P_{jk}^m / \tau_{jj} + \epsilon.$$

Upon replacing $\dfrac{1}{\tau_{jj}}$ by the limit of the mean in the asymptotic passage theorem with subscripts $h, j \in C'_t$, we obtain, by summing first over the j,

$$P_k^m \leqq p_t \lim_{n\to\infty} \frac{1}{n} \sum_{s=1}^{n} P_{hk}^{m+s} + \epsilon.$$

Hence

$$P_k^m \leqq \frac{p_t}{\tau_{kk}} + \epsilon,$$

so that, letting $\epsilon \to 0$, we have $P_k^m \leqq \dfrac{p_t}{\tau_{kk}}$.

If, for some k, the inequality is a strict one, then, since for null states $P_k^m = 0$, it follows, by summing over positive states k only, that

$$1 = \sum' P_k^m < \sum_t p_t \sum_{k \in C_t} \frac{1}{\tau_{kk}} = \sum_t p_t = 1.$$

Therefore, $P_k^m = \dfrac{p_t}{\tau_{kk}}$ for every m, and the "if" assertion is proved.

COMPLEMENTS AND DETAILS

I. Physical statistics. The problem is to determine the state of equilibrium of a physical system, of energy E, composed of a very large number N of "particles" of the same nature: electrons, protons, photons, mesons, neutrons, etc.

Hypotheses. There are g_1 microscopic states of energy e_1, g_2 of energy e_2, $\cdots$ and each particle is in one of these states. The macroscopic state, i.e., the state of the system, is specified by the number of particles at each energy level: ν_1 particles of energy e_1, ν_2 particles of energy e_2, $\cdots$. The set $\{\nu_1, \nu_2, \cdots\}$ is a set of random integers and the probability of a macroscopic state $\nu_1 = n_1$, $\nu_2 = n_2$, $\cdots$ is equal, up to a constant factor, to the number W of ways in which n_k particles can be distributed amongst g_k microscopic states of energy e_k, $k = 1, 2, \cdots$, provided

$$\sum_k n_k = N, \quad \sum_k n_k e_k = E.$$

The Maxwell-Boltzmann statistics (classical theory of gases) is that of distinguishable particles without exclusion, i.e., without any bound upon the possible number of particles in any of the microscopic states. The Bose-Einstein statistics (photons, mesons, deuterons, $\cdots$—particles with an integer "spin") is that of nondistinguishable particles without exclusion. The Fermi-Dirac statistics (electrons, protons, neutrons—particles with a semi-integer "spin") is that of nondistinguishable particles which obey the Pauli exclusion principle, that is, there cannot be more than one particle in any of the microscopic states.

Weights. Let w denote the weight of the macroscopic state $\{n_1, n_2, \cdots\}$ i.e., $W/N!$ in the distinguishable case and W in the nondistinguishable case. Prove that the combinatorial formulae give the following expressions for w, where it is assumed that $\sum_k n_k = N$, $\sum_k n_k e_k = E$ (in the case of photons N is not fixed and only the second condition remains):

	Distinguishable Particles	Nondistinguishable Particles
Without exclusion	$w = \prod g_i^{n_i}/\prod n_i!$ (Maxwell-Boltzmann)	$w = \prod \frac{(g_i + n_i - 1)!}{n_i!(g_i - 1)!}$ (Bose-Einstein)
With exclusion	$w = \prod \frac{g_i!}{n_i!(g_i - n_i)!}$ (corresponds to no physical reality)	$w = \prod \frac{g_i!}{n_i!(g_i - n_i)!}$ (Fermi-Dirac)

When $g_k \gg n_k$, then the expressions of the weights in B.-E. and F.-D. statistics are equivalent to w in M.-B. statistics. Assume distinguishability and let c be the "capacity" coefficient of the microscopic states, that is, if there are already n particles in the g_k states of energy $e_k (k = 1, 2, \cdots)$, the number of these g_k states which remains available for the $(n + 1)$th particle is $g_k - nc$—this is *Brillouin statistics.* The weights w of the macroscopic states, previously defined as $w = W/N!$, are given by

$$w = \prod_k \frac{1}{n_k!} g_k(g_k - c) \cdots [g_k - (n_k - 1)c]$$

and reduce to those of M.-B., B.-E., and F.-D. by giving to the parameter c the values 0, -1, $+1$ respectively.

Statistical equilibrium. For a very large N the equilibrium state of the macroscopic system is postulated to be the most probable one, that is, the one with the highest weight. Assume that Stirling's formula can be used for the factorials which figure in the table of weights above. Take the variation $\delta \log w$ which corresponds to the variation $\{\delta n_1, \delta n_2, \cdots\}$. Using the Lagrange multipliers method, the state which corresponds to the maximum of w is determined by solving the system (prove)

$$\delta \log w + \lambda \cdot \delta N + \mu \cdot \delta E = 0$$

$$\sum_k n_k = N, \quad \sum_k n_k e_k = E.$$

(In the case of photons take $\lambda = 0$ and suppress the second relation.) The equilibrium states for the various statistics are also obtained by replacing c by

0, −1, and 1 in the equilibrium state for Brillouin statistics, given by

$$n_k = g_k/(e^{\lambda+\mu e_k} + c)$$

where λ and μ are determined by the subsidiary conditions

$$N = \sum_k g_k/(e^{\lambda+\mu e_k} + c), \quad E = \sum_k g_k e_k/(e^{\lambda+\mu e_k} + c).$$

The Planck-Bose-Allard method. The macroscopic states can be described in a more precise manner. Instead of asking for the number n_k of particles in the states of energy e_k, we ask for the number g_{km} of states of energy e_k occupied by m particles. The particles are assumed to be nondistinguishable as required by modern physics. The combinatorial formulae give

$$w = \prod_k (g_k!/\prod_m g_{km}!) \text{ with } g_k = \sum_m g_{km}, \quad N = \sum_k \sum_m m g_{km}, \quad E = \sum_k \sum_m e_k m g_{km}.$$

To obtain the statistical equilibrium state use the procedure described above.

B.-E. statistics is obtained if no bounds are imposed upon the values of m. F.-D. statistics is obtained if m can take only the values 0 or 1; "intermediate" statistics is obtained if m can take only the values belonging to a fixed set of integers.

In the equilibrium state (with $c = -1$ or $+1$ when the statistics are B.-E.'s or F.-D.'s respectively), we have

$$g_{km} = g_k(1 + ca_k)^{-\frac{1}{c}} a_k^{m}, \text{ where } a_k = e^{-(\lambda+\mu e_k)}$$

and $g_k(u)$, determined by the usual subsidiary conditions, the generating function of the number of particles in a microscopic state of energy e_k, is

$$g_k(u) = (1 + ca_k)^{-g_k/c} \cdot (1 + ca_k u)^{g_k/c}.$$

II. The method of indicators.

1. Rule: In order to compute PB, $B = f(A_1, A_1^c, \cdots, A_m, A_m^c)$, take the following steps:

(a) Reduce the operations on events to complementations, intersections, and sums;

(b) Replace each event by its indicator, expand, and take the expectation.

In this way find

$$P(\bigcup_{j=1}^{m} A_j) \text{ and } P(\bigcap_{j=1}^{k} A_j \bigcup_{j=k+1}^{m} A_j^c) \text{ in terms of } P(\bigcap_{j=1}^{r} A_j)\text{'s}.$$

Notations. Let $I_{A_j} = I_j$ and let $R = \sum_{j=1}^{m} I_j$ be the "repetition" of A_j's, that is, the number of events A_j which occur. Let $J_o = 1$, $J_r = \sum I_{j_1} \cdots I_{j_r}$ where the summation is over all combinations $1 \leqq j_1 < j_2 < \cdots < j_r \leqq m$. Let $I_{[r]}$ and $I_{(r)}$ be indicators of the events exactly r A's occur and at least r A's occur, respectively; set

$$S_r = EJ_r, \quad P_{[r]} = EI_{[r]}, \quad P_{(r)} = EI_{(r)}.$$

2. Prove

(a) $$\sum_{r=0}^{m} u^r I_{[r]} = \sum_{s=0}^{m} (u-1)^s J_s = u^R,$$

and deduce

(b) $$P_{[r]} = \sum_{k=r}^{m} (-1)^{k-r} C_k^r S_k,$$

(c) $$S_k = \sum_{t=k}^{m} C_t^k P_{[t]} = \sum_{t=k}^{m} C_{t-1}^{k-1} P_{(t)},$$

(d) $$R(R-1)\cdots(R-k+1) = k!J_k.$$

3. Let $k \leqq r \leqq m$. Using *2*(c) and the relations

$$I_{(m)} \leqq \cdots \leqq I_{(r)} \leqq I_{(r-1)} \leqq \cdots \leqq I_{(k)},$$

prove that

$$(S_k - C_{r-1}^k)/(C_m^k - C_{r-1}^k) \leqq P_{(r)} \leqq S_k/C_r^k.$$

Examine the special case $r = m$; the left-hand side becomes Gumbel's inequality; the right-hand side becomes Fréchet's inequality.

Let

$$J(k) = 1 - J_k/C_m^k, \quad \Delta f(k) = f(k+1) - f(k).$$

4. Prove

(a) $$\Delta J(k) = \sum_{t=k}^{m-1} \frac{C_{m-k-1}^{t-k}}{C_m^t} I_{[t]},$$

(b) $$I_{[r]} \leqq \frac{C_m^r}{C_{m-k-1}^{r-k}} \Delta J(k), \quad k \leqq r \leqq m-1,$$

(c) $$\Delta J(k) \geqq 0;$$

deduce a scale of inequalities for the S_k's.

5.

(a) $$-m\Delta\left(\frac{J(k)}{k}\right) = \sum_{m=1}^{t=k} \frac{C_{m-k-1}^{t-k}}{C_{m-1}^{t-1}} (1 - I_{(t)}),$$

(b) $$1 - I_{(t)} \leqq \frac{C_{m-1}^{k-1}}{C_{m-k-1}^{r-k}} \left(-m\Delta \frac{J(k)}{k}\right),$$

(c) $$-m\Delta \frac{J(k)}{k} \geqq 0;$$

deduce another scale of inequalities for S_k's.

6. The general symbolic method. The events $B_1, \cdots, B_m$ are called exchangeable if $P(B_{i_1} \cdots B_{i_r} B_{i_{r+1}}^c \cdots B^c_{i_{r+s}})$ depends only on the number r of events B_i and on the number s of events B_i^c.

Let

$$J_{r/s} = \sum I(A_{i_1}) \cdots I(A_{i_r}) I(A_{i_{r+1}}^c) \cdots I(A_{i_{r+s}}^c)$$

$$S_{r/s} = E(J_{r/s}) = \sum P(A_{i_1} \cdots A_{i_r} A_{i_{r+1}}^c \cdots A_{i_{r+s}}^c)$$

$$p_{r/s} = P(B_{i_1} \cdots B_{i_r} B_{i_{r+1}}^c \cdots B_{i_{r+s}}^c).$$

If we choose the B_i's such that

$$\sum_i P(A_{i_1} \cdots A_{i_r} A_{i_{r+1}}{}^c \cdots A_{i_{r+s}}{}^c) = \sum_i P(B_{i_1} \cdots B_{i_r} B_{i_{r+1}}{}^c \cdots B_{i_{r+s}}{}^c),$$

then

$$p_{r/s} = S_{r/s}/C_m^r C_{m-r}^s.$$

If we further introduce symbolic independent events having the same probability p, the complementary events having probability $q = 1 - p$, then, symbolically,

$$p_{r/s} = p^r q^s.$$

The symbolic method consists of the following steps:

(a) In any given identity (or identical inequality) for p, q $(0 \leqq p \leqq 1$ $q = 1 - p)$ replace $p^r q^s$ by $p_{r/s}$.

(b) Replace $p_{r/s}$ by $S_{r/s}/C_m^r C_{m-r}^s$ and obtain an equality (or inequality resp.) for the $S_{r/s}$'s.

Examples:

(a) Starting from $p^r q^s = p^r(1 - p)^s$ obtain

$$\frac{S_{r/s}}{C_m^{r+s} C_{r+s}^r} = \sum_{i=0}^{p} (-1)^i \frac{C_s^i}{C_m^{r+i}} S_{r+i/0};$$

in the special case $r + s = m$, find

$$S_{r/m-r} = P_{[r]}.$$

(b) Starting from $p^r q^s = p^r q^s (p + q)^{m-r-s}$, obtain

$$S_{r/s} = \sum_{i=r}^{m-s} C_i^r C_{m-i}^s S_{i/m-i}.$$

In the special case $s = 0$, find

$$S_{r/0} = S_r = \cdots.$$

(c) Starting from $p^{r'} q^{s'} \geqq p^r q^s$, $r' \leqq r$, $s' \leqq s$, find

$$\frac{S_{r'/s'}}{C_m^{r'+s'} C_{r'+s'}^{r'}} \geqq \frac{S_{r/s}}{C_m^r C_{r+s}^r}, \quad r' \leqq r, \quad s' \leqq s,$$

and as a special case the scale of inequalities (4c).

(d) Starting from $1 \geqq \sum_{i=0}^{r}{}' C_i^n p^{r-i} q^i$ where $\sum'$ denotes a sum in which a certain number of terms is omitted, find

$$C_r^m \geqq \sum{}' S_{r-1/i}$$

and, taking only the terms $i = 0$ and $i = 1$, find the second scale of inequalities (5c).

7. *The classical problem of matching.* This problem (problème des rencontres) was studied first by Montmart (1708) and further treated by Lambert, Euler,

and others in different forms, all of which can be described by the following setup: given m distinct numbers $X_1, X_2, \cdots, X_m$, choose at random a first X_{i_1}, then a second X_{i_2} from the remaining ones, etc. A match (coincidence, rencontre) is an event A_i which consists in choosing exactly X_i at the ith draw.

In the following, assume that each permutation $(X_{i_1} \cdots X_{i_m})$ has the same probability of being chosen at random. Show that

(a) $$P(A_{i_1}, \cdots, A_{i_r}) = \frac{(m-r)!}{m!} \quad \text{and} \quad S_r = \frac{1}{r!},$$

(b) $$P_{[r]} = \frac{1}{r!} \sum_{s=0}^{m-r} \frac{(-1)^s}{s!},$$

(c) Find $\lim_{m \to \infty} P_{[r]}$; interpretation? Show that $P_{[m-1]} = 0$; interpretation?

(d) Show that $E(r) = \sum_{r=0}^{m} rP_{[r]} = S_1 = 1$ and $E[r - E(r)]^2 = 1$. (Use the generating function $\sum_{r=0}^{m} u^r P_{[r]}$.)

III. Random walk. A particle starting at some point of an m-dimensional space moves in such a way that its consecutive displacements can be represented by independent m-dimensional random vectors. Problems of the following type arise: find the probability that in time T or before time T the particle reaches a certain domain D, or that it reaches D without having reached previously a domain D', or find the expected time for the particle to reach D, etc.$\cdots$

We give a few examples which show the great variety of forms under which this problem occurs, questions which can be asked, and methods of solution. We restrict ourselves to the discontinuous case with every move taking one unit of time.

1. Game of "heads or tails" and combinatorial method. To n tosses of a coin with equal probabilities for heads and for tails we associate the score point whose coordinates are respectively the number of heads and the number of tails which occur. Thus, at every toss, the score point M moves by one unit either upwards or to the right, and the game is represented by a two-dimensional one-sided random walk on the lattice of points with integer coordinates.

The score points corresponding to the same number n of tosses lie on the line $x + y = n$. The total number of paths between 0 and $M = (a, b)$ is $\dfrac{(a+b)!}{a!b!}$.

(a) If A and B ran for office, A got a votes and B got $b < a$ votes, find the pr. P that in counting the votes A be always ahead of B.

(Equivalent to the pr. that the score point stays below the bisectrix until it reaches the point $M = (a, b)$. Compute the pr. of the complementary event by applying the symmetry principle of Désiré André as follows: the paths from 0 to M which intersect the bisectrix either go through (1, 0) or through (0, 1). By reason of symmetry both classes contain the same number of paths. The number of those which go from (0, 1) to M is $(a + b - 1)!/a!(b - 1)!$, and $P = \dfrac{a-b}{a+b}$.)

(b) The probability that there be neither gain nor loss in exactly $2n$ tosses is $\frac{1\cdot 3 \cdots (2n-3)}{2\cdot 4\cdot 6 \cdots 2n} \sim \frac{1}{2n\sqrt{\pi n}}$. (Start with the number of paths from 0 to $(n, n-1)$ which do not intersect the bisectrix.)

(c) The probability that the gambler who bets on heads and whose fortune is m times the stake loses his fortune in $m + 2n$ tosses is $m(m+n+1) \cdots (m+2n-1)/2^{m+2n}n!$. (Reduce to (a) by taking for origin the point $(n, m+n)$.)

2. *Gambler's ruin.*

(a) *Method of difference equations.* Consider a one-dimensional random walk on the lattice $x = 0, \pm 1, \pm 2, \cdots$. At each step the particle at y has probability p_k to move from y to $y + k$, $k = 0, \pm 1, \pm 2, \cdots$. Let P_x be the probability of ruin, that is, starting at x with $0 < x < a$ to arrive at $y \leqq 0$ before reaching $y \geqq a$. Then $P_x = \sum_y P_y p_{x-y}$ with boundary conditions $P_y = 1$ if $y \leqq 0$ and $P_y = 0$ if $y \geqq a$.

The gambler has x dollars and wins or loses one dollar with respective probabilities p and $q = 1 - p$. Find the probability P_x of his ruin. Find the probability P_{xn} of his ruin at the nth game.

(In the first case, $P_x = pP_{x+1} + qP_{x-1}$ with $P_o = 1$, $P_a = 0$. The solution is $P_x = \frac{(q/p)^a - (q/p)^x}{(q/p)^a - 1}$ for $p \neq q$ and $P_x = 1 - \frac{x}{a}$ for $p = q$.

In the second case $P_{x,n+1} = pP_{x+1,n} + qP_{x-1,n}$ with $P_{on} = P_{an} = 0$ and $P_{oo} = 1$, $P_{xo} = 0$. The solution is

$$P_{xn} = a^{-1}2^n p^{(n-x)/2} q^{(n+x)/2} \sum_{k=1}^{a-1} \cos^{n-1}\frac{\pi k}{a} \sin\frac{\pi k}{a} \sin\frac{\pi k x}{a}.$$

(b) *Method of matrices.* Same random walk but with $p_1 = p_{-1} = 1/2$. The particle starts from 0 and dies when it attains $a - 1 \leqq 0$ or $b = a + c \geqq -1$. Find the probability P_n that after n displacements the particle is still alive, as follows.

Set $g(k) = 1/2$ for $k = \pm 1$ and $g(k) = 0$ otherwise. Then $P_n = \sum g(k_1) \cdots g(k_n)$ where the sum is taken over all k's such that $a \leqq \sum_{j=1}^{h} k_j \leqq b$, $h = 1, 2, \cdots, n$. Set $d_j = k_1 + \cdots + k_j - a$. Then P_n is the sum of the elements of the $(1-a)$-th column or row of the matrix A^n where

$$A = (g(j-h)) = \begin{bmatrix} 0 & \frac{1}{2} & 0 & 0 & \cdots \\ \frac{1}{2} & 0 & \frac{1}{2} & 0 & \cdots \\ 0 & \frac{1}{2} & 0 & \frac{1}{2} & \cdots \\ \cdot & \cdot & \cdot & \cdot & \cdots \end{bmatrix}$$

The proper values λ_j of A are given by $\lambda_j = \cos\frac{\pi j}{c+2}$, the proper values of A^n are λ_j^n, and

$$P_n = \frac{2}{c+2} \sum_{j=1}^{c+1}{}' \cos\frac{\pi j}{c+2} \sin\frac{\pi j(1-a)}{c+2} \cot\frac{\pi j}{c+2},$$

where $\sum'$ denotes summation over the odd j's only.

IV. Geometric probabilities.

Elementary probabilities. Consider an n-dimensional space of points $U = (u_1, \cdots, u_n)$ and let G be a group of transformations of points into points. If there exists a differential element $d\mu = g(u_1, \cdots, u_n)du_1 \cdots du_n$ determined up to a constant factor by the property that its integral over domains is invariant with respect to the group G, $d\mu$ defines up to a constant factor an elementary probability. The constant factor is determined by fixing a domain D_0 within which all considered domains lie and by assigning to this domain the pr. one; that is, by setting $c\int_{D_0} d\mu = 1$. Then the points are said to be taken or thrown at random in D_0. To say that several points are taken or thrown at random means that the throws are stochastically independent; in other words, we make repeated trials.

Let M with or without affixes be points in an m-dimensional euclidean space and let $x_1, \cdots, x_m$ with same affixes, if any, be its cartesian coordinates with respect to a fixed orthogonal frame of reference. The group G which transforms points M into points M is the group of euclidean displacements (preserves euclidean lengths). This means that the probability is required to be independent of the choice of the frame of reference. Prove that $d\mu = c\, dx_1\, dx_2 \cdots dx_m$.

Let us now investigate straight lines in a euclidean plane determined by their equations $u_1x_1 + u_2x_2 = 1$ in rectangular coordinates, and let G_e be the group of euclidean displacements in the plane. Prove that $d\mu = c(u_1^2 + u_2^2)^{-3/2}\, du_1\, du_2$ or, using the normal equations: $x_1 \cos\theta + x_2 \sin\theta - p = 0$, $d\mu = cdp\, d\theta$.

(The transformations of the group G_e are of the form $x'_1 = a_1 + x_1 \cos\alpha - x_2 \sin\alpha$, $x'_2 = a_2 + x_1 \sin\alpha + x_2 \cos\alpha$ and induce transformations of a group G on the plane (u_1, u_2) defined by

$$u_1 = (u'_1 \cos\alpha + u'_2 \sin\alpha)/(a_1u'_1 + a_2u'_2 + 1),$$

$$u_2 = (-u'_1 \sin\alpha + u'_2 \cos\alpha)/(a_1u'_1 + a_2u'_2 + 1).$$

The invariance condition yields

$$g(u'_1, u'_2) = g(u_1, u_2)\frac{D(u_1, u_2)}{D(u'_1, u'_2)} \quad \text{with} \quad \frac{D(u_1, u_2)}{D(u'_1, u'_2)} = \frac{(u_1^2 + u_2^2)^{3/2}}{(u'^2_1 + u'^2_2)^{3/2}}.$$

With the same group G_e there is no elementary probability for circles in the plane. But there is one for circles of fixed radius.)

Points on a line. The elementary probability for a point M on a segment $[0, l]$ is dx/l. Throw n points at random on the segment. The probability, say, that there be no thrown points on $[0, x]$ is $\left(1 - \frac{x}{l}\right)^n$. What is the expected distance of the nearest to 0 of the thrown points? What is the probability that k out of the n thrown points lie on a fixed subinterval of length a? Find what happens as $l \to \infty$ with $n/l \to \lambda > 0$. Denote then by $M_1, M_2, \cdots$ the points in the nondecreasing order of their distance to 0. What is the elementary probability for the length M_jM_{j+1} to be between x and $x + dx$ and what is the expectation of this length?

Lines in a plane. The elementary probability of a straight line $x \cos\theta + y \sin\theta - p = 0$ thrown on a plane is $d\mu = cdp\, d\theta$. The integral $\int dp\, d\theta$ over a domain induced by a family of straight lines is said to be the measure of the family. The measure of the secants of a segment of length l is $2l$. (θ varies from $-\frac{\pi}{2}$ to $+\frac{\pi}{2}$ while p varies from 0 to $l \cos \theta$ for every fixed θ.) The measure of the secants of a polygonal line of length l is $2l$, provided every secant is counted as many times as there are points of intersections of the secant with the polygonal line; in particular, the measure of the secants of a closed convex polygon is its perimeter. The same is true for the secants of a curve formed by a finite number of analytic arcs. Prove it directly for the secants of a circle.

Let C and C_0 be two closed convex curves of respective lengths l and l_0 with C being interior to C_0. The probability that a secant of C_0 be secant of C is l/l_0.

Application to the needle problem. If C_0 is a circumference of radius r and C is a segment of length l, then $p = 2l/\pi r$. Throw the figure formed by the circumference and the segment on a plane with parallel equidistant straight lines with common distance r. The probability that one of these lines intersects the segment is $2l/\pi r$. Prove it directly by throwing a needle of length l on this plane.

(The position of the needle AB is determined by the coordinates x, y of A and the angle α that AB makes with Ox, one of the equidistant lines. The elementary probability is $dx\, dy\, d\alpha$. It is not a restriction to assume θ between 0 and $\pi/2$, $x = 0$, and y between 0 and r. Then $p = \dfrac{2l}{\pi r} \displaystyle\int_0^{\pi/2} \sin \alpha \, d\alpha$.)

A differential method. Let D_0 be a domain of the plane on which are thrown at random n points. Intrinsic properties of the figure formed by the points are defined independently of D_0; for example, $M_1M_2 < l$, triangle $M_1M_2M_3$ has acute angles, $\cdots$.

The probability of an intrinsic property is given by $P = a/s^n$ where s is the area of D_0 and a represents the measure of the set of favorable cases. Let D'_0 be a new domain containing D_0 and let $P + \Delta P = (a + \Delta a)/(s + \Delta s)^n$ be the new probability of the same property. If P_k is the probability of the property when $n - k$ points are in D_0 and k points are in $D'_0 - D_0$, then

$$a + \Delta a = a + a_1 + \cdots + a_n, \quad a_k = \frac{n!}{k!(n-k)!} P_k s^{n-k} (\Delta s)^k$$

and

$$(s + \Delta s)^n \Delta P = n(P_1 - P) s^{n-1} \Delta s + \cdots + \frac{n!}{k!(n-k)!} (P_k - P) s^{n-k} (\Delta s)^k + \cdots + (P_n - P)(\Delta s)^n.$$

Keeping infinitesimals of first order, we have

$$\delta P = n(P_1 - P) \frac{\delta s}{s}$$

where n is the number of points thrown at random on D_0, P is the probability of the property, P_1 is the probability of the same property when 1 point is thrown at random on an increment of D_0 of area δs, and $n - 1$ points are thrown at random on D_0. More generally,

$$\delta m = n(m_1 - m)\frac{\delta s}{s}$$

where m is the expectation of a function of the thrown points, and the other quantities are defined similarly to what precedes. The method and the formulae apply whatever be the number of dimensions of the space.

Application. Two points M_1 and M_2 are thrown at random on a segment of length l. The probability that $M_1M_2 < x$ is $\frac{2x}{l} - \frac{x^2}{l^2}$. What happens when the segment is replaced by a circle of radius r? Find $E\overline{M_1M_2{}^p}$ in both cases.

Part One

NOTIONS OF MEASURE THEORY

No rigorous presentation of probability theory is possible without using the notions of sets, measures, measurable functions, and integrals. Their first lineaments are already apparent in elementary probability theory. These notions are introduced and investigated systematically in this part.

The presentation is self-contained, and the material will suffice for later parts. It is organized—at the cost of a few repetitions—so as to make the unstarred portions independent of the starred ones and, at the same time, to make the sections on measurable functions, convergence, and integration independent of the remainder except for 1.1 to 1.5. This permits a reorganization of the course so as to proceed from the less abstract notions toward more abstract and more involved ones. The following order is possible: 1.1 to 1.5 with 5.1 to 7.2, then 3.1, 3.2 with 8.1, suffice for practically all of the unstarred portions of Parts II, III, then IV.

Chapter I

SETS, SPACES, AND MEASURES

§ 1. SETS, CLASSES, AND FUNCTIONS

1.1 Definitions and notations. A *set* is a collection of arbitrary elements. By an abuse of language, an *empty set* is a "set with no elements."

Unless otherwise stated, all sets will be sets of elements of a fixed non empty set Ω, to be called a *space*. Elements of Ω will be called *points* and denoted by ω, with or without affixes (such as subscripts, superscripts, primes, etc.). Capitals $A, B, C, \cdots$, with or without affixes, will denote sets of points, $\{\omega\}$ will denote a set consisting of the one point ω, and $\emptyset$ will denote the empty set, that is, the set "containing no points." If ω is a point of A, we write $\omega \in A$ and, if ω is not a point of A we write $\omega \notin A$.

A set of sets is called a *class* and classes will be denoted by $\mathcal{A}, \mathcal{B}, \mathcal{C}, \cdots$, with or without affixes. The class of all the sets in Ω is called the *space of sets* in Ω and will be denoted by $S(\Omega)$. Thus a class of sets in Ω is a set in $S(\Omega)$ and all set notions and operations apply to classes considered as sets in the corresponding space of sets.

A is said to be a *subset of* B, or *included in* B, or *contained in* B, if all points of A are points of B; we then write $A \subset B$ or, equivalently, $B \supset A$. In symbols, if $\omega \in A$ implies $\omega \in B$, then $A \subset B$, and conversely. Clearly, for every set A,

$$\emptyset \subset A \subset \Omega,$$

and *the relation of inclusion is reflexive and transitive:*

$$A \subset A; \quad A \subset B \quad \text{and} \quad B \subset C \quad \text{imply} \quad A \subset C.$$

A and B are said to be *equal* if $A \subset B$ and $B \subset A$; we then write $A = B$.

Clearly, *the relation of equality is reflexive, transitive, and symmetric:*

$$A = A; \quad A = B \text{ and } B = C \text{ imply } A = C;$$

$$A = B \text{ implies } B = A.$$

1.2 Differences, unions, and intersections. The *difference* $A - B$ is the set of all points of A which do not belong to B; in symbols, if $\omega \in A$ and $\omega \notin B$, then $\omega \in A - B$, and conversely. The particular difference $\Omega - A$, that is, the set of all points which do not belong to A, is called the *complement of* A and is denoted by A^c.

The *intersection* $A \cap B$, or simply AB, is the set of all points common to A and B; in symbols, if $\omega \in A$ and $\omega \in B$, then $\omega \in AB$ and conversely. The *union* $A \cup B$ is the set of all points which belong to at least one of the sets A or B; in symbols, if $\omega \in A$ or $\omega \in B$, then $\omega \in A \cup B$ and conversely. If $AB = \emptyset$, then A and B are said to be *disjoint*, and their union is then denoted by $A + B$ and called a *sum*.

It follows from the definitions that *the operations of intersection and union are associative, commutative, and distributive:*

$$(A \cup B) \cup C = A \cup (B \cup C), \quad (AB)C = A(BC);$$

$$A \cup B = B \cup A, \quad AB = BA;$$

$$(A \cup B)C = AC \cup BC, \quad (A \cup B)(A \cup C) = A \cup BC.$$

Moreover, the operation of complementation has the following properties:

$$A \subset B \text{ implies } A^c \supset B^c;$$

$$\Omega^c = \emptyset, \quad \emptyset^c = \Omega, \quad AA^c = \emptyset, \quad A + A^c = \Omega, \quad (A^c)^c = A;$$

$$A - B = AB^c, \quad (A \cup B)^c = A^cB^c, \quad (AB)^c = A^c \cup B^c.$$

The notions of intersection and union extend at once to arbitrary classes. Let T be a set, not necessarily in Ω, and to every $t \in T$ assign a set $A_t \subset \Omega$. The class $\{A_t, t \in T\}$ of all these sets, or simply $\{A_t\}$ if there is no confusion possible, is a class assigned to the *index set* T.

The *intersection*, or *infimum*, of all sets of $\{A_t\}$ is defined to be the set of all those points which belong to every A_t, and is denoted by $\bigcap_{t \in T} A_t$ or by $\inf_{t \in T} A_t$; we drop $t \in T$ if there is no confusion possible. In symbols, if $\omega \in A_t$ for every $t \in T$, then $\omega \in \bigcap A_t$ and conversely.

The *union*, or *supremum*, of all sets of the class $\{A_t\}$ is defined to be the set of all those points which belong to at least one A_t, and is denoted by $\bigcup_{t \in T} A_t$ or by $\sup_{t \in T} A_t$; we drop $t \in T$ if there is no confusion possible. In symbols, if $\omega \in A_t$ for at least one $t \in T$, then $\omega \in \bigcup A_t$ and conversely.

If all sets of $\{A_t\}$ are pairwise disjoint, $\{A_t\}$ is said to be a *disjoint class* and the union of its sets, denoted then by $\sum A_t$, is called a *sum*. Conversely, the term "sum" and the symbols $\sum$ and $+$ when used for sets of a class will imply that the class is disjoint.

If ω does not belong to at least one A_t, then it belongs to every A_t^c, and conversely; consequently (de Morgan rule),

$$(\bigcup A_t)^c = \bigcap A_t^c, \quad (\bigcap A_t)^c = \bigcup A_t^c.$$

When $\{A_t\}$ is empty, that is, T is empty, it is natural to make the convention that $\bigcup_{t \in \emptyset} A_t = \emptyset$. Then, in order to preserve the foregoing relations, we have to make the convention that $\bigcap_{t \in \emptyset} A_t = \Omega$. Thus, by convention,

$$\bigcup_{t \in \emptyset} A_t = \emptyset, \quad \bigcap_{t \in \emptyset} A_t = \Omega.$$

It is easily seen, collecting all the relations so far obtained, that the following *duality* rule holds:

Every valid relation between sets, obtained by taking complements, unions, and intersections, is transformed into a valid relation if, the symbols "$=$" and "^{c}" remaining unchanged, the symbols $\bigcap$, $\subset$, and $\emptyset$, are interchanged with the symbols $\bigcup$, $\supset$, and Ω, respectively.

Operations performed on elements of "countable" classes will play a prominent role later in connection with the notion of measure. A set, or a class, is said to be *finite*, or *denumerable*, according as its elements can be put in a one-to-one correspondence with the set $\{1, 2, \cdots, n\}$ of the first n positive integers, for some value of n, or with the set of all positive integers $\{1, 2, \cdots$ ad infinitum$\}$. It is said to be *countable* if it is either finite or denumerable. Similarly, operations performed on elements of finite, denumerable, or countable classes will be said to be *finite, denumerable, or countable operations*, respectively.

The following immediate transformation of countable unions into countable sums will prove useful in connection with the notion of measure:

$$\bigcup A_j = A_1 + A_1^c A_2 + A_1^c A_2^c A_3 + \cdots.$$

1.3 Sequences and limits. To every value of $n = 1, 2, \cdots$, assign a set A_n; these sets A_n, whether distinct or not, are distinguished by their indices. The ordered denumerable class $A_1, A_2, \cdots$, is called *sequence* A_n. The set of all those points which belong to almost all A_n (all but any finite number) is called the *inferior limit* of A_n, and is denoted by $\liminf A_n$. Clearly

$$\liminf A_n = \bigcup_{n=1}^{\infty} \bigcap_{k=n}^{\infty} A_k.$$

The set of all those points which belong to infinitely many A_n is called the *superior limit* of A_n and is denoted by $\limsup A_n$. Since every point which belongs to almost all A_n^c belongs to a finite number of A_n only, and conversely, it follows, by duality, that

$$\limsup A_n = (\bigcup_{n=1}^{\infty} \bigcap_{k=n}^{\infty} A_k^c)^c = \bigcap_{n=1}^{\infty} \bigcup_{k=n}^{\infty} A_k.$$

Every point which belongs to almost all A_n belongs to infinitely many A_n, so that

$$\liminf A_n \subset \limsup A_n.$$

Thus, if the reverse inclusion is true, $\liminf A_n$ and $\limsup A_n$ are equal to the same set A. Then A is called the *limit* of A_n and is denoted by $\lim A_n$; the sequence A_n is said to *converge* to A and we write $A_n \to A$. Clearly, limits (inferior or superior) of sequences of sets are formed by denumerable set operations.

Monotone sequences form a basic class of convergent sequences. A sequence A_n is said to be *monotone* if it is either *nondecreasing:* $A_1 \subset A_2 \subset \cdots$, and we then write $A_n \uparrow$; or if it is *nonincreasing:* $A_1 \supset A_2 \supset \cdots$, and we then write $A_n \downarrow$. From the expressions above of inferior and superior limits, it follows at once that

every monotone sequence is convergent, and $\lim A_n = \bigcup A_n$ *or* $\bigcap A_n$ *according as* $A_n \uparrow$ *or* $A_n \downarrow$.

Moreover, if we consider this proposition as a definition of limits of monotone sequences then, since for an arbitrary sequence B_n,

$$\bigcap_{k=n}^{\infty} B_k = \inf_{k \geqq n} B_k \uparrow \quad \text{and} \quad \bigcup_{k=n}^{\infty} B_k = \sup_{k \geqq n} B_k \downarrow,$$

it follows that its inferior and superior limits can be defined by

$$\liminf B_n = \lim_n (\inf_{k \geqq n} B_k) \quad \text{and} \quad \limsup B_n = \lim_n (\sup_{k \geqq n} B_k).$$

1.4 Indicators of sets. Set operations can be replaced by equivalent but more familiar ones, in the following manner. To every set A assign a function I_A of ω, to be called the *indicator of* A, defined by

$$I_A(\omega) = 1 \quad \text{or} \quad 0 \quad \text{according as} \quad \omega \in A \quad \text{or} \quad \omega \notin A.$$

Conversely, every function of ω which can take only the values 0 and 1 is the indicator of the set for the points of which it takes the value 1. The one-to-one correspondences (denoted by $\Leftrightarrow$) and relations listed below are immediate.

$$I_A \leqq I_B \Leftrightarrow A \subset B, \quad I_A = I_B \Leftrightarrow A = B, \quad I_{AB} = 0 \Leftrightarrow AB = \emptyset,$$

$$I_\emptyset = 0, \quad I_\Omega = 1, \quad I_A + I_{A^c} = 1,$$

$$I_{\inf A_t} = \inf I_{A_t}, \quad I_{\sup A_t} = \sup I_{A_t},$$

$$I_{\bigcap A_n} = \prod I_{A_n}, \quad I_{\sum A_n} = \sum I_{A_n},$$

$$I_{\bigcup A_n} = I_{A_1} + (1 - I_{A_1})I_{A_2} + (1 - I_{A_1})(1 - I_{A_2})I_{A_3} + \cdots$$

$$I_{\liminf A_n} = \liminf I_{A_n}, \quad I_{\limsup A_n} = \limsup I_{A_n}, \quad I_{\lim A_n} = \lim I_{A_n}.$$

1.5 Fields and σ-fields. Classes of sets in Ω are sets in the space $S(\Omega)$ of all sets in Ω and thus what precedes applies to classes. However, there is a notion specific to classes—that of closure under one or more set operations. A class $\mathcal{C}$ is said to be *closed* under a set operation if the sets obtained by performing this operation on sets of $\mathcal{C}$ are sets of $\mathcal{C}$. In particular, the class $S(\Omega)$ of all sets in Ω is closed under every set operation.

In connection with the notions of measurability and of measure, two species of classes play a prominent role—fields and σ-fields. A field is a (nonempty) class closed under all finite set operations; clearly, every field contains $\emptyset$ and Ω. A *σ-field* is a (nonempty) class closed under all countable set operations; clearly every σ-field is a field. We observe that, because of the duality rule, closure under complementations and finite (countable) intersections implies closure under finite (countable) unions. Also we can interchange in this property "intersections" and "unions."

Let $\mathcal{S}$-classes be species of classes closed under set operations $\mathcal{S}$; for example, the species of fields or the species of σ-fields. We observe that $S(\Omega)$ is an $\mathcal{S}$-class, whatever be the set operations $\mathcal{S}$.

a. *Arbitrary intersections of $\mathcal{S}$-classes are $\mathcal{S}$-classes. In particular, arbitrary intersections of fields or of σ-fields are fields or σ-fields, respectively.*

For the intersection of a collection of $\mathcal{S}$-classes belongs to every one of these classes. Therefore, performing operations $\mathcal{S}$ on sets of the intersection, we obtain sets belonging to every one of these classes, that is, to the intersection.

This property gives rise to the notion of a "minimal" $\mathcal{S}$-class over a given class. An $\mathcal{S}$-class $\mathcal{C}'$ containing $\mathcal{C}$ is a *minimal class over* $\mathcal{C}$ or the $\mathcal{S}$-class *generated* by $\mathcal{C}$ if every $\mathcal{S}$-class containing $\mathcal{C}$ contains $\mathcal{C}'$.

b. *There is one, and only one, minimal $\mathcal{S}$-class over a class $\mathcal{C}$. In particular, there is one, and only one, minimal field and one, and only one, minimal σ-field over $\mathcal{C}$.*

For the intersection of all $\mathcal{S}$-classes containing $\mathcal{C}$ contains $\mathcal{C}$ and is contained in every $\mathcal{S}$-class containing $\mathcal{C}$.

A space Ω in which is selected a fixed σ-field $\mathcal{A}$ is called a *measurable space* $(\Omega, \mathcal{A})$. If there is no confusion possible, the sets of $\mathcal{A}$ are said to be *measurable.*

1.6 Monotone classes. We shall need the notion of monotone classes in connection with the problem of extending measures on a field to its minimal σ-field. A *monotone* class is a class closed under formation of limits of monotone sequences.

a. *A σ-field is a monotone field and conversely.*

The first assertion is obvious and the second follows from the fact that every countable intersection $\bigcap_n A_n$ and union $\bigcup_n A_n$ is a monotone limit of sequences $\bigcap_{k=1}^{n} A_k$ and $\bigcup_{k=1}^{n} A_k$ of finite intersections and unions. The property we shall require is as follows:

A. *The minimal monotone class $\mathcal{M}$ and the minimal σ-field $\mathcal{A}$ over the same field $\mathcal{C}$ coincide.*

Proof. On account of **a** and minimality of $\mathcal{M}$ and $\mathcal{A}$, it suffices to prove that $\mathcal{M}$ is a field; for, a monotone field $\mathcal{M}$ is a σ-field so that $\mathcal{M} \supset \mathcal{A}$, and the σ-field $\mathcal{A}$ is monotone so that $\mathcal{M} \subset \mathcal{A}$. Since $\mathcal{M} \supset \mathcal{C} \ni \Omega$ and unions are reducible to intersections (by means of complementations), it suffices to prove that, if A and B belong to $\mathcal{M}$, so do AB, A^cB, and AB^c.

For every fixed $A \in \mathcal{M}$, let $\mathcal{M}_A$ be the class of all $B \in \mathcal{M}$ with the asserted property. Every $\mathcal{M}_A$ is monotone for, if the sequence $B_n \in \mathcal{M}_A$ is monotone, then $B = \lim B_n$ belongs to $\mathcal{M}$ and so do the limits of monotone sequences

$$AB = \lim AB_n, \quad A^cB = \lim A^cB_n, \quad AB^c = \lim AB_n{}^c.$$

It follows that, for every $A \in \mathcal{C}$, the class $\mathfrak{M}_A$ coincides with $\mathfrak{M}$. For $\mathcal{C}$ being a field, every $B \in \mathcal{C}$ is $\in \mathfrak{M}_A$, so that $\mathcal{C} \subset \mathfrak{M}_A \subset \mathfrak{M}$ and, hence, $\mathfrak{M}$ being minimal over $\mathcal{C}$, $\mathfrak{M}_A = \mathfrak{M}$. In fact, $\mathfrak{M}_B = \mathfrak{M}$ for every $B \in \mathfrak{M}$. For, the conditions imposed upon pairs A, B being symmetric, $B \in \mathfrak{M}(=\mathfrak{M}_A$ for $A \in \mathcal{C})$ is equivalent to $A \in \mathfrak{M}_B$ for every $A \in \mathcal{C}$ so that $\mathcal{C} \subset \mathfrak{M}_B$ and hence as above, $\mathfrak{M}_B = \mathfrak{M}$. But this last property means that $\mathfrak{M}$ is a field, and the proof is complete.

***1.7 Product sets.** We introduce now a different type of set operation and corresponding notions, for which we shall have need later. Let A_1 and A_2 be two arbitrary sets with elements ω_1 and ω_2, respectively. By the *product set* $A_1 \times A_2$ we shall mean the set of all ordered pairs $\omega = (\omega_1, \omega_2)$ where $\omega_1 \in A_1$ and $\omega_2 \in A_2$. If $A_1, B_1, \cdots$ are sets in a space Ω_1 and $A_2, B_2, \cdots$ are sets in a space Ω_2, then $A_1 \times A_2$, $B_1 \times B_2, \cdots$ are sets in the *product space* $\Omega_1 \times \Omega_2$, called *intervals* or *rectangles* in $\Omega_1 \times \Omega_2$ and the properties below follow readily from the definition:

$$(A_1 \times A_2) \cap (B_1 \times B_2) = (A_1 \cap B_1) \times (A_2 \cap B_2)$$

$$(A_1 \times A_2) - (B_1 \times B_2) = (A_1 - B_1) \times (A_2 - B_2) + (A_1 - B_1) \times (A_2 \cap B_2) + (A_1 \cap B_1) \times (A_2 - B_2)$$

In turn, it follows at once from these relations that

a. *If $\mathcal{C}_1$ and $\mathcal{C}_2$ are fields of sets in Ω_1 and Ω_2 respectively, then the class of all finite sums of intervals $A_1 \times A_2$, where $A_1 \in \mathcal{C}_1$ and $A_2 \in \mathcal{C}_2$, is a field of sets in $\Omega_1 \times \Omega_2$.*

This field will be called the *product field* of $\mathcal{C}_1$ and $\mathcal{C}_2$.

Yet, if $\mathcal{A}_1$ and $\mathcal{A}_2$ are σ-fields of sets in Ω_1 and Ω_2, respectively, then the product field of $\mathcal{A}_1$ and $\mathcal{A}_2$ is not necessarily a σ-field. The minimal σ-field over it will be called the *product σ-field* $\mathcal{A}_1 \times \mathcal{A}_2$. If $(\Omega_1, \mathcal{A}_1)$ and $(\Omega_2, \mathcal{A}_2)$ are measurable spaces, then their *product measurable space* is, by definition, $(\Omega_1 \times \Omega_2, \mathcal{A}_1 \times \mathcal{A}_2)$.

Let $\Omega = \Omega_1 \times \Omega_2$ and $\mathcal{A} = \mathcal{A}_1 \times \mathcal{A}_2$. If $A \subset \Omega$ is measurable and $\omega_1 \in \Omega_1$ is a fixed point, then the set $A(\omega_1)$ of all points $\omega_2 \in \Omega_2$ such that $\omega = (\omega_1, \omega_2) \in A$ is called the *section of A at ω_1*; similarly for the section $A(\omega_2)$ at $\omega_2 \in \Omega_2$; by the definition, $A(\omega_1) \subset \Omega_2$ and $A(\omega_2) \subset \Omega_1$.

b. *Every section of a measurable set is measurable.*

For let $\mathcal{C}$ be the class of all measurable sets in Ω whose sections are measurable. It is easily seen that $\mathcal{C}$ is a σ-field. On the other hand, if $A = A_1 \times A_2$ is a measurable interval, that is, A_1 and A_2 are measurable, then every section of A is either empty or is A_1 or A_2, so that $A_1 \times A_2 \in \mathcal{C}$. Therefore, $\mathcal{A} = \mathcal{A}_1 \times \mathcal{A}_2$, being the minimal σ-field over all measurable intervals, is contained in $\mathcal{C}$, and the assertion is proved.

The foregoing definitions and properties extend at once to any finite number of sets and of measurable spaces. However, in the nonfinite case, some of these definitions have to be modified in order to preserve these properties.

Let $\{A_t, t \in T\}$ be an arbitrary collection of arbitrary sets A_t in arbitrary spaces Ω_t of points ω_t. The *product set* $A_T = \prod_{t \in T} A_t$ is the set of all the new elements $\omega_T = (\omega_t, t \in T)$ such that $\omega_t \in A_t$ for every $t \in T$. The product set A_T is in the *product space* $\Omega_T = \prod_{t \in T} \Omega_t$; we drop "$t \in T$" if there is no confusion possible. It follows from the foregoing definition that, for any set B, when the Ω_t are identical

$$(\bigcap A_t) \times B = \bigcap (A_t \times B), \quad (\bigcup A_t) \times B = \bigcup (A_t \times B).$$

Let $T_N = (t_1, \cdots, t_N)$ be a finite index subset and let A_{T_N} be a set in the product space Ω_{T_N}. The set $A_{T_N} \times \Omega_{T-T_N}$ is a *cylinder* in Ω_T with *base* A_{T_N}. If the base is a product set $\prod_{t \in T_N} A_t$, the cylinder becomes a *product cylinder* or an *interval* in Ω_T with *sides* $A_t, t \in T_N$. Let $\mathcal{C}_t$ be fields in Ω_t. It is easily seen that, as in the finite case,

A. *The class of all finite sums of all the intervals in Ω_T with sides $A_t \in \mathcal{C}_t$, is a field of sets in Ω_T.*

This field is the *product field* of the fields $\mathcal{C}_t$.

Let $(\Omega_t, \mathcal{A}_t)$ be measurable spaces. The minimal σ-field over the product field of the $\mathcal{A}_t$ is the *product σ-field* $\mathcal{A}_T = \prod \mathcal{A}_t$ *of measurable sets in* Ω_T, and the measurable space $(\Omega_T, \mathcal{A}_T)$ is the *product measurable space* $(\prod \Omega_t, \prod \mathcal{A}_t)$ of the measurable spaces $(\Omega_t, \mathcal{A}_t)$. It is easily seen, as in the finite case, that **b** remains valid:

B. *Sections at ω_{T_N} of measurable sets in Ω_T are measurable sets in Ω_{T-T_N}.*

***1.8 Functions and inverse functions.** Perhaps the most important notion of mathematics is that of function (or transformation, or mapping, or correspondence). We have already encountered functions defined on an index set T whose "values" are sets in Ω. In general, a *function X on* a space Ω—the *domain of X*—to a space Ω'—the *range space of X*—is defined by assigning to every point $\omega \in \Omega$ a point $\omega' \in \Omega'$ called the *value of X at ω* and denoted by $X(\omega)$. Sets and classes of sets in Ω' will be denoted by $A', B', \cdots$, and $\mathcal{A}', \mathcal{B}', \cdots$, respective It will be assumed, once and for all, that functions are *single-value* that is, to every given $\omega \in \Omega$ corresponds one, and only one, value $X(\omega)$.

The set of values of X for all $\omega \in A$ is called the *image* $X(A)$ of A (by X) and the class of images $X(A)$ for all $A \in \mathcal{C}$ is called the *image* $X(\mathcal{C})$ of $\mathcal{C}$ (by X); in particular $X(\Omega)$ is the *range* (of all values) of X. Thus, a function X on Ω to Ω' determines a function on $S(\Omega)$ to $S(\Omega')$. While this new function is of no great interest, such is not the case for the inverse function that we shall introduce now.

By $[\omega; \cdots]$ where $\cdots$ stands for expressions and/or relations involving functions on Ω, we denote the set of points $\omega \in \Omega$ for which these expressions are defined and/or these relations are valid; if there is no confusion possible we drop "ω;". Thus, $[X = \omega']$, or *inverse image* of ω', is the set of all points ω for which $X(\omega) = \omega'$; $[X \in A']$, or *inverse image of* A', is the set of all points ω for which $X(\omega) \in A'$; and $[A; X(A) \in \mathcal{C}']$, or *inverse image of* $\mathcal{C}'$, is the class of inverse images of all sets $A' \in \mathcal{C}'$. We observe that the inverse image of an ω' which does not belong to the range of X is the empty set $\emptyset$ in Ω.

The *inverse function* X^{-1} of X is defined by assigning to every A' its inverse image $[X \in A']$. In other words, X^{-1} is a function on $S(\Omega')$ to $S(\Omega)$ with values $X^{-1}(A') = [X \in A']$; if $A' = \{\omega'\}$, then we write $X^{-1}(\omega')$ for $X^{-1}(\{\omega'\}) = [X = \omega']$. Since X is single-valued, X^{-1} generates a partition of Ω into disjoint inverse images of points $\omega' \in \Omega'$. It follows readily that

$$X^{-1}(A' - B') = X^{-1}(A') - X^{-1}(B'),$$

$$X^{-1}(\bigcup A'_t) = \bigcup X^{-1}(A'_t), \quad X^{-1}(\bigcap A'_t) = \bigcap X^{-1}(A'_t), \cdots$$

Therefore,

A. Basic property of inverse functions: *Inverse functions preserve all set and class inclusions and operations.*

It follows at once that

If $\mathcal{C}'$ is closed under a set operation so is $X^{-1}(\mathcal{C}')$. In particular, the inverse image of a σ-field is a σ-field, and the inverse image of the minimal σ-field over $\mathcal{C}'$ is the minimal σ-field over $X^{-1}(\mathcal{C}')$.

Moreover,

If $\mathcal{A}$ is a σ-field so is the class of all sets whose inverse images belong to $\mathcal{A}$.

The notion of function can be "iterated" as follows. Let X be a function on Ω to Ω' and let X' be a function on Ω' to Ω''. Then, the *function of function* $X'X$ defined by $(X'X)(\omega) = X'(X(\omega))$ is a function on Ω to Ω''. Clearly, its inverse function $(X'X)^{-1}$ is a function on $S(\Omega'')$ to $S(\Omega)$ such that, for every set $A'' \subset \Omega''$,

$$(X'X)^{-1}(A'') = X^{-1}(X'^{-1}(A''))$$

or, in a condensed form,

$$(X'X)^{-1} = X^{-1}X'^{-1}.$$

***1.9 Measurable spaces and functions.** So far, we did not consider particular species of functions. There are two species which play a basic role in abstract analysis. We shall introduce them now. But first we examine, in more detail, the class of inverse images of points of the range space.

Let X be a function on Ω to Ω'. The partition of Ω formed by the inverse images $X^{-1}(\omega')$ of all points $\omega' \in \Omega'$ is said to be *induced* (or *determined*) *by* X and X is said to be *constant* $(=\omega')$ on $X^{-1}(\omega')$. Since the class of values $X^{-1}(A')$ of X^{-1} is the inverse image of the σ-field of all sets A' in Ω', it is a σ-field. If the partition induced by X is finite, or denumerable, or countable, then X is said to be *finitely*, or *denumerably*, or *countably valued*, respectively; in other words, X is, say, countably valued if the set of its values is countable. Setting $A_j = [X = \omega'_j]$, we can write every countably valued function X as a countable combination of indicators:

$$X = \sum_j \omega'_j I_{A_j}.$$

Conversely, we make the convention that every time such a "sum" is written, the sets A_j form a partition of the domain of the function X. If the ω'_j are distinct, then this partition is the one induced by the function represented by the "sum."

Now, let $\mathcal{A}$ be a fixed σ-field in Ω. Ω, together with $\mathcal{A}$, is called a *measurable space* $(\Omega, \mathcal{A})$, and the sets of $\mathcal{A}$ are then said to be *measurable* (although this terminology derives from the notion of measure, we emphasize that, nowadays, the notion of measurability is independent of that of measure). A countably valued function $X = \sum \omega'_j I_{A_j}$, where the sets A_j are measurable, is called a countably valued measurable function—for short, an *elementary function;* if X is finitely valued, then this elementary function is also called a *simple function.* Clearly

the sets of the σ-field induced by an elementary function are measurable.

We are now in a position to introduce the general notion of measurable functions. However, there are several ways for doing so, and the classes of measurable functions so defined are, in general, not the same.

One way of defining measurable functions is to extend a basic property of inverse functions of elementary functions, as follows: Let $(\Omega, \mathfrak{A})$ and $(\Omega', \mathfrak{A}')$ be two measurable spaces. The inverse images by elementary functions on Ω to Ω' of measurable sets are measurable. Extending this property, we say that a function X on Ω to Ω' is *measurable* if the inverse images by X of measurable sets ($\in \mathfrak{A}'$) are measurable ($\in \mathfrak{A}$). If, moreover, $(\Omega'', \mathfrak{A}'')$ is a measurable space and X' on Ω' to Ω'' is a measurable function, then $X'X$ is measurable, for

$$(X'X)^{-1}(\mathfrak{A}'') = (X^{-1}X'^{-1}(\mathfrak{A}'')) \subset X^{-1}(\mathfrak{A}') \subset \mathfrak{A}.$$

Thus, *with this definition, a measurable function of a measurable function is measurable.*

Another way of defining measurable functions is as follows: Let $(\Omega, \mathfrak{A})$ be a measurable space on which are defined simple (elementary) functions to a space Ω' (there are no measurable sets in Ω'). A notion of limit is introduced on Ω', and *measurable functions in the sense of this limit* are then defined to be limits of convergent sequences of simple (elementary) functions. This approach is particularly suited for the introduction of integrals of measurable functions. Later we shall see cases in which measurable sets and the notion of limit are selected in such a manner that the two definitions are equivalent.

*§ 2. TOPOLOGICAL SPACES

The selections of measurable sets and of concepts of limit in range-spaces are rooted in the properties of the euclidean line: real line $R = (-\infty, +\infty)$ with euclidean distance $|x - y|$ of points (numbers, reals) x, y. Species of spaces vary according to the preserved amount of these properties, an amount which increases as we pass from separated spaces to metric spaces, then to Banach spaces and to Hilbert spaces. We examine here the basic properties of these spaces and shall encounter them in various guises throughout the book. At the same time, the few notions of topology which follow are a recapitulation of the properties of the euclidean line and, more generally, of euclidean spaces. We urge the reader to keep this fact constantly in mind by illustrating the concepts and their relationships in terms of euclidean spaces; for this reason, we denote here the points by x, y, z, with or without affixes.

Points, sets, and classes will be those of the space $\mathfrak{X}$ under consideration, unless otherwise stated.

We use without comment the *axiom of choice*: given a nonempty class of nonempty sets, there exists a function which assigns to every set of the class a point belonging to this set; in other words, we can always "choose" a point from every one of the sets of the class.

2.1 Topologies and limits. A class $\mathfrak{O}$ is a *topology* or the class of *open sets* if it is closed under formation of arbitrary unions and finite intersections and contains $\emptyset$ and Ω (the last property follows from the closure property by the conventions relative to intersections and unions of sets of an empty class). The dual class of complements of open sets is the class of *closed sets;* hence it is closed under formation of arbitrary intersections and finite unions and contains Ω and $\emptyset$.

A topological space $(\mathfrak{X}, \mathfrak{O})$ is a space $\mathfrak{X}$ in which is selected a topology $\mathfrak{O}$; from now on, all spaces under consideration will be topological and we shall frequently drop "$\mathfrak{O}$." A *topological subspace* thereof $(A, \mathfrak{O}_A)$ is a set A in which is selected its *induced topology* $\mathfrak{O}_A$ which consists of all the intersections of open sets with A and is, clearly, a topology in A. It is important to distinguish the properties of A considered as a set in $(\mathfrak{X}, \mathfrak{O})$ from those of A considered as a topological subspace of $(\mathfrak{X}, \mathfrak{O})$.

To every set A there are assigned an open set A^o and a closed set $\bar{A}$, as follows. The *interior* A^o of A is the maximal open set contained in A, that is, the union of all open sets in A; in particular, if A is open, then $A^o = A$. The *adherence* $\bar{A}$ of A is the minimal closed set containing A, that is, the intersection of all closed sets containing A; in particular, if A is closed, then $\bar{A} = A$. The definitions of interiors and adherences of A and A^c are clearly dual, so that

$$(A^o)^c = (\overline{A^c}), \quad (A^c)^o = (\bar{A})^c.$$

In topological spaces relations between sets and points are described in terms of neighborhoods. Every set containing a nonempty open set is a *neighborhood* of any point x of this open set; the symbol V_x will denote a neighborhood of x. The points of the interior A^o of A are "interior" to A; in other words, x is *interior* to A if A is a V_x. The points of the adherence $\bar{A}$ of A are adherent to A; in other words, x is *adherent* to A if no V_x is disjoint from A, that is, $x \notin (A^c)^o = (\bar{A})^c$.

Classical analysis is concerned primarily with continuous functions on euclidean lines to euclidean lines. In general, a function X on a topological domain Ω to a topological range space $\mathfrak{X}$ is *continuous at*

$\omega \in \Omega$ if the inverse images of neighborhoods of $x = X(\omega)$ are neighborhoods of ω; X is *continuous* (on Ω) if it is continuous at every $\omega \in \Omega$. Since taking inverse images preserves all set operations, it follows readily that we can limit ourselves to open (closed) sets. Thus X is continuous if, and only if, the inverse images of open (closed) sets are open (closed) and, hence, a continuous function induces on its domain a topology contained in (no "finer" than) that of the domain. Therefore, *if in topological spaces the σ-fields of measurable sets are selected to be the minimal σ-fields over the topologies, then continuous functions are measurable.* The importance of the concept of continuity is emphasized by the fact that two spaces $\mathfrak{X}$ and $\mathfrak{X}'$ are considered to be "topologically equivalent" if, and only if, there exists a one-to-one correspondence X on $\mathfrak{X}$ to $\mathfrak{X}'$ such that X and X^{-1} are continuous.

The basic concept which distinguishes classical analysis from classical algebra and which gave rise to the various concepts examined in this section is that of limit of sequences of numbers. In a topological space it becomes: x is *limit of a sequence* x_n or the sequence x_n *converges to* x if, for every V_x, there exists an integer $n(V_x)$ such that $x_n \in V_x$ for all $n \geqq n(V_x)$. However, the need for a more general concept of limit is already apparent in the classical theory of integration where the partitions of the interval of integration form a "direction" and the Riemann sums form a "directed set" of numbers of which the Riemann integral, if it exists, is the "limit." It so happens that this type of limit is precisely the one required for general topological spaces, and we now define the foregoing terms; the role of sequences in some species of spaces (including the euclidean ones) will be better understood when considered within the general setup.

Let T be a set of points t, with or without indices. T is *partially ordered* if a partial ordering is defined on it. A *partial ordering* "$<$," to be read "precedes," is a binary relation which is transitive ($t < t'$ and $t' < t''$ imply $t < t''$), reflexive ($t < t$), and such that, if $t < t'$ and $t' < t$, then $t = t'$; upon writing $t' > t$ when $t < t'$, the relation "$>$," to be read "follows," is also a partial ordering. T is a *direction* if it is partially ordered and if every pair t, t' is followed by some t'' ($t < t''$, $t' < t''$). T is *linearly ordered*, and *a fortiori* is a direction, if every pair t, t' is ordered (either $t < t'$ or $t' < t$). For example, the sets in a space are partially ordered by the relation of inclusion and the neighborhoods of a point x form a direction (this is the root of the definition of limit as given below); the finite partitions of an interval of integration form a direction when ordered by the relation of refinement; integers and, in general, sets of numbers are linearly ordered by the relation "$\leq$," etc.

A function X on T to $\mathfrak{X}$ can be represented by the indexed set $\{x_t\}$ of its values which may or may not be distinct but which are always distinguished by their indices t. The indexed set $\{x_t\}$ is *directed* if T is a direction; sequences $\{x_n\}$ are special directed sets representing functions on the (linearly ordered) set of positive integers. We are now ready to define the general concept of limit.

The point x is the *limit* of a directed set $\{x_t\}$ and we write $x = \lim x_t$, or, equivalently, x_t *converges to* x and we write $x_t \to x$, if, for every V_x, there exists an index $t(V_x)$ such that $x_t \in V_x$ for all those indices which follow $t(V_x)$. However, the concept of limit is of use only if, when the limit exists, it is unique; this requirement leads to the introduction of "separated" or "Hausdorff" space as follows:

A. SEPARATION THEOREM. *The following three definitions are equivalent. A topological space is separated if*

(S_1) *every directed set has at most one limit,*
(S_2) *every pair of distinct points has disjoint neighborhoods,*
(S_3) *the intersection of all closed neighborhoods of a point reduces to this point.*

The term "separated" expresses property (S_2).

We observe that, according to (S_3), in a separated space every set reduced to a point is closed.

Proof. (S_1) *and* (S_2) *are equivalent.* Let $x \neq y$. If $x_t \to x$ and $x_t \to y$, then $x_t \in V_x \cap V_y$ for all those t which follow both $t(V_x)$ and $t(V_y)$; since T is a direction such t exist so that no pair V_x, V_y is disjoint.

Conversely, if no pair V_x, V_y is disjoint, then there exist points $z(V_x, V_y) \in V_x \cap V_y$ and, since these pairs form a direction when ordered by the relation $(V_x, V_y) < (V'_x, V'_y)$ if $V_x \supset V'_x$ and $V_y \supset V'_y$, these points form a directed set converging to both x and y.

(S_2) *and* (S_3) *are equivalent.* If for every $y \neq x$ there exists a V_x such that $y \notin \overline{V}_x$, then the intersection of all $\overline{V}_x$ reduces to x. Conversely, if the intersection of all $\overline{V}_x$ reduces to the set formed by x, then, for every $y \neq x$, there exists a V_x such that $y \notin \overline{V}_x$, and the open set $(\overline{V}_x)^c$ is a neighborhood of y disjoint from V_x. The proof is terminated.

From now on, all spaces will be separated spaces.

2.2 Limit points and compact spaces. Analysis of concepts or properties leads to the introduction of "weaker" ones. A property $\mathcal{P}$ is *weaker* than a property $\mathcal{P}'$ if $\mathcal{P}'$ implies $\mathcal{P}$; $\mathcal{P}$ is a necessary condition for $\mathcal{P}'$ and $\mathcal{P}'$ is a sufficient condition for $\mathcal{P}$.

Perhaps even more basic than the concept of limit is the weaker one of limit point. A point x is a *limit point* of the directed set $\{x_t\}$ if, for every pair t, V_x, there exists *some* $t' > t$ such that $x_{t'} \in V_x$. The definitions of limit and of limit point yield at once (i) and (ii) of the proposition below, and then (iii) follows.

a. *Let the sets A_t be formed by all those points $x_{t'}$ for which t' follows t: $A_t = \{x_{t'}, t' > t\}$.*

(i) *$x_t \to x$ if, and only if, for every V_x there exists an $A_t \subset V_x$.*
(ii) *x is a limit point of $\{x_t\}$ if, and only if, no pair A_t, V_x is disjoint.*
(iii) *the set of all limit points of $\{x_t\}$ coincides with the intersection of all $\bar{A}_t$, and if $x_t \to x$ then this set reduces to the single point x.*

The reason for the somewhat confusing terminology above is that every limit point of $\{x_t\}$ is *the* limit of some subset of $\{x_t\}$, in the following sense. A direction S of elements s, s', $\cdots$ is a *subdirection* of the direction T when there exists a function f on S to T with the property that, for every t, there is an s such that, if s' follows s, then $t' = f(s')$ follows t. The set $\{x_{f(s)}\}$ directed by the subdirection S of T is a *subdirected set.* Clearly, if $x_t \to x$, then every subdirected set $x_{f(s)} \to x$.

b. *A point x is a limit point of a directed set $\{x_t\}$ if, and only if, the set contains a subdirected set which converges to x.*

Proof. The "if" assertion follows at once from the definitions. As for the "only if" assertion, it suffices for every pair $s' = (t, V_x)$ to take $f(s') = t' > t$ such that $x_{t'} \in V_x$ and direct the pairs by $(t_1, V_x^1) > (t_2, V_x^2)$ when $t_1 > t_2$ and $V_x^1 \subset V_x^2$.

Compact spaces are separated spaces in which every directed set has at least one limit point; a set is *compact* if it is compact in its induced topology. Compactness plays a prominent role in analysis and it is important to have equivalent characterizations of compact spaces. We shall use repeatedly the following terminology: a subclass of open sets is an *open covering* of a set if every point of the set belongs to at least one of the sets of the subclass.

A. COMPACTNESS THEOREM. *The following three properties of separated spaces are equivalent:*

(C_1) BOLZANO-WEIERSTRASS PROPERTY: *every directed set has at least one limit point.*

(C_2) HEINE-BOREL PROPERTY: *every open covering of the space contains a finite covering of the space.*

(C_3) INTERSECTION PROPERTY: *every class of closed sets such that all its finite subclasses have nonempty intersections has itself a nonempty intersection.*

If *some* class has the property described in (C_3), we say that it has the *finite intersection property.*

Proof. The intersection property means by contradiction that every class of closed sets whose intersection is empty contains a finite subclass whose intersection is empty. Thus, it is the dual of the Heine-Borel property, and it suffices to show that it is equivalent to the Bolzano-Weierstrass one.

Let $\{x_t\}$ be a directed set and, for every $t_0 \in T$, consider the adherence of the set of all the x_t with t following t_0. Since T is a direction, these adherences form a class of closed sets with finite intersection property. Thus, if the intersection property is true, then there exists an x common to all these adherences and it follows that x is a limit point of $\{x_t\}$.

Conversely, consider a class of closed sets with the finite intersection property and adjoin all finite intersections to the class. The class so obtained is directed by inclusion so that, by selecting a point from every set of this class, we obtain a directed set. If the Bolzano-Weierstrass property is true, then this set has a limit point and this point belongs to every set of the class; hence the intersection of the class is not empty. This completes the proof.

COMPACTNESS PROPERTIES. 1° *In a compact space, a directed set* $x_t \to x$ *if, and only if,* x *is its unique limit point.*

Proof. We use **a** and its notations. The "only if" assertion holds by **a**(iii). As for the "if" assertion, if $x_t \not\to x$ then, by **a**(i), there exists a V_x such that no A_t is disjoint from V_x^c; thus, for every t we can select a $t' > t$ such that $x_{t'} \in A_t \cap V_x^c$. Since the space is compact, the subdirected set $\{x_{t'}\}$, hence, by **b**, the directed set $\{x_t\}$, has a limit point $x' \in V_x^c$. Therefore, $x \neq x'$ and x cannot be the unique limit point of $\{x_t\}$.

2° *Every compact set is closed, and in a compact space the converse is true.*

Proof. Let A be compact and let V_x, $V_y(x)$ be a disjoint pair of open neighborhoods of $x \in A$ and $y \in A^c$. By the Heine-Borel property, the open covering $\{V_x\}$ of A where x ranges over A contains a finite subcovering $\{V_{x_k}\}$, and the disjoint open sets $V = \bigcup_k V_{x_k}$, $V' = \bigcap_k V_y(x_k)$ are such that $A \subset V$ and $y \in V'$. Thus, the open neighborhood V' of y contains no points of A; hence $y \notin \bar{A}$. Since $y \in A^c$ is arbitrary, it follows that A^c and $\bar{A}$ are disjoint, and the first assertion is proved. The second assertion follows readily from the intersection property.

3° *The intersection of a nonincreasing sequence of nonempty compact sets is not empty.*

Apply the intersection property.

4° *The range of a continuous function on a compact domain is compact.*

Proof. Because of continuity of the function, the inverse image of every open covering of the range is an open covering of the compact domain; hence it contains a finite open subcovering which is the inverse image of a finite open subcovering of the range. Thus, the range has the Heine-Borel property, and the assertion is proved.

The euclidean line $R = (-\infty, +\infty)$ is not compact but, according to the Bolzano-Weierstrass or Heine-Borel theorems, every closed interval $[a, b]$ is compact. These theorems become valid for the whole line if it is "extended"—that is, if points $-\infty$ and $+\infty$ are added. Thus, the extended euclidean line $\bar{R} = [-\infty, +\infty]$ is compact. In fact, R is locally compact and every locally compact space can be compactified by adding one point only, as below.

A separated space is *locally compact* if every point has a compact neighborhood; it is easily shown that every neighborhood then contains a compact one. The one-point *compactification* of a separated space $(\mathfrak{X}, \mathfrak{O})$ is as follows. Adjoin to the points of $\mathfrak{X}$ an arbitrary point $\infty \notin \mathfrak{X}$ and adjoin to the open sets all sets obtained by adjoining to the point ∞ those open sets whose complements are compact. Denote the topological space so obtained by $(\mathfrak{X}_\infty, \mathfrak{O}_\infty)$.

5° *The one-point compactification of a locally compact but not compact space is a compact space, and the induced topology of the original space is its original topology.*

Proof. The last assertion follows at once from the definition of $\mathcal{O}_\infty$. As for the first assertion, observe that the new space is separated, since two distinct points belonging to the separated original space are separated and the point ∞ is separated from any $x \in \mathfrak{X}$ by taking a compact and hence closed $V_x \subset \mathfrak{X}$, so that $\infty \in V_x^c$. Also, the new space has the Heine-Borel property, since an open covering of it has a member $O + \{\infty\}$ with O^c compact and hence contains a finite subcovering of O^c which, together with $O + \{\infty\}$, is a finite subcovering of the new space.

2.3 Countability and metric spaces. The euclidean line possesses many countability properties, among them separability (the countable set of rationals is dense in it) and a countable base (the countable class of all intervals with rational extremities); this permits us to define limits in terms of sequences only. In general topological spaces, a set A is *dense in* B if $\bar{A} \supset B$; in other words, taking for simplicity $B = \mathfrak{X}$, A is dense in $\mathfrak{X}$ if no neighborhood is disjoint from A; and B is *separable* if there exists a countable set A dense in B. A *countable base at* x is a countable class $\{V_x(j)\}$ of neighborhoods of x such that every neighborhood of x contains a $V_x(j)$; and the space has a *countable base* $\{V(j)\}$ if, for every point x, a subclass of $V(j)$'s is a base at x.

a. *A space has a countable base only if it is separable and has a countable base at every point. Then every open covering of the space contains a countable covering of the space.*

Note that if a countable set $\{x_j\}$ is dense in a metric space, then at every x_j there is a countable base of spheres of rational radii, and the countable union of all these countable bases is a base for the space.

Proof. If the space has a countable base $\{V(j)\}$, then it has a countable base at every point. Moreover, if A is a set formed by selecting a point x_j from every $V(j)$, then, since any neighborhood of any point contains a $V(j)$, it contains the corresponding point x_j, so that no neighborhood is disjoint from A.

Finally, given an open covering of the space, every one of its sets contains a $V(j)$ so that, for every $V(j)$, we can select one set O_j of the covering containing it. The countable class $\{O_j\}$ is an open covering of the space, and the proof is terminated.

A basic type of space with a countable base at every point is that of metric spaces. In fact, topologies in euclidean spaces are determined by means of distances; this approach characterizes metric spaces. A *metric space* is a space with a *distance* (or *metric*) d on $\mathfrak{X} \times \mathfrak{X}$ to R such that, whatever be the points x, y, z, this function has

the triangle property: $d(x, y) + d(x, z) \geqq d(y, z)$,
the identification property: $d(x, y) = 0 \Leftrightarrow x = y$.

Upon replacing z by x and interchanging x and y, it follows that

$$d(x, y) = d(y, x), \quad d(x, y) \geqq 0.$$

It happens frequently, and we shall encounter repeatedly such cases, that, for some space, a function d with the two foregoing properties can be defined—except for the property $d(x, y) = 0 \Rightarrow x = y$. Then the usual procedure is to identify all points x, y such that $d(x, y) = 0$; the space is replaced by the space of "classes of equivalence" so obtained, and this new space is metrized by d.

The topology of a metric space $(\mathfrak{X}, d)$ is defined as follows: Let the *sphere* $V_x(r)$ with "center" x and "radius" $r(>0)$ be the set of all points y such that $d(x, y) < r$. A set A is *open* if, for every $x \in A$, there exists a sphere $V_x(r) \subset A$; it follows, by the triangle property, that every sphere is open. Clearly, the class of open sets so defined is a topology. Since, by the identification property, $d(x, y) > 0$ when $x \neq y$ and the spheres $V_x(r)$ and $V_y(s)$ are disjoint for $0 < r, s \leqq \frac{1}{2}d(x, y)$, it follows that with the *metric topology* so defined, the space is separated; we observe that $x_n \to x$ means that $d(x_n, x) \to 0$.

A basic property of the metric topology is that at every point x there is a countable base, say, the sequence of spheres $V_x\left(\frac{1}{n}\right)$, $n = 1, 2, \cdots$, and it is to be expected that properties of metric spaces can be characterized in countable terms. To begin with:

1. *Sequences can converge to at most one point.*

2. *A point $x \in \bar{A}$ if, and only if, A contains a sequence $x_n \to x$, so that a set is closed if, and only if, limits of all convergent sequences of its points belong to it.*

3. *Every closed (open) set is a countable intersection (union) of open (closed) sets.*

4. *A metric space has a countable base if, and only if, it is separable.*

5. *If X is a function on a metric domain (Ω, ρ) to a metric space $(\mathfrak{X}, d)$, then $X(\omega') \to X(\omega)$ as $\omega' \to \omega$ if, and only if, $X(\omega_n) \to X(\omega)$ whatever be the sequence $\omega_n \to \omega$.*

Proof. The first assertion follows from the separation theorem. The "if" part of the second assertion is immediate, and for the "only if" part it suffices to take $x_n \in A \cap V_x\left(\frac{1}{n}\right)$.

For the third assertion, form the open sets $O_n = \bigcup_{x \in A} V_x\left(\frac{1}{n}\right)$; those sets contain A, so that $A \subset \bigcap O_n$. On the other hand, for every $x \in \bigcap O_n$ there exist points $x_n \in O_n$ such that $x \in V_{x_n}\left(\frac{1}{n}\right)$, and hence $x_n \to x$; since A is closed, it follows by the second assertion that $x \in A$, and hence $A \supset \bigcap O_n$. Thus, closed $A = \bigcap O_n$ and the dual assertion for open sets follows by complementations.

The fourth assertion follows from **a**.

Finally, if $X(\omega') \to X(\omega)$ as $\omega' \to \omega$, then, clearly, $X(\omega_n) \to X(\omega)$ as $\omega_n \to \omega$. Since $X(\omega') \nrightarrow X(\omega)$ as $\omega' \to \omega$ implies that there exist points $\omega_n \in V_\omega\left(\frac{1}{n}\right)$ such that $X(\omega_n) \nrightarrow X(\omega)$, while $\omega_n \to \omega$, the last assertion follows.

Metric completeness and compactness. The basic criterion for convergence of numerical sequences is the (Cauchy) *mutual convergence criterion:* a sequence x_n is mutually convergent, that is, $d(x_m, x_n) \to 0$ as $m, n \to \infty$ if, and only if, the sequence x_n converges. In a metric space, if $x_n \to x$, then, by the triangle inequality, $d(x_m, x_n) \leqq d(x, x_m) + d(x, x_n) \to 0$ as $m, n \to \infty$, but the converse is not necessarily true (take the space of all rationals with euclidean distance); if it is true, that is, if $d(x_m, x_n) \to 0$ implies that $x_n \to$ some x, then the mutual convergence criterion is valid, and we say that the space is *complete.* Complete metric spaces have many important properties, which follow.

Call $\Delta(A) = \sup_{x,y \in A} d(x, y)$ the *diameter* of A; A is *bounded* if $\Delta(A)$ is finite.

A. CANTOR'S THEOREM. *In a complete metric space, every nonincreasing sequence of closed nonempty sets A_n such that the sequence of their diameters $\Delta(A_n)$ converges to* 0 *has a nonempty intersection consisting of one point only.*

Proof. Take $x_n \in A_n$ and $m \geqq n$. Since $d(x_m, x_n) \leqq \Delta(A_n) \to 0$, it follows that $x_n \to$ some x. Since $x_m \in A_m \subset A_n$ for all $m \geqq n$ and the set A_n is closed, x belongs to every A_n; hence $x \in \bigcap A_n$. If now $d(x, x') > 0$, then, from some k on, $d(x, x') > \Delta(A_k)$ so that $x' \notin A_k \supset \bigcap A_n$. The assertion is proved.

A set A is *nowhere dense* if the complement of $\bar{A}$ is dense in the space, or, equivalently, if $\bar{A}$ contains no spheres, that is, if the interior of $\bar{A}$ is empty. A set is of the *first category* if it is a countable union of nowhere dense sets, and it is of the *second category* if it is not of the first category.

B. Baire's category theorem. *Every complete metric space is of the second category.*

Proof. Let $A = \bigcup A_n$ where the A_n are nowhere dense sets. There exist a point $x_1 \notin \bar{A}_1$ and a positive $r_1 < 1$ such that the adherence of $V_{x_1}(r_1)$ is disjoint from A_1. Proceeding by recurrence, we form a decreasing sequence of spheres $V_{x_n}(r_n)$ such that $\overline{V_{x_n}(r_n)}$ is disjoint from A_n and $r_n < \dfrac{1}{n} \to 0$. Therefore, by Cantor's theorem, there exists a point $x \in \bigcap \overline{V_{x_n}(r_n)}$ and, because of the foregoing disjunction, $x \notin \bigcup A_n$. Thus $A \neq \mathfrak{X}$, and the theorem follows.

We investigate now compact metric spaces and require the two following propositions.

b. *If every mutually convergent sequence contains a convergent subsequence, then the space is complete.*

This follows from the fact that if a sequence x_n is mutually convergent and contains a convergent subsequence $x_{n'} \to x$, then, by the triangle inequality, $d(x_n, x) \leqq d(x_{n'}, x_n) + d(x_{n'}, x) \to 0$ as $n, n' \to \infty$, so that $x_n \to x$.

A set is *totally bounded* if, for every $\epsilon > 0$, it can be covered by a finite number of spheres of radii $\leqq \epsilon$. Clearly, a totally bounded set is bounded, and a subset of a totally bounded set is totally bounded.

c. *A metric space is totally bounded if, and only if, every sequence of points contains a mutually convergent subsequence. A totally bounded metric space has a countable base.*

Proof. Let the space be not totally bounded; there exists an $\epsilon > 0$ such that the space cannot be covered by finitely many spheres of radii $\leqq \epsilon$. We can select by recurrence a sequence of points x_n whose mutual distances are $\geqq \epsilon$; for, if there is only a finite number of points $x_1, \cdots, x_m$ with this property, then the spheres of radius ϵ centered at these points cover the space. Clearly, this sequence cannot contain a mutually convergent subsequence.

Conversely, let the space be totally bounded, so that every set is totally bounded. Then any sequence of points belonging to a set contains a subsequence contained in a sphere of radius $\leqq \epsilon$—member of a finite covering of the set by spheres of radii $\leqq \epsilon$. Thus, given a sequence $\{x_n\}$, setting $\epsilon = \frac{1}{2}, \frac{1}{3}, \cdots$, and proceeding by recurrence, we obtain subsequences such that each is contained in the preceding one and the kth one is formed by points $x_{1k}, x_{2k}, \cdots$ belonging to a sphere of radius $\leqq \frac{1}{k}$. The "diagonal" subsequence $\{x_{nn}\}$ is such that, from the kth term on, the mutual distances are $\leqq \frac{1}{k}$; hence this subsequence is mutually convergent.

The last assertion follows from the fact that given a totally bounded space, the class formed by all finite coverings by spheres of radii $\leqq \frac{1}{n}$, $n = 1, 2, \cdots$ is a countable base.

C. Metric compactness theorem. *The three following properties of a metric space are equivalent:*

(MC$_1$) *every sequence of points contains a convergent subsequence;*
(MC$_2$) *every open covering of the space contains a finite covering of the space (Heine-Borel property);*
(MC$_3$) *the space is totally bounded and complete.*

Proof. It suffices to show that (MC$_2$) $\Rightarrow$ (MC$_1$) $\Rightarrow$ (MC$_3$) $\Rightarrow$ (MC$_2$).

(MC$_2$) $\Rightarrow$ (MC$_1$). Apply the compactness theorem.

(MC$_1$) $\Rightarrow$ (MC$_3$). Let every sequence of points contain a convergent (hence mutually convergent) subsequence. Then, by **b**, the space is complete and by **c**, it is also totally bounded.

(MC$_3$) $\Rightarrow$ (MC$_2$). According to **a**, an open covering of a totally bounded space contains a countable covering $\{O_j\}$ of the space. If no finite union of the O_j covers the space, then, for every n, there exists a point $x_n \not\subset \bigcup_{j=1}^{n} O_j$, and, according to **c**, the sequence of these points contains a mutually convergent subsequence. Therefore, when the totally bounded space is also complete, this sequence has a limit point x which necessarily belongs to some set O_{j_0} of the open countable covering of the space. Since x is a limit point of the sequence $\{x_n\}$, there exists some $n > j_o$ such that $x_n \in O_{j_o} \subset \bigcup_{j=1}^{n} O_j$, and we reach a contradiction. Thus, there exists a finite subcovering of the space.

COROLLARY 1. *A compact metric space is bounded and separable.*

COROLLARY 2. *A continuous function X on a compact metric space (Ω, ρ) to a metric space $(\mathfrak{X}, d)$ is uniformly continuous.*

By definition, X is *uniformly continuous* if for every $\epsilon > 0$ there exists a $\delta = \delta(\epsilon) > 0$, which depends only upon ϵ, such that $d(X(\omega), X(\omega')) < \epsilon$ for $\rho(\omega, \omega') < \delta$.

Proof. Let $\epsilon > 0$. Since X is continuous, for every $\omega \in \Omega$ there exists a δ_ω such that $d(X(\omega), X(\omega')) < \epsilon/2$ for $\rho(\omega, \omega') < 2\delta_\omega$. Since the domain is compact, it is covered by a finite number of spheres $V_{\omega_k}(\delta_{\omega_k})$, $k = 1, 2, \cdots, n$; let δ be the smallest of their radii. Any ω belongs to one of these spheres, say, $V_{\omega_k}(\delta_{\omega_k})$, and if $\rho(\omega, \omega') < \delta$, then $\rho(\omega_k, \omega') < 2\delta_{\omega_k}$. It follows, by the triangle inequality, that

$$d(X(\omega), X(\omega')) \leqq d(X(\omega_k), X(\omega)) + d(X(\omega_k), X(\omega')) < \frac{\epsilon}{2} + \frac{\epsilon}{2} = \epsilon$$

whenever $\rho(\omega, \omega') < \delta$, and the corollary is proved.

Let us indicate how a noncomplete metric space $(\mathfrak{X}, d)$ can be *completed*, that is, can be put in a one-to-one isometric correspondence with a set in a complete metric space—in fact, with a set dense in the latter space. The elementary computations will be left to the reader.

Consider all mutually convergent sequences $s = (x_1, x_2, \cdots)$, $s' = (x'_1, x'_2, \cdots)$, $\cdots$. The function ρ defined by $\rho(s, s') = \lim d(x_n, x'_n)$ exists and is finite and satisfies the triangular inequality. Let s, s' be *equivalent* if $\rho(s, s') = 0$; this notion is symmetric, transitive, and reflexive. It follows that the space (S, ρ) of all such equivalence classes is a metric space, and it is easily seen that it is complete. The one-to-one correspondence between $\mathfrak{X}$ and the set S' of classes of equivalence of all "constant sequences," defined by $x \leftrightarrow (x, x \cdots)$, preserves the distances. Moreover, S' is dense in S. Thus S may be considered as a "minimal completion" of $\mathfrak{X}$.

2.4 Linearity and normed spaces. Euclidean spaces are not only metric and complete but are also normed and linear as defined below. Unless specified, the "scalars" a, b, c, with or without subscripts, are *either* arbitrary real numbers *or* arbitrary complex numbers, and x, y, z, with or without subscripts, are arbitrary points in a space $\mathfrak{X}$.

A space $\mathfrak{X}$ is *linear* if a "linear operation" consisting of operations of "addition" and "multiplication by scalars" is defined on $\mathfrak{X}$ to $\mathfrak{X}$ with the properties:

$$\text{(i)} \qquad x + y = y + x, \quad x + (y + z) = (x + y) + z,$$

$$x + z = y + z \Rightarrow x = y;$$

$$\text{(ii)} \qquad 1 \cdot x = x, \quad a(x + y) = ax + ay, \quad (a + b)x = ax + bx,$$

$$a(bx) = (ab)x.$$

By setting $-y = -1 \cdot y$, "subtraction" is defined by $x - y = x + (-y)$. Elementary computations show that (i) and (ii) imply uniqueness of the "zero point" or "null point" or "origin" θ, defined by $\theta = 0 \cdot x$, and with the property $x + \theta = x$. A set in a linear space generates a *linear subspace*—the *linear closure* of the set—by adding to its points $x, y, \cdots t$ all points of the form $ax + by + \cdots lt$.

A *metric linear space* is a linear space with a metric d which is invariant under translations and makes the linear operations continuous:

$$\text{(iii)} \qquad d(x, y) = d(x - y, \theta), \quad x_n \to \theta \Rightarrow ax_n \to \theta,$$

$$a_n \to 0 \Rightarrow a_n x \to \theta.$$

If

$$\text{(iv)} \qquad d(x, y) = d(x - y, \theta), \quad d(ax, \theta) = |a| d(x, \theta),$$

then (iii) holds, $d(x, \theta)$ is called norm of x and is denoted by $\|x\|$, and the metric linear space is then a "normed linear space."

Equivalently, a *normed linear space* is a linear space on which is defined a *norm* with values $\|x\| \geqq 0$ such that

$$\text{(v)} \qquad \|x + y\| \leqq \|x\| + \|y\|, \quad \|x\| = 0 \Leftrightarrow x = \theta,$$

$$\|ax\| = |a| \cdot \|x\|,$$

and the metric d is determined by the norm by setting

$$d(x, y) = \|x - y\|.$$

A *Banach space* is a normed linear space complete in the metric determined by the norm. For example, the space of all bounded continuous functions f on a topological space $\mathfrak{X}$ to the euclidean line is a Banach space with a norm defined by $\|f\| = \sup_x |f(x)|$. Real spaces with points $x = (x_1, \cdots, x_N)$ and norms $\|x\| = (|x_1|^r + \cdots + |x_N|^r)^{1/r}$, $r \geqq 1$, are Banach spaces, and we shall encounter similar but more general spaces L_r. If $r = 2$, then these (euclidean) spaces are Hilbert spaces.

A *Hilbert space* is a Banach space whose norm has the parallelogram property: $\|x + y\|^2 + \|x - y\|^2 = 2\|x\|^2 + 2\|y\|^2$; such a norm determines a scalar product. It is simpler to determine the Hilbert norm by means of a scalar product (corresponding to the scalar product defined by $(x, y) = \sum_{k=1}^{N} x_k y_k$ in a euclidean space R^N) as follows:

A *scalar product* is a function on the product of a linear space by itself to its space of scalars, with values (x, y) such that

$$\text{(vi)} \qquad (ax + by, z) = a(x, z) + b(y, z), \quad (x, y) = \overline{(y, x)},$$

$$x \neq \theta \Rightarrow (x, x) > 0.$$

Clearly (x, x) is real and nonnegative. The function with values $\|x\| = (x, x)^{1/2} \geqq 0$ is the *Hilbert norm* determined by the scalar product. For, obviously, it has the two last properties (v) of a norm. And it also has the first property (v). This follows by using in the expansion of $(x + y, x + y)$ the Schwarz inequality

$$|(x, y)| \leqq \|x\| \cdot \|y\|;$$

when $(x, y) = 0$ this inequality is trivially true, and when $(x, y) \neq 0$ it is obtained by expanding $(x - ay, x - ay) \geqq 0$ and setting $a = (x, x)/(y, x)$. Finally, the parallelogram property is immediate.

Linear functionals. The basic concept in the investigation of Banach spaces is the analogue of $f(x) = cx$—the simplest of nontrivial functions of classical analysis. A *functional* f on a normed linear space has for range space the space of the scalars (the scalars and the points below are arbitrary, unless specified). f is

linear if $f(ax + by) = af(x) + bf(y)$;

continuous if $f(x_n) \to f(x)$ as $x_n \to x$; if this property holds only for a particular x, then f is *continuous at* this x;

normed or *bounded* if $|f(x)| \leqq c\|x\|$ where $c < \infty$ is independent of x; the *norm* of f is then the finite number $\|f\| = \sup_{x \neq \theta} \dfrac{|f(x)|}{\|x\|}$.

For example, a scalar product (x, y) is a linear continuous and normed functional in x for every fixed y. Clearly, if f is linear, then $f(\theta) = 0$, and a linear functional continuous at θ is continuous.

a. *Let f be a linear functional on a normed linear space. If f is normed, then it is continuous; and conversely.*

Proof. If f is normed, then it is continuous, since

$$|f(x_n) - f(x)| = |f(x_n - x)| \leqq c\|x_n - x\| \to 0 \text{ as } \|x_n - x\| \to 0.$$

If f is not normed, then it is not continuous, since whatever be n there exists a point x_n such that $|f(x_n)| > n\|x_n\|$, and, setting $y_n = x_n/n\|x_n\|$, we have $|f(y_n)| > 1$ while $\|y_n\| = \dfrac{1}{n} \to 0$.

b. *The space of all normed linear functionals f on a normed linear space is a Banach space with norm $\|f\|$.*

Proof. Clearly the space is normed and linear and it remains to prove that it is complete.

Let $\|f_m - f_n\| \to 0$ as $m, n \to \infty$. For every $\epsilon > 0$ there exists an n_ϵ such that $\|f_m - f_n\| < \epsilon$ for $m, n \geqq n_\epsilon$; hence $|f_m(x) - f_n(x)| < \epsilon\|x\|$ whatever be x. Since the space of scalars is complete, it follows that there exists a function f of x such that $f_n(x) \to f(x)$ and, clearly, f is linear and normed. By letting $m \to \infty$, we have, for $n \geqq n_\epsilon$, $|f(x) - f_n(x)| \leqq \epsilon\|x\|$ whatever be x, that is, $\|f_n - f\| \leqq \epsilon$. Hence $f_n \to f$ and the proposition is proved.

What precedes applies word for word to more general functions (mappings, transformations) on a normed linear space to a normed linear space with the same scalars, and the foregoing proposition remains valid, provided the range space is complete; it suffices to replace every $|f(x)|$ by $\|f(x)\|$.

The Banach space of normed linear functionals on a Banach space is said to be its *adjoint;* a Hilbert space is adjoint to itself. However, *a priori,* the adjoint space may consist only of the trivial null functional f with $\|f\| = 0$. That it is not so will follow (see Corollary 1) from the basic Hahn-Banach

A. EXTENSION THEOREM. *If f is a normed linear functional on a linear subspace A of a normed linear space, then f can be extended to a normed linear functional on the whole space without changing its norm.*

Proof. 1° We begin by showing that we can extend the domain of f point by point. Let $x_0 \notin A$ and let $\|f\| = 1$—this does not restrict the generality. First assume that the scalars, hence f, are real.

The linearity condition determines $f(x + ax_0)$, $x \in A$, by setting it equal to $f(x) + af(x_0)$, so that it suffices to show that there exists a number $f(x_0)$ such that $|f(x) + af(x_0)| \leqq \| x + ax_0 \|$ for every $x \in A$ and every number a. Since A is a linear subspace, we can replace x by ax and, by letting x vary, the condition becomes

$$\sup_x \{-\| x + x_0 \| - f(x)\} \leqq f(x_0) \leqq \inf_x \{\| x + x_0 \| - f(x)\}.$$

Therefore, acceptable values of $f(x_0)$ exist if the above supremum is no greater than the above infimum, that is, if whatever be $x', x'' \in A$

$$-\| x' + x_0 \| - f(x') \leqq \| x'' + x_0 \| - f(x'')$$

or

$$f(x'') - f(x') \leqq \| x'' + x_0 \| + \| x' + x_0 \|.$$

Since by linearity of f and the triangle inequality

$$f(x'') - f(x') = f(x'' - x') \leqq \| x'' - x' \| \leqq \| x'' + x_0 \| + \| x' + x_0 \|,$$

acceptable values of $f(x_0)$ exist.

We can pass from real scalars to complex scalars, as follows: From $f(ix) = if(x)$ it follows that $f(x) = g(x) - ig(ix)$, $x \in A$, where $g = \Re f$ is a real-valued linear functional with $\| g \| \leqq 1$; g extends first for all points $x + ax_0$ then for all points $(x + ax_0) + b \cdot ix_0 = x + (a + ib)x_0$, a, b real, and f extends by the foregoing relation. Now observe that f is linear on the so extended domain and that, for any given point x, upon setting $f(x) = re^{i\alpha}$, $r \geqq 0$, α real, we obtain $|f(x)| = g(e^{-i\alpha}x) \leqq \| x \|$.

2° We can extend the domain of f point by point. The family of all possible extensions of f to linear functionals without change of norm is partially ordered by inclusion of their domains. Any linearly ordered subfamily of extensions has a supremum in the family—the extension on the union of the domains. According to a consequence of the axiom of choice (Zorn's theorem), it follows that the whole family has a supremum which is a member of the family. It must have for domain the whole space, for otherwise, by 1°, it could be extended further. The theorem is proved.

COROLLARY 1. *Let x_0 be a nonzero point of a normed linear space, and let A be a closed linear subspace. There exist linear functionals f, f' on the space such that*

$$\| f \| = 1 \quad \textit{and} \quad f(x_0) = \| x_0 \|,$$

$$f' = 0 \textit{ on } A \quad \textit{and} \quad f'(x_0) = d(x_0, A) = \inf_{x \in A} d(x_0, x).$$

Set $f(ax_0) = a\| x_0 \|$, $f'(ax_0 + x) = ad(x_0, A)$, $x \in A$, and extend.

COROLLARY 2. *A functional f on a set A in a normed linear space extends to a normed linear functional on the whole space with norm bounded by $c(<\infty)$ if, and only if,*

$$\left| \sum_k a_k f(x_k) \right| \leqq c \left\| \sum_k a_k x_k \right\|$$

whatever be the finite number of arbitrary points $x_k \in A$ and of arbitrary scalars a_k.

Proof. The "only if" assertion is immediate. As for the "if" assertion, assume that the inequality is true, and observe that the linear closure of A consists of all points of the form $x = \sum_k a_k x_k$. Linearity of f on this closure implies that we must set $f(x) = \sum_k a_k f(x_k)$. Then, on the closure, $|f(x)| \leqq c\|x\|$, and f is uniquely determined, since, for $x = \sum_k a_k x_k = \sum_{k'} a'_{k'} x'_{k'}$, we have

$$\left| \sum_k a_k f(x_k) - \sum_{k'} a'_{k'} f(x'_{k'}) \right| \leqq c \left\| \sum_k a_k x_k - \sum_{k'} a'_{k'} x'_{k'} \right\| = 0.$$

The assertion follows by the extension theorem.

This corollary permits us to solve various moment problems as well as to find conditions for existence of solutions of systems of linear equations with an infinity of unknowns.

§ 3. ADDITIVE SET FUNCTIONS

3.1 Additivity and continuity. *A set function* φ is defined on a nonempty class $\mathcal{C}$ of sets in a space Ω by assigning to every set $A \in \mathcal{C}$ a single number $\varphi(A)$, finite or infinite, the *value of* φ *at* A. If all values of φ are finite, φ is said to be *finite*, and we write $|\varphi| < \infty$. If every set in $\mathcal{C}$ is a countable union of sets in $\mathcal{C}$ at which φ is finite, φ is said to be *σ-finite*. To avoid trivialities, we assume that every set function has at least one finite value. Unless otherwise stated, *φ denotes a set function and all sets considered are sets of the class on which this function is defined, so that the properties below are valid as long as φ is defined for the sets which appear there.*

φ is said to be *additive* if

$$\varphi(\sum A_j) = \sum \varphi(A_j)$$

either for every countable or only for every finite class of disjoint sets. In the first case φ is said to be *countably additive* or *σ-additive*, and in the second case φ is said to be *finitely additive*. In order that sums $\sum \varphi(A_j)$ be always meaningful we have to exclude the possibility of expressions of the form $+\infty - \infty$. In fact, if the sums always exist, φ is defined on a field, and $\varphi(A) = +\infty$ and $\varphi(B) = -\infty$, then $\varphi(\Omega) = \varphi(A) + \varphi(A^c) = +\infty$ and $\varphi(\Omega) = \varphi(B) + \varphi(B^c) = -\infty$, while the function φ is single-valued. Thus, by definition,

an additive set function has the additivity property above, and one of the values $+\infty$ or $-\infty$ is not allowed.

To fix ideas we assume that the value $-\infty$ is excluded, unless otherwise stated.

A nonnegative additive set function is called a *content* or a *measure* according as it is finitely additive or σ-additive. Let φ be additive. If $A \supset B$, then, by additivity,

$$\varphi(A) = \varphi(B) + \varphi(A - B).$$

It follows, upon taking $A = B + \emptyset = B$ with $\varphi(B)$ finite, that $\varphi(\emptyset) = 0$.

A convergent series of terms, which are not necessarily of constant sign, may depend upon the order of the terms. This possibility is excluded in our case by

a. *If φ is σ-additive and $|\varphi(\sum A_n)| < \infty$, then the series $\sum \varphi(A_n)$ is absolutely convergent.*

Proof. Set $A_n^+ = A_n$ or $\emptyset$ according as $\varphi(A_n) \geqq 0$ or $\varphi(A_n) < 0$, and set $A_n^- = A_n$ or $\emptyset$ according as $\varphi(A_n) \leqq 0$ or $\varphi(A_n) > 0$. Then

$$\varphi(\sum A_n^+) = \sum \varphi(A_n^+), \quad \varphi(\sum A_n^-) = \sum \varphi(A_n^-),$$

and the terms of each series are of constant sign. Since the value $-\infty$ is excluded, the last series converges. Since the sum of both series converges, so does the first series. The assertion follows.

b. *If $\varphi(A)$ is finite and $A \supset B$, then $\varphi(B)$ is finite; in particular, if $\varphi(\Omega)$ is finite, then φ is finite. If $\varphi \geqq 0$, then φ is nondecreasing: $\varphi(A) \geqq \varphi(B)$ for $A \supset B$, and subadditive: $\varphi(\bigcup A_j) \leqq \sum \varphi(A_j)$.*

Only the very last assertion needs verification and follows from

$$\begin{aligned}\varphi(\bigcup A_j) &= \varphi(A_1 + A_1^cA_2 + A_1^cA_2^cA_3 + \cdots)\\ &= \varphi(A_1) + \varphi(A_1^cA_2) + \varphi(A_1^cA_2^cA_3) + \cdots\\ &\leqq \varphi(A_1) + \varphi(A_2) + \varphi(A_3) + \cdots.\end{aligned}$$

We intend to show that the difference between finite additivity and σ-additivity lies in continuity properties. φ is said to be *continuous from below* or *from above* according as

$$\varphi(\lim A_n) = \lim \varphi(A_n)$$

for every sequence $A_n \uparrow$, or for every sequence $A_n \downarrow$ such that $\varphi(A_n)$ is finite for some value n_0 of n (hence, by **b**, for all $n \geqq n_0$). If φ is continuous from above and from below, it is said to be *continuous*. Continuity might hold at a fixed set A only, that is, for all monotone sequences which converge to A; continuity at $\emptyset$ reduces to continuity from above at $\emptyset$.

A. CONTINUITY THEOREM FOR ADDITIVE SET FUNCTIONS. *A σ-additive set function is finitely additive and continuous. Conversely, if a set function is finitely additive and, either continuous from below, or finite and continuous at $\emptyset$, then the set function is σ-additive.*

Proof. Let φ be σ-additive and, *a fortiori*, additive. φ is continuous from below, for, if $A_n \uparrow$, then

$$\lim A_n = \bigcup A_n = A_1 + (A_2 - A_1) + (A_3 - A_2) + \cdots$$

so that

$$\begin{aligned}\varphi(\lim A_n) &= \lim \{\varphi(A_1) + \varphi(A_2 - A_1) + \cdots + \varphi(A_n - A_{n-1})\}\\ &= \lim \varphi(A_n).\end{aligned}$$

φ is continuous from above, for, if $A_n \downarrow$ and $\varphi(A_{n_0})$ is finite, then $A_{n_0} - A_n \uparrow$ for $n \geqq n_0$, the foregoing result for continuity from below applies and, hence,

$$\begin{aligned}\varphi(A_{n_0}) - \varphi(\lim A_n) &= \varphi(\lim (A_{n_0} - A_n)) = \lim \varphi(A_{n_0} - A_n)\\ &= \varphi(A_{n_0}) - \lim \varphi(A_n)\end{aligned}$$

or

$$\varphi(\lim A_n) = \lim \varphi(A_n).$$

Conversely, let φ be finitely additive. If φ is continuous from below, then

$$\varphi(\sum A_n) = \varphi(\lim \sum_{k=1}^{n} A_k) = \lim \varphi(\sum_{k=1}^{n} A_k) = \lim \sum_{k=1}^{n} \varphi(A_k) = \sum \varphi(A_n),$$

so that φ is σ-additive. If φ is finite and continuous at $\emptyset$, then σ-additivity follows from

$$\varphi(\sum A_n) = \varphi(\sum_{k=1}^{n} A_k) + \varphi(\sum_{k=n+1}^{\infty} A_k) = \sum_{k=1}^{n} \varphi(A_k) + \varphi(\sum_{k=n+1}^{\infty} A_k)$$

and

$$\varphi(\sum_{k=n+1}^{\infty} A_k) \to \varphi(\emptyset) = 0.$$

The proof is complete.

The continuity properties of a σ-additive set function φ acquire their full significance when φ is defined on a σ-field. Then, not only is φ defined for all countable sums and monotone limits of sets of the σ-field but, moreover, φ attains its extrema at some sets of this σ-field. More precisely

c. *If φ on a σ-field $\mathcal{A}$ is σ-additive, then there exist sets C and D of $\mathcal{A}$ such that $\varphi(C) = \sup \varphi$ and $\varphi(D) = \inf \varphi$.*

Proof. We prove the existence of C; the proof of the existence of D is similar. If $\varphi(A) = +\infty$ for some $A \in \mathcal{A}$, then we can set $C = A$ and the theorem is trivially true. Thus, let $\varphi < \infty$, so that, since the value $-\infty$ is excluded, φ is finite.

There exists a sequence $\{A_n\} \subset \mathcal{A}$ such that $\varphi(A_n) \to \sup \varphi$. Let $A = \bigcup A_n$ and, for every n, consider the partition of A into 2^n sets A_{nm} of the form $\bigcap_{k=1}^{n} A'_k$ where $A'_k = A_k$ or $A - A_k$; for $n < n'$, every A_{nm} is a finite sum of sets $A_{n'm'}$. Let B_n be the sum of all those A_{nm} for which φ is nonnegative; if there are none, set $B_n = 0$. Since, on the one hand, A_n is the sum of some of the A_{nm} and, on the other hand, for $n' > n$, every $A_{n'm'}$ is either in B_n or disjoint from B_n, we have

$$\varphi(A_n) \leqq \varphi(B_n) \leqq \varphi(B_n \cup B_{n+1} \cup \cdots \cup B_{n'}).$$

Letting $n' \to \infty$, it follows, by continuity from below, that

$$\varphi(A_n) \leqq \varphi(B_n) \leqq \varphi(\bigcup_{k=n}^{\infty} B_k).$$

Letting now $n \to \infty$ and setting $C = \lim \bigcup_{k=n}^{\infty} B_k$, it follows, by continuity from above (φ is finite), that $\sup \varphi \leqq \varphi(C)$. But $\varphi(C) \leqq \sup \varphi$ and, thus, $\varphi(C) = \sup \varphi$. The proof is complete.

COROLLARY. *If φ on a σ-field $\mathcal{A}$ is σ-additive (and the value $-\infty$ is excluded), then φ is bounded below.*

3.2 Decomposition of additive set functions. We shall find later that the "natural" domains of σ-additive set functions are σ-fields. We intend to show that on such domains σ-additive set functions coincide with *signed measures*, that is, differences of two measures of which one at least is finite. Clearly, a signed measure is σ-additive so that we need only to prove the converse.

Let φ be an additive function on a field $\mathcal{C}$ and define φ^+ and φ^- on $\mathcal{C}$ by

$$\varphi^+(A) = \sup_{B \subset A} \varphi(B), \quad \varphi^-(A) = -\inf_{B \subset A} \varphi(B), \quad A, B \in \mathcal{C}.$$

The set functions φ^+, φ^- and $\bar{\varphi} = \varphi^+ + \varphi^-$ are called the *upper*, *lower*, and *total variation* of φ on $\mathcal{C}$, respectively. Since $\varphi(\emptyset) = 0$, these variations are nonnegative.

A. JORDAN-HAHN DECOMPOSITION THEOREM. *If φ on a σ-field $\mathcal{A}$ is σ-additive, then there exists a set D such that, for every $A \in \mathcal{A}$,*

$$-\varphi^-(A) = \varphi(AD), \quad \varphi^+(A) = \varphi(AD^c).$$

φ^+ and φ^- are measures and $\varphi = \varphi^+ - \varphi^-$ is a signed measure.

Proof. According to 3.1c, there exists a set $D \in \mathcal{A}$ such that $\varphi(D) = \inf \varphi$; since the value $-\infty$ is excluded, we have

$$-\infty < \varphi(D) = \inf \varphi \leqq 0.$$

For every set $A \in \mathcal{A}$, $\varphi(AD) \leqq 0$ and $\varphi(AD^c) \geqq 0$, since $\varphi \geqq \varphi(D)$ while, if $\varphi(AD) > 0$, then

$$\varphi(D - AD) = \varphi(D) - \varphi(AD) < \varphi(D),$$

and if $\varphi(AD^c) < 0$, then

$$\varphi(D + AD^c) = \varphi(D) + \varphi(AD^c) < \varphi(D).$$

It follows that, for every $B \subset A$, $(A, B \in \mathcal{A})$,

$$\varphi(B) \leqq \varphi(BD^c) \leqq \varphi(BD^c) + \varphi((A - B)D^c) = \varphi(AD^c),$$

and, hence, $\varphi^+(A) \leqq \varphi(AD^c)$. Since AD^c is one of the B's, the reverse inequality is also true. Therefore, for every $A \in \mathcal{A}$, $\varphi^+(A) = \varphi(AD^c)$ and, similarly, $-\varphi^-(A) = \varphi(AD)$, so that

$$\varphi(A) = \varphi(AD^c) + \varphi(AD) = \varphi^+(A) - \varphi^-(A).$$

Moreover, φ^+ on $\mathcal{A}$ is a measure since $\varphi^+ \geqq 0$ and

$$\varphi^+(\sum A_j) = \varphi(\sum A_j D^c) = \sum \varphi(A_j D^c) = \sum \varphi^+(A_j).$$

Similarly φ^- on $\mathcal{A}$ is a measure and, furthermore, it is bounded by $-\varphi(D)$ which is finite. Thus, $\varphi = \varphi^+ - \varphi^-$ is a signed measure, and the proof is complete.

Jordan decomposition. *If $\mathcal{A}$ is only a field but φ is also bounded, then it is still a signed measure.* Prove, proceeding directly from the definitions, showing first that $\varphi^\pm$ are bounded measures.

*§ 4. CONSTRUCTION OF MEASURES ON σ-FIELDS

4.1 Extension of measures. If two set functions φ on $\mathcal{C}$ and φ' on $\mathcal{C}'$ take the same values at sets of a common subclass $\mathcal{C}''$, we say that φ and φ' *agree* or *coincide* on $\mathcal{C}''$. If $\mathcal{C} \subset \mathcal{C}'$ and φ and φ' agree on $\mathcal{C}$, we say that φ is a *restriction* of φ' on $\mathcal{C}$, and φ' is an *extension* of φ on $\mathcal{C}'$. The general extension problem can be stated as follows: find extensions of φ which preserve some specified properties. If, given $\mathcal{C}' \supset \mathcal{C}$, there is one, and only one, such extension on $\mathcal{C}'$, we say that this extension is *determined.*

Here, we are concerned with the extension of measures to measures and shall denote extensions and restrictions of a measure μ by the same letter; as long as their domains are specified, there is no confusion possible. While any restriction of a measure is determined and is a measure, an extension of a measure to a measure on a given class may not exist, and if one exists it may not be unique. Our aim is to produce classes on which such extensions exist, and cases where they are determined. The results of the investigation are summarized by the Carathéodory

A. Extension theorem. *A measure μ on a field $\mathcal{C}$ can be extended to a measure on the minimal σ-field over $\mathcal{C}$. If, moreover, μ is σ-finite, then the extension is determined and is σ-finite.*

We prove the extension theorem by means of an intermediate weaker extension which preserves a part only of the properties characterizing a measure. We shall need various notions that we collect here.

A set function μ^o on the class $S(\Omega)$ of all sets in the space Ω is called an *outer measure* if it is sub σ-additive, nondecreasing, and takes the value 0 at $\emptyset$:

$$\mu^o(\bigcup A_j) \leqq \sum \mu^o(A_j) \text{ for every countable class } \{A_j\},$$

$$\mu^o(A) \leqq \mu^o(B) \text{ for } A \subset B, \quad \mu^o(\emptyset) = 0.$$

A set A is called *μ^o-measurable* if, for every set $D \subset \Omega$,

$$\mu^o(D) \geqq \mu^o(AD) + \mu^o(A^cD).$$

Since the relation is always true when $\mu^o(D) = \infty$, it suffices to consider sets D with $\mu^o(D) < \infty$. Since μ^o is sub σ-additive, the reverse inequality is always true and, hence, A is μ^o-measurable if, and only if,

$$\mu^o(D) = \mu^o(AD) + \mu^o(A^cD).$$

The class of all μ^o-measurable sets will be denoted by $\mathcal{A}^o$ and, clearly, contains $\emptyset$ and Ω. The *outer extension* of a measure μ given on a field $\mathcal{C}$ is defined for all sets $A \subset \Omega$ by

$$\mu^o(A) = \inf \sum \mu(A_j),$$

where the infimum is taken over all countable classes $\{A_j\} \subset \mathcal{C}$ such that $A \subset \bigcup A_j$—*coverings* in $\mathcal{C}$ of A, for short. Since $\Omega \in \mathcal{C}$, there is at least one covering (consisting of Ω) in $\mathcal{C}$ of every A so that the definition of an outer extension is justified. The use of the same symbol μ^o both for an outer measure and an outer extension is due to the property, to be proved first, that the outer extension of the measure μ on $\mathcal{C}$ is an extension of μ to an outer measure. Next we shall prove that the restriction to $\mathcal{A}^o$ of μ^o is a measure and that $\mathcal{A}^o$ is a σ-field, and the extension theorem will follow.

a. *The outer extension μ^o of a measure μ on a field $\mathcal{C}$ is an extension of μ to an outer measure.*

Proof. We prove first that μ^o is an extension of μ.

If $A \in \mathcal{C}$, then $\mu^o(A) \leqq \mu(A)$. On the other hand, since μ is a measure, $\mu(A) \leqq \sum \mu(A_j)$ for every covering $\{A_j\}$ in $\mathcal{C}$ of A, so that $\mu(A) \leqq \mu^o(A)$ and, hence, $\mu^o(A) = \mu(A)$ for $A \in \mathcal{C}$. It remains to prove that μ^o is an outer measure.

To begin with, $\mu^o(\emptyset) = 0$ since $\emptyset \in \mathcal{C}$. Furthermore, $\mu^o(A) \leqq \mu^o(B)$ for $A \subset B$, since every covering in $\mathcal{C}$ of B is also a covering of A. Finally, we prove that μ^o is sub σ-additive.

Let $\epsilon > 0$ and let $\{A_j\}$ be an arbitrary countable class. For every A_j there is a covering $\{A_{jk}\}$ in $\mathcal{C}$ such that

$$\sum_k \mu(A_{jk}) \leqq \mu^o(A_j) + \frac{\epsilon}{2^j}.$$

Since $\bigcup_j A_j \subset \bigcup_{j,k} A_{jk}$, it follows that

$$\mu^o(\bigcup_j A_j) \leqq \sum_{j,k} \mu(A_{jk}) \leqq \sum_j \mu^o(A_j) + \epsilon,$$

and, $\epsilon > 0$ being arbitrarily close to zero, sub σ-additivity is proved.

b. *If μ^o is an outer measure, then the class $\mathcal{A}^o$ of μ^o-measurable sets is a σ-field and μ^o on $\mathcal{A}^o$ is a measure.*

Proof. We prove first that $\mathcal{A}^o$ is a field and μ^o on $\mathcal{A}^o$ is a content.

If $A \in \mathcal{A}^o$, then $A^c \in \mathcal{A}^o$, since the definition of μ^o-measurability is symmetric in A and A^c. If $A, B \in \mathcal{A}^o$, then $AB \in \mathcal{A}^o$, since

$$\begin{aligned}\mu^o(D) &= \mu^o(AD) + \mu^o(A^cD)\\ &= \mu^o(ABD) + \mu^o(AB^cD) + \mu^o(A^cBD) + \mu^o(A^cB^cD)\\ &\geqq \mu^o(ABD) + \mu^o(AB^cD \cup A^cBD \cup A^cB^cD)\\ &= \mu^o(ABD) + \mu^o(AB)^cD.\end{aligned}$$

Thus $\mathcal{A}^o$ is closed under complementations and finite intersections and, hence, under finite unions, so that $\mathcal{A}^o$ is a field.

μ^o is finitely additive on $\mathcal{A}^o$ since, if $A, B \in \mathcal{A}^o$ and are disjoint,

$$\mu^o(A + B) = \mu^o((A + B)A) + \mu^o((A + B)A^c) = \mu^o(A) + \mu^o(B).$$

Since $\mu^o(A) \geqq \mu^o(\emptyset) = 0$, μ^o on $\mathcal{A}^o$ is a content.

To complete the proof, it suffices to show that, if the $A_n \in \mathcal{A}^o$ are disjoint, then $A = \sum A_n \in \mathcal{A}^o$ and $\mu^o(A) = \sum \mu^o(A_n)$.

Since $B_n = \sum_{k=1}^{n} A_k \in \mathcal{A}^o$, we have

$$\mu^o(D) = \mu^o(B_nD) + \mu^o(B_n{}^cD) \geq \sum_{k=1}^{n} \mu^o(A_kD) + \mu^o(A^cD)$$

and, letting $n \to \infty$,

$$\mu^o(D) \geqq \sum \mu^o(A_nD) + \mu^o(A^cD) \geq \mu^o(AD) + \mu^o(A^cD).$$

The inequality between the extreme sides shows that $A \in \mathcal{A}^o$. The first inequality with D replaced by A becomes

$$\mu^o(A) \geqq \sum \mu^o(A_n)$$

while the reverse inequality is always true.

Thus

$$\mu^o(A) = \sum \mu^o(A_n),$$

and the proof is complete.

REMARK. Most frequently, a measure μ is given on a class $\mathcal{D}$ whose closure under finite summations or under countable summations is a field $\mathcal{C}$. Then the requirement of σ-additivity determines the unique extension of μ on $\mathcal{C}$.

We are now in a position to prove the extension theorem.

1° For every $A \in \mathcal{C}$ and every D there is, for every $\epsilon > 0$, a covering $\{A_j\}$ in $\mathcal{C}$ of D such that

$$\mu^o(D) + \epsilon \geqq \sum \mu(A_j) = \sum \mu(AA_j) + \sum \mu(A^cA_j) \geqq \mu^o(AD) + \mu^o(A^cD).$$

Thus, $A \in \mathcal{A}^o$ and, hence, since the field $\mathcal{C}$ is contained in the σ-field $\mathcal{A}^o$, the minimal σ-field $\mathcal{A}$ over $\mathcal{C}$ is contained in $\mathcal{A}^o$. It follows, according to **a** and **b**, that the contraction on $\mathcal{A}$ of the measure μ^o on $\mathcal{A}^o$ is an extension of μ to a measure on $\mathcal{A}$. This proves the first part of the theorem.

2° Let μ on $\mathcal{C}$ be finite, let μ_1 and μ_2 be two extensions of μ to measures on $\mathcal{A}$, and let $\mathfrak{M} \subset \mathcal{A}$ be the class on which μ_1 and μ_2 agree. Since Ω belongs to $\mathcal{C}$, $\mu_1(\Omega) = \mu_2(\Omega) = \mu(\Omega) < \infty$; hence μ_1 and μ_2 are finite. Since $\mathfrak{M}$ contains $\mathcal{C}$ and, for every monotone sequence $A_n \in \mathfrak{M}$,

$$\mu_1(\lim A_n) = \lim \mu_1(A_n) = \lim \mu_2(A_n) = \mu_2(\lim A_n),$$

$\mathfrak{M}$ is a monotone class. It follows, by 1.6**A**, that $\mathfrak{M}$ contains the minimal σ-field $\mathcal{A}$ over the field $\mathcal{C}$ and, therefore, μ_1 and μ_2 agree on $\mathcal{A}$.

Let now μ on $\mathcal{C}$ be σ-finite so that there is a countable class $\{A_j\} \subset \mathcal{C}$ with μA_j finite which covers Ω. Thus, the foregoing result applies to every subspace A_j, and the second part of the theorem follows.

Generalization. The extension theorem is valid for σ-finite *signed* measures $\varphi = \mu' - \mu''$. Extend μ' and μ'' and observe that 2° applies with φ instead of μ.

Completion. Given a measure μ on a σ-field $\mathcal{A}$, it is always possible to extend μ to a larger σ-field obtained as follows: For every $A \in \mathcal{A}$ and an arbitrary subset N of a *null* set of $\mathcal{A}$, that is, a set of measure zero, set $\mu(A \cup N) = \mu(A)$. Clearly, the class of all sets $A \cup N$ is a σ-field $\mathcal{A}_\mu \supset \mathcal{A}$ and μ on $\mathcal{A}_\mu$ is an extension of μ to a measure on $\mathcal{A}_\mu$. $\mathcal{A}_\mu$ is called the *completion* of $\mathcal{A}$ for μ and μ on $\mathcal{A}_\mu$ is called a *complete* measure. It is easily seen that $\mathcal{A}_\mu \subset \mathcal{A}^o$, so that the extension theorem provides us automatically with extensions to complete measures.

4.2 Product probabilities. A measure on a class containing the space is called a *normed measure* or a *probability* when its value for the whole space is one; we reserve the symbol P, with or without affixes, for such measures.

Let $(\Omega_t, \mathcal{A}_t, P_t)$, $t \in T$, be *probability spaces*, that is, triplets consisting of a space Ω_t of points ω_t, a σ-field $\mathcal{A}_t$ of measurable sets A_t (with or without superscripts) in Ω_t, and a probability P_t on $\mathcal{A}_t$. Let $\mathcal{C}_T$ be the class of all measurable cylinders of the form $\prod_{t \in T_N} A_t \times \prod_{t \in T - T_N} \Omega_t$ in the product measurable space $(\prod \Omega_t, \prod \mathcal{A}_t)$. The class $\mathcal{B}_T$ of all finite sums of these cylinders is a field, and the minimal σ-field $\mathcal{A}_T$ over $\mathcal{B}_T$ is, by definition, the product σ-field $\prod \mathcal{A}_t$. The *product probability* $P_T = \prod P_t$ on the class $\mathcal{C}_T$ is defined by assigning to every interval cylinder the product of the probabilities of its sides: in symbols,

$$P_T\Big(\prod_{t \in T_N} A_t \times \prod_{t \in T-T_N} \Omega_t\Big) = \prod_{t \in T_N} P_t A_t \cdot \prod_{t \in T-T_N} P_t \Omega_t = \prod_{t \in T_N} P_t A_t.$$

Clearly, $P_T \Omega_T = 1$ and P_T on $\mathcal{C}_T$ is finitely additive and determines its extension to a finitely additive set function P_T on $\mathcal{B}_T$. The defining term "product-probability" is justified by the following theorem (Andersen and Jessen).

A. PRODUCT PROBABILITY THEOREM. *The product probability P_T on $\mathcal{B}_T$ is σ-additive and determines its extension to a probability P_T on the product σ-field $\mathcal{A}_T$.*

Thus, the triplet $(\Omega_T, \mathcal{A}_T, P_T)$ is a probability space, to be called the *product probability space.*

Proof. 1° On account of the extension theorem, it suffices to prove that P_T on $\mathcal{B}_T$ is σ-additive. Since it is obviously finitely additive on $\mathcal{B}_T$, on account of the continuity theorem for additive set functions it suffices to prove that P_T on $\mathcal{B}_T$ is continuous at $\emptyset$. *Ab contrario*, given $\epsilon > 0$ arbitrarily close to 0, it suffices to prove that, for every nonincreasing sequence of measurable cylinders $A^n \downarrow A$ with $P_T A^n > \epsilon$ for every n, the limit set A is not empty. Since every cylinder A^n depends only upon a finite subset of indices, the set of all indices involved in defining the sequence A^n is countable. By interchanging, if necessary, the indices, we can restrict ourselves to the product space $\Omega = \prod \Omega_n$ and set $\Omega'_n = \Omega_{n+1} \times \Omega_{n+2} \times \cdots$.

If the set of all indices is finite, then there is an integer N such that, for every n, all the factors which follow the Nth one reduce to Ω_N, and the argument below applies with corresponding modifications.

2° Let $P'_1, P'_2, \cdots$ be the set functions defined on the fields $\mathcal{B}_1'$, $\mathcal{B}'_2, \cdots$ of all measurable cylinders in $\Omega'_1, \Omega'_2, \cdots$, as P_T is defined on $\mathcal{B}_T$. Let $A^n(\omega_1), A^n(\omega_1, \omega_2), \cdots$ be the sections of A^n at $\omega_1 \in \Omega_1$, $(\omega_1, \omega_2) \in \Omega_1 \times \Omega_2$, etc. Clearly, $A^n(\omega_1) \in \mathcal{B}'_1$. It is easily seen that, if $B_1{}^n$ is the set of all ω_1 such that

$$P'_1 A^n(\omega_1) > \frac{\epsilon}{2},$$

then

$$P_1B_1^n + \frac{\epsilon}{2}(1 - P_1B_1^n) \geq P_TA^n > \epsilon$$

and, hence,

$$P_1B_1^n > \frac{\epsilon}{2}.$$

Since $A^n \downarrow$ implies that $B_1^n \downarrow$, it follows that, for $B_1 = \lim B_1^n$, $P_1B_1 \geq \frac{\epsilon}{2}$. Thus, B_1 is not empty; hence, there is a point $\bar{\omega}_1 \in \Omega_1$ common to all B_1^n and, for every n, $P'_1(A^n(\bar{\omega}_1)) > \frac{\epsilon}{2}$. The same argument applied to $A^n(\bar{\omega}_1) \downarrow$ yields a point $\bar{\omega}_2 \in \Omega_2$ such that $P'_2(A^n(\bar{\omega}_1, \bar{\omega}_2)) > \frac{\epsilon}{2^2}$, and so on. Therefore, the point $\bar{\omega} = (\bar{\omega}_1, \bar{\omega}_2, \cdots)$ is common to all A^n, so that the limit set A is not empty, and the proof is complete.

We pass now to Borel spaces.

4.3 Consistent probabilities on Borel fields. We introduce the following terminology. The set $R = (-\infty, +\infty)$ of all finite numbers x is a *real line*, the minimal σ-field over the class of all intervals is the *Borel field* $\mathcal{B}$ in R, the elements of $\mathcal{B}$ are *Borel sets* in R, and the measurable space $(R, \mathcal{B})$ is a *Borel line*. Similarly, the product space $R_T = \prod R_t$, where every R_t is a real line with points x_t, is a *real space* with points $x_T = (x_t)$, the product σ-field $\mathcal{B}_T = \prod \mathcal{B}_t$, where every $\mathcal{B}_t$ is the Borel field in R_t, is the *Borel field* in R_T whose elements are *Borel sets* in R_T, and the measurable space $(R_T, \mathcal{B}_T)$ is a *Borel space*. If T is a finite set, we say that R_T is a finite product space. Cylinders with Borel bases are *Borel cylinders* and, clearly, the Borel field $\mathcal{B}_T$ is the minimal σ-field over the class of all Borel cylinders or, equivalently, over the class of all cylinders whose bases are product Borel sets.

Given a finite measure on $\mathcal{B}_T$ we can assume, by dividing it by its value for R_T, that it is a probability P_T. Let $T_N = \{t_1, \cdots t_N\}$ be a finite subset of indices and let $(R_{T_N}, \mathcal{B}_{T_N})$ be the corresponding Borel space. We define on $\mathcal{B}_{T_N}$ the *marginal probability* P_{T_N}, or *projection of P on R_{T_N}*, by assigning to every Borel set B_{T_N} in R_{T_N} the measure of the cylinder with basis B_{T_N}; in symbols

$$P_{T_N}(B_{T_N}) = P_T(B_{T_N} \times R'_{T_N}), \quad R'_{T_N} = \prod_{t \notin T_N} R_t.$$

Marginal probabilities are *consistent* in the following sense. If R' and R'' are two finite product subspaces of R_{T_N} with marginal measures P'

and P'', respectively, then the projections of P' and P'' on their common subspace, if any, coincide (with the projection of P_T on this subspace). We want to prove that the converse is true (Daniell, Kolmogorov).

A. Consistency theorem. *Consistent probabilities P_{T_N} on Borel fields of all finite product subspaces R_{T_N} of R_T determine a probability P_T on the Borel field in R_T such that every P_{T_N} is the projection of P_T on R_{T_N}.*

Proof. To every Borel cylinder with Borel base B_{T_N} in R_{T_N} we assign the probability value

$$P_T(B_{T_N} \times R'_{T_N}) = P_{T_N}(B_{T_N}).$$

It is easily seen that P_T on the class $\mathcal{C}_T$ of all Borel cylinders is finitely additive, and the theorem will follow from the extension theorem if we prove that P_T on $\mathcal{C}_T$ is continuous at $\emptyset$.

As in the proof of the product probability theorem, it suffices to prove that, given $\epsilon > 0$ arbitrarily close to zero, if a sequence $A_n \downarrow A$ of Borel cylinders with bases B_n formed by finite sums of intervals in $R_1 \times \cdots \times R_n$ is such that, for every n,

$$P_T(A_n) = P_{12\cdots n}(B_n) > \epsilon,$$

then A is not empty. To simplify the writing, set $P = P_T$ and $P_n = P_{12\cdots n}$. Since P_n is bounded and continuous from below, in every interval in $R_1 \times \cdots \times R_n$ we can find a bounded closed interval whose P_n-measure is as close as we wish to that of the original interval. Therefore, in every B_n, we can find a bounded closed Borel set B'_n—formed by a finite sum of bounded closed intervals—such that $P_n(B_n - B'_n) < \frac{\epsilon}{2^{n+1}}$ and, hence, if A'_n is the Borel cylinder with basis B'_n, then

$$P(A_n - A'_n) = P_n(B_n - B'_n) < \frac{\epsilon}{2^{n+1}}.$$

It follows, setting $C_n = A'_1 \cap \cdots \cap A'_n$, that $P(A_n - C_n) < \frac{\epsilon}{2}$ or, since $C_n \subset A'_n \subset A_n$,

$$P(C_n) > P(A_n) - \frac{\epsilon}{2} > \frac{\epsilon}{2}.$$

Thus every C_n is nonempty and we can select in it a point $x^{(n)} = (x_1^{(n)}, x_2^{(n)}, \cdots)$. It follows from $C_1 \supset C_2 \supset \cdots$ that for every $p = 0, 1, \cdots, x^{(n+p)} \in C_n \subset A'_n$ and hence $(x_1^{(n+p)}, \cdots, x_n^{(n+p)}) \in B'_n$.

Since every B'_n is bounded, we can select a subsequence n_{1k} of integers such that $x_1^{(n_{1k})} \to x_1$ as $k \to \infty$, then within it a subsequence n_{2k} such that $x_2^{(n_{2k})} \to x_2$, and so on. The diagonal subsequence of points $x^{(n_{kk})} = (x_1^{(n_{kk})}, x_2^{(n_{kk})}, \cdots)$ converges to the point $x = (x_1, x_2, \cdots)$ and $(x_1^{(n_{kk})}, \cdots, x_m^{(n_{kk})}) \to (x_1, \cdots, x_m) \in B'_m$ for every m. Therefore, $x \in A'_m \subset A_m$ whatever be m so that $x \in \bigcap_{m=1}^{\infty} A_m$. Thus this intersection is not empty, and the assertion is proved.

Extensions. The foregoing theorem can be extended, as follows: Let $\mathfrak{A}_n$ be the σ-field of Borel cylinders with bases in $R_1 \times \cdots \times R_n$, and let $\mathfrak{A}_\infty$ be the Borel field in $\prod R_n$.

1° *If uniformly bounded measures μ_n on $\mathfrak{A}_n$ form a nondecreasing sequence, in the sense that $\mu_n A_n \leqq \mu_{n+1} A_n \leqq \cdots$ and hence $\mu_p A_n \uparrow \mu A_n$ as $p \to \infty$ whatever be n and $A_n \in \mathfrak{A}_n$, then μ extends to a bounded measure on $\mathfrak{A}_\infty$.*

The proof reduces to the previous one as follows. The set function μ so defined on the field $\bigcup \mathfrak{A}_n$ of all Borel cylinders in $\prod R_n$ is, clearly, finitely additive and bounded. Therefore, it suffices to prove that on this field μ is continuous at $\emptyset$. Given $\epsilon > 0$ and $A_n \in \mathfrak{A}_n$, we can find p sufficiently large so that $\mu_p A_n + \frac{\epsilon}{2^{n+2}} > \mu A_n$. Then we can select a Borel cylinder $A'_n \subset A_n$ whose basis is a closed and bounded Borel set in $R_1 \times \cdots \times R_n$ such that $\mu_p(A_n - A'_n) < \frac{\epsilon}{2^{n+2}}$. It follows that

$$\mu A'_n + \frac{\epsilon}{2^{n+1}} \geqq \mu_p A_n + \frac{\epsilon}{2^{n+1}} > \mu A_n$$

so that $\mu(A_n - A'_n) < \frac{\epsilon}{2^{n+1}}$. From here on, the end of the preceding proof applies word for word.

If φ_n on $\mathfrak{A}_n$, $n = 1, 2, \cdots$, are such that $\varphi_n(A_n) = \varphi_{n+1}(A_n) = \cdots$, $A_n \in \mathfrak{A}_n$, we say that the φ_n are *consistent.*

2° *If the uniformly bounded σ-additive set functions φ_n on $\mathfrak{A}_n$ are consistent, hence $\varphi_p(A_n) \to \varphi(A_n)$ as $p \to \infty$ whatever be n and $A_n \in \mathfrak{A}_n$, then φ extends to a σ-additive bounded set function on $\mathfrak{A}_\infty$.*

The assertion follows from what precedes. For, clearly, the total variations $\bar{\varphi}_n$ on $\mathfrak{A}_n$ form a nondecreasing bounded sequence on $\bigcup \mathfrak{A}_n$, in the sense of 1°. Hence lim $\bar{\varphi}_n$ is continuous at $\emptyset$ on $\bigcup \mathcal{C}_n$ and, *a fortiori*, so is φ. Now use Jordan decomposition and generalization in **4.1.**

4.4 Lebesgue-Stieltjes measures and distribution functions. Complete measures on the Borel field in a real line $R = (-\infty, +\infty)$ did, and still do, play a prominent role. However, being set functions, they are not easy to handle with the tools of classical analysis, for methods of analysis were developed to deal primarily with finite point functions on R. It is, therefore, of the greatest methodological importance to establish a link between the modern notion of measure and the classical notions. This will be done by showing that there is a class of point functions on R which can be placed in a one-to-one correspondence with a very wide class of measures. In this manner, investigations of measures (and, thereafter, of integrals) will be reduced to investigations of the corresponding point functions and, thus, the familiar methods of analysis will apply. Whatever be these point functions they will be said *to represent* the corresponding measure.

Among possible representations of measures there are two which are fundamental: "distribution functions" which represent measures assigning finite values to finite intervals, to be called *Lebesgue-Stieltjes* (*L.S.*) *measures*, that we shall introduce now, and "characteristic functions" which represent the subclass of finite Lebesgue-Stieltjes measures required in connection with probability problems—that we shall introduce in Part II. Let $\mathcal{B}$ be the Borel field in R and let μ be a Lebesgue-Stieltjes measure. The completion of $\mathcal{B}$ for μ will be denoted by $\mathcal{B}_\mu$, and called a *Lebesgue-Stieltjes field* in R, and its elements will be called *Lebesgue-Stieltjes sets* in R.

A function on R which is finite, nondecreasing, and continuous from the left is called a *distribution function* (d.f.). Two d.f.'s will be said to be *equivalent* if they differ by some fixed but arbitrary constant. This notion of equivalence has the usual properties of equivalence—it is reflexive, transitive, and symmetric. Thus, the class of all d.f.'s splits into equivalence classes. As the correspondence theorem below (Lebesgue, Radon) shows, the one-to-one correspondence between L.S.-measures and d.f.'s is not a correspondence between L.S.-measures and individual d.f.'s but a correspondence between L.S.-measures and classes of equivalent d.f.'s, each class to be represented by one of its elements, arbitrarily chosen.

Let F, with or without affixes, denote a d.f. and define its *increment function* by

$$F[a, b) = F(b) - F(a), \quad -\infty < a \leqq b < +\infty.$$

Since two equivalent d.f.'s have the same increment function and conversely, it follows that every class of equivalent d.f.'s is characterized by its increment function. Moreover, the defining properties of d.f.'s are equivalent to the following:

$$\text{(i)}\quad 0 \leqq F[a, b) < \infty, \qquad \text{(ii)}\quad F[a, b) \to 0 \text{ as } a \uparrow b,$$

and

$$\text{(iii)}\quad \sum_{k=1}^{n} F[a_k, b_k) + \sum_{k=1}^{n-1} F[b_k, a_{k+1}) = F[a_1, b_n)$$

where $a < b$, $a_1 \leqq b_1 \leqq a_2 \leqq \cdots \leqq a_n \leqq b_n$ are arbitrary.

A. CORRESPONDENCE THEOREM. *The relation*

$$\mu[a, b) = F[a, b), \quad -\infty < a \leqq b < +\infty$$

establishes a one-to-one correspondence between L.S.-measures μ and d.f.'s F defined up to an equivalence.

Proof. Let $\mathcal{B}_I$ be the class of all intervals $[a, b)$, $-\infty < a \leq b < +\infty$. $\mathcal{B}_I$ is closed under formation of finite intersections. The minimal field $\mathcal{B}_0$ over $\mathcal{B}_I$ is the class of all finite sums of elements of $\mathcal{B}_I$ and of intervals of the form $(-\infty, a)$, $[b + \infty)$, and the minimal σ-field over $\mathcal{B}_0$ is the Borel field $\mathcal{B}$.

The proof of the correspondence theorem is summarized by the diagram below, where c represents an arbitrary constant:

$$F + c \text{ on } R \Leftrightarrow \mu \text{ on } \mathcal{B}_I \Leftrightarrow \mu \text{ on } \mathcal{B}_0 \Leftrightarrow \mu \text{ on } \mathcal{B} \Leftrightarrow \mu \text{ on } \mathcal{B}_\mu.$$

1° *μ on $\mathcal{B}_\mu \Rightarrow F + c$ on R.* For, μ on $\mathcal{B}_\mu$ determines its restriction to $\mathcal{B}_I$ and, from properties of L.S.-measures it follows that the relation

$$F[a, b) = \mu[a, b)$$

determines an increment function with properties (i), (ii), and (iii) given above.

2° *μ on $\mathcal{B}_0 \Rightarrow \mu$ on $\mathcal{B}_\mu$.* For, R being a denumerable sum of finite intervals, the measure μ on $\mathcal{B}_0$ is σ-finite and the extension theorem applies followed by completion.

3° *μ on $\mathcal{B}_I \Rightarrow \mu$ on $\mathcal{B}_0$.* It suffices to prove that if $A = \sum_k I_k \in \mathcal{B}_0$, $I_k \in \mathcal{B}_I$, then $\mu(A)$ is determined by the σ-additivity requirement $\mu(A) = \sum_k \mu(I_k)$, that is, if A can also be written as $\sum_j I'_j$, where I'_j

$\in \mathcal{B}_I$, then $\sum_j \mu(I'_j) = \sum_k \mu(I_k)$. Since μ on $\mathcal{B}_I$ is additive and

$$I'_j = AI'_j = \sum_k I_k I'_j, \quad I_k = AI_k = \sum_j I'_j I_k,$$

it follows that

$$\sum_j \mu(I'_j) = \sum_j \sum_k \mu(I_k I'_j) = \sum_k \sum_j \mu(I'_j I_k) = \sum_k^n \mu(I_k),$$

and the assertion is proved.

4° $F + c \Rightarrow \mu$ *on* $\mathcal{B}_I$. We have to prove that the relation $\mu[a, b) = F[a, b)$ determines a measure μ on $\mathcal{B}_I$, that is, if $I = \sum I_n$, where $I = [a, b)$ and $I_n = [a_n, b_n)$, then $\mu I = \sum \mu I_n$. By interchanging, if necessary, the subscripts, we can assume that, for every n,

$$a \leqq a_1 \leqq b_1 \leqq \cdots \leqq a_n \leqq b_n \leqq b.$$

It follows that

$$\sum_{k=1}^n \mu(I_k) = \sum_{k=1}^n F[a_k, b_k) \leqq \sum_{k=1}^n F[a_k, b_k) + \sum_{k=1}^{n-1} F[b_k, a_{k+1})$$
$$= F[a_1, b_n) \leqq F[a, b) = \mu(I),$$

and, letting $n \to \infty$, we get $\sum \mu(I_n) \leqq \mu(I)$.

It remains for us to prove the reverse inequality. We exclude the trivial case $a = b$, select $\epsilon > 0$ such that $\epsilon < b - a$ and set $I^\epsilon = [a, b - \epsilon]$. Because of the continuity from the left, for every n there is an $\epsilon_n > 0$ such that $F[a_n - \epsilon_n, a_n) < \frac{\epsilon}{2^n}$. If $I_n^\epsilon = (a_n - \epsilon_n, b_n)$, then, from $I^\epsilon \subset \bigcup I_n^\epsilon$ it follows, by the Heine-Borel lemma, that there is an n_0 finite such that $I^\epsilon \subset \bigcup_{k=1}^{n_0} I_k^\epsilon$. Let $k_1 \leqq n_0$ be such that $a \in I_{k_1}^\epsilon$ and, if $b_{k_1} < b$, then let $k_2 \leqq n_0$ be such that $b_{k_1} \in I_{k_2}$. Continue in this manner until some $b_{k_m} \geqq b - \epsilon$—the process necessarily stops for some $m \leqq n_0$. Omitting intervals that were not selected and, if necessary, changing the subscripts, it follows that $I^\epsilon \subset \bigcup_{k=1}^m I_k^\epsilon$ and

$$a_1 - \epsilon_1 < a < b_1, \quad a_{k+1} - \epsilon_{k+1} < b_k < b_{k+1}$$

for

$$k = 1, 2, \cdots m - 1, \quad a_m - \epsilon_m < b - \epsilon \leqq b_m.$$

Therefore,

$$F[a, b - \epsilon) \leqq F[a_1 - \epsilon_1, b_m) = F[a_1 - \epsilon_1, b_1) + \sum_{k=1}^{m-1} F[b_k, b_{k+1})$$
$$\leqq \sum_{k=1}^{m} F[a_k - \epsilon_k, b_k) \leqq \sum_{k=1}^{\infty} F[a_k, b_k) + \epsilon$$

and, letting $\epsilon \to 0$,

$$F[a, b) \leqq \sum F[a_n, b_n), \text{ that is, } \mu(I) \leqq \sum \mu(I_n),$$

which completes the proof of the final assertion and, hence, of the correspondence theorem.

Particular case. If F is defined, up to an additive constant, by $F(x) = x$, $x \in R$, then the corresponding measure of an interval is its "length." The extension of "length" to a measure μ on $\mathcal{B}$ and the completed measure μ on $\mathcal{B}_\mu$ are called *Lebesgue measure* on $\mathcal{B}$ or $\mathcal{B}_\mu$, respectively, and $\mathcal{B}_\mu$ will be called *Lebesgue field.* The Lebesgue measure is at the root of the general notion of measure.

REMARK. We can define a L.S.-measure on the Borel field $\bar{\mathcal{B}}$-minimal σ-field over the class of all intervals in $\bar{R} = [-\infty, +\infty]$ and, hence, on $\bar{\mathcal{B}}_\mu$, by adjoining to a L.S.-measure on $\mathcal{B}$, arbitrary measures for the sets $\{-\infty\}$ and $\{+\infty\}$.

Extension. The preceding definitions, proofs, and results, remain valid, word for word, if Borel lines are replaced by finite-dimensional Borel spaces $R^N = R_1 \times \cdots \times R_N$, provided the following interpretation of symbols is used: $a, b, x, \cdots$ are points in R^N, say, $a = (a_1, \cdots, a_N)$; $a < b (a \leqq b)$ means that $a_k < b_k (a_k \leqq b_k)$ for $k = 1, \cdots, N$. F on R^N is a function with values $F(a) = F(a_1, \cdots, a_N)$ and increments $F[a, b)$ are defined by

$$F[a, b) = \Delta_{b-a}F(a) = \Delta_{b_1-a_1} \cdots \Delta_{b_N-a_N} F(a_1, a_2, \cdots a_N)$$

where, for every k, $\Delta_{b_k-a_k}$ denotes the difference operator of step $b_k - a_k$ acting on a_k. For instance, if $N = 2$,

$$\Delta_{b-a}F(a) = \Delta_{b_1-a_1}\Delta_{b_2-a_2}F(a_1, a_2) = \Delta_{b_1-a_1}\{F(a_1, b_2) - F(a_1, a_2)\}$$
$$= F(b_1, b_2) - F(a_1, b_2) - F(b_1, a_2) + F(a_1, a_2)$$

and, in particular, if $F(a_1, a_2) = a_1a_2$ is the area of the rectangle with sides 0 to a_1 and 0 to a_2, then $\Delta_{b-a}F(a) = (b_1 - a_1)(b_2 - a_2)$ is the area of the rectangle with sides a_1 to b_1 and a_2 to b_2.

The defining properties of a d.f. F on R^N become:

$$-\infty < F < +\infty, \quad F[a, b) = \Delta_{b-a}F(a) \geqq 0, \quad F[a, b) \to 0$$

as $a \uparrow b$, that is, $a_1 \uparrow b_1, \cdots, a_N \uparrow b_N$.

Product-d.f.'s and product-measures. A very important particular case is that of *product-d.f.'s:*

$$F(x_1, \cdots, x_N) = \prod_{k=1}^{N} F_k(x_k), \quad x_k \in R_k$$

where the F_k on R_k are d.f.'s. Then F on R^N is a d.f., for,

$$\Delta_{b-a}F(a) = \prod_{k=1}^{N} \Delta_{b_k-a_k}F_k(a_k) \geqq 0$$

and the other defining properties are clearly satisfied.

Every d.f. F_k determines a measure μ_k on the Borel field in R_k, by means of the relation $\mu_k[a_k, b_k) = F_k[a_k, b_k)$, and the measure μ on the product Borel field determined by means of the relation $\mu[a, b) = F[a, b)$, is clearly the product-measure $\prod_{k=1}^{N} \mu_k$.

Let now F_n be d.f.'s with $F_n(+\infty) - F_n(-\infty) = 1$, so that the measures μ_n are probabilities. Then, by the product-probability theorem or by the consistency theorem,

B. *A sequence F_n of d.f.'s corresponding to probabilities on R_n determines a product-probability on the Borel field in the product space $\prod R_n$.*

This result extends at once to any set $\{F_t, t \in T\}$, of such d.f.'s.

COMPLEMENTS AND DETAILS

In one guise or another, and especially when they are indefinite integrals, signed measures on a fixed σ-field are in constant use in measure theory and probability theory. Many of the properties established in this book are but properties of such set functions.

Notation. The measurable sets belong to a fixed σ-field on which the set functions and limits of their sequences are defined. Unless otherwise stated and with or without affixes, $A, B, \cdots$ denote sets, μ denotes a measure, φ denotes a signed measure.

1. If φ is σ-finite, then there are only countably many disjoint sets for which $\varphi \neq 0$ in every class.

2. For every A there exists a $B \subset A$ such that $\bar{\varphi}(A) \leqq 2|\varphi(B)|$.

3. If $\varphi_1 \leqq \varphi_2$, then $\varphi_1^+ \leqq \varphi_2^+$, $\varphi_1^- \geqq \varphi_2^-$. If $\varphi = \varphi_1 \pm \varphi_2$, then $\varphi^\pm \leqq \varphi_1^\pm + \varphi_2^\pm$.

4. Minimality of the Jordan-Hahn decomposition. If $\varphi = \mu^+ - \mu^-$, then $\varphi^\pm \leqq \mu^\pm$.

We say that A is a *φ-null set*, if $\varphi = 0$ on $\{AA', A' \in \mathfrak{A}\}$. We say that A and B are *φ-equivalent*, if they coincide up to a φ-null set. We say that a nonempty set is a *φ-atom*, if every measurable subset of A is φ-equivalent either to $\emptyset$ or to A.

5. The φ-null sets form a σ-ring; the φ-null sets of φ and of $\bar{\varphi}$ are the same. The φ-equivalence is an equivalence relation (reflexive, transitive, and symmetric), and $\mathcal{A}$ splits into φ-equivalence classes.

6. Every φ-null set and every measurable set consisting of one point is a φ-atom; $\bar{\varphi}(A) = |\varphi(A)|$ for every φ-atom A. Atoms of φ and $\bar{\varphi}$ are the same; atoms of φ are atoms of φ^+ and φ^-, but the converse is not necessarily true.

If A is a φ-atom, then $\varphi = 0$ or $\varphi(A)$ on $A \cap \mathcal{A}$; if φ is finite, then the converse is true. What if φ is σ-finite? What about $\varphi = \infty$ except for $\emptyset$?

7. If μ is finite, then $\Omega = \sum A_j + A$ where the A_j or A may be absent but, if present, then the A_j are μ-atoms of positive measure and, for every $B \subset A$ of positive measure, μ takes every value c between 0 and μB for measurable subsets of B. This decomposition of Ω is determined up to μ-null sets. Can μ be replaced by φ?

(There is only a countable number of μ-equivalence classes of such A_j's. Select representatives A_j of these classes and let $B \subset A = \Omega - \sum A_j$. Select inductively sets $C_n \in \mathcal{C}_n$ such that $\mu C_n > \sup \mu C - \dfrac{1}{n}$ for all $C \in \mathcal{C}_n$, where $\mathcal{C}_n$ is the class of all $C \subset B - (C_1 \cup C_2 \cdots \cup C_{n-1})$ for which $\mu C \leqq c - \mu(C_1 \cup C_2 \cup \cdots \cup C_{n-1})$. Then $\mu C = c$ for $C = \bigcup C_n$.)

8. If φ is finitely additive, μ is finite, and $\mu A_n \to 0$ implies $\varphi A_n \to 0$, then φ is σ-additive.

We say that φ is *φ_0-continuous* if $\varphi_0 A = 0$ implies $\varphi A = 0$.

9. If $\mu A_n \to 0$ implies $\varphi A_n \to 0 (\bar{\varphi} A_n \to 0)$, then φ is μ-continuous. If φ is finite, then the converse is true.

(Assume the contrary of the converse; there exist $\epsilon > 0$ and A_n such that $\mu A_n < \dfrac{1}{2^n}$ and $\bar{\varphi} A_n \geqq \epsilon$. Then $\mu B = 0$ and $\bar{\varphi} B \geqq \epsilon$ for $B = \limsup A_n$.)

What if φ is σ-finite? What about $\mathcal{A}$ consisting of all subsets of a denumerable space of points ω_n, and $\mu\{\omega_n\} = \dfrac{1}{2^n}$, $\varphi\{\omega_n\} = n$. What about μ replaced by φ_0?

10. If the μ_j are finite measures, then there exists a μ such that all the μ_j are μ-continuous. (Take $\mu = \sum \mu_j / 2^j \mu_j \Omega$.) What about μ_j's replaced by φ_j's?

Let $\mathcal{B} \subset \mathcal{A}$ be a σ-ring such that the measurable subsets of elements of $\mathcal{B}$ belong to $\mathcal{B}$. Let $\mathcal{B}(\varphi)$ be the class of sets such that their subsets which belong to $\mathcal{B}$ are φ-null. Call the sets of $\mathcal{B}$ "singular," and the sets of $\mathcal{B}(\varphi)$ "regular." Call φ regular (singular) if every singular (regular) set is φ-null.

Let $\varphi_r = \varphi_r^+ - \varphi_r^-$, $\varphi_s = \varphi_s^+ - \varphi_s^-$, defined by

$$\varphi_r^{\pm}(A) = \sup \varphi^{\pm}(B) \quad \text{for all regular} \quad B \subset A,$$

$$\varphi_s^{\pm}(A) = \sup \varphi^{\pm}(B) \quad \text{for all singular} \quad B \subset A.$$

11. Decomposition theorem. φ_r is regular, φ_s is singular, and $\varphi = \varphi_r + \varphi_s$. If φ is finite, then the decomposition of φ into a regular and a singular part is unique. What if φ is σ-finite? What if $\mathcal{A}$ consists of all subsets of a noncountable space, and $\varphi(A)$ equals the number of points of A? (Proceed as follows:

(i) $\mathcal{B}(\varphi) = \mathcal{B}(\bar{\varphi}) = \mathcal{B}(\varphi^+) \cap \mathcal{B}(\varphi^-)$ is a σ-field.

(ii) $\varphi_r(\varphi_s)$ is a regular (singular) signed measure.

(iii) Every A contains disjoint A_r regular and A_s singular such that $\varphi_r^{\pm}(A) = \varphi^{\pm}(A_r)$, $\varphi_s^{\pm}(A) = \varphi^{\pm}(A_s)$.

(iv) If $A = A'_r + A'_s$ with A'_r regular and A'_s singular, then we can take $A_r = A'_r$ and $A_s = A'_s$.

(v) If φ is finite, every A can be so decomposed.)

12. We can take for singular sets:

(i) the μ-null sets—regular (singular) becomes μ-continuous (μ-discontinuous);
(ii) the countable measurable sets—regular (singular) becomes continuous (purely discontinuous);
(iii) the countable sums of atoms—regular (singular) becomes nonatomic (atomic).

In each case investigate the regular and singular parts.

13. Intermediate-value theorem (compare with continuous function on a connected set). If A is nonatomic and $A_n \uparrow A$ with φA_n finite, then φ takes every value between $-\varphi^- A$ and $+\varphi^+ A$ for measurable subsets in A. (See *7.*) What if $\mathfrak{A}$ consists of all sets in a noncountable space, $\varphi(A) = 0$ or ∞ according as A is countable or not?

In what follows, the φ_n are σ-additive but, unless otherwise stated, lim φ_n is not assumed to be σ-additive.

14. If $\varphi_n \to \varphi$ σ-additive, then $\varphi^{\pm} \leqq \liminf \varphi_n^{\pm}$. If, moreover, $\varphi_n \uparrow$ or $\varphi_n \downarrow$, then $\varphi^{\pm} = \lim \varphi_n^{\pm}$.

15. If $\varphi_n \uparrow (\downarrow)$ and $\varphi_1 > -\infty (< +\infty)$, then $\varphi_n \to \varphi$ σ-additive.

16. If $\varphi_n \to \varphi$ uniformly on $\mathfrak{A}$ and $\varphi > -\infty$ or $\varphi < +\infty$, then φ is σ-additive.

17. To a measure space $(\Omega, \mathfrak{A}, \mu)$ associate a complete metric space $(\mathfrak{X}, d)$ as follows: $\mathfrak{X}$ is the space of all sets A, B of finite measure, d is a metric defined by $d(A, B) = \mu(AB^c + A^cB)$. Prove that the metric space is complete.

(If A_n is a mutually convergent sequence in $\mathfrak{X}$, then the sequence I_{A_n} mutually converges in measure and hence converges in measure—see 6.3.)

If ν on $\mathfrak{A}$ is a finite μ-continuous measure, then ν is defined and continuous on $(\mathfrak{X}, d)$.

We say that the φ_n are *uniformly μ-continuous* if $\mu A_m \to 0$ implies $\varphi_n A_m \to 0$ uniformly in n, as $m \to \infty$.

18. Let μ be σ-finite. If the finite φ_n are μ-continuous and lim φ_n exists and is finite, then the φ_n are uniformly μ-continuous and $\lim \varphi_n = \varphi$ is μ-continuous and σ-additive. (For every $\epsilon > 0$, set $A_k = \bigcap_{m=k}^{\infty} \bigcap_{n=k}^{\infty} \left[A \in \mathfrak{X}; |\varphi_m A - \varphi_n A| \leqq \frac{\epsilon}{3}\right]$. By (*17*), every A_k is closed. By Baire's category theorem, there exists k_0, d_0 and $A_0 \in \mathfrak{X}$ such that $[A \in \mathfrak{X}; d(A, A_0) < d_0] \subset A_{k_0}$. Let $0 < \delta_0 < d_0$ such that $|\varphi_n A| < \epsilon$ whenever $\mu A < \delta_0$ and $n \leqq k_0$. If $\mu A < \delta_0$, then $d(A_0 - A, A_0) < d_0$, $d(A_0 \cup A, A_0) < d_0$, and $|\varphi_n A| \leqq |\varphi_{k_0} A| + |\varphi_n(A_0 \cup A) - \varphi_{k_0}(A_0 \cup A)| + |\varphi_n(A_0 - A) - \varphi_{k_0}(A_0 - A)|$.)

19. If finite $\varphi_n \to \varphi$ finite, then φ is σ-additive. (If $|\varphi_n| \leqq c_n$, set $\mu A = \sum \frac{1}{2^n c_n} |\varphi_n A|$ and apply *18.*)

Chapter II

MEASURABLE FUNCTIONS AND INTEGRATION

§ 5. MEASURABLE FUNCTIONS

5.1 Numbers. Spaces built with numbers are prototypes of all spaces, and functions whose values are numbers are prototypes of all functions.

By a *number* x we mean either a usual real number—*finite number*—or one of the symbols $+\infty$ and $-\infty$—*infinite numbers*. These symbols are defined by the following properties:

$$-\infty \leqq x \leqq +\infty,$$

$$\pm\infty = (\pm\infty) + x = x + (\pm\infty), \quad \frac{x}{\pm\infty} = 0 \quad \text{if} \quad -\infty < x < +\infty,$$

$$x(\pm\infty) = (\pm\infty)x = \begin{cases} \pm\infty & \text{if} \quad 0 < x \leqq +\infty \\ 0 & \text{if} \quad x = 0 \\ \mp\infty & \text{if} \quad -\infty \leqq x < 0. \end{cases}$$

The expression $+\infty - \infty$ is meaningless, so that, when speaking of a "sum" of two numbers, we assume that, if one of them is $\mp\infty$, the other one is not $\pm\infty$; then the sum exists.

The reason for the introduction of infinite numbers lies in the fact that, then, $\sup x_t$ and $\inf x_t = -\sup(-x_t)$, where t varies over an arbitrary set T, always exist (but may be infinite). Moreover, if inclusion, union, and intersection of numbers are defined by $x \leqq y$, $\sup x_t$ and $\inf x_t$ respectively, then these operations have properties of the corresponding set operations; in particular, limits of monotone sequences of numbers always exist, but may be infinite.

If, as $n \to \infty$, the limit x of a sequence x_n of numbers exists, we write $x = \lim x_n$ or $x_n \to x$ and say that x_n *converges to* x; if x is infinite, say, $+\infty$, one also says that x_n *diverges to* $+\infty$. The Cauchy mutual convergence criterion is valid only for finite limits: *x_n converges to some finite x if, and only if, $x_m - x_n \to 0$ (as $m, n \to \infty$) or, equivalently, if $x_{n+\nu} - x_n \to 0$ uniformly in ν.* On the other hand, the Bolzano-Weierstrass lemma remains valid without the usual restriction of boundedness: *every sequence of numbers is compact,* that is, contains a convergent subsequence, but if the sequence is not bounded then the limits may be infinite.

The set of all finite numbers is a *real line* $R = (-\infty, +\infty)$ and the set of all numbers is an *extended real line* $\bar{R} = [-\infty, +\infty]$. The basic class of sets in R is the class of *intervals;* there are four types of finite intervals of respective form:

$[a, b)$: set of all points x such that $a \leqq x < b$;

$(a, b]$: set of all points x such that $a < x \leqq b$;

(a, b): set of all points x such that $a < x < b$;

$[a, b]$: set of all points x such that $a \leqq x \leqq b$.

The minimal σ-field over the class of all intervals in R is the *Borel field* in R and its elements are *Borel sets* in R. The Borel field in R coincides with the minimal σ-field over the subclass of all intervals of one of the foregoing four types, since countable operations performed upon elements of one of these subclasses yield any element of the other subclasses; for example, $(a, b) = \bigcup \left[a + \frac{1}{n}, b\right)$, $[a, b] = \bigcap \left[a, b + \frac{1}{n}\right)$, etc. Similarly, the Borel field in R is the minimal σ-field over the subclass of all infinite intervals of the form $(-\infty, x)$, $-\infty \leqq x \leqq +\infty$, since any finite interval $[a, b)$ is obtainable as a difference $\Delta_{b-a}(-\infty, a) = (-\infty, b) - (-\infty, a)$. The Borel field in $\bar{R}$ can be defined similarly by means of any of the foregoing types where $-\infty \leqq a \leqq b \leqq +\infty$, or by means of the intervals $[-\infty, x)$, $-\infty \leqq x \leqq +\infty$; but, frequently, the most convenient way is to take the minimal σ-field over the class formed by the Borel field in R and the two sets $\{-\infty\}$, $\{+\infty\}$.

Extension. The preceding notions extend at once to finite-dimensional real spaces. The set of all ordered N-uples $x = (x_1, \cdots, x_N)$ of finite numbers is the *N-dimensional real space* R^N or, equivalently, the *product space* $\prod_{\nu=1}^{N} R_\nu$ of N real lines $R_\nu = (-\infty < x_\nu < +\infty)$. If every R_ν is replaced by $\overline{R}_\nu = [-\infty \leqq x_\nu \leqq +\infty]$, then we have the *extended N-dimensional real space* $\overline{R}^N$. If $a, b \in R^N$, then $a \leqq b$ means that $a_\nu \leqq b_\nu$ for $\nu = 1, 2, \cdots, N$, and, similarly, for $a < b$, $a = b$.

An interval, say $[a, b)$, will also be written more explicitly as $[a_1, a_2, \cdots, a_N; b_1, b_2, \cdots, b_N)$, and

$$[a, b) = \Delta_{b-a}(-\infty, a) = \Delta_{b_1-a_1}\Delta_{b_2-a_2}\cdots \Delta_{b_N-a_N}(-\infty, -\infty, \cdots -\infty; a_1, a_2, \cdots a_N)$$

where $\Delta_{b_\nu - a_\nu}$ is the difference operator of step $b_\nu - a_\nu$ acting on a_ν. For example, if $N = 2$, then

$$\begin{aligned}[a_1, a_2; b_1, b_2) &= \Delta_{b_1-a_1}\Delta_{b_2-a_2}(-\infty, -\infty; a_1, a_2)\\ &= \Delta_{b_1-a_1}\{(-\infty, -\infty; a_1, b_2) - (-\infty, -\infty; a_1, a_2)\}\\ &= (-\infty, -\infty; b_1, b_2) - (-\infty, -\infty; a_1, b_2) - (-\infty, -\infty;\\ &\qquad b_1, a_2) + (-\infty, -\infty; a_1, a_2).\end{aligned}$$

With this interpretation, the foregoing definitions of types of intervals and, thereafter, of Borel fields, remain the same.

5.2 Numerical functions. A *numerical function* X on a space Ω is a function on Ω to $\overline{R}$, defined by assigning to every point $\omega \in \Omega$ a single number $x = X(\omega)$, the *value* of X at ω. If infinite values are excluded, X is a *finite function* or, equivalently, a function on Ω to R. Ω is called the *domain* of X and $\overline{R}$ (or R) is called the *range space* of X. The functions $X^+ = XI_{[X \geqq 0]}$ and $X^- = -XI_{[X<0]}$ will be called the *positive part* and the *negative part* of X, respectively, and we have

$$X = X^+ - X^-, \quad |X| = X^+ + X^-.$$

Unless otherwise stated, all functions will be numerical functions and, in general, will be denoted by $X, Y, \cdots$, with or without affixes.

If definitions or relations between values of given functions hold for every ω belonging to a set $A \subset \Omega$, we say that these definitions or relations hold *on A* and drop "on A" if $A = \Omega$. For example,

$|X| < \infty$ means that X is finite;
$X \geqq 0$ on A means that $X(\omega) \geqq 0$ for every $\omega \in A$;
$X = \inf X_n$ means that $X(\omega) = \inf X_n(\omega)$ for every $\omega \in \Omega$;
$X_n \to X$ on A means that $X_n(\omega) \to X(\omega)$ for every $\omega \in A$, etc.

Conversely, the set of *all* $\omega \in \Omega$ on which definitions or relations hold is denoted by $[\omega; \cdots]$ or, if there is no confusion possible, by $[\cdots]$ where $\cdots$ stand for the definitions or relations. For example,

$[X]$ is the set on which X is defined;
$[X \geqq Y]$ is the set of all $\omega \in \Omega$ for which $X(\omega) \geqq Y(\omega)$;
$[X \in S]$, where $S \subset \bar{R}$, is the set of all $\omega \in \Omega$ for which the values $X(\omega)$ belong to the set S.

The set $[X = x]$ is called the inverse image of the set $\{x\}$ which consists of x only or, simply, of x. Since X is single-valued, the inverse images of distinct numbers x are disjoint, and the partition of Ω into inverse images of all $x \in \bar{R}$ is called the partition of the domain *induced* by X; we sometimes write $X = \sum_{x \in \bar{R}} x I_{[X=x]}$ where $I_{[X=x]}$ is the indicator of $[X = x]$. In particular, if X is *countably valued*, that is, takes only a countable number of values x_j, then, and only then,

$$X = \sum_j x_j I_{[X=x_j]}.$$

More generally, the set $[X \in S]$ is called *inverse image* of S and is also denoted by $X^{-1}(S)$. The symbol X^{-1}, which can be considered as representing a mapping of sets in $\bar{R}$ onto sets in Ω, is called the *inverse function* of X. Since inverse images of disjoint sets of $\bar{R}$ are disjoint, it follows easily that

X^{-1} *and set operations commute:*

$$X^{-1}(S - S') = X^{-1}(S) - X^{-1}(S'), \quad X^{-1}(\bigcup S_t) = \bigcup X^{-1}(S_t),$$
$$X^{-1}(\bigcap S_t) = \bigcap X^{-1}(S_t).$$

Similarly, $X^{-1}(\mathcal{C})$ or the *inverse image* of $\mathcal{C}$, where $\mathcal{C}$ is a class of sets in R, is the class of all inverse images of elements of $\mathcal{C}$. Since set operations commute with inverse functions, it follows that

a. *The inverse image of a σ-field is a σ-field, the inverse image of the minimal σ-field over a class is the minimal σ-field over the inverse image of the class, the class of all sets whose inverse images belong to a σ-field is a σ-field.*

The foregoing definitions and properties extend at once to functions $X = (X_1, \cdots, X_N)$ on Ω to an N-dimensional real space $\bar{R}^N$ (or R^N) or, equivalently, to N-uples of numerical functions $X_1, \cdots, X_N$. Classical analysis is concerned with functions from a real line to a real line or, more generally, from a finite-dimensional real space R^N to a finite-dimensional real space R^N. Still more generally, let X be a function on Ω to $\bar{R}^N$ and let g be a function on $\bar{R}^N$ to $\bar{R}^{N'}$. *The function of func-*

tion gX defined by $(gX)(\omega) = g(X(\omega))$ is a function on Ω to $\bar{R}^{N'}$. Clearly, its inverse function $(gX)^{-1}$ is a mapping of sets S' in $\bar{R}^{N'}$ onto sets in Ω such that

$$(gX)^{-1}(S') = X^{-1}(g^{-1}(S'))$$

or, in a condensed form,

$$(gX)^{-1} = X^{-1}g^{-1}.$$

5.3 Measurable functions. Classical analysis is concerned primarily with continuous functions on R to R' or, more generally, on R^N to $R^{N'}$. However, passages to the limit, which play such a basic role in analysis, do not, in general, preserve continuity (and also they cause the appearance of $\mp\infty$). The essential achievement of modern analysis, due to Borel, Baire, and Lebesgue, is the introduction of a wider class of functions which is closed under the "usual" operations of analysis: arithmetic operations and formation of infima, suprema, and limits of sequences. Those are the functions we intend to define now.

In the domain Ω of our functions we select a σ-field $\mathcal{A}$ of sets, to be called *$\mathcal{A}$-sets* or, if there is no confusion possible, *measurable sets;* the doublet $(\Omega, \mathcal{A})$ is called a *measurable space.* In the range space $\bar{R}$ of our functions we select the σ-field $\bar{\mathcal{B}}$ of Borel sets—the Borel field in $\bar{R}$; the doublet $(\bar{R}, \bar{\mathcal{B}})$ is an *(extended) Borel line.* Thus, our functions are defined on a measurable space $(\Omega, \mathcal{A})$ to the Borel line $(\bar{R}, \bar{\mathcal{B}})$. More generally, if the range space is $\bar{R}^N$, then we select the Borel field $\bar{\mathcal{B}}^N$, and the doublet $(\bar{R}^N, \bar{\mathcal{B}}^N)$ is an *extended Borel space;* then the functions are defined on a measurable space $(\Omega, \mathcal{A})$ to the Borel space $(\bar{R}^N, \bar{\mathcal{B}}^N)$.

A countably valued function $X = \sum x_j I_{A_j}$ where the sets A_j are measurable is called an elementary measurable function or, simply, an *elementary function;* if the number of distinct values of X is finite, then X is also called a *simple function.*

(C) *Limits of convergent sequences of simple functions are called measurable functions.*

This is a constructive definition and, because of that, will play an essential role in the constructive definition of integrals. However, general properties of measurable functions are easier to discover and to prove when using the descriptive definition which follows.

(D) *Functions such that inverse images of all Borel sets are measurable sets are called measurable functions.*

Yet this definition is not the most economical one, since

(D′) *In* (D), *it suffices to require measurability of inverse images of elements of any fixed class* $\mathcal{C}$ *such that the minimal* σ*-field over* $\mathcal{C}$ *is the Borel field.*

For example, we can take $\mathcal{C}$ to be the class of all intervals, or the class of all intervals $[-\infty, x]$, etc.

The proof is immediate. Since a mapping X^{-1} preserves all sets operations and the measurable sets form a σ-field, it follows that the class of all sets whose inverse images are measurable is a σ-field. Therefore, if, according to (D′), it contains $\mathcal{C}$, then it contains the minimal σ-field over $\mathcal{C}$ which, by assumption, is the Borel field.

Similarly, the constructive definition (C) is not the most economical one as we shall find in proving the basic theorem below.

A. MEASURABILITY THEOREM. *The constructive and descriptive definitions are equivalent, and the class of measurable functions is closed under the usual operations of analysis.*

Proof. 1° Let X_n be functions measurable (D), that is, measurable according to (D) or, equivalently, (D′). Then all sets

$$[\inf X_n < x] = \bigcup [X_n < x], \quad [-X_n < x] = [X_n > -x]$$

are measurable and, hence, the functions

$$\sup X_n = -\inf(-X_n), \quad \liminf X_n = \sup_n (\inf_{k \geq n} X_k),$$
$$\limsup X_n = -\liminf(-X_n)$$

are measurable (D). Thus, the class of functions measurable (D) is closed under formation of infima, suprema, and limits. But every simple function $X = \sum x_j I_{A_j}$ is measurable (D), since all sets $[X \leqq x] = \sum_{x_j \leqq x} A_j$ are measurable. Therefore, limits of convergent sequences of simple functions are measurable (D); in particular, functions measurable (C) are measurable (D).

2° Conversely, let X be measurable (D) so that the functions

$$X_n = -nI_{[X<-n]} + \sum_{-n2^n+1}^{n2^n} \frac{k-1}{2^n} I_{\left[\frac{k-1}{2^n} \leqq X < \frac{k}{2^n}\right]} + nI_{[X \geqq n]},$$
$$n = 1, 2, \cdots$$

are simple. Since

$$|X_n(\omega) - X(\omega)| < \frac{1}{2^n} \quad \text{for} \quad |X(\omega)| < n$$

and

$$X_n(\omega) = \pm n \quad \text{for} \quad X(\omega) = \pm\infty,$$

it follows that $X_n \to X$ and this, together with what precedes, completes the proof of the equivalence of the two definitions of measurability.

We observe that if X is nonnegative, then the foregoing functions X_n become

$$X_n = \sum_{k=1}^{n2^n} \frac{k-1}{2^n} I_{\left[\frac{k-1}{2^n} \leqq X < \frac{k}{2^n}\right]} + nI_{[X \geqq n]}$$

and we have $0 \leqq X_n \uparrow X$. Also, if

$$X'_n = \sum_{k=-\infty}^{+\infty} \frac{k-1}{2^n} I_{\left[\frac{k-1}{2^n} \leqq X < \frac{k}{2^n}\right]} + (-\infty)I_{[X=-\infty]} + (+\infty)I_{[X=+\infty]},$$

then $|X'_n - X| < \frac{1}{2^n}$ on $[|X| < \infty]$ and $X'_n = X$ on $[|X| = \infty]$, so that $X'_n \to X$ uniformly.

3° It remains to prove closure under the arithmetic operations. Using definition (C) and the fact that arithmetic operations commute with passages to the limit by convergent sequences, it suffices to show that the class of simple functions is closed under the arithmetic operations. But much more is true, for if g on $\bar{R}^N$ is an arbitrary function and $X_k = \sum_j x_{kj} I_{A_{kj}}$, $k = 1, \cdots, N$, are simple (elementary) functions, then the function of functions

$$g(X_1, \cdots, X_N) = \sum g(x_{1j_1}, \cdots, x_{Nj_N}) I_{A_{1j_1}} \cdots I_{A_{Nj_N}}$$

is simple (elementary). This completes the proof.

According to this proof we have new *equivalent constructive definitions* of measurable functions that we state now.

(C′) *A nonnegative function is measurable if it is the limit of a nondecreasing sequence of nonnegative simple functions. A function X is measurable if its positive and negative parts X^+ and X^- are measurable.*

(C″) *A function is measurable if it is the limit of a uniformly convergent sequence of elementary functions. In particular, every bounded measurable function is limit of a uniformly convergent sequence of simple functions.*

Definition (C′) will play a central role in the theory of integration.

Closure under the arithmetic operations is a very particular case of

a. *A Baire function of measurable functions is measurable.*

Proof. Let us recall a (constructive) definition of Baire functions (we consider only finite-dimensional Borel spaces). *Baire functions* are elements of the smallest class closed under passages to the limit containing all continuous functions. Therefore, since the class of measurable functions is closed under passages to the limit, it suffices to prove that

A continuous function of measurable functions is measurable.

Thus, let g on $\bar{R}^N$ be continuous; that is, for every point $(x_1, \cdots, x_N) \in \bar{R}^N$,

$$g(x'_1, \cdots, x'_N) \to g(x_1, \cdots, x_N) \quad \text{as} \quad x'_1 \to x_1, \cdots, x'_N \to x_N.$$

Let X_k, $k = 1, 2, \cdots, N$, be measurable and let X_{nk} be sequences of simple functions such that $X_{nk} \to X_k$ for every k. We found (in 3°) that the functions $g(X_{n1}, \cdots, X_{nN})$(that we assumed tacitly to have meaning) are measurable and hence, by continuity and closure under passages to the limit, the function

$$g(X_1, \cdots, X_N) = \lim g(X_{n1}, \cdots, X_{nN})$$

is measurable. This completes the proof.

All the foregoing definitions and properties extend at oncc, and word for word, to functions on a measurable space to any finite-dimensional Borel space, provided we replace $\bar{R}$ by $\bar{R}^N$ and leave out the operations of multiplication and division that we do not define (at least here) for such functions. For example,

functions such that inverse images of Borel sets in their range space are measurable sets in their domain are called measurable functions.

This extension is useful but, in fact, brings nothing new, for

b. *A function $X = (X_1, \cdots, X_N)$ is measurable if, and only if, its components $X_1, \cdots, X_N$ are measurable.*

In other words such a function is merely an N-uple of numerical measurable functions.

Proof. If $X = (X_1, \cdots, X_N)$ is measurable, then, for every $k \leqq N$, the sets

$$[X_k \leqq x_k] = X_k^{-1}[-\infty, x_k]$$
$$= X^{-1}[-\infty, \cdots, -\infty; +\infty, \cdots, +\infty, x_k, +\infty, \cdots, +\infty]$$

are measurable, so that X_k is measurable.

Conversely, if all X_k are measurable, then the sets

$$[X \leqq x] = [X_1 \leqq x_1, \cdots, X_N \leqq x_N] = \bigcap_{k=1}^{N} [X_k \leqq x_k]$$

are measurable, so that $X = (X_1, \cdots, X_N)$ is measurable.

We give another (descriptive) definition of Baire functions. With this definition, it is customary to call these functions Borel functions. A measurable function on a finite-dimensional Borel space to a finite-dimensional Borel space is called a *Borel function*. In other words, g on $\overline{R}^N$ to $\overline{R}^{N'}$ is a Borel function if, and only if, the inverse images of Borel sets S' in $\overline{R}^{N'}$ are Borel sets S in $\overline{R}^N$. The proof of **a** in this more general case is then immediate and we have

a'. *A Borel function of a measurable function is measurable.*

For, if X is a measurable function (not necessarily numerical) and g is a Borel function on the range space of X, then, for every Borel set S' in the range space of g, the set $(gX)^{-1}(S') = X^{-1}(g^{-1}(S'))$ is measurable and, hence, gX is a measurable function.

§ 6. MEASURE AND CONVERGENCES

6.1 Definitions and general properties. The notions of "measurable" sets and "measurable" functions are two out of a triplet of notions, due essentially to Lebesgue, the third being the notion of "measure" which gave its name to the two others, and which we shall introduce now.

A function φ on a σ-field $\mathcal{A}$ is said to be *σ-additive* if, for every countable disjoint class $\{A_j\} \subset \mathcal{A}$,

$$\varphi(\sum A_j) = \sum \varphi(A_j).$$

To avoid trivialities, it is assumed that at least one value of φ, say, $\varphi(A_0)$, $A_0 \in \mathcal{A}$, is finite. Since

$$\varphi(A_0 + \emptyset) = \varphi(A_0) = \varphi(A_0) + \varphi(\emptyset),$$

this assumption is equivalent to $\varphi(\emptyset) = 0$. To avoid meaningless expressions of the form $+\infty - \infty$, it is assumed that at least one of the possible values $-\infty$ or $+\infty$ is excluded.

φ is said to be *finite* if its values are finite, and it is said to be *σ-finite* if the space in which $\mathcal{A}$ is defined can be partitioned into a countable number of sets in $\mathcal{A}$ for which the values of φ are finite.

A *measure* μ on a σ-field $\mathcal{A}$ is a nonnegative and σ-additive function. In other words, μ is defined by the three following properties:

(i) $\mu(\sum A_j) = \sum \mu(A_j)$ for every countable disjoint class $\{A_j\} \subset \mathcal{A}$;
(ii) $\mu(A) \geqq 0$ for every $A \in \mathcal{A}$;
(iii) $\mu(\emptyset) = 0$.

The value $\mu(A)$ of μ at A is called the *measure of A* and, if there is no confusion possible, we drop the bracket following the symbol μ.

A *measure space* $(\Omega, \mathcal{A}, \mu)$ is formed by the space Ω, the σ-field $\mathcal{A}$ of measurable sets in this space, and the measure μ defined on this σ-field. Unless otherwise stated all sets under consideration will be measurable sets in our measure space. A set of measure 0 is said to be a *μ-null set* or, if there is no confusion possible, a *null set*, and definitions or relations valid outside a μ-null set are said to be valid *almost everywhere* (*a.e.*). The following properties of the measure μ are immediate:

a. *μ is nondecreasing, and μ is bounded if the space Ω is of finite measure.*

This follows from

$$\mu B = \mu A + \mu(B - A) \geqq \mu A \quad \text{for} \quad B \supset A.$$

b. *μ is sub σ-additive:* $\mu \bigcup A_j \leqq \sum \mu A_j$.

This follows from

$$\begin{aligned}\mu \bigcup A_j &= \mu(A_1 + A_1{}^c A_2 + \cdots) \\ &= \mu A_1 + \mu A_1{}^c A_2 + \cdots \leqq \mu A_1 + \mu A_2 + \cdots.\end{aligned}$$

A. SEQUENCES THEOREM. *If $A_n \uparrow A$, then $\mu A_n \uparrow \mu A$ and, in general,*

$$\liminf \mu A_n \geqq \mu(\liminf A_n).$$

If μ is finite, then, moreover,

$$A_n \downarrow A \textit{ implies } \mu A_n \downarrow \mu A, \quad \limsup \mu A_n \leqq \mu(\limsup A_n),$$

$$A_n \to A \textit{ implies } \mu A_n \to \mu A.$$

Proof. If $A_n \uparrow A$, then, by σ-additivity,

$$\begin{aligned}\mu A &= \mu A_1 + \mu(A_2 - A_1) + \cdots \\ &= \lim \{\mu A_1 + \mu(A_2 - A_1) + \cdots + \mu(A_n - A_{n-1})\} \\ &= \lim \mu A_n.\end{aligned}$$

f A_n is an arbitrary sequence, then, since $B_n = \bigcap_{k \geqq n} A_k \uparrow \liminf A_n$ and $\mu A_n \geqq \mu B_n$, it follows that

$$\liminf \mu A_n \geqq \lim \mu B_n = \mu(\liminf A_n),$$

and the first assertion is proved.

Let now μ be finite and use the proved assertion. If $A_n \downarrow A$, then $A_1 - A_n \uparrow A_1 - A$ and, hence,

$$\mu A_1 - \mu A_n = \mu(A_1 - A_n) \uparrow \mu(A_1 - A) = \mu A_1 - \mu A,$$

so that $\mu A_n \downarrow \mu A$. If A_n is an arbitrary sequence, then $\mu\Omega - \limsup \mu A_n = \liminf \mu A_n^c \geqq \mu(\liminf A_n^c) = \mu\Omega - \mu(\limsup A_n)$ and, hence, $\limsup \mu A_n \leqq \mu(\limsup A_n)$. Finally, if $A_n \to A$, then the two inequalities proved above yield $\mu A_n \to \mu A$, and the proof is complete.

The introduction of measures yields new types of convergence founded upon the notion of measure and unknown in classical analysis. Before we introduce them, we recall the classical types of convergence; unless otherwise stated, we consider sequences X_n of measurable functions on a fixed measure space $(\Omega, \mathfrak{A}, \mu)$ and limits taken as $n \to \infty$.

If X_n converges to X on A according to a definition "c" of convergence, we say that X_n converges "c" on A and write $X_n \xrightarrow{c} X$ on A. The Cauchy convergence criterion leads to the corresponding notion of mutual convergence: if $X_{n+\nu} - X_n$ converges "c" to 0 on A uniformly in ν (or $X_m - X_n$ converges "c" to 0 on A as $m, n \to \infty$), we say that X_n *mutually converges* "c" on A and write $X_{n+\nu} - X_n \xrightarrow{c} 0$ (or $X_m - X_n \xrightarrow{c} 0$). In defining mutual convergence, we naturally must assume that the differences exist, that is, meaningless expressions $+\infty - \infty$ do not occur. We drop "on A" if $A = \Omega$ and drop "c" if the convergence is ordinary pointwise convergence.

We recall that $X_n \to X$ on A means that, for every $\omega \in A$ and every $\epsilon > 0$, there is an integer $n_{\epsilon,\omega}$ such that, for $n \geqq n_{\epsilon,\omega}$,

if $X(\omega)$ is finite, then $|X(\omega) - X_n(\omega)| < \epsilon$,

if $X(\omega) = -\infty$, then $X_n(\omega) < -\dfrac{1}{\epsilon}$,

if $X(\omega) = +\infty$, then $X_n(\omega) > +\dfrac{1}{\epsilon}$.

If $n_{\epsilon,\omega} = n_\epsilon$ is independent of $\omega \in A$, then the convergence is *uniform* and, according to the preceding conventions, we write $X_n \xrightarrow{u} X$ on A. According to the closure property of measurable functions, if a sequence of measurable functions $X_n \to X$, then X is measurable. According to the Cauchy criterion, if X_n are finite, then

$X_n \to X$ finite if, and only if, $X_m - X_n \to 0$ or, equivalently $X_{n+\nu} - X_n \to 0$

$X_n \xrightarrow{u} X$ finite if, and only if, $X_m - X_n \xrightarrow{u} 0$ or, equivalently, $X_{n+\nu} - X_n \xrightarrow{u} 0.$

6.2 Convergence almost everywhere. A sequence X_n is said to *converge* a.e. to X, and we write $X_n \xrightarrow{\text{a.e.}} X$, if $X_n \to X$ outside a null set; it *mutually converges* a.e., and we write $X_m - X_n \xrightarrow{\text{a.e.}} 0$ or $X_{n+\nu} - X_n \xrightarrow{\text{a.e.}} 0$, if it mutually converges outside a null set. It follows, by the Cauchy criterion and the fact that a countable union of null sets is a null set, that

a. *A sequence of a.e. finite functions converges a.e. to an a.e. finite function if, and only if, the sequence mutually converges a.e.*

Let $X_n \xrightarrow{\text{a.e.}} X$. Since X_n are taken to be measurable, X is a.e. measurable, that is, X is the a.e. limit of a sequence of simple functions. Also, if X' is such that $X_n \xrightarrow{\text{a.e.}} X'$, then $X = X'$ a.e., for X can differ from X' only on the null set on which X_n converges neither to X nor to X'. Thus, the limit of the sequence X_n is a.e. determined and a.e. measurable. Moreover, if every X_n is modified arbitrarily on a null set N_n, then the whole sequence is modified at most on the null set $\bigcup N_n$ and, therefore, the so modified sequence still converges a.e. to X.

These considerations lead to the introduction of the notion of "equivalent" functions: X and X' are *equivalent* if $X = X'$ a.e. Since the notion has the usual properties of an equivalence—it is reflexive, transitive, and symmetric—it follows that the class of all functions on our measure space splits into equivalence classes, and the discussion which precedes can be summarized as follows.

b. *Convergence a.e. is a type of convergence of equivalence classes to an equivalence class.*

In other words, as long as we are concerned with convergence a.e. of *sequences* of functions, these functions as well as the limit functions are to be considered as defined up to an equivalence. In particular, we can replace an a.e. finite and a.e. measurable function by a finite and measurable function, and conversely, without destroying convergence a.e.

Let us investigate in more detail the set on which a given sequence converges. To simplify, we restrict ourselves to the most important case of *finite* measurable functions, the study of the general case being similar. By definition of ordinary convergence, the set of convergence $[X_n \to X]$ of finite X_n to a finite measurable X is the set of all points $\omega \in \Omega$ at which, for every $\epsilon > 0$, $|X(\omega) - X_n(\omega)| < \epsilon$ for $n \geqq n_{\epsilon,\omega}$ sufficiently large. Since, moreover, the requirement "for every $\epsilon > 0$" is equivalent to "for every term of a sequence $\epsilon_k \downarrow 0$ as $k \to \infty$," say, the sequence $\frac{1}{k}$, we have

$$[X_n \to X] = \bigcap_{\epsilon>0} \bigcup_n \bigcap_\nu [|X_{n+\nu} - X| < \epsilon]$$

$$= \bigcap_k \bigcup_n \bigcap_\nu \left[|X_{n+\nu} - X| < \frac{1}{k}\right],$$

so that the set $[X_n \to X]$ is measurable. Similarly for the set of mutual convergence, since the set

$$[X_{n+\nu} - X_n \to 0] = \bigcap_{\epsilon>0} \bigcup_n \bigcap_\nu [|X_{n+\nu} - X_n| < \epsilon]$$

$$= \bigcap_k \bigcup_n \bigcap_\nu \left[|X_{n+\nu} - X_n| < \frac{1}{k}\right]$$

is measurable. Thus

c. *The sets of convergence (to a finite measurable function) and of mutual convergence of a sequence of finite measurable functions are measurable.*

In other words, to every sequence we can assign a "measure of convergence" and, the sets of divergence $[X_n \nrightarrow X]$ and $[X_{n+\nu} - X_n \nrightarrow 0]$ being complements of those of convergence and, hence, measurable, to every sequence we can assign a "measure of divergence." In particular, the definitions of a.e. convergence of a sequence X_n mean that

$$\mu[X_n \nrightarrow X] = 0 \quad \text{or} \quad \mu[X_{n+\nu} - X_n \nrightarrow 0] = 0.$$

Upon applying repeatedly the sequences theorem to the above-defined sets, we obtain the following

A. Convergence a.e. criterion. *Let X, X_n be finite measurable functions.*

$X_n \xrightarrow{\text{a.e.}} X$ *if, and only if, for every* $\epsilon > 0$,

$$\mu \bigcap_n \bigcup_\nu [|X_{n+\nu} - X| \geqq \epsilon] = 0$$

and, if μ *is finite, this criterion becomes*

$$\mu \bigcup_\nu [|X_{n+\nu} - X| \geqq \epsilon] \to 0.$$

$X_{n+\nu} - X_n \xrightarrow{\text{a.e.}} 0$ *if, and only if, for every* $\epsilon > 0$,

$$\mu \bigcap_n \bigcup_\nu [|X_{n+\nu} - X_n| \geqq \epsilon] = 0$$

and, if μ *is finite, this criterion becomes*

$$\mu \bigcup_\nu [|X_{n+\nu} - X_n| \geqq \epsilon] \to 0.$$

6.3 Convergence in measure. A sequence X_n of finite measurable functions is said to *converge in measure* to a measurable function X and we write $X_n \xrightarrow{\mu} X$ if, for every $\epsilon > 0$,

$$\mu[|X_n - X| \geqq \epsilon] \to 0.$$

The limit function X is then necessarily a.e. finite, since

$$\mu[|X| = \infty] = \mu[|X_n - X| = \infty] \leqq \mu[|X_n - X| \geqq \epsilon] \to 0.$$

Similarly, $X_{n+\nu} - X_n \xrightarrow{\mu} 0$ if, for every $\epsilon > 0$,

$$\mu[|X_{n+\nu} - X_n| \geqq \epsilon] \to 0 \text{ (uniformly in } \nu).$$

All considerations about equivalence classes in the case of convergence a.e. remain valid for convergence in measure. In particular, if $X_n \xrightarrow{\mu} X$ and $X_n \xrightarrow{\mu} X'$, then X and X' are equivalent, for

$$\mu[|X - X'| \geqq \epsilon] \leqq \mu\left[|X - X_n| \geqq \frac{\epsilon}{2}\right] + \mu\left[|X_n - X'| \geqq \frac{\epsilon}{2}\right] \to 0$$

and, hence,

$$\mu[X \neq X'] = \mu \bigcup_k \left[|X - X'| \geqq \frac{1}{k}\right] = 0.$$

We compare now convergence in measure and convergence a.e.

A. COMPARISON OF CONVERGENCES THEOREM. *Let X_n be a sequence of finite measurable functions.*

If X_n converges or mutually converges in measure, then there is a subsequence X_{n_k} which converges in measure and a.e. to the same limit function. If μ is finite, then convergence a.e. to an a.e. finite function implies convergence in measure to the same limit function.

Proof. The second assertion is an immediate consequence of the a.e. convergence criterion, since μ finite and $X_n \xrightarrow{\text{a.e.}} X$ imply that, for every $\epsilon > 0$,

$$\mu[|X_{n+1} - X| \geqq \epsilon] \leqq \mu \bigcup_{\nu} [|X_{n+\nu} - X| \geqq \epsilon] \to 0.$$

As for the first assertion, let $X_{n+\nu} - X_n \xrightarrow{\mu} 0$. Then, for every integer k there is an integer $n(k)$ such that, for $n \geqq n(k)$ and all ν,

$$\mu\left[|X_{n+\nu} - X_n| \geqq \frac{1}{2^k}\right] < \frac{1}{2^k}.$$

Let $n_1 = n(1)$, $n_2 = \max(n_1 + 1, n(2))$, $n_3 = \max(n_2 + 1, n(3))$, etc., so that $n_1 < n_2 < n_3 < \cdots \to \infty$. Let $X'_k = X_{n_k}$ and

$$A_k = \left[|X'_{k+1} - X'_k| \geqq \frac{1}{2^k}\right], \quad B_n = \bigcup_{k \geqq n} A_k,$$

so that

$$\mu A_k < \frac{1}{2^k}, \quad \mu B_n \leqq \sum_{k \geqq n} \mu A_k < \frac{1}{2^{n-1}}.$$

Thus, for a given $\epsilon > 0$, n large enough so that $\frac{1}{2^{n-1}} < \epsilon$, and all ν, we have on B_n^c

$$|X'_{n+\nu} - X'_n| \leqq \sum_{k \geqq n} |X'_{k+1} - X'_k| < \frac{1}{2^{n-1}} < \epsilon.$$

Therefore,

$$\mu \bigcap_n \bigcup_\nu [|X'_{n+\nu} - X'_n| \geqq \epsilon] \leqq \mu \bigcup_\nu [|X'_{n+\nu} - X'_n| \geqq \epsilon]$$
$$\leqq \mu B_n < \frac{1}{2^{n-1}} \to 0$$

and, hence, by the convergence a.e. criterion, $X'_{n+\nu} - X'_n \xrightarrow{\text{a.e.}} 0$. Thus, by **6.2a**, there is a finite X' such that $X'_n \xrightarrow{\text{a.e.}} X'$. Since on

B_n^c we have $| X'_{n+\nu} - X'_n | < \epsilon$ for all ν it follows, upon letting $\nu \to \infty$, that on B_n^c we also have $| X' - X'_n | < \epsilon$ outside perhaps a null subset. Therefore, upon taking complements,

$$\mu[| X' - X'_n | \geqq \epsilon] \leqq \mu B_n < \frac{1}{2^{n-1}} \to 0,$$

so that $X'_n \xrightarrow{\mu} X'$. A similar argument shows that $X_n \xrightarrow{\mu} X$ implies $X'_n \xrightarrow{\text{a.e.}} X$. This completes the proof.

Corollary. *Convergence and mutual convergence in measure imply one another.*

Proof. If $X_n \xrightarrow{\mu} X$, then, for every $\epsilon > 0$ and all ν,

$$\mu[| X_{n+\nu} - X_n | \geqq \epsilon] \leqq \mu\left[| X_{n+\nu} - X | \geqq \frac{\epsilon}{2}\right] + \mu\left[| X - X_n | \geqq \frac{\epsilon}{2}\right] \to 0,$$

so that $X_{n+\nu} - X_n \xrightarrow{\mu} 0$. Conversely, if $X_{n+\nu} - X_n \xrightarrow{\mu} 0$, then, upon taking the subsequence X_{n_k} of the foregoing theorem, we obtain, for every $\epsilon > 0$, by letting $n_k, n \to \infty$,

$$\mu[| X - X_n | \geqq \epsilon] \leqq \mu\left[| X - X_{n_k} | \geqq \frac{\epsilon}{2}\right] + \mu\left[| X_{n_k} - X_n | \geqq \frac{\epsilon}{2}\right] \to 0,$$

so that $X_n \xrightarrow{\mu} X$, and the corollary is proved.

§ 7. INTEGRATION

The concepts of σ-field, measure, and measurable function are born from the efforts, made in the nineteenth and the beginning of the twentieth centuries, to extend the concept of integration to wider and wider classes of functions. The decisive extension was accomplished by Lebesgue, after Borel opened the way. Lebesgue worked with the special "Lebesgue" measure. Radon applied the same approach working with Lebesgue-Stieltjes measures. Finally, Fréchet, still using Lebesgue's approach, got rid of the restrictions on the measure space on which the numerical functions to be integrated were defined.

Lebesgue had two equivalent definitions of the integral, a descriptive one and a constructive one. We shall use a constructive definition of the integral of which there are many variants, but the basic ideas are always the same and, in general, the integral is first defined for simple functions. Although infinite values are not excluded, nevertheless, the expression $+\infty - \infty$, being meaningless, must be avoided. Therefore, it behooves us to start with integrals of functions of constant sign, say, nonnegative ones. Furthermore, the central property of the integral, called "the monotone convergence theorem," says that for a nondecreasing sequence of nonnegative functions integration and passage to the limit can be interchanged. Therefore, we give here the approach aimed directly at this theorem, an approach which requires a minimum of notions and of effort. The reader will recognize in the central definition 2° below, a particular form of the monotone convergence theorem.

7.1 Integrals. We consider a fixed measure space $(\Omega, \mathcal{A}, \mu)$; A, B, $\cdots$, and X, Y, $\cdots$, with or without affixes, will denote measurable sets and (numerical) measurable functions, respectively.

DEFINITIONS 1° The *integral on Ω of a nonnegative simple function* $X = \sum_{j=1}^{m} x_j I_{A_j}$ is defined by

$$\int_\Omega X\,d\mu = \sum_{j=1}^{m} x_j \mu A_j.$$

2° The *integral on Ω of a nonnegative measurable function* X is defined by

$$\int_\Omega X\,d\mu = \lim \int_\Omega X_n\,d\mu,$$

where X_n is a nondecreasing sequence of nonnegative simple functions which converges to X.

3° The *integral on Ω of a measurable function* X is defined by

$$\int_\Omega X\,d\mu = \int_\Omega X^+\,d\mu - \int_\Omega X^-\,d\mu,$$

where $X^+ = XI_{[X \geqq 0]}$ and $X^- = -XI_{[X<0]}$, are the positive and negative parts of X respectively, provided the defining difference exists, that is, provided at least one of the terms of this difference is finite. If $\int_\Omega X\,d\mu$ is finite, that is, if both of the terms of the difference are finite, X is said to be *integrable* on Ω.

Finally, if X is a.e. determined and measurable, that is, there exists a measurable function X' such that $X = X'$ outside a μ-null set, we set $\int X = \int X'$, provided the right-hand side exists.

Upon replacing, in the preceding definitions, Ω by a measurable set A (hence replacing, in 1°, every $A_j = \Omega A_j$ by AA_j), they become definitions of the *integral of* X *on* A, to be denoted by $\int_A X\,d\mu$. Since, for $X = \sum_{j=1}^{m} x_j I_{A_j} \geqq 0$, we have

$$\int_\Omega XI_A\,d\mu = \sum_{j=1}^{m} x_j\mu AA_j \leqq \int_\Omega X\,d\mu,$$

it follows immediately that

$$\textit{if } \int_\Omega X\,d\mu \textit{ exists so does } \int_A X\,d\mu, \quad \textit{and} \quad \int_A X\,d\mu = \int_\Omega XI_A\,d\mu.$$

To simplify the writing, we drop $d\mu$ and Ω in the foregoing symbols, unless confusion is possible; thus, the symbols $\int_\Omega X\,d\mu$ and $\int_A X\,d\mu$ will be replaced by $\int X$ and $\int_A X$, respectively.

Justification and additivity. We have to justify the three definitions 1°, 2°, 3°, that is, we have to show that the concepts as defined exist and are uniquely determined. In the course of the justification we shall have use for the elementary properties below; the first one is called the *additivity property* of the operation of integration.

A. ELEMENTARY PROPERTIES. *Let* $\int X, \int Y, \int X + \int Y$ *exist.*

I *Linearity:*

$$\int (X+Y) = \int X + \int Y, \quad \int_{A+B} X = \int_A X + \int_B X, \quad \int cX = c\int X.$$

II *Order-preservation:*

$$X \geqq 0 \Rightarrow \int X \geqq 0, \quad X \geqq Y \Rightarrow \int X \geqq \int Y,$$

$$X = Y \text{ a.e.} \Rightarrow \int X = \int Y.$$

III *Integrability:*

$$X \text{ integrable} \Leftrightarrow |X| \text{ integrable} \Rightarrow X \text{ a.e. finite;}$$

$$|X| \leqq Y \text{ integrable} \Rightarrow X \text{ integrable;}$$

$$X \text{ and } Y \text{ integrable} \Rightarrow X + Y \text{ integrable.}$$

Assume that the additivity property is proved. Then the second of properties I follows by replacing in the first one X by XI_A and Y by XI_B. The third one follows directly by successive use of the definitions.

The successive use of the definitions also proves directly the first and third of properties II, and the second one follows by the additivity property upon setting $X = Y + Z$, where $Z \geqq 0$.

Similarly for properties III, except for $|X|$ integrable $\Rightarrow X$ a.e. finite. But, if $\mu A > 0$ where $A = [|X| = \infty]$, then, on account of II, $\int |X| \geqq \int |X| I_A \geqq c\mu A$ whatever be $c > 0$. It follows, by letting $c \to \infty$, that $\int |X| = \infty$, and the property is proved *ab contrario*. Thus

For each of the successive definitions, the elementary properties hold as soon as the additivity property is proved.

We use this fact repeatedly in proceeding to the successive justifications of the definitions and to the proof of the additivity property.

1° *Nonnegative simple functions.* Since $X = \sum_{j=1}^{m} x_j I_{A_j}$ is nonnegative, the defining sum in

$$\int X = \sum_{j=1}^{m} x_j \mu A_j \geqq 0$$

exists; it may be infinite. Its value is independent of the way in which X is written. For, if X is written in some other form $\sum_{k=1}^{n} y_k I_{B_k}$, then $x_j = y_k$ if $A_j B_k \neq \emptyset$ and, from $\sum_{j=1}^{m} A_j = \sum_{k=1}^{n} B_k = \Omega$, it follows that

$$\sum_{j=1}^{m} x_j \mu A_j = \sum_{j,k} x_j \mu A_j B_k = \int \sum_{j,k} x_j I_{A_j B_k} = \int \sum_{j,k} y_k I_{A_j B_k}$$

$$= \sum_{j,k} y_k \mu A_j B_k = \sum_{k=1}^{n} y_k \mu B_k.$$

Thus, $\int X$ is unambiguously defined.

Let now $X = \sum_{j=1}^{m} x_j I_{A_j}$ and $Y = \sum_{k=1}^{n} y_k I_{B_k}$ be two nonnegative simple functions, so that $X + Y = \sum_{jk} (x_j + y_k) I_{A_j B_k}$. Proceeding as above, we have

$$\int (X + Y) = \sum_{jk} (x_j + y_k) \mu A_j B_k = \sum_{jk} x_j \mu A_j B_k + \sum_{jk} y_k \mu A_j B_k$$

$$= \sum_{j=1}^{m} x_j \mu A_j + \sum_{k=1}^{n} y_k \mu B_k = \int X + \int Y,$$

and the additivity property is proved.

2° *Nonnegative measurable functions.* In definition 2°, the sequence of simple functions $X_n \geqq 0$ is nondecreasing, so that, by **AII** for simple functions, the sequence $\int X_n$ is nondecreasing and, hence, has a limit, finite or not. Moreover, for every nonnegative measurable function X there exists such a sequence $X_n \uparrow X$. Therefore, to justify the definition, it suffices to show that the defining limit is independent of the particular choice of the sequence X_n. In other words

a. *If two nondecreasing sequences X_n and Y_n of nonnegative simple functions have the same limit, then*

$$\lim \int X_n = \lim \int Y_n.$$

Proof. It suffices to prove that $0 \leqq X_n \uparrow X$ and $\lim X_n \geqq Y$, where Y is a nonnegative simple function, imply $\lim \int X_n \geqq \int Y$. For, then, it follows from the assumptions that, for every integer p,

$$\lim \int X_n \geqq \int Y_p, \quad \lim \int Y_n \geqq \int X_p,$$

and the asserted equality is obtained by letting $p \to \infty$.

First, we prove the asserted inequality under the supplementary restrictions

$$\mu\Omega < \infty, \quad m = \min Y > 0, \quad M = \max Y < \infty.$$

Let $\epsilon > 0$ be less than m. Since $\lim X_n \geqq Y$, it follows that $A_n = [X_n > Y - \epsilon] \uparrow \Omega$. But, on account of the validity of **A** for simple functions and the finiteness of μ and Y, we have

$$\int X_n \geqq \int X_n I_{A_n} \geqq \int (Y - \epsilon) I_{A_n} = \int Y - \int Y I_{A_n^c} - \epsilon\mu A_n$$

$$\geqq \int Y - M\mu A_n^c - \epsilon\mu A_n$$

and, hence, by letting $n \to \infty$ and then $\epsilon \to 0$, the asserted inequality follows. Now, we get rid of the supplementary restrictions.

If $\mu\Omega = \infty$, then

$$\int X_n \geqq \int X_n I_{A_n} \geqq \int (Y - \epsilon) I_{A_n} \geqq (m - \epsilon)\mu A_n \to \infty,$$

and the asserted inequality is trivially true.

If $M = \infty$, then, the inequality being valid with X_n and $YI_{[Y<\infty]} + cI_{[Y=+\infty]}$ where c is an arbitrary finite number, we have

$$\lim \int X_n \geqq \int Y I_{[Y<\infty]} + c\mu[Y = +\infty]$$

and, letting $c \to \infty$, the right-hand side becomes $\int Y$.

Finally, if $m = 0$, then, since the functions X_n and Y are nonnegative and, by what precedes, the inequality is true for integrals on $[Y > 0]$, we have

$$\lim \int X_n \geqq \lim \int_{[Y>0]} X_n \geqq \int_{[Y>0]} Y = \int Y.$$

This completes the proof and the definition of the integral of a nonnegative measurable function is justified.

Since the additivity property was proved for nonnegative simple functions X_n, Y_n, and $0 \leqq X_n \uparrow X$, $0 \leqq Y_n \uparrow Y$ imply $0 \leqq X_n + Y_n \uparrow X + Y$, it follows, by letting $n \to \infty$ in

$$\int (X_n + Y_n) = \int X_n + \int Y_n,$$

that

$$\int (X + Y) = \int X + \int Y.$$

Thus, the additivity property remains valid for nonnegative measurable functions.

3° *Measurable functions.* The decomposition $X = X^+ - X^-$ of a measurable function into its positive and negative parts is unique, so that $\int X = \int X^+ - \int X^-$ is unambiguously defined, provided $\int X^+$ or $\int X^-$ is finite.

Finally, if X is determined and measurable outside a μ-null set N, then let X' be any measurable function such that $X = X'$ on N^c. The integral of X is defined by setting $\int X = \int X'$, provided $\int X'$ exists. By **AII** for nonnegative measurable functions, the integrals of such functions which coincide on N^c are equal. It follows, by definition 3°, that the same is true when the functions are not of constant sign. Therefore, $\int X$ is unambiguously defined.

It remains to prove the additivity property.

Since we assume that not only $\int X$ and $\int Y$ exist but also that $\int X + \int Y$ exists, that is, is not of the form $+\infty - \infty$, it follows that (excluding the trivial case of the three integrals infinite of the same sign) at least one of the functions, say Y, is integrable and, hence, by **AIII**, is a.e. finite. Therefore, $X + Y$ is a.e. determined, and we do not restrict the generality by taking determined X and Y, and changing Y to 0 on the μ-null event on which it is infinite and $X + Y$ may be not determined.

We decompose Ω into the six sets on each of which X, Y, and $X + Y$ are of constant sign ($\geqq 0$ or < 0). Because of definition 3° and property **AI** for nonnegative functions, it suffices to prove the additivity property on each of these sets, say $A = [X \geqq 0,\ Y < 0,\ X + Y \geqq 0]$. But, on account of definition 3° and the additivity property for nonnegative functions $(X + Y)I_A$ and $-YI_A$, we have

$$\int_A X = \int_A (X + Y) + \int_A (-Y) = \int_A (X + Y) - \int_A Y$$

and, $\int_A Y$ being finite,

$$\int_A X + \int_A Y = \int_A (X + Y).$$

Similarly for the other sets, and the additivity property follows.

This completes the justification of the definitions and the proof of the elementary properties.

7.2 Convergence theorems. The central convergence property is as follows:

A. MONOTONE CONVERGENCE THEOREM. *If* $0 \leqq X_n \uparrow X$, *then* $\int X_n \uparrow \int X$.

Proof. Choose nonnegative simple functions $X_{km} \uparrow X_k$ as $m \to \infty$. The sequence $Y_n = \max_{k \leqq n} X_{kn}$ of nonnegative simple functions is nondecreasing, and

$$X_{kn} \leqq Y_n \leqq X_n, \quad \int X_{kn} \leqq \int Y_n \leqq \int X_n.$$

It follows, by letting $n \to \infty$, that

$$X_k \leqq \lim Y_n \leqq X, \quad \int X_k \leqq \int \lim Y_n \leqq \lim \int X_n$$

and, by letting $k \to \infty$, we obtain

$$X \leqq \lim Y_n \leqq X, \quad \lim \int X_n \leqq \int \lim Y_n \leqq \lim \int X_n.$$

Thus $\lim Y_n = X$ and $\int X = \lim \int X_n$. The assertion is proved.

COROLLARY 1. *The integral is σ-additive on the family of nonnegative measurable functions.*

This means that, if the X_n are nonnegative, then $\int \sum X_n = \sum \int X_n$. and follows by $0 \leqq \sum_{k=1}^{n} X_k \uparrow \sum X_n$.

COROLLARY 2. *If X is integrable, then* $\int_A |X| \to 0$ *as* $\mu A \to 0$.

For, if $X_n = X$ or n according as $|X| < n$ or $|X| \geqq n$, then $\int |X_n| \uparrow \int |X|$, so that, given $\epsilon > 0$, there exists an n_0 such that $\int |X| < \int |X_{n_0}| + \frac{\epsilon}{2}$. It follows that, for A with $\mu A < \epsilon/2n_0$,

$$\int_A |X| = \int_A |X_{n_0}| + \int_A (|X| - |X_{n_0}|) < \frac{\epsilon}{2} + \int |X| - \int |X_{n_0}| < \epsilon.$$

The monotone convergence theorem extends as follows:

B. Fatou-Lebesgue theorem. *Let Y and Z be integrable functions. If $Y \leqq X_n$ or $X_n \leqq Z$, then*

$$\int \liminf X_n \leqq \liminf \int X_n, \quad \textit{resp.} \limsup \int X_n \leqq \int \limsup X_n.$$

If $Y \leqq X_n \uparrow X$, or $Y \leqq X_n \leqq Z$ and $X_n \xrightarrow{\text{a.e.}} X$, then $\int X_n \to \int X$.

Proof. If the X_n are nonnegative, then

$$X_n \geqq Y_n = \inf_{k \geqq n} X_k \uparrow \liminf X_n,$$

so that, by the monotone convergence theorem,

$$\liminf \int X_n \geqq \lim \int Y_n = \int \liminf X_n.$$

The asserted inequalities follow, by the additivity property, upon applying this result to the sequences $X_n - Y$ and $Z - X_n$ of nonnegative measurable functions, and the asserted equalities are immediate consequences.

Clearly, if the assumptions of this theorem hold only a.e., the conclusions continue to hold. In fact, the last assertion, frequently called the dominated convergence theorem, extends as follows:

C. Dominated convergence theorem. *If $|X_n| \leqq Y$ a.e. with Y integrable and if $X_n \xrightarrow{\text{a.e.}} X$ or $X_n \xrightarrow{\mu} X$, then $\int X_n \to \int X$. In fact, $\int_A X_n - \int_A X \to 0$ uniformly in A or, equivalently, $\int |X_n - X| \to 0$.*

Proof. Since

$$\left|\int_A (X_n - X)\right| \leqq \int |X_n - X| = \int (X_n - X)^+ + \int (X_n - X)^-,$$

it follows that the last two assertions are equivalent and imply the first one. Thus, it suffices to prove that $\int |X_n - X| \to 0$. Set $Y_n = |X_n - X|$ and observe that $Y_n \leqq 2Y$ a.e. and that the $\int Y_n$ remain the same when the Y_n are modified on null events. Therefore, it suffices to prove that, if $0 \leqq Y_n \leqq Z$ integrable and $Y_n \xrightarrow{\text{a.e.}} 0$ or $Y_n \xrightarrow{\mu} 0$, then $\int Y_n \to 0$.

The case $Y_n \xrightarrow{\text{a.e.}} 0$ follows from the last assertion in **B**. It implies the case $Y_n \xrightarrow{\mu} 0$, since, by selecting a subsequence $Y_{n'}$ ($\xrightarrow{\mu} 0$) such that $\int Y_{n'} \to \limsup \int Y_n$ and, within this subsequence, a sequence $Y_{n''} \xrightarrow{\text{a.e.}} 0$, it follows that $\int Y_{n''} \to 0$ and $\limsup \int Y_n = 0$. Hence $\int Y_n \to 0$, and the proof is complete.

Extension. In all the preceding convergence theorems the parameter $n \to \infty$ can be replaced by a parameter $t \to t_0$ along an arbitrary set $T \subset \bar{R}$ of values, the reason for this being that $a_t \to a$ as $t \to t_0$ along T is equivalent to $a_{t_n} \to a$ for every sequence t_n in T converging to t_0.

Applications I. We assume all functions X_t to be integrable.

The dominated convergence theorem yields at once

1° *If* $|X_t| \leqq Y$ *integrable and* $X_t \to X_{t_0}$ *as* $t \to t_0 (t \in T)$, *then* $\int X_t \to \int X_{t_0}$.

This proposition yields, by applying the definition of derivative,

2° *If, on* T, $\frac{dX_t}{dt}$ *exists at* t_0 *and* $\left|\frac{X_t - X_{t_0}}{t - t_0}\right| \leqq Y$ *integrable, then*

$$\left(\frac{d}{dt}\int X_t\right)_{t_0} = \int\left(\frac{dX_t}{dt}\right)_{t_0}.$$

In turn, this proposition yields

3° *If, on a finite interval* $[a, b]$, $\frac{dX_t}{dt}$ *exists and* $\left|\frac{dX_t}{dt}\right| \leqq Y$ *integrable, then, on* $[a, b]$,

$$\frac{d}{dt}\int X_t = \int \frac{dX_t}{dt}.$$

This follows from

$$X_t - X_{t'} = (t - t')\left(\frac{dX_t}{dt}\right)_{t''}$$

where t'' lies between t and t'. And in its turn, this proposition yields

4° *If, on a finite interval* $[a, b]$, X_t *is continuous and* $|X_t| \leqq Y$ *integrable then*, for every $t \in [a, b]$,

$$\int_a^t \left(\int X_{t'}\right) dt' = \int\left(\int_a^t X_{t'}\, dt'\right).$$

Moreover, if the foregoing assumptions hold for every finite interval and $\int_{-\infty}^{+\infty} |X_t|\, dt \leqq Z$ *integrable, then*

$$\int_{-\infty}^{+\infty} \left(\int X_t\right) dt = \int\left(\int_{-\infty}^{+\infty} X_t\, dt\right).$$

The integrals with respect to t are Riemann integrals.

The first assertion follows from the fact that the derivative of a Riemann integral $\int_a^t g(t)\, dt$ where g is continuous is $g(t)$ which is bounded on $[a, b]$, so that, upon applying 3° to the asserted equality, it follows that derivatives of both sides are equal and, since both sides vanish for $t = a$, the equality is proved. The second assertion follows by 1° from the first one, by letting $a \to -\infty$ and $t \to +\infty$.

II. *Integrals over the Borel line.* Let $\mathcal{B}$ be the Borel field in $R = (-\infty, +\infty)$ and let μ be *a measure on* $\mathcal{B}$ *which assigns finite values to finite intervals.* Let $\mathcal{B}_\mu$ be the class of all sets which are unions of a Borel set and a subset of a μ-null Borel set. $\mathcal{B}_\mu$ is closed under formation of complements and countable unions and, hence, is a σ-field. By assigning to every set of $\mathcal{B}_\mu$ the measure of the Borel set from which it differs by a subset of a μ-null set, μ is extended to a σ-finite measure on $\mathcal{B}_\mu$, that we continue to denote by μ. $\mathcal{B}_\mu$ will be called a *Lebesgue-Stieltjes field* in R and μ on $\mathcal{B}_\mu$ will be called a *Lebesgue-Stieltjes measure.* The relation

$$F(b) - F(a) = F[a, b) = \mu[a, b)$$

determines, up to an additive constant, a function F on R which is clearly finite, nondecreasing, and continuous from the left, called *a distribution function* corresponding to μ. (It was proved that, conversely, such a function *determines* a Lebesgue-Stieltjes measure μ.)

Let g be a $\mathcal{B}_\mu$-measurable function. If g is integrable, the integral $\int g\, d\mu$ is called a *Lebesgue-Stieltjes integral.* If F is a distribution function corresponding to μ, this integral is also denoted by $\int g\, dF$, and the integral $\int_{[a,b)} g\,d\mu$ is also denoted by $\int_a^b g\,dF$. If $F(x) = x$, $x \in R$, the corresponding measure is called the *Lebesgue measure;* it assigns to every interval its "length" and, thus, is a direct extension of the notion of length. The corresponding σ-field, or *Lebesgue field,* is formed by *Lebesgue sets* and the corresponding integrals, say $\int g\,dx$, $\int_a^b g\,dx$, are called *Lebesgue integrals.* Lebesgue field, measure, and integral are prototypes of general σ-fields, measures, and integrals. One may say that the basic ideas and methods relative to measure spaces and integrals belong to Lebesgue.

Let g be continuous on $[a, b]$. The Lebesgue-Stieltjes integral $\int_a^b g\,dF$ becomes then a Riemann-Stieltjes integral and the Lebesgue integral $\int_a^b g\,dx$ becomes then a Riemann integral.

The proof is easy. We have to show that, g being continuous on $[a, b]$, $\int_a^b g\,dF$ is limit of Riemann-Stieltjes sums. This is possible because a continuous function on a closed interval is bounded and is the (uniform) limit of any sequence of step-functions

$$g_n = \sum g(x'_{nk}) I_{[x_{nk}, x_{n,k+1})}, \quad a = x_{n1} < \cdots < x_{n,k_n+1} = b, \quad x_{nk} \leqq x'_{nk} < x_{n,k+1},$$

such that $\max\limits_{k \leqq k_n} (x_{n,k+1} - x_{nk}) \to 0$. Therefore, by the dominated convergence theorem or, more specifically, by the last assertion of the Fatou-Lebesgue theorem,

$$\int_a^b g\,d\mu = \lim \int_a^b g_n\,d\mu = \lim \sum_{k=1}^{k_n} g(x'_{nk})\,\mu[x_{nk}, x_{n,k+1}),$$

that is,

$$\int_a^b g\,dF = \lim \sum_{k=1}^{k_n} g(x'_{nk}) F[x_{nk}, x_{n,k+1}),$$

where the right-hand side sums are precisely the usual Riemann-Stieltjes sums. Thus, in the case of g continuous on $[a, b]$, the integral $\int_a^b g\,dF$ can be defined directly in terms of F, or of measures assigned to intervals only.

However, when g is continuous on R, its Lebesgue-Stieltjes integral over R and its *improper* Riemann-Stieltjes integral do not necessarily coincide. In fact, the last integral is defined by

$$\int g\,dF = \lim_{\substack{a \to -\infty \\ b \to +\infty}} \int_a^b g\,dF,$$

provided the limit exists and is finite. It may happen that at the same time

$$\int |g|\,dF = \lim_{\substack{a \to -\infty \\ b \to +\infty}} \int_a^b |g|\,dF$$

is infinite so that $|g|$ not being Lebesgue-Stieltjes integrable, g is not Lebesgue-Stieltjes integrable. Such examples are familiar; one of the most classical ones is that of the improper Riemann-integral of $g(x) = \sin x/x$. However, if g is Lebesgue-Stieltjes integrable then, clearly, both integrals coincide. Thus, the class of continuous functions whose improper Riemann-Stieltjes integrals with respect to a distribution function F exist (and are finite) contains the class of continuous functions which are Lebesgue-Stieltjes integrable with respect to F.

§ 8. INDEFINITE INTEGRALS; ITERATED INTEGRALS

8.1 Indefinite integrals and Lebesgue decomposition. We characterize now the indefinite integrals by using repeatedly the monotone convergence theorem. Let X be a measurable function whose integral exists—say, $\int X^-$ is finite. Then the indefinite integral φ on $\mathcal{A}$ defined by

$$\varphi(A) = \int_A X = \int XI_A$$

exists, for $\int X^- I_A$ is finite and $\int X^+ I_A$ exists. Since the integral of a function which vanishes a.e. is 0, *the indefinite integral is μ-continuous*, that is, vanishes for μ-null sets. Since for a countable measurable partition $\{A_j\}$, $X^\pm I_A = \sum X^\pm I_{AA_j}$, it follows that, by the monotone convergence theorem,

$$\int_{\sum A_j} X = \sum \int_{A_j} X,$$

and the *indefinite integral is σ-additive.*

If X is integrable, then it is a.e. finite, and the indefinite integral is finite. If X is not integrable but, still, *X is a.e. finite and μ is σ-finite, then the indefinite integral is σ-finite.* For, by decomposing Ω into sets A_n of finite measure, we have

$$\int X = \sum_{m=-\infty}^{+\infty} \sum_{n=1}^{\infty} \int_{A_n[m \leqq X < m+1]} X$$

and every term of the double sum is finite.

The problem which arises is whether the foregoing properties characterize indefinite integrals and the answer lies in the celebrated Lebesgue (-Radon-Nikodym) decomposition theorem that we shall establish now. But first we introduce a notion in opposition to that of μ-continuity. A set function φ_s on $\mathfrak{a}$ is said to be *μ-singular* if it vanishes outside a μ-null set; in symbols, there is a μ-null set N such that

$$\varphi_s(AN^c) = 0, \quad A \in \mathfrak{a}.$$

A. LEBESGUE DECOMPOSITION THEOREM. *If, on $\mathfrak{a}$, the measure μ and the σ-additive function φ are σ-finite, then there exists one, and only one, decomposition of φ into a μ-continuous and σ-additive set function φ_c and a μ-singular and σ-additive set function φ_s,*

$$\varphi = \varphi_c + \varphi_s,$$

and φ_c is the indefinite integral of a finite measurable function X determined up to a μ-equivalence.

φ_c and φ_s are called μ-continuous and μ-singular *parts* of φ, and X is called the *derivative $d\varphi/d\mu$ with respect* to μ; we emphasize that $d\varphi/d\mu$ is determined up to μ-equivalence.

Proof. 1° Since Ω is a countable sum of sets for which μ and φ are finite and since, by the Hahn decomposition theorem, φ is a difference of two measures, it suffices to prove the theorem for finite measures μ and φ. Furthermore, if there are two decompositions of φ into a μ-continuous and a μ-singular part:

$$\varphi = \varphi_c + \varphi_s = \varphi'_c + \varphi'_s,$$

then

$$\varphi_c - \varphi'_c = \varphi'_s - \varphi_s = 0,$$

for the μ-continuous function $\varphi_c - \varphi'_c$ vanishes for all μ-null sets while the μ-singular function $\varphi'_s - \varphi_s$ vanishes outside a μ-null set. Finally, an indefinite integral determines the integrand up to an equivalence:

if, for every $A \in \mathcal{A}$,

$$\varphi_c(A) = \int_A X = \int_A X'$$

then $X = X'$ a.e.; for, if, say, $\mu A = \mu[X - X' > \epsilon] > 0$, then

$$\int_A (X - X') > 0.$$

Thus the uniqueness assertions hold if we prove the existence assertions under the assumption that μ and φ are finite measures.

2° Let Φ be the class of all nonnegative integrable functions X whose indefinite integrals are majorized by φ:

$$\int_A X \leqq \varphi(A), \quad A \in \mathcal{A}.$$

Φ is not empty, since $X = 0$ belongs to it; and there is a sequence $\{X_n\} \subset \Phi$ such that

$$\int X_n \to \sup_{X \in \Phi} \int X = \alpha \leqq \varphi(\Omega) < \infty.$$

Let $X'_n = \sup_{k \leqq n} X_k$, so that $0 \leqq X'_n \uparrow X = \sup X_n$. Let

$$A_k = [X_k = X'_n], \quad A'_k = A_1{}^c A_2{}^c \cdots A_{k-1}{}^c A_k, \quad A'_1 = A_1,$$

so that

$$\sum_{k=1}^{n} A'_k = \bigcup_{k=1}^{n} A_k = \Omega$$

and, for every A,

$$\int_A X'_n = \sum_{k=1}^{n} \int_{AA'_k} X'_n = \sum_{k=1}^{n} \int_{AA'_k} X_k \leqq \sum_{k=1}^{n} \varphi(AA'_k) = \varphi(A).$$

Upon letting $n \to \infty$ and applying the monotone convergence theorem, we get

$$\int_A X \leqq \varphi(A), \quad \int X = \alpha.$$

Therefore X is a "maximal" element of Φ. This property will allow us to show that

$$\varphi_s = \varphi - \varphi_c \geqq 0,$$

where φ_c is the indefinite integral of X, is μ-singular, and the proof will be complete.

3° Let $D_n + D_n^c$ be a Hahn decomposition for the finite and σ-additive set function $\varphi_n = \varphi_s - \frac{1}{n}\mu$, that is, $\varphi_n(AD_n) \leqq 0$ and $\varphi_n(AD_n^c) \geqq 0$ for every A. Let $D = \bigcap D_n$ (whence $D^c = \bigcup D_n^c$), so that, for every A and all n,

$$0 \leqq \varphi_s(AD) \leqq \frac{1}{n}\mu(AD).$$

Upon letting $n \to \infty$, it follows that $\varphi_s(AD) = 0$ and, hence, $\varphi_s(A) = \varphi_s(AD^c)$. Since

$$\varphi_c(A) = \varphi(A) - \varphi_s(AD^c) \leqq \varphi(A) - \varphi_s(AD_n^c),$$

it follows that

$$\int_A \left(X + \frac{1}{n} I_{D_n^c}\right) = \varphi_c(A) + \frac{1}{n}\mu(AD_n^c) \leqq \varphi(A) - \varphi_n(AD_n^c) \leqq \varphi(A),$$

so that $X + \frac{1}{n} I_{D_n^c} \in \Phi$. But this conclusion is contradicted by

$$\int \left(X + \frac{1}{n} I_{D_n^c}\right) = \alpha + \frac{1}{n}\mu D_n^c > \alpha$$

unless $\mu D_n^c = 0$. Therefore, all sets D_n^c are μ-null sets and so is their countable union D^c. Since $\varphi_s(A) = \varphi_s(AD^c)$, it follows that φ_s is μ-singular, and the proof is complete.

In the particular case of a μ-continuous φ, the foregoing theorem reduces to

B. RADON-NIKODYM THEOREM. *If, on $\mathfrak{A}$, the measure μ and the σ-additive set function φ are σ-finite and φ is μ-continuous, then φ is the indefinite integral of a finite function determined up to an equivalence.*

We are now in a position to characterize indefinite integrals of finite functions on a σ-finite measure space.

C. *A set function φ on $\mathfrak{A}$ is the indefinite integral on a σ-finite measure space of a finite function X determined up to an equivalence, if, and only if, φ is σ-finite, σ-additive, and μ-continuous; and X is integrable if, and only if, this φ is finite.*

The "if" assertion is the Radon-Nikodym theorem and the "only if" assertion is contained in the discussion at the beginning of this subsection.

COROLLARY. *Let λ and μ be σ-finite measures on $\mathfrak{A}$. If μ is λ-continuous and X is a measurable function whose integral $\int X\,d\mu$ exists, then, for every $A \in \mathfrak{A}$,*

$$\int_A X\,d\mu = \int_A X\frac{d\mu}{d\lambda}\,d\lambda.$$

Proof. If $X = I_B$, $B \in \mathfrak{A}$, then the equality is valid, since

$$\int_A I_B\,d\mu = \mu AB = \int_{AB}\frac{d\mu}{d\lambda}\,d\lambda = \int_A I_B\frac{d\mu}{d\lambda}\,d\lambda.$$

It follows that the equality is valid for nonnegative simple functions and hence, by the monotone convergence theorem, for nonnegative measurable functions and, consequently, for measurable functions whose integral exists.

Extension. The indefinite integral of a measurable function X which is not necessarily finite is still σ-additive and μ-continuous, but it is not necessarily σ-finite. The question arises whether the Radon-Nikodym theorem can be extended to this case. The answer is in the affirmative.

D. *The Radon-Nikodym theorem remains valid if finiteness of X and σ-finiteness of φ are simultaneously suppressed therein.*

Proof. As usual, it suffices to consider a finite measure μ and a μ-continuous measure φ on $\mathfrak{A}$.

Let $\mathfrak{B}$ be the class of all measurable sets such that φ on $\mathfrak{B}$ is σ-finite, and let s be the supremum of μ on $\mathfrak{B}$.

There exists a sequence $B_n \in \mathfrak{B}$ such that $s = \lim \mu B_n$ and, hence, $B = \bigcup B_n \in \mathfrak{B}$ with $\mu B = s$. If there exists a $C \in \{B^cA,\ A \in \mathfrak{A}\}$ such that $0 < \varphi(C) < \infty$, then $B + C \in \mathfrak{B}$, $\mu C > 0$, and

$$s \geqq \mu(B + C) = \mu B + \mu C > s.$$

Therefore, while φ on $\{BA,\ A \in \mathfrak{A}\}$ is σ-finite, φ on $\{B^cA,\ A \in \mathfrak{A}\}$ can take values 0 and ∞ only.

Furthermore, whatever be $C \in \{B^cA, A \in \mathfrak{A}\}$, it is impossible to have $\mu C > 0$ and $\varphi(C) = 0$ since then $B + C \in \mathfrak{B}$ and, as above, $s > s$. Since φ is μ-continuous, it is also impossible to have $\mu C = 0$ and $\varphi(C) > 0$. Thus, for every $C \in \{B^cA,\ A \in \mathfrak{A}\}$, either $\mu C > 0$ and $\varphi(C) = \infty \cdot \mu C = \infty$ or $\mu C = 0$ and $\varphi(C) = 0$. In other words, φ on $\{B^cA,\ A \in \mathfrak{A}\}$ is the indefinite integral of a function $X = \infty$ on B^c, determined up to an equivalence. On the other hand, by **B**, φ on $\{BA, A \in \mathcal{A}\}$ is the indefinite integral of a function X on B, determined up to an equivalence. These values of X on B and on B^c determine it on Ω, up to an equivalence and, for every $A \in \mathcal{A}$,

$$\int_A X = \int_{AB} X + \int_{AB^c} X = \varphi(AB) + \varphi(AB^c) = \varphi(A).$$

The extension follows.

8.2 Product measures and iterated integrals. Let $(\Omega_i, \mathcal{A}_i, \mu_i)$, $i = 1, 2$, be two measure spaces. A space $(\Omega, \mathcal{A}, \mu)$ is their *product-measure space* if

$\Omega = \Omega_1 \times \Omega_2$ is the space of all points $\omega = (\omega_1, \omega_2)$, $\omega_i \in \Omega_i$;
$\mathcal{A} = \mathcal{A}_1 \times \mathcal{A}_2$ is the minimal σ-field over the class of all measurable "rectangles" $A_1 \times A_2$, $A_i \in \mathcal{A}_i$, where $A_1 \times A_2$ is the set of all points ω with $\omega_i \in A_i$;
$\mu = \mu_1 \times \mu_2$ is the "product-measure" on $\mathcal{A}$, provided it exists, that is, is a measure on $\mathcal{A}$ uniquely determined by the relations $\mu(A_1 \times A_2) = \mu_1 A_1 \times \mu_2 A_2$ for all measurable rectangles $A_1 \times A_2$.

We intend to find conditions under which the product-measure exists and conditions under which integrals with respect to this measure can be expressed in terms of integrals with respect to the factor measures μ_i. In what follows the subscripts 1 and 2 can be interchanged. We shall also frequently proceed to the usual abuse of notation which consists in the use of the same symbol for a function and for its values.

For every set $A \subset \Omega$, the *section* A_{ω_1} of A at ω_1 is the set of all points ω_2 such that $(\omega_1, \omega_2) \in A$. For every function X on Ω, the *section* X_{ω_1} of X at ω_1 is the function defined on Ω_2 by $X_{\omega_1}(\omega_2) = X(\omega_1, \omega_2)$.

a. *Every section of a measurable set or function is measurable.*

If $\mathcal{C}$ is the class of all the sets in Ω whose every section is measurable, then it is readily seen that $\mathcal{C}$ is a σ-field. But every section of a measurable rectangle $A_1 \times A_2$ is measurable, since it is either empty or is one of the sides. Therefore, $\mathcal{C} \supset \mathcal{A}$ and the first assertion is proved. If X on Ω is measurable and $S \subset \bar{R}$ is an arbitrary Borel set, the second assertion follows by

$$X_{\omega_1}^{-1}(S) = [\omega_2; X_{\omega_1}(\omega_2) \in S] = [\omega_2; X(\omega_1, \omega_2) \in S]$$
$$= [\omega_2; (\omega_1, \omega_2) \in X^{-1}(S)] = (X^{-1}(S))_{\omega_1}.$$

A. Product-measure theorem. *If μ_1 on $\mathcal{A}_1$ and μ_2 on $\mathcal{A}_2$ are σ-finite, then, for every $A \in \mathcal{A}_1 \times \mathcal{A}_2$, the functions with values $\mu_1 A_{\omega_2}$ and $\mu_2 A_{\omega_1}$ are measurable, and the set function μ with values*

$$\mu A = \int (\mu_1 A_{\omega_2})\, d\mu_2 = \int (\mu_2 A_{\omega_1})\, d\mu_1,$$

is a σ-finite measure μ on $\mathcal{A}_1 \times \mathcal{A}_2$ uniquely determined by the relation

$$\mu(A_1 \times A_2) = \mu_1 A_1 \times \mu_2 A_2, \quad A_i \in \mathcal{A}_i.$$

In other words, μ is the product-measure $\mu_1 \times \mu_2$.

Proof. The proof is based upon the fact that, by the monotone convergence theorem, the class $\mathfrak{M}$ of all those sets A for which the integrals are equal is closed under formation of countable sums.

Since the measures μ_1 and μ_2 are σ-finite, the product space is decomposable into a countable sum of rectangles with sides of finite measure. It follows that, without restricting the generality, we can suppose that these measures are finite. If $A = A_1 \times A_2$ is a measurable rectangle, then $\mu_1 A_{\omega_2} = \mu_1 A_1 \times I_{A_2}(\omega_2)$ and similarly by interchanging the subscripts 1 and 2. Thus, the functions with these values are measurable and both integrals reduce to $\mu_1 A_1 \times \mu_2 A_2$. The last asserted equality is proved and $\mathfrak{M}$ contains all measurable rectangles. It follows that $\mathfrak{M}$ contains the field of finite sums of these rectangles. But, $\mathfrak{M}$ is closed under nondecreasing passages to the limit, on account of the monotone convergence theorem, and, under nonincreasing ones, on account of the dominated convergence theorem and the finiteness of measures. Therefore, by 1.6, it contains the product σ-field $\mathcal{A}_1 \times \mathcal{A}_2$, and the equality of the integrals is proved. The finite set function μ on $\mathcal{A}$ so defined is a measure, on account of the monotone convergence theorem, and it is uniquely determined by the stated relation, on account of the extension theorem. This terminates the proof.

Corollary. *$A \in \mathcal{A}_1 \times \mathcal{A}_2$ is a $(\mu_1 \times \mu_2)$-null set if, and only if, almost every section A_{ω_1} is a μ_2-null set.*

For the integral of a nonnegative function vanishes if, and only if, the integrand vanishes a.e.

We are now in a position to answer the second stated question. The result is due to Lebesgue and Fubini and is generally called the Fubini theorem.

B. ITERATED INTEGRALS THEOREM. *Let* $(\Omega_1, \mathcal{A}_1, \mu_1)$ *and* $(\Omega_2, \mathcal{A}_2, \mu_2)$ *be* σ*-finite measure spaces.*

If the $\mathcal{A}_1 \times \mathcal{A}_2$*-measurable function* X *on* $\Omega_1 \times \Omega_2$ *is nonnegative or* $\mu_1 \times \mu_2$*-integrable, then*

$$\int_{\Omega_1 \times \Omega_2} X d(\mu_1 \times \mu_2) = \int_{\Omega_1} d\mu_1 \int_{\Omega_2} X_{\omega_1} d\mu_2 = \int_{\Omega_2} d\mu_2 \int_{\Omega_1} X_{\omega_2} d\mu_1,$$

and in the integrability case almost every section of X *is integrable.*

The iterated integrals are to be read from right to left.

Proof. For $X = I_A$, the asserted equality reduces to that of the product-measure theorem. It follows that it holds for simple functions and hence holds for nonnegative measurable functions because of the monotone convergence theorem, since, if $0 \leqq X_n \uparrow X$, then $0 \leqq (X_n)_{\omega_i} \uparrow (X)_{\omega_i}$. If $X \geqq 0$ is integrable, then the function $\int X_{\omega_1} d\mu_2$ of ω_1 is integrable and hence a.e. finite, so that the functions X_{ω_1} of ω_2 are almost all integrable. Therefore, if $X = X^+ - X^-$ is integrable, that is, X^+ and X^- are integrable, then $(X)_{\omega_i} = (X^+)_{\omega_i} - (X^-)_{\omega_i}$ are almost all integrable and a.e. finite. This terminates the proof.

Finite-dimensional case. What precedes extends in an obvious manner to the product of an arbitrary but finite number of measure spaces. The interesting case is the infinitely dimensional one, and we shall now investigate it from a somewhat more general point of view.

***8.3 Iterated integrals and infinite product spaces.** In what follows we push the abuse of notation to its extreme.

We consider a sequence of measurable spaces $(\Omega_n, \mathcal{A}_n)$ and denote by ω_n points of Ω_n and by A_n measurable sets in Ω_n (sets of $\mathcal{A}_n$). The product measurable space $(\Omega_1 \times \cdots \times \Omega_n, \mathcal{A}_1 \times \cdots \times \mathcal{A}_n)$ is the space of points $(\omega_1, \cdots, \omega_n)$ together with the minimal σ-field over the *intervals* $A_1 \times \cdots \times A_n$. The product measurable space $(\prod \Omega_n, \prod \mathcal{A}_n)$ is the space of points $(\omega_1, \omega_2, \cdots)$ and the minimal σ-field over all cylinders of the form $A_1 \times \cdots \times A_n \times \prod_{k=n+1}^{\infty} \Omega_k$ or, equivalently, over all cylinders of the form $C(B_n) = B_n \times \prod_{k=n+1}^{\infty} \Omega_k$ where the *base* B_n is a measurable set in $\Omega_1 \times \cdots \times \Omega_n$.

In the infinitely dimensional case, we must, for reasons of "consistency" (to be made clear later), limit ourselves to probabilities, that is, to measures which assign value one to the space, to be denoted by P, Q, $\cdots$, with or without affixes. Furthermore, in probability theory, the following more general concept plays a basic role (at least when "independence"—see Part III—is not assumed). Every function—to be denoted by $P_n(\omega_1, \cdots, \omega_{n-1}; A_n)$—which is a probability in A_n for every fixed point $(\omega_1, \cdots, \omega_{n-1})$ and a measurable function in this point for every fixed A_n will be called a *regular conditional probability*. For $n = 1$ it reduces to a probability P_1 on $\mathfrak{A}_1$ but for $n > 1$ it reduces to a probability on $\mathfrak{A}_n$ only when it is constant in $(\omega_1, \cdots, \omega_{n-1})$ for every fixed A_n, provided the ordered T has a first element. We observe that the functions $\mu_2 A_{\omega_1} = \mu_2(\omega_1; A_{\omega_1})$ are regular conditional probabilities when $\mu_2(\omega_1; \Omega_2) = 1$. On account of the monotone convergence theorem, iterated integrals of the form

$$Q_n B_n = \int P_1(d\omega_1) \int P_2(\omega_1; d\omega_2) \cdots \int P_n(\omega_1, \cdots, \omega_{n-1}; d\omega_n) I_{B_n}(\omega_1, \cdots, \omega_n)$$

define probabilities Q_n on $\mathfrak{A}_1 \times \cdots \times \mathfrak{A}_n$. It follows by the same theorem that if a measurable function X on $\Omega_1 \times \cdots \times \Omega_n$ is nonnegative or Q_n-integrable, then

$$\int_{\Omega_1 \times \cdots \times \Omega_n} X\, dQ_n = \int P_1(d\omega_1) \int P_2(\omega_1; d\omega_2) \cdots \int P_n(\omega_1, \cdots, \omega_{n-1}; d\omega_n) X(\omega_1, \cdots, \omega_n).$$

A. ITERATED REGULAR CONDITIONAL PROBABILITIES THEOREM. *The iterated integrals*

$$QC(B_n) = \int P_1(d\omega_1) \int P_2(\omega_1; d\omega_2) \cdots \int P_n(\omega_1, \cdots, \omega_{n-1}; d\omega_n) I_{B_n}(\omega_1, \cdots, \omega_n),$$

determine a probability Q on $\prod \mathfrak{A}_n$.

This extension of the product-probability theorem is due to Tulcea and, proceeding as therein (in 1°), permits one to determine Q on an arbitrary $\prod_{t \in T} \mathfrak{A}_t$ under obvious consistency conditions on the regular conditional pr.'s $P_{t_{n+1}}(\omega_{t_1}, \cdots, \omega_{t_n}; A_{t_{n+1}})$.

Proof. To begin with, the definition of Q_n on the class $\mathfrak{C}$ of all cylinders of the form $C(B_n)$ is consistent. For, if $C(B_n) = C(B_m)$, $m < n$, then integrations with respect to the ω_k which do not belong to the product subspace where B_m lies yield factors one.

Since Q on $\mathfrak{C}$ is finitely additive, the assertion will follow by the extension theorem if we prove that Q on $\mathfrak{C}$ is continuous at $\emptyset$. We have to consider nonincreasing sequences of cylinders which converge to $\emptyset$.

Upon renumbering the indices, we can suppose that the sequences are of the form $C(B_n) \downarrow \emptyset$ with nonempty bases $B_n \in \mathcal{A}_1 \times \cdots \times \mathcal{A}_n$. We can write

$$(1) \qquad QC(B_n) = \int P(d\omega_1) Q^{(1)} C(B_n)_{\omega_1}$$

where $(B_n)_{\omega_1}$ is the section of B_n at ω_1 and

$$Q^{(1)}C(B_n)_{\omega_1} = \int P_2(\omega_1; d\omega_2) \cdots \int P_n(\omega_1, \cdots, \omega_{n-1}; d\omega_n) I_{B_n}(\omega_1, \cdots, \omega_n).$$

In (1) the left-hand side is nonincreasing in n, and the integrand converges nonincreasingly to a certain limit $X_1(\omega_1) \geqq 0$. By the dominated convergence theorem, the limit of the left-hand side is $\int P(d\omega_1) X_1(\omega_1)$.

Assume that this integral is positive. Then there exists a point $\bar{\omega}_1$ such that $X_1(\bar{\omega}_1) > 0$. It follows that we find ourselves in the same situation but with the sequence $Q^{(1)}C(B_n)_{\bar{\omega}_1}$ instead of $QC(B_n)$. Repeating the argument over and over again, we obtain a sequence $\bar{\omega} = (\bar{\omega}_1, \bar{\omega}_2, \cdots)$ such that $\bar{\omega}_n \in \Omega_n$ and $Q^{(n)}C(B_n)_{\bar{\omega}_1, \cdots, \bar{\omega}_n} \downarrow X_n(\bar{\omega}_n) > 0$. Therefore, every $C(B_n)$ contains at least one point of the form $(\bar{\omega}_1, \cdots, \bar{\omega}_n, \omega_{n+1}, \cdots)$. Since $C(B_n) = B_n \times \prod_{k=n+1}^{\infty} \Omega_k$, it contains the point $\bar{\omega}$ and, hence, $\bar{\omega} \in \bigcap C(B_n)$. Thus, when $QC(B_n) \nrightarrow 0$ the intersection is not empty, and the theorem follows *ab contrario*.

Particular cases. 1° If $P_n(\omega_1, \cdots, \omega_{n-1}; A_n) = P_n A_n$ are constant for every fixed A_n, then we write $Q = \prod P_n$ and call it a *product-probability*. Then the theorem reduces to the *product-probability theorem* in the denumerable case (4.2**A**).

2° If the factor spaces are finite-dimensional Borel spaces, then, it follows from 27.2, Application 1, that the theorem yields the *consistency theorem*.

COMPLEMENTS AND DETAILS

Notation. Unless otherwise stated, the measure space $(\Omega, \mathcal{A}, \mu)$ is fixed, the (measurable) sets $A, B, \cdots$, with or without affixes, belong to $\mathcal{A}$, and the functions $X, Y, \cdots$, with or without affixes, are finite measurable functions.

1. The set C of convergence of a sequence X_n (to a finite or infinite limit function) is measurable.

$$(C = [\liminf X_n = \limsup X_n].)$$

2. If μ is finite, then given X, for every $\epsilon > 0$ there exists A such that $\mu A < \epsilon$ and X is bounded on A^c. If X is bounded, then there exists a sequence of simple functions which converges uniformly to X. Combine both propositions.

We say that a sequence X_n converges *almost uniformly* (a.u.) to X, and write $X_n \xrightarrow{\text{a.u.}} X$, if, for every $\epsilon > 0$, there exists a set A with $\mu A < \epsilon$ such that $X_n \xrightarrow{\text{u}} X$ on A^c.

3. If $X_n \xrightarrow{\text{a.u.}} X$, then $X_n \xrightarrow{\text{a.e.}} X$ and $X_n \xrightarrow{\mu} X$. (For the first assertion, form A_n where A_n is the A of the foregoing definition with $\epsilon = \frac{1}{n}$.)

4. If $X_n \xrightarrow{\mu} X$, then there exists a subsequence $X_{n'} \xrightarrow{\text{a.u.}} X$.

5. Egoroff's theorem. If μ is finite, then $X_n \xrightarrow{\text{a.e.}} X$ implies that $X_n \xrightarrow{\text{a.u.}} X$. Compare with *3.* (Neglect the null set of divergence, and form $A = \bigcup_{m=1}^{\infty} A_m$ with $A_m = \bigcup_{k \geqq n(m)} \left[\,|X_k - X| \geqq \frac{1}{m}\right]$ and $n(m)$ such that $\mu A_m < \frac{\epsilon}{2^m}$.)

6. Lusin's theorem. If μ is σ-finite, then $X_n \xrightarrow{\text{a.e.}} X$ implies that $X_n \xrightarrow{\text{u}} X$ on every element A_j of some countable partition of Ω-N where N is some null set. (Neglect the null set of divergence, and start with μ finite. Use Egoroff's theorem to select inductively sets A_k such that $\mu \bigcap_{k=1}^{n} A_k < \frac{1}{n}$ and $X_n \xrightarrow{\text{u}} X$ on A_k^c for every k.)

7. If μ is finite, then $X_n \xrightarrow{\text{a.e.}} X$ implies existence of a set of positive measure on which the X_n are uniformly bounded. What if μ is σ-finite?

8. If μ is finite, then $X_{mn} \xrightarrow{\text{a.e.}} X_m$ as $n \to \infty$ and $X_m \xrightarrow{\text{a.e.}} X$ as $m \to \infty$ imply that there exists subsequences m_k, n_k such that $X_{m_k n_k} \xrightarrow{\text{a.e.}} X$ as $k \to \infty$. What if μ is σ-finite?

(Neglect the null sets of divergence. Select A_k and m_k such that $\mu A_k < \frac{1}{2^k}$ and $|X_{m_k} - X| < \frac{1}{2^k}$ on A_k^c. Select $B_k \subset A_k$ and n_k such that $\mu B_k < \frac{1}{2^k}$ and $|X_{m_k n_k} - X_{m_k}| < \frac{1}{2^k}$ on $A_k - B_k$.)

9. Let $X_n \xrightarrow{\mu} X, Y_n \xrightarrow{\mu} Y$. Then $aX_n + bY_n \xrightarrow{\mu} aX + bY, |X_n| \xrightarrow{\mu} |X|$, $X_n^2 \xrightarrow{\mu} X^2, X_nY_n \xrightarrow{\mu} XY$. What about $1/X_n$? Let μ be finite and let g on R or on $R \times R$ be continuous. What about the sequences $g(X_n)$ and $g(X_n, Y_n)$?

10. Let the functions X_n, X on the measure space be complex-valued or vector-valued or, more generally, let them take their values in some fixed Banach space. Denote the norm of X by $|X|$, and denote $|X_n - X| \to 0$ by $X_n \to X$.

Transpose the constructive definitions of measurability and the definitions of various types of convergence. Investigate the validity of the transposed of the corresponding properties established in the text, as well as of those stated above.

11. Examples and counterexamples of mutual implications of types of convergence. Investigate convergences of the sequences defined below:

(i) The measure space is the Borel line with Lebesgue measure, $X_n = 1$ on $[n, n+1]$ and $X_n = 0$ elsewhere.

(ii) The measure space is the Borel interval (0, 1) with Lebesgue measure, $X_n = 1$ on $\left(0, \frac{1}{n}\right)$ and $X_n = 0$ elsewhere.

(iii) The measure space is the Borel interval [0, 1] with Lebesgue measure, the sequence is $X_{11}, X_{21}, X_{22}, X_{31}, X_{32}, X_{33}, \cdots$ with $X_{nk} = 1$ on $\left[\frac{k-1}{n}, \frac{k}{n}\right]$ and $X_{nk} = 0$ elsewhere.

(iv) $\mathfrak{A}$ consists of all subsets of the set of positive integers, μA is the number of points of A, X_n is indicator of the set of the n first integers.

12. If X is integrable, then the set $[X \neq 0]$ is of σ-finite measure. What if $\int X$ exists? ($\mu[|X| \geqq c] \leqq \frac{1}{c}\int |X|$.)

13. Let $(T, \mathfrak{I}, \tau)$ be a measure space, to every point t of which is assigned a measure μ_t on $\mathfrak{A}$. Let the function on T defined by $\mu_t A$ for any fixed A be $\mathfrak{I}$-measurable.

The relation $\mu A = \int_T \mu_t A \, d\tau(t)$ defines a measure μ on $\mathfrak{A}$. If $\int_\Omega X(\omega) \, d\mu(\omega)$ exists, then the function defined on T by $U(t) = \int_\Omega X(\omega) \, d\mu_t(\omega)$ exists and is $\mathfrak{I}$-measurable, and $\int_\Omega X(\omega) \, d\mu(\omega) = \int_T U(t) \, d\tau(t)$.

14. Let φ be the indefinite integral of X. Express $\varphi^+, \varphi^-, \bar{\varphi}$ in terms of X.

15. If $\int_A X_n \to 0$ uniformly in n as $\mu A \to 0$ or as $A \downarrow \emptyset$, then the same is true of $\int_A |X_n|$; and conversely. Interpret in terms of signed measures.

$$\left(\int_A |X_n| = \int_{A[X_n \geqq 0]} X_n - \int_{A[X_n < 0]} X_n.\right)$$

16. If finite $\int_A X_n \to \int_A X$ finite, uniformly in $A \,(\in \mathfrak{A})$, then $\int_\Omega |X_n - X| \to 0$; and conversely.

17. If $0 \leqq X_n \xrightarrow{\mu} X$, then finite $\int_\Omega X_n \to \int_\Omega X$ finite implies that $\int_A X_n \to \int_A X$ uniformly in A (also if $\xrightarrow{\mu}$ is replaced by $\xrightarrow{\text{a.e.}}$)

$$(0 \leqq (X - X_n)^+ \leqq X \text{ integrable, and } \int (X - X_n)^+ - \int (X - X_n) \to 0.)$$

18. Rewrite in terms of integrals as many as possible of the complements and details of Chapter I.

19. If the X_n are integrable and $\lim \int_A X_n$ exists and is finite for every A, then the $\int |X_n|$ are uniformly bounded, $\int_A |X_n| \to 0$ uniformly in n as $\mu A \to 0$ and as $A \downarrow \emptyset$, and there exists an integrable X, determined up to an equivalence, such that $\int_A X_n \to \int_A X$ for every A. (Use *18.*)

20. If integrable $X_n \to X$ integrable, then existence and finiteness of $\lim \int_A X_n$ for every A are equivalent to the following properties:

(i) $$\int_A X_n \to \int_A X \text{ uniformly in } A;$$

(ii) $$\int_A X_n \to 0 \text{ uniformly in } n \text{ as } \mu A \to 0 \text{ and as } A \downarrow \emptyset.$$

If μ is finite, then "as $A \downarrow \emptyset$" can be suppressed. (Use the preceding propositions and the relations

$$\int_A |X_n| \leqq \int_A |X_n - X| + \int_A |X|,$$

$$\int_A |X_n - X| \leqq \epsilon + \int_{A[|X_n - X| \geqq \epsilon]} (|X_n| + |X|).)$$

21. The differential formalism applies to Radon-Nikodym derivatives: Let μ, ν be finite measures on $\mathfrak{A}$ and φ, φ' be σ-finite signed measures on $\mathfrak{A}$. Let φ be ν-continuous and ν, φ, φ' be μ-continuous. Then

$$\frac{d(\varphi + \varphi')}{d\mu} = \frac{d\varphi}{d\mu} + \frac{d\varphi'}{d\mu} \quad \mu\text{-a.e.}$$

$$\frac{d\varphi}{d\mu} = \frac{d\varphi}{d\nu}\frac{d\nu}{d\mu} \quad \mu\text{-a.e.}$$

(For the second assertion, it suffices to consider $\varphi \geqq 0$, $X = \dfrac{d\varphi}{d\nu} \geqq 0$ $Y = \dfrac{d\nu}{d\mu} \geqq 0$. Take simple X_n with $0 \leqq X_n \uparrow X$ so that

$$\int_A X\,d\nu \leftarrow \int_A X_n\,d\nu = \int_A X_n Y\,d\mu \to \int_A XY\,d\mu.)$$

Let $\{\mu_t, t \in T\}$ and $\{\mu'_{t'}, t' \in T'\}$ be two families of measures on $\mathfrak{A}$; we drop $t \in T$ and $t' \in T'$ unless confusion is possible. We say that $\{\mu_t\}$ is $\{\mu'_{t'}\}$-continuous if every set null for all $\mu'_{t'}$ is null for all μ_t. If the converse is also true, we say that the two families are mutually continuous.

22. If $\{\mu_j\}$ is a countable family of finite measures, then there exists a finite measure μ such that $\{\mu_t\}$ and μ are mutually continuous. (Take $\mu = \sum \mu_j/2^j\mu_j\Omega$.)

23. Let the μ_t and μ be finite measures. If $\{\mu_t\}$ is μ-continuous, then there exists a finite measure μ' such that $\{\mu_t\}$ and μ' are mutually continuous. (Select sets $A_t = \left[\dfrac{d\mu_t}{d\mu} > 0\right]$. Denote by B, with or without affixes, sets such that, for some t, $B \subset A_t$ and $\mu_t B > 0$. Denote countable sums of sets B up to μ-null sets by C, with or without affixes. Every subset $C' \subset C$ with $\mu_t C' > 0$ is a set C; every countable union of sets C is a set C. Let $\mu C_n \to s$ where s is the supremum of values of μ over all the sets C. Then $s = \mu \bigcup C_n = \mu \bigcup B_m$ and to every m there corresponds a μ_t, say μ_m, such that $B_m \subset A_m$ and $\mu_m B_m > 0$. The families $\{\mu_t\}$ and $\{\mu_m\}$ are mutually continuous.)

24. Let $\bar{\mu}_n = \sum_{k=1}^{n} \mu_k \to \bar{\mu}$ and $\bar{\nu}_n = \sum_{k=1}^{n} \nu_k \to \bar{\nu}$, all the μ and ν with various affixes being finite measures on $\mathfrak{A}$ and every $\bar{\nu}_n$ being $\bar{\mu}_n$-continuous.

(i) $\frac{d\mu_1}{d\bar{\mu}_n} \to \frac{d\mu_1}{d\bar{\mu}}$ μ-a.e.

(ii) if $\{\mu_n\}$ is ν-continuous, then $\frac{d\bar{\mu}_n}{d\nu} \to \frac{d\bar{\mu}}{d\nu}$ ν-a.e.

(iii) $\bar{\nu}$ is $\bar{\mu}$-continuous and $\frac{d\bar{\nu}_n}{d\bar{\mu}_n} \to \frac{d\bar{\nu}}{d\bar{\mu}}$ $\bar{\mu}$-a.e.

(For the last assertion, if $\bar{\mu}_n A_n = 0$ for all n, then $\bar{\mu}$ (lim sup A_n) $= 0$. It follows that it suffices to consider a particular choice of the $\frac{d\bar{\nu}_n}{d\bar{\mu}_n} = \sum_{k=1}^{n} X_k / \sum_{k=1}^{n} Y_k$ where $X_k = \frac{d\nu_k}{d\bar{\mu}}$, $Y_k = \frac{d\mu_k}{d\bar{\mu}}$. But $\sum X_n = \frac{d\bar{\nu}}{d\bar{\mu}}$ and $\sum Y_n = 1\bar{\mu}$-a.e.)

The propositions which follow correspond to various definitions of the concept of integration. We shall assume that the measures and the functions are finite. Besides proving the statements, the reader should also examine removal of the restriction of finiteness as well as of other restrictions which may be introduced.

25. Set

$$\int X\,d\varphi = \int X\,d\varphi^+ - \int X\,d\varphi^-, \quad \int (X + iY)\,d\mu = \int X\,d\mu + i\int Y\,d\mu,$$

$$\int X d(\mu + i\nu) = \int X\,d\mu + i\int X\,d\nu$$

and investigate existence and properties of integrals so defined.

26. Descriptive approach. The Radon-Nikodym theorem characterizes an indefinite integral but not that of a given function. The following proposition answers this requirement.

φ on $\mathfrak{A}$ is indefinite integral of X on Ω if, and only if, φ is σ-additive and, for every set $A = [a \leqq X \leqq b]B$, $B \in \mathfrak{A}$,

$$a\mu A \leqq \varphi(A) \leqq b\mu A.$$

27. In the definition of the integral given in the text, start with (nonnegative) elementary functions instead of simple ones. The integral so defined coincides with the initial one.

28. Lebesgue's approach. The Cauchy-Riemann approach starts with arbitrary finite partitions of the interval of integration into intervals. The Lebesgue approach consists in partitioning the set of integration according to the function to be integrated so that the integral is tailored to order as opposed to the ready-to-wear Cauchy-Riemann one.

Set

$$\sum_n(X) = \sum_{k=-\infty}^{+\infty} \frac{k-1}{2^n} \mu\left[\frac{k-1}{2^n} \leqq X < \frac{k}{2^n}\right].$$

If X is bounded, these sums correspond to finite partitions and $\int X =$

$\lim \sum_n(X)$. If X is not bounded, set $X_{mn} = X$ if $-m \leqq X \leqq n$ and $X_{mn} = 0$ otherwise. If X is integrable, then $\int X_{mn} \to \int X$ as $m, n \to \infty$.

If X is not bounded, the series $\sum_n(X)$ correspond to countable partitions and $\int X = \lim \sum_n(X)$, in the sense that if X is integrable, then these series are absolutely convergent and the equality holds and, conversely, if one of these series is absolutely convergent, so are all of them and the equality holds.

(For the last assertion, it suffices to consider nonnegative elementary functions $X_n = \sum_1^\infty \frac{k-1}{2^n} I_{\left[\frac{k-1}{2^n} \leqq X < \frac{k}{2^n}\right]}$. For the converse, use the relation $X \leqq 2X_n + \mu\Omega$.)

29. Darboux-Young approach. Let X be measurable or not and set

$$\underline{\int} X = \sup \sum_{k=1}^{n} \inf_{\omega \in A_k} X(\omega)\mu A_k, \quad \overline{\int} X = \inf \sum_{k=1}^{n} \sup_{\omega \in A_k} X(\omega)\mu A_k$$

where the extrema of sums are taken over all finite measurable partitions $\sum_{k=1}^{n} A_k = \Omega$. If X is measurable and bounded, then

$$\int X = \underline{\int} X = \overline{\int} X.$$

If $\underline{\int} X$ and $\overline{\int} X$ exist and are equal, we say that $\int X$ exists and equals their common value.

We can also set

$$\underline{\int}' X = \sup \int Y, \quad \overline{\int}' X = \inf \int Z$$

where the extrema are taken over all integrable (and measurable) Y and Z such that $Y \leqq X \leqq Z$ and define $\int X$ as above. Compare the two definitions.

30. Completion approach. The Meray-Cantor method for completion of metric spaces adjoins to the given metric space elements which represent mutually convergent (in distance) sequences of its points. This method permits (Dunford) to define and study the integral of functions with values in an arbitrary Banach space (Bochner), as follows:

(i) Define the indefinite integral of a simple function as in the text. Since nonnegativity and infinite values may be meaningless, all simple functions under consideration are integrable.

(ii) Adjoin to the space of these integrable functions $X_m, X_n, \cdots$ all functions X such that $\int | X_m - X_n | \to 0$ and $X_n \to X$, by defining the indefinite integral of X as the limit of the indefinite integrals of the X_n. To justify this definition, prove for simple functions those elementary properties of integrals which continue to have content for an arbitrary Banach space: $\int | X_m - X_n | \to 0$ if, and only if, $X_n \xrightarrow{\mu} X$ where X is some measurable function, and $\int_A | X_n | \to 0$ uniformly in n as $\mu A \to 0$; $\int | X_m - X_n | \to 0$ implies that $\varphi_n \to \varphi$ where φ is σ-additive.

(iii) Extend the foregoing properties to all integrable functions and obtain the dominated convergence theorem.

31. Kolmogorov's approach. Let $\mathfrak{C}$ be a class closed under intersections. Let $\mathfrak{D}$, with or without affixes, be finite disjoint subclasses of $\mathfrak{C}$. Order them by the relation $\mathfrak{D}_1 < \mathfrak{D}_2$ if every set of $\mathfrak{D}_2$ is contained in some set of $\mathfrak{D}_1$. Fix $A \in \mathfrak{C}$ and consider all the $\mathfrak{D}$ which are partitions of A. They form a "direction" Δ in the sense that, if $\mathfrak{D}_1$ and $\mathfrak{D}_2$ are such partitions, then there exists such a partition which "follows" both, namely, $\mathfrak{D}_1 \cap \mathfrak{D}_2$.

Let φ on $\mathfrak{C}$ be a function, additive or not, single-valued or not. By definition,

$$\hat{\varphi}(A) = \int_A d\varphi = \lim \sum \varphi(A_j)$$

where the A_j are elements of partitions $\mathfrak{D}$ of A and the limit $\hat{\varphi}(A)$, if it exists, is "along the direction Δ," that is, to every $\epsilon > 0$ there corresponds a $\mathfrak{D}_\epsilon$ such that $| \hat{\varphi}(A) - \sum \varphi(A_j) | < \epsilon$ for all $\mathfrak{D} > \mathfrak{D}_\epsilon$ and all values of the $\varphi(A_j)$—if φ is multivalued. If $\hat{\varphi}(A)$ exists, it is unique. If $\hat{\varphi}$ on $\mathfrak{C}$ exists, then it is finitely additive.

Compare this integral to the Riemann-Stieltjes integral by selecting conveniently φ.

Compare $\int_{\alpha\beta} d\varphi$ with the length (if it exists) of the arc $\alpha\beta$ of a plane curve, by taking $\varphi(\alpha_{k-1}, \alpha_k) = \overline{\alpha_{k-1}\alpha_k}$, the length of the cord α_{k-1} to α_k, the $\alpha = \alpha_1, \cdots \alpha_{k-1}, \alpha_k, \cdots, \alpha_n = \beta$ being consecutive points on the arc $\alpha\beta$.

We say that φ and φ' on $\mathfrak{C}$ are "differentially equivalent" on A if, for every $\epsilon > 0$, there exists a partition $\mathfrak{D}_\epsilon$ of A such that $\sum | \varphi(A_j) - \varphi'(A_j) | < \epsilon$ for all $\mathfrak{D} > \mathfrak{D}_\epsilon$. If φ is finitely additive, then $\int_A d\varphi = \varphi(A)$. If not, then $\hat{\varphi}$ on $A \cap \mathfrak{C}$ (if it exists) is the unique additive function differentially equivalent on A to φ. Proceed as follows:

(i) φ and φ' are differentially equivalent on A if, and only if, $\hat{\varphi} = \hat{\varphi}'$.

(ii) φ and $\hat{\varphi}$ are differentially equivalent on A.

(iii) If finitely additive functions φ and φ' are differentially equivalent on A, then they coincide on A.

In all which precedes replace "finite" by "countable" and investigate the validity of the propositions so obtained. Compare the various definitions of the integral, by selecting conveniently φ.

Finally, take φ with values in a fixed but arbitrary Banach space, and go over what precedes.

32. A structure of the concept of integration. The concept of integration is constructed by means of the concepts of summations and of passage to the limit along a direction or, more generally, a cut-direction. A bipartition $\underline{\overline{\Delta}} = \underline{\Delta} + \overline{\Delta}$

of a set Δ with an order relation $\prec$ is a "cut-direction" if $\underline{\Delta}$ and $\overline{\Delta}$ are directions and every element of $\overline{\Delta}$ follows every element of $\underline{\Delta}$.

Let φ be a function, single-valued or not, on a direction Δ to a real line or a plane or, more generally, a Banach space. The element φ_Δ of the range space is "limit of φ along Δ" if, for every $\epsilon > 0$, there exists an $\alpha_\epsilon \in \Delta$ such that $|\varphi_\Delta - \varphi(\alpha)| < \epsilon$ for all $\alpha \succ \alpha_\epsilon$ and for all values of $\varphi(\alpha)$. If the direction Δ is replaced by a cut-direction $\underline{\overline{\Delta}}$, then $\varphi_{\underline{\overline{\Delta}}}$ is "limit of φ along $\underline{\overline{\Delta}}$ if, for every $\epsilon > 0$ there exist $\underline{\alpha}_\epsilon \in \Delta$ and $\bar{\alpha}_\epsilon \in \overline{\Delta}$ such that $|\varphi_{\underline{\overline{\Delta}}} - \varphi(\alpha)| < \epsilon$ for all α such that $\underline{\alpha}_\epsilon \prec \alpha \prec \bar{\alpha}_\epsilon$ and for all values of $\varphi(\alpha)$. If φ_Δ or $\varphi_{\underline{\overline{\Delta}}}$ exist, they are unique.

To every $\alpha \in \Delta$ assign some finite collection of points α_j of a Banach space, not necessarily distinct and not necessarily uniquely determined. Form $\varphi(\alpha) = \sum \varphi(\alpha_j)$. By definition, $\int_\Delta d\varphi$ is the limit, if it exists, of φ along Δ. If Δ is replaced by $\underline{\overline{\Delta}}$, the definition continues to apply.

Investigate all definitions of the integral you know of from this structural point of view, that is, the selections of Δ or $\underline{\overline{\Delta}}$, and of the functions φ.

33. Daniell approach. Let S be a family of bounded real-valued functions on Ω, closed under finite linear combinations and lattice operations $f \cup g = \max(f, g)$, $f \cap g = \min(f, g)$. Then $f \in L \Rightarrow |f| = f \cup 0 - f \cap 0 \in S$. Suppose that on S is defined an *integral* $\int$: a nonnegative linear functional continuous under monotone limits: $f \geqq 0 \Rightarrow \int f \geqq 0, \int (af + bg) = a\int f + b\int g, f_n \downarrow 0 \Rightarrow \int f_n \downarrow 0.$

a) Let U be the family of limits (not necessarily finite) of nondecreasing sequences in S. U contains S and is closed under addition, multiplication by nonnegative constants, and lattice operations. Extend the integral on U, setting $\int f = \lim \int f_n$ when $S \ni f_n \uparrow f$ (infinite values being permitted).
The definition is justified, for if the nondecreasing sequences f_n and g_n in S are such that $\lim f_n \leqq \lim g_n$, then $\lim \int f_n \leqq \lim \int g_n$.

If $U \ni f_n \uparrow f$ then $f \in U$ and $\int f_n \uparrow \int f$.

b) Let $-U$ be the family of functions f such that $-f \in U$, and set $\int f = -\int(-f)$.

If $g \in -U, h \in U$ and $g \leqq h$, then $h - g \in U$ and $\int h - \int g = \int (h - g) \geqq 0$.

By definition, f is *integrable* if, for every $\epsilon > 0$, there exist $g_\epsilon \in -U$ and $h_\epsilon \in U$ such that $g_\epsilon \leqq f \leqq h_\epsilon$, $\int g_\epsilon$ and $\int h_\epsilon$ are finite, and $\int h_\epsilon - \int g_\epsilon < \epsilon$. Then $\inf_\epsilon \int h_\epsilon = \sup_\epsilon \int g_\epsilon$ and $\int f$ is defined to be this common value.

Let L be the family of integrable functions. L and the integral on L have all the properties of S and of the integral on S.

If $L \ni f_n \uparrow f$ and $\lim \int f_n < \infty$, then $f \in L$ and $\int f_n \uparrow \int f$.

Let $\mathfrak{F}$ be the smallest monotone family over S (closed under monotone passages to the limit by sequences). $\mathfrak{F}$ is closed under algebraic and lattice operations.

Let $L_1 = L \cap \mathfrak{F}$. $f \in L_1$ if and only if $f \in \mathfrak{F}$ and there exists $g \in L_1$ such that $|f| \leqq g$.

e) Let $\mathfrak{F}^+$ be the smallest monotone family over S^+ (consisting of all nonnegative functions of S). Set $\int f = \infty$ if $f \in \mathfrak{F}^+$ is not integrable. By definition, for $f \in \mathfrak{F}, \int f = \int f^+ - \int f^-$ *exists* if f^+ or f^- is integrable.

If $\int f$ and $\int g$ exist and they are not infinite with opposite sign, then $\int (f+g)$ exists and equals $\int f + \int g$.

If $\int f_n$ exist, $\int f_1 > -\infty$, and $f_n \uparrow f$, then $\int f$ exists and $\int f_n \uparrow \int f$.

f) If $I_A \in \mathfrak{F}$, then, by definition, the *measure* of A is $\mu A = \int I_A$.

If I_A, $I_B \in \mathfrak{F}$, then $I_{A \cup B}$, $I_{A \cap B}$, $I_{A-B} \in \mathfrak{F}$ and if the $I_{A_n} \in \mathfrak{F}$, then $I_{\Sigma I A_n} \in \mathfrak{F}$ and $\mu \sum A_n = \sum \mu A_n$.

g) Suppose that $f \in S \Rightarrow f \cap 1 \in S$. Then $f \in \mathfrak{F} \Rightarrow f \cap 1 \in \mathfrak{F}$ and if $a > 0$, then $I_{[f>a]} \in \mathfrak{F}$.

If $f \geqq 0$, $I_{[f>a]} \in \mathfrak{F}$ for every $a > 0$, then $f \in \mathfrak{F}$.

h) Suppose that $1 \in S$. Then $f \in \mathfrak{F}^+ \Rightarrow \int f = \int f d\mu$ where the right side is taken in the customary sense. What if $f \in \mathfrak{F}$?

i) The family S is a real linear normed space with the uniform norm $\|f\| = \sup f$. Every bounded linear functional $\varphi(f)$ on this space is difference of two bounded nonnegative linear functionals $\varphi(f) = \varphi^+(f) - \varphi^-(f)$: Take $\varphi^+(f) = \sup \{\varphi(f'), 0 \leqq f' \leqq f\}$ on S^+, then extend to S by linearity.

34. Riesz representation. Let $\mathfrak{X}$ be a locally compact space with points x, compacts K, and the σ-field $\mathcal{S}$ of topological Borel sets S, with or without subscripts. Let C be the space of bounded continuous functions g, with or without affixes, with the uniform norm $\|g\| = \sup g$. $C_0 \subset C$ consists of those g which vanish or infinity: Given $\epsilon > 0$ there exists a K_ϵ such that $|g| < \epsilon$ on K_ϵ^c. $C_{00} \subset C$ consists of those g which vanish off compacts and $C_K \subset C_{00}$ of those g which vanish off K. If $\mathfrak{X}$ is compact, then $C_{\mathfrak{X}} = C_{00} = C_0 = C$.

a) *Dini.* If $g_n \in C_{00}$ and $g_n \downarrow 0$, then $g_n \downarrow 0$ uniformly, that is, $\|g\| \downarrow 0$.

b) Nonnegative linear functionals $\mu(g)$ on C_{00} are bounded on every C_K and are integrals on C_{00}: Bounded, since there exists $g_0 \in C_{00}{}^+$ with $g_0 \geqq 1$ on C_K, hence $g \in C_K$ implies $|g| \leqq g_0 \|g\|$ and $|\mu(g)| \leqq \mu(g_0) \|g\|$. Integrals, since $g_1 \in C_K$ and $g_n \downarrow 0$ imply $g_n \in C_K$, $\|g\| \downarrow 0$, hence $|\mu(g_n)| \leqq \mu(g_0) \|g_n\| \downarrow 0$.

c) There is a one-to-one correspondence between nonnegative linear functionals $\mu(g)$ on C_{00} and measures $\mu(S)$ bounded on compacts, given by $\mu(g) = \int \mu(dx) g(x)$: By b) and 33, $\mu(g)$ determines the measure $\mu(S)$.

d) There is a one-to-one correspondence between bounded linear functionals $\varphi(g)$ on C_{00} and bounded signed measures $\varphi(S)$ on $\mathcal{S}$ given by $\varphi(g) = \int \varphi(dx) g(x)$ with $\|g\| = \operatorname{Var} \varphi$: Apply c) and 35i).

e) There is a one-to-one correspondence between bounded linear functionals on C_0 and bounded signed measures on $\mathcal{S}$. Compactify and apply d.

Part Two

GENERAL CONCEPTS AND TOOLS OF PROBABILITY THEORY

Probability concepts can be defined in terms of measure-theoretic concepts. Since probability is a normed measure and random variables are finite measurable functions, the properties of sequences of random variables are more precise than those of measurable functions on a general measure space. Since in probability theory probability spaces are but frames of reference for families of random variables, probability properties are to be expressed in terms of the laws of the families only. These laws are expressed in terms of distributions which are set functions on the Borel fields in the range spaces. The distributions are expressed in terms of distribution functions which are point functions on the range spaces. In turn, to distribution functions correspond their Fourier-Stieltjes transforms (called characteristic functions) which are easier to deal with.

The following Parts utilize the tools so developed to investigate probability problems. These problems are centered about the concepts of independence and of conditioning introduced in Parts III and IV, respectively. The corresponding sections 15 and 24 may be read immediately after section 9.

Chapter III

PROBABILITY CONCEPTS

§ 9. PROBABILITY SPACES AND RANDOM VARIABLES

9.1 Probability terminology. Probability theory has its own terminology, born from and directly related and adapted to its intuitive background; for the concepts and problems of probability theory are born from and evolve with the analysis of random phenomena. As a branch of mathematics, however, probability theory partakes of and contributes to the whole domain of mathematics and, at present, its general set-up is expressible in terms of measure spaces and measurable functions. We give below a first table of correspondences between the probability and measure theoretic terms. Within parentheses appear the abbreviations to be used throughout this book.

probability space (pr. space)	normed measure space
elementary event	point belonging to the space
event	measurable set
sure event	whole space
impossible event	empty set
probability (pr.)	normed measure
almost sure, almost surely (a.s.)	almost everywhere
random variable (r.v.)	finite numerical measurable function
expectation E	integral $\int$

We shall use the pr. theory terms or the measure theory terms according to our convenience. We summarize below in pr. terms the properties which are specializations of those established in Part I.

I. A *pr. space* $(\Omega, \mathcal{A}, P)$ consists of the *sure event* Ω, the (nonempty) σ-field $\mathcal{A}$ of *events* and the *pr.* P on $\mathcal{A}$. Unless otherwise stated, the pr. space $(\Omega, \mathcal{A}, P)$ is fixed and $A, B, \cdots$, with or without affixes, represent events. If so required, the pr. space can always be *completed*, so that every subset of a null event becomes an event—necessarily null.

1° *$\mathcal{A}$ is a σ-field: for all A's, A^c, $\bigcup_{j=1}^{\infty} A_j$, $\bigcap_{j=1}^{\infty} A_j$ are events.*
It follows that, for every sequence A_n, $\liminf A_n$, $\limsup A_n$, *and* $\lim A_n$ *(if it exists) are events.*

2° P *is defined on $\mathcal{A}$ and, for all A's,*

$$PA \geqq 0, \quad P(\sum A_j) = \sum PA_j, \quad P\Omega = 1.$$

It follows that

$$P\emptyset = 0, \quad PA \leqq PB \quad when \quad A \subset B, \quad P(\bigcup A_j) \leqq \sum PA_j,$$

$$P(\liminf A_n) \leqq \liminf PA_n \leqq \limsup PA_n \leqq P(\limsup A_n),$$

and, if $\lim A_n$ *exists, then* $P(\lim A_n) = \lim PA_n$.

II. A r.v. X is a function on Ω to $R = (-\infty, +\infty)$ such that the inverse images under X of all Borel sets in R are events; it suffices to require the same of all intervals, or of all intervals $[a, b)$, or of all intervals $(-\infty, b)$, etc.

An *elementary r.v.* is a function on Ω to R of the form $X = \sum x_j I_{A_j}$ where x_j's are finite numbers, A_j's are disjoint events, and $\sum A_j = \Omega$; if there is only a finite number of distinct A_j's, then X is a *simple r.v.*

1° *Every r.v. is the finite limit of a sequence of simple r.v.'s and the finite uniform limit of a sequence of elementary r.v.'s; and conversely.*

Every nonnegative r.v. is the finite limit of a nondecreasing sequence of nonnegative simple r.v.'s; and conversely.

2° *The class of all r.v.'s is closed under the usual operations of analysis, provided these operations yield finite functions.*

3° *Every finite Borel function of a finite number of r.v.'s is a r.v.*

A *random function* is a family of r.v.'s; if the family is finite, it is a *random vector*, and, if the family is denumerable, it is a *random sequence*, that is, a sequence of r.v.'s.

III. Unless otherwise stated, $X, Y, \cdots$, with or without affixes, will represent r.v.'s and, as usual, limits will be taken for $n \to \infty$.

X_n *converges in pr.* to X, and we write $X_n \xrightarrow{P} X$, if, for every $\epsilon > 0$,

$$P[|X_n - X| \geqq \epsilon] \to 0.$$

X_n *converges a.s.* to X, and we write $X_n \xrightarrow{\text{a.s.}} X$, if $X_n \to X$, except perhaps on a null event (event of pr. 0) or, *equivalently*, if for every $\epsilon > 0$,

$$P \bigcup_{k \geqq n} [|X_k - X| \geqq \epsilon] \to 0.$$

Mutual convergence in pr. ($X_n - X_m \xrightarrow{P} 0$) and a.s. ($X_n - X_m \xrightarrow{\text{a.s.}} 0$) are defined by replacing above $X_n - X$ by $X_n - X_m$ and $X_k - X$ by $X_k - X_l$ with $k, l \geqq n$, and taking limits as $m, n \to \infty$.

1° $X_n \xrightarrow{P} X$ *if, and only if,* $X_n - X_m \xrightarrow{P} 0$. $X_n \xrightarrow{\text{a.s.}} X$ *if, and only if,* $X_n - X_m \xrightarrow{\text{a.s.}} 0$.

2° *If* $X_n \xrightarrow{\text{a.s.}} X$ *then* $X_n \xrightarrow{P} X$. *If* $X_n \xrightarrow{P} X$, *then there is a subsequence* $X_{n_k} \xrightarrow{\text{a.s.}} X$ *as* $k \to \infty$, *with*

$$\sum_{k=1}^{\infty} P\left[|X_{n_k} - X| \geqq \frac{1}{2^k}\right] < \infty.$$

The terms "integral" and "expectation" and the notations $\int$ and E will be considered as equivalent. In the case of r.v.'s, we have

IV. The *expectation of a simple r.v.* $X = \sum_{k=1}^{n} x_k I_{A_k}$ is defined by

$$EX = \sum_{k=1}^{n} x_k P A_k.$$

The *expectation of a nonnegative r.v.* $X \geqq 0$ is the limit of expectations of nonnegative simple r.v.'s X_n which converge nondecreasingly to X:

$$EX = \lim EX_n, \quad 0 \leqq X_n \uparrow X.$$

The *expectation of a r.v.* $X = X^+ - X^-$ is given by

$$EX = EX^+ - EX^-,$$

provided the right-hand side is not of the form $+\infty - \infty$, and if EX exists and is finite, X is *integrable*.

1° *X is integrable if, and only if, $|X|$ is integrable.*

If X_1 and X_2 are integrable and a_1 and a_2 are finite numbers, then $a_1X_1 + a_2X_2$ is integrable and $E(a_1X_1 + a_2X_2) = a_1EX_1 + a_2EX_2$; if, moreover, $X_1 \leqq X_2$, then $EX_1 \leqq EX_2$.

If $|X_1| \leqq X_2$ and X_2 is integrable, then X_1 is integrable; in particular, every bounded r.v. is integrable, and if X degenerates at a ($X = a$ a.s.), then $EX = a$.

The *indefinite expectation* φ_X of a r.v. X whose expectation exists is defined on the σ-field $\mathfrak{A}$ of events A by $\varphi_X(A) = EXI_A$.

2° *φ_X on $\mathfrak{A}$ is σ-finite, σ-additive, and* P*-continuous; if X is integrable, then φ_X is bounded* by $E|X|$, *and $\varphi_X(A) \to 0$ as $PA \to 0$.*

3° MONOTONE CONVERGENCE THEOREM. *If $0 \leqq X_n \uparrow X$ finite or not, then $EX_n \uparrow EX$; if EX is finite, then the measurable function X is a.s. a r.v.*

DOMINATED CONVERGENCE THEOREM. *If $X_n \xrightarrow{P} X$ and $|X_n| \leqq Y$ integrable, then X is integrable, and $EX_n \to EX$.*

FATOU-LEBESGUE THEOREM. *If Y and Z are integrable r.v.'s and $Y \leqq X_n$ or $X_n \leqq Z$, then*

$$E(\liminf X_n) \leqq \liminf EX_n \quad or \quad \limsup EX_n \leqq E(\limsup X_n).$$

If, moreover, $\liminf EX_n$ *or* $\limsup EX_n$ *is finite, then, respectively,* $\liminf X_n$ *or* $\limsup X_n$ *is a.s. a r.v.*

EQUIVALENCE. Two functions on Ω are *equivalent* if they agree outside a null event. Convergences in pr. and a.s., integrals and integrability are, in fact, defined for equivalence classes and not for individual functions. Therefore, as long as we are concerned with a sequence of r.v.'s we can consider every r.v. of the sequence as defined up to an equivalence. In particular, we can then extend the notion of a r.v. as follows: a r.v. is an a.s. defined, a.s. finite and a.s. measurable function.

Let us observe, once and for all, that when the measurable functions under consideration are *by definition* $\mathfrak{B}$-measurable where $\mathfrak{B}$ is a sub σ-field of events, then almost sure relations are $P_{\mathfrak{B}}$-equivalences, that is, valid up to null $\mathfrak{B}$-measurable sets.

THE COMPLEX-VALUED CASE. A *complex r.v.* X is of the form $X = X' + iX''$ where X' and X'' are "ordinary" or "real-valued" r.v.'s as defined at the beginning of this section and where $i^2 = -1$; X takes its values in the complex plane of points $x' + ix''$, that is, in the plane $R \times R$, and its expectation is the point $EX = EX' + iEX''$. In other words, a complex r.v. X is a representation of the random vector $\{X', X''\}$. Similarly, a complex Borel function $g = g' + ig''$ is a representation of the Borel vector $\{g', g''\}$. The definitions and properties given below of random vectors, random sequences and, in general, random functions extend at once to the complex case where the components instead of being ordinary r.v.'s are complex-valued r.v.'s or, equivalently, two-dimensional random vectors. The relation $|EX| \leqq E|X|$ is still true; it suffices to use polar coordinates, setting $X = \rho e^{i\alpha}$, $EX = re^{it}$, and observe that

$$r = e^{-it}E\rho e^{i\alpha} = E\rho\cos(\alpha - t) \leqq E\rho$$

***9.2 Random vectors, sequences, and functions.** A *random vector* $X = (X_1, \cdots, X_n)$ is a finite family of r.v.'s called *components* of the random vector. Every component X_k induces a sub σ-field $\mathcal{B}(X_k)$ of events—inverse image of the Borel field in the range-space R_k of X_k. The random vector has for range space the n-dimensional real space $R^n \doteq \prod_{k=1}^{n} R_k$ with points $x = (x_1, \cdots, x_n)$ and it induces a σ-field $\mathcal{B}(X) = \mathcal{B}(X_1, X_2, \cdots, X_n)$—inverse image of the Borel field in R^n. The inverse images of intervals $(-\infty, x) \subset R^n$ are events

$$[X < x] = [X_1 < x_1, \cdots, X_n < x_n] = \bigcap_{k=1}^{n} [X_k < x_k]$$

and, hence, are intersections of events belonging to the $\mathcal{B}(X_k)$. Since the Borel field $\mathcal{B}^n$ in R^n is the minimal σ-field over the class of these intervals, the σ-field $\mathcal{B}(X)$ is the minimal σ-field over these intersections or, equivalently, over the union of the $\mathcal{B}(X_k)$—a *compound or union σ-field* $\mathcal{B}(X_1, \cdots X_n)$ with *component* σ-fields $\mathcal{B}(X_k)$. Thus, the elements of $\mathcal{B}(X)$ are events and the random vector X can be defined as a measurable function on the pr. space to the n-dimensional Borel space $(R^n, \mathcal{B}^n)$. We define EX to be $(EX_1, EX_2, \cdots, EX_n)$—a point in the space R^n.

A *random sequence* $X = (X_1, X_2, \cdots)$ is a sequence of r.v.'s called its *components;* it takes its values in the space $R^\infty = \prod_{n=1}^{\infty} R_n$ of points $x = (x_1, x_2, \cdots)$, that is, the space of numerical sequences. To every point x with an arbitrary but *finite number of finite coordinates* $x_{k_1}, \cdots x_{k_n}$ there corresponds the interval $(-\infty, x)$ of all points y such that $y_{k_1} < x_{k_1}, \cdots y_{k_n} < x_{k_n}$, and the minimal σ-field over the class of these intervals is the Borel field $\mathcal{B}^\infty$ in R^∞. Exactly as for random vectors, it follows that the inverse image under X of $\mathcal{B}^\infty$ is the minimal σ-field over the class of all finite intersections of events $A_n \in \mathcal{B}(X_n)$—the *compound* or *union σ-field* $\mathcal{B}(X)$ with *component σ-fields* $\mathcal{B}(X_n)$—then we write $\mathcal{B}(X) = \mathcal{B}(X_1, X_2, \cdots)$ and the random sequence can be defined as a measurable function on the pr. space to the *Borel space* $(R^\infty, \mathcal{B}^\infty)$. Similarly, the definition of the expectation of the random sequence is $EX = \{EX_1, EX_2, \cdots\}$—when $EX_1, EX_2, \cdots$ exist.

A random function $X_T = (X_t, t \in T)$ is a family of r.v.'s X_t where t varies over an arbitrary but fixed index set T. Exactly as above, the range space of X_T is the *real space* $R_T = \prod_{t \in T} R_t$ of points $x_T = (x_t, t \in T)$—the space of numerical functions; *intervals* $(-\infty, x_T)$ are defined for points x_T with an arbitrary but finite number of finite coordinates to be sets of all points $y_T < x_T$, that is, $y_t < x_t$, $t \in T$; the *Borel field* $\mathcal{B}_T$ is the minimal σ-field over the class of these intervals. The random function X_T induces the *compound* or *union σ-field* $\mathcal{B}(X_T)$ with *component σ-fields* $\mathcal{B}(X_t)$—the minimal σ-field over the class of all finite intersections of events $A_t \in \mathcal{B}(X_t)$ as t varies on T or, equivalently, the inverse image under X_T of the Borel field $\mathcal{B}_T$; and the random function X_T can be defined as a measurable function on the pr. space to the *Borel space* $(R_T, \mathcal{B}_T)$. By definition, $EX_T = \{EX_t, t \in T\}$ is a numerical function—when the EX_t exist.

A *Borel function* $g_{T'}$ is a function on a Borel space $(R_T, \mathcal{B}_T)$ to a Borel space $(R_{T'}, \mathcal{B}_{T'})$ such that the inverse image under $g_{T'}$ of the Borel field in the range space is contained in the Borel field $\mathcal{B}_T$ in the domain R_T. Therefore, if X_T is a random function to R_T, then the function of function $g_{T'}(X_T)$ on the pr. space to the Borel space $(R_{T'}, \mathcal{B}_{T'})$ induces a sub σ-field of events—inverse image under X_T of the inverse image under $g_{T'}$ of the Borel field $\mathcal{B}_{T'}$. Thus, $\mathcal{B}(g_{T'}(X_T)) \subset \mathcal{B}(X_T)$; in other words, $g_{T'}(X_T)$ is $\mathcal{B}(X_T)$-measurable and, hence, is a random function. We state this conclusion as a theorem.

A. BOREL FUNCTIONS THEOREM. *A Borel function of a random function is a random function which induces a sub σ-field of events contained in the one induced by the original random function.*

Loosely speaking, a Borel function of a random function induces a "coarser" sub σ-field of events and has "fewer" values.

9.3 Moments, inequalities, and convergences. Expectations of powers of r.v.'s are called *moments* and play an essential role in the investigations of pr. theory. They appear in the simple but powerful Markov inequality and in the definition of the very useful notion of convergence "in the rth mean," that we shall introduce in this subsection. They appear in the expansions of "characteristic functions" that we shall examine in the next chapter. They play a basic role in the study of sums of "independent" r.v.'s to which the next part is devoted. Furthermore, the powerful "truncation" method—to be used extensively in the following parts—expands tremendously the domain of applicability of the methods of investigation based upon the use of moments.

EX^k ($k = 1, 2, \cdots$) and $E|X|^r$ ($r > 0$) are called, respectively, the kth *moment* and the rth *absolute moment* of the r.v. X. We may also consider 0th moments but, for all r.v.'s, the 0th moments are 1, and we shall limit ourselves to kth moments where k is a positive integer, and to rth absolute moments where r is a positive number, unless otherwise stated.

We establish now a few simple properties of moments. While a kth moment may not exist, absolute moments always exist but may be infinite. Since integrability is equivalent to absolute integrability, if the kth absolute moment of X is finite, then its kth moment exists and is finite; and conversely. More generally, since $|X|^{r'} \leqq 1 + |X|^r$ for $0 < r' < r$, we have

a. *If $E|X|^r < \infty$, then $E|X|^{r'}$ is finite for $r' \leqq r$ and EX^k exists and is finite for $k \leqq r$.*

In other words, finiteness of a moment of X implies existence and finiteness of all moments of X of lower order.

Upon applying the elementary inequality

$$|a + b|^r \leqq c_r|a|^r + c_r|b|^r, \quad r > 0,$$

where $c_r = 1$ or 2^{r-1} according as $r \leqq 1$ or $r \geqq 1$, replacing a by X, b by Y and, taking expectations of both sides, we obtain the

c_r-INEQUALITY. *$E|X + Y|^r \leqq c_rE|X|^r + c_rE|Y|^r$, where $c_r = 1$ or 2^{r-1} according as $r \leqq 1$ or $r \geqq 1$.*

This inequality shows that if the rth absolute moments of X and Y exist and are finite, so is the rth absolute moment of $X + Y$.

Similarly, excluding the trivial case of vanishing $E|X|^r$ or $E|Y|^s$ (in which case the Hölder inequality below is trivially true), and replacing a by $X/E^{\frac{1}{r}}|X|^r$, b by $Y/E^{\frac{1}{s}}|Y|^s$ in the elementary inequality

$$|ab| \leqq \frac{|a|^r}{r} + \frac{|b|^s}{s}, \quad r > 1, \quad \frac{1}{r} + \frac{1}{s} = 1,$$

we obtain the

Hölder inequality. *$E|XY| \leqq E^{\frac{1}{r}}|X|^r \cdot E^{\frac{1}{s}}|Y|^s$, where $r > 1$ and $\frac{1}{r} + \frac{1}{s} = 1$.*

From this inequality follows the

Minkowski inequality. *If $r \geqq 1$, then*

$$E^{\frac{1}{r}}|X + X'|^r \leqq E^{\frac{1}{r}}|X|^r + E^{\frac{1}{r}}|X'|^r.$$

In fact, upon excluding the trivial case $r = 1$, and applying the Hölder inequality with $Y = |X + X'|^{r-1}$ to the right-hand side terms in the obvious inequality

$$E|X + X'|^r \leqq E(|X| \cdot |X + X'|^{r-1}) + E(|X'| \cdot |X + X'|^{r-1}),$$

we find

$$E|X + X'|^r \leqq (E^{\frac{1}{r}}|X|^r + E^{\frac{1}{r}}|X'|^r)E^{\frac{1}{s}}|X + X'|^{(r-1)s},$$

where $\frac{1}{r} + \frac{1}{s} = 1$. Upon excluding the trivial case of vanishing $E|X + X'|^r$, noticing that $(r - 1)s = r$, and dividing both sides by $E^{\frac{1}{s}}|X + X'|^r$, the asserted inequality follows.

Hölder's inequality with $r = s = 2$, is called the

Schwarz inequality: $E^2|XY| \leqq E|X|^2 \cdot E|Y|^2$.

Replacing X by $|X|^{\frac{r-r'}{2}}$ and Y by $|X|^{\frac{r+r'}{2}}$, with $r' \leqq r$, and, taking logarithms of both sides, we obtain the inequality

$$\log E|X|^r \leqq \tfrac{1}{2} \log E|X|^{r-r'} + \tfrac{1}{2} \log E|X|^{r+r'}$$

b. *$\log E|X|^r$ is a convex function of r.*

Hölder's inequality with X, Y, r, s replaced respectively by $|X|^p, 1^p, p/r, q/r$ $\left(\text{hence } \frac{1}{r} = \frac{1}{p} + \frac{1}{q}\right)$ becomes $E^{1/r}|X|^r \leqq E^{1/p}|X|^p$ for $r < p$.

Hence,

c. *$E^{1/r}|X|^r$ is nondecreasing in r.*

In fact, $E^{1/r}|X|^r \uparrow E^{1/p}|X|^p$ as $r \uparrow p$. For, if $E|X|^p < \infty$ then $|X|^r \leqq \max(1, |X|^p)$ and the dominated convergence theorem applies. If $E|X|^p = \infty$ apply what precedes to $Y_n = |X| I_{[|X|<n]}$ then let $n \uparrow \infty$.

We introduce now convergence in the rth mean. Let X_n and X be r.v.'s with finite rth absolute moments, so that, by the c_r-inequality, the same is true of $X_n - X$. We say that the sequence X_n converges to X *in the* rth *mean*, and write $X_n \xrightarrow{r} X$, if $E|X_n - X|^r \to 0$.

Let $X_n \xrightarrow{r} X$. If $r \leq 1$ then it follows, by the c_r-inequality, that

$$|E|X_n|^r - E|X|^r| \leq E|X_n - X|^r \to 0,$$

and, if $r > 1$, then it follows, by the Minkowski inequality, that

$$|E^{\frac{1}{r}}|X_n|^r - E^{\frac{1}{r}}|X|^r| \leq E^{\frac{1}{r}}|X_n - X|^r \to 0.$$

This proves that

d. *If* $X_n \xrightarrow{r} X$, *then* $E|X_n|^r \to E|X|^r$.

We conclude this subsection with a simple but basic inequality and a few of its applications.

A. Basic inequality. *Let X be an arbitrary r.v. and let g on R be a nonnegative Borel function.*

If g is even and is nondecreasing on $[0, +\infty)$ *then, for every* $a \geq 0$

$$\frac{Eg(X) - g(a)}{\text{a.s.}\sup g(X)} \leq P[|X| \geq a] \leq \frac{Eg(X)}{g(a)}.$$

If g is nondecreasing on R, then the middle term is replaced by $P[X \geq a]$, *where a is an arbitrary number.*

The proof is immediate. Since g is a Borel function on R, it follows that $g(X)$ is a measurable function on Ω and, since g is nonnegative on R, its integral exists. If g is even and is nondecreasing on $[0, +\infty)$, then, setting $A = [|X| \geq a]$, from the obvious relations

$$Eg(X) = \int_A g(X) + \int_{A^c} g(X)$$

and

$$g(a)PA \leq \int_A g(X) \leq \text{a.s.}\sup g(X) \cdot PA, \quad 0 \leq \int_{A^c} g(X) \leq g(a),$$

it follows that

$$g(a)PA \leq Eg(X) \leq \text{a.s.}\sup g(X) \cdot PA + g(a).$$

This proves the first assertion and the second is similarly proved.

Applications. (1) Upon taking $g(x) = e^{rx}(r > 0)$, we obtain

$$\frac{Ee^{rX} - e^{ra}}{\text{a.s. sup } e^{rX}} \leqq P[X \geqq a] \leqq e^{-ra}Ee^{rX}$$

(2) Upon taking $g(x) = |x|^r(r > 0)$ we obtain

$$\frac{E|X|^r - a^r}{\text{a.s. sup } |X|^r} \leqq P[|X| \geqq a] \leqq \frac{E|X|^r}{a^r};$$

the right-hand side inequality is called the *Markov inequality*, and for $r = 2$ it reduces to the celebrated *Tchebichev inequality*.

Upon applying Markov's inequality with X replaced by $X_n - X$, it follows that

If $X_n \xrightarrow{r} X$, then $X_n \xrightarrow{P} X$, and if the X_n are a.s. uniformly bounded, then conversely, $X_n \xrightarrow{P} X$ implies that $X_n \xrightarrow{r} X$.

(3) Upon taking $g(x) = \dfrac{|x|^r}{1 + |x|^r}$ $(r > 0)$, we obtain

$$E\frac{|X|^r}{1 + |X|^r} - \frac{a^r}{1 + a^r} \leqq P[|X| \geqq a] \leqq \frac{1 + a^r}{a^r} E\frac{|X|^r}{1 + |X|^r};$$

replacing X by $X_n - X$ and by $X_m - X_n$, it follows that, as $m, n \to \infty$,

$$X_n \xrightarrow{P} X \quad \textit{if, and only if,} \quad E\frac{|X_n - X|^r}{1 + |X_n - X|^r} \to 0;$$

$$X_m - X_n \xrightarrow{P} 0 \quad \textit{if, and only if,} \quad E\frac{|X_m - X_n|^r}{1 + |X_m - X_n|^r} \to 0.$$

REMARK. Observe that the function defined by $d(X, Y) = E\dfrac{|X - Y|}{1 + |X - Y|}$ has the triangular and identification properties of a distance, except that $d(X, Y) = 0$ implies only that $X = Y$ a.s. It follows from the foregoing proposition that

The space of the equivalence classes of the r.v.'s defined in a pr. space is a complete metric space with distance d defined by

$$d(X, Y) = E\frac{|X - Y|}{1 + |X - Y|},$$

and convergence in distance is equivalent to convergence in pr.

*CONVEX FUNCTIONS. The relations between moments established at the beginning of this subsection are essentially convexity properties. Let us recall a few classical properties of convex functions.

Let g be a (numerical) Borel function defined on a finite or an infinite *open* interval $I \subset R$. g is said to be *convex* if, for every pair of points x, x' of I,

$$g\left(\frac{x + x'}{2}\right) \leqq \frac{1}{2}g(x) + \frac{1}{2}g(x');$$

if g is twice differentiable on I, then the convexity property is equivalent to $g'' \geqq 0$ on I. The same definition applies to g on an N-dimensional interval I^N and is equivalent to the convexity of the function $g(x + ux')$ of the numerical argument u for all values of u for which $x + ux' \in I^N$, so that it suffices to consider convex functions on $I \subset R$. A convex function on I is either continuous on I or is not a Borel function. Thus, from now on, a convex function will be assumed to be continuous on its domain. In that case, g is convex on I if, and only if, to every $x_0 \in I$ there corresponds a number $\lambda(x_0)$ such that, for all $x \in I$,

$$\lambda(x_0)(x - x_0) \leqq g(x) - g(x_0).$$

Let X be a r.v. whose values lie a.s. in I and whose expectation EX exists and is finite. Replacing x_0 by EX and x by X, and taking the expectation of both sides of the foregoing inequality, it follows that

e. *If g is convex and EX is finite, then*

$$g(EX) \leqq Eg(X).$$

If g is strictly monotone, then this relation can be written

$$EX \leqq g^{-1}(Eg(X)).$$

For example, for $r \geqq 1$, $g(x) = x^r (x \in (0, +\infty))$ being convex, we have $E|X| \leqq E^{1/r}|X|^r$.

More generally, let G_1 and G_2 be two continuous and strictly monotone functions such that $g = G_2G_1^{-1}$ is convex; we say then that G_2 *is convex in* G_1. Since $Y = G_1(X)$ implies that $X = G_1^{-1}(Y)$, it follows by **e**, upon assuming that EX and EY are finite, that

$$G_2G_1^{-1}(EY) \leqq EG_2G_1^{-1}(Y)$$

and, hence,

e'. *If G_2 is convex in G_1, then*

$$G_1^{-1}(EG_1(X)) \leqq G_2^{-1}(EG_2(X)).$$

For example, since on $(0, +\infty)$, x^{r_2} is convex in x^{r_1} for $r_2 \geqq r_1$, that is, the function x^{r_2/r_1} is convex, we have

$$E^{\frac{1}{r_1}}|X|^{r_1} \leqq E^{\frac{1}{r_2}}|X|^{r_2} \quad \text{for} \quad r_2 \geqq r_1.$$

***9.4 Spaces L_r.** The r.v.'s whose rth absolute moments are finite are said to form the *space* L_r over the pr. space $(\Omega, \mathcal{A}, P)$; in symbols, $X \in L_r$ if $E|X|^r < \infty$; we drop r if $r = 1$. We shall find later that the space L_2 is a very important tool in the investigation of pr. problems, especially those relative to sums of "independent" r.v.'s. It will be convenient to introduce two boundary cases. The first is the trivial space L_0 of all r.v.'s X since $E|X|^0 = 1$ is finite. The second is the space L_∞ of all a.s. bounded r.v.'s. Since $\lim_{r \to \infty} E|X|^r < \infty$ if, and only if, $|X| \leqq 1$ a.s., it seems that only the subspace $L'_\infty \subset L_\infty$ of r.v.'s a.s. bounded by 1 ought to be introduced. However, for $r \to \infty$ it is $\lim E^{\frac{1}{r}}|X|^r$ which counts, and this limit is finite if, and only if, X is a.s. bounded. In fact, let s be the a.s. supremum of $|X|$, defined by $P[|X| > s] = 0$ and $P[|X| \geqq c] > 0$ for every $c < s$; we have $s \leqq \infty$. The foregoing assertion is implied by

a. $E^{\frac{1}{\infty}}|X|^\infty \equiv \lim_{r \to \infty} E^{\frac{1}{r}}|X|^r = \text{a.s.} \sup |X| \equiv s.$

For

$$s \geqq E^{\frac{1}{r}}|X|^r \geqq E^{\frac{1}{r}}(|X|^r I_{[|X| \geqq c]}) \geqq cP^{\frac{1}{r}}[|X| \geqq c] \to s$$

as $r \to \infty$, then $c \uparrow s$.

The foregoing definitions permit us to state 9.3**a** as follows:

b. $L_0 \supset L_r \supset L_s \supset L_\infty \supset L'_\infty$, $0 \leqq r \leqq s \leqq \infty$.

Let us observe that the space of all simple r.v.'s is a subspace of L_∞ and, hence, of all the spaces L_r.

Since, by the c_r- and Minkowski inequalities and by **a**,

$$E|X + Y|^r \leqq E|X|^r + E|Y|^r, \quad 0 < r < 1,$$

$$E^{\frac{1}{r}}|X + Y|^r \leqq E^{\frac{1}{r}}|X|^r + E^{\frac{1}{r}}|Y|^r, \quad 1 \leqq r \leqq \infty,$$

and $E|X - Y|^r = 0$ if, and only if, X and Y are equivalent, we have, according to the definitions relative to metric and normed spaces, the following theorem.

A. *The spaces L_r are linear metric spaces with metric defined by*

$$d(X, Y) = E|X - Y|^r \quad \textit{for} \quad 0 < r < 1$$

and norm

$$\|X\| = E^{\frac{1}{r}}|X|^r \quad \textit{for} \quad 1 \leqq r \leqq \infty,$$

provided equivalent r.v.'s are identified.

The problem arises whether the spaces L_r are complete and what are the convergence theorems in these spaces. Unless otherwise stated, *from now on* $0 < r < \infty$ (the reader is invited to examine in each case the boundary spaces L_0 and L_∞).

First we observe that on account of **A** and 9.3**d** we have

c. *Convergence in distance $d(X_n, X) \to 0$ in L_r is equivalent to convergence in the* rth *mean $X_n \xrightarrow{r} X$ and implies convergence of distances $d(X_n, X_0) \to d(X, X_0)$ to any fixed $X_0 \in L_r$.*

Also, if $X_n \in L_r$, then, for a r.v. X, $E|X_n - X|^r$, which always exists, can converge to 0 only if, from some value of n on, $E|X_n - X|^r$ is finite and, hence, only if $X \in L_r$, so that

d. *If X_n is a sequence in L_r and $E|X_n - X|^r \to 0$, then $X \in L_r$.*

We are now in a position to prove the

B. L_r-COMPLETENESS THEOREM. *Let the $X_n \in L_r$. Then $X_n \xrightarrow{r}$ some X if, and only if, $X_m - X_n \xrightarrow{r} 0$, as $m, n \to \infty$.*

Proof. If $X_n \xrightarrow{r} X$, then $X_m - X_n \xrightarrow{r} 0$, since, by the c_r-inequality,

$$E|X_m - X_n|^r \leqq c_r E|X_m - X|^r + c_r E|X - X_n|^r \to 0.$$

Conversely, if $X_m - X_n \xrightarrow{r} 0$, then, by the Markov inequality, for every $\epsilon > 0$,

$$P[|X_m - X_n| \geqq \epsilon] \leqq \frac{1}{\epsilon^r} E|X_m - X_n|^r \to 0 \quad \text{as} \quad m, n \to \infty,$$

so that $X_m - X_n \xrightarrow{P} 0$. Therefore, there is a subsequence $X_{n'} \xrightarrow{\text{a.s.}}$ some X as $n' \to \infty$ and, for every fixed m, $X_m - X_{n'} \xrightarrow{\text{a.s.}} X_m - X$ as $n' \to \infty$. Since $E|X_m - X_{n'}|^r \to 0$ as $m, n' \to \infty$, it follows, by the Fatou-Lebesgue theorem and the hypothesis, that

$$E|X_m - X|^r \leqq \liminf_{n'} E|X_m - X_{n'}|^r \to 0 \quad \text{as} \quad m \to \infty.$$

Thus, $X_n \xrightarrow{r} X$, and the proof is complete.

If a r.v. X is integrable, then the (indefinite) integral of X is P-absolutely continuous: $\int_A |X| \to 0$ as $PA \to 0$. Let $B = [|X| \geqq a]$. Since $PB \to 0$ as $a \to \infty$, it follows that $\int_B |X| \to 0$ as $a \to \infty$. Conversely, this implies that

$$\int_A |X| = \int_{AB} |X| + \int_{AB^c} |X| \leqq \int_B |X| + aPA \to 0$$

as $PA \to 0$ then $a \to \infty$, and thus implies that X is integrable, since, given $\epsilon > 0$,

$$\int |X| \leqq \int_B |X| + a < \epsilon + a_\epsilon < \infty$$

for $a = a_\epsilon$ sufficiently large.

The integrals of r.v.'s X_n are *uniformly P-absolutely continuous* or simply *uniformly continuous* if $\int_A |X_n| \to 0$ uniformly in n as $PA \to 0$; in other words, for every $\epsilon > 0$ there exists a δ_ϵ independent of n such that $\int_A |X_n| < \epsilon$ for any set A with $PA < \delta_\epsilon$. Let $B_n = [|X_n| \geqq a]$. The r.v.'s $|X_n|$ are *uniformly integrable*, if $\int_{B_n} |X_n| \to 0$ uniformly in n, as $a \to \infty$. Observe that if the $\int |X_n|$ are uniformly bounded, say, by $c(< \infty)$, then, by Markov's inequality, $PB_n \leqq c/a \to 0$ as $a \to \infty$. Upon replacing X by X_n and B by B_n in the foregoing discussion, it follows that

e. *The r.v.'s X_n are uniformly integrable if, and only if, their integrals are uniformly bounded and uniformly continuous.*

Let $X_n \xrightarrow{r} X$ hence $X_n I_A \xrightarrow{r} X I_A$. It follows, by 9.3d and the above lemma (take $A = \Omega$, and take A such that $PA \to 0$)

f. *If $X_n \xrightarrow{r} X$, then the $|X_n|^r$ are uniformly integrable.*

For use on the forthcoming theorem, note that (Young)

The Fatou-Lebesgue theorem and the dominated convergence theorem remain valid if therein Y and Z are replaced by U_n and V_n with $U_n \xrightarrow{a.e.} U$, $V_n \xrightarrow{a.e.} V$ and $\int U_n \to \int U$ finite, $\int V_n \to \int V$ finite.

For then, the argument pp. 125–6 remains valid. Furthermore, by selecting $\{n''\}$ p. 126 so that also $U_{n''} \xrightarrow{\text{a.e.}} U$, we have

If $|X_n| \leq U_n$ *with* $U_n \xrightarrow{\mu} U$ *and* $\int U_n \to \int U$ *finite, then* $X_n \xrightarrow{\mu} X$ *implies that* $\int X_n \to \int X$ *in fact* $\int |X_n - X| \to 0$.

C. L_r-CONVERGENCE THEOREM. *Let the* $X_n \in L_r$. *Then*
(i) $X_n \xrightarrow{r} X$ *if and only if* (ii) $X_n \xrightarrow{P} X$
and one of the following conditions holds:

(iii) $\int |X_n|^r \to \int |X|^r < \infty$; (iv) *the* $|X_n|^r$ *are uniformly integrable;* (v) *the* $|X_n|^r$, *or* (vi) *the* $|X_n - X|^r$, *have uniformly continuous integrals.*

Proof. Let $\epsilon > 0$ be arbitrary, set $A_n = [|X_n - X| \geq \epsilon]$, $A_{mn} = [|X_m - X_n| \geq \epsilon]$, and let $m, n \to \infty$. We use the c_r-inequality without further comment. Note that (iv) implies $X_n \epsilon L_r$.

Condition (i) implies (ii) by Markov inequality ($PA_n \leq E|X_n - X|^r/\epsilon^r \to 0$) and implies (iii) by 9.3**d**. Conversely, (ii) and (iii) imply (i), since then $|X_n - X|^r \leq c_r|X_n|^r + c_r|X|^r = U_n$ with $U_n \xrightarrow{P} 2c_r|X|^r$ and $\int U_n \to 2c_r \int |X|^r < \infty$.

As for the remaining assertions, (i) implies (iv) by **f**, and (iv) implies (v) by **e** applied to the $|X_n|^r$ in lieu of the X_n. Also, clearly (i) implies (vi), and (vi) implies (v), since it implies integrability of $|X_n - X|^r$ hence of $|X|^r$ (because $X_n \in L_r$) so that $\int_A |X_n| \leq c_r \int_A |X_n - X|^r + c_r \int_A |X|^r < \epsilon$ for PA sufficiently small.

Thus, to complete the proof, it suffices to show that (ii) and (v) imply (i). Since convergence in pr. (in the rth mean) is equivalent to mutual convergence in pr. (in the rth mean) and $X_n \xrightarrow{P} X$, $X_n \xrightarrow{r} Y$ imply that $Y = X$ a.s., we can replace (i) and (ii) by (i′) $E|X_m - X_n|^r \to 0$ and (ii′) $PA_{mn} \to 0$. The assertion follows since, upon integrating $|X_m - X_n|^r$ on A_{mn} and on $A_{mn}{}^c$, (ii′) and (v) imply that as $m, n \to 0$ then $\epsilon \to 0$, $E|X_m + X_n|^r \leq c^r \int A_{mn} |X_m|^r + c_r \int A_{mn} |X_n|^r + \epsilon^r \to 0$.

Corollary 1. *$X_n \xrightarrow{r} X$ implies $X_n \xrightarrow{r'} X$ for $r' < r$.*

Set $A_n = [|X_n - X| \geqq 1]$ and observe that

$$\int_A |X_n - X|^{r'} = \int_{AA_n} |X_n - X|^{r'} + \int_{AA_n^c} |X_n - X|^{r'} \leqq \int_A |X_n - X|^r + PA.$$

Corollary 2. *If $\sup E|X_n|^r = c < \infty$, then $X_n \xrightarrow{P} X$ implies $X_n \xrightarrow{r'} X$ for $r' < r$.*

Let $A_n = [|X_n| \geqq a]$ and observe that

$$\int_A |X_n|^{r'} = \int_{AA_n} |X_n|^{r'} + \int_{AA_n^c} |X_n|^{r'} \leqq ca^{r'-r} + a^r PA < \epsilon$$

by taking a sufficiently large to have $ca^{r'-r} < \frac{\epsilon}{2}$ and, then, PA sufficiently small to have $a^r PA < \frac{\epsilon}{2}$.

Corollary 3. *If $|X_n| \leqq Y \in L_r$ for large n, then $X_n \xrightarrow{P} X$ implies $X_n \xrightarrow{r} X \in L_r$.*

Observe that for large n, $\int_A |X_n|^r \leqq \int_A Y^r$.

We proved in 9.3 a particular case of this corollary, with $Y = c < \infty$.

We summarize below the relations between various types of convergence:

$$X_n \xrightarrow{\text{a.s.}} X \Rightarrow X_n \xrightarrow{P} X \Rightarrow X_{n_k} \xrightarrow{\text{a.s.}} X \text{ with } \sum_k P\left[|X_{n_k} - X| \geqq \frac{1}{2^k}\right] < \infty$$

$$\Uparrow$$

$$X_n \xrightarrow{r} X \Rightarrow X_n \xrightarrow{r'} X, \quad r' < r.$$

The operation of integration on the complete normed linear space L_r with $r \geqq 1$ can be characterized as a functional of the integrand, as follows:

D. Integral representation theorem. *Let $\frac{1}{r} + \frac{1}{s} = 1$ with $1 \leqq r < \infty$.*

A functional f on L_r is linear and continuous if, and only if, there exists a r.v. $Y \in L_s$ such that $f(X) = EXY$ for every $X \in L_r$; then f determines Y up to an equivalence and $||f|| = E^{\frac{1}{s}}|Y|^s$.

Proof. Since $\frac{1}{r} + \frac{1}{s} = 1$ and $1 \leqq r < \infty$, it follows that $1 < s \leqq \infty$, and we apply repeatedly Hölder's inequality $E|XY| \leqq ||X||_r||Y||_s$, where $||X||_r = E^{\frac{1}{r}}|X|^r$ and $||Y||_s = E^{\frac{1}{s}}|Y|^s$ with $||Y||_\infty = \lim_{s\to\infty} E^{\frac{1}{s}}|Y|^s = \text{a.s.}\sup|Y|$.

If $||X||_r||Y||_s$ is finite, then $f(X) = EXY$ exists, is finite, and defines a normed functional f on L_r with $||f|| \leqq ||Y||_s$. Since EXY is linear in $X \in L_r$, so is $f(X)$. Being normed and linear, f is continuous.

Conversely, let a functional f on L_r be continuous and linear; linearity implies additivity and additivity implies $f(\theta) = 0$, where θ is the zero-point of L_r, that is, the class of r.v.'s degenerate at 0. Therefore, the set function φ on $\mathcal{a}$ defined by $\varphi(A) = f(I_A)$ is continuous and additive, hence σ-additive, and vanishes for null events, hence is P-continuous. Thus, the Radon-Nikodym theorem applies and φ on $\mathcal{a}$ determines up to an equivalence a r.v. Y such that

$$f(I_A) = \varphi(A) = EI_AY.$$

Since $f(X)$ and EXY are both linear in X, it follows that $f(X) = EXY$ for all simple finite $X(\in L_r)$. If $Y \in L_s$ and $L_r \ni X_n \xrightarrow{r} X$ hence $X_nY \xrightarrow{1} XY$, then, by continuity of f and of E on L_r, this equality extends to all $X \in L_r$. Since f has finite norm $||f|| \leqq ||Y||_s$, to complete the proof it suffices to show that the reverse inequality $||f|| \geqq ||Y||_s$ is true.

Let $r > 1$. If the X_n are simple finite and $0 \leqq X_n \uparrow |Y|$, then

$$E|X_n|^s \leqq E(X_n^{s-1}\,\text{sign}\,Y)Y \leqq ||f||E^{\frac{1}{r}}|X_n|^{(s-1)r}$$

yields

$$||Y||_s \leftarrow ||X_n||_s \leqq ||f||.$$

Let $r = 1$. If there exists an $\epsilon > 0$ such that $||Y||_\infty \geqq ||f|| + 2\epsilon$ and we set $A = [|Y| \geqq ||f|| + \epsilon]$, then $PA > 0$ while

$$(||f|| + \epsilon)PA \leqq E|I_AY| = E(I_A\,\text{sign}\,Y)Y \leqq ||f||PA,$$

and we reach a contradiction. This completes the proof.

Remark. The definitions and results of this subsection extend at once to complex-valued r.v.'s.

§ 10. PROBABILITY DISTRIBUTIONS

10.1 Distributions and distribution functions. Let X be a r.v. on our pr. space $(\Omega, \mathcal{A}, P)$. The nonnegative set function P_X defined on the Borel field $\mathcal{B}$ in R by

$$P_X S = P[X \in S], \quad S \in \mathcal{B}$$

is called the *pr. distribution* or, simply, *distribution* of X. Since X is finite, the inverse image under X of R is Ω and, since the inverse image of a sum of Borel sets is the sum of their inverse images, we have

$$P_X R = 1, \quad P_X(\sum S_j) = \sum P_X S_j, \quad S_j \in \mathcal{B}.$$

Therefore, P_X on $\mathcal{B}$ is a probability. Thus, the r.v. X induces on its range space a new pr. space $(R, \mathcal{B}, P_X)$, to be called a pr. space *induced* by X on its range space or the *sample pr. space* of X. Moreover,

a. *The distribution P_X of X determines the distributions of all r.v.'s $g(X)$ where g is a finite Borel function on R; and $Eg(X) = \int_R g\, dP_X$ in the sense that, if either side of this expression exists, so does the other, and then they are equal.*

Proof. Every finite Borel function $g(X)$ of a r.v. X is a r.v. and, by definition,

$$[g(X) \in S] = [X \in g^{-1}(S)]$$

where S and $g^{-1}(S)$ are Borel sets. Therefore

$$P_{g(X)}(S) = P_X g^{-1}(S), \quad S \in \mathcal{B},$$

and the first assertion is proved.

The second assertion will follow if we prove it for nonnegative functions g. Because of the monotone convergence theorem, it suffices to prove it for nonnegative simple functions g and, because of the additivity property of integrals, it suffices to prove the assertion for indicators. Thus, let $g = I_S$, so that $g(X) = I_{[X \in S]}$. But, then, the left-hand side of the asserted equality becomes

$$\int_\Omega I_{[X \in S]} dP = P[X \in S],$$

while the right-hand side becomes $\int_R I_S\, dP_X = P_X S$. Therefore, by definition of P_X, the asserted equality holds, and the proof is complete.

Distributions are *set* functions and are not easy to handle by means of classical analysis developed primarily to deal with point functions. Thus, in order to be able to use analytical methods and tools, it is of the greatest importance to find, and learn to use, *point* functions which "represent" distributions, that is, which are in a one-to-one correspondence with distributions. Such functions are obtained by the correspondence theorem according to which, to the finite measure P_X corresponds one, and only one, interval function defined by

$$F_X[a, b) = P_X[a, b) = P[a \leqq X < b), \quad [a, b) \subset R.$$

In turn, to this interval function corresponds one, and only one, class of point functions on R defined up to an additive constant, by

$$F_X(b) - F_X(a) = F_X[a, b), \quad a < b \in R.$$

Recalling that P_X is the distribution of a r.v. X, we select among all those functions the function F_X defined on R by

$$F_X(x) = P_X(-\infty, x) = P[X < x], \quad x \in R,$$

and call it the *distribution function* (*d.f.*) *of* X. Then, according to the usual notational convention, the equality in **a** can be written $Eg(X) = \int_R g\,dF_X$ and, if g is integrable and continuous on R, then the right-hand side L.-S.-integral becomes an improper R.-S.-integral.

b. *The d.f.* F_X *of a r.v.* X *is nondecreasing and continuous from the left on* R, *with* $F_X(-\infty) = 0$ *and* $F_X(+\infty) = 1$. *Conversely, every function* F *with the foregoing properties is the d.f. of a r.v. on some pr. space.*

Proof. The first assertion follows from the fact that $P[X < x]$ does not decrease as x increases, approaches $P[X < x']$ as $x \uparrow x'$, and approaches $P[X = -\infty] = 0$ or $P[X < +\infty] = 1$ according as $x \to -\infty$ or $x \to +\infty$. The converse follows by taking, say, for pr. space (R, $\mathcal{B}$, P) where P is the pr. determined, according to the correspondence theorem, by F. Then F is the d.f. of the r.v. X defined on this pr. space by $X(x) = x$, $x \in R$.

REMARK. *There are pr. spaces on which there can be defined r.v.'s for every function* F *with the stated properties.*

For example, take for the space Ω the interval (0, 1), for the σ-field of events the σ-field of all Borel sets in this interval, and for pr. the Lebesgue measure on this σ-field. Then any function F with the stated properties is the d.f. of an inverse function X of F.

The weakest type of convergence of sequences of r.v.'s considered so far is convergence in pr. In turn, it implies a type of convergence of d.f.'s, as follows:

c. *If* $X_n \xrightarrow{P} X$, *then* $F_{X_n} \to F_X$ *on the continuity set* $C(F_X)$ *of* F_X.

Proof. Since

$$[X < x'] = [X_n < x, X < x'] + [X_n \geqq x, X < x']$$
$$\subset [X_n < x] + [X_n \geqq x, X < x'],$$

we have

$$P[X < x'] \leqq F_{X_n}(x) + P[X_n \geqq x, X < x'].$$

If $X_n - X \xrightarrow{P} 0$, then, for $x' < x$,

$$P[X_n \geqq x, X < x'] \leqq P[|X_n - X| \geqq x - x'] \to 0$$

and, hence,

$$F_X(x') \leqq \liminf F_{X_n}(x), \quad x' < x.$$

Similarly, interchanging X and X_n, x and x', we obtain

$$\limsup F_{X_n}(x) \leqq F_X(x''), \quad x < x''.$$

Therefore, for $x' < x < x''$,

$$F_X(x') \leqq \liminf F_{X_n}(x) \leqq \limsup F_{X_n}(x) \leqq F_X(x'')$$

and, if $x \in C(F_X)$, it follows, letting $x' \uparrow x$ and $x'' \downarrow x$, that

$$F_X(x) = \lim F_{X_n}(x).$$

The same argument with X'_n in lieu of X and $x', x'' \in C(F_X)$ yields

d. If $X_n - X'_n \xrightarrow{P} 0$ and $F_{X'_n} \to F_X$ on $C(F_X)$, then $F_{X_n} \to F_X$ on $C(F_X)$.

Particular case. There is an important case in which convergence in pr. and convergence of d.f.'s are equivalent:

$X_n \xrightarrow{P} c$ *if, and only if,* $F_{X_n} \to 0$ *or* 1 *according as* $x < c$ *or* $x > c$.

Follows by **c** and **d**.

First Extension. Let $X = (X_1, \cdots, X_N)$ be a random vector or, equivalently, a finite class of r.v.'s $X_1, \cdots, X_N$. The *distribution* of X

is defined on the Borel field $\mathcal{B}^N$ in the N-dimensional space $R^N = \prod_{k=1}^{N} R_k$, by

$$P_X(S) = P[X \in S], \quad S \in \mathcal{B}^N.$$

As for a r.v., P_X is a pr. and the *induced* pr. space is $(R^N, \mathcal{B}^N, P_X)$. Proposition **a**, with its proof, continues to be valid: the first part holds for every finite Borel function g on R^N to some $R^{N'}$ and the second part holds for every component of g.

The *distribution function* (d.f.) F_X on R^N of X is still defined by

$$F_X(x) = P_X(-\infty, x) = P[X < x], \quad x \in R^N,$$

or, more explicitly, by

$$F_{X_1,\cdots,X_N}(x_1, \cdots, x_N) = P[X_1 < x_1, \cdots, X_N < x_N].$$

P_X determines the increment function of F_X and, conversely, by

$$P_X[a, b) = F_X[a, b) = \Delta_{b-a}F_X(a), \quad a < b \in R^N$$

or, more explicitly, by

$$P[a_1 \leqq X_1 < b_1, \cdots, a_N \leqq X_N < b_N]$$
$$= \Delta_{b_1-a_1} \cdots \Delta_{b_N-a_N}F_{X_1,\cdots,X_N}(a_1, \cdots, a_N),$$

where $\Delta_{b_k-a_k}$, $k = 1, \cdots, N$, is the difference operator of step $b_k - a_k$ operating on a_k.

Proposition **b** and its proof, as well as the remark, remain valid, provided F_X "nondecreasing" means that $\Delta_h F_X \geqq 0$ for $h > 0$, that is, $h_1 > 0, \cdots, h_N > 0$, and $x \to -\infty$ or $x \to +\infty$ means that one at least of the $x_k \to -\infty$ or that all the $x_k \to +\infty$, respectively.

Proposition **c** and its proof remain valid, provided $X_n \xrightarrow{P} X$ means that every one of the components $X_{nk} \xrightarrow{P} X_k$, $k = 1, \cdots, N$.

*Let $X = \{X_t, t \in T\}$ be an arbitrary random function or, equivalently, an arbitrary class of r.v.'s X_t, $t \in T$. Then X *induces* the pr. space $(R^T, \mathcal{B}^T, P_X)$—its *sample pr. space*—where $R^T = \prod_{t \in T} R_t$ is the range space of X, $\mathcal{B}^T$ is the Borel field in R^T, and P_X is the *distribution* of X defined by

$$P_X(S) = P[X \in S], \quad S \in \mathcal{B}^T.$$

According to the consistency theorem, P_X determines the consistent family of the distributions $P_{X_{t_1},\cdots,X_{t_N}}$ of all finite subfamilies $(X_{t_1}, \cdots, X_{t_N})$ of the family X and, conversely, a consistent family of distributions on Borel fields of all finite subspaces $R_{t_1,\ldots,t_N}$ of R^T determines a distribution on $\mathcal{B}^T$. Similarly, the d.f. F_X on R^T is defined by the consistent family of the d.f.'s $F_{X_{t_1},\ldots,X_{t_N}}$ of all finite subfamilies of the family X and, conversely, a consistent family of d.f.'s on all finite subspaces of R^T defines a d.f. on R^T.

Remark. So far, the numerical functions under consideration were r.v.'s, that is, finite (or a.s. finite) measurable functions. However, the preceding definitions remain valid for nonfinite measurable functions, provided the range-spaces are extended, that is, R, R_k, $R_t = (-\infty, +\infty)$ are replaced by $\bar{R}$, $\bar{R}_k$, $\bar{R}_t = [-\infty, +\infty]$. Thus, say, R^N is replaced by $\bar{R}^N = \prod_{k=1}^{N} \bar{R}_k$ and, at the same time, $\mathcal{B}^N$ is replaced by $\bar{\mathcal{B}}^N$—the Borel field in $\bar{R}^N$, and P_X on $\mathcal{B}^N$ is replaced by P_X on $\bar{\mathcal{B}}^N$.

To fix the ideas, let X be a numerical measurable function, not necessarily finite. Since $\bar{\mathcal{B}}$ is determined by $\mathcal{B}$ and the sets $\{-\infty\}$ and $\{+\infty\}$, P_X on $\bar{\mathcal{B}}$ is determined by P_X on $\mathcal{B}$ and the values

$$P_X(-\infty) = P[X = -\infty], \quad P_X(+\infty) = P[X = +\infty].$$

In fact, P_X on $\bar{\mathcal{B}}$ is determined by the d.f. F_X of X, defined by

$$F_X(x) = P[X < x] = P_X[-\infty, x), \quad x \in R,$$

since

$$F_X(-\infty) = \lim_{x \to -\infty} F_X(x) = P[X = -\infty] \geqq 0$$

and

$$F_X(+\infty) = \lim_{x \to +\infty} F_X(x) = P[X < +\infty] = 1 - P[X = +\infty] \leqq 1.$$

10.2 The essential feature of pr. theory. We are now in a position to describe the essential feature of pr. theory as distinct from measure theory.

While pr. concepts are born from experience and, in their rough form, are perhaps older than the measure-theoretic ones, yet their rigorous formulation was given in this chapter in terms of and by specializing the measure-theoretic concepts. Thus, it looks as if, nowadays, pr. theory were a part of measure theory or, conversely, as if measure theory were a generalized and rigorous pr. theory. Therefore, it is important to point out the basic distinction between these two interlocking branches of mathematics. The fact is that the distinction does not lie in the greater or lesser generality of the concepts, but in the properties investigated in these branches of mathematics.

Let us start with an analogy. Geometry, say, euclidean plane geometry, appears to be a part of algebra and analysis, since we can consider a point in a plane as an ordered pair (x, y) of reals or as a complex number, a straight line as a linear equation in x and y, etc. Yet, geometry remains a science *per se*, not because it has its own terminology or is older than algebra and analysis, but because geometry studies those properties of sets of points that remain invariant under all the transformations which, say, preserve the distances; for example, euclidean displacements in the case of the euclidean geometry. And geometric terminology developed, frequently unconsciously, for this specific purpose is, on the whole, well adapted to the geometrical intuition, problems, and methods.

Now, measure theory investigates families of functions on a measure space to other spaces, distinct or not from the first. On the other hand, pr. theory has developed and continues to develop the intuition, problems, and methods of its own in exploring those properties of families of functions which remain invariant under all the transformations which preserve their joint distributions—the reason being that the primary datum in random phenomena is not the pr. space but the joint distributions of the families of r.v.'s which describe the characteristics of the phenomena. Since the measurable characteristics are finite, pr. theory limited itself to r.v.'s (which, by definition, are finite). This explains the historical reason for the restrictions imposed on the measure-theoretic setup of pr. theory. However, today pr. theory is sufficiently mature mathematically to show signs of getting rid of those restrictions, by considering more general families of functions on measure spaces (normed or not) to more and more abstract spaces. We can summarize the essential feature of pr. theory as follows:

A PROPERTY IS PR.-THEORETICAL IF, AND ONLY IF, IT IS DESCRIBABLE IN TERMS OF A DISTRIBUTION.

In other words,

A property of a family of functions on a measure space is pr.-theoretical if, and only if, the property remains the same when the family is replaced by any other family with the same distribution.

In particular, since in the numerical case a distribution is represented by the corresponding d.f.'s, we can say that

—*the pr.-theoretic properties of a r.v. X are those which can be expressed in terms of its d.f. F_X,*

—*the pr.-theoretic properties of a finite family* $(X_1, X_2, \cdots, X_N)$ *of r.v.'s are those which can be expressed in terms of the joint d.f.* $F_{X_1, X_2, \cdots, X_N}$,

—*the pr.-theoretic properties of any family* $(X_t, t \in T)$ *of r.v.'s are those which can be expressed in terms of the joint d.f.'s of its finite subfamilies.*

More generally, consider a function X on a pr. space $(\Omega, \mathcal{A}, P)$ to some abstract space Ω'. The class of all sets in Ω' whose inverse images under X are events is a σ-field $\mathcal{A}'$ in Ω'; assign to $A' \in \mathcal{A}'$ the number $P'A' = P(X^{-1}A')$. This defines the induced pr. space $(\Omega', \mathcal{A}', P')$. *The pr.-theoretic properties of X are those which can be expressed in terms of P' on $\mathcal{A}'$.* If we limit ourselves to these properties only, we can speak of a "*stochastic variable*" X described by a "*pr. law*" represented by P'. Those are the mathematical beings we are concerned with, and the function X, the measure P' (or the d.f.'s in the preceding cases) are only various ways of talking about those beings in various languages. It is important to realize fully that measurements of a stochastic variable are relative to the induced pr. space; the original pr. space is but a mathematical fiction. Yet it is basic, for it permits the use of a "common frame of reference" for the families of stochastic variables we investigate—the families of sub σ-fields of events they induce on the original pr. space. However, precisely because of the existence of a common frame of reference in the present setup, modern physics forces us to introduce a different setup that we shall see in the next volume.

COMPLEMENTS AND DETAILS

Notation. Unless otherwise stated, the pr. space $(\Omega, \mathcal{A}, P)$ is fixed, the spaces L_r, $L_s (r, s > 0)$ are defined over the pr. space, and, with or without affixes, $A, B, \cdots$ denote events, while $X, Y, \cdots$ denote r.v.'s.

1. Rewrite in pr. terms as many as possible of the complements and details of Part I.

2. The convex function $\log E|X|^r$ of r is linear if, and only if, X is a degenerate r.v.

3. Liapounov's inequality. Let $\mu_r = E|X|^r$. If $r \geqq s \geqq t \geqq 0$, then $\mu_r^{s-t}\mu_s^{t-r}\mu_t^{r-s} \geqq 1$. When does this inequality become an equality? Prove Hölder's inequality by means of properties of convex functions. When does this inequality become an equality?

4. Investigate the possible behaviors of $E^{\frac{1}{r}}|X|^r$ as r varies from $-\infty$ to 0.

5. Apply Markov's inequality to $X - \dfrac{a+b}{2}$ to obtain a bound for $P[a \leqq X \leqq b]$. Also use the method of proof of the basic inequalities to obtain various bounds for this pr.

6. If g_0 on $[0, +\infty)$ is a nonnegative Borel function such that $g_0(x) \geqq g_0(\epsilon)$ for $x \geqq \epsilon$, then $P[|X| \geqq \epsilon] \leqq Eg_0(|X|)/g_0(\epsilon)$. Construct a function g on $[0, +\infty)$ with $g(0) = 0$, $g(\epsilon) = g_0(\epsilon)$, which is nondecreasing, continuous where g_0 is continuous, and such that $Eg(|X|) \leqq Eg_0(|X|)$. Then the above bound is at least as sharp with g instead of g_0.

(Form $g_i(x) = \inf g(x')$ for $x' \geqq x$ and $g(x) = \min\left(g_i(x), \frac{x}{\epsilon} g_0(x)\right)$.)

7. Let g with $g(0) = 0$ be a continuous and nondecreasing function on $[0, +\infty)$. If there exists an $h = h(Eg(|X|), \epsilon)$ such that $P[|X| \geqq \epsilon] \leqq h \leqq Eg(|X|)/g(\epsilon)$ for all r.v.'s X, then $h = Eg(|X|)/g(\epsilon)$ for those $\epsilon > 0$ for which the bound is of interest, that is, for which $Eg(|X|) < g(\epsilon)$. Loosely speaking, the bound $Eg(|X|)/g(\epsilon)$ is the sharpest of all bounds which depend upon $Eg(|X|)$ and ϵ.

(Take $|X| = \epsilon$ or 0 with pr. p and $q = 1 - p$ $(pq \neq 0)$, respectively.)

8. For $\epsilon > 0$ sufficiently small, the bound $E|X|^r/\epsilon^r$ is at least as sharp as the bound $E|X|^s/\epsilon^s$ with $s > r$.

9. Let

$d_0(X, Y) = \inf\{P[|X - Y| \geqq \epsilon] + \epsilon\}$ for all $\epsilon > 0$;

$d_1(X, Y) = \inf \epsilon$ such that $P[|X - Y| \geqq \epsilon] < \epsilon$;

$d_2(X, Y) = Eg(|X - Y|)$, g on $[0, +\infty)$ is bounded continuous and increasing with $g(0) = 0$ and $g(x + x') \leqq g(x) + g(x')$; for instance, take $g(x) = \dfrac{cx}{1 + cx}$ with $c > 0$, $g(x) = 1 - e^{-x}$, or $g(x) = \tanh x$.

Each of the three functions d_0, d_1, d_2 is a metric on the space of all r.v.'s, provided equivalent r.v.'s are identified. Convergence in pr. is equivalent to convergence in any of the corresponding metric spaces.

10. (a) $\sum |X_n| < \infty$ a.s. if, and only if, the sequence of d.f.'s of consecutive sums converges to the d.f. of a r.v.

(b) If $E \sum |X_n|^r < \infty$, then $\sum |X_n|^r < \infty$ a.s.

(c) Let $s = 1$ or $\frac{1}{r}$ according as $r < 1$ or $r \geqq 1$. If $\sum E^s|X_n|^r < \infty$, then $\sum |X_n| < \infty$ a.s.

11. $X_n \xrightarrow{P} X$ if, and only if, given $\epsilon > 0$ and $\delta > 0$, there exists $n(\epsilon, \delta)$ such that $P[|X_n - X| \geqq \epsilon] < \delta$ for $n \geqq n(\epsilon, \delta)$.

(a) $X_n \xrightarrow{\text{a.s.}} X$ if, and only if, given $\epsilon > 0$ and $\delta > 0$, there exists $n(\epsilon, \delta)$ such that $P[|X_n - X| \geqq \epsilon$ for some $n \geqq n(\epsilon, \delta)] < \delta$.

(b) $X_n \xrightarrow{\text{u}} X$ except on a null event if, and only if, given $\epsilon > 0$ there exists $n(\epsilon)$ such that $P[|X_n - X| \geqq \epsilon] = 0$ for $n \geqq n(\epsilon)$ or, equivalently, $P[|X_n - X| \geqq \epsilon$ for some $n \geqq n(\epsilon)] = 0$.

12. $P[X_n \nrightarrow X] = \lim_{\epsilon \to 0} \lim_{n \to \infty} P \bigcup_{k \geqq n} [|X_k - X| \geqq \epsilon]$.

(a) If $\sum P[|X_n - X| \geqq \epsilon] < \infty$ for every $\epsilon > 0$, then $X_n \xrightarrow{\text{a.s.}} X$.

(b) If $\sum E|X_n - X|^r < \infty$ for some $r > 0$, then $X_n \xrightarrow{\text{a.s.}} X$.

13. $X_n \xrightarrow{\text{a.s.}} X$ if, and only if, there exists a sequence $\epsilon_n \to 0$ such that $P \bigcup_{k \geqq n} [|X_k - X| \geqq \epsilon_k] \to 0$. (For the "only if" assertion select $n_m \uparrow \infty$ by $P \bigcup_{k \geqq n_m} \left[|X_k - X| \geqq \frac{1}{m}\right] < \frac{1}{2^m}$ and take $\epsilon_n = \frac{1}{m}$ for $n_m \leqq n < n_{m+1}$.) Let D be the set where the sequence X_n does not converge to a finite function.

$$PD = \lim_{\epsilon \to 0} \lim_{m \to \infty} \lim_{n \to \infty} P \bigcup_{k=m}^{n} [|X_k - X| \geqq \epsilon]$$

$$PD = \lim_{\epsilon \to 0} \lim'_{m \to \infty} \lim_{n \to \infty} P \bigcup_{k=m}^{n} [|X_k - X_m| \geqq \epsilon]$$

where lim′ denotes lim inf or lim sup indifferently. Can lim′ be replaced by lim?

14. (a) If $\sum P[X_{n+1} - X_n| \geqq \epsilon_n] < \infty$ and $\sum \epsilon_n < \infty$, then the sequence X_n converges a.s. to a r.v.

(b) If $\sum \sup_p P[|X_{n+p} - X_n| \geqq \epsilon] < \infty$ for every $\epsilon > 0$, or

$$\sup_p P[|X_{n+p} - X_n| \geqq \epsilon] \to 0 \quad \text{and} \quad \sum \liminf_p P[|X_{n+p} - X_n| \geqq \epsilon] < \infty$$

for every $\epsilon > 0$, then the sequence X_n converges a.s. to a r.v. (In the last two cases, $X_n \xrightarrow{P}$ some r.v. X and $P[|X_n - X| \geqq 2\epsilon]$ is bounded by the corresponding term of each of the two series.)

15. Take $X_n = n^c$ or 0 with pr. $\frac{1}{n}$ and $1 - \frac{1}{n}$, respectively, and investigate convergences of the sequences X_n and $E|X_n|^r$ according to the choice of c and of r.

16. If $F_{X_n} \to F_X$ on $C(F_X)$ and $Y_n \xrightarrow{P} c$, then $F_{X_n+Y_n} \to F_{X+c}$ on $C(F_{X+c})$ (*Slutsky*).
What about X_nY_n, X_n/Y_n and in general $g(X_n, Y_n)$ where g is continuous? (Use 10.1d.)

17. Take $X_{2n-1} = \frac{1}{n}$, $X_{2n} = -\frac{1}{n}$ and investigate the sequences X_n and F_{X_n}.

Take $X_n = 0$ or 1, each with pr. $\frac{1}{2}$, and $X = 1$ or 0, each with pr. $\frac{1}{2}$. Then $|X_n - X| = 1$ but $F_{X_n} = F$. To what converse is it a counterexample?

18. If the sequence X_n converges a.s. to a nonfinite function, what can be said about the sequence F_{X_n}?

19. Let $\{F_n\}$ be a denumerable family of d.f.'s with $F_n(-\infty) = 0$ and $F_n(+\infty) = 1$. The family of all functions $F_{n_1,\cdots,n_m} = F_{n_1} \times \cdots \times F_{n_m}$ is a consistent family of d.f.'s. Construct as many pr. spaces as you can, on which are defined r.v.'s X_n such that $F_{X_{n_1},\ldots,X_{n_m}} = F_{n_1,\cdots,n_m}$ for all finite index sets.

Extend what precedes to a family $\{F_t\}$ where t ranges over an arbitrarily given set T.

20. There is no universal pr. space for all possible r.v.'s on all possible pr. spaces.

21. Extend as much as possible of this chapter and of the foregoing complements and details to complex-valued r.v.'s and to complex vectors, by suitably interpreting the symbols used.

Chapter IV

DISTRIBUTION FUNCTIONS AND CHARACTERISTIC FUNCTIONS

§ 11. DISTRIBUTION FUNCTIONS

11.1 Decomposition. In pr. theory, a *distribution function* (*d.f.*), to be denoted by F, with or without affixes, is a nondecreasing function, continuous from the left and bounded by 0 and 1 on R. This definition entails at once that the quantities,

$$F(-\infty) = \lim_{x \to -\infty} F(x) = \inf F, \quad F(+\infty) = \lim_{x \to +\infty} F(x) = \sup F,$$

$$F(x) = F(x-0) = \lim_{x_n \uparrow x} F(x_n) = \sup_{x' < x} F(x'),$$

$$F(x+0) = \lim_{x_n \downarrow x} F(x_n) = \inf_{x' > x} F(x'),$$

exist and are bounded by 0 and 1, and x is a continuity or a discontinuity point of F according as $F(x+0) - F(x-0) = 0$ or > 0. As we have seen, a d.f. is always the d.f. of a measurable function on a pr. space, and if $F(-\infty) = 0$, $F(+\infty) = 1$, then it is the d.f. of a r.v.

The requirement of continuity from the left is of no importance, since every nondecreasing function F_1 on R bounded by 0 and 1 determines a d.f. F by setting $F(x) = F_1(x)$ or $F(x) = F_1(x-0)$ according as x is a continuity or a discontinuity point of F_1. In fact, even less is necessary to determine a d.f.

Let D denote a set dense in R (for example, the set of all rationals) and let F_D denote a nondecreasing function on D bounded by 0 and 1. We can assume, without loss of generality, that it is continuous from the left on D. Since, for every $x \in R$, there exists a sequence $\{x_n\} \subset D$

such that $x_n \uparrow x$, $x_n < x$, it follows easily that, according to the definition of d.f.'s,

a. *The function F defined on R by*

$$F(x) = \lim_{x_n \uparrow x} F_D(x_n), \quad x_n \in D, \quad x_n < x$$

is a d.f.

It follows that, if two d.f.'s coincide on a set dense in R, they coincide everywhere. Furthermore, monotoneity of d.f.'s leads to the

A. Decomposition theorem. *Every d.f. F has a countable set of discontinuity points and determines two d.f.'s F_c and F_d such that F_c is continuous, F_d is a step-function, and $F = F_c + F_d$.*

Proof. If F has at least n discontinuity points x_k

$$a \leqq x_1 < x_2, \cdots, < x_n < b$$

in a finite interval $[a, b)$, then, from

$$F(a) \leqq F(x_1) < F(x_1 + 0) \leqq \cdots \leqq F(x_n) < F(x_n + 0) \leqq F(b),$$

it follows, setting $p(x_k) = F(x_k + 0) - F(x_k)$, that

$$\sum_{k=1}^{n} p(x_k) = \sum_{k=1}^{n} \{F(x_k + 0) - F(x_k)\} \leqq F(b) - F(a).$$

Therefore, the number of discontinuity points x in $[a, b)$ with jumps $p(x) > \epsilon > 0$ is bounded by $\frac{1}{\epsilon}\{F(b) - F(a)\}$. Thus, for every integer m, the number of discontinuity points with jumps greater than $\frac{1}{m}$ is finite and, hence, there is no more than a countable set of discontinuity points in every finite interval $[a, b)$. Since R is a denumerable sum of such intervals, the same is true of the set of all discontinuity points, and the first assertion is proved. Furthermore, denoting the discontinuity set by $\{x_n\}$, we have, for every interval $[a, b)$, finite or not,

$$\sum_{a \leqq x_n < b} p(x_n) \leqq F(b) - F(a).$$

Upon defining F_d by

$$F_d(x) = \sum_{x_n < x} p(x_n), \quad x \in R,$$

and setting $F_c = F - F_d$, it follows at once that F_d and F_c are d.f.'s.

But, for $x < x'$,

$$F_c(x') - F_c(x) = F(x') - F(x) - \sum_{x \leq x_n < x'} p(x_n)$$
$$= F(x') - F(x+0) - \sum_{x < x_n < x'} p(x_n),$$

so that, letting $x' \downarrow x$, we obtain

$$F_c(x+0) - F_c(x) = 0;$$

thus F_c is also continuous from the right and hence continuous.

Finally, if there are two such decompositions of F,

$$F = F_c + F_d = F'_c + F'_d,$$

then $F_c - F'_c = F'_d - F_d$, and both sides must vanish since the left-hand side is continuous while the right-hand side is discontinuous, except when it vanishes identically. This completes the proof.

Remark. Since the discontinuity set of a d.f. is countable, its continuity set is *always* dense in R. However, the discontinuity set *can* also be dense in R. For example, let $\{r_n\}$ be the set of all rationals in R (it is dense in R); if $p(r_n) = \frac{6}{\pi^2} \cdot \frac{1}{n^2}$, then the function F defined by

$$F(x) = \sum_{r_n < x} p(r_n), \quad x \in R,$$

is a d.f. and, in fact, is the d.f. of a r.v., since $F(-\infty) = 0$ and $F(+\infty) = \frac{6}{\pi^2}\sum_1^\infty \frac{1}{n^2} = 1.$

Further decomposition. F_c determines, by $\mu_c(-\infty, x) = F_c(x) - F_c(-\infty)$, a finite measure μ_c on the Borel field $\mathfrak{B}$ in R. Upon applying to μ_c the Lebesgue decomposition theorem with respect to the Lebesgue measure on $\mathfrak{B}$ we obtain

$$\mu_c = \mu_{ac} + \mu_s, \quad \mu_{ac}(S) = \int_S g(x)\, dx, \quad S \in \mathfrak{B},$$

where $g \geqq 0$ is a Borel function and $\mu_s = 0$ on the complement of some Lebesgue-null set N_s. It follows that there are d.f.'s F_{ac} and F_s which correspond to the measures μ_{ac} and μ_s, respectively, such that

$$F_c = F_{ac} + F_s, \quad F_{ac}(x) = \int_{-\infty}^{x} g(x)\, dx, \quad g \geqq 0,$$

and F_s is a continuous d.f. whose points of increase all lie in N_s. Thus

A′. *Every d.f. F determines three d.f.'s of which F is the sum:*
—the step part F_d which is a step function,
—the absolutely continuous part F_{ac} such that

$$F_{ac}(x) = \int_{-\infty}^{x} g(x)\,dx, \quad g \geqq 0, \quad x \in R,$$

—the singular part F_s which is a continuous function with points of increase all belonging to a Lebesgue-null set.

11.2 Convergence of d.f.'s. As 10.1**c** and 11.1**a** suggest, convergence of d.f.'s to a d.f. F ought to be defined without taking into account what happens on the discontinuity set of F.

We say that a sequence F_n of d.f.'s converges *weakly* to a d.f. F and write $F_n \xrightarrow{w} F$, if $F_n \to F$ on the continuity set $C(F)$ of F. This definition is justified—that is, the weak limit, if it exists, is unique, since $F_n \xrightarrow{w} F$ and $F_n \xrightarrow{w} F'$ imply $F = F'$ on the set $C(F) \cap C(F')$ and, on the remaining set, which, by 11.1**A**, is countable, $F = F'$ by continuity from the left.

We say that a sequence F_n of d.f.'s converges *completely* and write $F_n \xrightarrow{c} F$, if $F_n \xrightarrow{w} F$ and $F_n(\mp\infty) \to F(\mp\infty)$. Weak convergence does not imply complete convergence. For example, given a d.f. F_0 with at least one point of increase so that $F_0(-\infty) \neq F_0(+\infty)$, let $F_n(x) = F_0(x + n)$. Then $F_n \to F_0(+\infty)$ and the weak convergence holds but not the complete convergence. However, in the case of weak convergence we have

a. Let $F_n \xrightarrow{w} F$. *Then*

$$\limsup F_n(-\infty) \leqq F(-\infty) \leqq F(+\infty) \leqq \liminf F_n(+\infty),$$

$$\operatorname{Var} F \leqq \liminf \operatorname{Var} F_n$$

and $F_n \xrightarrow{c} F$ *if, and only if,* $\operatorname{Var} F_n \to \operatorname{Var} F$ *or* $\operatorname{Var} F_n - F_n[-a, +a) \to 0$ *uniformly in n as $a \to \infty$.*

For, from

$$F_n(-\infty) \leqq F_n(x) \leqq F_n(+\infty),$$

it follows that, for $x \in C(F)$,

$$\limsup F_n(-\infty) \leqq F(x) \leqq \liminf F_n(+\infty)$$

and, letting $x \to \mp\infty$ along $C(F)$, the first inequalities are proved.

Thus

$$\operatorname{Var} F = F(+\infty) - F(-\infty) \leqq \liminf (F_n(+\infty) - F_n(-\infty))$$
$$= \liminf \operatorname{Var} F_n,$$

and the second assertion follows from the same inequalities.

We still have to find a way to recognize whether a given sequence F_n of d.f.'s converges, weakly or completely.

b. *A sequence F_n of d.f.'s converges weakly if, and only if, it converges on a set D dense in R.*

Proof. The "only if" assertion follows from the fact that the continuity set of a d.f. is dense in R. As for the "if" assertion, let $F_D = \lim F_n$ on D. The relation of 11.1**a** determines a d.f. F on R. Since, for $x' < x < x''$,

$$F_n(x') \leqq F_n(x) \leqq F_n(x''),$$

it follows that, for $x', x'' \in D$

$$F_D(x') \leqq \liminf F_n(x) \leqq \limsup F_n(x) \leqq F_D(x'').$$

Taking $x \in C(F)$ and letting $x' \uparrow x$ and $x'' \downarrow x$ along D, we obtain

$$F(x) = \lim F_n(x), \quad x \in C(F),$$

and the "if" assertion is proved.

We are now in a position to prove the basic Helly

A. WEAK COMPACTNESS THEOREM. *Every sequence of d.f.'s is weakly compact.*

We recall that (at least here) a set is *compact* in the sense of a type of convergence if every infinite sequence in the set contains a subsequence which converges in the same sense.

Proof. It suffices to show that, if F_n is a sequence of d.f.'s, then there is a subsequence which converges weakly. According to **b**, it suffices to prove that there is a subsequence which converges on a set D dense in R.

Let $D = \{x_n\}$ be an arbitrary countable set dense in R, say, the set of all rationals. All terms of the numerical sequence $F_n(x_1)$ lie between 0 and 1 and, therefore, by the Bolzano-Weierstrass compactness lemma, this sequence contains a convergent subsequence $F_{n1}(x_1)$. Similarly, the numerical sequence $F_{n1}(x_2)$ contains a convergent subsequence $F_{n2}(x_2)$ and the sequence $F_{n2}(x_1)$ converges, and so on. It follows that the "diagonal" sequence F_{nn} of d.f.'s, contained in all the subsequences $\{F_{n1}\}, \{F_{n2}\}, \cdots$, converges on D, and the proof is complete.

B. Complete compactness criterion. *A sequence F_n of d.f.'s is completely compact if, and only if, it is equicontinuous at infinity*: $\operatorname{Var} F_n - F_n[-a, +a) \to 0$ *uniformly in n as $a \to +\infty$.*

Proof. The "if" assertion is immediate. As for the "only if" assertion, if the F_n are not equicontinuous at infinity, then, by **a** and **A**, there exists a subsequence $F_{n'}$ which converges weakly but not completely.

11.3 Convergence of sequences of integrals. Let g denote a function continuous on R and let F, with or without affixes, denote a d.f. We intend to investigate conditions under which weak or complete convergence of a sequence F_n implies convergence of the corresponding sequence of integrals $\int g\, dF_n$, when these integrals exist. Let us observe that these integrals do not change if arbitrary constants are added to the d.f.'s. The investigation is centered upon the basic

a. Helly-Bray lemma. *If $F_n \xrightarrow{w} F$ up to additive constants, then, for every pair $a < b$ such that $F_n(a) \to F(a)$ and $F_n(b) \to F(b)$,*

$$\int_a^b g\, dF_n \to \int_a^b g\, dF.$$

Proof. Setting $g_m = \sum_{k=1}^{k_m} g(x_{mk}) I_{[x_{mk},\, x_{m,k+1})}$, where

$$a = x_{m1} < x_{m2} < \cdots < x_{m,k_m+1} = b$$

and $\Delta_m = \sup_k (x_{m,k+1} - x_{mk}) \to 0$ as $m \to \infty$, we have, according to the definition of R.-S. integrals,

$$\int_a^b g_m\, dF_n \to \int_a^b g\, dF_n, \quad \int_a^b g_m\, dF \to \int_a^b g\, dF, \quad m \to \infty.$$

Upon selecting all subdivision points x_{mk} to be continuity points of F, it follows from $F_n \xrightarrow{w} F$ that, for every m and every k, as $n \to \infty$

$$F_n[x_{mk}, x_{m,k+1}) \to F[x_{mk}, x_{m,k+1}),$$

and, hence,

$$\int_a^b g_m\, dF_n = \sum_{k=1}^{k_m} g(x_{mk})F_n[x_{mk}, x_{m,k+1}) \to \sum_{k=1}^{k_m} g(x_{mk})F[x_{mk}, x_{m,k+1})$$
$$= \int_a^b g_m\, dF.$$

Since

$$\int_a^b g\, dF_n - \int_a^b g\, dF$$
$$= \int_a^b (g - g_m)\, dF_n + \int_a^b g_m\, dF_n - \int_a^b g_m\, dF + \int_a^b (g_m - g)\, dF$$

and the first and last integrals on the right-hand side are bounded by $\sup_{a \leqq x \leqq b} | g(x) - g_m(x) | \to 0$ as $m \to \infty$, the assertion follows by letting $n \to \infty$ and then $m \to \infty$.

The extensions of this lemma will be based upon the obvious inequality

$$\text{(I)} \quad \left|\int g\, dF_n - \int g\, dF\right| \leqq \left|\int g\, dF_n - \int_a^b g\, dF_n\right|$$
$$+ \left|\int_a^b g\, dF - \int_a^b g\, dF_n\right| + \left|\int_a^b g\, dF - \int g\, dF\right|$$

with a and b continuity points of F, provided the integrals exist and are finite.

A. EXTENDED HELLY-BRAY LEMMA. *If* $g(\mp\infty) = 0$, *then* $F_n \xrightarrow{w} F$ *up to additive constants, implies* $\int g\, dF_n \to \int g\, dF$.

Proof. Since g is continuous and its limits as $x \to \mp\infty$ exist and are finite, g is bounded on R and the integrals $\int g\, dF_n$ and $\int g\, dF$ exist and are finite. Letting $n \to \infty$ and then $a \to -\infty$, $b \to +\infty$, it follows that, out of the three right-hand side terms in (I), the second converges to 0 by the Helly-Bray lemma, whereas the first and the third ones are bounded by $\sup_{x \notin (a,b)} | g(x) | \to 0$. The assertion is proved.

B. HELLY-BRAY THEOREM. *If g is bounded on R, then $F_n \xrightarrow{c} F$ up to additive constants implies $\int g\,dF_n \to \int g\,dF$.*

Proof. Since $|g| \leqq c < \infty$, the integrals exist and are finite. Letting $n \to \infty$ and then $a \to -\infty$, $b \to +\infty$, it follows that, out of the three terms on the right-hand side of (I), the second converges to 0 by the Helly-Bray lemma, whereas the first and the third ones are bounded, respectively, by

$$c\{\operatorname{Var} F_n - F_n[a, b)\} \to 0 \quad \text{and} \quad c\{\operatorname{Var} F - F[a, b)\} \to 0;$$

and the assertion follows.

REMARK. All the results of these subsections extend, without further ado, to d.f.'s F on R^N and continuous functions g on R^N, with the usual conventions for the symbols used above.

***11.4 Final extension and convergence of moments.** Let g on R be continuous and F on R, with or without affixes, be a d.f. The integrals we are interested in, are finite Lebesgue-Stieltjes integrals of the form $\int g\,dF$, that is, such that $\int |g|\,dF < \infty$; they are, therefore, absolutely convergent improper Riemann-Stieltjes integrals.

We say that $|g|$ is *uniformly integrable in F_n* if, as $a \to -\infty$, $b \to +\infty$, $\int_a^b |g|\,dF_n \to \int |g|\,dF_n < \infty$ uniformly in n; in other words, given $\epsilon > 0$,

$$\int |g|\,dF_n - \int_a^b |g|\,dF_n < \epsilon$$

for $a \leqq a_\epsilon$ and $b \geqq b_\epsilon$ independent of n. Since $\int_a^b |g|\,dF_n$ does not decrease as $a \downarrow -\infty$ and/or $b \uparrow +\infty$, it suffices to require the foregoing conditions for some set of values of $|a|$ and b going to infinity; for example, that $\int_{|x| \geqq c_m} |g|\,dF_n \to 0$ uniformly in n as $c_m \to \infty$ with $m \to \infty$.

We consider now properties of the foregoing integrals which follow from the weak convergence of d.f.'s F_n; they contain the extensions of the Helly-Bray lemma of the preceding subsection (we leave the verification to the reader).

A. Convergence theorem. *If $F_n \xrightarrow{w} F$ up to additive constants, then*

$$\text{(i)} \qquad \liminf \int |g|\, dF_n \geqq \int |g|\, dF$$

$$\text{(ii)} \qquad |g| \text{ is uniformly integrable in } F_n \Rightarrow \int g\, dF_n \to \int g\, dF$$

$$\text{(iii)} \qquad \int |g|\, dF_n \to \int |g|\, dF < \infty \Leftrightarrow |g| \text{ is uniformly integrable in } F_n.$$

Proof. Let $\pm c$ be continuity points of F, and use repeatedly the Helly-Bray lemma.

(i) follows, by letting $n \to \infty$ and then $c \to +\infty$, from

$$\int |g|\, dF_n \geqq \int_{-c}^{+c} |g|\, dF_n \to \int_{-c}^{+c} |g|\, dF \to \int |g|\, dF.$$

(ii) is proved as follows:

Given $\epsilon > 0$, let $\int_{|x| \geqq c} |g|\, dF_n < \epsilon$ for $c \geqq c_\epsilon$ whatever be n. By the Helly-Bray lemma, if $c' > c$ and $\pm c'$ (like $\pm c$) are continuity points of F, then $\int_{c \leqq |x| < c'} |g|\, dF < \epsilon$ and, letting $c' \to \infty$, we have $\int_{|x| \geqq c} |g|\, dF < \epsilon$ and hence $\int |g|\, dF < \infty$. Furthermore, by taking $c \geqq c_\epsilon$ and letting $n \to \infty$ and then $\epsilon \to 0$,

$$\left| \int g\, dF_n - \int g\, dF \right| \leqq \int_{|x| \geqq c} |g|\, dF_n + \left| \int_{-c}^{+c} g\, dF_n - \int_{-c}^{+c} g\, dF \right| + \int_{|x| \geqq c} |g|\, dF \to 0.$$

(iii) $\Rightarrow$ follows from

$$\int_{|x| \geqq c} |g|\, dF_n \leqq \left| \int |g|\, dF_n - \int |g|\, dF \right| + \int_{|x| \geqq c} |g|\, dF + \left| \int_{-c}^{+c} |g|\, dF - \int_{-c}^{+c} |g|\, dF_n \right|$$

by taking $c = c_0$ such that the second right-hand side term is less than $\epsilon/3$, then $n \geqq n_0$ such that the first and the third right-hand side terms are less than $\epsilon/3$, and finally $c_\epsilon = \max(c_0, c_1, \cdots, c_{n_0-1})$ where c_k $(k = 1, \cdots n_0 - 1)$ are such that $\int_{|x| \geqq c_k} |g|\, dF_k < \epsilon$; thus,

$$\int_{|x| \geqq c} |g|\, dF_n < \epsilon$$

for $c \geqq c_\epsilon$ whatever be n.

(iii) $\Leftarrow$ follows by (ii) where g is replaced by $|g|$.
This proves the last assertion and terminates the proof.

Application. Let

$$m^{(k)} = \int x^k\, dF(x), \quad k = 0, 1, 2, \cdots, \quad \mu^{(r)} = \int |x|^r\, dF(x), \quad r \geqq 0$$

define, respectively, the kth *moment* (if it exists) and the rth *absolute moment* of the d.f. F or, equivalently, of the finite part of a measurable function X with d.f. F; if X is a r.v., then this definition coincides with that given in 9.3. If F possesses subscripts, we affix the same subscripts to its moments.

B. MOMENT CONVERGENCE THEOREM. *If, for a given $r_0 > 0$, $|x|^{r_0}$ is uniformly integrable in F_n, then the sequence F_n is completely compact and, for every subsequence $F_{n'} \xrightarrow{c} F$ and all $k, r \leqq r_0$,*

$$m_{n'}^{(k)} \to m^{(k)} \text{ finite}, \quad \mu_{n'}^{(r)} \to \mu^{(r)} \text{ finite}.$$

Proof. According to the weak compactness theorem, there is a subsequence $F_{n'}$ and a d.f. F such that $F_{n'} \xrightarrow{w} F$. On the other hand, the uniformity condition for $|x|^{r_0}$ implies that, for every $r \leqq r_0$,

$$\int_{|x| \geqq c} |x|^r\, dF_{n'}(x) \leqq c^{r-r_0} \int_{|x| \geqq c} |x|^{r_0}\, dF_{n'}(x) \to 0 \quad \text{as} \quad c \to +\infty$$

uniformly in n', so that the uniformity condition holds for $|x|^r$. Therefore, the preceding convergence theorem applies to every sequence $m_{n'}^{(k)}$ and $\mu_{n'}^{(r)}$ with $k, r \leqq r_0$. In particular, taking $r = 0$, we obtain $\operatorname{Var} F_{n'} \to \operatorname{Var} F$, so that $F_{n'} \xrightarrow{c} F$. The theorem is proved.

COROLLARY. *If the sequence $\mu_n^{(r_0+\delta)}$ is bounded for some $\delta > 0$, then the conclusion of the foregoing theorem holds.*

For $\mu_n^{(r_0+\delta)} \leqq a < \infty$ implies that, as $c \to +\infty$,

$$\int_{|x| \geqq c} |x|^{r_0}\, dF_n(x) \leqq c^{-\delta} \int_{|x| \geqq c} |x|^{r_0+\delta}\, dF_n(x) \leqq c^{-\delta} a \to 0,$$

so that the uniformity condition holds for $|x|^{r_0}$.

This corollary yields at once the following solution of the celebrated "moment convergence problem" (Fréchet and Shohat).

C. *If, for $k \geq k_0$ arbitrary but fixed, the sequences $m_n^{(k)} \to m^{(k)}$ finite, then these sequences converge for every value of k, and their limits $m^{(k)}$ are finite and are the moments of a d.f. F such that there exists a subsequence $F_{n'} \xrightarrow{c} F$.*

If, moreover, these limits determine F up to an additive constant, then $F_n \xrightarrow{c} F$ up to an additive constant.

It suffices to apply the foregoing corollary and to observe that, if the $m^{(k)}$ determine F up to an additive constant, then all completely convergent subsequences $F_{n'}$ have the same limit d.f. F up to additive constants.

§ 12. CHARACTERISTIC FUNCTIONS AND DISTRIBUTION FUNCTIONS

Pr. properties are properties describable in terms of distributions—and those are set functions. The introduction of d.f.'s makes it possible to describe pr. properties in terms of point functions, easier to handle with the tools of classical analysis. Yet, to a distribution corresponds not a single d.f. F but the family of all functions $F + c$ where c is an arbitrary constant. The selection of one of them is somewhat arbitrary, and we have constantly to bear this fact in mind. The introduction of characteristic functions (ch.f.) assigned to the family $F + c$ by the relation

$$f(u) = \int e^{iux}\, dF(x), \quad u \in R$$

obviates this difficulty and, moreover, is of the greatest practical importance for the following reasons.

1° To the family $F + c$ corresponds a unique ch.f., and conversely. Therefore, there is a one-to-one correspondence between distributions and ch.f.'s.

2° The methods and results of classical analysis are particularly well suited to the handling of ch.f.'s. In fact, ch.f.'s are continuous and uniformly bounded (by 1) functions. Moreover, to complete and weak convergence of d.f.'s (defined up to additive constants) correspond, respectively, ordinary convergence of ch.f.'s and ordinary convergence of their indefinite integrals.

3° The oldest and, until recent years, almost the only general problem of pr. theory is the "Central Limit Problem," concerned with the asymptotic behavior of d.f.'s of sequences of sums of independent r.v.'s. Much of Part III will be devoted to this problem. The d.f.'s of such sums are obtained by "composition" of the d.f's of their summands, and this "composition" involves repeated integrations and results in unwieldly expressions, whereas the ch.f.'s of these sums are simply the products of the ch.f.'s of the summands. The Central Limit Problem was satisfactorily solved in the 15 years (1925–1940) which followed the establishment by P. Lévy of the properties of ch.f.'s.

12.1 Uniqueness. *The characteristic function (ch.f.) f of a d.f. F is defined on R by*

$$f(u) = \int e^{iux}\, dF(x) = \int \cos ux\, dF(x) + i\int \sin ux\, dF(x), \quad u \in R.$$

Since, for every $u \in R$, the function of x with values e^{iux} is continuous and bounded by 1, f exists and is continuous and bounded by 1 on R. Moreover, to all functions $F + c$, where c is an arbitrary constant, corresponds the same function f. The converse (and, thus, the one-to-one correspondence between distributions and ch.f.'s) follows from the formula below.

A. Inversion formula.

$$F[a, b) = \lim_{U \to \infty} \frac{1}{2\pi} \int_{-U}^{+U} \frac{e^{-iua} - e^{-iub}}{iu} f(u)\, du,$$

provided $a < b$ are continuity points of F.

The inversion formula holds for all $a < b \in R$, provided F is normalized.

We say that F is *normalized* if the values of F at its discontinuity points x are taken to be $\dfrac{F(x-0) + F(x+0)}{2}$. Normalization destroys the continuity from the left of F at its discontinuity points. However, according to 11.1, the normalized d.f. determines the original one, so that nothing is lost by normalization.

We observe that, in the integral which figures on the right-hand side of the inversion formula, the integrand is defined at $u = 0$ by continuity, so that it is continuous on R; also it is bounded on R by its value $(b - a)$ $f(0)$ at $u = 0$. Thus, for every finite U, this integral is an ordinary

Riemann integral and, in proving the inversion formula, we shall find that the limit of this integral, as $U \to \infty$, exists.

Proof. The proof uses repeatedly the dominated convergence theorem applied to an interchange of integrations and is based on the classical Dirichlet formula

$$\frac{1}{\pi}\int_a^b \frac{\sin v}{v}\,dv \to 1 \quad \text{as} \quad a \to -\infty, \quad b \to +\infty,$$

so that the left-hand side is bounded uniformly in a and b. Let

$$I_U = \frac{1}{2\pi}\int_{-U}^{+U} \frac{e^{-iua} - e^{-iub}}{iu} f(u)\,du, \quad a < b \in R,$$

and replace $f(u)$ by its defining integral $\int e^{iux}\,dF(x)$. We can interchange the integrations, so that, by elementary computations,

$$I_U = \int J_U(x)\,dF(x),$$

where

$$J_U(x) = \frac{1}{\pi}\int_{U(x-b)}^{U(x-a)} \frac{\sin v}{v}\,dv.$$

Since J_U is bounded uniformly in U, integration and passage to the limit as $U \to \infty$ can be interchanged in

$$\lim_{U\to\infty} I_U = \lim_{U\to\infty} \int J_U(x)\,dF(x).$$

Therefore

$$\lim_{U\to\infty} I_U = \int J(x)\,dF(x)$$

where

$$J(x) = \lim_{U\to\infty} J_U(x) = \begin{cases} 1 & \text{for} \quad a < x < b \\ \frac{1}{2} & \text{for} \quad x = a, \quad x = b \\ 0 & \text{for} \quad x < a, \quad x > b, \end{cases}$$

and, hence,

$$\begin{aligned} \lim_{U\to\infty} I_U &= \tfrac{1}{2}\{F(a+0) - F(a-0)\} + \{F(b-0) - F(a+0)\} \\ &\quad + \tfrac{1}{2}\{F(b+0) - F(b-0)\} \\ &= \frac{F(b-0) + F(b+0)}{2} - \frac{F(a-0) + F(a+0)}{2}. \end{aligned}$$

Thus, if F is normalized or if $a < b \in C(F)$, then

$$\lim_{U \to \infty} I_U = F[a, b),$$

and the inversion formula is proved.

REMARK. If an improper Riemann integral

$$\int_{-\infty}^{+\infty} g\,dx = \lim_{\substack{a \to -\infty \\ b \to +\infty}} \int_a^b g\,dx$$

exists and is finite, then

$$\lim_{U \to \infty} \int_{-U}^{+U} g\,dx = \int_{-\infty}^{+\infty} g\,dx.$$

However, the left-hand side limit may exist and be finite (as in the inversion formula), whereas the right-hand side improper integral does not exist. Yet the inversion formula can be written in terms of an improper Riemann integral as follows:

$$F[a, b) = \frac{1}{\pi} \int_0^\infty \frac{\mathcal{I}\{(e^{-iua} - e^{-iub})f(u)\}}{u}\,du$$

where $\mathcal{I}$ stands for "imaginary part of," so that

$$\mathcal{I}\{(e^{-iua} - e^{-iub})f(u)\} =$$
$$(\cos ua - \cos ub)\mathcal{I}f(u) - (\sin ua - \sin ub)\mathcal{R}f(u).$$

It suffices to write $\int_{-U}^{+U} = \int_{-U}^{0} + \int_0^U$, change u into $-u$ in the first right-hand side integral, and take into account the fact that then the integrand changes into its complex-conjugate.

COROLLARY. *F is differentiable at a and its derivative $F'(a)$ at a is given by*

$$(1) \qquad F'(a) = \lim_{h \to 0} \lim_{U \to \infty} \frac{1}{2\pi} \int_{-U}^{+U} \frac{1 - e^{-iuh}}{iuh} e^{-iua} f(u)\,du$$

if, and only if, the right-hand side exists.

In particular, if f is absolutely integrable on R, then F' exists and is bounded and continuous on R and, for every $x \in R$,

$$(2) \qquad F'(x) = \frac{1}{2\pi} \int_{-\infty}^{+\infty} e^{-iux} f(u)\,du.$$

Proof. The first assertion follows directly from the inversion formula by the definition of the derivative. The second assertion follows from the first and from the assumption that $\int |f|\, du < \infty$ since, the integrand in (1) being bounded by $|f|$, we have, in (1),

$$\lim_{U\to\infty}\int_{-U}^{+U} = \int_{-\infty}^{+\infty} \quad\text{and}\quad \lim_{h\to 0}\int_{-\infty}^{+\infty} = \int_{-\infty}^{+\infty}\lim_{h\to 0}.$$

REMARK. Thus, if the ch.f.'s f_n of d.f.'s F_n are uniformly Lebesgue-integrable on R, and if $f_n \to f$ ch.f. of F, then f is Lebesgue-integrable on R, and $F'_n \to F'$.

B. *For every* $x \in R$,

$$F(x+0) - F(x-0) = \lim_{U\to\infty}\frac{1}{2U}\int_{-U}^{+U} e^{-iux} f(u)\, du.$$

For we can interchange below the integrations and the passage to the limit, so that

$$\lim_{U\to\infty}\frac{1}{2U}\int_{-U}^{+U} e^{-iux} f(u)\, du = \lim_{U\to\infty}\frac{1}{2U}\int_{-U}^{+U} du\left\{\int e^{iu(y-x)}\, dF(y)\right\}$$

$$= \lim_{U\to\infty}\int \frac{\sin U(y-x)}{U(y-x)}\, dF(y)$$

$$= F(x+0) - F(x-0).$$

12.2 Convergences. Since there is a one-to-one correspondence between d.f.'s defined up to additive constants and ch.f.'s, it has to be expected that a one-to-one correspondence also exists between the weak and complete convergence, up to additive constants, of sequences of d.f.'s and certain types of convergence—to be found—of ch.f.'s. For this purpose we introduce the *integral ch.f.* $\hat{f}$ of F defined on R by

$$\hat{f}(u) = \int_0^u f(v)\, dv = \int \frac{e^{iux} - 1}{ix}\, dF(x).$$

The last integral is obtained upon replacing $f(v)$ by its defining integral and noting that the interchange of integrations is permissible. Since there is a one-to-one correspondence between $\hat{f}$ and its continuous derivative f, it follows, by 12.1, that there is a one-to-one correspondence between $\hat{f}$ and F defined up to an additive constant.

We are now in a position to show that the weak and the complete convergence up to additive constants of sequences of d.f.'s correspond to the ordinary convergence of the corresponding sequences of integral ch.f.'s and of ch.f.'s, respectively. Unless otherwise stated, a d.f., its ch.f., and its integral ch.f. will be denoted by $F, f, \hat{f}$ respectively, with the same affixes if any.

A. WEAK CONVERGENCE CRITERION. *If $F_n \xrightarrow{w} F$ up to additive constants, then $\hat{f}_n \to \hat{f}$. Conversely, if $\hat{f}_n$ converges to some function $\hat{g}$, then there exists a d.f. F with $F_n \xrightarrow{w} F$ up to additive constants and $\hat{f} = \hat{g}$.*

Proof. Since $\dfrac{e^{iux} - 1}{ix} \to 0$ as $x \to \mp\infty$, the first assertion follows at once, by the extended Helly-Bray lemma, from the definition of the integral ch.f.'s.

Conversely, let $\hat{f}_n \to \hat{g}$. According to the weak compactness theorem, there is a d.f. F and a subsequence $F_{n'} \xrightarrow{w} F$ as $n' \to \infty$. Therefore, by the extended Helly-Bray lemma, for every $u \in R$,

$$\hat{g}(u) = \lim_{n'} \hat{f}_{n'}(u) = \lim_{n'} \int \frac{e^{iux} - 1}{ix}\, dF_{n'}(x) = \int \frac{e^{iux} - 1}{ix}\, dF(x) = \hat{f}(u).$$

Since $\hat{f}$ determines F up to an additive constant, it follows that weakly convergent subsequences of the sequence F_n have the same limit F up to additive constants, with $\hat{f} = \hat{g}$. This proves the second assertion.

COROLLARY 1. *Every sequence $\hat{f}_n$ of integral ch.f.'s is compact in the sense of ordinary convergence on R.*

For, in view of the above criterion, this statement is equivalent to the weak compactness theorem for d.f.'s.

COROLLARY 2. *If $f_n \to g$ a.e., then $F_n \xrightarrow{w} F$ up to additive constants, with $f = g$ a.e.*

Here "a.e." is taken with respect to the Lebesgue measure on R.

Proof. Since $f_n \to g$ a.e. and the f_n are continuous and uniformly bounded by 1, it follows that g is measurable and bounded a.e. so that, by the dominated convergence theorem, $\hat{f}_n \to \hat{g}$ where $\hat{g}$ is defined on R by the Lebesgue integral

$$\hat{g}(u) = \int_0^u g(v)\, dv, \quad u \in R.$$

Therefore, by the foregoing criterion, $F_n \xrightarrow{w} F$ up to additive constants, and $\hat{f} = \hat{g}$. Since the derivative of $\hat{f}$ is f, whereas that of the indefinite Lebesgue integral $\hat{g}$ exists and equals g a.e., it follows that $f = g$ a.e.

B. COMPLETE CONVERGENCE CRITERION. *If $F_n \xrightarrow{c} F$ up to additive constants, then $f_n \to f$. Conversely, if $f_n \to g$ continuous at $u = 0$, then $F_n \xrightarrow{c} F$ up to additive constants, and $f = g$.*

When the F_n and f_n are d.f.'s and ch.f.'s of r.v.'s, the converse becomes the celebrated P. Lévy's *continuity theorem* for ch.f.'s.

Proof. Let $F_n \xrightarrow{c} F$ up to additive constants. Then, by the Helly-Bray theorem, for every $u \in R$,

$$f_n(u) = \int e^{iux}\, dF_n(x) \to \int e^{iux}\, dF(x) = f(u).$$

Conversely, let $f_n \to g$ continuous at $u = 0$. Then, for every $u \in R$,

$$\hat{f}_n(u) = \int_0^u f_n(v)\, dv \to \int_0^u g(v)\, dv = \hat{g}(u),$$

and, hence, by the weak convergence criterion, for some d.f. F with ch.f. f, $F_n \xrightarrow{w} F$ up to additive constants, and $\hat{f} = \hat{g}$. Therefore,

$$\frac{1}{u}\int_0^u f(v)\, dv = \frac{1}{u}\int_0^u g(v)\, dv$$

and, letting $u \to 0$, we obtain $f(0) = g(0)$ on account of continuity of f and of g at the origin. Thus,

$$\operatorname{Var} F_n = f_n(0) \to g(0) = f(0) = \operatorname{Var} F,$$

and the proof is completed by taking into account the direct assertion.

C. UNIFORM CONVERGENCE THEOREM. *If a sequence f_n of ch.f.'s converges to a ch.f. f, then the convergence is uniform on every finite interval $[-U, +U]$.*

Proof. On account of **B**, $F_n \xrightarrow{c} F$ up to additive constants. Let $\epsilon > 0$ and $U > 0$ be arbitrarily fixed. We have

$$|f_n(u) - f(u)| \leq \left| \int_a^b e^{iux}\, dF_n(x) - \int_a^b e^{iux}\, dF(x) \right|$$
$$+ \operatorname{Var} F_n - F_n[a, b) + \operatorname{Var} F - F[a, b)$$

where we take a, b to be continuity points of F. Let $|a|, b$ and then n be so large that $\text{Var}\, F - F[a, b)\,| < \frac{\epsilon}{6}$,

$$\text{Var}\, F_n - F_n[a, b) < \text{Var}\, F - F[a, b) + \frac{\epsilon}{6} < \frac{\epsilon}{3}.$$

It suffices to show that, for n sufficiently large and all $u \in [-U, +U]$,

$$\Delta_n = \left| \int_a^b e^{iux}\, dF_n(x) - \int_a^b e^{iux}\, dF(x) \right| < \frac{\epsilon}{2}.$$

Let

$$a = x_1 < x_2, \cdots < x_{N+1} = b$$

where the subdivision points are continuity points of F and $\alpha = \max\limits_{k \leqq N} (x_{k+1} - x_k) < \epsilon/8U$. Since, by the mean value theorem,

$$|e^{iux} - e^{iux'}| \leqq |x - x'|U \quad \text{for} \quad |u| \leqq U,$$

it follows that, upon replacing x by x_k in every interval $[x_k, x_{k+1})$, Δ_n is modified by at most

$$\alpha U \int_a^b dF_n(x) + \alpha U \int_a^b dF(x) \leqq 2\alpha U < \frac{\epsilon}{4}.$$

Thus, it remains to show that, for n sufficiently large,

$$\left|\sum_{k=1}^{N} e^{iux_k}\{F_n[x_k, x_{k+1}) - F[x_k, x_{k+1})\}\right| \leqq \sum_{k=1}^{N} |F_n[x_k, x_{k+1}) - F[x_k, x_{k+1})| < \frac{\epsilon}{4}.$$

Since $F_n[x_k, x_{k+1}) \to F[x_k, x_{k+1})$ for every $k \leqq N$, the last assertion follows and the proof is complete.

REMARK. In fact, we proved, with a supplementary detail, the first assertion of the complete convergence criterion without using the Helly-Bray theorem.

COROLLARY 1. *If $f_n \to f$ and $u_n \to u$ finite, then $f_n(u_n) \to f(u)$.*

This follows, by **C** and continuity of f, from

$$|f_n(u_n) - f(u)| \leqq |f_n(u_n) - f(u_n)| + |f(u_n) - f(u)|.$$

COROLLARY 2. *A set $\{F_t\}$ of d.f.'s is completely compact (up to additive constants) if, and only if, the corresponding set $\{f_t\}$ of ch.f.'s is equicontinuous at $u = 0$.*

Proof. By 12.4**B** equicontinuity of $\{f_t\}$ at $u = 0$ is equivalent to equicontinuity on R.

On the other hand, Ascoli's theorem and its converse say that a set of continuous functions is compact in the sense of uniform convergence on a finite closed interval if, and only if, it is uniformly bounded and equicontinuous on this interval. Since the f_t are uniformly bounded, the assertion follows by **B** and **C**.

REMARK. If the d.f.'s F_n, F of r.v.'s, are differentiable and $F_n' \to F'$ on R, then $f_n \to f$ *uniformly on* R. It suffices to use 17 in Complements and Details of Ch. II.

12.3 Composition of d.f.'s and multiplication of ch.f.'s. A function F on $R = (-\infty, +\infty)$ is said to be *composed* of d.f.'s F_1 and F_2, and written $F_1 * F_2$, if

$$F(x) = \int F_1(x - y)\, dF_2(y), \quad x \in R$$

where we assume, for simplicity, that $F_1(-\infty) = F_2(-\infty) = 0$; otherwise, to avoid trivial complications, we would have to replace F_1 by $F_1 - F_1(-\infty)$.

Since, for every fixed y, $F_1(x - y)$ are values of a d.f., nondecreasing, continuous from the left and bounded by $F_1(-\infty) = 0$ and $F_1(+\infty) \leqq 1$, it follows, upon applying the dominated convergence theorem, that F has the same properties and that $\mathrm{Var}\, F = \mathrm{Var}\, F_1 \cdot \mathrm{Var}\, F_2$.

A. COMPOSITION THEOREM. *If $F = F_1 * F_2$, then $f = f_1 f_2$, and conversely.*

Proof. Let $F = F_1 * F_2$ and let $a = x_{n1} < \cdots < x_{n,k_n+1} = b$ with $\sup_k (x_{n,k+1} - x_{nk}) \to 0$ as $n \to \infty$. Since, for every $u \in R$,

$$\int_a^b e^{iux}\, dF(x) = \lim \sum_k e^{iux_{nk}} F[x_{nk}, x_{n,k+1})$$

$$= \lim \int \sum_k e^{iu(x_{nk}-y)} F_1[x_{nk} - y, x_{n,k+1} - y) e^{iuy}\, dF_2(y),$$

it follows that

$$\int_a^b e^{iux}\, dF(x) = \int \left\{ \int_{a-y}^{b-y} e^{iux}\, dF_1(x) \right\} e^{iuy}\, dF_2(y)$$

and, letting $a \to -\infty$ and $b \to +\infty$,

$$\int e^{iux}\, dF(x) = \int e^{iux}\, dF_1(x) \int e^{iuy}\, dF_2(y),$$

so that $f = f_1 f_2$ and the first assertion is proved.

Conversely, according to the first assertion, $f_1 f_2$ is the ch.f. of $F_1 * F_2$ and, hence, on account of the one-to-one correspondence between f and $F + c$, $F = F_1 * F_2$ up to an additive constant. The converse is proved.

Corollary 1. *A product of ch.f.'s is a ch.f. and, in particular, if f is a ch.f. so is $|f|^2$.*

For $f = f_1 f_2$ is the ch.f. of the d.f. $F = F_1 * F_2$, and the particular case follows from the fact that, if f is a ch.f., so is its complex-conjugate $\bar{f}$ which corresponds to the d.f. $F(+\infty) - F(-x + 0)$.

Corollary 2. *Composition of d.f.'s is commutative and associative.*

For the corresponding multiplication of ch.f.'s has these properties.

12.4 Elementary properties of ch.f.'s and first applications. In the sequel, the elementary properties we establish now will play an important ancillary role, and the first applications will be used, improved, and generalized.

We denote by F and f, with same subscripts if any, corresponding d.f.'s and ch.f.'s; in general, the corresponding d.f.'s F are defined up to additive constants, but if f is ch.f. of a r.v., then, as usual, we take $F(-\infty) = 0$, $F(+\infty) = 1$. We say that a r.v. X is *symmetric* if X and $-X$ have the same d.f., that is, for every $x \in R$, $P[X < x] = P[X > -x]$.

A. General properties. *Every ch.f. f is uniformly continuous and*

$$|f| \leqq f(0) = \operatorname{Var} F \leqq 1, \quad f(-u) = \bar{f}(u).$$

If f is the ch.f. of a r.v. X, then the function with values $e^{iua}f(bu)$ is the ch.f. of the r.v. $a + bX$. In particular, $\bar{f}$ is the ch.f. of $-X$ and f is real if, and only if, X is symmetric.

Proof. The first assertion follows from $f(u) = \int e^{iux}\, dF(x)$. The second assertion follows from $Ee^{iu(a+bX)} = e^{iua}Ee^{ibuX}$. Finally, if X is symmetric, then $f(u) = Ee^{iuX} = Ee^{-iuX} = f(-u) = \bar{f}(u)$ so that f is real; conversely, if f is real, then changing the signs of a and b in the inversion formula is equivalent to taking the complex-conjugate of the integrand and changing its sign, so that $F[a, b) = F[-b, -a)$ and, hence, by letting $a \to -\infty$ and $b \uparrow x$, we have $P[X < x] = F(x) = 1 - F(-x + 0) = P[X > -x]$.

B. INCREMENTS INEQUALITY: *for any* $u, h \in R$

$$|f(u) - f(u + h)|^2 \leqq 2f(0)\{f(0) - \mathcal{R}f(h)\}.$$

INTEGRAL INEQUALITY: *for* $u > 0$ *there exist functions* $0 < m(u) < M(u) < \infty$ *such that*

$$m(u)\int_0^u \{f(0) - \mathcal{R}f(v)\}\, dv \leqq \int \frac{x^2}{1 + x^2}\, dF(x)$$

$$\leqq M(u)\int_0^u \{f(0) - \mathcal{R}f(v)\}\, dv;$$

if $f(0) = 1$, *then, for* u *sufficiently close to* 0,

$$\int \frac{x^2}{1 + x^2}\, dF(x) \leqq -M(u)\int_0^u (\log \mathcal{R}f(v))\, dv.$$

Proof. The increments inequality follows, by Schwarz's inequality, from

$$|f(u) - f(u + h)|^2 = \left|\int e^{iux}(1 - e^{ihx})\, dF(x)\right|^2$$

$$\leqq \int dF(x) \int |1 - e^{ihx}|^2\, dF(x)$$

$$= 2f(0)\int (1 - \cos hx)\, dF(x)$$

$$= 2f(0)\{f(0) - \mathcal{R}f(h)\}.$$

The integral inequality follows, by the elementary inequality with $u \neq 0$

$$0 < M^{-1}(u) \leqq |u| \left(1 - \frac{\sin ux}{ux}\right)\frac{1 + x^2}{x^2} \leqq m^{-1}(u) < \infty, \quad x \in R,$$

from

$$\int_0^u dv \int (1 - \cos vx)\, dF(x) = u\int \left(1 - \frac{\sin ux}{ux}\right)\frac{1 + x^2}{x^2} \cdot \frac{x^2}{1 + x^2}\, dF(x).$$

The case $f(0) = 1$ follows then from the elementary inequality $1 - a \leqq -\log a$ for $a \geqq 0$.

The integral inequality permits us in turn to find bounds for $\int_{|x|<c} x^2\, dF(x)$ and $\int_{|x|\geqq c} dF(x)$, $(c > 0)$, by

$$\text{(I)}\quad \frac{1}{1+c^2}\int_{|x|<c} x^2\, dF(x) + \frac{c^2}{1+c^2}\int_{|x|\geqq c} dF(x)$$

$$\leqq \int \frac{x^2}{1+x^2}\, dF(x)$$

$$\leqq \int_{|x|<c} x^2\, dF(x) + \int_{|x|\geqq c} dF(x).$$

However, it is sometimes more convenient to use the direct

B′. TRUNCATION INEQUALITY: *for* $u > 0$:

$$\int_{|x|<1/u} x^2\, dF(x) \leqq \frac{3}{u^2}\{f(0) - \mathcal{R}f(u)\},$$

$$\int_{|x|\geqq 1/u} dF(x) \leqq \frac{7}{u}\int_0^u \{f(0) - \mathcal{R}f(v)\}\, dv.$$

If $f(0) = 1$ *and* u *is sufficiently close to* 0, *then we can replace* $1 - \mathcal{R}f$ *in the foregoing by* $-\log \mathcal{R}f$.

These inequalities follow, respectively, from

$$\int (1 - \cos ux)\, dF(x) \geqq \int_{|x|<1/u} \frac{u^2x^2}{2}\left(1 - \frac{u^2x^2}{12}\right) dF(x)$$

$$\geqq \frac{11u^2}{24}\int_{|x|<1/u} x^2\, dF(x)$$

and from

$$\frac{1}{u}\int_0^u dv \int (1 - \cos vx)\, dF(x) = \int\left(1 - \frac{\sin ux}{ux}\right) dF(x)$$

$$\geqq (1 - \sin 1)\int_{|x|\geqq 1/u} dF(x).$$

The case $f(0) = 1$ follows as in **B**.

Applications. 1° *If* $f_n \to g$ *continuous at* $u = 0$, *then* g *is continuous on* R.

This follows from the fact that the increments inequality with f_n becomes, as $n \to \infty$, the same inequality with g.

2° *If the sequence* f_n *is equicontinuous at* $u = 0$, *then it is equicontinuous at every* $u \in R$.

For, then, as $h \to 0$,

$$|f_n(u) - f_n(u+h)|^2 \leqq 2\{f_n(0) - \Re f_n(h)\} \to 0$$

uniformly in n.

3° *If* $f_n \to 1$ *on* $(-U, +U)$, *then* $f_n \to 1$ *on* R.

This follows by induction as $f_n(2u) \to 1$ for $|u| < U$ follows from

$$|f_n(u) - f_n(2u)|^2 \leqq 2\{f_n(0) - \Re f_n(u)\} \to 0 \quad \text{for} \quad |u| < U.$$

If we take into account the fact that the set of all differences of numbers belonging to a set of positive Lebesgue measure contains a nondegenerate interval $(-U, +U)$, this proposition can be improved as follows:

If $f_n \to 1$ *on a set* A *of positive Lebesgue measure, then* $f_n \to 1$ *on* R.

For, we can assume that the set A is symmetric with respect to the origin and contains it, since, for $u \in A$,

$$f_n(-u) = \bar{f}_n(u) \to 1, \quad 1 \geqq f_n(0) \geqq |f_n(u)| \to 1,$$

and, then, $f_n(u - u') \to 1$ for $u, u' \in A$ on account of

$$|f_n(u) - f_n(u - u')|^2 \leqq 2\{f_n(0) - \Re f_n(-u')\} \to 0.$$

4° We shall now prove an elegant proposition (slightly completed) due to Kawata and Ugakawa. We use repeatedly Corollary 2 of the weak convergence criterion which says that, if a sequence of ch.f.'s $g_n \to g$ a.e., then the corresponding sequence of d.f.'s $G_n \xrightarrow{w} G$ up to additive constants and the ch.f. of G coincides a.e. with g.

Let $g_n = \prod_{k=1}^{n} f_k \to g$ *a.e. Either* $g = 0$ *a.e., and then* $G_n \xrightarrow{w} 0$ *up to additive constants. Or* $g \neq 0$ *on a set* A *of positive Lebesgue measure, and then* $G_n \xrightarrow{c} G$ *up to additive constants.*

Proof. In both cases $G_n \xrightarrow{w} G$ up to additive constants. The first case follows from the recalled proposition. In the second case, we have to prove that $\operatorname{Var} G_n \to \operatorname{Var} G$. Since $\operatorname{Var} F^s = \operatorname{Var}(F * F) = (\operatorname{Var} F)^2$ and $f^s = |f|^2$, it suffices to consider real-valued nonnegative ch.f.'s. But then $\lim_{m\to\infty} \prod_{n+1}^{m} f_k$ exists on R and coincides a.e. with a ch.f., while, for m, n sufficiently large, $g_m g_n \neq 0$ a.e. on A, and, as $m \to \infty$ and then $n \to \infty$,

$$\prod_{k=n+1}^{m} f_k = g_m/g_n \to g/g_n = \prod_{k=n+1}^{\infty} f_k \to 1 \quad \text{a.e. on } A.$$

It follows, by 3°, that $\prod_{k=n+1}^{\infty} f_k \to 1$ a.e. on R. Therefore, if H_n is the d.f. whose ch.f. coincides a.e. with $\prod_{k=n+1}^{\infty} f_k$, then $\operatorname{Var} H_n \to 1$. But, by 11.2**a** and the composition theorem 12.3**A**,

$$\liminf \operatorname{Var} G_n \geqq \operatorname{Var} G = \operatorname{Var} G_n \cdot \operatorname{Var} H_n.$$

It follows, by letting $n \to \infty$, that $\operatorname{Var} G_n \to \operatorname{Var} G$. The proof is completed.

5° Let F_{nk} be d.f.'s of r.v.'s, $k = 1, \cdots, k_n \to \infty$, $\gamma_n = \sum_k (1 - f_{nk})$. Set $\Psi_n(x) = \sum_k \int_{-\infty}^{x} \frac{y^2}{1+y^2}\, dF_{nk}(y)$ and $\alpha(c) = \sup_n \sum_k \int_{|x|\geq c} dF_{nk}(x)$, $\beta(c) = \sup_n \sum_k \int_{|x|<c} x^2\, dF_{nk}(x)$, $c > 0$ finite.

If $f_n = \prod_k f_{nk}$ with f_{nk} real-valued, then the following properties are equivalent:

(C_1) *the sequence F_n is completely compact.*
(C_2) *the sequence γ_n is equicontinuous at $u = 0$.*
(C_3) $\alpha(c) \to 0$ *as* $c \to \infty$ *and* $\alpha(c) + \beta(c) < \infty$ *for every (some)* c.
(C_4) *the sequence Ψ_n is bounded and completely compact.*

Proof. (C_1) $\Leftrightarrow$ (C_2) by 12.2 **C** Cor. 2 and the inequality $1 - \sum a_k \leqq \prod(1 - a_k) \leqq \exp\{-\sum a_k\}$, $0 \leq a_k \leq 1$. (C_2) $\Rightarrow$ (C_3) by **B′** and (C_3) $\Rightarrow$ (C_2) by $\gamma_n(u) \leq 2\alpha(c) + \beta(c)u^2/2$. Finally, ($C_3$) $\Leftrightarrow$ (C_4) and "some c" $\Leftrightarrow$ "every c" by (I), $\alpha(c)c^2/(1 + c^2) \leq \int_{|x|\geq c} d\Psi_n(x) \leq \alpha(c)$ and 11.2B.

Let

$$m^{(k)} = \int x^k \, dF(x), \quad \mu^r = \int |x|^r \, dF(x), \quad k = 0, 1, 2, \cdots, \quad r \geqq 0,$$

be, respectively, the kth moments and the rth absolute moments of F. Let $f^{(k)}$ be the kth derivative of $f(f^{(0)} = f)$ and, as usual, let θ, θ' be quantities with modulus bounded by 1.

C. DIFFERENTIABILITY PROPERTIES. *If $f^{(2n)}(0)$ exists and is finite, then $\mu^{(r)} < \infty$ for $r \leqq 2n$.*

If $\mu^{(n+\delta)} < \infty$ for a $\delta \geqq 0$, then for every $k \leqq n$

$$f^{(k)}(u) = i^k \int e^{iux} x^k \, dF(x), \quad u \in R,$$

and $f^{(k)}$ is continuous and bounded by $\mu^{(k)}$; moreover

$$f(u) = \sum_{k=0}^{n-1} m^{(k)} \frac{(iu)^k}{k!} + \rho_n(u), \quad u \in R$$

where

$$\rho_n(u) = u^n \int_0^1 \frac{(1-t)^{n-1}}{(n-1)!} f^{(n)}(tu) \, dt = m^{(n)} \frac{(iu)^n}{n!} + o(u^n) = \theta \mu^{(n)} \frac{|u|^n}{n!},$$

and if $0 < \delta \leqq 1$, then

$$\rho_n(u) = m^{(n)} \frac{(iu)^n}{n!} + 2^{1-\delta} \theta' \mu^{(n+\delta)} \frac{|u|^{n+\delta}}{(1+\delta)(2+\delta)\cdots(n+\delta)}.$$

Proof. To begin with, we observe that, since $|x|^{r'} \leqq 1 + |x|^r$ for $r' < r$, finiteness of $\mu^{(r)}$ implies that of $\mu^{(r')}$.

The first assertion follows from the existence and finiteness of the $2n$th symmetric derivative by using the Fatou-Lebesgue theorem in

$$|f^{(2n)}(0)| = \lim_{h \to 0} \int \left(\frac{\sin hx}{hx} \right)^{2n} x^{2n} \, dF(x) \geqq \int x^{2n} \, dF(x).$$

The second assertion follows from the fact that, by differentiating $\int e^{iux} \, dF(x)$ k times under the integral sign, the integral so obtained is absolutely convergent and, hence, this differentiation and the integration can be interchanged.

The limited expansions follow by integrating the limited expansions of e^{iux} with corresponding forms of its remainder term. The last and less usual corresponding form of its remainder is obtained upon observing that $|e^{ia} - 1| \leqq 2|a/2|^\delta$ (since, for $0 < \delta \leqq 1$, if $|a/2| < 1$, then $|e^{ia} - 1| \leqq |a| \leqq 2|a/2|^\delta$ and, if $|a/2| \geqq 1$, then $|e^{ia} - 1| \leqq 2 \leqq 2|a/2|^\delta$), and using successive integrations by parts in

$$\left|\int_0^1 \frac{(1-t)^{n-1}}{(n-1)!}(e^{itux} - 1)\,dt\right| \leqq 2^{1-\delta}|ux|^\delta \int_0^1 \frac{(1-t)^{n-1}}{(n-1)!} t^\delta\,dt$$

$$= \frac{2^{1-\delta}|ux|^\delta}{(1+\delta)(2+\delta)\cdots(n+\delta)}.$$

COROLLARY. *If all moments of F exist and are finite, then $f^{(k)}(0) = i^k m^{(k)}$ for every k, and*

$$f(u) = \sum_{n=0}^{\infty} m^{(n)} \frac{(iu)^n}{n!}$$

in the interval of convergence of the series.

Applications. We consider d.f.'s F and ch.f.'s f of r.v.'s X, with the same subscripts if any. If $m^{(1)} = EX = 0$, we write σ^2 instead of $m^{(2)} = EX^2$.

1° NORMAL DISTRIBUTION. A "reduced normal" d.f. is defined by $F'(x) = e^{-x^2/2}/\sqrt{2\pi}$. It is the d.f. of a r.v., since $m^{(0)} = 1$ by

$$\left(\frac{1}{\sqrt{2\pi}}\int e^{-x^2/2}\,dx\right)\left(\frac{1}{\sqrt{2\pi}}\int e^{-y^2/2}\,dy\right) = \frac{1}{2\pi}\iint e^{-(x^2+y^2)/2}\,dx\,dy$$

$$= \frac{1}{2\pi}\int_0^{2\pi} d\theta \int_0^{\infty} e^{-\rho^2/2}\rho\,d\rho = 1$$

Since $F'(-x) = F'(x)$, it follows at once that the odd moments vanish, while, by integration by parts, we obtain

$$m^{(2n)} = (2n-1)m^{(2n-2)} = \cdots = (2n)!/2^n n!.$$

Therefore, by the foregoing corollary, the "reduced normal" ch.f. is

$$f(u) = \sum_{n=0}^{\infty} \frac{(-u^2/2)^n}{n!} = e^{-u^2/2}, \quad u \in R.$$

2° BOUNDED LIAPOUNOV THEOREM. *Let $|X_n| \leqq c < \infty$ and $EX_n = 0$.*

If $s_n^2 = \sum_{k=1}^{n} \sigma_k^2 \to \infty$, then $\prod_{k=1}^{n} f_k(u/s_n) \to e^{-u^2/2}$ for every $u \in R$.

Since $E|X_n|^3 \leqq cEX_n^2$ and $\sigma_n^2 = EX_n^2 \leqq c^2$, it follows, upon fixing u arbitrarily, that

$$f_k\left(\frac{u}{s_n}\right) = 1 - \frac{\sigma_k^2}{2s_n^2}u^2 + \theta_{nk}\frac{c\sigma_k^2}{6s_n^3}|u|^3 \to 1$$

uniformly in $k \leqq n$. Therefore, for n sufficiently large,

$$\sum_{k=1}^{n} \log f_k\left(\frac{u}{s_n}\right) = -\frac{u^2}{2}(1 + o(1)) + \theta_n \frac{c|u|^3}{6s_n}(1 + o(1)) \to -\frac{u^2}{2},$$

and the assertion is proved.

§ 13. PROBABILITY LAWS AND TYPES OF LAWS

13.1 Laws and types; the degenerate type. Since there is a one-to-one correspondence between distributions, d.f.'s defined up to an additive constant, and ch.f.'s, they are different but equivalent "representations" of the same mathematical concept which we shall call *pr. law* or, simply, *law*. Moreover, to a given distribution on the Borel field $\mathcal{B}$ we can always make correspond the finite part of a measurable function X on some pr. space $(\Omega, \mathcal{A}, P)$, and the restriction of P to $X^{-1}(\mathcal{B})$ with values $P[X \in S]$, $S \in \mathcal{B}$, is still another representation of the law defined by the given distribution; there are many such measurable functions and many such spaces. Nevertheless, the various representations of a given law have their own intuitive value. Thus, for every law we have a multiplicity of representations and we shall use them according to convenience.

A law will be denoted by the symbol $\mathcal{L}$, with the same affixes if any as the d.f. or the ch.f. which represents this law, and the terminology and notations for operations on laws will be those introduced for d.f.'s; in particular, if $F_n \xrightarrow{w} F$ we write $\mathcal{L}_n \xrightarrow{w} \mathcal{L}$, and if $F_n \xrightarrow{c} F$ we write $\mathcal{L}_n \xrightarrow{c} \mathcal{L}$. The case of laws of r.v.'s (with d.f.'s of variation 1) is by far the most important. The law of a r.v. X will be denoted by $\mathcal{L}(X)$, and if a sequence $\mathcal{L}(X_n)$ of laws of r.v.'s converges completely—necessarily to the law $\mathcal{L}(X)$ of a r.v. X—we shall drop "complete" and write $\mathcal{L}(X_n) \to \mathcal{L}(X)$. *From now on a law will be law of a r.v., unless otherwise stated.*

The origin and the scale of values of measured quantities, say a r.v. X, are more or less arbitrarily chosen. By modifying them we modify linearly the results of measurements, that is, we replace X by $a + bX$

where a and $b > 0$ are finite numbers. If, moreover, the orientation of values can be modified, then the only restriction on the finite numbers a and b is that $b \neq 0$. This leads us to assign to a law $\mathcal{L}(X)$ the family $\mathfrak{I}(X) = \{\mathcal{L}(a + bX)\}$ of all laws obtainable by changes of origin, scale, and orientation, to be called a *type* of laws. If b is restricted either to positive or to negative values, the corresponding families of laws will be called *positive*, resp. *negative* types of laws.

Letting $b \to 0$ we encounter a boundary case—the simplest and at the same time the everywhere pervading *degenerate type* $\{\mathcal{L}(a)\}$ of laws of r.v.'s which degenerate at some arbitrary but finite value a, that is, such that $X = a$ a.s. The corresponding family of "degenerate" d.f.'s is that of d.f.'s with one, and only one, point of increase $a \in R$ with $F(a + 0) - F(a - 0) = 1$. The corresponding family of "degenerate" ch.f.'s is that of all ch.f.'s of the form $f(u) = e^{iua}$, $u \in R$, so that their moduli reduce to 1. The converse is also true and, more precisely,

a. *A ch.f. is degenerate if, and only if, its modulus equals* 1 *for two values $h \neq 0$ and $\alpha h \neq 0$ of the argument whose ratio α is irrational. In particular, a ch.f. f is degenerate if $|f(u)| = 1$ in a nondegenerate interval.*

Proof. Since $|f(h)| = 1$, there is a finite number a such that $f(h) = e^{iha}$ and, hence,

$$e^{-iha}f(h) = \int e^{ih(x-a)}\, dF(x) = 1.$$

Thus

$$\int [1 - \cos h(x - a)]\, dF(x) = 0$$

and, since the integrand is nonnegative, it follows that, for points $\bar{x}$ of increase of F, $\cos h(\bar{x} - a) = 1$ so that $\bar{x}' - \bar{x}''$ is a multiple of $\frac{2\pi}{h}$ when the points of increase $\bar{x}', \bar{x}''$ are distinct. Replacing h by αh, we find that $\bar{x}' - \bar{x}''$ is also a multiple of $\frac{2\pi}{\alpha h}$, which is impossible when α is irrational unless there is only one point of increase. The particular case follows.

REMARK. The foregoing argument proves that, if $|f(h)| = 1$ for an $h \neq 0$, then $f(u) = \sum_{k=0}^{\infty} p_k e^{iux_k}$, $u \in R$, where $p_k \geqq 0$, $\sum_{k=0}^{\infty} p_k = 1$ and $x_k = a + k \cdot \frac{2\pi}{h}$; the converse is immediate.

13.2 Convergence of types. If $\mathfrak{L}(X_n) \to \mathfrak{L}(X)$, then, for every a, $b \neq 0$, $\mathfrak{L}(a + bX_n) \to \mathfrak{L}(a + bX)$, since $f_n \to f$ implies that $e^{iua}f_n(bu) \to e^{iua}f(bu)$, $u \in R$. Thus, we may say that convergence of sequences of laws to a law is, in fact, convergence of sequences of types to a type. It may even happen that, given a sequence $\mathfrak{L}(X_n)$ convergent or not, we can proceed to changes of origin and of scale varying with n and giving rise to a convergent sequence $\mathfrak{L}(a_n + b_nX_n)$. In the particular case of consecutive sums X_n of "independent" r.v.'s, a special form of the problem of finding the sequences of laws which converge for given changes of origin and of scale is the oldest and, until recently, was the only limit problem of pr. theory; we shall investigate it in Part III. Meanwhile there is an immediate question to answer: given a sequence $\mathfrak{L}(X_n)$ of laws, do all the limit laws of convergent sequences of the form $\mathfrak{L}(a_n + b_nX_n)$ belong to a same type? The answer, due to Khintchine for positive types, is as follows:

A. Convergence of types theorem. *If $\mathfrak{L}(X_n) \to \mathfrak{L}(X)$ nondegenerate and $\mathfrak{L}(a_n + b_nX_n) \to \mathfrak{L}(X')$ nondegenerate, then the laws $\mathfrak{L}(X)$ and $\mathfrak{L}(X')$ belong to the same type. More precisely, $\mathfrak{L}(X') = \mathfrak{L}(a + bX)$ with $|b_n| \to |b|$, and if $b_n > 0$ then $b_n \to b$, $a_n \to a$.*

However, for every finite a and for every sequence $\mathfrak{L}(X_n)$ of laws, there exist numbers a_n and $b_n \neq 0$ such that $\mathfrak{L}(a_n + b_nX_n) \to \mathfrak{L}(a)$.

In other words, given a sequence of laws, the changes of origin, scale, and orientation can yield in the limit no more than one nondegenerate type and can always yield in the limit the degenerate type. This shows once more that the degenerate type is to be considered as the "degenerate part" of every type.

Proof. The second assertion is immediate. For, by taking the numbers c_n sufficiently large so as to have $P[|X_n| \geq c_n] < \frac{1}{n} \to 0$, we obtain

$$P\left[\frac{|X_n|}{nc_n} \geq \frac{1}{n}\right] < \frac{1}{n} \to 0$$

and, it follows at once, that $\mathfrak{L}\left(\frac{X_n}{nc_n}\right) \to \mathfrak{L}(0)$, so that $\mathfrak{L}\left(a + \frac{X_n}{nc_n}\right) \to \mathfrak{L}(a)$.

The first assertion means that $f_n \to f$ nondegenerate and $e^{iua_n}f_n(b_nu) \to f'(u)$ nondegenerate, $u \in R$, imply existence of two finite numbers a

and $b \neq 0$ such that $f'(u) = e^{iua}f(bu)$, $u \in R$. We can always select from the sequence b_n a convergent subsequence $b_{n'}$, but its limit b may be 0 or $\pm\infty$. If $b = 0$, then, since the convergence of ch.f.'s to a ch.f. is uniform in every finite interval, we have, for every fixed $u \in R$,

$$|f'(u)| = \lim_{n'} |f_{n'}(b_{n'}u)| = |f(0)| = 1,$$

so that, by **13.1a**, f' is degenerate and this contradicts the assumption. Similarly, if $b_{n'} \to \pm\infty$, then, replacing u by $\frac{u}{b_{n'}}$, it follows that

$$|f(u)| = \lim_{n'} \left|f'_{n'}\left(\frac{u}{b_{n'}}\right)\right| = |f'(0)| = 1,$$

so that f is degenerate and this contradicts the assumption. Thus $b_{n'} \to b$ finite and different from 0. On the other hand, for all u sufficiently close to 0, the continuous functions $f(bu)$ and $f'(u)$ (with values 1 for $u = 0$) differ from 0; and we have, for n' sufficiently large,

$$e^{iua_{n'}} = \frac{e^{iua_{n'}} f_{n'}(b_{n'}u)}{f_{n'}(b_{n'}u)} \to \frac{f'(u)}{f(bu)} \neq 0, \quad n' \to \infty,$$

so that $a_{n'} \to a = \frac{1}{iu} \log \frac{f'(u)}{f(bu)}$. Therefore $f'(u) = e^{iua}f(bu)$, $u \in R$.

Clearly, it remains only to prove that $|b_n| \to |b|$. Let $b_{n'} \to b$ and $b_{n''} \to b'$ hence $a_{n'} \to a$ and $a_{n''} \to a'$; it suffices to prove that if, for every u, $e^{iua}f(bu) = e^{iua'}f(b'u)$, then $|b| = |b'|$. Upon replacing $b'u$ by u and $\frac{b}{b'}$ by c, it suffices to prove that, if $|c| \leqq 1$ and, for every u, $|f(u)|^2 = |f(cu)|^2$, then $|c| = 1$. But $|c| < 1$ entails, upon replacing repeatedly u by cu,

$$|f(u)|^2 = |f(cu)|^2 = \cdots = \lim |f(c^n u)|^2 = 1.$$

Thus, the nondegeneracy assumption excludes the possibility $|c| < 1$, so that $|c| = 1$ and the proof is complete.

REMARK. It is immediately seen that if we limit ourselves to, say, positive types only, then, under the foregoing assumptions, $a_n \to a$ and $b_n \to b$. We leave to the reader to find conditions under which this property remains valid for types.

COROLLARY. *If, for every u,*

$$e^{iua_n}f_n(b_nu) \to f(u) \quad and \quad e^{iua'_n}f_n(b'_nu) \to f(u)$$

where f is a nondegenerate ch.f. and $b_nb'_n > 0$ *for every n, then*

$$\frac{a_n - a'_n}{b'_n} \to 0 \quad and \quad \frac{b_n}{b'_n} \to 1.$$

Replace in the theorem X_n by $a'_n + b'_nX_n$.

13.3 Extensions. The results and terminology of this chapter extend at once to families of r.v.'s, and we shall content ourselves with a few generalities.

The *law of a random vector* $X = \{X_1, \cdots, X_N\}$ with d.f. F_X on R^N is represented by the ch.f. f_X on R^N defined by the N-uple integral

$$f_X(u) = \int e^{iux}\, dF_X(x), \quad ux = u_1x_1 + \cdots + u_Nx_N$$

or, explicitly, by

$$f_X(u_1, \cdots, u_N) = \overbrace{\int \cdots \int}^{N\text{-uple}} e^{i(u_1x_1 + \cdots + u_Nx_N)}\, d_1d_2 \cdots d_NF_X(x_1, \cdots, x_N).$$

The integral which appears in the inversion formula becomes an N-uple Riemann-Stieltjes integral $\int_{-U_1}^{+U_1} \cdots \int_{-U_N}^{+U_N}$ and the "kernel" $\dfrac{e^{-iua} - e^{-iub}}{iu}$ becomes $\prod_{k=1}^{N} \dfrac{e^{-iu_ka_k} - e^{-iu_kb_k}}{iu_k}$.

We observe that there is a one-to-one correspondence between the law of the random vector $X = \{X_1, \cdots, X_N\}$ and the laws of the r.v.'s $uX = u_1X_1 + \cdots + u_NX_N$, *where u varies over* R^N, since

$$f_X(tu) = f_{uX}(t), \quad t \in R$$

and, in particular, $f_X(u) = f_{uX}(1)$.

Finally, the *law of a random function* $X = \{X_t, t \in T\}$ is the set of joint laws of all its finite subfamilies.

§ 14. NONNEGATIVE-DEFINITENESS; REGULARITY

14.1 Ch.f.'s and nonnegative-definiteness. The class of ch.f.'s has been defined to be the class of Fourier-Stieltjes transforms of d.f.'s. Conversely, given a continuous function g on R, we can recognize whether or not it is a ch.f. by applying the inversion formula: if the right-hand side of the inversion formula exists and is nonnegative for all pairs $a < b$ of finite numbers, then g is a ch.f. up to a multiplicative constant. If g is absolutely integrable on R, then it suffices to apply Corollary 1 of the inversion formula and verify that the function F' is nonnegative. A very important criterion of a different type is that of nonnegative-definiteness that we investigate now.

Let g be a real or complex-valued function on a set $D_S \subset R$ obtained by forming all differences of the elements of a set $S \ni 0$; for example, $S = [0, U)$ and $D_S = (-U, +U)$ S = set of all nonnegative integers and D_S = set of all integers. Sets D_S are necessarily symmetric with respect to the origin $u = 0$ and contain it. We say that g on D_S is *nonnegative-definite* if for every finite set $S_n \subset S$ and every real or complex-valued function h on S_n

$$\sum_{u,v \in S_n} g(u - v)h(u)\bar{h}(v) \geqq 0;$$

we shall omit mention of D_S when $D_S = R$.

a. *If g on D_S is nonnegative-definite, then, for every $u \in D_S$,*

$$g(0) \geqq 0, \quad g(-u) = \bar{g}(u), \quad |g(u)| \leqq g(0).$$

If, moreover, $D_S \supset (-U, +U)$ and g is continuous at the origin, then g is continuous at every limit point of D_S.

Proof. We apply the defining relation with

$$S_1 = \{0\}, \quad S_2 = \{0, u\}, \quad S_3 = \{0, u, u'\}.$$

With S_1 we obtain $g(0) \geqq 0$. It follows with S_2 that $g(u)h(u) + g(-u)\bar{h}(u)$ is real and hence $g(-u) = \bar{g}(u)$ (take $h(u) = 1$ and $h(u) = i$). We use these two properties below.

The discriminant of a nonnegative quadratic form being nonnegative, elementary computations with S_2 yield $|g(u)| \leqq g(0)$. For the last assertion we exclude the trivial case $g(0) = 0$ which implies $g = 0$, and, to simplify the writing, assume that $g(0) = 1$ (it suffices to replace g by $g/g(0)$). The same discriminant property but with S_3 yields, by elementary computations,

$$|g(u) - g(u')|^2 \leqq 1 - |g(u - u')|^2 - 2\mathcal{R}\{\bar{g}(u)g(u')(1 - g(u - u'))\}$$

Therefore, if g is continuous at the origin, that is, if $g(u - u') \to g(0) = 1$ as $u' \to u$, then $g(u') \to g(u)$. The proof is complete.

The foregoing proposition shows that a nonnegative-definite function g on R continuous at the origin has properties similar to those of ch.f.'s. In fact, g coincides on R with a ch.f.—up to a multiplicative constant; and this is what we intend to prove now. According to **a**, if $g(0) = 0$, then $g = 0$ so that, by excluding this trivial case and dividing by $g(0)$ *we can and will assume from now on that* $g(0) = 1$.

b. HERGLOTZ LEMMA. *A function g on the set $D_S = \{\cdots -2c, -c, 0, +c, +2c, \cdots\}$ is nonnegative-definite if, and only if, it coincides on this set with a ch.f.* $f(u) = \int_{-\pi/c}^{+\pi/c} e^{iux}\, dF(x)$.

Proof. We can assume that $c > 0$. If g on D_S is nonnegative-definite, then, for every integer n and every finite number x,

$$G'_n(x) = \frac{1}{2\pi} \sum_{k=-n+1}^{n-1} \left(1 - \frac{|k|}{n}\right) g(kc)e^{-ikx}$$

$$= \frac{1}{2\pi n} \sum_{j=1}^{n} \sum_{h=1}^{n} g((j-h)c)e^{-i(j-h)x} \geqq 0.$$

Upon multiplying by e^{ikx} with some fixed value of k and integrating over $[-\pi, +\pi)$, we obtain

$$\left(1 - \frac{|k|}{n}\right) g(kc) = \int_{-\pi}^{+\pi} e^{ikx} G'_n(x)\, dx = \int_{-\pi/c}^{+\pi/c} e^{i(kc)x}\, dF_n(x)$$

where F_n is a d.f. with $F_n(-\pi/c) = 0$, $F_n(+\pi/c) = g(0) = 1$. The "only if" assertion follows, on account of the weak compactness and Helly-Bray lemma, by letting $n \to \infty$ along a suitable subsequence of integers. The "if" assertion is immediate (as below).

A. BOCHNER'S THEOREM. *A function g on R is nonnegative-definite and continuous if, and only if, it is a ch.f.*

Proof. The "if" assertion (Mathias) is immediate, since, if g is a ch.f. with d.f. G, then, letting u and v range over an arbitrary but finite set in R,

$$\sum_{u,v} g(u-v)h(u)\bar{h}(v) = \int \Big\{ \sum_{u,v} e^{i(u-v)x} h(u)\bar{h}(v) \Big\}\, dG(x)$$

$$= \int \Big| \sum_{u} e^{iux} h(u) \Big|^2 dG(x) \geqq 0.$$

Conversely, let g on R be nonnegative-definite and continuous. It coincides on R with a ch.f. if it does so on the set S_r (dense in R) of all rationals of the form $k/2^n$, $k = 0, \pm 1, \pm 2, \cdots, n = 1, 2, \cdots$. For every integer n, let S_n be the corresponding subset of all rationals of the form $k/2^n$ so that $S_n \uparrow S_r$. Since g is nonnegative-definite on R, it is nonnegative-definite on every S_n. Therefore, by **b**, there exist ch.f.'s f_n such that $g(k/2^n) = f_n(k/2^n)$ whatever be k and n. Since $S_n \uparrow S_r$, it follows that $f_n \to g$ on S_r. Let $0 \leqq \theta, \theta_n \leqq 1$, so that, by **b**,

$$1 - \mathcal{R}f_n(\theta/2^n) = \int_{-\pi}^{+\pi} (1 - \cos \theta x)\, dF_n(2^n x)$$

$$\leqq \int_{-\pi}^{+\pi} (1 - \cos x)\, dF_n(2^n x) = 1 - \mathcal{R}g(1/2^n).$$

Therefore, by the elementary inequality $|a + b|^2 \leqq 2|a|^2 + 2|b|^2$ and the increments inequality, for every fixed $h = (k_n + \theta_n)/2^n$,

$$|1 - f_n(h)|^2 \leqq 2|1 - f_n(k_n/2^n)|^2 + 4(1 - \mathcal{R}f_n(\theta_n/2^n))$$

$$\leqq 2|1 - g(k_n/2^n)| + 4(1 - \mathcal{R}g(1/2^n)).$$

Since g is continuous at the origin, it follows by 12.4, 2°, that the sequence f_n of ch.f.'s is equicontinuous. Hence, by Ascoli's theorem, it contains a subsequence converging to a continuous function f, so that $g = f$ on S_r and hence on R. Since by the continuity theorem f is a ch.f., the proof is complete.

The "only if" assertion can be proved directly, and this direct proof will extend to a more general case: For every $T > 0$ and $x \in R$

$$p_T(x) = \frac{1}{T}\int_0^T \int_0^T g(u - v)e^{-i(u-v)x}\, du\, dv \geqq 0,$$

since, g on R being nonnegative-definite and continuous, the integral can be written as a limit of nonnegative Riemann sums. Let $u = v + t$, integrate first with respect to v and set $g_T(t) = \left(1 - \frac{|t|}{T}\right) g(t)$ or 0 according as $|t| \leqq T$ or $|t| \geqq T$. The above relation becomes

$$p_T(x) = \int e^{-itx} g_T(t)\, dt \geqq 0.$$

Now multiply both sides by $\frac{1}{2\pi}\left(1 - \frac{|x|}{X}\right) e^{iux}$ and integrate with respect to x on $(-X, +X)$. The relation becomes

$$\frac{1}{2\pi}\int_{-X}^{+X}\left(1-\frac{|x|}{X}\right)p_T(x)e^{iux}\,dx = \frac{1}{2\pi}\int \frac{\sin^2 \frac{1}{2}X(t-u)}{\frac{1}{4}X(t-u)^2}\,g_T(t)\,dt.$$

The left-hand side is a ch.f. (since its integrand is a product of e^{iux} by a nonnegative function) and the right-hand side converges to $g_T(u)$ as $X \to \infty$. Therefore, g_T is the limit of a sequence of ch.f.'s. Since it is continuous at the origin, the continuity theorem applies and g_T is a ch.f. Since $g_T \to g$ as $T \to \infty$, the same theorem applies, and the assertion is proved.

Extension* 1. The question arises whether in **A continuity at the origin is necessary. Let g on R be nonnegative-definite and Lebesgue-measurable.

By integrating

$$\sum_{u_j, u_k \in S_n} g(u_j - u_k)e^{i(u_j - u_k)x} \geqq 0, \quad x \in R$$

with respect to every $u \in S_n$ over $(0, T)$, we obtain

$$nT^n + n(n-1)T^{n-2}\int_0^T\int_0^T g(u-v)e^{i(u-v)x}\,du\,dv \geqq 0.$$

Dividing by $n(n-1)T^{n-2}$ and letting $n \to \infty$, it follows that

$$\int_0^T\int_0^T g(u-v)e^{i(u-v)x}\,du\,dv \geqq 0.$$

Therefore, the direct proof of the "only if" assertion in **A** continues to apply, but instead of the continuity theorem we have to use its Corollary 2, and we obtain $g = f$ ch.f. almost everywhere (in Lebesgue measure). The "if" assertion is modified accordingly. Thus (F. Riesz)

A′. *A function g on R is nonnegative-definite and Lebesgue-measurable if, and only if, it coincides a.e. with a ch.f.*

Extension* 2. It can be shown that Herglotz lemma remains valid with $D_S = \{-Nc, -(N-1)c, \cdots, 0, \cdots (N-1)c, Nc\}$ whatever be the fixed integer N. Then, replacing S_r and S_n by their intersections with $(-U, +U)$ whatever be the fixed U, the proof of **A remains valid. Thus (Krein)

A″. *A function g on $(-U, +U)$ is nonnegative-definite and continuous if, and only if, it coincides on $(-U, +U)$ with a ch.f.*

REMARK 1. The proofs of **A** and **A''** use only the fact that g is continuous at the origin, so that these theorems imply the last assertion in **a.**

REMARK 2. The foregoing proofs show that in the definition of a nonnegative-definite g it suffices to take $h(u) = e^{iux}$ where x runs over R. Also if g is Lebesgue-measurable, then the definition can be taken to be

$$\int_0^{T_n}\int_0^{T_n} g(u - v)e^{i(u-v)x}\,du\,dv \geqq 0$$

for every $x \in R$ and a sequence $T_n \to \infty$.

According to the second extension, a function which coincides with a ch.f. on $(-U, +U)$ can be extended to a ch.f. on R. The problem which arises is under what conditions this extension is unique. This is part of the problem we investigate in the following subsection.

***14.2 Regularity and extension of ch.f.'s.** According to 13.1**a**, if $f = 1$ on an interval $(-U, +U)$, then $f = 1$ on R. Also, according to 12.4, 3°, if $f_n \to 1$ on $(-U, +U)$ then $f_n \to 1$ on R. Thus, in these cases a ch.f. is determined by its values on an interval, and convergence of a sequence of ch.f.'s on R follows from its convergence on an interval. We intend to investigate more general conditions under which these properties hold. To simplify the writing, we assume that the ch.f.'s are those of r.v.'s, that is, take the value 1 at $u = 0$.

a. *If $\hat{f}$ is the integral ch.f. corresponding to the ch.f. f, then*

$$\left|\frac{\hat{f}(u+h) - \hat{f}(u-h)}{2h}\right|^2 \leqq \frac{1}{2}\{1 + \mathcal{R}f(h)\}.$$

For, from

$$\frac{\sin^2 x}{x^2} = \frac{\sin^2 2\frac{x}{2}}{4\sin^2\frac{x}{2}}\cdot\frac{\sin^2\frac{x}{2}}{\left(\frac{x}{2}\right)^2} \leqq \cos^2\frac{x}{2} = \frac{1 + \cos x}{2},$$

it follows, upon applying the Schwarz inequality, that

$$\left|\frac{\hat{f}(u+h) - \hat{f}(u-h)}{2h}\right|^2 = \left|\int e^{iux}\frac{\sin hx}{hx}\,dF(x)\right|^2$$

$$\leqq \int \frac{1 + \cos hx}{2}\,dF(x) = \frac{1}{2}\{1 + \mathcal{R}f(h)\}.$$

We extend now the uniform convergence theorem 12.2**C**. Let f_n be ch.f.'s.

b. *If* $f_n \to g$ *on* $(-U, +U)$ *and* g *is continuous at* $u = 0$, *then the* f_n *are equicontinuous and the convergence is uniform.*

Proof. Because of 12.4 (1°,2°) and Ascoli's theorem, it suffices to prove that the f_n are equicontinuous at $u = 0$. If this conclusion is not true, then there exist an $\epsilon > 0$, a sequence $n' \to \infty$, and a sequence $u_{n'} \to 0$, such that $|f_{n'}(u_{n'})| < 1 - \epsilon$ for all n'; given a positive $h \in (-U, +U)$, we take $m_{n'} = \left[\frac{h}{u_{n'}}\right]$, so that $m_{n'}u_{n'} \to h$. Upon applying **a** with $u = kh$ and summing over $k = -m + 1, -m + 3, \cdots, m - 1$, we obtain by the elementary inequality $|a_1 + \cdots + a_m|^2 \leqq m|a_1|^2 + \cdots + m|a_m|^2$

$$\left|\frac{\hat{f}(mh) - \hat{f}(-mh)}{2mh}\right|^2 \leqq \frac{1}{2}\{1 + \Re f(h)\}.$$

It follows that

$$\left|\frac{1}{2m_{n'}u_{n'}}\int_{-m_{n'}u_{n'}}^{+m_{n'}u_{n'}} f_{n'}(v)\,dv\right|^2 \leqq \frac{1}{2}\{1 + \Re f_{n'}(u_{n'})\} < 1 - \frac{\epsilon}{2}$$

and, letting $n' \to \infty$, we have

$$\left|\frac{1}{2h}\int_{-h}^{+h} g(v)\,dv\right|^2 \leqq 1 - \frac{\epsilon}{2}.$$

Since $1 = f_n(0) \to g(0)$ and g is continuous at $u = 0$, it follows, letting $h \to 0$, that $1 \leqq 1 - \frac{\epsilon}{2}$. Therefore, *ab contrario*, the f_n are equicontinuous at $u = 0$, and the assertion is proved.

A. CONTINUITY THEOREM ON AN INTERVAL. *If* $f_n \to f_U$ *on* $(-U, +U)$ *and* f_U *is continuous at* $u = 0$, *then* f_U *extends to a ch.f.* f *on* R; *if the extension* f *is unique, then* $f_n \to f$ *on* R.

Proof. According to **b**, the f_n are equicontinuous. Therefore, by Ascoli's theorem, the sequence f_n is compact in the sense of uniform convergence and, since $f_n \to f_U$ on $(-U, +U)$, all its limit ch.f.'s coincide with f_U on $(-U, +U)$. It follows that, if there is only one ch.f. f which coincides with f_U on $(-U, +U)$, then $f_n \to f$ on R.

The second part of the problem raised above is reduced to its first part: find ch.f.'s determined by their values on an interval $(-U, +U)$. A partial answer is given by the following theorem (Marcinkiewicz).

B. Extension theorem for ch.f.'s. *If the restriction f_U of a ch.f. f to an interval $(-U, +U)$ is regular or is the boundary function of a regular function, then f_U determines f.*

This theorem follows, by the unicity of analytic continuation, from the three propositions below of independent interest. Let $f(z) = \int e^{izx}\, dF(x)$, where $z = u + iv$ is a point of the complex plane $R_u \times R_v$.

a. *$f(z)$ is regular in a circle $|z| < R$ if, and only if, for every positive $r < R$, $\int e^{r|x|}\, dF(x)$ is finite.*

Proof. The "if" assertion is immediate and it suffices to prove the "only if" assertion.

Let

$$m^{(n)} = \int x^n\, dF(x) \quad \text{and} \quad \mu^{(n)} = \int |x|^n\, dF(x).$$

If $f(z)$ is regular for $|z| < R$, then, for every positive $r < R$,

$$\sum \frac{1}{n!} |m^{(n)}| r^n < \infty,$$

and, in particular,

$$\sum \frac{1}{(2n)!} \mu^{(2n)} r^{2n} < \infty.$$

Since

$$(\mu^{(2n-1)})^{\frac{1}{2n-1}} \leqq (\mu^{(2n)})^{\frac{1}{2n}},$$

it follows that

$$\sum \frac{1}{(2n-1)!} \mu^{(2n-1)} r^{2n-1} < \infty$$

and, hence,

$$\int e^{r|x|}\, dF(x) = \sum \frac{1}{n!} \mu^{(n)} r^n < \infty.$$

This proves the assertion.

b. *If $f(z)$ is regular in the circle $|z| < R$ or in the rectangle $|\Re z| < U$, $|\Im z| < R$, then $f(z)$ is regular in the strip $|\Im z| < R$.*

Proof. The first assertion follows at once from **a**. As for the second assertion, let V be the largest number such that $f(z)$ is regular in the circle $|z| < V$ and assume that $V < R$. According to **a**, $f(z)$ is regular in the strip $|\Im z| < V$. But it is also regular in the rectangle $|\Re z| < U$, $|\Im z| < R$ and, hence, in the circle whose radius equals min $(R, \sqrt{U^2 + V^2})$. Therefore V cannot be less than R and the proof is concluded.

For every ch.f. f, we have $f(z) = f^+(z) + f^-(z)$ where

$$f^+(z) = \int_0^\infty e^{izx}\, dF(x) \quad \text{and} \quad f^-(z) = \int_{-\infty}^0 e^{izx}\, dF(x)$$

are regular for $\Im z > 0$ and $\Im z < 0$, respectively. Therefore, if, say, $f^+(z)$ is regular for $0 > \Im z > -R$, then $f(z)$ is regular for $0 > \Im z > -R$, so that the ch.f. with values $f(x)$ is the boundary function of a regular function. Thus, the following proposition completes the proof of the foregoing extension theorem.

c. *$f^+(z)$ is regular for $0 > \Im z > -R$ if, and only if, for every positive $r < R$, $\int_0^\infty e^{rx}\, dF(x)$ is finite.*

Proof. The "if" assertion is immediate. As for the "only if" assertion, we observe that, since $f^+(z)$ is regular for $\Im z < 0$ and continuous on the real axis, regularity for $0 > \Im z > -R$ implies, by a well-known symmetry property, regularity for $|\Im z| < R$ and, hence, according to **a**, $\int_0^\infty e^{rx}\, dF(x)$ is finite for $0 < r < R$.

PARTICULAR CASES. Upon applying what precedes, we have

1° *If $f_n(u) \to e^{iua}$ on $(-U, +U)$, then $f_n(u) \to e^{iua}$ for every $u \in R$.*

2° *If $f_n(u) \to e^{-\frac{u^2}{2}}$ on $(-U, +U)$, then $f_n(u) \to e^{-\frac{u^2}{2}}$ for every $u \in R$.*

3° *If $f_n \to f$ on $(-U, +U)$ and f is ch.f. of a r.v. bounded either above or below, then $f_n \to f$ on R.*

***14.3 Composition and decomposition of regular ch.f.'s.** Let F denote the composed $F_1 * F_2$ of d.f.'s F_1 and F_2. In the case of f or f_1, f_2 regular, the composition theorem 12.4A can be completed as follows:

A. COMPOSITION THEOREM FOR REGULAR CH.F.'S. *$f(z)$ is regular in the strip $|\Im z| < R$ if, and only if, $f_1(z)$ and $f_2(z)$ are regular in $|\Im z| < R$.*

This theorem follows at once, by 14.2**a** and **b**, from the

Composition lemma. *If* $F = F_1 * F_2$ *then, for every* v,

$$\int e^{vx}\, dF(x) = \int e^{vx}\, dF_1(x) \int e^{vx}\, dF_2(x),$$

and there exist finite numbers $\alpha_j > 0$, $\beta_j \geqq 0$ *such that*

$$\int e^{vx}\, dF(x) \geqq \alpha_j e^{-\beta_j |v|} \int e^{vx}\, dF_j(x), \quad j = 1, 2.$$

Proof. We exclude the trivial case of degenerate F_1 or F_2. The first assertion follows, using Fatou's lemma, in a way similar to that of the proof of the composition theorem 12.3**A**, whether the integrals are finite or not.

As for the second assertion, for every b, either

$$\int e^{vx}\, dF_1(x) \geqq \int_b^{\infty} e^{vx}\, dF_1(x) \geqq e^{bv} F_1[b, +\infty)$$

or

$$\int e^{vx}\, dF_1(x) \geqq \int_{-\infty}^{b} e^{vx}\, dF_1(x) \geqq e^{bv} F_1(b),$$

according as $v \geqq 0$ or $v < 0$. Let β_2 be the larger of two finite numbers $|b_1|$ and $|b_2|$ such that

$$a_1 = F_1[b_1, +\infty) > 0 \quad \text{and} \quad a_2 = F_1(b_2) > 0$$

and let α_2 be the smaller of a_1 and a_2. Then the inequalities above and the first assertion yield

$$\int e^{vx}\, dF(x) \geqq \alpha_2 e^{-\beta_2 |v|} \int e^{vx}\, dF_2(x)$$

and the proof is complete.

COMPLEMENTS AND DETAILS

Unless otherwise stated, functions F, with or without affixes, are d.f.'s of r.v.'s: $F(-\infty) = 0$, $F(+\infty) = 1$, and functions f, with same affixes if any, are corresponding ch.f.'s.

1. If F is purely discontinuous and the discontinuity set is dense in R, then the nondecreasing inverse function is singular.

2. If $F_{X_n} \xrightarrow{c} F_X$ and μ is any limit point of the sequence $\mu(X_n)$ of medians of the X_n, then μ is a median of X. In particular, if $\mu(X)$ is the unique median of X, then $\mu(X_n) \to \mu(X)$. (Take $x' < \mu < x''$ to be continuity points of F, then $F(x') \leqq \frac{1}{2}$.)

3. P. Lévy's space. Let $\mathfrak{F}$ be the space of all d.f.'s F of r.v.'s. Set $d(F, F')$ to be the infimum of all those h for which $F(x - h) - h \leqq F'(x) \leqq F(x + h) + h$ whatever be $x \in R$.

(a) Draw a graph and interpret $d(F, F')$ geometrically by considering lengths of segments intercepted by the graphs of F and F' on parallels to the second bisector.

(b) The function d so defined is a distance, and $(\mathfrak{F}, d)$ is a complete metric space.

(c) The following three assertions are equivalent:

$$F_n \xrightarrow{c} F, \quad d(F_n, F) \to 0, \quad \int g\, dF_n \to \int g\, dF$$

for every function g continuous and bounded on R.

(d) A set S in $\mathfrak{F}$ is compact if, and only if, $F(x) \to 0$ as $x \to -\infty$ and $F(x) \to 1$ as $x \to +\infty$, uniformly on S.

4. Establish the following correspondences for laws.

Binomial: $p_k = C_n^k p^k q^{n-k}$, $k \leqq n$, $f(u) = (pe^{iu} + q)^n$.

Poissonian: $p_k = \dfrac{\lambda^k}{k!} e^{-\lambda}$, $k = 0, 1, \cdots$, $f(u) = e^{\lambda(e^{iu} - 1)}$.

Uniform: $F'(x) = \dfrac{1}{b - a}$ in (a, b), and 0 outside, $f(u) = \dfrac{e^{ibu} - e^{iau}}{i(b - a)u}$.

Cauchy: $F'(x) = \dfrac{1}{\pi} \dfrac{a}{a^2 + (x - b)^2}$, $a > 0$, $f(u) = e^{-a|u| + ibu}$.

Laplace: $F'(x) = \dfrac{1}{2a} e^{-|x-b|/a}$, $a > 0$, $f(u) = (1 + a^2u^2)^{-1} e^{ibu}$.

Normal: $F'(x) = \dfrac{1}{\sigma\sqrt{2\pi}} e^{-(x-m)^2/2\sigma^2}$, $\sigma > 0$, $f(u) = e^{imu - \frac{\sigma^2 u^2}{2}}$.

Squared Normal: $(m = 0, \sigma = 1)$: $F'(x) = \dfrac{1}{\sqrt{2\pi x}} e^{-x/2}$ for $x > 0$ and $= 0$ for $x \leqq 0$, $f(u) = (1 - 2iu)^{-1/2}$.

Γ-type: $F'(x) = \dfrac{c^\gamma}{\Gamma(\gamma)} x^{\gamma - 1} e^{-cx}$ for $x > 0$, $c > 0$, $\gamma > 0$, 0 for $x \leqq 0$, $f(u) = \left(1 - \dfrac{iu}{c}\right)^{-\gamma}$.

5. The composed $\bar{F}$ of F with the uniform distribution on $(-h, +h)$ is given by

$$\bar{F}(x) = \frac{1}{2h} \int_{x-h}^{x+h} F(y)\, dy, \quad \bar{f}(u) = \frac{\sin hu}{hu} f(u).$$

An absolutely convergent inversion integral follows:

$$\frac{1}{2h} \int_x^{x+2h} F(y)\, dy - \frac{1}{2h} \int_{x-2h}^{x} F(y)\, dy = \frac{1}{\pi} \int_{-\infty}^{\infty} \left(\frac{\sin u}{u}\right)^2 e^{-iux/h} f\left(\frac{u}{h}\right) du.$$

Deduce the continuity theorem.

6. Let $M_h f = \frac{2}{\pi h}\int_0^\infty |f(u)|^2 \frac{\sin^2 hu}{u^2}\,du$, $h > 0$, and let

$$Mf = \lim_{u\to\infty} \frac{1}{2u}\int_{-u}^{+u} |f(v)|^2\,dv.$$

(a) $M_h f$ is nondecreasing in h and converges to 1 or Mf according as $h \to \infty$ or $h \to 0$. $\lim_{m\to\infty}\lim_{n\to\infty} M_h(\prod_{k=m+1}^{n} f_k)$ is either 0 or 1 (identically in h).

(b) $Mf = \sum p_k^2$ where the p_k are jumps of F; $Mf_1f_2 \geqq Mf_1\cdot Mf_2$; $M_h f = 2\int_0^{2h}\left(1 - \frac{x}{2h}\right) dF^s(x)$ where F^s is d.f. with ch.f. $f^s = |f|^2$. (The sum is the jump at 0 of $\mathfrak{L}(X) * \mathfrak{L}(-X)$ where X is a r.v. with d.f. F.)

(c) If $f_n \to f$ with $Mf = 0$, then $Mf_n \to 0$; the converse is not necessarily true. If $\prod_{k=1}^{n} f_k \to f$, then $M(\prod_{k=1}^{n} f_k) \to Mf$.

7. A law is a "lattice" law if the only possible values are of form $a + ns$ only, $s > 0$; $n = 0, \pm 1, \cdots$; if s is the largest possible, then s is the "step" of the law. The step is well determined.

(a) A law is a lattice law if, and only if, $|f(u_0)| = 1$ for an $u_0 \neq 0$. The step s is given by the property that $|f(u)| < 1$ in $0 < |u| < 2\pi/s$ and $f(2\pi/s) = 1$.

(b) Let $p_n = P[X = a + ns]$ where X has a lattice law with step s. Then

$$p_n = \frac{s}{2\pi}\int_{-\pi/s}^{+\pi/s} e^{-iau - insu} f(u)\,du,$$

$$F(x_2) - F(x_1) = \frac{s}{2\pi}\int_{-\pi/s}^{+\pi/s} \frac{e^{-iux_1} - e^{-iux_2}}{2i\sin\frac{su}{2}} f(u)\,du$$

where $x_1 = a + ms - \frac{1}{2}s$, $x_2 = a + ns + \frac{1}{2}s$, $n \geqq m$.

8. If the moment m_k exists and is finite, then

$$\log f(u) = \sum_{k=1}^{n} \frac{a_k}{k!}(iu)^k + o(u^k).$$

The a_k are called semi-invariants; formally

$$\sum_{n=1}^{\infty} \frac{a_n}{n!} z^n = \log \sum_{n=0}^{\infty} \frac{m_n}{n!} z^n.$$

Deduce the expression of a few first semi-invariants in terms of moments, and conversely. Prove that

$$|a_k| \leqq k^k \mu_k.$$

($\log \sum_{k=1}^{n} \frac{m_k}{k!} z^k$ is majorized by $\sum_{k=1}^{\infty} \frac{1}{k}(e^{\mu_k^{1/k}z} - 1)^k$.)

9. If the derivative F' on R exists and is finite, then $f(u) \to 0$ as $|u| \to \infty$. (Use Riemann-Lebesgue lemma.)

If the nth derivative $F^{(n)}$ on R exists, is finite, and is absolutely integrable, then $f(u) = o(|u|^{1-n})$ for $|u| \to \infty$. (Integrate by parts.)

10. Let X be a r.v. with d.f. F.

(a) If $P[|X| \geqq x] \to 0$ as $x \to \infty$ faster than any power of x^{-1}, then all moments exist and are finite. (Integrate by parts $\int |x|^n \, dF(x)$.)

A pr. law is determined by the sequence of moments assumed finite if the series $\sum_{n=0}^{\infty} \frac{m_n}{n!} u^n$ has a nonnull radius of convergence ρ. (Use Schwarz's inequality to show that the series with the m_n replaced by μ_n majorizes the expansion of about any value of u, and then use analytic continuation.)

(b) Formally, by integration by parts,

$$f(z) = 1 - iz\int_{-\infty}^{0} e^{izx}F(x)\,dx + iz\int_{0}^{\infty} e^{izx}(1 - F(x))\,dx.$$

If $P[|X| \geqq x] \to 0$ as $x \to \infty$ faster than e^{-rx} for every positive $r < \rho$, then $f(z)$ is analytic in the strip $|\mathfrak{J}z| < \rho$. If $\rho = \infty$, then $f(z)$ is an entire function.

(c) If $e^{|x|^r}F'(x) \geqq c > 0$ on R for an $r < \frac{1}{2}$, then the pr. law is not determined by its moments.

11. If f' exists and is finite on R, $\int |x| \, dF(x)$ may be infinite: take $f(u) = c\sum_{n=2}^{\infty} \frac{\cos nu}{n^2 \log n}$. (The differentiated series converges uniformly but $\sum 1/n \log n = \infty$.) Let $m' = \lim_{a \to +\infty} \int_{-a}^{+a} x\,dF(x)$ be the "symmetric" first moment. If m' exists and is finite, $f'(0)$ may not exist: take a Weierstrass non-differentiable function $c \sum a^n \cos b^n u$.

If the derivative at $u = 0$ of $\mathfrak{R}f$ exists, then

$$\frac{f(h) - 1}{h} = o(1) + i\int_{-1/h}^{+1/h} x\,dF(x), \quad 0 < h \to 0.$$

(Set $G(x) = F(x) - G(x)$, $H(x) = F(x) + F(-x)$, so that $|\Delta H| \leqq \Delta G$. Show that $\int \frac{\sin^2 hx}{h}\,dG(x) \to 0$ as $h \to 0$, $\int_{1/h}^{\infty} \frac{\sin hxq}{x}\,dH(x) = o(1)$.)

Under the foregoing condition, f' exists and is finite if, and only if, $m' = \lim_{a \to \infty} \int_{-a}^{+a} x\,dF(x)$ exists and is finite, and then $f'(0) = im'$. Extend to any derivative of odd order. What about those of even order?

12. If g on R is not constant and $g(u) = 1 + o(u) + o(u^2)$ near $u = 0$ with $o(u)$ an odd function, then g is not a ch.f. (Observe that $g(u)g(-u) = 1 + o(u^2)$.) Examples: e^{-u^4}, $e^{-|u|^r}$ for $r > 2$, $e^{-u^4-u^6}$, $1/(1 + u^4)$.

13. Let g on R be real, even and continuous, with $g(0) = 1$, $g(u) \to 0$ as $u \to \infty$.

If g is convex from below, on $[0, +\infty)$, then g is a ch.f. (To prove $\int_0^{\infty} g(u) \cos xu\,du \geqq 0$ for $x > 0$; observe that on $[0, \infty)$, say, the left-hand side derivative g' exists and is nondecreasing, with $g'(u) \leqq 0$ and $g'(u) \to 0$ as $u \to \infty$. Set $h = -g'$ so that, by integration by parts,

$$x\int_0^\infty g(u)\cos xu\,du = \int_0^\infty h(u)\sin xu\,du.$$

For $x > 0$, the last integral is

$$\int_0^{\pi/x}\left\{h(u) - h\left(u+\frac{\pi}{x}\right) + h\left(u+\frac{2\pi}{x}\right) - h\left(u+\frac{3\pi}{x}\right)+\cdots\right\}\sin xu\,du \geqq 0.)$$

Examples: $e^{-|u|}$, $1/(1+|u|)$, $1-|u|$ for $|u| \leqq 1$ and 0 for $|u| > 1$.

14. (a) Two ch.f.'s may coincide on intervals without being identical.

Take $F'_1(x) = \dfrac{1-\cos x}{\pi x^2}$ hence $f_1(u) = 1 - |u|$ for $|u| \leqq 1$ and 0 for $|u| > 1$, and take F_2 defined by $p_0 = \frac{1}{2}$, $p_{\pm\pi(2k+1)} = \dfrac{2}{\pi^2(2k+1)^2}$; $f_2(u)$ is periodic of period two and coincides with f_1 on $[-1, +1]$. Or, take f to be a ch.f. of the type described in *13* with f' continuous and strictly increasing on $[0, \infty)$. Replace two arbitrarily small arcs of the graph of f which are symmetric with respect to the y-axis by their chords, and compare the function so defined with f.

(b) The compositions of a law with either one of two distinct laws may coincide ($f_1f_1 = f_1f_2$).

(c) If $f_n \to f$ on $[-U, +U]$, the same may not be true on R.

15. f on R is a ch.f. if, and only if, there exists a sequence g_n such that $\int |g_n(v)|^2\,dv \to 1$ and $\int g_n(u+v)\bar{g}_n(v)\,dv \to f(u)$ uniformly in every finite interval.

(For the "if" assertion, observe that every integral is positive-definite. For the "only if" assertion, divide $[-n, +n]$ into n^2 equal subintervals, set $F_n(-n) = 0$, $F_n(n) = 1$, $F_n = F$ at the subdivision points, and linear inside every subinterval; set $c_ng_n(u) = \int_{-n}^{+n} \sqrt{F'_n(x)}\,e^{iux}\,dx$ with $g_n(0) = 1$. Compute f_n and observe that $f_n \to f$.)

16. Let g and h be bounded and continuous on R, with $\bar{g}(u) = g(-u)$, and let $\lambda(u)$ be an arbitrary finite function on R.

If for every finite set A of values of u

$$\left|\sum_{u\in A}\sum_{v\in A} g(u-v)\lambda(u)\bar{\lambda}(v)\right| \leqq \sum_{u\in A}\sum_{v\in A} h(u-v)\lambda(u)\bar{\lambda}(v),$$

then

$$h(u) = \int e^{iux}\,dH(x)$$

where H is a d.f. up to a multiplicative constant.

The foregoing inequalities represent a necessary and sufficient condition for g to be of the form

$$g(u) = \int e^{iux}\,dG(x)$$

with $|\Delta G| \leqq \Delta H$. Find the relation between discontinuity and continuity points of G and H.

17. The uniqueness and composition properties determine "essentially" the form of ch.f.'s. Let K on $R \times R$ be bounded and continuous. If the functions g on R are defined by $g(u) = \int K(u, x)\, dF(x)$ for every d.f. F, and the uniqueness and composition properties hold, then $K(u, x) = e^{ixh(u)}$ and $f(h(u)) = g(u)$.

18. Normal vectors. A normal vector $X = (X_k,\, k \leqq n)$ is so defined that all r.v.'s of the form $\sum_k u_k X_k$ are normal. Let the X_k be centered at expectations.

A ch.f. f on R^n is that of a normal vector (centered at its expectation) if, and only if,

$$\log f(u_1, \cdots, u_n) = Q(u_1, \cdots, u_n) = -\tfrac{1}{2} \sum_{jk} m_{jk} u_j u_k \geqq 0$$

where $m_{jk} = EX_j X_k$.

If the inequality is strict, then the normal d.f. is defined by

$$\frac{\partial^n}{\partial_{x_1} \cdots \partial_{x_n}} F(x_1, \cdots, x_n) = \frac{1}{(2\pi)^{n/2} D^{1/2}} e^{-\frac{1}{2} q(x_1, \cdots, x_n)}$$

where $D = \| m_{jk} \| > 0$ and $q(x_1, \cdots, x_n) = \frac{1}{D} \sum_{jk} D_{jk} x_j x_k$ is the reciprocal form of $Q(u_1, \cdots, u_n)$ with the variables x_k. What if $Q \geqq 0$?

19. If (X, Y) is a normal pair centered at expectations, then $EXY/\sigma X \sigma Y = \cos p\pi$ where $p = P[XY < 0]$. (Compute $P[XY < 0]$ using the d.f.)

Part Three

INDEPENDENCE

Until very recently, probability theory could have been defined to be the investigation of the concept of independence. This concept continues to provide new problems. Also it has originated and continues to originate most of the problems where independence is not assumed.

The main model is that of sequences of sums of independent random variables. The main problems are the Strong Central Limit Problem and the (Laws) Central Limit Problem. The first is concerned with almost sure convergence and stability properties. The second one is concerned with convergence of laws. All general results were obtained since 1900.

Chapter V

SUMS OF INDEPENDENT RANDOM VARIABLES

Two properties play a basic role in the study of independent r.v.'s: the Borel zero-one law and the multiplication theorem for expectations. Two general a.s. limit problems for sums of independent r.v.'s have been investigated: the a.s. convergence problem and the a.s. stability problem. Both of them took their present form in the second quarter of this century.

§ 15. CONCEPT OF INDEPENDENCE

CONVENTION. To avoid endless repetitions, we make the convention that, unless otherwise stated,

—r.v.'s, random vectors and, in general, random functions are defined on a fixed but otherwise arbitrary pr. space $(\Omega, \mathcal{A}, P)$.

—indices t vary on a fixed but otherwise arbitrary index set T, and events of a class have the index of the class.

15.1 Independent classes and independent functions. Events A_t are said to be *independent* if, for every finite subset $(t_1, \cdots, t_n)$,

$$\text{(I)} \qquad P\bigcap_{k=1}^{n} A_{t_k} = \prod_{k=1}^{n} PA_{t_k}.$$

In fact, the concept of independence is relative to families of classes (see Application 1° below).

Classes $\mathcal{C}_t$ of events are said to be *independent* if their events are independent; in other words, if events selected arbitrarily one from each class are independent. Clearly, if the $\mathcal{C}_t$ are independent so are the $\mathcal{C}'_{t'} \subset \mathcal{C}_{t'}$, $t' \in T' \subset T$. Because of its constant use, we state this fact as a theorem.

A. *Subclasses of independent classes are independent.*

Let X_t be r.v.'s or random vectors or, in general, random functions. Let $\mathcal{B}(X_t)$ be the sub σ-field of events induced by X_t, that is, the inverse image under X_t of the Borel field in the range space of X_t.

The X_t are said to be *independent* if they induce independent σ-fields $\mathcal{B}(X_t)$. Then classes $\mathcal{B}_t \subset \mathcal{B}(X_t)$ are independent. Since a Borel function of X_t induces a sub σ-field $\mathcal{B}_t$ of events contained in $\mathcal{B}(X_t)$, it follows that

A′. BOREL FUNCTIONS THEOREM. *Borel functions of independent random functions are independent.*

Independent classes can be enlarged, to some extent, without destroying independence. More precisely

Let $\mathcal{C}_t$ be independent classes. Independence is preserved if to every $\mathcal{C}_t$ we adjoin

1° *the null and the a.s. events*, for (I) is trivially true—both sides reducing to 0—when at least one of the events which figure in it is null, while (I) with n indices reduces to (I) with fewer indices when at least one of the events which figure in it is a.s.;

2° *the proper differences of its elements and, in particular, their complements* (because of 1°), for if $A_{t_1} \supset A'_{t_1}$, then

$$\begin{aligned} P(A_{t_1} - A'_{t_1})A_{t_2} \cdots A_{t_n} &= PA_{t_1}A_{t_2} \cdots A_{t_n} - PA'_{t_1}A_{t_2} \cdots A_{t_n} \\ &= (PA_{t_1} - PA'_{t_1})PA_{t_2} \cdots PA_{t_n} \\ &= P(A_{t_1} - A'_{t_1})PA_{t_2} \cdots PA_{t_n}; \end{aligned}$$

3° *the countable sums of its elements*, for

$$\begin{aligned} P(\sum_j A_{t_1}^j)A_{t_2} \cdots A_{t_n} &= \sum_j PA_{t_1}^j A_{t_2} \cdots A_{t_n} \\ &= (\sum_j PA_{t_1}^j)PA_{t_2} \cdots PA_{t_n} \\ &= P(\sum_j A_{t_1}^j)PA_{t_2} \cdots PA_{t_n}; \end{aligned}$$

4° *the limits of sequences of its elements*, for if $A_{t_1}^m \to A_{t_1}$ as $m \to \infty$, then

$$\begin{aligned} PA_{t_1}A_{t_2} \cdots A_{t_n} &\leftarrow PA_{t_1}^m A_{t_2} \cdots A_{t_n} \\ &= PA_{t_1}^m PA_{t_2} \cdots PA_{t_n} \to PA_{t_1}PA_{t_2} \cdots PA_{t_n}. \end{aligned}$$

It follows easily that

B. Extension theorem. *Minimal σ-fields over independent classes $\mathcal{C}_t$ closed under finite intersections are independent.*

Applications. 1° If the events A_t are independent, so are the σ-fields $(A_t, A_t^c, \emptyset, \Omega)$.

2° If the inverse images $\mathcal{C}_t$ of the classes of all intervals $(-\infty, x_t)$ in Borel spaces R_t are independent, so are the inverse images $\mathcal{B}_t$ of the Borel fields in the R_t. For, every $\mathcal{C}_t$ is closed under finite intersections and $\mathcal{B}_t$ is the minimal σ-field over $\mathcal{C}_t$.

*3° Let $\mathcal{B}_t$ be σ-fields (or fields) of events and let T_s be a subset of the index set T. The compound σ-field $\mathcal{B}_{T_s}$ with components $\mathcal{B}_t$, $t \in T_s$, is the minimal σ-field over the class $\mathcal{C}_{T_s}$ of all finite intersections of events A_t, $t \in T_s$, and contains all its components; since the $\mathcal{B}_t$ are closed under finite intersections so is $\mathcal{C}_{T_s}$. $\mathcal{B}_{T_s}$ is a compound sub σ-field of $\mathcal{B}_T$ and, if T_s is finite, then $\mathcal{B}_{T_s}$ is a "finitely compound" sub σ-field.

If compound σ-fields are independent, then, by **A**, their finitely compound sub σ-fields are independent. Conversely, if the finitely compound sub σ-fields are independent, then, by the extension theorem, the compound σ-fields are independent. We state these facts as a theorem.

C. Compounds theorem. *Compound σ-fields are independent if, and only if, their finitely compound sub σ-fields are independent.*

In particular, if the $\mathcal{B}_t$ are independent, so are the $\mathcal{B}_{T_s}$ for every partition of T into set T_s.

Families $X_{T_s} = \{X_t, t \in T_s\}$ of r.v.'s induce sub σ-fields $\mathcal{B}(X_{T_s})$ of events. Every $\mathcal{B}(X_{T_s})$ is the minimal σ-field over the class $\mathcal{C}(X_{T_s})$ of inverse images of all intervals in the range space R_{T_s} of X_{T_s}. But the intervals in the Borel space R_{T_s} are products of intervals in the factor spaces R_t, with only a finite number of factor intervals different from the whole factor spaces, and the inverse image of any factor space is Ω. Therefore the elements of $\mathcal{C}(X_{T_s})$ are all the finite intersections of elements of the $\mathcal{B}(X_t)$. It follows that the σ-field $\mathcal{B}(X_{T_s})$ is a compound of the σ-fields $\mathcal{B}(X_t)$, and theorem **C** becomes

C′. Families theorem. *Families of random variables are independent if, and only if, their finite subfamilies are mutually independent.*

Thus, in the last analysis, independence of random functions reduces to independence of random vectors.

To conclude this investigation of the definition of independence, let us observe that all which precedes applies to complex r.v.'s, to complex random vectors, and, in general, to complex random functions $X_t = X'_t + iX''_t$ considered as vector random functions (X'_t, X''_t), $t \in T$.

15.2 Multiplication properties. The direct definition of independent r.v.'s is as follows:

Random variables X_t, $t \in T$, are *independent* if, for every finite class $(S_{t_1}, \cdots, S_{t_n})$ of Borel sets in R,

$$P \bigcap_{k=1}^{n} [X_{t_k} \in S_{t_k}] = \prod_{k=1}^{n} P[X_{t_k} \in S_{t_k}].$$

The basic expectation property of independent r.v.'s is expressed by

a. Multiplication lemma. *If $X_1, \cdots, X_n$ are independent nonnegative r.v.'s, then $E \prod_{k=1}^{n} X_k = \prod_{k=1}^{n} EX_k$.*

Proof. It suffices to prove the assertion for two independent r.v.'s X and Y, for then the general case follows by induction. First, let $X = \sum_j x_j I_{A_j}$ and $Y = \sum_k y_k I_{B_k}$ be nonnegative simple (or elementary) r.v.'s; we can always take the x_j, and, similarly, the y_k, to be all distinct, so that $A_j = [X = x_j]$, $B_k = [Y = y_k]$. Since X and Y are independent, $PA_jB_k = PA_jPB_k$ and, hence,

$$EXY = \sum_{j,k} x_j y_k PA_j PB_k = \sum_j x_j PA_j \cdot \sum_k y_k PB_k = EXEY.$$

Now, let X and Y be nonnegative r.v.'s and set

$$A_{nj} = \left[\frac{j-1}{2^n} \leqq X < \frac{j}{2^n}\right], \quad B_{nk} = \left[\frac{k-1}{2^n} \leqq Y < \frac{k}{2^n}\right].$$

Since X and Y are independent so are these events and, hence, so are the simple r.v.'s

$$X_n = \sum_{j=1}^{n2^n} \frac{j-1}{2^n} I_{A_{nj}}, \quad Y_n = \sum_{k=1}^{n2^n} \frac{k-1}{2^n} I_{B_{nk}}.$$

But $0 \leqq X_n \uparrow X$, $0 \leqq Y_n \uparrow Y$, so that $0 \leqq X_n Y_n \uparrow XY$ and, by what precedes, $EX_nY_n = EX_nEY_n$. Therefore, by the monotone convergence theorem, $EXY = EXEY$, and the lemma is proved.

A. Multiplication theorem. *Let $X_1, \cdots, X_n$ be independent r.v.'s. If these r.v.'s are integrable so is their product, and $E \prod_{k=1}^{n} X_k = \prod_{k=1}^{n} EX_k$. Conversely, if their product is integrable and none is degenerate at 0, then they are integrable.*

Proof. It suffices to prove the assertion for two independent r.v.'s X and Y. We observe that independence of X and Y implies that the nonnegative r.v.'s $X' = X^+$ or X^- or $|X|$ and $Y' = Y^+$ or Y^- or $|Y|$ are independent, so that, by **a**, $EX'Y' = EX'EY'$. Now, if X and Y are integrable so are X' and Y' and, by the foregoing equality, so is $X'Y'$. Therefore $|XY|$ and hence XY are integrable and, by the same equality,

$$\begin{aligned} EXY &= E(X^+ - X^-)(Y^+ - Y^-) \\ &= EX^+EY^+ - EX^+EY^- - EX^-EY^+ + EX^-EY^- \\ &= EXEY. \end{aligned}$$

Conversely, if XY is integrable so that $E|X|E|Y| = E|XY| < \infty$, and neither X nor Y degenerates at 0 so that $E|X|$ and $E|Y|$ do not vanish, then $E|X|$ and $E|Y|$ are finite, and the proof is concluded.

Extension. The multiplication theorem remains valid for independent complex r.v.'s $X_k = X'_k + iX''_k$, since it applies to every term of the expansion of $\prod_{k=1}^{n} (X'_k + iX''_k)$. In particular, according to the Borel functions theorem, if the X_k are independent so are the e^{iuX_k} and, hence,

$$Ee^{iu\sum_{k=1}^{n} X_k} = E\prod_{k=1}^{n} e^{iuX_k} = \prod_{k=1}^{n} Ee^{iuX_k}.$$

In other words,

COROLLARY. *Ch.f.'s of sums of independent r.v.'s are products of ch.f.'s of the summands.*

This proposition, to be used extensively in the following chapter, is but a special case of a property which can serve as an equivalent definition of independent r.v.'s, as follows:

Let F_t and f_t, $F_{t_1\cdots t_n}$ and $f_{t_1\cdots t_n}$ be the d.f.'s and ch.f.'s of the r.v. X_t and of the random vector $(X_{t_1} \ldots X_{t_n})$, respectively.

B. EQUIVALENCE THEOREM. *The three following definitions of independence of the r.v.'s X_t are equivalent.*

For every finite class of Borel sets S_t and of points x_t, $u_t \in R$

$$\text{(I}_1\text{)} \qquad P\bigcap_{k=1}^{n} [X_{t_k} \in S_{t_k}] = \prod_{k=1}^{n} P[X_{t_k} \in S_{t_k}],$$

$$\text{(I}_2\text{)} \qquad F_{t_1\cdots t_n}(x_{t_1}, \cdots, x_{t_n}) = F_{t_1}(x_{t_1}) \cdots F_{t_n}(x_{t_n}),$$

$$\text{(I}_3\text{)} \qquad f_{t_1\cdots t_n}(u_{t_1}, \cdots, u_{t_n}) = f_{t_1}(u_{t_1}) \cdots f_{t_n}(u_{t_n}).$$

Proof. (I_1) implies (I_2) by taking $S_t = (-\infty, x_t)$. Conversely, (I_2) implies (I_1) with $S_t = (-\infty, x_t)$ and, on account of 15.1, Application 2°, this implies (I_1) for all S_t.

(I_2) implies (I_3), for (I_2) implies (I_1) which implies (I_3) exactly as the multiplication theorem implies its corollary. Conversely, (I_3) implies (I_2), for the inversion formula for one- and multi-dimensional ch.f.'s shows at once that if (I_3) is true, then, for all continuity intervals,

$$F_{t_1\cdots t_n}[a_{t_1}, \cdots, a_{t_n}; b_{t_1}, \cdots, b_{t_n}) = F_{t_1}[a_{t_1}, b_{t_1}) \cdots F_{t_n}[a_{t_n}, b_{t_n}),$$

and (I_2) follows by letting the $a_t \to -\infty$ and $b_t \uparrow x_t$. This completes the proof.

Extension. The equivalence theorem is valid when the X_t are random vectors, for the proof applies word by word, provided R is replaced by the range space R_t of X_t.

15.3 Sequences of independent r.v.'s. At the root of known a.s. limit properties of sequences of independent r.v.'s lies the celebrated

A. BOREL ZERO-ONE CRITERION. *If the events A_n are independent, then* $P(\limsup A_n) = 0$ *or* 1 *according as* $\sum PA_n < \infty$ *or* $= \infty$.

Proof. Since

$$P(\limsup A_n) = \lim_m \lim_n P \bigcup_{k=m}^{n} A_k = \lim_m \lim_n \left(1 - P \bigcap_{k=m}^{n} A_k^c\right)$$

and the events A_n and hence A_n^c are independent, the assertion follows by passing to the limit in the elementary inequality

$$1 - \exp\left[-\sum_{k=m}^{n} PA_k\right] \leqq 1 - \prod_{k=m}^{n} (1 - PA_k) \leqq \sum_{k=m}^{n} PA_k.$$

Since, whatever be the events A_n, $\sum PA_n < \infty$ implies that

$$\lim_m \lim_n P \bigcup_{k=m}^{n} A_k \leqq \lim_m \lim_n \sum_{k=m}^{n} PA_k = 0,$$

the "zero" part of this criterion is valid *with no assumption of independence:*

a. BOREL-CANTELLI LEMMA. *If* $\sum PA_n < \infty$, *then* $P(\limsup A_n) = 0$.

COROLLARY 1. *If the events A_n are independent and $A_n \to A$, then* $PA = 0$ *or* 1.

COROLLARY 2. *If the r.v.'s X_n are independent and $X_n \xrightarrow{\text{a.s.}} 0$, then* $\sum P[|X_n| \geqq c] < \infty$ *whatever be the finite number* $c > 0$.

For $X_n \xrightarrow{\text{a.s.}} 0$ implies that, if $A_n = [|X_n| \geqq c]$, then $P(\limsup A_n) = 0$, and independence of the X_n implies that of the A_n.

Because of its intuitive appeal, instead of "lim sup A_n" we shall sometimes write "A_n i.o."; to be read "A_n's occur infinitely often" or "infinitely many A_n occur." This terminology corresponds to the fact that lim sup A_n is the set of all those elementary events which belong to infinitely many A_n or, equivalently, to some of "the $A_n, A_{n+1}, \cdots$ however large be n"—the "tail" of the sequence A_n. To the "tail" of the sequence A_n of events corresponds the "tail" of the sequence I_{A_n} of their indicators. More generally, the "tail" of a sequence X_n of r.v.'s is "the sequence $X_n, X_{n+1}, \cdots$ however large be n."

To be precise, let $X_1, X_2, \cdots$ be a sequence of r.v.'s and let $\mathcal{B}(X_n)$, $\mathcal{B}(X_n, X_{n+1}), \cdots, \mathcal{B}(X_n, X_{n+1}, \cdots), \mathcal{B}(X_{n+1}, X_{n+2}, \cdots), \cdots$ be sub σ-fields of events induced by the random functions within the brackets. We give a precise meaning to lim sup $\mathcal{B}(X_n)$, as follows: The sequence $\mathcal{B}(X_n), \mathcal{B}(X_n, X_{n+1}), \cdots$ is a nondecreasing sequence of σ-fields, its supremum or union is a field, and the minimal σ-field over this field is $\mathcal{B}(X_n, X_{n+1}, \cdots)$ or, writing loosely, "$\sup\limits_{m \geqq n} \mathcal{B}(X_m)$." In turn, the sequence $\mathcal{B}(X_n, X_{n+1}, \cdots), \mathcal{B}(X_{n+1}, X_{n+2}, \cdots), \cdots$ is a nonincreasing sequence of σ-fields and its limit or intersection is a σ-field $\mathcal{C}$ contained in $\mathcal{B}(X_n, X_{n+1}, \cdots)$ however large be n or, writing loosely, "lim sup $\mathcal{B}(X_n)$." The σ-field $\mathcal{C}$ will be called the *tail σ-field* of the sequence X_n or "the sub σ-field of events induced by the tail of the sequence X_n." Let us observe that all the foregoing σ-fields and, in particular, the tail σ-field, are contained in the σ-field $\mathcal{B}(X_1, X_2, \cdots)$ induced by the whole sequence X_n. The elements of the tail σ-field $\mathcal{C}$ are *tail events* and the numerical (finite or not) $\mathcal{C}$-measurable functions, that is, those functions which induce sub σ-fields of events contained in $\mathcal{C}$ are *tail functions*—they are defined on the "tail" of the sequence. For example, the limits inferior and superior of the sequence X_n and of the sequence $(X_1 + X_2 + \cdots + X_n)/b_n$, where $b_n \to \infty$, are tail functions (not necessarily finite), while the sets of convergence of these sequences, as well as the set of convergence of the series $\sum X_n$, are tail events.

To Borel's result corresponds the basic Kolmogorov's

B. ZERO-ONE LAW. *On a sequence of independent r.v.'s, the tail events have for pr. either* 0 *or* 1 *and the tail functions are degenerate.*

In other words, the tail σ-field of a sequence of independent r.v.'s is equivalent to $\{\emptyset, \Omega\}$.

Proof. We observe that an event A is independent of itself if, and only if, $PAA = PA \cdot PA$, that is, if $PA = 0$ or 1—and such events are mutually independent. Thus, the first assertion means that the tail σ-field $\mathcal{C}$ of the sequence X_n of independent r.v.'s is independent of itself. Since $\mathcal{C} \subset \mathcal{B}(X_{n+1}, X_{n+2}, \cdots)$ whatever be n and, because of the independence assumption, $\mathcal{B}(X_1, \cdots, X_n)$ is independent of $\mathcal{B}(X_{n+1}, X_{n+2}, \cdots)$, it follows that $\mathcal{C}$ is independent of $\mathcal{B}(X_1, X_2, \cdots, X_n)$ whatever be n. Therefore, $\mathcal{C}$ is independent of $\mathcal{B}(X_1, X_2, \cdots)$ and, being contained in $\mathcal{B}(X_1, X_2, \cdots)$, it is independent of itself. This proves the first assertion and the second follows, since, if X is a tail function, then it is a.s. $\{\emptyset, \Omega\}$-measurable hence degenerates.

COROLLARY. *If X_n are independent r.v.'s, then the sequence X_n either converges a.s. or diverges a.s.; and similarly for the series $\sum X_n$. Moreover, the limits of the sequences X_n and $(X_1 + \cdots + X_n)/b_n$ where $b_n \uparrow \infty$, are degenerate.*

***15.4 Independent r.v.'s and product spaces.** Let X_t, where t runs over an index set T, be independent r.v.'s with d.f.'s F_{X_t} on R_t. Because of the correspondence theorem, every F_{X_t} determines a pr. P_{X_t} on the Borel field $\mathcal{B}_t$ in R_t. On account of the product-measure theorem, the P_{X_t} determine a product-measure $\prod P_{X_t}$ on the product Borel field $\prod \mathcal{B}_t$ in the product space $\prod R_t$. On the other hand, the law of the family $X = \{X_t, t \in T\}$, represented by the family of d.f.'s $\{F_{X_{t_1}, \cdots, X_{t_N}}\}$ of all finite subfamilies of X determines, by the correspondence theorem, a family $\{P_{X_{t_1}, \cdots, X_{t_N}}\}$ of consistent measures on the product Borel fields $\prod_{k=1}^{N} \mathcal{B}_{t_k}$. Owing to the consistent measures theorem, this family of pr.'s determines a pr. P_X on $\prod \mathcal{B}_t$.

Since the X_t are independent,

$$F_{X_{t_1} \cdots X_{t_N}} = F_{X_{t_1}} \times \cdots \times F_{X_{t_N}}$$

so that

$$P_{X_{t_1} \cdots X_{t_N}} = P_{X_{t_1}} \times \cdots \times P_{X_{t_N}}$$

and, therefore, P_X coincides with $\prod P_{X_t}$. In other words,

A. *The pr. space induced on its range space by a family of independent r.v.'s is the product of pr. spaces induced on their respective range spaces by the r.v.'s of the family.*

Let us observe that this reduces the multiplication theorem to the Fubini theorem.

The question arises whether the converse is true: Given a product pr. space $(\prod R_t, \prod \mathcal{B}_t, \prod P_t)$, is there a family $\{X_t, t \in T\}$ of independent r.v.'s on some pr. space $(\Omega, \mathcal{A}, P)$ which induces this product pr. space? Equivalently, given a family $\{F_t, t \in T\}$ of d.f.'s with variation 1, is there a family $\{X_t, t \in T\}$ of independent r.v.'s with $F_{X_t} = F_t$?

If the pr. space on which the r.v.'s have to be defined is fixed, then, in general, the answer is in the negative, since on a fixed pr. space even one r.v. with a given d.f. might not exist. However, if we are at liberty to select the pr. space on which to define r.v.'s, and *we shall always do so*, then the answer is in the affirmative, as follows:

Let the pr. space be the product pr. space $(\prod R_t, \prod \mathcal{B}_t, \prod P_t)$ where, if the F_t are given, the P_t are determined upon applying the correspondence theorem. The r.v.'s X_t, defined on this pr. space by $X_t(x) = x_t$, $x = \{x_t, t \in T\}$, are then independent, since their pr.d.'s are P_t and their d.f.'s are F_t. Thus

B. *The relation $X_t(x) = x_t$, $x = \{x_t, t \in T\}$ establishes a one-to-one correspondence between families $\{X_t\}$ of independent r.v.'s and product pr. spaces on $\prod R_t$.*

REMARK. There exist pr. spaces on which can be defined all possible families of independent r.v.'s with a given index set T. For example, take the pr. space $(\Omega, \mathcal{A}, P)$ where $\Omega = \prod \Omega_t$ with $\Omega_t = (0, 1)$ and $P = \prod P_t$ on the Borel field $\mathcal{A}$ in Ω, with P_t being the Lebesgue measure on the Borel field in Ω_t (class of Borel sets in Ω_t). Then the r.v.'s X_t—inverse functions of arbitrarily given d.f.'s F_t—are independent and $F_{X_t} = F_t$.

Extension. The preceding considerations apply, word for word, to random vectors. They also apply to arbitrary random functions, provided we consider that the d.f. of a random function is defined in terms of its "finite sections," that is, the family of d.f.'s of projections of the random function on finite subspaces.

§ 16. CONVERGENCE AND STABILITY OF SUMS; CENTERING AT EXPECTATIONS AND TRUNCATION

This section and the following one are devoted to the investigation of sums $S_n = \sum_{k=1}^{n} X_k$ of independent r.v.'s $X_1, X_2, \cdots$ and, especially, of their limit properties—convergence to r.v.'s and stability.

Given two numerical sequences a_n and $b_n \uparrow \infty$, we say that the sequence S_n is *stable* in pr. or a.s. if $\frac{S_n}{b_n} - a_n \xrightarrow{P} 0$ or $\frac{S_n}{b_n} - a_n \xrightarrow{\text{a.s.}} 0$. In fact, a stability property is at the root of the whole development of pr. theory. If $X_1, X_2, \cdots$ are independent and identically distributed indicators with $P[X_n = 1] = p$ and $P[X_n = 0] = q = 1 - p$, we have the *Bernoulli case.* The first stability property is the

BERNOULLI LAW OF LARGE NUMBERS: *In the Bernoulli case* $\frac{S_n}{n} - p \xrightarrow{P} 0.$

The Central Limit Problem, to which the following chapter is devoted, is the direct descendant of its sharpening by de Moivre and by Laplace. On the other hand, the following strengthening

BOREL STRONG LAW OF LARGE NUMBERS: *In the Bernoulli case*

$$\frac{S_n}{n} - p \xrightarrow{\text{a.s.}} 0,$$

is at the origin of the results given in this chapter. Perhaps the importance of the methods overwhelms that of the results and emphasis will be laid upon the methods. These methods are (1) centering at expectations and truncation and (2) centering at medians and symmetrization.

16.1 Centering at expectations and truncation. We say that we *center* X at c if we replace X by $X - c$. If X is integrable, then we can center it at its expectation EX and, thus, X is replaced by $X - EX$. In other words, a r.v. is *centered at its expectation* if, and only if, its expectation exists and equals 0.

Let X be integrable. The second moment of $X - EX$ is called *variance* of X; it exists but may be infinite and will be denoted by $\sigma^2 X$. Thus

$$\sigma^2 X = E(X - EX)^2 = EX^2 - (EX)^2.$$

Since, for every finite c, we have

$$\sigma^2(X - c) = E(X - c - E(X - c))^2 = E(X - EX)^2,$$

centerings do not modify variances.

The importance of variances is due to the fact that we have at our disposal bounds, in terms of variances of summands, of pr.'s of events defined in terms of sums S_n of independent r.v.'s; we shall find and use such bounds in this section. However, variances can be introduced only when the summands are integrable. Moreover, the bounds mentioned above are nontrivial only when the variances are finite. This seems to limit the use of such bounds to square-integrable summands. Yet this obstacle can be overcome by means of the *truncation method.*

We *truncate* X at $c > 0$ (finite) when we replace X by $X^c = X$ or 0 according as $|X| < c$ or $|X| \geqq c$, and X^c is *X truncated at c.* It follows that, if F is the d.f. of X, then all moments of X^c

$$EX^c = \int_{|x|<c} x\,dF, \quad E(X^c)^2 = \int_{|x|<c} x^2\,dF, \text{ etc.,}$$

exist and are finite. We can always select c sufficiently large so as to make $P[X \neq X^c] = P[|X| \geqq c]$ arbitrarily small. Furthermore, we can always select the c_j sufficiently large so as to make $P \bigcup [X_j \neq X_j^{c_j}]$ arbitrarily small, since, given $\epsilon > 0$, we have

$$P \bigcup [X_j \neq X_j^{c_j}] \leqq \sum P[|X_j| \geqq c_j] < \epsilon$$

if, say, the c_j are selected so as to make $P[|X_j| \geqq c_j] < \frac{\epsilon}{2^j}$. Thus, to every countable family of r.v.'s we can make correspond a family of bounded r.v.'s which differs from the first on an event of arbitrarily small pr. Moreover, if we are interested primarily in limit properties there is no need for arbitrarily small pr., for the following reasons.

Let two sequences X_n and X'_n of r.v.'s be called *tail-equivalent* if they differ a.s. only by a finite number of terms; in other words, if for a.e. $\omega \in \Omega$ there exists a finite number $n(\omega)$ such that for $n \geqq n(\omega)$ the two sequences $X_n(\omega)$ and $X'_n(\omega)$ are the same; in symbols $P[X_n \neq X'_n \text{ i.o.}] = 0$. If the sequences X_n and X'_n only converge on the same event except for a null subset, then we say that they are *convergence equivalent.*

Let $S_n = \sum_{k=1}^{n} X_k$ and $S'_n = \sum_{k=1}^{n} X'_k$. Since

$$P[X_n \neq X'_n \text{ i.o.}] = \lim_n P \bigcup_{k=n}^{\infty} [X_k \neq X'_k] \leqq \lim_n \sum_{k=n}^{\infty} P[X_k \neq X'_k]$$

it follows that

a. EQUIVALENCE LEMMA. *If the series $\sum P[X_n \neq X'_n]$ converges, then the sequences X_n and X'_n are tail-equivalent and, hence, the series $\sum X_n$ and $\sum X'_n$ are convergence-equivalent and the sequences $\frac{S_n}{b_n}$ and $\frac{S'_n}{b_n}$, where $b_n \uparrow \infty$, converge on the same event and to the same limit, excluding a null event.*

16.2 Bounds in terms of variances. To avoid repetitions, we make the convention that, unless otherwise stated, $S_0 = 0$, $S_n = \sum_{k=1}^{n} X_k$, $n = 1, 2, \cdots$, and the summands $X_1, X_2, \cdots$ are independent r.v.'s.

Let $X_1, X_2, \cdots$, be integrable. Since centerings do not modify the variances, we can assume, when computing variances, that these r.v.'s are centered at expectations. Then

$$\sigma^2 S_n = ES_n^{\ 2} = \sum_{k=1}^{n} EX_k^{\ 2} + \sum_{j\neq k=1}^{n} EX_jX_k = \sum_{k=1}^{n} \sigma^2 X_k,$$

since independence of X_j and X_k entails, by 15.2,

$$EX_jX_k = EX_j \cdot EX_k = 0.$$

Thus, we obtain the classical

BIENAYMÉ EQUALITY. *If the r.v.'s X_n are independent and integrable, then*

$$\sigma^2 S_n = \sum_{k=1}^{n} \sigma^2 X_k.$$

The basic inequalities 9.3**A** become

a. $$\frac{\sum_{k=1}^{n} \sigma^2 X_k - \epsilon^2}{\text{a.s. sup } (S_n - ES_n)^2} \leqq P[|S_n - ES_n| \geqq \epsilon] \leqq \frac{1}{\epsilon^2} \sum_{k=1}^{n} \sigma^2 X_k.$$

The right-hand side inequality is the celebrated BIENAYMÉ-TCHEBICHEV INEQUALITY. Applied to $(S_{n+k} - ES_{n+k}) - (S_n - ES_n)$ and to $S_n - ES_n$ with ϵ replaced by ϵb_n, it yields, by passage to the limit,

b. *If the series $\sum \sigma^2 X_n$ converges, then the series $\sum (X_n - EX_n)$ converges in pr. If $\frac{1}{b_n^{\ 2}} \sum_{k=1}^{n} \sigma^2 X_k \rightarrow 0$, then $\frac{S_n - ES_n}{b_n} \xrightarrow{P} 0$.*

This last property is due to Tchebichev (when $b_n = n$). In the Bernoulli case, where $b_n = n$, $EX_n = p$, $\sigma^2 X_n = pq$, it reduces to the Bernoulli law of large numbers. It is of some interest to observe that Borel's strengthening can also be obtained by means of the Bienaymé-Tchebichev inequality (see Introductory part).

So far, the assumption of independence was used only to establish that the summands were *orthogonal*, that is, $EX_jX_k = 0 (j \neq k)$ when

X_j and X_k are centered at expectations. In fact, the foregoing results remain valid under the even less restrictive assumption of orthogonality of S_{n-1} and X_n, $n = 1, 2, \cdots$, since, then,

$$\sigma^2 S_n = \sigma^2 S_{n-1} + \sigma^2 X_n,$$

and the Bienaymé equality follows by induction.

But, in the case of independence, the r.v.'s $S_{n-1} I_{A_{n-1}}$ and X_n are orthogonal, not only for $A_{n-1} = \Omega$ but also for *every* event A_{n-1} defined in terms of $X_1, X_2, \cdots, X_{n-1}$. Therefore, it is to be expected that the foregoing results can be strengthened by using more completely the properties of independence, in particular the orthogonality property just mentioned.

A. KOLMOGOROV INEQUALITIES. *If the independent r.v.'s X_k are integrable and the $|X_k| \leqq c$ finite or not, then, for every $\epsilon > 0$,*

$$1 - \frac{(\epsilon + 2c)^2}{\sum_{k=1}^{n} \sigma^2 X_k} \leqq P[\max_{k \leqq n} |S_k - ES_k| \geqq \epsilon] \leqq \frac{1}{\epsilon^2} \sum_{k=1}^{n} \sigma^2 X_k.$$

If one of the variances is infinite, then the right-hand side inequality is trivial and the left-hand side inequality has no content (for, then, $c = \infty$), so that we assume that all variances are finite. In that case, the left-hand side inequality is trivial when c is infinite and therefore we assume, in proving this inequality, that, moreover, c is finite.

Proof. We can assume, without restricting the generality, that the X_n and hence the S_n are centered at expectations, provided we note that $|X| \leqq c$ implies $|EX| \leqq c$ and, hence, $|X - EX| \leqq 2c$.

Let

$$A_k = [\max_{j \leqq k} |S_j| < \epsilon],$$

$$B_k = A_{k-1} - A_k = [|S_1| < \epsilon, \cdots, |S_{k-1}| < \epsilon, |S_k| \geqq \epsilon]$$

so that

$$A_0 = \Omega, \quad A_n^c = \sum_{k=1}^{n} B_k, \quad B_k \subset [|S_{k-1}| < \epsilon, |S_k| \geqq \epsilon].$$

1° Since $S_k I_{B_k}$ and $S_n - S_k$ are orthogonal, it follows that

$$\int_{B_k} S_n^2 = E(S_n I_{B_k})^2$$
$$= E(S_k I_{B_k})^2 + E((S_n - S_k) I_{B_k})^2 \geqq E(S_k I_{B_k})^2 \geqq \epsilon^2 P B_k.$$

Summing over $k = 1, \cdots, n$, we obtain

$$\sum_{k=1}^{n} \sigma^2 X_k = ES_n^2 \geqq \int_{A_n^c} S_n^2 = \sum_{k=1}^{n} \int_{B_k} S_n^2 \geqq \epsilon^2 \sum_{k=1}^{n} PB_k = \epsilon^2 PA_n^c,$$

and the right-hand side inequality is proved.

2° Since

$$S_{k-1}I_{A_{k-1}} + X_k I_{A_{k-1}} = S_k I_{A_{k-1}} = S_k I_{A_k} + S_k I_{B_k}$$

and $S_{k-1}I_{A_{k-1}}$ and X_k are orthogonal while $I_{A_k}I_{B_k} = 0$, it follows that

$$E(S_{k-1}I_{A_{k-1}})^2 + \sigma^2 X_k \cdot PA_{k-1} = E(S_k I_{A_k})^2 + E(S_k I_{B_k})^2.$$

Since $PA_{k-1} \geqq PA_n$ and $| X_k | \leqq 2c$, and hence

$$| S_k I_{B_k} | \leqq | S_{k-1} I_{B_k} | + | X_k I_{B_k} | \leqq (\epsilon + 2c) I_{B_k},$$

it follows that

$$E(S_{k-1}I_{A_{k-1}})^2 + \sigma^2 X_k \cdot PA_n \leqq E(S_k I_{A_k})^2 + (\epsilon + 2c)^2 PB_k.$$

Summing over $k = 1, \cdots, n$, we obtain

$$(\sum_{k=1}^{n} \sigma^2 X_k) PA_n \leqq E(S_n I_{A_n})^2 + (\epsilon + 2c)^2 \sum_{k=1}^{n} PB_k$$

$$\leqq \epsilon^2 PA_n + (\epsilon + 2c)^2 PA_n^c \leqq (\epsilon + 2c)^2,$$

and the left-hand side inequality follows.

16.3 Convergence and stability. We apply now Kolmogorov inequalities and the truncation method to convergence and stability problems for consecutive sums S_n of independent r.v.'s $X_1, X_2, \cdots$.

I. CONVERGENCE. In this Chapter, convergence means convergence to a *finite* number or to a *finite* function (r.v.).

a. *If $\sum \sigma^2 X_n$ converges, then $\sum (X_n - EX_n)$ converges a.s. If $\sum \sigma^2 X_n$ diverges and the X_n are uniformly bounded, then $\sum (X_n - EX_n)$ diverges a.s. Thus, if the X_n are uniformly bounded, then $\sum (X_n - EX_n)$ converges a.s. if, and only if, $\sum \sigma^2 X_n$ converges.*

This follows, by letting $m, n \to \infty$ in Kolmogorov's inequalities with S_k replaced by $S_{m+k} - S_m$.

b. *If the X_n are uniformly bounded and $\sum X_n$ converges a.s., then $\sum \sigma^2 X_n$ and $\sum EX_n$ converge.*

Proof. To the r.v.'s X_n we associate r.v.'s X'_n such that X_n and X'_n are identically distributed for every n and $X_1, X'_1, X_2, X'_2, \cdots$ is a sequence of independent r.v.'s. We form the "symmetrized" sequence $X_n^s = X_n - X'_n$ of independent r.v.'s, and have

$$|X_n^s| \leqq |X_n| + |X'_n| \leqq 2c, \quad EX_n^s = EX_n - EX'_n = 0,$$

$$\sigma^2 X_n^s = \sigma^2 X_n + \sigma^2 X'_n = 2\sigma^2 X_n.$$

Since $\sum X_n$ converges a.s., so does $\sum X'_n$ and hence $\sum X_n^s$ $(= \sum X_n - \sum X'_n)$. It follows, by **a**, that $\sum \sigma^2 X_n^s$ and hence $\sum \sigma^2 X_n$ converge and, again by **a**, $\sum (X_n - EX_n)$ converges a.s., so that $\sum EX_n = \sum X_n - \sum (X_n - EX_n)$ converges. The assertion is proved.

Let X^c be X truncated at (a finite) $c > 0$. We have Kolmogorov's

A. THREE-SERIES CRITERION. *The series $\sum X_n$ of independent summands converges a.s. to a r.v. if, and only if, for a fixed $c > 0$, the three series*

$$\text{(i) } \sum P[|X_n| \geqq c], \quad \text{(ii) } \sum \sigma^2 X_n^c, \quad \text{(iii) } \sum EX_n^c,$$

converge.

Proof. Convergence of (i) entails, by the equivalence lemma, convergence-equivalence of $\sum X_n$ and $\sum X_n^c$, and convergence of (ii) and (iii) entails, by **a**, a.s. convergence of $\sum X_n^c$. This proves the "if" assertion.

Conversely, let $\sum X_n$ converge a.s. so that $X_n \xrightarrow{\text{a.s.}} 0$. By 15.3**A**, (i) converges, so that, by the equivalence lemma, $\sum X_n^c$ converges a.s. and, by **b**, (ii) and (iii) converge. This proves the "only if" assertion.

COROLLARY. *If at least one of the three series in* **A** *does not converge, then $\sum X_n$ diverges a.s.*

For, by **15.3B** (Corollary), $\sum X_n$ either converges a.s. or diverges a.s.

REMARK. In the proof of **b** we introduced a "symmetrized" sequence. This is an application of the "symmetrization method," to be expounded in the next section.

II. A.S. STABILITY. We seek conditions under which $\dfrac{S_n}{b_n} - a_n \xrightarrow{\text{a.s.}} 0$ when $b_n \uparrow \infty$, and require the following elementary proposition.

Toeplitz lemma. *Let a_{nk}, $k = 1, 2, \cdots, k_n$, be numbers such that, for every fixed k, $a_{nk} \to 0$ and, for all n, $\sum_k |a_{nk}| \leqq c < \infty$; let $x'_n = \sum_k a_{nk}x_k$.*

Then, $x_n \to 0$ entails $x'_n \to 0$ and, if $\sum_k a_{nk} \to 1$, then $x_n \to x$ finite entails $x'_n \to x$. In particular, if $b_n = \sum_{k=1}^{n} a_k \uparrow \infty$, then $x_n \to x$ finite entails $\frac{1}{b_n} \sum_{k=1}^{n} a_k x_k \to x$.

The proof is immediate. If $x_n \to 0$ then, for a given $\epsilon > 0$ and $n \geqq n_\epsilon$ sufficiently large, $|x_n| < \frac{\epsilon}{c}$ so that

$$|x'_n| \leqq \sum_{k<n_\epsilon} |a_{nk}x_k| + \epsilon.$$

Letting $n \to \infty$ and then $\epsilon \to 0$, it follows that $x'_n \to 0$. The second assertion follows, since then

$$x'_n = \sum_k (a_{nk})x + \sum_k a_{nk}(x_k - x) \to x.$$

And setting $a_{nk} = \frac{a_k}{b_n}$, $k \leq n$, the particular case is proved.

The particular case yields the powerful

Kronecker lemma. *If $\sum x_n$ converges to s finite and $b_n \uparrow \infty$, then $\frac{1}{b_n} \sum_{k=1}^{n} b_k x_k \to 0$.*

For, setting $b_0 = 0$, $a_k = b_k - b_{k-1}$, $s_{n+1} = \sum_{k=1}^{n} x_k$, we have

$$\frac{1}{b_n} \sum_{k=1}^{n} b_k x_k = \frac{1}{b_n} \sum_{k=1}^{n} b_k(s_{k+1} - s_k) = s_{n+1} - \frac{1}{b_n} \sum_{k=1}^{n} a_k s_k \to s - s = 0.$$

We are now in a position to prove Kolmogorov's proposition below.

A. *If the integrable r.v.'s X_n are independent, then $\sum \frac{\sigma^2 X_n}{b_n^2} < \infty$, $b_n \uparrow \infty$, entails $\frac{S_n - ES_n}{b_n} \xrightarrow{\text{a.s.}} 0$.*

For, by **Ia**, convergence of $\sum \frac{\sigma^2 X_n}{b_n^2}$ entails a.s. convergence of $\sum \frac{X_n - EX_n}{b_n}$, and the Kronecker lemma applies.

We can now prove an extension of Borel's strong law of large numbers.

B. KOLMOGOROV STRONG LAW OF LARGE NUMBERS. *If the independent r.v.'s X_n are identically distributed with a common law $\mathcal{L}(X)$, then $\frac{X_1 + \cdots + X_n}{n} \xrightarrow{\text{a.s.}} c$ finite if, and only if, $E|X| < \infty$; and then $c = EX$.*

Proof. We set $A_n = [|X| \geqq n]$, $A_0 = \Omega$, and observe that, for every n, $PA_n = P[|X_n| \geqq n]$, while

$$\sum PA_n = \sum (n-1)(PA_{n-1} - PA_n) \leqq \sum E|X| I_{A_{n-1} - A_n}$$
$$\leqq \sum n(PA_{n-1} - PA_n) \leqq 1 + \sum PA_n$$

or

$$\sum PA_n \leqq E|X| \leqq 1 + \sum PA_n.$$

If $\frac{S_n}{n} \xrightarrow{\text{a.s.}} c$ finite, then $\frac{X_n}{n} = \frac{S_n}{n} - \frac{n-1}{n}\frac{S_{n-1}}{n-1} \xrightarrow{\text{a.s.}} 0$ and, hence, by 15.3**a**, $\sum PA_n < \infty$. This proves the "only if" assertion and it remains to prove that, if $E|X| < \infty$, then $\frac{S_n}{n} \xrightarrow{\text{a.s.}} EX$.

Let $E|X| < \infty$ and set $\bar{S}_n = \sum_{k=1}^{n} \bar{X}_k$, where $\bar{X}_k$ represents X_k truncated at k. Since

$$\sum P[|X_n| \geqq n] = \sum PA_n \leqq E|X| < \infty,$$

it follows that the sequences S_n/n and $\bar{S}_n/n$ have same limit, and it suffices to prove that $\frac{\bar{S}_n}{n} \xrightarrow{\text{a.s.}} EX$. Since, by the dominated convergence theorem,

$$E\bar{X}_n = EXI_{A_n^c} \rightarrow EX$$

and, hence, by the Toeplitz lemma, $\frac{E\bar{S}_n}{n} \rightarrow EX$, it suffices to prove that $\frac{\bar{S}_n - E\bar{S}_n}{n} \xrightarrow{\text{a.s.}} 0$. But

$$\sum \frac{\sigma^2 \bar{X}_n}{n^2} \leqq \sum \frac{E\bar{X}_n^2}{n^2} = E \sum \frac{X^2}{n^2} I_{A_n^c} \leqq 2 + E|X| < \infty,$$

since, setting $B_m = [\, m-1 \leqq |X| < m \,]$, we have $A_n^c B_m = \emptyset$ or B_m according as $n < m$ or $n \geqq m$ and, hence,

$$\sum_n \frac{X^2}{n^2} I_{A_n^c B_m} \leqq m^2 \left(\frac{1}{m^2} + \frac{1}{(m+1)^2} + \cdots\right) I_{B_m}$$

$$\leqq \left(1 + m^2 \int_m^\infty \frac{dx}{x^2}\right) I_{B_m} \leqq (2 + |X|) I_{B_m},$$

so that, by summing over m, we obtain the bound $2 + E|X|$. Thus, theorem **A** applies, and the proof is complete.

***16.4 Generalization.** Let c, with or without affixes, be finite positive numbers and let g_n be continuous and nondecreasing functions on $[0, +\infty]$ such that $g_n(0) = 0$ *and* $g_n(x) \geqq cx^2$ *or* $\geqq c'$ *according as* $0 < x < c_n$ *or* $x \geqq c_n$.

a. *If the series* (i) $\sum P[|X_n| \geqq c_n]$ *and* (ii) $\sum Eg_n(|X_n^{c_n}|)$ *converge, then* $\sum (X_n - EX_n^{c_n})$ *converges a.s.*

For convergence of (i) entails, by the equivalence lemma, convergence-equivalence of $\sum (X_n - EX_n^{c_n})$ and $\sum (X_n^{c_n} - EX_n^{c_n})$ and, by **Ia**, this last series converges a.s., since convergence of (ii) entails

$$\sum \sigma^2 X_n^{c_n} \leqq \sum E|X_n^{c_n}|^2 \leqq \frac{1}{c} \sum Eg_n(|X_n^{c_n}|) < \infty.$$

b. *If the series* (i) $\sum Eg_n(|X_n|)$ *or* (ii) $\sum \int_0^{c_n} P[|X_n| \geqq x]\, dg_n(x)$ *converges, then* $\sum (X_n - EX_n^{c_n})$ *converges a.s.*

For convergence of (i) entails

$$\sum P[|X_n| \geqq c_n] \leqq \frac{1}{c'} \sum Eg_n(|X_n|) < \infty$$

and

$$\sum Eg_n(|X_n^{c_n}|) \leqq \sum Eg_n(|X_n|) < \infty,$$

so that **a** applies.

Similarly, convergence of (ii) entails, by integration by parts,

$$\infty > \sum \int_0^{c_n} P[|X_n| \geqq x]\, dg_n(x) = \sum g_n(c_n) P[|X_n| \geqq c_n] + \int_0^{c_n} g_n(x)\, dP[|X_n| < x]$$

$$\geqq c' \sum P[|X_n| \geqq c_n] + \sum Eg_n(|X_n^{c_n}|),$$

so that **a** applies.

A. *If the series* (i) $\sum Eg_n\left(\frac{|X_n|}{b_n}\right)$ *or* (ii) $\sum \int_0^{c_n} P[|X_n| \geqq b_n x]\, dg_n(x)$ *converges, then*

$$\sum \frac{X_n - EX_n^{b_n c_n}}{b_n} \text{ converges a.s. and } \frac{1}{b_n}\sum_{k=1}^{n}(X_k - EX_k^{b_n c_n}) \xrightarrow{\text{a.s.}} 0.$$

Moreover, if (i) *converges and* $g_n(x) \geqq c''x$ *for* $0 < x \leqq c_n$ *or for* $x \geqq c_n$, *then* $EX_n^{b_n c_n}$ *can be replaced by* 0 *or by* EX_n, *respectively.*

Proof. The first assertion follows from **b** and the Kronecker lemma. As for the second assertion, if $\sum'$ and $\sum''$ denote summations over those values of n for which the first, respectively, the second, assumption about g_n holds, then

$$\begin{aligned}\sum{}' \frac{1}{b_n} E|X_n^{b_n c_n}| &= \sum{}' \int_0^{b_n c_n} \frac{x}{b_n}\, dP[|X_n| < x]\\ &\leqq \frac{1}{c''}\sum{}' \int_0^{b_n c_n} g_n\left(\frac{x}{b_n}\right) dP[|X_n| < x]\\ &\leqq \frac{1}{c''}\sum Eg_n\left(\frac{|X_n|}{b_n}\right) < \infty,\end{aligned}$$

and

$$\begin{aligned}\left|\sum{}'' \frac{1}{b_n} EX_n - \sum{}'' \frac{1}{b_n} EX_n^{b_n c_n}\right| &\leqq \sum{}'' \frac{1}{b_n} E|X_n - X_n^{b_n c_n}|\\ &= \sum{}'' \int_{b_n c_n}^{\infty} \frac{x}{b_n}\, dP[|X_n| < x]\\ &\leqq \frac{1}{c''}\sum{}'' \int_{b_n c_n}^{\infty} g_n\left(\frac{x}{b_n}\right) dP[|X_n| < x]\\ &\leqq \frac{1}{c''}\sum Eg_n\left(\frac{|X_n|}{b_n}\right) < \infty.\end{aligned}$$

This completes the proof.

Particular cases. 1° Let $g_n(x) = |x|^{r_n}$ with $0 < r_n \leqq 2$. Theorem **A** yields

If $b_n \uparrow \infty$ *and* $\sum \frac{E|X_n|^{r_n}}{b_n^{r_n}} < \infty$, *then* $\frac{1}{b_n}\sum_{k=1}^{n}(X_k - a_k) \xrightarrow{\text{a.s.}} 0$ *where* $a_k = 0$ *or* EX_k *according as* $0 < r_n < 1$ *or* $1 \leqq r_n \leqq 2$.

For $r_n \equiv 2$, we find 16.3II**A**.

2° Let $g_n(x) = x^2$ for $0 \leqq x \leqq 1$ and, to simplify the writing, set $q(x) = P[|X| \geqq x]$, $q_n(x) = P[|X_n| \geqq x]$. Theorem **A** yields

If $\int_0^1 x\{\sum q_n(b_n x)\}\, dx < \infty$, *then* $\sum \frac{1}{b_n}(X_n - EX_n^{b_n})$ *converges a.s. and* $\frac{1}{b_n}\sum_{k-1}^{n}(X_k - EX_k^{b_k}) \xrightarrow{\text{a.s.}} 0.$

We require the following

Moments lemma. *For every* $r > 0$ *and* $x > 0$

$$x^r \sum q(n^{\frac{1}{r}}x) \leqq E|X|^r \leqq 1 + x^r \sum q(n^{\frac{1}{r}}x).$$

This follows from

$$E|X|^r = -\int_0^\infty t^r\, dq(t) = -\sum \int_{(n-1)^{\frac{1}{r}}x}^{n^{\frac{1}{r}}x} t^r\, dq(t)$$

and

$$(n-1)x^r\{q((n-1)^{\frac{1}{r}}x) - q(n^{\frac{1}{r}}x)\}$$

$$\leqq -\int_{(n-1)^{\frac{1}{r}}x}^{n^{\frac{1}{r}}x} t^r\, dq(t) \leqq nx^r\{q((n-1)^{\frac{1}{r}}x) - q(n^{\frac{1}{r}}x)\},$$

by summing the inequalities over $n = 1, 2, \cdots$ and rearranging the terms.

3° If $b_n = n^{\frac{1}{r}}$ and the laws of the r.v.'s X_n are uniformly bounded by the law of a r.v. X, that is, $q_n \leqq q$, then $E|X|^r < \infty$ entails

$$\int_0^1 x(\sum q_n(n^{\frac{1}{r}}x))\, dx \leqq \int_0^1 x(\sum q(n^{\frac{1}{r}}x))\, dx \leqq E|X|^r \int_0^1 \frac{dx}{x^{r-1}},$$

so that the left-hand side is finite for $r < 2$. Therefore, on account of 2°,

If $q_n \leqq q$ *and* $E|X|^r < \infty$ *with* $r < 2$, *then* $\frac{1}{n^{\frac{1}{r}}}\sum_{k=1}^{n}(X_k - a_k) \xrightarrow{\text{a.s.}} 0$ *where* $a_k = 0$ *or* EX_k *according as* $r < 1$, *or* $\geqq 1$.

4° If $F_n = F$, then the converse is also true. More precisely (Kolmogorov: $r = 1$; Marcinkiewicz: $r \neq 1$),

Let the independent r.v.'s X_n be identically distributed with common law $\mathfrak{L}(X)$, and let $0 < r < 2$.

If $E|X|^r < \infty$, then $\frac{1}{n^{1/r}} \sum_{k=1}^{n} (X_k - a_k) \xrightarrow{\text{a.s.}} 0$ with $a_k = 0$ or EX according as $r < 1$ or $r \geqq 1$.

Conversely, if $\frac{1}{n^{1/r}} \sum_{k=1}^{n} (X_k - a_k) \xrightarrow{\text{a.s.}} 0$, then $E|X|^r < \infty$.

Proof. The first assertion is a particular case of the preceding proposition. As for the converse proposition, we use the symmetrization method expounded in the following section.

Let X'_n be a sequence independent of the sequence X_n and with same distribution, and let X' be independent of X and with same distribution; set $X_n^s = X_n - X'_n$ and $X^s = X - X'$. Then, on account of the assumption,

$$Y_n = \frac{1}{n^{1/r}} \sum_{k=1}^{n} X_k^s = \frac{1}{n^{1/r}} \sum_{k=1}^{n} (X_k - a_k) - \frac{1}{n^{1/r}} \sum_{k=1}^{n} (X'_k - a_k) \xrightarrow{\text{a.s.}} 0$$

and, hence,

$$\frac{X_n^s}{n^{1/r}} = Y_n - \left(\frac{n-1}{n}\right)^{1/r} Y_{n-1} \xrightarrow{\text{a.s.}} 0.$$

Since the X_n^s are independent r.v.'s, it follows that, for every $x > 0$,

$$\sum q^s(n^{1/r}x) = \sum P[|X_n^s| \geqq n^{1/r}x] < \infty.$$

Therefore, by the moments lemma, $E|X^s|^r < \infty$ so that, by **17.1A**, Corollary 2,

$$E|X - \mu X|^r \leqq 2E|X^s|^r < \infty$$

and, hence, by the c_r-inequality,

$$E|X|^r \leqq c_r E|X - \mu X|^r + c_r|\mu X|^r < \infty.$$

The proof is complete.

*§ 17. CONVERGENCE AND STABILITY OF SUMS; CENTERING AT MEDIANS AND SYMMETRIZATION

While centering at expectations goes back to Bernoulli and use of bounds in terms of variances goes back to Tchebichev, centering at medians and symmetrization are relatively recent. Yet, not only do they complete the first ones, but they also tend to replace them altogether. Moreover, medians always exist and the ch.f.'s of symmetrized r.v.'s, being real-valued, are much easier to handle than complex-valued ones.

*17.1 **Centering at medians and symmetrization.** Let F be the d.f. of a r.v. X. There exists at least one finite number μX called a *median* of X, such that

$$P[X \geqq \mu X] \geqq \tfrac{1}{2} \leqq P[X \leqq \mu X]$$

or, equivalently,

$$F(\mu X) \leqq \tfrac{1}{2}, \quad F(\mu X + 0) \geqq \tfrac{1}{2}.$$

For, F being nondecreasing on R with $F(-\infty) = 0$, $F(+\infty) = 1$, the graph of $y = F(x)$ completed at its discontinuity points by the segments $(x, F(x))$ to $(x, F(x + 0))$ has either a point or a segment parallel to the x-axis, in common with the line $y = \frac{1}{2}$. According to the foregoing definition, the abscissae of the common point or of the common segment are medians of X so that either X has a unique median or it has for medians all points of a closed interval on R—the *median segment* of X.

It follows from the definition of medians that, for every finite number c, we can set $\mu(cX) = c\mu X$. Furthermore, there is a relation between μX, EX, and $\sigma^2 X$, namely,

a. *If X is integrable, then* $|\mu X - EX| \leqq \sqrt{2\sigma^2 X}$.

For, by Tchebichev's inequality,

$$P[|X - EX| \geqq \sqrt{2\sigma^2 X}] \leqq \tfrac{1}{2},$$

so that

$$EX - \sqrt{2\sigma^2 X} \leqq \mu X \leqq EX + \sqrt{2\sigma^2 X}.$$

A r.v. X and its law as well as its d.f. F and ch.f. f are said to be *symmetric* if, for every x,

$$P[X \leqq x] = P[X \geqq -x]; \tag{1}$$

equivalently,

$$F(-x + 0) = 1 - F(x), \tag{2}$$

or, for every pair $a < b$ of continuity points of F,

$$F[a, b) = F[-b, -a), \tag{3}$$

or

$$f = \bar{f} \text{ is real.} \tag{4}$$

The *symmetrization* procedure consists in assigning to a r.v. X the *symmetrized* r.v. $X^s = X - X'$, where X' is independent of X and has the same distribution. More generally, if $X = \{X_t, t \in T\}$ is a family of r.v.'s, then the *symmetrized family* is $X^s = \{X_t - X'_t, t \in T\}$ where the family X' is independent of X and has same distribution. If X has affixes we affix them to X^s as well as to its d.f. and ch.f. Clearly

b. *To a r.v. X with ch.f. f, there corresponds a symmetric r.v. $X^s = X - X'$ where X and X' are independent and identically distributed, and $f^s = |f|^2$ is the ch.f. of X^s.*

We arrive now at inequalities which are the basic reason for centering at medians.

A. WEAK SYMMETRIZATION INEQUALITIES. *For every ϵ and every a,*

(i)
$$\tfrac{1}{2}P[X - \mu X \geqq \epsilon] \leqq P[X^s \geqq \epsilon]$$

and

(ii)
$$\tfrac{1}{2}P[|X - \mu X| \geqq \epsilon] \leqq P[|X^s| \geqq \epsilon] \leqq 2P\left[|X - a| \geqq \frac{\epsilon}{2}\right].$$

Proof. Since $X^s = X - X'$ where X and X' are independent and identically distributed, it follows that to a median $\mu = \mu X$ corresponds an equal median $\mu = \mu X'$ and

$$\begin{aligned}P[X^s \geqq \epsilon] &= P[(X - \mu) - (X' - \mu) \geqq \epsilon] \geqq P[X - \mu \geqq \epsilon, X' - \mu \leqq 0]\\ &= P[X - \mu \geqq \epsilon]\cdot P[X' - \mu \leqq 0] \geqq \tfrac{1}{2}P[X - \mu \geqq \epsilon].\end{aligned}$$

This proves inequality (i) which, together with the inequality obtained by changing in (i) X into $-X$, entails the left-hand side inequality in (ii). The right-hand side inequality in (ii) follows from the identical distribution of X and X' only, by

$$\begin{aligned}P[|X^s| \geqq \epsilon] &= P[|(X - a) - (X' - a)| \geqq \epsilon]\\ &\leqq P\left[|X - a| \geqq \frac{\epsilon}{2}\right] + P\left[|X' - a| \geqq \frac{\epsilon}{2}\right]\\ &= 2P\left[|X - a| \geqq \frac{\epsilon}{2}\right].\end{aligned}$$

COROLLARY 1. *If $X_n - a_n \xrightarrow{P} 0$, then $X_n^s \xrightarrow{P} 0$ and $a_n - \mu X_n \to 0$, and conversely.*

This follows by letting $n \to \infty$ in (ii) where X is replaced by X_n.

COROLLARY 2. *For $r > 0$ and every a,*

$$\tfrac{1}{2}E\,|\,X - \mu X\,|^r \leqq E\,|\,X^s\,|^r \leqq 2c_rE\,|\,X - a\,|^r$$

where $c_r = 1$ or 2^{r-1} according as $r \leqq 1$ or $r \geqq 1$.

Proof. The right-hand side inequality follows, by the c_r-inequality, from

$$E|\,X^s\,|^r = E\,|\,(X - a) - (X' - a)\,|^r \leqq c_rE\,|\,X - a\,|^r + c_r\,E\,|\,X' - a\,|^r$$
$$= 2c_rE\,|\,X - a\,|^r.$$

As for the left-hand side inequality, it is trivial when $E\,|\,X^s\,|^r = \infty$ and then, according to the inequality just proved (with $a = \mu X$), $E\,|\,X - \mu X\,|^r = \infty$; thus, we can assume that $E|\,X^s\,|^r$ is finite. Let

$$q(t) = P[|\,X - \mu X\,| \geqq t] \quad \text{and} \quad q^s(t) = P[|\,X^s\,| \geqq t]$$

so that, by **A**(ii),

$$q(t) \leqq 2q^s(t).$$

It follows, upon integrating by parts, that

$$E|\,X - \mu X\,|^r = -\int_0^\infty t^r\,dq(t) = \int_0^\infty q(t)\,d(t^r) \leqq 2\int_0^\infty q^s(t)\,d(t^r)$$
$$= -\,2\int_0^\infty t^r\,dq^s(t) = 2E|\,X^s\,|^r,$$

and the proof is concluded.

This corollary was used at the end of the preceding section.

We pass now to symmetrized families and recall that, if two families $\{X_t,\, t \in T\}$ and $\{X'_t,\, t \in T\}$ are independent, then events defined in terms of the X_t and in terms of the X'_t, respectively, are independent. We require the following

c. LEMMA FOR EVENTS. *Let events with subscript 0 be empty. If, for every integer $j \geqq 1$, $A_jA_{j-1}^c \cdots A_0^c$ and B_j are independent, then*

$$P \bigcup A_jB_j \geqq \alpha P \bigcup A_j,\ \alpha = \inf PB_j.$$

More generally, if $(A_j + A_j')(A_{j-1} + A_{j-1}')^c \cdots (A_0 + A_0')^c$ are independent of B_j and of B'_j, then

$$P \bigcup (A_jB_j + A'_jB'_j) \geqq \alpha P \bigcup (A_j + A'_j),\ \alpha = \inf (PB_j, PB'_j).$$

Proof. The same method applies to both cases. For instance

$$\begin{aligned} P\bigcup A_jB_j &= PA_1B_1 + P(A_1B_1)^cA_2B_2 + P(A_1B_1)^c(A_2B_2)^cA_3B_3 + \cdots \\ &\geqq PA_1B_1 + PA_1^cA_2B_2 + PA_1^cA_2^cA_3B_3 + \cdots \\ &\geqq PA_1\cdot PB_1 + PA_1^cA_2\cdot PB_2 + PA_1^cA_2^cA_3\cdot PB_3 + \cdots \\ &\geqq \alpha(PA_1 + PA_1^cA_2 + PA_1^cA_2^cA_3 + \cdots) = \alpha P\bigcup A_j. \end{aligned}$$

B. SYMMETRIZATION INEQUALITIES. *For every* ϵ *and every* $a_j, j \leqq n$,

(i) $$\tfrac{1}{2}P[\sup_j (X_j - \mu X_j) \geqq \epsilon] \leqq P[\sup_j X_j^s \geqq \epsilon]$$

and

(ii) $$\tfrac{1}{2}P[\sup_j | X_j - \mu X_j | \geqq \epsilon] \leqq P[\sup_j | X_j^s | \geqq \epsilon]$$

$$\leqq 2P\left[\sup_j | X_j - a_j | \geqq \frac{\epsilon}{2}\right].$$

Proof. Since $X_j^s = X_j - X'_j$ and the families $\{X_j\}$ and $\{X'_j\}$ are independent and identically distributed, it follows that to medians $\mu_j = \mu X_j$ correspond equal medians $\mu_j = \mu X'_j$; setting

$$A_j = [X_j - \mu_j \geqq \epsilon'], \quad B_j = [X'_j - \mu_j \leqq 0], \quad C_j = [X_j^s \geqq \epsilon'],$$

so that $A_jB_j \subset C_j$, the lemma for events applies, with $\alpha = \frac{1}{2}$, and

$$\tfrac{1}{2}P\bigcup A_j \leqq P\bigcup A_jB_j \leqq P\bigcup C_j.$$

This proves (i) by letting $\epsilon' \uparrow \epsilon$, and (ii) follows by arguments similar to those used in the proof of **A** and by the lemma for events.

COROLLARY. *If* $X_n - a_n \xrightarrow{\text{a.s.}} 0$, *then* $X_n^s \xrightarrow{\text{a.s.}} 0$ *and* $a_n - \mu X_n \to 0$; *and conversely.*

By centering sums of independent r.v.'s at suitable medians, we obtain inequalities which can play the role of Kolmogorov's inequalities.

C. P. LÉVY INEQUALITIES. *If* $X_1, \cdots, X_n$ *are independent r.v.'s and* $S_k = \sum_{j=1}^{k} X_j$, *then, for every* ϵ,

(i) $$P[\max_{k \leqq n} (S_k - \mu(S_k - S_n)) \geqq \epsilon] \leqq 2P[S_n \geqq \epsilon]$$

and

(ii) $$P[\max_{k \leqq n} | S_k - \mu(S_k - S_n) | \geqq \epsilon] \leqq 2P[| S_n | \geqq \epsilon].$$

Proof. Let $S_0 = 0$, $S^*_k = \max_{j \leqq k} (S_j - \mu(S_j - S_n))$ and set

$$A_k = [S^*_{k-1} < \epsilon, \quad S_k - \mu(S_k - S_n) \geqq \epsilon],$$
$$B_k = [S_n - S_k - \mu(S_n - S_k) \geqq 0]$$

where $\mu(S_n - S_k) = -\mu(S_k - S_n)$. Since

$$[S^*_n \geqq \epsilon] = \sum_{k=1}^{n} A_k, \quad [S_n \geqq \epsilon] \supset \sum_{k=1}^{n} A_k B_k, \quad PB_k \geqq \tfrac{1}{2},$$

(i) follows upon applying the lemma for events or, directly, by

$$P[S_n \geqq \epsilon] \geqq \sum_{k=1}^{n} PA_k PB_k \geqq \tfrac{1}{2} \sum_{k=1}^{n} PA_k = \tfrac{1}{2} P[S^*_n \geqq \epsilon].$$

By changing the signs of all r.v.'s which figure in (i) and combining with (i), inequality (ii) follows, and the proof is complete.

REMARK. Let $X_1, \cdots, X_n$ be independent, square-integrable, and centered at expectations. Since, by **a**,

$$|\mu(S_k - S_n)| \leqq \sqrt{2\sigma^2(S_n - S_k)} \leqq \sqrt{2\sigma^2 S_n}$$

inequality (i) remains valid if $\mu(S_k - S_n)$ is replaced by $-\sqrt{2\sigma^2 S_n}$ and, hence, changing ϵ into $\epsilon - \sqrt{2\sigma^2 S_n}$,

$$P[\max S_k \geqq \epsilon] \leqq 2P[S_n \geqq \epsilon - \sqrt{2\sigma^2 S_n}].$$

***17.2 Convergence and stability.** We are now in possession of the basic tools and shall apply them to the investigation of convergence and stability of sums $S_n = \sum_{k=1}^{n} X_k$ of independent r.v.'s. We recall that here we say that a sequence of r.v.'s converges a.s. if it converges a.s. to a r.v., and their sequence of laws converges if it converges to the law of a r.v., that is, converges completely.

I. CONVERGENCE. Whatever be the sequence of r.v.'s, we have the comparison table of convergences below:

$$\begin{array}{ccccc} \text{convergence a.s.} & \Rightarrow & \text{convergence in pr.} & \Rightarrow & \text{convergence of laws} \\ & & \Uparrow & & \\ & & \text{convergence in q.m.} & & \end{array}$$

("in q.m." means "in the 2nd mean" and reads "in quadratic mean").

For series of independent r.v.'s, reverse implications are also true, either with no restriction or under a uniform boundedness restriction. More precisely

a. IMPROVED CONVERGENCE LEMMA. *For series of independent r.v.'s:*

(i) *Convergence a.s. and convergence in pr. are equivalent.*

(ii) *If the summands are uniformly bounded and centered at expectations, then convergence a.s., convergence in pr., convergence in q.m., and convergence of laws, are equivalent.*

Proof. 1° Let $S_n \xrightarrow{P} S$, so that, by 6.3**A**, there exists a subsequence $S_{n_k} \xrightarrow{\text{a.s.}} S$ with $\sum_k P\left[|S_{n_{k+1}} - S_{n_k}| \geqq \frac{1}{2^k}\right] < \infty$. Let $n_k < n \leqq n_{k+1}$ and set $T_k = \max_n | S_n - S_{n_k} - \mu(S_n - S_{n_{k+1}}) |$, so that, by P. Lévy's inequality (ii),

$$\sum_k P\left[T_k \geqq \frac{1}{2^k}\right] \leqq 2 \sum_k P\left[| S_{n_{k+1}} - S_{n_k} | \geqq \frac{1}{2^k}\right] < \infty$$

and, hence, $T_k \xrightarrow{\text{a.s.}} 0$ as $k \to \infty$. Therefore,

$$\begin{aligned} | S_n - S - \mu(S_n - S_{n_{k+1}}) | &\leqq | S_n - S_{n_k} - \mu(S_n - S_{n_{k+1}}) | + | S_{n_k} - S | \\ &\leqq T_k + | S_{n_k} - S | \xrightarrow{\text{a.s.}} 0, \end{aligned}$$

that is, $S_n - \mu(S_n - S_{n_{k+1}}) \xrightarrow{\text{a.s.}} S$ and, *a fortiori*, $S_n - \mu(S_n - S_{n_{k+1}}) \xrightarrow{P} S$. Since $S_n \xrightarrow{P} S$, it follows that $\mu(S_n - S_{n_{k+1}}) \to 0$ and, hence, $S_n \xrightarrow{\text{a.s.}} S$. Thus, convergence in pr. of the series $\sum X_n$ entails its convergence a.s. and, the converse being always true, the first assertion is proved.

2° Let $| X_n | \leqq c < \infty$ and $EX_n = 0$. The series $\sum X_n$ converges in q.m. if, and only if, as $m, n \to \infty$

$$E(S_m - S_n)^2 = \sum_{m+1}^{n} \sigma^2 X_k \to 0$$

or, equivalently, $\sum \sigma^2 X_n < \infty$; then it converges in pr. and, hence, by the first assertion, it converges a.s. But if $\mathcal{L}(S_n) \xrightarrow{c} \mathcal{L}(S)$, so that for all u in some neighborhood of the origin

$$-\sum \log |f_n| = -\log |f_S| < \infty,$$

then, by 12.4**B**′, for u belonging to the intersection of this neighborhood with $(-1/c, +1/c)$,

$$2\sum \sigma^2 X_n = \sum \sigma^2 X_n^s \leqq -\frac{3}{u^2}\sum \log |f_n(u)|^2 < \infty,$$

and the second assertion follows.

The three-series criterion follows from this improved convergence lemma exactly as it followed from the convergence lemma in section 16.

REMARK. A better insight into the behavior of the series is provided by the Liapounov theorem for the bounded case, according to which, if $s_n^2 = \sum_{k=1}^{n} \sigma^2 X_k \to \infty$ and $ES_n = 0$, then, for any fixed $a > 0$ and $\epsilon > 0$ and n large enough to have $\epsilon s_n > a$, we have

$$(1) \qquad P[|S_n| \geqq a] \geqq P[|S_n| \geqq \epsilon s_n] \to \frac{1}{\sqrt{2\pi}}\int_{|x|\geqq\epsilon} e^{-x^2/2}\,dx.$$

Thus, as $\epsilon \to 0$, $P[|S_n| \geqq a] \to 1$ for any fixed but arbitrarily large a, and the sequence $\mathcal{L}(S_n)$ of laws diverges to a law degenerate at infinity. The second assertion follows *ab contrario*, and we see that when the sequence of laws does not converge, then, as $n \to \infty$, the distribution of S_n escapes to infinity in the fashion described by (1).

So far we have been concerned with convergence of a given series. Yet various auxiliary centering constants appeared during the investigation, and the problem arises whether, given the series $\sum X_n$ of independent r.v.'s, there exist centering constants a_n such that the series $\sum (X_n - a_n)$ converges. If $\sum (X_n - a_n)$ converges a.s. for some numerical constants a_n, we say that the series $\sum X_n$ is *essentially convergent;* otherwise, we say that it is *essentially divergent*, since, then, by the corollary of the zero-one law, $\sum (X_n - a_n)$ diverges a.s. whatever be the a_n. As above, our problem is to find criteria for this dichotomy and to find the suitable centering constants when the series is essentially convergent; at the same time, we shall be able to improve the preceding results (see also 37.1).

b. ESSENTIAL CONVERGENCE LEMMA. *The series $\sum X_n$ is essentially convergent if, and only if, the symmetrized series $\sum X_n^s$ converges a.s.*

Proof. If $\sum X_n^s$ converges a.s., then, for every finite $c > 0$, using 17.1**A**, by the three series criterion,

$$\sum P[|X_n - \mu X_n| \geqq c] \leqq \sum 2P[|X_n^s| \geqq c] < \infty$$

and, upon integrating by parts,

$$\tfrac{1}{2}\sum \sigma^2(X_n - \mu X_n)^c \leqq \sum \sigma^2(X_n^s)^c + c^2 \sum P[|X_n^s| \geqq c] < \infty.$$

Therefore, the series $\sum \{X_n - \mu X_n - E(X_n - \mu X_n)^c\}$ converges a.s. and the "if" assertion is proved while the "only if" assertion is immediate.

From this proof follows the

A. TWO-SERIES CRITERION. *The series $\sum X_n$ is essentially convergent if, and only if, for some arbitrarily fixed $c > 0$, the two series $\sum P[|X_n - \mu X_n| \geqq c]$ and $\sum \sigma^2(X_n - \mu X_n)^c$ converge; then the centered series $\sum \{X_n - \mu X_n - E(X_n - \mu X_n)^c\}$ converges a.s.*

The essential convergence lemma permits us to improve further the convergence lemma.

B. EQUIVALENCE THEOREM. *For series of independent r.v.'s, convergence of laws, convergence in pr. and a.s. convergence are equivalent.*

Proof. It suffices to prove that convergence of laws implies a.s. convergence. Let f_n be the ch.f. of X_n so that $|f_n|^2$ is ch.f. of X_n^s. If $\prod_{k=1}^n f_k \to f$ ch.f., then $\prod_{k=1}^n |f_k|^2 \to |f|^2$ and, by 12.4 B', the two series $\sum P[|X_n^s| \geqq c]$ and $\sum \sigma^2(X_n^s)^c$ converge. Since $E(X_n^s)^c = 0$, it follows, by the three series criterion, that the symmetrized series $\sum X_n^s$ converges a.s. Therefore, by the essential convergence lemma, there exist constants a_n such that the series $\sum (X_n - a_n)$ converges a.s. to a r.v. and *a fortiori* its law converges completely, so that, for every u, $\prod_{k=1}^n e^{-ia_k u} f_k(u) \to f'(u)$, where f' is a ch.f. By taking u close enough to 0 so that $f(u)f'(u) \neq 0$, it follows that the series $\sum a_n$ converges and, hence, the series $\sum X_n$ converges a.s. This completes the proof.

COROLLARY 1. *A series $\sum X_n$ of independent r.v.'s converges a.s. if, and only if, $\prod_{k=1}^n f_k \to f$ and f is continuous at the origin or $f \neq 0$ on a set of positive Lebesgue measure.*

This follows by the continuity theorem or 12.4, 4°.

COROLLARY 2. *A series $\sum X_n$ of independent r.v.'s is essentially convergent or divergent according as*

$$\lim \prod_{k=1}^n |f_k| \neq 0 \quad \textit{on a set of positive Lebesgue measure or}$$

$$\lim \prod_{k=1}^n |f_k| = 0 \quad \textit{a.e.}$$

This follows by 12.4, 4° and **b**.

II. STABILITY. Given sequences a_n and $b_n \uparrow \infty$, we seek conditions for a.s. stability of sequences S_n of sums of independent r.v.'s. On account of the corollary to the symmetrization lemma, a first condition is that $a_n = \mu\left(\frac{S_n}{b_n}\right) + o(1)$. Thus, it suffices to take $a_n = \mu\left(\frac{S_n}{b_n}\right)$ and investigate conditions under which $\frac{S_n}{b_n} - \mu\left(\frac{S_n}{b_n}\right) \xrightarrow{\text{a.s.}} 0$.

We have $b_n \uparrow \infty$ and, moreover, assume that there exists a subsequence b_{n_k} and finite numbers c, c' such that, for all k sufficiently large, $1 < c' \leqq \frac{b_{n_{k+1}}}{b_{n_k}} \leqq c < \infty$. Roughly speaking, this assumption means that the sequence b_n does not increase too fast, and it is always satisfied (with an arbitrary $c > 1$) when $\frac{b_{n+1}}{b_n} \to 1$. Let $S_{n_0} = 0$ and $T_k = \frac{S_{n_k} - S_{n_{k-1}}}{b_{n_k}}$.

A. A.S. STABILITY CRITERION. (i) $\frac{S_n}{b_n} - \mu\left(\frac{S_n}{b_n}\right) \xrightarrow{\text{a.s.}} 0$ *if, and only if,* (ii) $T_k - \mu T_k \xrightarrow{\text{a.s.}} 0$ *as* $k \to \infty$ *or, equivalently,* (ii') *for every* $\epsilon > 0$, $\sum P[|T_k - \mu T_k| \geqq \epsilon] < \infty$.

Proof. Since the T_k are nonoverlapping sums of independent r.v.'s, it follows, by 15.3A, that conditions (ii) and (ii') are equivalent. And, on account of the symmetrization lemma, it suffices to prove equivalence of (i) and (ii) for symmetric summands; then the medians which figure in these conditions vanish.

If $\frac{S_n}{b_n} \xrightarrow{\text{a.s.}} 0$, then

$$\frac{S_{n_k}}{b_{n_k}} \xrightarrow{\text{a.s.}} 0 \text{ as } k \to \infty, \quad T_k = \frac{S_{n_k} - S_{n_{k-1}}}{b_{n_k}} = \frac{S_{n_k}}{b_{n_k}} - \frac{b_{n_{k-1}}}{b_{n_k}} \frac{S_{n_{k-1}}}{b_{n_{k-1}}} \xrightarrow{\text{a.s.}} 0,$$

and the "only if" assertion is proved.

Conversely, if $T_k \xrightarrow{\text{a.s.}} 0$, then, by the Toeplitz lemma,

$$\frac{S_{n_k}}{b_{n_k}} = \frac{1}{b_{n_k}} \sum_{j=1}^{k} b_{n_j} T_j \xrightarrow{\text{a.s.}} 0.$$

Furthermore, upon setting $U_k = \max_{n_{k-1} < n \leqq n_k} \frac{|S_n - S_{n_{k-1}}|}{b_{n_k}}$ and applying P. Lévy's inequality we obtain, for every $\epsilon > 0$,

$$\sum P[U_k \geqq \epsilon] \leqq 2 \sum P[|T_k| \geqq \epsilon] < \infty,$$

so that $U_k \xrightarrow{\text{a.s.}} 0$. Therefore, for $n_{k-1} < n \leqq n_k$,

$$\left|\frac{S_n}{b_n}\right| = \left|\frac{S_n - S_{n_{k-1}}}{b_n} + \frac{S_{n_{k-1}}}{b_n}\right| \leqq \frac{b_{n_k}}{b_n} U_k + \frac{b_{n_{k-1}}}{b_n}\left|\frac{S_{n_{k-1}}}{b_{n_{k-1}}}\right|$$

$$\leqq cU_k + \frac{S_{n_k}}{b_{n_k}} \xrightarrow{\text{a.s.}} 0,$$

and the "if" assertion is proved.

COROLLARY 1. *If* $|X_n| < b_n$, *then* $\frac{S_n - ES_n}{b_n} \xrightarrow{\text{a.s.}} 0$ *if, and only if,* $T_k - ET_k \xrightarrow{\text{a.s.}} 0$ *as* $k \to \infty$ *or, equivalently, for every* $\epsilon > 0$,

$$\sum_k P[|T_k - ET_k| \geqq \epsilon] < \infty.$$

Proof. The "only if" assertion is proved as that of the foregoing criterion. As for the "if" assertion, set $X_{nk} = (X_n - EX_n)/b_{n_k}$, $n_{k-1} < n \leq n_k$, so that $\sum_n X_{nk} = T_k - ET_k \xrightarrow{\text{a.s.}} 0$. Note that $|X_{nk}| < 2$ and apply 22.5, 3° and 17.1 **a**. It follows that

$$|\mu T_k - ET_k| \leqq \sqrt{2t_k^2} \to 0,$$

so that $T_k - \mu T_k \xrightarrow{\text{a.s.}} 0$ and, by the foregoing criterion, $\frac{S_n}{b_n} - \mu\left(\frac{S_n}{b_n}\right) \xrightarrow{\text{a.s.}} 0$. But

$$\left|\mu\left(\frac{S_n}{b_n}\right) - E\left(\frac{S_n}{b_n}\right)\right| \leqq \sqrt{2\frac{s_n^2}{b_n^2}} \to 0, \quad (s_n^2 = \sigma^2 S_n),$$

since, for $n_{k-1} < n \leqq n_k$,

$$\frac{1}{c^2}\frac{s_n^2}{b_n^2} \leqq \frac{s_{n_k}^2}{b_{n_k}^2} = \frac{1}{b_{n_k}^2}\sum_{j=1}^{k} b_{n_j}^2 t_j^2 \to 0.$$

Therefore, $\frac{S_n - ES_n}{b_n} \xrightarrow{\text{a.s.}} 0$, and the proof is concluded.

COROLLARY 2. *If the* X_n *are centered at expectations and* $\sum \frac{\sigma^2 X_n}{b_n^2} < \infty$, *then* $\frac{S_n}{b_n} \xrightarrow{\text{a.s.}} 0$.

Let $\overline{X}_n$ be X_n truncated at b_n, and set $\overline{S}_n = \sum_{k=1}^{n} \overline{X}_k$, $\overline{T}_k = \frac{\overline{S}_{n_k} - \overline{S}_{n_{k-1}}}{b_{n_k}}$. Since, by Tchebichev's inequality,

$$\sum P[|X_n \neq \overline{X}_n] = \sum P[|X_n| \geqq b_n] \leqq \sum \frac{\sigma^2 X_n}{b_n^2} < \infty,$$

it follows, by the equivalence lemma, that the sequences $\frac{S_n}{b_n}$ and $\frac{\overline{S}_n}{b_n}$ are tail-equivalent. But

$$\epsilon^2 P[|\overline{T}_k| \geqq \epsilon] \leqq \sum_{n>n_{k-1}}^{n_k} \frac{\sigma^2 \overline{X}_n}{b_{n_k}^2} \leqq \sum_{n>n_{k-1}}^{n_k} \frac{\sigma^2 \overline{X}_n}{b_n^2} \leqq \sum_{n>n_{k-1}}^{n_k} \frac{\sigma^2 X_n}{b_n^2},$$

so that

$$\epsilon^2 \sum P[|\overline{T}_k| \geqq \epsilon] \leqq \sum \frac{\sigma^2 X_n}{b_n^2} < \infty,$$

Corollary 1 applies, $\frac{\overline{S}_n}{b_n} \xrightarrow{\text{a.s.}} 0$ and, therefore, $\frac{S_n}{b_n} \xrightarrow{\text{a.s.}} 0$.

*§ 18. EXPONENTIAL BOUNDS AND NORMED SUMS

In this section, the r.v.'s X_n, $n = 1, 2, \cdots$, are independent and centered at expectations with variance $\sigma_n^2 = \sigma^2 X_n = EX_n^2$; and $S_n = \sum_{k=1}^{n} X_k$ are their consecutive sums, so that $ES_n = 0$, $s_n^2 = \sigma^2 S_n = \sum_{k=1}^{n} \sigma_k^2$. We exclude the trivial case of degenerate summands.

18.1 Exponential bounds. Kolmogorov's inequalities led, in Section 16, to asymptotic properties of sums S_n. His inequalities below, where to simplify the writing we drop the subscript n, will lead to deeper results but under more restrictive assumptions.

A. EXPONENTIAL BOUNDS. *Let* $c = \max_{k \leqq n} \left| \frac{X_k}{s} \right|$ *and let* $\epsilon > 0$.

(i) *If* $\epsilon c \leqq 1$, *then* $P\left[\frac{S}{s} > \epsilon\right] < \exp\left[-\frac{\epsilon^2}{2}\left(1 - \frac{\epsilon c}{2}\right)\right]$ *and, if* $\epsilon c \geqq 1$, *then* $P\left[\frac{S}{s} > \epsilon\right] < \exp\left[-\frac{\epsilon}{4c}\right]$.

(ii) *Given* $\gamma > 0$, *if* $c = c(\gamma)$ *is sufficiently small and* $\epsilon = \epsilon(\gamma)$ *is sufficiently large, then* $P\left[\frac{S}{s} > \epsilon\right] > \exp\left[-\frac{\epsilon^2}{2}(1 + \gamma)\right]$.

Proof. 1° Let $t > 0, |X| \leqq c < \infty$, $EX = 0$ and $\sigma^2 = \sigma^2 X$. Since

$$|EX^n| \leqq c^n, \quad Ee^{tX} = 1 + \frac{t^2}{2!}EX^2 + \frac{t^3}{3!}EX^3 + \cdots,$$

$$e^{t(1-t)} < 1 + t < e^t,$$

it follows that, for $tc \leqq 1$,

$$Ee^{tX} < 1 + \frac{t^2\sigma^2}{2}\left(1 + \frac{tc}{3} + \frac{t^2c^2}{3.4} + \cdots\right) < 1 + \frac{t^2\sigma^2}{2}\left(1 + \frac{tc}{2}\right)$$

$$< \exp\left[\frac{t^2\sigma^2}{2}\left(1 + \frac{tc}{2}\right)\right]$$

and

$$Ee^{tX} > 1 + \frac{t^2\sigma^2}{2}\left(1 - \frac{tc}{3} - \frac{t^2c^2}{3.4} - \cdots\right) > 1 + \frac{t^2\sigma^2}{2}\left(1 - \frac{tc}{2}\right)$$

$$> \exp\left[\frac{t^2\sigma^2}{2}(1 - tc)\right].$$

Replacing X by $\frac{X_k}{s}$, setting $S' = \frac{S}{s}$, and taking into account that

$$Ee^{tS'} = \prod_{k=1}^{n} E\exp\left[\frac{tX_k}{s}\right]$$

we obtain

$$\text{(1)} \qquad \exp\left[\frac{t^2}{2}(1 - tc)\right] < Ee^{tS'} < \exp\left[\frac{t^2}{2}\left(1 + \frac{tc}{2}\right)\right], \quad tc \leqq 1.$$

Inequalities (i) follow then from

$$P[S' > \epsilon] \leqq e^{-t\epsilon}Ee^{tS'} < \exp\left[-t\epsilon + \frac{t^2}{2}\left(1 + \frac{tc}{2}\right)\right]$$

where t is replaced by ϵ or $\frac{1}{c}$ according as $\epsilon c \leqq 1$ or $\geqq 1$.

2° The proof of inequality (ii) is much more involved. Let α and β be two positive numbers less than 1; they will be selected later in terms of the given number γ. According to (1), we can take c sufficiently small so as to have

$$\text{(2)} \qquad Ee^{tS'} > \exp\left[\frac{t^2}{2}(1 - \alpha)\right].$$

On the other hand, setting $q(x) = P[S' > x]$ and integrating by parts, we have

$$Ee^{tS'} = -\int e^{tx}\,dq(x) = t\int e^{tx}q(x)\,dx.$$

We decompose the interval $(-\infty, +\infty)$ of integration into the five intervals $I_1 = (-\infty, 0]$, $I_2 = (0, t(1-\beta)]$, $I_3 = (t(1-\beta), t(1+\beta)]$, $I_4 = (t(1+\beta), 8t]$ and $I_5 = (8t, +\infty)$ and search for upper bounds of the integral over I_1 and I_5 and over I_2 and I_4. We have

$$J_1 = t\int_{-\infty}^{0} e^{tx}q(x)\,dx < t\int_{-\infty}^{0} e^{tx}\,dx = 1.$$

On account of (i), we have on I_5, for $8tc < 1$,

$$q(x) < \exp\left[-\frac{x}{4c}\right] < \exp[-2tx] \quad \text{for} \quad x \geqq \frac{1}{c}$$

$$q(x) < \exp\left[-\frac{x^2}{2}\left(1-\frac{xc}{2}\right)\right] \leqq \exp\left[-\frac{x^2}{4}\right] < \exp[-2tx] \quad \text{for} \quad x < \frac{1}{c}.$$

Therefore, for c sufficiently small

$$J_5 = t\int_{8t}^{\infty} e^{tx}q(x)dx < t\int_{8t}^{\infty} e^{-tx}\,dx < 1$$

and

(3) $$J_1 + J_5 < 2.$$

On the intervals I_2 and I_4 we have $x < \frac{1}{c}$ for c sufficiently small and, by (i),

$$e^{tx}q(x) < \exp\left[tx - \frac{x^2}{2}\left(1-\frac{xc}{2}\right)\right] \leqq \exp\left[tx - \frac{x^2}{2}(1-4tc)\right] = e^{g(x)}.$$

The quadratic expression $g(x)$ attains its maximum for $x = \dfrac{t}{1-4tc}$ which, for c sufficiently small, lies in I_3. Therefore, for c sufficiently small and $x \in I_2$,

$$g(x) \leqq g(t(1-\beta)) = \frac{t^2}{2}(1-\beta)(1+\beta+4tc-4tc\beta) < \frac{t^2}{2}\left(1-\frac{\beta^2}{2}\right)$$

and, then,

$$J_2 = t\int_0^{t(1-\beta)} e^{tx}q(x)\,dx < t\int_0^{t(1-\beta)} e^{g(x)}\,dx < t^2\exp\left[\frac{t^2}{2}\left(1-\frac{1}{2}\beta^2\right)\right];$$

similarly,

$$J_4 = t\int_{t(1+\beta)}^{8t} e^{tx}q(x)\,dx < t\int_{t(1+\beta)}^{8t} e^{g(x)}\,dx < 8t^2 \exp\left[\frac{t^2}{2}\left(1 - \frac{1}{2}\beta^2\right)\right].$$

We set now $\alpha = \dfrac{\beta^2}{4}$ and $t = \dfrac{\epsilon}{1-\beta}$ so that, by (2),

$$(4)\quad J_2 + J_4 < 9t^2 \exp\left[\frac{t^2}{2}\left(1 - \frac{1}{2}\beta^2\right)\right]$$

$$< \frac{9\epsilon^2}{(1-\beta)^2}\exp\left[-\frac{\epsilon^2\beta^2}{8(1-\beta)^2}\right] E\exp\left[\frac{\epsilon}{1-\beta}S'\right].$$

Since the last expectation and the inverse of its coefficient increase indefinitely as $\epsilon \to \infty$, it follows, by (3) and (4), that for ϵ sufficiently large

$$J_1 + J_5 < 2 < \tfrac{1}{4}Ee^{tS'}, \quad J_2 + J_4 < \tfrac{1}{4}Ee^{tS'}.$$

Then

$$J_3 = t\int_{t(1-\beta)}^{t(1+\beta)} e^{tx}q(x)\,dx > \tfrac{1}{2}Ee^{tS'},$$

a fortiori,

$$2t^2\beta e^{t^2(1+\beta)}q(\epsilon) > \frac{1}{2}\exp\left[\frac{t^2}{2}(1-\alpha)\right]$$

and, since as $\epsilon \to \infty$, $\dfrac{1}{4t^2}\exp\left[\dfrac{t^2}{2}\alpha\right] \to \infty$, replacing t by its value, it follows that, for ϵ sufficiently large,

$$q(\epsilon) > \frac{1}{4t^2\beta}\exp\left[\frac{t^2}{2}\alpha\right]\exp\left[-\frac{t^2}{2}(1 + 2\alpha + 2\beta)\right]$$

$$> \exp\left[-\frac{\epsilon^2}{2}\frac{1 + 2\alpha + 2\beta}{(1-\beta)^2}\right]$$

But, given $\gamma > 0$, we can select $\beta > 0$ so as to have

$$\frac{1 + 2\beta + \dfrac{\beta^2}{2}}{(1-\beta)^2} \leqq 1 + \gamma.$$

Therefore, for $c = c(\gamma)$ sufficiently small and $\epsilon = \epsilon(\gamma)$ sufficiently large,

$$q(\epsilon) > \exp\left[-\frac{\epsilon^2}{2}(1+\gamma)\right],$$

and (ii) is proved.

18.2 Stability. The a.s. stability criterion (which is due to Prokhorov for $b_n = n$) is a criterion in the sense that it is both necessary and sufficient. Yet, it is not satisfactory, since, because of the independence of the summands, it has to be expected that a satisfactory criterion ought to be expressed in terms of individual summands and not in terms of nonoverlapping sums. The nearest to this requirement is a criterion in terms of variances (due also to Prokhorov for $b_n = n$), valid when the summands are suitably bounded, and whose proof is based upon the exponential bounds.

Let $b_n \uparrow \infty, 0 < \delta' \leq \frac{b_{n_{k+1}}}{b_{n_k}} \leqq c < \infty$ and set $T_k = \frac{S_{n_k} - S_{n_{k-1}}}{b_{n_k}}$, $t_k^2 = \sigma^2 T_k = \frac{1}{b_{n_k}^2} \sum_{n_{k-1} < n \leq n_k} \sigma^2 X_k$. We write $\log_2$ for loglog.

A. *If* $\frac{|X_n|}{b_n} = o(\log_2^{-1} b_n)$ *then* $\frac{S_n}{b_n} \xrightarrow{\text{a.s.}} 0$ *if, and only if, for every* $\epsilon > 0$, *the series* (i) $\sum \exp\left[-\frac{\epsilon^2}{t_k^2}\right]$ *converges.*

Proof. For n sufficiently large $\frac{|X_n|}{b_n} < 1$, so that corollary 1 of the a.s. stability criterion applies: for every $\epsilon > 0$

$$\text{(ii)} \qquad \sum P[|T_k| > \epsilon] < \infty.$$

We have to prove that convergence of series (i) for some ϵ implies that of series (ii) for the same or distinct ϵ; and conversely. On the other hand, elementary computations show that, setting $c_k = \max_{n_{k-1} < n \leqq n_k} \frac{|X_n|}{b_n}$, the assumption made implies that $c_k = \frac{a_k}{\log k}$ with $a_k \to 0$ as $k \to \infty$.

We use now the upper exponential bounds and observe that for $c_k \frac{\epsilon}{t_k} \geqq 1$ and k sufficiently large

$$P[|T_k| > \epsilon] < 2 \exp\left[-\frac{\epsilon}{4a_k} \log k\right] = 2\left(\frac{1}{k}\right)^{\frac{\epsilon}{4a_k}} < \frac{2}{k^2}$$

and

$$\exp\left[-\frac{\epsilon^2}{4t_k^2}\right] \leqq \exp\left[-\frac{\epsilon^2}{4t_k \epsilon c_k}\right] = \exp\left[-\frac{\epsilon}{4a_k} \log k\right] < \frac{1}{k^2}$$

so that the corresponding sums in (i) and (ii) converge and we can neglect all those terms for which $c_k \frac{\epsilon}{t_k} \geqq 1$.

Since for $c_k \frac{\epsilon}{t_k} < 1$

$$P[|T_k| > \epsilon] < 2 \exp\left[-\frac{\epsilon^2}{t_k^2}\left(1 - \frac{c_k\epsilon}{2t_k}\right)\right] < 2 \exp\left[-\frac{\epsilon^2}{4t_k^2}\right],$$

it follows that convergence of series (i) for every $\epsilon > 0$ entails that of series (ii) for every $\epsilon > 0$. Conversely, if series (ii) converges, then $T_k \xrightarrow{\text{a.s.}} 0$ and $t_k^2 \to 0$, so that, for k sufficiently large, $\frac{\epsilon}{t_k^2}$ is as large as we please and $c_k < \frac{t_k}{\epsilon}$ is as small as we please. Therefore, the exponential bound is valid with, say, $\gamma = 1$, and

$$P[|T_k| > \epsilon] > 2 \exp\left[-\frac{\epsilon^2}{t_k^2}\right],$$

so that convergence of series (ii) for every $\epsilon > 0$ entails that of series (i) for every $\epsilon > 0$, and the proof is concluded.

COROLLARY. *If, for an* $r \geqq 1$, $\sum \frac{E|X_n|^{2r}}{n^{r+1}} < \infty$, *then* $\frac{S_n}{n} \xrightarrow{\text{a.s.}} 0$.

For $r = 1$, this proposition coincides with Corollary 2 of the a.s. stability criterion, so that it suffices to consider the case $r > 1$ (due to Brunk).

Proof. Let $\bar{X}_n = X_n$ or 0 according as $|X_n| < n^{\frac{r+1}{2r}}$ or $\geqq n^{\frac{r+1}{2r}}$, so that

$$\frac{|\bar{X}_n|}{n} = O(\log_2^{-1} n), \quad \sum \frac{E|\bar{X}_n|^{2r}}{n^{r+1}} \leqq \sum \frac{E|X_n|^{2r}}{n^{r+1}} < \infty$$

and, by Tchebichev's inequality,

$$\sum P[X_n \neq \bar{X}_n] = \sum P[|X_n| \geqq n^{\frac{r+1}{2r}}] \leqq \sum \frac{E|X_n|^{2r}}{n^{r+1}} < \infty.$$

Therefore, on account of the equivalence lemma, it suffices to prove that the assertion holds for r.v.'s X_n which satisfy the assumption made in **A**.

But, upon applying for $r > 1$ the inequality $E^r|X| \leqq E|X|^r$, setting $n_k = 2^k$, and applying the c_r-inequality with $n_k - n_{k-1}$ summands, we have, summing over $n = n_{k-1} + 1, \cdots, n_k$,

$$t_k^{2r} = \frac{1}{n_k^{2r}}(E\sum X_n^2)^r \leqq \frac{1}{n_k^{2r}}E(\sum X_n^2)^r$$

$$\leqq \frac{1}{n_k^{r+1}}\sum E|X_n|^{2r} \leqq \sum \frac{E|X_n|^{2r}}{n^{r+1}}.$$

Therefore,

$$\sum_{k=1}^{\infty} t_k^{2r} \leqq \sum_{n=1}^{\infty} \frac{E|X_n|^{2r}}{n^{r+1}}$$

and, since we have $\exp\left[-\frac{\epsilon}{t_k^2}\right] < t_k^{2r}$ for k sufficiently large, criterion **A** is satisfied, and the proof is concluded.

18.3 Law of the iterated logarithm. We say that a numerical sequence b_n belongs to the *upper class* or to the *lower class* of a sequence S_n of r.v.'s, according as $P[S_n > b_n \text{ i.o.}] = 0$ or 1. *A priori*, there may be sequences b_n which belong to neither of these two classes. However, if S_n is an essentially divergent sequence of consecutive sums of independent r.v.'s, then every sequence b_n belongs to one of the foregoing two classes. The problem which arises is that of corresponding criteria. Relatively little is known about its general solution (in the case of unbounded summands), and the proofs of what is known are quite involved; the best results are due to Feller. The basic known result was first obtained by Khintchine (also P. Lévy) in the Bernoulli case as a strengthening of consecutive improvements of Borel's strong law of large numbers and, then, was extended by Kolmogoroff (also Cantelli) to more general cases, as follows:

A. Law of the iterated logarithm. *If*

$$s_n^2 \to \infty \quad \textit{and} \quad \frac{|X_n|}{s_n} = o(\log_2^{-1/2} s_n^2), \quad t_n = (2\log_2 s_n^2)^{1/2},$$

then

$$P\left[\limsup \frac{S_n}{s_n t_n} = 1\right] = 1.$$

In other words, for every $\delta > 0$, the sequence $(1+\delta)s_n t_n$ belongs to the upper class of the sequence S_n while the sequence $(1-\delta)s_n t_n$ belongs to the lower class; clearly, it suffices to prove these assertions for δ arbitrarily small.

We observe that, since the assumptions remain valid if every X_n is replaced by $-X_n$, the conclusion yields

$$P\left[\liminf \frac{S_n}{s_n t_n} = -1\right]$$

and, therefore, it holds for both sequences S_n and $|S_n|$ if it holds for the first one.

Proof. Since $s_n^2 \to \infty$ and $\frac{s_{n+1}^2}{s_n^2} = 1 + o(\log_2^{-1} s_n^2) \to 1$, it follows that, for every $c > 1$, there exists a sequence $n_k = n_k(c) \uparrow \infty$ as $k \to \infty$, such that $s_{n_k} \sim c^k$. Let $\delta, \delta', \delta''$ be positive numbers.

1° We prove that the sequences $(1 + \delta)s_n t_n$ belong to the upper class of the sequence S_n by proving the same for the sequence $S^*_{n_k} = \max_{n \leq n_k} S_n$. For

$$P[S_n > (1 + \delta)s_n t_n \text{ i.o.}] \leq P[S^*_{n_k} > (1 + \delta)s_{n_{k-1}} t_{n_{k-1}} \text{ i.o.}]$$

where

$$(1 + \delta)s_{n_{k-1}} t_{n_{k-1}} \sim \frac{1 + \delta}{c} s_{n_k} t_{n_k},$$

hence, taking $\delta' < \delta$, we can select $c > 1$ so that $\frac{1 + \delta}{c} > 1 + \delta'$ and, for k sufficiently large,

$$P[S^*_{n_k} > (1 + \delta)s_{n_{k-1}} t_{n_{k-1}} \text{ i.o.}] \leq P[S^*_{n_k} > (1 + \delta')s_{n_k} t_{n_k} \text{ i.o.}].$$

Thus, the assertion will follow from the Cantelli lemma if we prove that

$$\sum P[S^*_{n_k} > (1 + \delta')s_{n_k} t_{n_k}] < \infty.$$

But, by the remark at the end of 17.1, the general term of this series is bounded by $2P\left[S_{n_k} > \left(1 + \delta' - \frac{\sqrt{2}}{t_{n_k}}\right) s_{n_k} t_{n_k}\right]$, where $1 + \delta' - \frac{\sqrt{2}}{t_{n_k}} \to 1 + \delta'$. Therefore, for $\delta'' < \delta'$ and k sufficiently large,

$$P\left[S_{n_k} > \left(1 + \delta' - \frac{\sqrt{2}}{t_{n_k}}\right) s_{n_k} t_{n_k}\right] \leq P[S_{n_k} > (1 + \delta'')s_{n_k} t_{n_k}],$$

and it suffices to prove that the right-hand side is general term of a convergent series. This follows by applying the first upper exponential bound with $\epsilon_k = (1 + \delta'')t_{n_k}$ and $c_k = \max |X_j|/s_{n_k}$, valid for k sufficiently large since $c_k t_{n_k} \to 0$, so that

$$P[S_{n_k} > (1 + \delta'')s_{n_k}t_{n_k}] \leqq \exp[-\tfrac{1}{2}(1 - \epsilon)(1 + \delta'')^2 t_{n_k}^2]$$

$$\leqq \exp[-(1 + \delta'')\log_2 s_{n_k}^2] \sim \frac{1}{(2k \log c)^{1+\delta''}},$$

and the assertion is proved. Furthermore, according to the considerations which follow the statement of the theorem, this assertion entails that $P[|S_n| > (1 + \delta)s_n t_n \text{ i.o.}] = 0$.

2° It remains to prove that the sequences $(1 - \delta')s_n t_n$ belong to the lower class of the sequence S_n where we will take $1 > \delta' > \delta$. This assertion will be *a fortiori* true if we prove that it holds for a sequence S_{n_k}. Let

$$u_k^2 = s_{n_k}^2 - s_{n_{k-1}}^2 \sim s_{n_k}^2\left(1 - \frac{1}{c^2}\right),$$

$$v_k = (2 \log_2 u_k^2)^{1/2} \sim (2 \log_2 s_{n_k}^2) = t_{n_k}$$

and set

$$A_k = [S_{n_k} - S_{n_{k-1}} > (1 - \delta)u_k v_k].$$

We prove first that $P[A_k \text{ i.o.}] = 1$, as follows: The sums $S_{n_k} - S_{n_{k-1}}$, being nonoverlapping sums of independent r.v.'s, are independent and, by the Borel criterion, it suffices to prove that $\sum PA_k = \infty$. But, $\epsilon_k = (1 - \delta)v_k \to \infty$ while $c_k = \max\limits_{n_{k-1} < n \leqq n_k} (X_n/u_k) \to 0$ as $k \to \infty$; hence the lower exponential bound for PA_k applies with $1 + \gamma = \dfrac{1}{1 - \delta}$. Therefore,

$$PA_k > \exp[-\tfrac{1}{2}(1 + \gamma)(1 - \delta)^2 v_k^2] = \exp[-(1 - \delta)\log_2 u_k^2]$$

$$\sim \frac{1}{(2k \log c)^{1-\delta}},$$

the series $\sum PA_k$ diverges, and $P[A_k \text{ i.o.}] = 1$.

On the other hand, if $B_k = [|S_{n_{k-1}}| \leqq 2s_{n_{k-1}}t_{n_{k-1}}]$, then, according, to the end of 1°, $P[B_k^c \text{ i.o.}] = 0$; thus, from some value $n = n(\omega)$ on $|S_{n_{k-1}}(\omega)| \leqq 2s_{n_{k-1}}t_{n_{k-1}}$ except for ω belonging to the null event $[B_k^c \text{ i.o.}]$. Therefore, $P[A_k B_k \text{ i.o.}] = 1$, and this entails the assertion. For,

$$A_kB_k \subset [S_{n_k} > (1-\delta)u_kv_k - 2s_{n_{k-1}}t_{n_{k-1}}],$$

$$(1-\delta)u_kv_k - 2s_{n_{k-1}}t_{n_{k-1}} \sim \left\{(1-\delta)\left(1-\frac{1}{c^2}\right)^{\frac{1}{2}} - \frac{2}{c}\right\} s_{n_k}t_{n_k}$$

and, if we take c sufficiently large so that for $\delta' > \delta$

$$(1-\delta)\left(1-\frac{1}{c^2}\right)^{\frac{1}{2}} - \frac{2}{c} > 1 - \delta',$$

then

$$P[A_kB_k \text{ i.o.}] \leqq P[S_{n_k} > (1-\delta')s_{n_k}t_{n_k} \text{ i.o.}].$$

The proof is terminated.

COMPLEMENTS AND DETAILS

As throughout this chapter, $S_n = \sum_{k=1}^{n} X_k$ and a.s. convergence is to a r.v.

1. If the ch.f. of a sum of two r.v.'s is the product of the ch.f.'s of the summands, the summands may not be independent. Construct examples. Here is one: X is a Cauchy r.v.—with ch.f. $e^{-|u|}$; consider $X + Y$ where $Y = cX$, $c > 0$.

2. Let X, Y be independent r.v.'s and let $r \geqq 1$.

If X and Y are centered at expectations, then $E|X+Y|^r$ majorizes $E|X|^r$ and $E|Y|^r$. More generally, if, say, A is an event defined on X, then $E|X+Y|^r I_A \geqq E|X|^r I_A$.

If $E|X+Y|^r$ is finite, so are $E|X|^r$ and $E|Y|^r$. (Since $|x|^r = |E(x+Y)|^r \leqq E|x+Y|^r$, it follows that

$$E|X+Y|^r I_A = \int_A dF_X(x)\left\{\int |x+y|^r\, dF_Y(y)\right\} \geqq \int_A |x|^r\, dF_X(x) = E|X|^r I_A.$$

For $r > 1$ the first assertion implies the second one. For $r = 1$, set $A = [|X| < a]$ and observe that $E|X+Y| \geqq E(|Y| - a)I_A = (E|Y| - a)PA$.)

3. Generalized Kolmogorov inequality. Let $X_1, X_2, \cdots$ be independent r.v.'s centered at expectations, and let $r \geqq 1$. Set $C = [\sup_{k \leqq n} |S_k| \geqq c]$ and prove that

$$c^r PC \leqq E|S_n|^r I_C \leqq E|S_n|^r.$$

Apply to the same problems to which Kolmogorov's inequality was applied. For example, if $S_n \xrightarrow{r} S$, then $S_n \xrightarrow{\text{a.s.}} S$. (Set $C_k = [\sup_{j<k} |S_j| < c, |S_k| \geqq c]$, $S_0 = 0$. By *2*, $E|S_n|^r I_C = \sum_{k=1}^{n} E|S_n|^r I_{C_k} \geqq \sum_{k=1}^{n} E|S_k|^r I_{C_k} \geqq c^r PC$.)

4. Let $X_1, X_2, \cdots$ be independent r.v.'s, and let $T_n{}^r = \sup_{k \leqq n} |S_k|^r$, $r \geqq 1$.

If the X_k are symmetric, then $ET_n{}^r \leqq 2E|S_n|^r$.

If the X_k are centered at expectations, then $ET_n{}^r \leqq 2^{2r+1}E|S_n|^r$.

Extend to $n = \infty$ when $S_n \xrightarrow{\text{a.s.}} S_\infty$. (If symmetric, then

$$ET_n^r = \int_0^\infty P[T_n^r \geqq t]\,dt \leqq 2\int_0^\infty P[|S_n|^r \geqq t]\,dt = 2E|S_n|^r.$$

If centered at expectations symmetrize; then

$$|S_k|^r \leqq 2^{r-1} \sup_{k \leqq n} |S_k - S'_k|^r + 2^{r-1}|S'_k|^r.$$

Integrate over $X'_1, \cdots, X'_n$, take sup, integrate over $X_1, \cdots, X_n$, and apply the first assertion.)

5. Let $X_1, X_2, \cdots$ be independent r.v.'s centered at expectations, and let $r \geqq 1$. If $\sum E|X_n|^{2r}/n^{r+1} < \infty$, then $\frac{S_n}{n} \xrightarrow{\text{a.s.}} 0$. (Apply *4* and the elementary inequality $(\sum_{k=1}^n a_k^2)^r \leqq n^{r-1} \sum_{k=1}^n |a_k|^{2r}$ to obtain $E|S_n|^{2r} \leqq cn^{r-1} \sum_{k=1}^n E|X_k|^{2r}$. By Tchebichev's inequality,

$$P[|S_{2^{k+1}} - S_{2^k}| \geqq 2^k\epsilon] \leqq c2^{r+1}\epsilon^{-2r} \sum_{j=2^k+1}^{2^{k+1}} E|X_j|^{2r}/j^{r+1}.$$

Apply the a.s. stability criterion with $n_k = 2^k$.)

6. The series $\sum c_n e^{i\theta_n}$ where the θ_n are independent r.v.'s with $Ee^{i\theta_n} = 0$, converges or diverges a.s., according as the series $\sum c_n^2$ converges or diverges.

7. If a series $\sum X_n$ of independent r.v.'s converges a.s., then by centering the summands at the terms of some convergent series, the a.s. convergence and the limit are preserved under all changes of the order of the summands. (Start with a series which converges in q.m. Use the centering in the two series criterion.)

8. A series $\sum X_n$ of independent r.v.'s with ch.f.'s f_n converges a.s. whatever be the order of summands if, and only if, $\sum |f_n - 1| < \infty$.

9. If a series $\sum X_n$ of independent r.v.'s is essentially divergent, then it degenerates at infinity: $P[|S_n| < c] \to 0$ however large be $c > 0$. State the dual form for essential convergence. (This is true for the symmetrized series. Prove and apply: if X and X' are independent and identically distributed, then $P^2[|X| < c] \leqq P[|X - X'| < 2c]$.)

10. Let $\sum X_n$ be a series of independent r.v.'s with ch.f.'s f_n.

If for a subsequence of integers $m \to \infty$ there exist r.v.'s Y_m with ch.f. g_m such that S_m and $Y_m - S_m$ are independent and $|g_m|^2 \to |g|^2$ continuous at the origin, then $\sum X_n$ is essentially convergent. (This follows from $\prod_{k=1}^m |f_k| \geqq |g_m| \to |g| > \epsilon > 0$ in a neighborhood of the origin.)

11. Smoothing by addition. Loosely speaking, a sum of independent r.v.'s is at least as "smooth" as any of its summands. More precisely, continuity or analyticity properties of the law of one of the summands continue to hold for the law of the sum. Examples:

(a) If one of the summands has a continuous law so does the sum. (Introduce the "concentration" C_X defined by $C_X(l) = \max_{x \in R} P[x \leqq X \leqq x + l]$, $l \geqq 0$. Observe that $C_X(0) = 0$ if, and only if, F_X is continuous. By the composition theorem for independent r.v.'s X and Y, $C_{X+Y} \leqq C_X$, $C_{X+Y} \leqq C_Y$.)

(b) If one of the summands has an absolutely continuous law, so does the sum. (In defining the concentration replace translates of segments of length l by translates of Lebesgue sets of measure l.)

(c) If one of the summands has a strictly increasing d.f., so does the sum. What about unicity of medians?

12. The symmetrization method reduces medians to zero and transforms essentially convergent series into a.s. convergent ones. However, only centering at medians does not yield a.s. convergence. In fact, let $\sum_{n=0}^{\infty} X_n$ be an a.s. convergent series of independent summands. The sequence $\mu(S_n)$ of medians may not converge. However, if $\sum_{n=0}^{\infty} X_n$ is essentially convergent and the r.v. Y is independent of all the X_n and has a strictly increasing d.f., then, after centering the $S_n + Y$ at medians, the series converges a.s.

(For the counterexample, take $X_0 = -1$ or $+1$ with same pr. 1/2; let $0 < p_n < 1$ with $\sum p_n < \infty$ and, for $n \geqq 1$, take X_{2n-1} and X_{2n} with values $2(-1)^n$ of pr. p_n and 0 of pr. $1 - p_n$. The sequence S_n converges a.s., yet the S_n are odd integers with $\mu(S_{4n-1}) \geqq 1$ and $\mu(S_{4n+1}) \leqq 1$. For the last assertion use (c) and Ch. V, *CD*.)

13. The X_n are not assumed to be independent. If $\frac{S_{n^2}}{n^2} \xrightarrow{\text{a.s.}} U$ and the X_n are uniformly bounded, then $\frac{S_n}{n} \xrightarrow{\text{a.s.}} U$. What if n^2 is replaced by n^k where k is a fixed integer? What if n^2 is replaced by $[q^n]$ with $q > 1$ arbitrarily close to 1?

More generally, let $\sum P[|U_n - U| > \epsilon]/n^\alpha < \infty$ for every $\epsilon > 0$, $\sum P[|X_n| > cn^\beta] < \infty$ for some $c > 0$, $0 < \alpha \leqq 1$, $\beta > 0$. If $\gamma \geqq \alpha + \beta$, then $U_n \xrightarrow{\text{a.s.}} U$, where $U_n = S_n/n^\gamma$

(For the first assertion, the second part of the proof of Borel's strong law of large numbers (see Introductory Part) applies. For the second assertion, use the following property of series: if $\sum |p_n|/n^\alpha < \infty$ with $0 < \alpha \leqq 1$, then $\sum_k |p_{n_k}| < \infty$ for $n_{k+1} - n_k = o(n_k^\alpha)$).

In what follows, *the r.v.'s* $X_1, X_2, \cdots$, *are independent and identically distributed* with common d.f. F, and ch.f. f of a r.v. X; the trivial case of $X = 0$ a.s. is excluded. In other words, repeated trials are performed on X.

14. Random selection. Let $\nu_1 < \nu_2 < \cdots$ be integer-valued r.v.'s such that every $[\nu_j = n]$ is defined on $X_1, \cdots, X_{n-1}$. The r.v.'s $X_{\nu_1}, X_{\nu_2}, \cdots$, are independent and identically distributed—as X. (Proceed as in

$$\begin{aligned} P[X_{\nu_1} < x_1, X_{\nu_2} < x_2] &= \sum_{1 \leqq n_1 < n_2 < \infty} P[\nu_1 = n_1, X_{n_1} < x_1; \nu_2 = n_2, X_{n_2} < x_2] \\ &= \sum_{1 \leqq n_1 < n_2 < \infty} P[\nu_1 = n_1, X_{n_1} < x_1; \nu_2 = n_2] P[X_{n_2} < x_2] \\ &= P[X_1 < x_1] P[X_2 < x_2].) \end{aligned}$$

15. Deviations from the median. If X is centered at a median, then

$$E\left|\sum_{k=1}^{n} X_k\right| \geqq \frac{g(n)}{n}\sum_{k=1}^{n} E|X_k|,\quad g(2n+1) = g(2n+2) = \frac{(2n+1)!}{(2^n n!)^2}.$$

This inequality is not necessarily true when X is centered at its expectation. Extend to nonidentically distributed X_k's. (Divide R^n into its 2^n "octants" and consider the corresponding parts of the left-hand side. For a counter-example, take $n = 3$, $X = 1$ with pr. 2/3 and -2 with pr. 1/3.)

16. Equidistribution of sums. If X is a lattice r.v. with step h—only possible values kh, $k = 0, \pm 1, \cdots$—set $M(g) = \lim\limits_{n\to\infty} \frac{1}{2n+1}\sum\limits_{k=-n}^{+n} g(kh)$ and otherwise set $M(g) = \lim\limits_{h\to\infty} \frac{1}{2h}\int_{-h}^{+h} g(x)\,dx$ for those functions g on R for which either of the foregoing limits exists and is finite.

(a) In the first case $M(e^{iux}) = 1$ or 0, according as $u = 0 \left(\bmod \frac{2\pi}{h}\right)$ or $u \neq 0 \left(\bmod \frac{2\pi}{h}\right)$. In the second case $M(e^{iux}) = 1$ or 0 according as $u = 0$ or $u \neq 0$.

(b) For every $u \in R$,

$$Y_n = \frac{1}{n}\sum_{k=1}^{n} e^{iuS_k} \xrightarrow{\text{a.s.}} M(e^{iux}).$$

(This is immediate in the lattice case and if $u = 0$. Otherwise $f(u) \neq 1$ and

$$E|Y_n|^2 = \frac{1}{n} + \frac{2}{n^2}\Re\sum_{j>k} f^{j-k}(u) \leqq \frac{c}{n}$$

where c is finite. Use *13*.)

(c) The family G of functions g on R such that $\frac{1}{n}\sum\limits_{k=1}^{n} g(S_k) \xrightarrow{\text{a.s.}} M(g)$ contains all almost periodic functions and functions with period p Riemann-integrable on $[o, p]$. (G contains all functions $g(x) = e^{iux}$. It is closed under additions, multiplications by complex numbers, conjugations, and uniform passages to the limit. M is a linear monotone operation on G.)

If $g_n \in G$ and $g_n \xrightarrow{u} g$, then $M(g_n) \to M(g)$. If $g'_n, g''_n \in G$ and $M(g'_n) - M(g''_n) \to 0$, then for every g such that $g'_n \leqq g \leqq g''_n$ whatever be n, $g \in G$ and $M(g) = \lim M(g'_n) = \lim M(g''_n)$.

(d) For X degenerate at an irrational a, the classical equidistribution (modulo 1) of the fractional parts of na follows: for g bounded with $g(x) \to c$ finite as $x \to \pm\infty$,

$$\frac{1}{n}\sum_{k=1}^{n} g(S_k) \xrightarrow{\text{a.s.}} c.$$

For every finite segment I, (no. of $S_1, \cdots, S_n$ in I)$/n \xrightarrow{\text{a.s.}} 0$.

17. Normal r.v.'s. Let X be normal with $EX = 0$, $EX^2 = 1$, let g on R^n be a finite Borel function, and set $\bar{X} = S_n/n$.

(a) If $g(x_1 + c, \cdots, x_n + c) = g(x_1, \cdots, x_n)$ for all $x_k, c \in R$, then the ch.f. of the pair $\bar{X}$, $g(X_1, \cdots, X_n)$ is $f(u, v) = f_1(u)f_2(u, v)$ where $f_1(u) = e^{-u^2/2}$ is ch.f. of $\bar{X}$ and $f_2(u, v) = (2\pi)^{-n/2}\int h(x_1, \cdots, x_n)\, dx_1, \cdots, dx_n$ with

$$\log h(x_1, \cdots, x_n) = -\frac{1}{2}\sum_k \left(x_k - \frac{iu}{n}\right)^2 + ivg(x_1, \cdots, x_n).$$

(b) If f_2 is analytic in u, then $\bar{X}$ and $g(X_1, \cdots, X_n)$ are independent. In particular, $\bar{X}$ is independent of $\max_{j,k} |X_j - X_k|$ and of $\sum_{k=1}^{n} |X_k - \bar{X}|^r$, $r > 0$. (f_2 is independent of u: set $u = inc$ and use the translation property of g.)

(c) Let p with or without affixes denote a pr. density with respect to the Lebesgue measure. Let $p(x) = \frac{1}{b\sqrt{2\pi}} \exp[-(x - a)^2/2b^2]$ be the pr. density of X_k, and set

$$S^2 = \frac{1}{n}\sum_{k=1}^{n} (X_k - \bar{X})^2, \quad \bar{S} = S/\sqrt{n}, \quad Y = \frac{\sqrt{n}}{b}(\bar{X} - a), \quad Z = \frac{\sqrt{2n}}{b}(\bar{S} - b).$$

Then the pr. density of Y is $\frac{1}{\sqrt{2\pi}} e^{-x^2/2}$, the pr. density of Z converges to $\frac{1}{\sqrt{2\pi}} e^{-x^2/2}$, the pr. density of (Y, Z) converges to $\frac{1}{\sqrt{2\pi}} e^{-(x^2+y^2)/2}$, and

$$E\bar{S} = b\left(1 + O\left(\frac{1}{n}\right)\right), \quad \sigma^2\bar{S} = \frac{\sigma^2}{2n}\left(1 + O\left(\frac{1}{n}\right)\right).$$

Chapter VI

CENTRAL LIMIT PROBLEM

The Central Limit Problem of probability theory is the problem of convergence of laws of sequences of sums of r.v.'s.

For more than two centuries a particular case—the Classical Limit Problem—has been the limit problem of probability theory. The precise formulation of this case and its solution were obtained in the second quarter of this century. At the very time that this particular problem was receiving its definite answer, the much more general Central Limit Problem appeared, and was solved almost at once, thanks to the powerful ch.f.'s tool and to the truncation and symmetrization methods.

§ 19. DEGENERATE, NORMAL, AND POISSON TYPES

19.1 First limit theorems and limit laws. Three limit theorems and corresponding limit laws are at the origin of the classical limit problem. Let S_n be the number of occurrences of an event of pr. p in n independent and identical trials; to avoid trivialities we assume that $pq \neq 0$, where $q = 1 - p$. If X_k denotes the indicator of the event in the kth trial, then $S_n = \sum_{k=1}^{n} X_k$, $n = 1, 2, \cdots$, where the summands are independent and identically distributed indicators—this is the Bernoulli case. Since $EX_k = p$, $EX_k^2 = p$ and, hence, $\sigma^2 X_k = p - p^2 = pq$, it follows that

$$ES_n = \sum_{k=1}^{n} EX_k = np, \quad \sigma^2 S_n = \sum_{k=1}^{n} \sigma^2 X_k = npq.$$

The first limit theorem of pr. theory, published in 1713, says that $\frac{S_n}{n} \xrightarrow{P} p$. Bernoulli found it by a direct but cumbersome analysis of the asymptotic behavior of the "binomial pr.'s" $P[S_n = k] = C_n^k p^k q^{n-k}$, $k = 0, 1, 2, \cdots, n$.

Sharpening this analysis, de Moivre obtained the second limit theorem which, in its integral form due to Laplace, says that

$$P\left[\frac{S_n - np}{\sqrt{npq}} < x\right] \to \frac{1}{\sqrt{2\pi}} \int_{-\infty}^{x} \exp\left[-\frac{1}{2}y^2\right] dy, \quad -\infty \leqq x \leqq \infty.$$

The third limit theorem was obtained by Poisson, who modified the Bernoulli case by assuming that the pr. $p = p_n$ depends upon the total number n of trials in such a manner that $np_n \to \lambda > 0$. Thus, writing now X_{nk} and S_{nn} instead of X_k and S_n, the *Poisson case* corresponds to sequences of sums $S_{nn} = \sum_{k=1}^{n} X_{nk}$, $n = 1, 2, \cdots$, where, for every fixed n, the summands X_{nk} are independent and identically distributed indicators with $P[X_{nk} = 1] = \frac{\lambda}{n} + o\left(\frac{1}{n}\right)$. By a direct analysis of the asymptotic behavior of the binomial pr.'s, much easier to carry than the preceding ones, Poisson proved that

$$P[S_{nn} = k] \to \frac{\lambda^k}{k!} e^{-\lambda}, \quad k = 0, 1, 2, \cdots.$$

Thus are born the three basic laws of pr. theory.

1° *The degenerate law* $\mathfrak{L}(0)$ of a r.v. degenerate at 0 with d.f. having one point of increase only at $x = 0$ and ch.f. reduced to 1.

2° *The normal law* $\mathfrak{N}(0, 1)$ of a *normal r.v.* with d.f. defined by

$$F(x) = \frac{1}{\sqrt{2\pi}} \int_{-\infty}^{x} \exp\left[-\frac{1}{2}y^2\right] dy$$

and ch.f. given by

$$f(u) = \frac{1}{\sqrt{2\pi}} \int \exp\left[iux - \frac{x^2}{2}\right] dx$$

$$= \exp\left[-\frac{u^2}{2}\right] \cdot \frac{1}{\sqrt{2\pi}} \int_{-\infty - iu}^{+\infty - iu} \exp\left[-\frac{z^2}{2}\right] dz = \exp\left[-\frac{u^2}{2}\right].$$

The well-known value of the last integral is obtained by using Cauchy contour integration theorem.

3° *The Poisson law* $\mathcal{P}(\lambda)$ of a *Poisson r.v.* with d.f. defined by

$$F(x) = e^{-\lambda} \sum_{k=0}^{[x]} \frac{\lambda^k}{k!},$$

and ch.f. given by

$$f(u) = e^{-\lambda} \sum_{k=0}^{\infty} e^{iuk} \frac{\lambda^k}{k!} = e^{-\lambda} \sum_{k=0}^{\infty} \frac{(\lambda e^{iu})^k}{k!} = e^{\lambda(e^{iu}-1)}.$$

While the first two limit laws played a central role in the development of pr. theory, Poisson's law long stood isolated and ignored. We shall see later that there was a deep reason for this isolation and also that, unexpectedly enough, Poisson's law is, in a sense to be made precise, more fundamental for the central limit problem than the two others. With the notation introduced above, the three first limit theorems can be summarized as follows:

A. FIRST LIMIT THEOREMS. *In the Bernoulli case* $\mathcal{L}\left(\frac{S_n - ES_n}{n}\right) \to \mathcal{L}(0)$ *and* $\mathcal{L}\left(\frac{S_n - ES_n}{\sigma S_n}\right) \to \mathfrak{N}(0, 1)$, *while in the Poisson case* $\mathcal{L}(S_{nn}) \to \mathcal{P}(\lambda)$.

The proof by means of ch.f.'s reduces to elementary computations. We have, taking limited expansions of exponentials,

$$\begin{aligned} E \exp\left[iu \frac{S_n - np}{n}\right] &= \prod_{k=1}^{n} E \exp\left[iu \frac{X_k - p}{n}\right] \\ &= \left(p \exp\left[\frac{iuq}{n}\right] + q \exp\left[-\frac{iup}{n}\right]\right)^n \\ &= \left(1 + o\left(\frac{u}{n}\right)\right)^n \to 1; \end{aligned}$$

$$\begin{aligned} E \exp\left[iu \frac{S_n - np}{\sqrt{npq}}\right] &= \prod_{k=1}^{n} E \exp\left[iu \frac{X_k - p}{\sqrt{npq}}\right] \\ &= \left(p \exp\left[\frac{iuq}{\sqrt{npq}}\right] + q \exp\left[\frac{-iup}{\sqrt{npq}}\right]\right)^n \\ &= \left(1 - \frac{u^2}{2n} + o\left(\frac{u^2}{n}\right)\right)^n \to \exp\left[-\frac{u^2}{2}\right]; \end{aligned}$$

$$E \exp [iuS_{nn}] = \prod_{k=1}^{n} E \exp [iuX_{nk}] = (p_n \exp [iu] + q_n)^n$$

$$= \left(1 + \frac{\lambda}{n} (\exp [iu] - 1) + o\left(\frac{1}{n}\right)\right)^n$$

$$\to \exp [\lambda(e^{iu} - 1)].$$

The three first limit laws give rise to the three first limit types:

the *degenerate type* of degenerate laws $\mathcal{L}(a)$ with $f(u) = e^{iua}$;

the *normal type* of normal laws $\mathfrak{N}(a, b^2)$ with $f(u) = \exp\left[iua - \frac{b^2}{2} u^2\right]$;

the *Poisson type* of Poisson laws $\mathcal{P}(\lambda; a, b)$ with

$$f(u) = \exp [iua + \lambda(e^{iub} - 1)].$$

The three first limit theorems extend at once by means of the convergence of types theorem; we leave the corresponding statements to the reader.

***19.2 Composition and decomposition.** The three first limit types possess an important closure property. Its deep parts are the normal and the Poisson "decompositions" discovered between 1935 and 1937. P. Lévy surmised and Cramer proved the first one and, then, Raikov proved the second one.

Let $\mathcal{L}(X)$, $\mathcal{L}(X_1)$, $\mathcal{L}(X_2)$ be laws of r.v.'s with corresponding ch.f.'s f, f_1, f_2. We say that $\mathcal{L}(X)$ is *composed* of $\mathcal{L}(X_1)$ and $\mathcal{L}(X_2)$ or that $\mathcal{L}(X_1)$ and $\mathcal{L}(X_2)$ are *components* of $\mathcal{L}(X)$ if, X_1 and X_2 being independent, $\mathcal{L}(X) = \mathcal{L}(X_1 + X_2)$ or, equivalently, if $f = f_1 f_2$.

A. COMPOSITION AND DECOMPOSITION THEOREM. *The degenerate and the normal types are closed under compositions and under decompositions. The same is true of every family of Poisson laws $\mathcal{P}(\lambda; a, b)$ with the same b.*

To avoid exceptions we consider degenerate laws as degenerate normal and as degenerate Poisson ones.

Proof. 1° Closure under compositions

$$\mathcal{L}(a_1) * \mathcal{L}(a_2) = \mathcal{L}(a_1 + a_2)$$

$$\mathfrak{N}(a_1, b_1^2) * \mathfrak{N}(a_2, b_2^2) = \mathfrak{N}(a_1 + a_2, b_1^2 + b_2^2)$$

$$\mathcal{P}(\lambda_1; a_1, b) * \mathcal{P}(\lambda_2; a_2, b) = \mathcal{P}(\lambda_1 + \lambda_2; a_1 + a_2, b)$$

follows at once by means of ch.f.'s, for

$$e^{iua_1}\cdot e^{iua_2} = e^{iu(a_1+a_2)}$$

$$\exp\left[iua_1 - \frac{b_1^2}{2}u^2\right]\cdot\exp\left[iua_2 - \frac{b_2^2}{2}u^2\right] = \exp\left[iu(a_1+a_2) - \frac{b_1^2+b_2^2}{2}u^2\right]$$

$$\exp\,[iua_1 + \lambda_1(e^{iub}-1)]\cdot\exp\,[iua_2 + \lambda_2(e^{iub}-1)] = \exp\,[iu(a_1+a_2) + (\lambda_1+\lambda_2)(e^{iub}-1)].$$

The decomposition property of the degenerate type is immediate. For, if for every $u \in R, f_1(u)f_2(u) = e^{iua}$, then $|f_1|\,|f_2| = 1$ and, since $|f_1| \leqq 1$, $|f_2| \leqq 1$, it follows that $|f_1| = |f_2| = 1$, so that by 13.1**a**

$$f_1(u) = e^{iua_1},\quad f_2(u) = e^{iua_2},\quad u \in R.$$

The proof in the normal and Poisson cases is much more involved. To begin with, we can, by a linear change of variable, make $a = 0$ and $b = 1$ in the laws to be decomposed. Thus, we have to seek ch.f.'s f_1 and f_2 such that, for every $u \in R$,

$$f_1(u)f_2(u) = e^{-\frac{u^2}{2}}$$

or

$$f_1(u)f_2(u) = e^{\lambda(e^{iu}-1)}.$$

2° We consider first the normal decomposition and apply 14.3**A**. Since $e^{-\frac{z^2}{2}}$ is an entire nonvanishing function in the complex plane, the same is true of $f_1(z)$ and $f_2(z)$, and there exists a constant $c > 0$ such that $|f_1(z)| \leqq e^{c|z|^2}$. Therefore, upon taking the principal branch of $\log f_1(z)$ (vanishing at $u = 0$), it follows from the Hadamard factorization theorem that $\log f_1(z)$ is a polynomial in z of, at most, second degree. Since $f_1(u)$ being a ch.f., reduces to 1 at $u = 0$, equals $\bar{f}_1(-u)$, and is bounded on R, it follows that

$$\log f_1(u) = iua_1 - \frac{b_1^2}{2}u^2,\quad u \in R,$$

where a and b are real numbers. Similarly for $f_2(u)$, and the normal decomposition is proved.

3° It remains for us to consider the Poisson decomposition. Let X_1 and X_2 be two independent r.v.'s with d.f.'s F_1 and F_2, and let F be the d.f. of their sum. Since

$$[a_1 \leqq X_1 < b_1][a_2 \leqq X_2 < b_2] \subset [a_1 + a_2 \leqq X_1 + X_2 < b_1 + b_2]$$

and X_1, X_2 are independent, we have

$$F_1[a_1, b_1)F_2[a_2, b_2) \leqq F[a_1 + a_2, b_1 + b_2) \tag{1}$$

and, letting $b_1, b_2 \to \infty$, it follows that

$$F(a_1 + a_2) \leqq F_1(a_1) + F_2(a_2). \tag{2}$$

Let now α_1 and α_2 be points of increase of F_1 and F_2, respectively. If $\alpha_1 \in (a_1, b_1)$ and $\alpha_2 \in (a_2, b_2)$ whence $\alpha_1 + \alpha_2 \in (a_1 + a_2, b_1 + b_2)$, then the left-hand side in (1) is positive and, hence, $\alpha_1 + \alpha_2$ is point of increase of F. Moreover, if α_1 and α_2 are first points of increase, then, taking $a_1 < \alpha_1$ and $a_2 < \alpha_2$ in (2), we have $F(a_1 + a_2) = 0$, and, hence, $\alpha_1 + \alpha_2$ is the first point of increase of F.

Now let F be the Poisson d.f. corresponding to $\mathcal{P}(\lambda)$; its only points of increase are $k = 0, 1, 2, \cdots$. Therefore, on account of what precedes, all points of increase α_1 and α_2 of its components F_1 and F_2 are such that $\alpha_1 + \alpha_2 =$ some k and the first points of increase are α and $-\alpha$ where α is some finite number. It follows, replacing $F_1(x)$ by $F_1(x - \alpha)$ and $F_2(x)$ by $F_2(x + \alpha)$ (this does not change F), that the new d.f.'s have $k = 0, 1, 2, \cdots$ as the only possible points of increase. Thus, we can set for the corresponding ch.f.'s

$$f_1(u) = \sum_{k=0}^{\infty} a_k e^{iuk}, \quad f_2(u) = \sum_{k=0}^{\infty} b_k e^{iuk}$$

with

$$a_0, b_0 > 0,\ a_k, b_k \geqq 0 \quad \text{for} \quad k > 0,\ \sum_{k=0}^{\infty} a_k = \sum_{k=0}^{\infty} b_k = 1.$$

Upon setting $z = e^{iu}$, $\varphi_1(z) = f_1(u)$, $\varphi_2(z) = f_2(u)$, we have to find nonvanishing functions φ_1 and φ_2 such that

$$\varphi_1(z)\varphi_2(z) = \sum_{k,l=0}^{\infty} a_k b_l z^{k+l} = \sum_{k=0}^{\infty} \frac{\lambda^k e^{-\lambda}}{k!} z^k.$$

Therefore,

$$a_0 b_k + a_1 b_{k-1} + \cdots + a_k b_0 = \frac{\lambda^k e^{-\lambda}}{k!}, \quad k = 0, 1, 2, \cdots,$$

and it follows that

$$a_k \leqq \frac{1}{b_0}\frac{\lambda^k e^{-\lambda}}{k!}, \quad |\varphi_1(z)| \leqq \frac{1}{b_0} e^{\lambda(|z|-1)}.$$

Thus, $\varphi_1(z)$ and similarly $\varphi_2(z)$ are nonvanishing entire functions at most of first order. It follows from the Hadamard factorization theorem that they are of the form $e^{cz+c'}$. Since $f_1(u)$ reduces to 1 at $u = 0$ and is bounded by 1, we have

$$\log f_1(u) = \lambda_1(e^{iu} - 1), \quad \lambda_1 \geqq 0.$$

Similarly for $f_2(u)$, and the Poisson decomposition is proved. This terminates the proof of the theorem.

§ 20. EVOLUTION OF THE PROBLEM

20.1 The problem and preliminary solutions. From the time of Laplace and until 1935, the limit problem aims at weakenings of the assumptions under which the *law of large numbers* (convergence to $\mathfrak{L}(0)$) and the *normal convergence* (convergence to $\mathfrak{N}(0, 1)$) hold. This classical problem can be stated as follows:

Let $S_n = \sum_{k=1}^{n} X_k$ be consecutive sums of independent r.v.'s. Find conditions under which

$$\mathfrak{L}\left(\frac{S_n - ES_n}{n}\right) \to \mathfrak{L}(0), \quad \mathfrak{L}\left(\frac{S_n - ES_n}{\sigma S_n}\right) \to \mathfrak{N}(0, 1).$$

It is implicitly assumed, in the first case, that the summands are integrable, and in the second case that their squares also are integrable. To simplify the writing, we shall center the summands at expectations, so that, in this section, $EX_k = 0$, $ES_n = 0$. We also set $f_k(u) = Ee^{iuX_k}$, $\sigma_k = \sigma X_k$ and $s_n = \sigma S_n$, and exclude the trivial case of all summands degenerate.

Although not the first historically, the solution of the extension of the Bernoulli case to independent and identically distributed summands (not necessarily indicators) is immediate—when ch.f.'s are used.

A. *If the summands are independent, identically distributed, and centered at expectations, then* $\mathfrak{L}\left(\frac{S_n}{n}\right) \to \mathfrak{L}(0)$ *and* $\mathfrak{L}\left(\frac{S_n}{s_n}\right) \to \mathfrak{N}(0, 1)$.

For, if f is the common ch.f. of the summands, then, by using its limited expansions, we have

$$E \exp\left[iu\frac{S_n}{n}\right] = \left(f\left(\frac{u}{n}\right)\right)^n = \left(1 + o\left(\frac{u}{n}\right)\right)^n \to 1,$$

and, since $s_n^2 = n\sigma^2 > 0$,

$$E \exp\left[iu\frac{S_n}{s_n}\right] = \left(f\left(\frac{u}{s_n}\right)\right)^n = \left(1 - \frac{\sigma^2}{2s_n^2}u^2 + o\left(\frac{\sigma^2}{s_n^2}u^2\right)\right)^n$$

$$= \left(1 - \frac{u^2}{2n} + o\left(\frac{u^2}{n}\right)\right)^n \to \exp\left[-\frac{u^2}{2}\right].$$

However, the first reasonably general conditions are the following.

B. *Let* $S_n = \sum_{k=1}^{n} X_k$ *and* $s_n = \sigma S_n$, *where the summands are independent r.v.'s centered at expectations.*

(i) *If* $\frac{1}{n^{1+\delta}} \sum_{k=1}^{n} E|X_k|^{1+\delta} \to 0$ *for a positive* $\delta \leqq 1$, *then*

$$\mathcal{L}\left(\frac{S_n}{n}\right) \to \mathcal{L}(0).$$

(ii) *If* $\frac{1}{s_n^{2+\delta}} \sum_{k=1}^{n} E|X_k|^{2+\delta} \to 0$ *for a positive* δ, *then*

$$\mathcal{L}\left(\frac{S_n}{s_n}\right) \to \mathfrak{N}(0, 1).$$

The assumptions imply finiteness of moments $E|X_k|^{1+\delta}$ and $E|X_k|^{2+\delta}$, respectively.

The first assertion is slightly more general than the classical ones. For $\delta = 1$, it becomes the celebrated *Tchebichev's theorem*. It also contains *Markov's theorem:* if $E|X_k|^{1+\delta} \leqq c < \infty$, then $\mathcal{L}\left(\frac{S_n}{n}\right) \to \mathcal{L}(0)$ (since, then, the asserted condition becomes $\frac{c}{n^\delta} \to 0$); since, for $\delta > 1$, $EX_k^2 \leqq (E|X_k|^{1+\delta})^{2/1+\delta}$ Markov's theorem is valid with any $\delta > 0$.

The second assertion is the celebrated *Liapounov's theorem* which has been the turning point for the entire Central Limit theorem. Moreover, while the ch.f.'s were known to and used by Laplace, the first continuity theorem for ch.f.'s:

$$\text{if } f_n(u) \to e^{-\frac{u^2}{2}}, \quad \text{then} \quad \mathcal{L}(X_n) \to \mathfrak{N}(0, 1),$$

is to be found, proved but not stated, in Liapounov's proof of his theorem. We observe that (ii) has content only when at least one of the r.v.'s is not degenerate at zero and, then, the hypothesis implies that $s_n \to \infty$.

Proof. 1° To begin with, let us reduce in (ii) the case $\delta > 1$ to $\delta = 1$, so that it will suffice to assume that $0 < \delta \leqq 1$.

Let Y be a r.v. whose d.f. is $\frac{1}{n}\sum_{k=1}^{n} F_k$ and, hence,

$$E|Y|^r = \frac{1}{n}\sum_{k=1}^{n} E|X_k|^r.$$

According to 9.3**b**. $\log E|Y|^r$ is a convex from below function of $r > 0$. Therefore, for $2 + \delta > 3$, we have

$$\delta \cdot \log E|Y|^3 \leqq (\delta - 1)\log E|Y|^2 + \log E|Y|^{2+\delta}$$

or, equivalently,

$$\frac{1}{s_n^3}\sum_{k=1}^{n} E|X_k|^3 \leqq \left(\frac{1}{s_n^{2+\delta}}\sum_{k=1}^{n} E|X_k|^{2+\delta}\right)^{1/\delta}.$$

It follows that, if the condition in (ii) holds for a $\delta > 1$, then it holds for $\delta = 1$. Thus, in what follows we can limit ourselves to $0 < \delta \leqq 1$.

2° We use limited expansions of ch.f.'s, the continuity theorem, and the expansion $\log(1 + z) = z + o(|z|)$ valid for $|z| < 1$. As usual, θ with or without affixes denotes quantities bounded by 1.

Condition (i) implies that

$$\max_{k \leqq n} \frac{E|X_k|^{1+\delta}}{n^{1+\delta}} \leqq \frac{1}{n^{1+\delta}}\sum_{k=1}^{n} E|X_k|^{1+\delta} \to 0,$$

so that, for u arbitrary but fixed,

$$f_k\left(\frac{u}{n}\right) = 1 + \frac{2^{1-\delta}}{1+\delta}\theta_{nk}|u|^{1+\delta}\frac{E|X_k|^{1+\delta}}{n^{1+\delta}} \to 1$$

uniformly in $k \leqq n$. Therefore, for n sufficiently large,

$$\sum_{k=1}^{n} \log f_k\left(\frac{u}{n}\right) = 2\theta_n|u|^{1+\delta} \cdot \frac{1}{n^{1+\delta}}\sum_{k=1}^{n} E|X_k|^{1+\delta} \to 0,$$

and the first assertion is proved.

Condition (ii) implies that

$$\max_{k \leqq n} \left(\frac{\sigma_k}{s_n}\right)^{2+\delta} \leqq \max_{k \leqq n} \frac{E|X_k|^{2+\delta}}{s_n^{2+\delta}} \leqq \frac{1}{s_n^{2+\delta}} \sum_{k=1}^{n} E|X_k|^{2+\delta} \to 0,$$

so that, for u arbitrary but fixed,

$$f_k\left(\frac{u}{s_n}\right) = 1 - \frac{u^2}{2} \cdot \frac{\sigma_k^2}{s_n^2} + \frac{2^{1-\delta}}{(1+\delta)(2+\delta)} \theta'_{nk} |u|^{2+\delta} \frac{E|X_k|^{2+\delta}}{s_n^{2+\delta}} \to 1$$

uniformly in $k \leqq n$. Therefore, for n sufficiently large,

$$\sum_{k=1}^{n} \log f_k\left(\frac{u}{s_n}\right) = -\frac{u^2}{2}(1 + o(1)) + 2\theta'_n |u|^{2+\delta} \frac{1}{s_n^{2+\delta}} \sum_{k=1}^{n} E|X_k|^{2+\delta} \to -\frac{u^2}{2},$$

and Liapounov's theorem is proved

BOUNDED CASE. *If the summands are uniformly bounded, then* $\mathcal{L}(S_n/n) \to \mathcal{L}(0)$. *If, moreover,* $s_n \to \infty$, *then* $\mathcal{L}(S_n/s_n) \to \mathfrak{N}(0, 1)$.

For, if $|X_k| \leqq c < \infty$, then $E|X_k|^{1+\delta} \leqq c^{1+\delta}$ and $E|X_k|^{2+\delta} \leqq c^\delta \sigma_k^2$, and, hence,

$$\frac{1}{n^{1+\delta}} \sum_{k=1}^{n} E|X_k|^{1+\delta} \leqq \frac{c^{1+\delta}}{n^\delta} \to 0,$$

$$\frac{1}{s_n^{2+\delta}} \sum_{k=1}^{n} E|X_k|^{2+\delta} \leqq \frac{c^\delta}{s_n^\delta} \to 0 \quad \text{as} \quad s_n \to \infty.$$

Tools for solution. The preceding theorem is not satisfactory since moments of higher order than those which figure in the formulation of the problem are used. Yet a restatement of this theorem with $\delta = 1$, together with the truncation method, will provide the stepping stone towards the solution.

a. BASIC LEMMA. *If* $S_{nn} = \sum_{k=1}^{n} X_{nk}$, *where the summands are independent r.v.'s (centered at expectations), then*

(i) *if* $\frac{1}{n^2} \sum_{k=1}^{n} E|X_{nk}|^2 \to 0$, *then* $\mathcal{L}\left(\frac{S_{nn}}{n}\right) \to \mathcal{L}(0)$

(ii) *if* $\frac{1}{s_{nn}^3} \sum_{k=1}^{n} E|X_{nk}|^3 \to 0$, *then* $\mathcal{L}\left(\frac{S_{nn}}{s_{nn}}\right) \to \mathfrak{N}(0, 1)$.

It suffices to replace in the proof of 20.2B subscripts k and n by double subscripts nk and nn, respectively.

In order to use the truncation method we shall require a weak form of the equivalence lemma. We say that two sequences $\mathfrak{L}(X_n)$ and $\mathfrak{L}(X'_n)$ of laws are *equivalent* if, for every subsequence $\mathfrak{L}(X_{n'}) \to \mathfrak{L}(X)$, we have $\mathfrak{L}(X'_{n'}) \to \mathfrak{L}(X)$, and conversely.

b. LAW-EQUIVALENCE LEMMA. *If* $X_n - X'_n \xrightarrow{P} 0$ *or* $P[X_n \neq X'_n] \to 0$, *then the sequences* $\mathfrak{L}(X_n)$ *and* $\mathfrak{L}(X'_n)$ *of laws are equivalent.*

For the second condition implies the first one which, by 10.1**d**, implies the asserted equivalence.

20.2 Solution of the Classical Limit Problem. We are now in a position to give a complete solution of the problem.

$X_1, X_2, \cdots$ are independent r.v.'s centered at expectations, with d.f.'s $F_1, F_2, \cdots$, ch.f.'s $f_1, f_2, \cdots$, and variances $\sigma_1^2, \sigma_2^2, \cdots$; $S_n = \sum_{k=1}^{n} X_k$ are their consecutive sums with variances $s_n^2 = \sum_{k=1}^{n} \sigma_k^2$. To simplify the writing, we make the convention that all summations are over $k = 1, \cdots, n$.

A. CLASSICAL DEGENERATE CONVERGENCE CRITERION. $\mathfrak{L}\left(\frac{S_n}{n}\right) \to \mathfrak{L}(0)$ *if, and only if,*

(i)
$$\sum \int_{|x| \geq n} dF_k \to 0,$$

(ii)
$$\frac{1}{n} \sum \int_{|x| < n} x\, dF_k \to 0,$$

(iii)
$$\frac{1}{n^2} \sum \left\{ \int_{|x| < n} x^2\, dF_k - \left(\int_{|x| < n} x\, dF_k \right)^2 \right\} \to 0.$$

Proof. 1° Let (i), (ii), and (iii) hold. We wish to prove that $\mathfrak{L}\left(\frac{S_n}{n}\right) \to \mathfrak{L}(0)$. In what follows we apply the law equivalence lemma and the first part of the basic lemma.

Let $S_{nn} = \sum X_{nk}$, where $X_{nk} = X_k$ or 0 according as $|X_k| < n$ or $|X_k| \geq n$. On account of (i)

$$P\left[\frac{S_{nn}}{n} \neq \frac{S_n}{n}\right] \leq \sum P[X_{nk} \neq X_k] = \sum \int_{|x| \geq n} dF_k \to 0,$$

so that it suffices to prove that $\mathfrak{L}\left(\frac{S_{nn}}{n}\right) \to \mathfrak{L}(0)$. But, on account of (ii),

$$\frac{1}{n} ES_{nn} = \frac{1}{n} \sum \int_{|x|<n} x\, dF_k \to 0,$$

so that it suffices to prove that $\mathfrak{L}\left(\frac{S_{nn} - ES_{nn}}{n}\right) \to \mathfrak{L}(0)$. But this follows, by Tchebichev inequality, from (iii) and

$$\frac{1}{n^2} \sum E|X_{nk} - EX_{nk}|^2 = \frac{1}{n^2} \sum \left\{ \int_{|x|<n} x^2\, dF_k - \left(\int_{|x|<n} x\, dF_k \right)^2 \right\} \to 0.$$

2° Conversely, let $\mathfrak{L}\left(\frac{S_n}{n}\right) \to \mathfrak{L}(0)$; equivalently, $\frac{S_n}{n} \xrightarrow{\text{P}} 0$ or $g_n(u) = \prod_{k=1}^{n} f_k(u/n) \to 1$ uniformly on every finite interval. Let n be sufficiently large so that $\log|g_n(u)|$ is bounded on $[-c, +c]$. By the weak symmetrization lemma and the second truncation inequality

$$\frac{1}{2} \sum_{k=1}^{n} P\left[\left|\frac{X_k - \mu X_k}{n}\right| \geqq c\right] \leqq \sum_{k=1}^{n} P\left[\left|\frac{X_k^s}{n}\right| \geqq c\right] \leqq 7\int_0^c \log|g_n(v)|^2\, dv \to 0.$$

Since

$$\frac{X_n}{n} = \frac{S_n}{n} - \frac{n-1}{n}\frac{S_{n-1}}{n-1} \xrightarrow{\text{P}} 0,$$

so that $\mu X_n/n \to 0$, it follows that the foregoing relation with $c < 1$ yields (i) and, hence, $\mathfrak{L}(S_{nn}/n) \to \mathfrak{L}(0)$. But, by the first truncation inequality,

$$(1) \qquad 2\sum_{k=1}^{n} \sigma^2(X_{nk}/n) = \sum_{k=1}^{n} \sigma^2(X_{nk}^s/n) \leqq -3 \log|g_n(1)|^2 \to 0,$$

so that (iii) holds, and, by Tchebichev inequality, $\frac{S_{nn} - ES_{nn}}{n} \xrightarrow{\text{P}} 0$. Therefore,

$$\frac{ES_{nn}}{n} = \frac{S_{nn}}{n} - \frac{S_{nn} - ES_{nn}}{n} \to 0$$

and (ii) holds. The proof is completed.

Observe that centering at, and in fact existence of, expectations were not required. Also, according to the proof,

$$\mathcal{L}\left(\frac{S_n - ES_{nn}}{n}\right) \to \mathcal{L}(0) \Leftrightarrow \text{(i) } \textit{and} \text{ (iii) } \textit{hold.}$$

B. Classical normal convergence criterion. $\mathcal{L}\left(\frac{S_n}{s_n}\right) \to \mathfrak{N}(0, 1)$ *and* $\max_{k \leqq n} \frac{\sigma_k}{s_n} \to 0$ *if, and only if, for every* $\epsilon > 0$,

$$g_n(\epsilon) = \frac{1}{s_n^2} \sum \int_{|x| \geqq \epsilon s_n} x^2 \, dF_k \to 0.$$

The "if" part is due to Lindeberg and the "only if" part is due to Feller.

Proof. 1° Let $g_n(\epsilon) \to 0$ for every $\epsilon > 0$. We apply the law equivalence lemma and the basic lemma.

Since $g_n(\epsilon) \to 0$ for every $\epsilon > 0$, there is a sufficiently slowly decreasing sequence $\epsilon_n \downarrow 0$ such that $\frac{1}{\epsilon_n^2} g_n(\epsilon_n) \to 0$ and, *a fortiori*, $\frac{1}{\epsilon_n} g_n(\epsilon_n) \to 0$, $g_n(\epsilon_n) \to 0$ (it suffices to select a sequence $n_k \uparrow \infty$ as $k \to \infty$ such that $g_n\left(\frac{1}{k}\right) < \frac{1}{k^3}$ for $n \geqq n_k$ and, then, take $\epsilon_n = \frac{1}{k}$ for $n_k \leqq n < n_{k+1}$). We have

$$\max_{k \leqq n} \frac{\sigma_k^2}{s_n^2} \leqq \max_{k \leqq n} \frac{1}{s_n^2} \int_{|x| \geqq \epsilon_n s_n} x^2 \, dF_k + \epsilon_n^2 \leqq g_n(\epsilon_n) + \epsilon_n^2 \to 0,$$

and the "if" assertion will be proved if we show that $\mathcal{L}\left(\frac{S_n}{s_n}\right) \to \mathfrak{N}(0, 1)$. Let $X_{nk} = X_k$ or 0 according as $|X_k| < \epsilon_n s_n$ or $|X_k| \geqq \epsilon_n s_n$. Since

$$P\left[\frac{S_{nn}}{s_n} \neq \frac{S_n}{s_n}\right] \leqq \sum P[X_{nk} \neq X_k] = \sum \int_{|x| \geqq \epsilon_n s_n} dF_k \leqq \frac{1}{\epsilon_n^2} g_n(\epsilon_n) \to 0,$$

it suffices to prove that $\mathcal{L}\left(\frac{S_{nn}}{s_n}\right) \to \mathfrak{N}(0, 1)$.

Since the X_k are centered at expectations, we have

$$|EX_{nk}| = \left|\int_{|x| < \epsilon_n s_n} x \, dF_k\right| = \left|\int_{|x| \geqq \epsilon_n s_n} x \, dF_k\right| \leqq \frac{1}{\epsilon_n s_n} \int_{|x| \geqq \epsilon_n s_n} x^2 \, dF_k.$$

Therefore,

$$\frac{1}{s_n}\sum |EX_{nk}| \leqq \frac{g_n(\epsilon_n)}{\epsilon_n} \to 0$$

and, setting $s_{nn}^2 = \sigma^2 S_{nn}$, we obtain

$$1 - \frac{s_{nn}^2}{s_n^2} = \frac{1}{s_n^2}\sum\int_{|x|\geqq\epsilon_n s_n} x^2\,dF_k + \frac{1}{s_n^2}(\sum |EX_{nk}|)^2 \leqq g_n(\epsilon_n) + \frac{g_n^2(\epsilon_n)}{\epsilon_n^2} \to 0.$$

Thus, it suffices to prove that $\mathfrak{L}\left(\frac{S_{nn} - ES_{nn}}{s_{nn}}\right) \to \mathfrak{N}(0, 1)$. But, this follows from

$$\frac{1}{s_{nn}^3}\sum E|X_{nk} - EX_{nk}|^3 \leqq \frac{2\epsilon_n s_n}{s_{nn}^3}\sum E(X_{nk} - EX_{nk})^2 \leqq 2\epsilon_n\frac{s_n}{s_{nn}} \to 0,$$

and the "if" assertion is proved.

2° It remains to prove the "only if" assertion.

Since $\max_{k\leqq n}\frac{\sigma_k}{s_n} \to 0$, it follows from

$$f_k\left(\frac{u}{s_n}\right) = 1 - \theta_k\frac{u^2}{2}\frac{\sigma_k^2}{s_n^2}$$

that

$$\max_{k\leqq n}\left|f_k\left(\frac{u}{s_n}\right) - 1\right| \to 0, \quad \sum\left|f_k\left(\frac{u}{s_n}\right) - 1\right|^2 \to 0.$$

Therefore, for n sufficiently large, $\log f_k\left(\frac{u}{s_n}\right)$ exists, so that

$$E\exp\left[iu\frac{S_n}{s_n}\right] = \prod_{k=1}^{n} f_k\left(\frac{u}{s_n}\right) \to \exp\left[-\frac{u^2}{2}\right]$$

becomes

$$\sum \log f_k\left(\frac{u}{s_n}\right) \to -\frac{u^2}{2}$$

and, since $\log z = z - 1 + \theta|z - 1|^2$,

$$\left|\frac{u^2}{2} - \sum\left\{1 - f_k\left(\frac{u}{s_n}\right)\right\}\right| \to 0.$$

Upon taking the real parts, we obtain

$$\frac{u^2}{2} - \sum \int_{|x|<\epsilon s_n} \left(1 - \cos\frac{ux}{s_n}\right) dF_k$$

$$= \sum \int_{|x| \geqq \epsilon s_n} \left(1 - \cos\frac{ux}{s_n}\right) dF_k + o(1).$$

Since

$$\sum \int_{|x|<\epsilon s_n} \left(1 - \cos\frac{ux}{s_n}\right) dF_k \leqq \frac{u^2}{2s_n^2} \sum \int_{|x|<\epsilon s_n} x^2\, dF_k$$

$$= \frac{u^2}{2s_n^2}\left(s_n^2 - \sum \int_{|x| \geqq \epsilon s_n} x^2\, dF_k\right) = \frac{u^2}{2}(1 - g_n(\epsilon))$$

and

$$\sum \int_{|x| \geqq \epsilon s_n} \left(1 - \cos\frac{ux}{s_n}\right) dF_k \leqq 2 \sum \int_{|x| \geqq \epsilon s_n} dF_k$$

$$\leqq \frac{2}{\epsilon^2 s_n^2} \sum \int_{|x| \geqq \epsilon s_n} x^2\, dF_k \leqq \frac{2}{\epsilon^2},$$

it follows that

$$\frac{u^2}{2} g_n(\epsilon) \leqq \frac{2}{\epsilon^2} + o(1).$$

Therefore, letting $n \to \infty$ and then $u \to \infty$ in

$$0 \leqq g_n(\epsilon) \leqq \frac{2}{u^2}\left(\frac{2}{\epsilon^2} + o(1)\right),$$

we obtain $g_n(\epsilon) \to 0$. This concludes the proof.

***20.3 Normal approximation.** In his celebrated investigation of normal convergence, Liapounov examined not only conditions for, but also the speed of, this convergence. His results were greatly improved by Berry (and, independently, by Esseen) and to present the basic one we shall proceed in steps.

Let F and G be d.f.'s of r.v.'s with corresponding ch.f.'s f and g, and let $H = F - G$, $h = f - g$. We exclude the trivial case of $\alpha = \sup |H| = 0$, that is, $H = h = 0$.

a. *If G is continuous on R, then there exists a finite number s such that either $H(s) = \mp\alpha$ or $H(s + 0) = \alpha$.*

Proof. Let x_n be a sequence such that $|H(x_n)| \to \alpha$. It contains a subsequence $x_{n'} \to s$ finite or infinite. Since $H(x) \to 0$ as $x \to \mp\infty$ and $\alpha > 0$, s must be finite.

The sequence $x_{n'}$ contains a subsequence $x_{n''}$ such that either $H(x_{n''}) \to -\alpha$ or $H(x_{n''}) \to +\alpha$. It suffices to consider one case only, say the first, for the same argument is valid for the other. Thus, let $x_{n''} \to s$, $H(x_{n''}) \to -\alpha$; we know that H is continuous from the left.

If the sequence $x_{n''}$ contains a subsequence converging to s from the left, then $-\alpha = \lim H(x_{n''}) = H(s)$, and the assertion is proved. Otherwise, this sequence contains a subsequence converging to s from the right, $-\alpha = H(s+0)$ and, G being continuous on R,

$$-\alpha \leqq H(s) \leqq F(s+0) - G(s) = F(s+0) - G(s+0) = -\alpha,$$

so that $-\alpha = H(s)$. The assertion is proved.

Let p be the derivative of a symmetric d.f. (of a r.v.) differentiable on R, so that $p(x) = p(-x)$, $x \in R$.

b. *If G has a derivative G' on R, then there exists a finite number "a" such that*

$$\left|\int H(x+a)p(x)\,dx\right| \geqq \frac{\alpha}{2}\Big(1 - 6\int_{\frac{\alpha}{2\beta}}^{\infty} p(x)\,dx\Big), \quad \beta = \sup|G'|.$$

Proof. If $\beta = \infty$, then $\dfrac{\alpha}{2\beta} = 0$, and the inequality is trivially true whatever be a. Thus, it suffices to prove it when $\beta < \infty$. Let $\gamma = \dfrac{\alpha}{2\beta} > 0$.

We have, for an arbitrary a,

$$(1)\qquad \left|\int H(x+a)p(x)\,dx\right| \geqq \left|\int_{|x|<\gamma} H(x+a)p(x)\,dx\right| - \left|\int_{|x|\geqq\gamma} H(x+a)p(x)\,dx\right|$$

and

$$(2)\qquad \left|\int_{|x|\geqq\gamma} H(x+a)p(x)\,dx\right| \leqq \alpha\int_{|x|\geqq\gamma} p(x)\,dx.$$

On the other hand, according to **a**, there exists a finite number s such that, say, $-\alpha = H(s)$. For $|x| < \gamma$, we have, setting $a = s - \gamma$ so that

$$s - 2\gamma < x + a < s, \quad x - \gamma < 0,$$

the relation

$$G(x + a) = G(s) + \theta(x - \gamma)G'(x'), \quad |\theta| \leqq 1, \quad s - 2\gamma < x' < s.$$

Thus, for $|x| < \gamma$,

$$\begin{aligned} H(x + a) &= F(x + a) - G(s) - \theta(x - \gamma)G'(x') \\ &\leqq F(s) - G(s) - \beta(x - \gamma) \\ &= -\alpha - \beta(x - \gamma) = -\beta(x + \gamma), \end{aligned}$$

and it follows that

$$(3) \qquad \begin{aligned} \int_{|x|<\gamma} H(x + a)p(x)\,dx &\leqq -\beta \int_{|x|<\gamma} (x + \gamma)p(x)\,dx \\ &= -\beta\gamma \int_{|x|<\gamma} p(x)\,dx \\ &= -\frac{\alpha}{2}\Big(1 - \int_{|x|\geqq\gamma} p(x)\,dx\Big). \end{aligned}$$

Upon substituting in (1) the bounds given by (2) and (3), we obtain

$$\left|\int H(x + a)p(x)\,dx\right| \geqq \frac{\alpha}{2}\Big(1 - 3\int_{|x|\geqq\gamma} p(x)\,dx\Big)$$

and the assertion follows. In the case $\alpha = H(s + 0)$, the argument is similar.

Let $\bar{\omega}$ be a real ch.f. with $\int |\bar{\omega}(u)|\,du < \infty$, so that the corresponding d.f. has a symmetric derivative continuous on R, given by

$$p(x) = \frac{1}{2\pi}\int e^{-iux}\bar{\omega}(u)\,du = \frac{1}{2\pi}\int \cos ux \cdot \bar{\omega}(u)\,du.$$

c. *For every* $a \in R$

$$\frac{1}{2\pi}\int \left|\frac{h(u)\bar{\omega}(u)}{u}\right| du \geqq \left|\int H(x + a)p(x)\,dx\right|.$$

Proof. We can assume $\frac{h(u)\bar{\omega}(u)}{u}$ to be integrable, for otherwise the inequality is trivially true. According to the composition theorem, $h\bar{\omega}$ is the Fourier-Stieltjes transform of $\bar{H}$ defined by

$$\bar{H}(x) = \int H(x-y)p(y)\,dy.$$

Since $\frac{h(u)\bar{\omega}(u)}{u}$ is integrable, the inversion formula yields

$$\bar{H}(x) - \bar{H}(x') = \frac{1}{2\pi}\int \frac{e^{-iux} - e^{-iux'}}{-iu} h(u)\bar{\omega}(u)\,du.$$

But, as $x' \to -\infty$, $H(x') \to 0$ and, by the Riemann-Lebesgue theorem, $\int e^{-iux'} \frac{h(u)\bar{\omega}(u)}{-iu}\,du \to 0$. Therefore,

$$\int H(x-y)p(y)\,dy = \frac{1}{2\pi}\int e^{-iux}\frac{h(u)\bar{\omega}(u)}{-iu}\,du$$

and, hence, replacing x by a, y by $-x$, and taking into account that p is symmetric, we obtain

$$\int H(x+a)p(x)\,dx = \frac{1}{2\pi}\int e^{-iua}\frac{h(u)\bar{\omega}(u)}{-iu}\,du.$$

The asserted inequality follows.

We are now in a position to establish the basic inequality below, of independent interest. We shall require a real integrable function $\bar{\omega}_0$ defined by $\bar{\omega}_0(u) = 1 - \frac{|u|}{U}$ or 0 according as $|u| < U$ or $|u| \geqq U$. Its Fourier-Stieltjes transform p_0 is given by

$$p_0(x) = \frac{1}{2\pi}\int_{-U}^{+U}\left(1 - \frac{|u|}{U}\right)\cos ux\,du = \frac{1 - \cos Ux}{\pi x^2 U}$$

and we have $p_0 \geqq 0$, $\int p_0(x)\,dx = 1$, so that $\bar{\omega}_0$ is a ch.f.

A. Basic inequality. *If G has a derivative G' on R, then, for every $U > 0$,*

$$\sup|H| \leqq \frac{2}{\pi}\int_0^U \left|\frac{h(u)}{u}\right| du + \frac{24}{\pi U}\sup|G'|.$$

Proof. Upon replacing $\bar{\omega}$ and p by $\bar{\omega}_0$ and p_0, the propositions **b** and **c** yield the inequality

$$\frac{1}{\pi}\int_{-U}^{U}\left|\frac{h(u)}{u}\right|du \geqq \frac{1}{2\pi}\int\left|\frac{h(u)\bar{\omega}_0(u)}{u}\right|du \geqq \frac{\alpha}{2}\left(1-6\int_{\gamma}^{\infty}p_0(x)\,dx\right)$$

where

$$\int_{\gamma}^{\infty}p_0(x)\,dx = \frac{1}{\pi}\int_{\gamma}\frac{1-\cos Ux}{x^2U}dx \leqq \frac{2}{\pi}\int_{\gamma U}^{\infty}\frac{dx}{x^2} = \frac{2}{\pi\gamma U} = \frac{4\beta}{\pi\alpha U}.$$

Therefore,

$$\frac{1}{\pi}\int_{0}^{U}\left|\frac{h(u)}{u}\right|du \geqq \frac{\alpha}{2} - \frac{12\beta}{\pi U}$$

and the asserted inequality follows.

In order to apply the basic inequality to the normal approximation problem, we have to bound the corresponding h. Let F^*_n and G^* be the d.f.'s of $\mathfrak{L}\left(\frac{S_n}{s_n}\right)$ and $\mathfrak{N}(0, 1)$ and let $h^*_n = f^*_n - e^{-\frac{u^2}{2}}$ denote the difference of the corresponding ch.f.'s. The summands X_n are independent r.v.'s centered at expectations, and we set $\gamma_n^3 = E|X_n|^3$, $g_n^3 = \frac{2^3}{s_n^3}\sum_{k=1}^{n}\gamma_k^3$. We exclude the case of one of the γ_n infinite, for then the normal approximation theorem below is trivially true.

d. *If* $|u| < \frac{2}{g_n^3}$, *then* $|h^*_n(u)| \leqq 2g_n^3|u|^3\exp\left[-\frac{u^2}{3}\right]$.

Proof. 1° First, we prove the assertion under the supplementary condition $|u| \geqq \frac{1}{g_n}$. Then $g_n^3|u|^3 \geqq 1$ and it suffices to prove that $|h^*_n(u)| \leqq 2\exp\left[-\frac{u^2}{3}\right]$. But, since

$$|h^*_n(u)| \leqq |f^*_n(u)| + \exp\left[-\frac{u^2}{2}\right] \leqq |f^*_n(u)| + \exp\left[-\frac{u^2}{3}\right],$$

it will suffice to prove that $|f^*_n(u)|^2 \leqq \exp\left[-\frac{2u^2}{3}\right]$.

Consider the symmetrized r.v. $X_k - X'_k$ where X_k and X'_k are independent and identically distributed, so that its ch.f. is $|f_k|^2$ and

$$E(X_k - X'_k)^2 = 2\sigma_k^2,\quad E|X_k - X'_k|^3 \leqq 2^3\gamma_k^3 < \infty.$$

Therefore,

$$|f_k(u)|^2 \leqq 1 - \sigma_k^2 u^2 + \frac{2^2}{3}\gamma_k^3|u|^3 \leqq \exp\left[-\sigma_k^2u^2 + \frac{2^2}{3}\gamma_k^3|u|^3\right]$$

and, replacing u by $\frac{u}{s_n}$ and summing over $k = 1, \cdots, n$, we obtain, using the fact that, by assumption, $g_n^3|u| < 2$,

$$|f^*_n(u)|^2 \leqq \exp\left[-u^2 + \frac{g_n^3|u|^3}{3.2}\right] \leqq \exp\left[-u^2 + \frac{u^2}{3}\right]$$

$$= \exp\left[-\frac{2}{3}u^2\right].$$

2° It remains to prove the assertion when $|u| < \frac{1}{g_n}$ and, hence,

$$\frac{\sigma_k}{s_n}|u| \leqq \frac{\gamma_k}{s_n}|u| \leqq \frac{g_n}{2}|u| < \frac{1}{2}.$$

Then, we have

$$f_k\left(\frac{u}{s_n}\right) = 1 - \frac{\sigma_k^2}{2s_n^2}u^2 + \theta\frac{\gamma_k^3}{6s_n^3}|u|^3 = 1 - r_k,$$

where $|r_k| < \frac{1}{2}$, so that

$$\log f_k\left(\frac{u}{s_n}\right) = -r_k + \theta'_k r_k^2.$$

On the other hand,

$$|r_k|^2 \leqq 2\left(\frac{\sigma_k^2u^2}{2s_n^2}\right)^2 + 2\left(\frac{\gamma_k^3|u|^3}{6s_n^3}\right)^2 \leqq \frac{\gamma_k^3}{3s_n^3}|u|^3.$$

so that

$$\log f_k\left(\frac{u}{s_n}\right) = -\frac{\sigma_k^2}{2s_n^2}u^2 + \theta''_k\frac{|\gamma_k|^3}{2s_n^3}|u|^3$$

and, summing over k, we obtain

$$\log f^*_n(u) = -\frac{u^2}{2} + \theta\frac{g_n^3}{2^4}|u|^3.$$

Since, for every number a, $e^a = 1 + \theta'_a e^{\theta' a}$, it follows, taking $a = \frac{g_n^3}{2^4}|u|^3 \leqq \frac{1}{2^4}$ so that $e^a < 2$, that

$$\left|f^*_n(u) - \exp\left[-\frac{u^2}{2}\right]\right| \leqq 2\frac{g_n^3}{2^4}|u|^3 \exp\left[-\frac{u^2}{2}\right]$$

$$\leqq 2g_n^3|u|^3 \exp\left[-\frac{u^2}{3}\right],$$

and the proof is complete.

B. NORMAL APPROXIMATION THEOREM. *There exists a numerical constant $c < \infty$ such that, for all x and all n, if F^*_n is d.f. of $\mathfrak{L}(S_n/s_n)$ and G^* is d.f. of $\mathfrak{N}(0, 1)$, then*

$$|F^*_n(x) - G^*(x)| \leqq \frac{c}{s_n^3}\sum_{k=1}^{n} E|X_k|^3.$$

For, upon replacing h^*_n by its bound obtained above in the basic inequality with $U = \frac{2}{g_n^3}$, $F = F^*_n$, and $G = G^*$ hence $\sup|G'| = \frac{1}{\sqrt{2\pi}}$, we obtain

$$\alpha \leqq \frac{4}{\pi}g_n^3\left(\int_0^\infty u^2 \exp\left[-\frac{u^2}{3}\right]du + \frac{3}{\sqrt{2\pi}}\right).$$

§ 21. CENTRAL LIMIT PROBLEM; THE CASE OF BOUNDED VARIANCES

21.1 Evolution of the problem. The classical limit problem deals with independent summands X_n with finite first moments and, in the normal convergence case, with finite second moments as well. Those moments are used for changing origins and scales of values of the consecutive sums $S_n = \sum_{k=1}^{n} X_k$ so as to avoid shifts of the pr. spreads towards infinite values. There is no reason for these choices of "norming" quantities except an historical one; they are a straightforward extension to more general cases of the norming quantities which appeared in the Bernoulli case. *A priori*, there is no reason to expect that these quantities will continue to play the same role in the general case. Furthermore, whether they are available (that is, exist and are finite) or not, other choices might achieve the same purpose. Thus, the problem becomes a search for conditions under which the law of large numbers and the normal convergence hold for normed sums $\frac{S_n}{b_n} - a_n$. The methods remain those of the classical problem, but the computations become more involved. However, remnants of the two first limit theorems in the Bernoulli case are still visible. For there is no other reason to expect or to look for limit laws which are either degenerate or normal.

The real liberation which gave birth to the Central Limit Problem came with a new approach due to P. Lévy. He stated and solved the following problem: Find the family of *all possible limit laws* of normed sums of independent and identically distributed r.v.'s. We saw that when these r.v.'s have a finite second moment, the limit law (with classical norming quantities) is normal. Thus, P. Lévy was concerned primarily with the novel case of infinite second moments and finite or infinite first moments.

Naturally, the question of all possible limit laws of normed sums with independent, but not necessarily identically distributed, r.v.'s arises at once. Yet, the Poisson limit theorem is still out, for it is relative to sequences of sums and not to sequences of normed consecutive sums. Moreover, as we shall find it later (end 23.4), under "natural" restrictions Poisson laws *cannot* be limit laws of sequences of normed sums—which explains their isolation. But sequences $\frac{S_n}{b_n} - a_n$ are a particular form of sequences $\sum_{k=1}^{n} X_{nk} \left(\text{set } X_{nk} = \frac{X_k}{b_n} - \frac{a_n}{n}\right)$, and this provides the final modification of the problem.

The general outline of the Central Limit Problem is now visible: Find *the* limit laws of sequences of sums of independent summands and find conditions for convergence to a specified one. Yet, so general a problem is without content. In fact, let Y_n be arbitrary r.v.'s, set $X_{n1} = Y_n$ and $X_{nk} = 0$ a.s. for $k > 1$ and every n. Then the sequence of laws becomes the sequence $\mathfrak{L}(Y_n)$, so that the family of possible limit laws contains *any* law $\mathfrak{L}$—take $\mathfrak{L}(Y_n) \equiv \mathfrak{L}$. Thus, some restriction is needed.

To find a "natural" one, let us consider the problems which led to this one. Their common feature is that the number of summands increases indefinitely and that the limit law remains the same if an arbitrary but finite number of summands is dropped. To emphasize this feature, we are led to the following "natural" restriction: the summands

X_{nk} are *uniformly asymptotically negligible* (*uan*), that is, $X_{nk} \xrightarrow{P} 0$ uniformly in k or, equivalently, for every $\epsilon > 0$,

$$\max_k P[|X_{nk}| \geqq \epsilon] \to 0.$$

Finally, the precise formulation of the problem is as follows:

Central limit problem. *Let* $S_{nk_n} = \sum_{k=1}^{k_n} X_{nk}$ *be sums of uan independent summands* X_{nk}, *with* $k_n \to \infty$.

1° *Find the family of all possible limit laws of these sums.*

2° *Find conditions for convergence to any specified law of this family.*

To simplify the writing, we make the following conventions valid for the whole chapter.

(i) $k = 1, \cdots, k_n$, $k_n \to \infty$, the summations $\sum_k$, the products $\prod_k$, the maxima $\max_k$, are over these values of k, and the limits are taken as usually for $n \to \infty$, unless otherwise stated.

(ii) F_{nk} and f_{nk} denote the d.f. and the ch.f. of r.v.'s X_{nk}, F_n and f_n denote the d.f. and the ch.f. of $\sum_k X_{nk}$. Thus, the uan condition becomes: $\max_k \int_{|x|\geqq\epsilon} dF_{nk} \to 0$ for every $\epsilon > 0$, and the assumption of independence becomes $f_n = \prod_k f_{nk}$. The problem becomes

Given sequences $f_n = \prod_k f_{nk}$ *of products of ch.f.'s of uan r.v.'s:* 1° *Find all ch.f.'s* f *such that* $f_n \to f$; 2° *Find conditions under which* $f_n \to f$ *given.* If these ch.f.'s have log's on $I = [-U, +U]$, we always select their principal branches—continuous and vanishing at $u = 0$, and then on I: $\log f_n = \sum_k \log f_{nk}$, $f_n \to f$ (uniformly) $\Leftrightarrow$ $\log f_n \to \log f$ (uniformly).

The solution of the problem is due to the introduction, by de Finetti, of the "infinitely decomposable" family of laws and to the discovery of their explicit representation by Kolmogorov in the case of finite second moments and by P. Lévy in the general case.

It has been obtained, with the help of the preceding family of laws, by the efforts of Kolmogorov, P. Lévy, Feller, Bawly, Khintchine, Marcinkiewicz, Gnedenko, and Doblin (1931–1938). The final form is essentially due to Gnedenko.

21.2 The case of bounded variances. As a preliminary to the investigation of the general problem, and independently of it, we examine here the particular "case of bounded variances"—a "natural" extension of the classical normal convergence problem. It is much less involved computationally than the general one, while the method of attack is essentially the same.

We consider sums $\sum_k X_{nk}$ of independent r.v.'s, centered at expectations, with d.f.'s F_{nk}, ch.f.'s f_{nk} and finite variances $\sigma_{nk}^2 = \sigma^2 X_{nk}$ such that

(C): $\max_k \sigma_{nk}^2 \to 0$ *and* $\sum_k \sigma_{nk}^2 \leqq c < \infty$, *where* c *is a constant independent of* n.

Since, for every $\epsilon > 0$,

$$\max_k P[|X_{nk}| \geqq \epsilon] \leqq \frac{1}{\epsilon^2} \max_k \sigma_{nk}^2 \to 0,$$

the uan condition is satisfied and the model is a particular case of that of the Central Limit Problem. The boundedness of the sequence of variances of the sums entails finiteness of the variance of the limit law.

a. COMPARISON LEMMA. *Under* (C), $\log f_{nk}(u)$ *exists and is finite for* $n \geqq n_u$ *sufficiently large and, for any fixed* u,

$$\sum_k \{\log f_{nk}(u) - (f_{nk}(u) - 1)\} \to 0.$$

Proof. Since $f_{nk}(u) = 1 - \theta_{nk} \dfrac{\sigma_{nk}^2}{2} u^2$, it follows from (C) that

$$\max_k |f_{nk}(u) - 1| \leqq \frac{u^2}{2} \max_k \sigma_{nk}^2 \to 0, \quad \sum_k |f_{nk}(u) - 1| \leqq \frac{c}{2} u^2.$$

Therefore, for $n \geqq n_u$ sufficiently large, $|f_{nk}(u) - 1| \leqq \frac{1}{2}$, so that the $\log f_{nk}(u)$ exist and are finite,

$$\log f_{nk}(u) = f_{nk}(u) - 1 + \theta'_{nk}|f_{nk}(u) - 1|^2,$$

and it follows that

$$\begin{aligned} \Big|\sum_k \{\log f_{nk}(u) - (f_{nk}(u) - 1)\}\Big| & \\ &\leqq \sum_k |f_{nk}(u) - 1|^2 \\ &\leqq \max_k |f_{nk}(u) - 1| \sum_k |f_{nk}(u) - 1| \to 0. \end{aligned}$$

The comparison lemma is proved.

Let

$$\psi_n(u) = \sum_k (f_{nk}(u) - 1) = \sum_k \int (e^{iux} - 1)\, dF_{nk}.$$

Since

$$\int x\, dF_{nk} = 0, \quad \sum_k \int x^2\, dF_{nk} \leqq c,$$

we have

$$\psi_n(u) = \sum_k \int (e^{iux} - 1 - iux)\frac{1}{x^2} \cdot x^2\, dF_{nk}$$

or

$$\psi_n(u) = \int (e^{iux} - 1 - iux)\frac{1}{x^2}\, dK_n,$$

where K_n on R is a continuous from the left nondecreasing function with $K_n(-\infty) = 0$, $\operatorname{Var} K_n \leqq c < \infty$, defined by

$$K_n(x) = \sum_k \int_{-\infty}^{x} y^2\, dF_{nk},$$

and the integrand, defined by continuity at $x = 0$, takes there the value $-u^2/2$. The comparison lemma becomes

a′. *Under* (C), $\log \prod_k f_{nk} - \psi_n \to 0$.

Functions of the foregoing type will be denoted in this subsection by ψ and K, with or without affixes. Thus, unless otherwise stated, ψ is a function defined on R by

$$\psi(u) = \int (e^{iux} - 1 - iux)\frac{1}{x^2}\, dK(x),$$

and K is a d.f.—up to a multiplicative constant—with $K(-\infty) = 0$, $\operatorname{Var} K \leqq c$; ψ and K will have same affixes if any.

b. *Every e^{ψ} is a ch.f. with null first moment and finite variance $\sigma^2 =$* $\operatorname{Var} K$, *and is a limit law under* (C).

Proof. The integrand is bounded in x and continuous in u (or x) for every fixed x (or u). It follows that ψ is continuous on R and is limit of Riemann-Stieltjes sums of the form $\sum_k \{iua_{nk} + \lambda_{nk}(e^{iub_{nk}} - 1)\}$ where

$$\lambda_{nk} = \frac{1}{x_{nk}^2} K[x_{nk}, x_{n,k+1}), \quad a_{nk} = -\lambda_{nk}x_{nk}, \quad b_{nk} = x_{nk};$$

we can and do take all subdivision points $x_{nk} \neq 0$. Since every summand is log of a (Poisson type) ch.f., the sums are log of ch.f.'s, and so is their limit ψ according to the continuity theorem. The second assertion follows, since, by elementary computations,

$$(e^{\psi})'_{u=0} = (\psi)'_{u=0} = 0, \quad (e^{\psi})''_{u=0} = (\psi'')_{u=0} = -\operatorname{Var} K.$$

Finally, let X_{nk}, $k = 1, \cdots, n$ be independent r.v.'s with common log of ch.f. being ψ/n. Since ψ/n corresponds to K/n, we have $\sigma^2 X_{nk} = \operatorname{Var} K/n$ while $EX_{nk} = 0$. Since $\sum_{k=1}^{n} X_{nk}$ has for ch.f. e^{ψ} whatever be n and condition (C) is fulfilled, the last assertion is proved.

c. Uniqueness lemma. *ψ determines K, and conversely.*

Proof. Since

$$-\psi''(u) = \int e^{iux}\, dK(x), \quad \operatorname{Var} K < \infty, \quad K(-\infty) = 0,$$

the inversion formula applies and K is determined by ψ by means of ψ''. The converse is obvious.

d. Convergence lemma. *Let* (C) *hold. If* $K_n \xrightarrow{w} K$, *then* $\psi_n \to \psi$. *Conversely, if* $\psi_n \to \log f$, *then* $K_n \xrightarrow{w} K$ *and* $\log f = \psi$ *determined by* K.

Proof. The first assertion follows at once from the extended Helly-Bray lemma. As for the converse, since the variations are uniformly bounded, the weak compactness theorem applies and there exists a K (with $\operatorname{Var} K \leqq c$) such that $K_{n'} \xrightarrow{w} K$ as $n' \to \infty$ along some subsequence of integers. Therefore, by the same lemma, $\psi_{n'} \to \psi = \log f$ since $\psi_n \to \log f$. But, by the uniqueness lemma, $\psi = \log f$ determines K, and it follows that $K_n \xrightarrow{w} K$. The proposition is proved.

Upon applying the foregoing lemmas, the answer to our problem follows:

A. Bounded variances limit theorem. *If independent summands X_{nk} are centered at expectations and $\max_k \sigma_{nk}^2 \to 0$, $\sum_k \sigma_{nk}^2 \leqq c < \infty$ for all n, then*

1° *the family of limit laws of sequences $\mathfrak{L}(\sum_k X_{nk})$ coincides with the family of laws of r.v.'s centered at expectations with finite variances and ch.f.'s of the form $f = e^{\psi}$, where ψ is of the form*

$$\psi(u) = \int (e^{iux} - 1 - iux) \frac{1}{x^2}\, dK(x),$$

with K continuous from the left and nondecreasing on R and $\operatorname{Var} K \leqq c < \infty$; *$\psi$ determines K and conversely.*

2° $\mathfrak{L}(\sum_k X_{nk}) \to \mathfrak{L}(X)$ *with ch.f. necessarily of the form e^{ψ} if, and only if, $K_n \xrightarrow{w} K$ where K_n are defined by*

$$K_n(x) = \sum_k \int_{-\infty}^{x} y^2 \, dF_{nk}.$$

If $\sum_k \sigma_{nk}^2 \leqq c < \infty$ is replaced by $\sum_k \sigma_{nk}^2 \to \sigma^2 X < \infty$, then $K_n \xrightarrow{w} K$ is to be replaced by $K_n \xrightarrow{c} K$.

Proof. 1° follows from **b**, the comparison lemma and the convergence lemma.

2° follows from 1° and the convergence lemma; and the particular case follows from the fact that the assumption made becomes

$$\operatorname{Var} K_n = \sum_k \sigma_{nk}^2 \to \sigma^2 X = \operatorname{Var} K.$$

EXTENSION. So far the r.v.'s under consideration were all centered at expectations. If we suppress this condition and set

$$a_{nk} = EX_{nk}, \quad \bar{F}_{nk}(x) = F_{nk}(x + a_{nk}), \quad \bar{f}_{nk}(u) = e^{-iua_{nk}} f_{nk}(u),$$

then the foregoing results continue to apply, provided F_{nk} and f_{nk} are replaced everywhere by $\bar{F}_{nk}$ and $\bar{f}_{nk}$; and then we write $\bar{\psi}$ instead of ψ. Going back to the noncentered r.v.'s, we have to introduce limit laws $\mathfrak{L}(X)$ with finite variances but not necessarily null expectations $a = EX$, whose log's of ch.f.'s are of the form $\psi(u) = iua + \bar{\psi}(u)$, so that $\left(\dfrac{d\psi(u)}{du}\right)_0 = ia$.

The uniqueness lemma becomes: ψ determines a and K, and conversely.

In the convergence lemma, $K_n \xrightarrow{w} K$ is replaced by $K_n \xrightarrow{w} K$ and $a_n \to a$.

The same is to be done in the limit theorem with $a_n = \sum_k a_{nk}$ and F_{nk} replaced by $\bar{F}_{nk}$.

Thus, the convergence criterion **A2°** becomes

EXTENDED CONVERGENCE CRITERION. *If independent summands X_{nk} are such that $\max_k \sigma_{nk}^2 \to 0$ and $\sum_k \sigma_{nk}^2 \leqq c < \infty$, then $\mathfrak{L}(\sum_k X_{nk}) \to$*

$\mathfrak{L}(X)$ with ch.f. necessarily of the form e^{ψ} if, and only if, $K_n \xrightarrow{w} K$ and $\sum_k a_{nk} \to a$ where

$$K_n(x) = \sum_k \int_{-\infty}^{x} y^2\, dF_{nk}(y + a_{nk}), \quad a_{nk} = EX_{nk}.$$

If $\sum_k \sigma_{nk}^2 \leqq c < \infty$ is replaced by $\sum_k \sigma_{nk}^2 \to \sigma^2 X < \infty$, then $K_n \xrightarrow{w} K$ is to be replaced by $K_n \xrightarrow{c} K$.

Particular cases:

1° NORMAL CONVERGENCE. The normal law $\mathfrak{N}(0, 1)$ corresponds to $\psi(u) = -\dfrac{u^2}{2}$ and, hence, to K defined by $K(x) = 0$ or 1 according as $x < 0$ or $x > 0$ (because of the uniqueness lemma, it suffices to verify that this K gives the above ψ).

NORMAL CONVERGENCE CRITERION. *Let the independent summands X_{nk}, centered at expectations, be such that $\sum_k \sigma_{nk}^2 = 1$ for all n:*

then $\mathfrak{L}(\sum_k X_{nk}) \to \mathfrak{N}(0, 1)$ and $\max_k \sigma_{nk}^2 \to 0$ if, and only if, for every $\epsilon > 0$,

$$g_n(\epsilon) = \sum_k \int_{|x| \geqq \epsilon} x^2\, dF_{nk} \to 0.$$

Proof. Since

$$\max_k \sigma_{nk}^2 = \max_k \int x^2\, dF_{nk}(x) \leqq \epsilon^2 + \max_k \int_{|x| \geqq \epsilon} x^2\, dF_{nk} \leqq \epsilon^2 + g_n(\epsilon),$$

it follows that $g_n(\epsilon) \to 0$ for every $\epsilon > 0$ implies (letting $n \to \infty$ and then $\epsilon \to 0$ in the foregoing relation) $\max_k \sigma_{nk}^2 \to 0$. Then, immediate computations show that the convergence criterion **A2°** is equivalent to $g_n(\epsilon) \to 0$ for every $\epsilon > 0$.

Upon setting $X_{nk} = \dfrac{X_k}{s_n}$, $k = 1, \cdots, n$, $EX_k = 0$, $s_n^2 = \sum_k \sigma^2 X_k$, we obtain the classical normal convergence criterion. Liapounov's theorem follows from

$$\int_{|x| \geqq \epsilon s_n} x^2\, dF_k \leqq \frac{1}{\epsilon^\delta s_n^\delta} \int |x|^{2+\delta}\, dF_k.$$

2° POISSON CONVERGENCE. The Poisson law $\mathcal{P}(\lambda)$ corresponds to $\psi(u) = iu\lambda + \lambda(e^{iu} - 1 - iu) = iu\lambda + \bar{\psi}(u)$ and, hence, the function

K which corresponds to ψ is defined by $K(x) = 0$ or λ according as $x < 1$ or $x > 1$. The extended convergence criterion yields, by immediate transformations, the following

Poisson convergence criterion. *If the independent summands* X_{nk} *are such that* $\max \sigma_{nk}^2 \to 0$ *and* $\sum_k \sigma_{nk}^2 \to \lambda$, *then* $\mathfrak{L}(\sum_k X_{nk}) \to \mathcal{P}(\lambda)$ *if, and only if,* $\sum_k EX_{nk} \to \lambda$ *and, for every* $\epsilon > 0$,

$$\sum_k \int_{|x-1|\geqq\epsilon} x^2\, dF_{nk}(x + EX_{nk}) \to 0.$$

*§ 22. SOLUTION OF THE CENTRAL LIMIT PROBLEM

We consider now the general problem. As was pointed out, the method of attack will be essentially the same as in the case of bounded variances. The computational difficulties will arise from two facts. (1) Even existence of first moments is not assumed, and the centerings, instead of being at expectations, will have to be at truncated expectations. (2) The functions K defined previously are not necessarily of bounded variation and, even when they are, they are not assumed to be of uniformly bounded variation. They will have to be replaced by functions of the form $\Psi_n(x) = \sum_k \int_{-\infty}^{x} \frac{y^2}{1+y^2}\, d\overline{F}_{nk}$ where $\overline{F}_{nk}$ will be d.f.'s of the summands centered at truncated expectations. This will lead to limit laws with log ch.f.'s of a more complicated form, which we investigate first.

22.1 A family of limit laws; the infinitely decomposable laws. A law $\mathfrak{L}$ and its ch.f. f are said to be *infinitely decomposable* (*i.d.*) if, for every integer n, there exist (on some pr. space) n independent and identically distributed r.v.'s X_{nk}, such that $\mathfrak{L} = \mathfrak{L}(\sum_{k=1}^{n} X_{nk})$; in other words, for every n there exists a ch.f. f_n such that $f = f_n^n$. If $f \neq 0$, then $\log f$ exists and is finite and $f_n = e^{(1/n)\log f}$; unless otherwise stated, we select for log of a ch.f. its principal branch (vanishing at $u = 0$) and for the nth root of f we take the function defined by the preceding equality.

Clearly, if a law is i.d., so is its type. The degenerate, normal, and Poisson type are i.d., since if $\log f(u) = iua$ or $iua - \sigma^2 \frac{u^2}{2}$ or $iua + \lambda(e^{iub} - 1)$, then $\frac{1}{n} \log f(u)$ has the same form whatever be n. More generally, the limit laws e^{ψ} obtained in the case of bounded variances are i.d., since the corresponding functions ψ are such that ψ/n is log of a ch.f. of the same form (with $a/n \in R$ and K/n d.f. up to a multiplicative constant). In fact

a. *The i.d. family belongs to that of limit laws of the Central Limit Problem.*

For, on the one hand, the uan condition for independent and identically distributed r.v.'s X_{nk} which figure in the definition of i.d. laws becomes convergence of their common law to the degenerate at 0, that is, $f_n \to 1$; on the other hand,

b. *If, for every n, $f = f_n^n$ where f_n is a ch.f., then $f_n \to 1$; and, moreover, $f \neq 0$.*

Proof. Since $|f| \leq 1$, we have $|f_n|^2 = |f|^{2/n} \to g$ with $g(u) = 0$ or 1 according as $f(u) = 0$ or $f(u) \neq 0$. Since f is continuous and $f(0) = 1$, there exists a neighborhood of the origin where $|f(u)| > 0$ and, hence, $g(u) = 1$, so that g is continuous in this neighborhood. Thus, the sequence $|f_n|^2$ of ch.f.'s converges to a function g continuous at the origin, the continuity theorem applies, and g is a ch.f. Therefore, g is continuous on R with $g(0) = 1$ and, since it takes at most two values 0 and 1, it reduces to 1. Consequently, $f \neq 0$, $\log f$ exists and is finite, and $f_n = e^{\frac{1}{n}\log f} \to 1$. The proposition is proved.

We shall see later that the family of limit laws of the problem *coincides* with the i.d. one. This explains the property below.

A. Closure theorem. *The i.d. family is closed under compositions and passages to the limit.*

Proof. If f and f' are i.d. ch.f.'s, then, for every n, there exist ch.f.'s f_n and f'_n such that $f = f_n^n$, $f' = f'^n_n$, so that $ff' = (f_n f'_n)^n$ where $f_n f'_n$ are ch.f.'s, and the first assertion is proved.

On the other hand, if a sequence f_n of i.d. ch.f.'s converges to a ch.f. f, then, for every integer m, $|f_n|^{\frac{2}{m}} \to |f|^{\frac{2}{m}}$ and, by the continuity theorem, $|f|^{\frac{2}{m}}$ is a ch.f. Therefore, $|f|^2$ is an i.d. ch.f. and, hence, by **b**, $f \neq 0$. Since $\log f$ exists and is finite, and

$$f_n^{\frac{1}{m}} = e^{\frac{1}{m}\log f_n} \to e^{\frac{1}{m}\log f} = f^{\frac{1}{m}},$$

it follows that $f^{1/m}$ is a ch.f., so that f is i.d. This concludes the proof.

The basic feature of i.d. laws (hence, as we shall see, of all the limit laws of the Central Limit Problem) is that they are constructed by means of Poisson type laws. This is made precise in the theorem below and explicited by the representation theorem which will follow.

B. Structure theorem. *A ch.f. is i.d. if, and only if, it is the limit of sequences of products of Poisson type ch.f.'s.*

In other words, the class of i.d. laws coincides with the limit laws of sequences of sums of independent Poisson type r.v.'s.

Proof. Products f_n of Poisson type ch.f.'s are defined by finite sums of the form

$$\log f_n(u) = \sum_k \{iua_{nk} + \lambda_{nk}(e^{iub_{nk}} - 1)\}, \quad \lambda_{nk} \geqq 0,$$

so that the functions $\frac{1}{m}\log f_n$ are log of ch.f.'s (of the same kind) whatever be the fixed integer m and the f_n are i.d.ch.f.'s. Thus, by **A**, if $f_n \to f$ ch.f., then f is i.d. This proves the "if" assertion.

Conversely, if f is i.d., then $\log f$ exists and is finite and

$$n(f^{\frac{1}{n}} - 1) \to \log f,\ f^{\frac{1}{n}}(u) - 1 = \int (e^{iux} - 1)\, dF_n(x)$$

where F_n are d.f.'s. By taking Riemann-Stieltjes sums which approximate $f^{1/n}(u) - 1$ by less than $1/n^2$, the "only if" assertion follows, and the proof is terminated.

In what precedes, $\psi_n(u) = \int (e^{iux} - 1) n dF_n(x) \to \log f(u)$ and ψ_n is itself log of an i.d.ch.f. Since Var $(nF_n) = n \to \infty$, brutal interchange of integration and passage to the limit is excluded. However, the integral inequality in 12.4 yields Var $\Psi_n \leqq c < \infty$ with $d\Psi_n(x) = (x^2/1 + x^2) n dF_n(x)$ so that the weak compactness theorem applies. But the integrand for $d\Psi_n(x)$ is undetermined at $x = 0$, and we have to modify it. This leads to the ψ-functions below:

Unless otherwise stated, ψ, with or without affixes, will denote a function defined on R by

$$\psi(u) = iu\alpha + \int \left(e^{iux} - 1 - \frac{iux}{1 + x^2}\right) \frac{1 + x^2}{x^2}\, d\Psi(x)$$

where $\alpha \in R$ and Ψ denoting a d.f.—up to a multiplicative constant, with $\Psi(-\infty) = 0$; the corresponding ψ, α, Ψ will have same affixes if any. The value of the integrand at $x = 0$, defined by continuity, is $-u^2/2$.

c. *Every e^ψ is an i.d. ch.f.*

Proof. We use repeatedly the fact that the class of log's of ch.f.'s is closed under additions.

The integrand is bounded in x and continuous in u (or x) for every fixed x (or u). It follows that the integral is continuous in u and is limit of Riemann-Stieltjes sums of the form

$$\sum_k \{iua_{nk} + \lambda_{nk}(e^{iub_{nk}} - 1)\}$$

where

$$\lambda_{nk} = \frac{1 + x_{nk}^2}{x_{nk}^2}\Psi[x_{nk}, x_{n,k+1}), \quad a_{nk} = -\lambda_{nk}\frac{x_{nk}}{1 + x_{nk}^2}, \quad b_{nk} = x_{nk}$$

we can and do take all $x_{nk} \neq 0$. Since every nonvanishing summand is log of a (Poisson type) ch.f., the sums are log's of ch.f.'s, and so is the integral according to the continuity theorem. Since $iu\alpha$ is log of a ch.f., so is ψ and, hence, so is every ψ/n corresponding to $\alpha/n \in R$ and Ψ/n—d.f. up to a multiplicative constant. The assertion is proved.

REMARK. If $\int x^2 d\Psi(x) < \infty$, then

$$\psi(u) = iua + \int (e^{iux} - 1 - iux)\frac{1}{x^2}\, dK(x)$$

where

$$a = \alpha + \int x\, d\Psi(x) \in R, \quad dK(x) = (1 + x^2)\, d\Psi(x),$$

and the i.d. ch.f. e^ψ has for first moment a and for variance Var $K < \infty$ (take the first two derivatives at $u = 0$). Conversely, if an i.d. ch.f. e^ψ has second (hence first) finite moment, then $\int x^2 d\Psi(x) < \infty$ (take the second symmetric derivative at $u = 0$). Thus, the family of all limit laws in the case of bounded variances coincides with the subfamily of i.d. laws with finite second moments.

We establish now two properties of functions ψ corresponding to the unicity and continuity theorems for ch.f.'s. They will be reduced to these theorems by making correspond to functions ψ functions φ and Φ, with same affixes as ψ if any. We define φ on R by

$$\varphi(u) = \psi(u) - \int_0^1 \frac{\psi(u + h) + \psi(u - h)}{2}\, dh.$$

We have, upon replacing ψ by its defining relation and interchanging the integrations,

$$\varphi(u) = \int_0^1 \left\{ \int e^{iux}(1 - \cos hx) \frac{1 + x^2}{x^2} d\Psi \right\} dh = \int e^{iux} \, d\Phi$$

with

$$\Phi(x) = \int_{-\infty}^{x} \left(1 - \frac{\sin y}{y}\right) \frac{1 + y^2}{y^2} d\Psi.$$

Since

$$0 < c' \leqq \left(1 - \frac{\sin x}{x}\right) \frac{1 + x^2}{x^2} \leqq c'' < \infty$$

where c' and c'' are independent of $x \in R$, it follows that Φ is nondecreasing on R with

$$c' \operatorname{Var} \Psi \leqq \operatorname{Var} \Phi \leqq c'' \operatorname{Var} \Psi < \infty$$

and

$$\Psi(x) = \int_{-\infty}^{x} d\Phi \Big/ \left(1 - \frac{\sin y}{y}\right) \frac{1 + y^2}{y^2}.$$

C. UNICITY THEOREM. *There is a one-to-one correspondence between functions ψ and couples (α, Ψ).*

For this reason we shall sometimes write $\psi = (\alpha, \Psi)$.

Proof. By definition, every couple (α, Ψ) determines a function ψ. Conversely, if ψ is given, then, by the foregoing considerations, ψ determines a function φ which is a ch.f. (up to a constant factor). By the inversion formula for ch.f.'s, φ determines Φ and, in its turn, Φ determines Ψ; furthermore, ψ and Ψ determine α, which completes the proof.

D. CONVERGENCE THEOREM. *If $\alpha_n \to \alpha$ and $\Psi_n \xrightarrow{c} \Psi$, then $\psi_n \to \psi$. Conversely, if $\psi_n \to g$ continuous at the origin, then $\alpha_n \to \alpha$ and $\Psi_n \xrightarrow{c} \Psi$ such that $g = \psi = (\alpha, \Psi)$.*

Proof. The first assertion follows at once by the Helly-Bray theorem. As for the converse, since the sequence e^{ψ_n} of i.d. ch.f.'s converges to e^g continuous at the origin, this convergence is uniform in every finite interval and, by 22.1**b** and **A**, e^g is an i.d. ch.f. with $e^g \neq 0$. Hence, g is finite and continuous on R, the sequence ψ_n converges to g uniformly on every finite interval, and

$$\varphi_n(u) \to g(u) - \int_0^1 \frac{g(u + h) + g(u - h)}{2} dh$$

continuous on R. In particular,

$$\operatorname{Var}\Phi_n = \varphi_n(0) \to \int_0^1 \frac{g(h) + g(-h)}{2}\,dh < \infty,$$

so that variations of the Φ_n are uniformly bounded. Thus, the continuity theorem applies to the sequence φ_n, and there exists a nondecreasing function Φ of bounded variation on R such that, upon applying the Helly-Bray theorem, at every continuity point x of Φ as well as for $x = +\infty$,

$$\Psi_n(x) = \int_{-\infty}^{x} d\Phi_n \Big/ \left(1 - \frac{\sin y}{y}\right)\frac{1 + y^2}{y^2}$$
$$\to \Psi(x) = \int_{-\infty}^{x} d\Phi \Big/ \left(1 - \frac{\sin y}{y}\right)\frac{1 + y^2}{y^2}.$$

Hence, $\Psi_n \xrightarrow{c} \Psi$ and, by the same theorem,

$$iu\alpha_n = \psi_n(u) - \int\left(e^{iux} - 1 - \frac{iux}{1 + x^2}\right)\frac{1 + x^2}{x^2}\,d\Psi_n$$
$$\to g(u) - \int\left(e^{iux} - 1 - \frac{iux}{1 + x^2}\right)\frac{1 + x^2}{x^2}\,d\Psi$$
$$= iu\alpha.$$

This terminates the proof.

E. Representation theorem. *The family of i.d. ch.f.'s coincides with the family of ch.f.'s of the form e^{ψ}.*

Proof. According to 22.1**c**, every e^{ψ} is an i.d. ch.f. Conversely if, for every $n, f = f_n^{\,n}$ where f_n is a ch.f. corresponding to a d.f. F_n, then, upon applying the preceding convergence theorem, we obtain

$$\log f(u) = \lim n(f^{\frac{1}{n}} - 1) = \lim n(f_n - 1) = \lim \int (e^{iux} - 1)n\,dF_n$$
$$= \lim\left(iu\int \frac{nx}{1 + x^2}\,dF_n\right.$$
$$\left. + \int\left(e^{iux} - 1 - \frac{iux}{1 + x^2}\right)\frac{1 + x^2}{x^2}\cdot\frac{x^2}{1 + x^2}\,n\,dF_n\right)$$
$$= \lim \psi_n = \text{some } \psi,$$

with

$$d\Psi_n(x) = n\frac{x^2}{1 + x^2}\,dF_n(x) \quad \text{and} \quad \Psi_n \xrightarrow{c} \Psi.$$

The theorem is proved.

22.2 The uan condition. The main computational difficulties arise in connection with the uan condition, and we have to investigate it in detail. We recall that given a sequence of sums $\sum_k X_{nk}$ of independent r.v.'s, the uan condition is that, for every $\epsilon > 0$,

$$\max_k P[|X_{nk}| \geqq \epsilon] = \max_k \int_{|x|\geqq\epsilon} dF_{nk} \to 0.$$

a. *The uan condition implies that*

$$\max_k |\mu X_{nk}| \to 0, \quad \max_k \int_{|x|<\tau} |x|^r \, dF_{nk} \to 0, \quad r > 0, \quad \tau > 0 \textit{ finite}.$$

Proof. The medians of a r.v. belong to any interval such that the pr. for the r.v. to be in the interval is greater than 1/2. Since under the uan condition $\min_k P[|X_{nk}| < \epsilon] > 1/2$ whatever be $\epsilon > 0$, provided $n \geqq n_\epsilon$ sufficiently large, it follows that $\max_k |\mu X_{nk}| < \epsilon$ for $n \geqq n_\epsilon$, and the first assertion is proved.

Under the same condition, by letting $n \to \infty$ and then $\epsilon \to 0$, we have

$$\max_k \int_{|x|<\tau} |x|^r \, dF_{nk} \leqq \epsilon^r + \max_k \int_{\epsilon \leqq |x| < \tau} |x|^r \, dF_{nk}$$

$$\leqq \epsilon^r + \tau^r \max_k \int_{|x|\geqq\epsilon} dF_{nk} \to 0,$$

and the second assertion is proved.

A. UAN CRITERIA. *The uan condition is equivalent to*

$$\max_k \int \frac{x^2}{1+x^2} \, dF_{nk} \to 0 \quad \textit{or} \quad \max_k |f_{nk} - 1| \to 0$$

uniformly on every finite interval.

Proof. Under the uan condition, by letting $n \to \infty$ and then $\epsilon \to 0$, we have

$$\max_k \int \frac{x^2}{1+x^2} \, dF_{nk} \leqq \epsilon^2 + \max_k \int_{|x|\geqq\epsilon} dF_{nk} \to 0$$

and for $|u| \leqq b < \infty$,

$$\max_k |f_{nk}(u) - 1|$$
$$\leq \max_k \left| \int_{|x|<\epsilon} (e^{iux} - 1)\, dF_{nk} \right| + \max_k \left| \int_{|x|\geq\epsilon} (e^{iux} - 1)\, dF_{nk} \right|$$
$$\leq b\epsilon + 2 \max_k \int_{|x|\geq\epsilon} dF_{nk} \to 0.$$

Conversely, if $\max_k \int \frac{x^2}{1+x^2}\, dF_{nk} \to 0$, then, for every $\epsilon > 0$,

$$\max_k \int_{|x|\geq\epsilon} dF_{nk} \leq \frac{1+\epsilon^2}{\epsilon^2} \max_k \int_{|x|\geq\epsilon} \frac{x^2}{1+x^2}\, dF_{nk} \to 0,$$

and the uan condition holds.

Since, upon replacing $f_{nk}(u)$ by $\int e^{iux}\, dF_{nk}$ and interchanging the integrations, we have

$$\max_k \int \frac{x^2}{1+x^2}\, dF_{nk} = \max_k \int_0^\infty e^{-u}(1 - \mathcal{R} f_{nk}(u))\, du$$
$$\leq \int_0^\infty e^{-u} \max_k |f_{nk}(u) - 1|\, du,$$

it follows, by the dominated convergence theorem, that $\max_k |f_{nk} - 1| \to 0$ implies the uan condition, and the proof is complete.

From now on, we fix a finite $\tau > 0$ and, for every d.f. F, with or without affixes, we set

$$a = \int_{|x|<\tau} x\, dF, \quad \bar{F}(x) = F(x + a), \quad \bar{f}(u) = \int e^{iux}\, d\bar{F}$$

with same affixes if any.

We observe that $|a| < \tau$ and that the "bar" does not mean "complex-conjugate."

COROLLARY 1. *Under the uan condition, $\max_k |\bar{f}_{nk} - 1| \to 0$ uniformly on every finite interval.*

Since, by **a**, $\max_k |a_{nk}| \leq \max_k \int_{|x|<\tau} |x|\, dF_{nk} \to 0$, the r.v.'s $\bar{X}_{nk} = X_{nk} - a_{nk}$ obey the uan condition, and the assertion follows by **A**.

COROLLARY 2. *Under the uan condition, given $b < \infty$, all $\log f_{nk}(u)$ exist and are finite for $|u| \leqq b$ and $n \geqq n_b$ sufficiently large, and*

$$\log f_{nk}(u) = f_{nk}(u) - 1 + \theta_{nk}|f_{nk}(u) - 1|^2, \quad |\theta_{nk}| \leqq 1;$$

similarly for the $\bar{f}_{nk}(u)$.

This follows from **A** and $\log z = (z-1) + |\theta|\, |z-1|^2$ for $|z-1| < \frac{1}{2}$.

From now on, if $b > 0$ is given, then we take $n \geqq n_b$ so that the foregoing relations hold.

We are now in a position to establish the inequalities which will lead almost at once to the solution of the Central Limit Problem.

B. CENTRAL INEQUALITIES. *Under the uan condition, for $n \geqq n_b$ sufficiently large, there exist two finite positive constants $c_1 = c_1(b, \tau)$ and $c_2 = c_2(b, \tau)$ such that*

$$c_1 \max_{|u| \leqq b} |\bar{f}_{nk}(u) - 1| \leqq \int \frac{x^2}{1+x^2}\, d\bar{F}_{nk} \leqq c_2 \int_0^b |\log|f_{nk}(u)||\, du.$$

The inequalities follow at once, upon applying **a**, from two inequalities, valid for arbitrary r.v.'s, that we establish now. We shall use repeatedly the two relations

$$\int g(x)\, dF(x+c) = \int g(x-c)\, dF(x),$$

and

$$\int_{|x|<\tau} (x-a)\, dF = a - a\int_{|x|<\tau} dF = a\int_{|x|\geqq\tau} dF.$$

B₁. LOWER BOUND. *There exists a finite positive number $c_1 = c_1(a, b, \tau)$ such that*

$$c_1 \max_{|u| \leqq b} |\bar{f}(u) - 1| \leqq \int \frac{x^2}{1+x^2}\, d\bar{F}.$$

Proof. Since, for $|u| \leqq b < \infty$,

$$|\bar{f}(u) - 1|$$

$$= \left|\int (e^{iu(x-a)} - 1)\, dF\right| \leqq 2\int_{|x|\geqq\tau} dF + b\left|\int_{|x|<\tau} (x-a)\, dF\right| + \frac{b^2}{2}\int_{|x|<\tau} (x-a)^2\, dF$$

$$= (2 + |a|b)\int_{|x|\geqq\tau} dF + \frac{b^2}{2}\int_{|x|<\tau} (x-a)^2\, dF$$

where

$$\int_{|x|\geq\tau} dF \leq \frac{1+(\tau+|a|)^2}{(\tau-|a|)^2}\int_{|x|\geq\tau}\frac{(x-a)^2}{1+(x-a)^2}\,dF$$

and

$$\int_{|x|<\tau}(x-a)^2\,dF \leq \{1+(\tau+|a|)^2\}\int_{|x|<\tau}\frac{(x-a)^2}{1+(x-a)^2}\,dF,$$

it follows that

$$|\bar{f}(u)-1| \leq \frac{1}{c_1}\int\frac{x^2}{1+x^2}\,d\bar{F}$$

where

$$\frac{1}{c_1} = \{1+(\tau+|a|)^2\}\left\{\frac{2+|a|b}{(\tau-|a|)^2}+\frac{b^2}{2}\right\},$$

and the asserted inequality is proved.

Under the uan assertion, for n sufficiently large, we have, according to **a**, $|a| < \frac{\tau}{2}$, and we can take for $c_1 = c_1(b, \tau)$ the value of c_1 obtained upon replacing $|a|$ by $\frac{\tau}{2}$. This proves the left-hand side central inequality.

$\mathbf{B}_2$. Upper bound. *For $\tau > |\mu|$, μ a median of F, there exists a finite positive number $c_2 = c(\mu, b, \tau)$ such that*

$$\int\frac{x^2}{1+x^2}\,d\bar{F} \leq c_2\int_0^b (1-|f(u)|^2)\,du.$$

If $f(u) \neq 0$ for $|u| \leq b$, then $1-|f(u)|^2$ can be replaced by $2|\log|f(u)||$.

Proof. On account of the elementary inequality

$$1-|f|^2 \leq -\log|f|^2 = 2|\log|f||,$$

the second assertion follows from the first one. To prove the first assertion, we shall use the symmetrization method and denote by F^s the d.f. of the symmetrized r.v. $X - X'$ where X and X' are independent and identically distributed, so that the corresponding ch.f. $f^s = |f|^2$.

From the elementary inequality

$$\left(1 - \frac{\sin bx}{bx}\right)\frac{1 + x^2}{x^2} \geqq c(b) > 0, \quad x \in R,$$

and the relation (obtained upon interchanging the integrations)

$$\int_0^b (1 - |f(u)|^2)\, du = \int\left\{\int_0^b (1 - \cos ux)\, du\right\} dF^s$$

$$= b\int\left(1 - \frac{\sin bx}{bx}\right)\frac{1 + x^2}{x^2}\cdot\frac{x^2}{1 + x^2}\, dF^s,$$

it follows that

$$(1) \qquad \int_0^b (1 - |f(u)|^2)\, du \geqq bc(b)\int\frac{x^2}{1 + x^2}\, dF^s.$$

We pass now from F^s to F^μ, the d.f. of $X - \mu$, and set

$$q^\mu(t) = P[|X - \mu| \geqq t], \quad q^s(t) = P[|X^s| \geqq t], \quad t \in [0, +\infty),$$

so that, upon applying the weak symmetrization lemma (which says that $q^\mu \leqq 2q^s$) and integrating by parts, we obtain

$$(2) \qquad \int\frac{x^2}{1 + x^2}\, dF^\mu = -\int_0^\infty \frac{t^2}{1 + t^2}\, dq^\mu = \int_0^\infty q^\mu(t) d\left(\frac{t^2}{1 + t^2}\right)$$

$$\leqq 2\int_0^\infty q^s(t)\, d\left(\frac{t^2}{1 + t^2}\right) = 2\int\frac{x^2}{1 + x^2}\, dF^s.$$

Now, we pass from F^μ to $\overline{F}$. From the elementary inequality

$$(x - a)^2 \leqq (x - \mu)^2 + 2(\mu - a)(x - a),$$

it follows that

$$\int_{|x|<\tau}(x - a)^2\, dF \leqq \int_{|x|<\tau}(x - \mu)^2\, dF + 2(\tau + |\mu|)\left|\int_{|x|<\tau}(x - a)\, dF\right|$$

$$\leqq \int_{|x|<\tau}(x - \mu)^2\, dF + 2\tau(\tau + |\mu|)\int_{|x|\geqq\tau} dF$$

and, hence,

$$\int\frac{x^2}{1 + x^2}\, d\overline{F} = \int\frac{(x - a)^2}{1 + (x - a)^2}\, dF \leqq \int_{|x|<\tau}(x - a)^2\, dF + \int_{|x|\geqq\tau} dF$$

$$\leqq \int_{|x|<\tau}(x - \mu)^2\, dF + \{1 + 2\tau(\tau + |\mu|)\}\int_{|x|\geqq\tau} dF.$$

Since

$$\int_{|x|<\tau}(x-\mu)^2\,dF \leqq \{1+(\tau+|\mu|)^2\}\int_{|x|<\tau}\frac{(x-\mu)^2}{1+(x-\mu)^2}\,dF$$

$$\leqq \{1+(\tau+|\mu|)^2\}\int\frac{x^2}{1+x^2}\,dF^{\mu}$$

and

$$\int_{|x|\geqq\tau}dF \leqq \frac{1+(\tau+|\mu|)^2}{(\tau-|\mu|)^2}\int_{|x|\geqq\tau}\frac{(x-\mu)^2}{1+(x-\mu)^2}\,dF$$

$$\leqq \frac{1+(\tau+|\mu|)^2}{(\tau-|\mu|)^2}\int\frac{x^2}{1+x^2}\,dF^{\mu},$$

it follows that

$$(3)\qquad \int\frac{x^2}{1+x^2}\,d\overline{F} \leqq c'\int\frac{x^2}{1+x^2}\,dF^{\mu}$$

where

$$c' = c'(\mu,\tau) = \{1+(\tau+|\mu|)^2\}\left\{1+\frac{1+2\tau(\tau+|\mu|)}{(\tau-|\mu|)^2}\right\}.$$

Together, the inequalities (1), (2), and (3) yield the inequality

$$\int\frac{x^2}{1+x^2}\,d\overline{F} \leqq c_2\int_0^b(1-|f(u)|^2)\,du$$

with $c_2 = \dfrac{2c'}{bc(b)}$, and the proof is concluded.

Under the uan condition, for $n \geqq n_\tau$ sufficiently large, $|\mu| < \dfrac{\tau}{2}$ and we can take for $c_2 = c_2(b,\mu,\tau)$ the value of c_2 obtained upon replacing $|\mu|$ by $\dfrac{\tau}{2}$. This proves the right-hand side central inequality.

22.3 Central Limit Theorem. We are ready for the solution of the Central Limit Problem and can follow the same approach as in the case of bounded variances, since

a. BOUNDEDNESS LEMMA. *Under the uan condition, if* $\prod_k |f_{nk}| \to |f|$ *continuous, then there exists a finite constant* $c > 0$ *such that*

$$\sum_k\int\frac{x^2}{1+x^2}\,d\overline{F}_{nk} \leqq c < \infty.$$

Proof. It suffices to prove the assertion for n sufficiently large so that, by 22.2**A**, Corollary 2, all $\log f_{nk}$ exist and are finite. Let $b > 0$ be sufficiently small so that, for $|u| \leqq b$, $|f(u)| > 0$, and $\log|f(u)|$ exists and is finite. Since $|f|^2$ is a ch.f., $\sum_k \log|f_{nk}| \to \log|f|$ uniformly on $[-b, +b]$, and, by the right-hand side central inequality,

$$\sum_k \int \frac{x^2}{1+x^2}\, d\bar{F}_{nk} \leqq -c_2 \sum_k \int_0^b \log|f_{nk}(u)|\, du$$

$$\to -c_2 \int_0^b \log|f(u)|\, du < \infty.$$

The assertion follows.

b. COMPARISON LEMMA. *Under the uan condition, if there exists a constant c such that whatever be n*

$$\sum_k \int \frac{x^2}{1+x^2}\, d\bar{F}_{nk} \leqq c < \infty,$$

then

$$\sum_k \{\log \bar{f}_{nk}(u) - (\bar{f}_{nk}(u) - 1)\} \to 0, \quad u \in R.$$

Proof. By 22.2**A**, Corollaries 1 and 2, $\max_k |\bar{f}_{nk} - 1| \to 0$ and, given $b > 0$, for $|u| \leqq b$ and n sufficiently large,

$$\log \bar{f}_{nk} = \bar{f}_{nk} - 1 + \frac{\theta_{nk}}{2}|\bar{f}_{nk} - 1|^2, \quad |\theta_{nk}| \leqq 1.$$

By the left-hand side central inequality

$$\sum_k |f_{nk}(u) - 1| \leqq \frac{1}{c_1} \sum_k \int \frac{x^2}{1+x^2}\, d\bar{F}_{nk} \leqq \frac{c}{c_1} < \infty.$$

It follows that by taking $b > |u|$, where $u \in R$ is arbitrarily fixed,

$$\left|\sum_k \{\log \bar{f}_{nk}(u) - (\bar{f}_{nk}(u) - 1)\}\right| \leqq \sum |\bar{f}_{nk}(u) - 1|^2$$

$$\leqq \frac{c}{c_1} \max_k |\bar{f}_{nk}(u) - 1| \to 0,$$

and the theorem is proved.

Since (omitting the subscripts)

$$\log \bar{f}(u) - (\bar{f}(u) - 1) = \log f(u) - \{iua + \int (e^{iux} - 1)\, d\bar{F}\}$$

and

$$\int (e^{iux} - 1)\, d\overline{F} = iu \int \frac{x}{1+x^2}\, d\overline{F}$$
$$+ \int \left(e^{iux} - 1 - \frac{iux}{1+x^2} \right) \frac{1+x^2}{x^2} \cdot \frac{x^2}{1+x^2}\, d\overline{F},$$

the sums which figure in the comparison lemma are

$$\log \prod_k f_{nk}(u) - \psi_n(u)$$

where

$$\psi_n(u) = iu\alpha_n + \int \left(e^{iux} - 1 - \frac{iux}{1+x^2} \right) \frac{1+x^2}{x^2}\, d\Psi_n(x)$$

with

$$\alpha_n = \sum_k \{a_{nk} + \int \frac{x}{1+x^2}\, d\overline{F}_{nk}\}, \quad d\Psi_n(x) = \sum_k \frac{x^2}{1+x^2}\, d\overline{F}_{nk}(x).$$

A. CENTRAL LIMIT THEOREM. *Let X_{nk} be uan independent summands.*

1° *The family of limit laws of sequences $\mathfrak{L}(\sum_k X_{nk})$ coincides with the family of i.d. laws or, equivalently, with the family of laws with log of ch.f. $\psi = (\alpha, \Psi)$ defined by*

$$\psi(u) = iu\alpha + \int \left(e^{iux} - 1 - \frac{iux}{1+x^2} \right) \frac{1+x^2}{x^2}\, d\Psi(x)$$

where $\alpha \in R$, and Ψ is a d.f. up to a multiplicative constant.

2° $\mathfrak{L}(\sum_k X_{nk}) \to \mathfrak{L}(X)$ *with log of ch.f. necessarily of the form $\psi = (\alpha, \Psi)$ if, and only if,*

$$\Psi_n \xrightarrow{c} \Psi, \quad \alpha_n \to \alpha,$$

where

$$\alpha_n = \sum_k \{a_{nk} + \int \frac{x}{1+x^2}\, d\overline{F}_{nk}\}, \quad \Psi_n(x) = \sum_k \int_{-\infty}^{x} \frac{y^2}{1+y^2}\, d\overline{F}_{nk}$$

and

$$a_{nk} = \int_{|x|<\tau} x\, dF_{nk}, \quad \overline{F}_{nk}(x) = F_{nk}(x + a_{nk}),$$

with $\tau > 0$ finite and arbitrarily fixed.

Proof. Every i.d. law is a limit law of the Central Limit Problem. Conversely, if, under the uan condition, $\prod_k f_{nk} \to f$ ch.f., then, on account of the boundedness lemma, the comparison lemma applies and $e^{\psi_n} \to f$. Thus, on the one hand, by the closure theorem for i.d. laws $f = e^{\psi}$ is i.d. and 1° is proved. On the other hand, $\psi_n \to \psi$ and hence, by the convergence theorem for i.d. laws, $\Psi_n \xrightarrow{c} \Psi$, $\alpha_n \to \alpha$, and the "only if" part of 2° is proved.

Conversely, if $\alpha_n \to \alpha$ and $\Psi_n \xrightarrow{c} \Psi$, so that

$$\operatorname{Var} \Psi_n = \sum_k \int \frac{x^2}{1+x^2}\, d\bar{F}_{nk} \to \operatorname{Var} \Psi < \infty$$

and the comparison lemma applies, then $\psi_n \to \psi$ hence $\prod_k f_{nk} \to e^{\psi}$, and the "if" part of 2° is proved. This terminates the proof.

Extension. It may happen that under the uan condition, the sequence $\mathfrak{L}(\sum X_{nk})$ does not converge, yet the sequence $\mathfrak{L}(\sum_k X_{nk} - a_n)$ converges for suitably chosen constants a_n; this is the situation in the Bernoulli case and, more generally, in the classical limit problem where $X_{nk} = X_k/b_n$ with $b_n = n$ or s_n. Then $\prod_k f_{nk}(u)$ is replaced by $e^{-iua_n} \prod_k f_{nk}(u)$ and the boundedness lemma can still be used, since it refers only to the moduli of products. On the other hand, the sums in the comparison lemma can be written $\log \{e^{-iua_n} \prod_k f_{nk}(u)\} - \{-iua_n + \psi_n(u)\}$. Since $-iua_n + \psi_n(u)$ is still a ψ-function, the Central Limit theorem remains valid, provided α_n is replaced by $\alpha_n - a_n$, and the theorem can be stated as follows:

B. Extended central limit theorem. *Let X_{nk} be uan independent summands.*

1° *The family of limit laws of sequences $\mathfrak{L}(\sum_k X_{nk} - a_n)$ coincides with the family of i.d. laws.*

2° *There exist constants a_n such that the sequence $\mathfrak{L}(\sum X_{nk} - a_n)$ converges if, and only if, $\Psi_n \xrightarrow{c}$ some Ψ, where*

$$\Psi_n(x) = \sum_k \int_{-\infty}^{x} \frac{y^2}{1+y^2}\, d\bar{F}_{nk}.$$

Then all admissible a_n are of the form $a_n = \alpha_n - \alpha + o(1)$ where α is an arbitrary finite number and $\alpha_n = \sum_k \left\{a_{nk} + \int \frac{x}{1+x^2}\, d\bar{F}_{nk}\right\}$, and all possible limit laws have for log of ch.f. $\psi = (\alpha, \Psi)$.

22.4 Central convergence criterion. The convergence criterion 22.3**A** 2° is expressed in terms of expressions twice removed from the primary datum—the d.f.'s of the summands, and the probabilistic meaning of these expressions is somewhat hidden. We transform it by unpleasant but elementary computations as follows:

A. CENTRAL CONVERGENCE CRITERION. *If X_{nk} are uan independent summands, then*

$$\prod_k f_{nk} \to f = e^{\psi}, \quad \psi = (\alpha, \Psi),$$

if, and only if,

(i) *at every continuity point $x \neq 0$ of Ψ*

$$\sum_k F_{nk}(x) \to \int_{-\infty}^{x} \frac{1+y^2}{y^2}\, d\Psi \quad \text{for} \quad x < 0,$$

$$\sum_k \{1 - F_{nk}(x)\} \to \int_{x}^{\infty} \frac{1+y^2}{y^2}\, d\Psi \quad \text{for} \quad x > 0$$

(ii) *as $n \to \infty$ and then $\epsilon \to 0$*

$$\sum_k \left\{ \int_{|x|<\epsilon} x^2\, dF_{nk} - \left(\int_{|x|<\epsilon} x\, dF_{nk} \right)^2 \right\} \to \Psi(+0) - \Psi(-0)$$

(iii) *for a fixed $\tau > 0$ such that $\pm\tau$ are continuity points of Ψ*

$$\sum_k \int_{|x|<\tau} x\, dF_{nk} \to \alpha + \int_{|x|<\tau} x\, d\Psi - \int_{|x|\geqq\tau} \frac{1}{x}\, d\Psi.$$

The iterated limit in (ii) is the generalized iterated limit $\lim_{\epsilon\to 0} \overline{\underline{\lim}}_n$.

Proof. We have to prove that the three stated conditions are equivalent to

(C) $$\Psi_n \xrightarrow{c} \Psi \quad \text{with} \quad d\Psi_n(x) = \frac{x^2}{1+x^2} \sum_k d\bar{F}_{nk}$$

and

(C′) $$\sum_k \left\{ a_{nk} + \int \frac{x}{1+x^2}\, d\bar{F}_{nk} \right\} \to \alpha, \quad \text{with} \quad a_{nk} = \int_{|x|<\tau} x\, dF_{nk},$$

$\bar{F}_{nk}(x) = F_{nk}(x + a_{nk})$.

1° Let x be continuity points of Ψ. It is readily seen that condition (C) can be written as follows:

$$\Psi_n(x) \to \Psi(x) \quad \text{for} \quad x < 0,$$

$$\Psi_n(+\infty) - \Psi_n(x) \to \Psi(+\infty) - \Psi(x) \quad \text{for} \quad x > 0$$

and, as $n \to \infty$ and then $\epsilon \to 0$,

$$\Psi_n(+\epsilon) - \Psi_n(-\epsilon) \to \Psi(+0) - \Psi(-0).$$

It follows, upon replacing Ψ_n by its defining expression and applying the Helly-Bray theorem, that (C) is equivalent to

$$(\mathrm{C}_1) \qquad \sum_k \overline{F}_{nk}(x) \to \int_{-\infty}^{x} \frac{1+y^2}{y^2}\, d\Psi \quad \text{for} \quad x < 0,$$

$$\sum_k \{1 - \overline{F}_{nk}(x)\} \to \int_x^{\infty} \frac{1+y^2}{y^2}\, d\Psi \quad \text{for} \quad x > 0$$

and

$$(\mathrm{C}_2) \qquad \sum_k \int_{|x|<\epsilon} \frac{x^2}{1+x^2}\, d\overline{F}_{nk} \to \Psi(+0) - \Psi(-0)$$

$$\text{as} \quad n \to \infty \quad \text{and then} \quad \epsilon \to 0.$$

Let $a_n = \max_k \int_{|x|<\tau} |x|\, dF_{nk}$ so that $|a_{nk}| \leqq a_n \to 0$. Since

$$\sum_k F_{nk}(x - a_n) \leqq \sum_k \overline{F}_{nk}(x) \leqq \sum_k F_{nk}(x + a_n),$$

and the continuity points x of Ψ are continuity points of the integrals in (C_1), it follows at once that the first parts of (i) and (C_1) are equivalent; similarly for the second parts. Thus (C_1) is equivalent to (i).

2° Since

$$\frac{1}{1+\epsilon^2} \sum_k \int_{|x|<\epsilon} x^2\, d\overline{F}_{nk} \leqq \sum_k \int_{|x|<\epsilon} \frac{x^2}{1+x^2}\, d\overline{F}_{nk} \leqq \sum_k \int_{|x|<\epsilon} x^2\, d\overline{F}_{nk},$$

condition (C_2) is equivalent to

$$\sum_k \int_{|x|<\epsilon} x^2\, d\overline{F}_{nk} \to \Psi(+0) - \Psi(-0) \quad \text{as} \quad n \to \infty \quad \text{and then} \quad \epsilon \to 0.$$

But, on account of (i), as $n \to \infty$ and then $\epsilon \to 0$,

$$|\sum_k \int_{|x|<\epsilon} x^2\, d\bar{F}_{nk} - \sum_k \int_{|x|<\epsilon} (x - a_{nk})^2\, dF_{nk}|$$

$$\leqq \sum_k \int_{|x|\geqq\epsilon,\, |x-a_{nk}|<\epsilon} (x - a_{nk})^2\, dF_{nk} \leqq \epsilon^2 \sum_k \int_{|x|\geqq\epsilon} dF_{nk} \to 0$$

and, since $a_n \to 0$, we have, for $\epsilon < \tau$,

$$|\sum_k \int_{|x|<\epsilon} (x - a_{nk})^2\, dF_{nk} - \sum_k \left\{ \int_{|x|<\epsilon} x^2\, dF_{nk} - \left(\int_{|x|<\epsilon} x\, dF_{nk} \right)^2 \right\}|$$

$$= |\sum_k \left(\int_{\epsilon\leqq|x|<\tau} x\, dF_{nk} \right)^2 - \sum_k a_{nk}^2 \int_{|x|\geqq\epsilon} dF_{nk}|$$

$$\leqq (\tau a_n + a_n^2) \sum_k \int_{|x|\geqq\epsilon} dF_{nk} \to 0.$$

Therefore, under (i) or its equivalent (C_1), condition (C_2) is equivalent to (ii). Thus, condition (C) is equivalent to (i) and (ii).

3° It remains to prove that, under (C) or its equivalent (i) and (ii), condition (C′) is equivalent to (iii). Since

$$\sum_k \int \frac{x}{1+x^2}\, d\bar{F}_{nk}$$

$$= \sum_k \int_{|x|<\tau} x\, d\bar{F}_{nk} - \sum_k \int_{|x|<\tau} \frac{x^3}{1+x^2}\, d\bar{F}_{nk} + \sum_k \int_{|x|\geqq\tau} \frac{x}{1+x^2}\, d\bar{F}_{nk}$$

and, $\pm\tau$ being continuity points of Ψ, we have, by the Helly-Bray theorem,

$$\sum_k \int_{|x|<\tau} \frac{x^3}{1+x^2}\, d\bar{F}_{nk} = \int_{|x|<\tau} x\, d\Psi_n \to \int_{|x|<\tau} x\, d\Psi$$

$$\sum_k \int_{|x|\geqq\tau} \frac{x}{1+x^2}\, d\bar{F}_{nk} = \int_{|x|\geqq\tau} \frac{1}{x}\, d\Psi_n \to \int_{|x|\geqq\tau} \frac{1}{x}\, d\Psi,$$

it suffices to prove that $\sum_k \int_{|x|<\tau} x\, d\bar{F}_{nk} \to 0$. This assertion follows from the fact that $a_n \to 0$ and $\pm\tau$, being continuity points of Ψ, are continuity points of integrals in (i), so that, by (i),

$$\left|\sum_k \int_{|x|<\tau} x\,d\overline{F}_{nk}\right| \leqq \left|\sum_k \int_{|x|<\tau} (x - a_{nk})\,dF_{nk}\right|$$

$$+ \left|\sum_k \left\{ \int_{|x-a_{nk}|<\tau} (x - a_{nk})\,dF_{nk} - \int_{|x|<\tau} (x - a_{nk})\,dF_{nk} \right\}\right|$$

$$\leqq a_n \sum_k \int_{|x|\geqq \tau} dF_{nk} + (\tau + a_n) \int_{\tau \leqq |x| < \tau + a_n} dF_{nk} \to 0.$$

This terminates the proof.

REMARK 1. In the course of the proof, it was found that condition (i) can be written with $\overline{F}_{nk}$ instead of F_{nk} and condition (ii) is equivalent to

(ii′) $$\sum_k \int_{|x|<\epsilon} x^2\,d\overline{F}_{nk} \to \Psi(+0) - \Psi(-0)$$

as $n \to \infty$ and then $\epsilon \to 0$.

REMARK 2. In conditions (ii) or (ii′), the passages to the limit can be taken indifferently to be $\lim_{\epsilon \to 0} \limsup_n$ or $\lim_{\epsilon \to 0} \liminf_n$, instead of the generalized iterated limit; we leave the verification to the reader.

Upon using the extended Central Limit theorem, the central convergence criterion extends at once to sums with variable origin, as follows:

B. EXTENDED CENTRAL CONVERGENCE CRITERION. *If X_{nk} are uan independent summands, then there exist constants a_n such that $e^{-iua_n} \prod_k f_{nk}(u) \to e^{\psi(u)}$ where $\psi = (\alpha, \Psi)$ if, and only if, conditions* (i) *and* (ii) *of the central convergence criterion hold. Then the admissible a_n are of the form*

$$a_n = \sum_k \int_{|x|<\tau} x\,dF_{nk} - \alpha - \int_{|x|<\tau} x\,d\Psi + \int_{|x|\geqq \tau} \frac{1}{x}\,d\Psi + o(1)$$

where $\pm\tau$ are fixed continuity points of Ψ.

This criterion implies properties of $\min_k X_{nk}$ and $\max_k X_{nk}$. In fact, it takes then a more intuitive form, as follows:

C. EXTREMA CRITERION. *Let X_{nk} be uan independent summands, and let $X_{nk}^{\epsilon} = X_{nk}$ or 0 according as $|X_{nk}| < \epsilon$ or $|X_{nk}| \geqq \epsilon$.*

The sequence $\mathfrak{L}(\sum_k X_{nk} - a_n)$ converges for suitable constants a_n if, and only if, the sequences $\mathfrak{L}(\min_k X_{nk})$, $\mathfrak{L}(\max_k X_{nk})$ and $\sum_k \sigma^2 X_{nk}^{\epsilon}$ converge as $n \to \infty$ and then $\epsilon \to 0$.

More precisely, $\mathfrak{L}(\sum_k X_{nk} - a_n) \to \mathfrak{L}(X)$ with $\mathfrak{L}(X)$ necessarily an i.d. law (α, Ψ) if, and only if, as $n \to \infty$ and then $\epsilon \to 0$,

$$\sum \sigma^2 X_{nk}^{\epsilon} \to \Psi(+0) - \Psi(-0)$$

and

$$\mathfrak{L}(\min_k X_{nk}) \to \mathfrak{L}(Y), \quad \mathfrak{L}(\max_k X_{nk}) \to \mathfrak{L}(Z)$$

with

$F_Y(x) = 1 - e^{-L(x)}$ *or* 1 *and* $F_Z(x) = 0$ *or* $e^{L(x)}$, *according as* $x < 0$ *or* $x > 0$,

where

$$L(x) = \int_{-\infty}^{x} \frac{1+y^2}{y^2}\, d\Psi(y),\ x < 0; \quad L(x) = -\int_{x}^{\infty} \frac{1+y^2}{y^2}\, d\Psi(y),\ x > 0.$$

Proof. Let G_n be the d.f. of $\min_{k \leqq k_n} X_{nk}$, so that $1 - G_n = \prod_{k=1}^{k_n} (1 - F_{nk})$. For every fixed $x > 0$, $F_{nk}(x) \to 1$ uniformly in k and, hence, $G_n(x) \to 1$. For every fixed $x < 0$, $F_{nk}(x) \to 0$ uniformly in k and, hence, for n sufficiently large,

$$\log(1 - G_n(x)) = \sum_k \log(1 - F_{nk}(x)) = -(1 + o(1)) \sum_k F_{nk}(x).$$

Therefore, the assertion relative to F_Y is equivalent to the first part of condition (i) of the central convergence criterion; similarly for the assertion relative to F_Z. The theorem follows.

22.5 Normal, Poisson, and degenerate convergence. We apply now the central convergence criterion to the three first-discovered limit types. We set

$$a_{nk}(\tau) = \int_{|x|<\tau} x\, dF_{nk}, \quad \sigma_{nk}^2(\tau) = \int_{|x|<\tau} x^2\, dF_{nk} - \left(\int_{|x|<\tau} x\, dF_{nk}\right)^2$$

1° A normal law $\mathfrak{N}(\alpha, \sigma^2)$ corresponds to $\psi(u) = iu\alpha - \frac{\sigma^2}{2} u^2$, that is, $\psi = (\alpha, \Psi)$ where $\Psi(x) = 0$ or σ^2 according as $x < 0$ or $x > 0$.

NORMAL CONVERGENCE CRITERION. *If X_{nk} are independent summands, then, for every $\epsilon > 0$,*

$$\mathfrak{L}(\sum_k X_{nk}) \to \mathfrak{N}(\alpha, \sigma^2) \quad and \quad \max_k P[|X_{nk}| \geqq \epsilon] \to 0$$

if, and only if, for every $\epsilon > 0$ and a $\tau > 0$,

(i) $$\sum_k P[|X_{nk}| \geqq \epsilon] \to 0$$

(ii) $$\sum_k \sigma_{nk}^2(\tau) \to \sigma^2, \quad \sum_k a_{nk}(\tau) \to \alpha.$$

Proof. We have, under (i),

$$\max_k P[|X_{nk}| \geqq \epsilon] \leqq \sum_k P[|X_{nk}| \geqq \epsilon] \to 0.$$

Furthermore, always under (i), if $\epsilon < \tau$, then

$$\left|\sum_k \sigma_{nk}^2(\tau) - \sum_k \sigma_{nk}^2(\epsilon)\right| \leqq \sum_k \int_{\epsilon \leqq |x| < \tau} x^2\, dF_{nk} + 2\tau \sum_k \left|\int_{\epsilon \leqq |x| < \tau} x\, dF_{nk}\right|$$

$$\leqq 3\tau^2 \sum_k \int_{\epsilon \leqq |x| < \tau} dF_{nk} \to 0$$

and the same is true of $\epsilon > \tau$; it suffices to interchange ϵ and τ in the foregoing chain of inequalities. Upon taking into account these consequences of (i), the foregoing criterion follows from the central convergence criterion applied to the limit law $\mathfrak{N}(\alpha, \sigma^2)$.

COROLLARY. *If X_{nk} are independent summands and the sequence $\mathfrak{L}(\sum_k X_{nk})$ converges, then the limit law is normal and the uan condition is satisfied if, and only if,* $\max_k |X_{nk}| \xrightarrow{P} 0$.

Upon setting $p_{nk} = P[|X_{nk}| \geqq \epsilon]$, it suffices to observe that, because of the independence of the summands,

$$P[\max_k |X_{nk}| \geqq \epsilon] = 1 - \prod_k (1 - p_{nk}).$$

For, upon applying the elementary inequality

$$1 - \exp[-\sum_k p_{nk}] \leqq 1 - \prod_k (1 - p_{nk}) \leqq \sum_k p_{nk},$$

it follows that the asserted condition is equivalent to condition (i) of the above criterion.

2° The Poisson law $\mathcal{P}(\lambda)$ corresponds to $\psi(u) = \lambda(e^{iu} - 1)$ and, consequently, to $\psi = \left(\frac{\lambda}{2}, \Psi\right)$ with $\Psi(x) = 0$ or $\frac{\lambda}{2}$ according as $x < 1$ or $x > 1$. Upon applying the central convergence criterion and observing that the condition relative to the $\sigma_{nk}^2(\epsilon)$ reduces exactly as in the normal case, we obtain the

POISSON CONVERGENCE CRITERION. *If X_{nk} are uan independent summands, then $\mathcal{L}(\sum_k X_{nk}) \to \mathcal{P}(\lambda)$ if, and only if, for every $\epsilon \in (0, 1)$ and a $\tau \in (0, 1)$,*

(i) $$\sum_k \int_{|x| \geq \epsilon, |x-1| \geq \epsilon} dF_{nk} \to 0 \quad \textit{and} \quad \sum_k \int_{|x-1| < \epsilon} dF_{nk} \to \lambda$$

(ii) $$\sum_k \sigma_{nk}^2(\tau) \to 0 \quad \textit{and} \quad \sum_k a_{nk}(\tau) \to 0.$$

3° The degenerate law $\mathcal{L}(0)$ can be considered as a degenerate normal $\mathfrak{N}(0, 0)$ so that the normal convergence criterion reduces to the

DEGENERATE CONVERGENCE CRITERION. *If X_{nk} are independent summands, then $\mathcal{L}(\sum_k X_{nk}) \to \mathcal{L}(0)$ and the uan condition is satisfied if, and only if, for every $\epsilon > 0$ and a $\tau > 0$*

(i) $$\sum_k \int_{|x| \geq \epsilon} dF_{nk} \to 0$$

(ii) $$\sum_k \sigma_{nk}^2(\tau) \to 0, \quad \sum_k a_{nk}(\tau) \to 0.$$

COROLLARY 1. *If X_k are independent summands and $b_n \uparrow \infty$, then $\mathcal{L}\left(\frac{S_n}{b_n}\right) \to 0$ if, and only if, for every $\epsilon > 0$*

(i) $$\sum_k \int_{|x| \geq \epsilon b_n} dF_k \to 0$$

(ii) $$\frac{1}{b_n^2} \sum_k \left\{ \int_{|x| < b_n} x^2 \, dF_k - \left(\int_{|x| < b_n} x \, dF_k \right)^2 \right\} \to 0,$$

$$\frac{1}{b_n} \sum_k \int_{|x| < b_n} x \, dF_k \to 0.$$

Because of the above criterion, taking $\tau = 1$ and observing that for $X_{nk} = \dfrac{X_k}{b_n}$, $F_{nk}(x) = F_k(b_n x)$, it remains only to prove that $\mathcal{L}\left(\dfrac{S_n}{b_n}\right) \to \mathcal{L}(0)$ implies the uan condition. This follows from the fact that $P\left[\left|\dfrac{S_n}{b_n}\right| < \epsilon\right] > 1 - \delta$, for $n \geqq n_{\epsilon,\delta}$ sufficiently large, implies that, for $n > n_{\epsilon,\delta}$,

$$P\left[\left|\frac{X_n}{b_n}\right| < 2\epsilon\right] = P\left[\left|\frac{S_n}{b_n} - \frac{b_{n-1}}{b_n}\frac{S_{n-1}}{b_{n-1}}\right| < 2\epsilon\right]$$

$$\geqq P\left[\left|\frac{S_n}{b_n}\right| < \epsilon\right]\left[\left|\frac{S_{n-1}}{b_{n-1}}\right| < \epsilon\right] \geqq 1 - 2\delta.$$

Remark. For the degenerate convergence criterion, (ii) and (i) with $\epsilon = \tau$ imply that $\mathcal{L}(\sum_k X_{nk}) \to \mathcal{L}(0)$. For, as in 20.2**A**, by Tchebichev inequality, (ii) implies that $\mathcal{L}(\sum_k X_{nk}^{\tau}) \to \mathcal{L}(0)$ and then, by 20.1**b**, (i) implies that $\mathcal{L}(\sum_k X_{nk}) \to \mathcal{L}(0)$.

In particular, in Corollary 1, we may take $\epsilon = 1$. Thus, for $b_n = n$, we have

Corollary 2. *If X_k are independent summands, then $\mathcal{L}\left(\dfrac{S_n}{n}\right) \to \mathcal{L}(0)$ if, and only if,*

(i) $$\sum_k \int_{|x| \geqq n} dF_k \to 0,$$

(ii) $$\frac{1}{n^2}\sum_k \left\{\int_{|x|<n} x^2\, dF_k - \left(\int_{|x|<n} x\, dF_k\right)^2\right\} \to 0,$$

(iii) $$\frac{1}{n}\sum_k \int_{|x|<n} x\, dF_k \to 0.$$

This is the classical degenerate convergence criterion.

The reader is invited to specialize 22.4**C** to the three foregoing cases.

In particular, it implies the corollary to the normal convergence criterion. As for the Poisson case, $dL(x) = 0$ or λ according as $x \neq 1$ or $x = 1$ so that

If $\mathcal{L}(\sum_k X_{nk}) \to \mathcal{L}(X)$, then $\mathcal{L}(X) = \mathcal{P}(\lambda)$ if and only if $\mathcal{L}\min_k(X_{nk}) \to \mathcal{L}(0)$ and $\mathcal{L}(\max_k X_{nk}) \to \mathcal{L}(0, 1)$ with two values 0 and 1 only of pr. $e^{-\lambda}$ and $1 - e^{-\lambda}$, respectively.

*§ 23. NORMED SUMS

23.1 The problem. Let $\frac{S_n}{b_n} - a_n$ be normed sums with d.f. G_n and ch.f. g_n, where $S_n = \sum_{k=1}^{n} X_k$ are consecutive sums of independent r.v.'s X_k with d.f.'s F_k and ch.f.'s f_k, and where a_n, $b_n > 0$ are finite numbers; thus

$$g_n(u) = e^{-iua_n} \prod_{k=1}^{n} f_k\left(\frac{u}{b_n}\right).$$

In what follows k runs over $1, \cdots, n$; $n = 1, \cdots$

If the $X_{nk} = X_k/b_n$ obey the uan condition:

$$\max_k P[|X_k| \geqq \epsilon b_n] \to 0 \quad \text{or} \quad \max_k \int \frac{x^2}{b_n^2 + x^2} dF_k(x) \to 0$$

$$\text{or} \quad \max_k \left| f_k\left(\frac{u}{b_n}\right) - 1 \right| \to 0,$$

then, according to the extended Central Limit Theorem, all possible limit laws of sequences $\frac{S_n}{b_n} - a_n$ of normed sums form a family $\mathfrak{N}$ of i.d. laws, and the extended central convergence criterion applies with $F_{nk}(x) = F_k(b_n x)$.

However, in the case of normed sums, new problems arise.

1° Given a sequence X_n of independent r.v.'s, find whether there exist sequences a_n and $b_n > 0$ such that the uan condition (for the X_k/b_n) is satisfied and $g_n \to f$ ch.f., necessarily of the form e^{ψ} with $\psi = (\alpha, \Psi)$; and if such sequences exist, then characterize them.

2° Characterize the family $\mathfrak{N}$; in other words, characterize those i.d. ch.f.'s e^{ψ} and the corresponding functions Ψ which represent limit laws of normed sums obeying the uan condition.

But on the one hand, according to the convergence of types theorem, there always exist sequences a_n and $b_n > 0$ such that the limit laws of $\frac{S_n}{b_n} - a_n$ are degenerate and, on the other hand, all degenerate laws belong to $\mathfrak{N}$: $e^{iua} = (e^{iua/n})^n$. Thus, whenever convenient, *we can and do exclude degenerate limit laws from our considerations.*

a. *If $g_n \to f$ nondegenerate ch.f., then the uan condition for the X_k/b_n implies that $b_n \to \infty$ and $b_{n+1}/b_n \to 1$.*

Proof. We have

$$g_n(u) = e^{-iua_n}\prod_k f_k\left(\frac{u}{b_n}\right) \to f(u) \quad \text{nondegenerate.}$$

If $b_n \not\to \infty$, then the sequence b_n contains a bounded subsequence and, by the Bolzano-Weierstrass lemma, this subsequence contains another sequence $b_{n'} \to b$ finite as $n' \to \infty$. Setting $u_{n'} = b_{n'}u$, the uan condition implies that for every $k, f_k(u) = f_k(u_{n'}/b_{n'}) \to 1$; hence, $f_k = 1$ and $f = 1$. This contradicts the nondegeneracy assumption so that, *ab contrario*, $b_n \to \infty$.

Since $X_{n+1}/b_{n+1} \xrightarrow{P} 0$, it follows by the law-equivalence lemma that the limit laws of the sequences $\dfrac{S_n}{b_n} - a_n$ and $\dfrac{S_n}{b_{n+1}} - a_{n+1} = \dfrac{S_{n+1}}{b_{n+1}} - a_{n+1} - \dfrac{X_{n+1}}{b_{n+1}}$ are the same. Thus $e^{-iua_n'} g_n(b_n'u) \to f(u)$ as $n' \to \infty$, with $b_n' = b_n/b_{n+1}$ and f nondegenerate. It follows, by the corollary to the convergence of types theorem, that $b_{n+1}/b_n \to 1$. The proof is complete.

23.2 Norming sequences. We have at our disposal the necessary tools to solve the problem of existence and determination of norming sequences a_n and $b_n > 0$. Given the summands, we know, according to the convergence of types theorem, that 1° all the limit laws belong—if they exist—to the positive type of *one* i.d. law and 2° it suffices to find one pair of such sequences. Furthermore, on account of the extended convergence criterion (with $X_{nk} = X_k/b_n$), 3° if there exists a limit i.d. positive type, then the a_n are determined by the expression given there 4° the uan condition is satisfied and $g_n \to e^{\psi}$ if, and only if,

$$\max_k \int \frac{x^2}{b_n{}^2 + x^2}\, dF_k(x) \to 0 \quad \text{and} \quad \Psi_n \xrightarrow{c} \Psi$$

where Ψ_n are defined on R by

$$\text{(D)} \quad \Psi_n(x) = \sum_k \int_{-\infty}^{b_n x} \frac{y^2}{b_n{}^2 + y^2}\, dF_k(y + \dot{b}_{nk}), \quad b_{nk} = \int_{|x| < b_n\tau} x\, dF_k(x)$$

with $\pm\tau \neq 0$ fixed continuity points of Ψ (we shall see later that any τ is admissible, so that we may set, say, $\tau = 1$). The theorem below completes the answer. As usual, the superscript "s" will denote the operation of symmetrization.

A. NORMING THEOREM. *There exist sequences* b_n *such that* $\mathcal{L}\left(\frac{S_n}{b_n} - a_n\right) \to \mathcal{L}(X)$ *for suitable* a_n *if, and only if, there exists a* Ψ *such that, upon setting in* (D), $b_n = b'_n > 0$ *determined by*

$$\frac{1}{2}\sum_{k=1}^{n}\int \frac{x^2}{b'^2_n + x^2}\, dF_k^s(x) = \Psi(+\infty).$$

we have

$$\text{(i)} \qquad \max_k \int \frac{x^2}{b'^2_n + x^2}\, dF_k(x) \to 0,$$

$$\text{(ii)} \qquad \Psi_n \xrightarrow{c} \Psi.$$

Proof. The "if" assertion follows by taking normed sums $\frac{S_n}{b'_n} - a_n$. Because of the corollary to the convergence of types theorem and of the extended central convergence criterion, the "only if" assertion will follow by proving that if $\mathcal{L}\left(\frac{S_n}{b_n} - a_n\right) \to \mathcal{L}(X)$ with ch.f. e^{ψ}, $\psi = (\alpha, \Psi)$, then $b'_n/b_n \to 1$.

Upon symmetrizing, the hypothesis becomes $\mathcal{L}(S_n^s/b_n) \to \mathcal{L}(X^s)$ and the corresponding Ψ^s is defined by

$$\Psi^s(x) = \Psi(x) + \Psi(+\infty) - \Psi(-x + 0).$$

Thus $\Psi_n^s \xrightarrow{c} \Psi^s$ where Ψ_n^s are defined by

$$\Psi_n^s(x) = \sum_{k=1}^{n}\int_{-\infty}^{b_n x} \frac{x^2}{b_n^2 + x^2}\, dF_k^s(x).$$

Upon using $\Psi^s(+\infty) = 2\Psi(+\infty)$, and (D) with b_n replaced by b'_n, it follows that

$$\Delta_n = \sum_{k=1}^{n}\left\{\int \frac{x^2}{b_n^2 + x^2}\, dF_k^s(x) - \int \frac{x^2}{b'^2_n + x^2}\, dF_k^s(x)\right\} \to 0.$$

On the other hand, since degenerate limit laws are excluded, Ψ^s does not reduce to a constant. Therefore, there exists an $a > 0$ such that $2\delta = \Psi^s(a) - \Psi^s(-a + 0) > 0$ and, hence, for $n \geqq n_a$ sufficiently large,

$$\sum_{k=1}^{n}\int_{-ab_n}^{+ab_n} \frac{x^2}{b_n^2 + x^2}\, dF_k^s(x) > \delta > 0.$$

It follows that

$$0 \leftarrow |\Delta_n| = |b_n^2 - b'^2_n| \sum_{k=1}^{n} \int \frac{x^2}{(b_n^2 + x^2)(b'^2_n + x^2)} dF_k^s(x)$$

$$\geqq \frac{|b_n^2 - b'^2_n|}{b_n^2 + a^2 b'^2_n} \cdot \sum_{k=1}^{n} \int_{-ab_n}^{+ab_n} \frac{x^2}{(b'^2_n + x^2)} dF_k^s(x)$$

$$\geqq \frac{|(b_n/b'_n)^2 - 1|}{1 + a^2 b_n^2/b'^2_n} \cdot \delta \geqq 0,$$

so that $b_n/b'_n \rightarrow 1$, and the proof is complete.

23.3 Characterization of $\mathfrak{N}$. We characterize $\mathfrak{N}$ by a decomposability property and, then, we characterize the corresponding functions Ψ.

In order to define the decomposability property we prove

a. *If to a ch.f. f there corresponds a number $c > 0$ and a nondegenerate ch.f. f_c such that, for every $u, f(u) = f(cu)f_c(u)$, then $c < 1$.*

Proof. If $c = 1$, then $f_c = 1$. If $c > 1$, then, replacing repeatedly in the assumed relation u by $\frac{u}{c}$ and $|f_c|$ by 1, we have

$$1 \geqq |f(u)| \geqq \left|f\left(\frac{u}{c}\right)\right| \geqq \left|f\left(\frac{u}{c^2}\right)\right| \geqq \cdots \geqq \lim \left|f\left(\frac{u}{c^n}\right)\right| = f(0) = 1$$

and f is degenerate, so that f_c is degenerate. The assertion follows *ab contrario.*

We say that a law and its ch.f. f are *self-decomposable* if, for every $c \in (0, 1)$, there exists a ch.f. f_c such that, for every $u, f(u) = f(cu)f_c(u)$. Clearly, a degenerate ch.f. is self-decomposable and all its components f_c are also degenerate.

b. *If f is self-decomposable, then $f \neq 0$.*

Proof. If $f(2a) = 0$ and $f(u) \neq 0$ for $0 \leqq u < 2a$, then $f_c(2a) = 0$. Upon replacing t and h by a in

$$|f_c(t + h) - f_c(t)|^2 \leqq 2\{1 - \mathcal{R}f_c(h)\},$$

we obtain

$$|f_c(a)|^2 \leqq 2\{1 - \mathcal{R}f_c(a)\}.$$

This leads to a contradiction since, by letting $c \rightarrow 1$, we obtain

$f_c(a) = \dfrac{f(a)}{f(ca)} \to 1$ and the inequality becomes $1 \leqq 0$. The assertion follows *ab contrario*.

A. SELF-DECOMPOSABILITY CRITERION. *A law belongs to $\mathfrak{N}$ if, and only if, it is self-decomposable.*

Proof. A degenerate law certainly belongs to $\mathfrak{N}$, so that it suffices to consider nondegenerate laws with ch.f. f.

1° If f is self-decomposable, then let $X_k (k = 1, \cdots, n)$ be independent r.v.'s, with ch.f. f_k defined by

$$f_k(u) = f_{\frac{k-1}{k}}(ku) = \frac{f(ku)}{f((k-1)u)}.$$

Since $f_k\left(\dfrac{u}{n}\right) \to 1$ uniformly in k and the ch.f. of $\dfrac{S_n}{n}$ is given by

$$\prod_k f_k\left(\frac{u}{n}\right) = f(u),$$

the "if" assertion follows.

2° Conversely, let f belong to $\mathfrak{N}$. There exist normed sums $\dfrac{S_n}{b_n} - a_n$ with ch.f. g_n such that, denoting by f_k the ch.f. of summands X_k,

$$g_n(u) = e^{-iua_n} \prod_k f_k\left(\frac{u}{b_n}\right) \to f(u)$$

and, by 23.1b, $b_n \to \infty$, $\dfrac{b_{n+1}}{b_n} \to 1$. Then, given $c \in (0, 1)$, we can make correspond to every integer n an integer $m < n$ such that $\dfrac{b_m}{b_n} \to c$ and $m, n - m \to \infty$ as $n \to \infty$. Since

$$(1) \quad g_n(u) = \left\{e^{iuca_m} \prod_{k=1}^{m} f_k\left(\frac{b_m}{b_n} \cdot \frac{u}{b_m}\right)\right\} \left\{e^{-iu(a_n - ca_m)} \prod_{k=m+1}^{n} f_k\left(\frac{u}{b_n}\right)\right\}$$

where $g_n(u) \to f(u)$, and the first bracket converges to $f(cu)$, it follows that the ch.f. $g_{m,n}$, whose values figure within the second bracket, converges to the continuous function f_c defined by $f_c(u) = \dfrac{f(u)}{f(cu)}$. Therefore, by the continuity theorem, f_c is a ch.f., and the proof is concluded.

COROLLARY. *A self-decomposable ch.f. f and its components f_c are i.d.*

Proof. Since f belongs to $\mathfrak{N}$, f is i.d. On the other hand, upon taking for f_k the ch.f. of r.v.'s X_k defined in 1° and $m < n$ such that $\frac{m}{n} \to c$, we have

$$f(u) = \prod_{k=1}^{m} f_k\left(\frac{m}{n} \cdot \frac{u}{m}\right) \prod_{k=m+1}^{n} f_k\left(\frac{u}{n}\right).$$

The first product converges to $f(cu)$; the second one converges to $f_c(u)$. Thus, f_c is ch.f. of the limit law of sums $\sum_{k=m+1}^{n} X_{nk}$ where the summands $X_{nk} = \frac{X_k}{n}$ obey the uan condition. Therefore, f_c is an id. ch.f., and the proof is concluded.

We express now the self-decomposability criterion in terms of functions Ψ which figure in the representation of the i.d. self-decomposable ch.f.'s.

B. Ψ-CRITERION. *Self-decomposable laws coincide with i.d. laws with functions Ψ such that on $(-\infty, 0)$ and on $(0, +\infty)$, their left and right derivatives, denoted indifferently by $\Psi'(x)$, exist and $\frac{1+x^2}{x}\Psi'(x)$ do not increase.*

Proof. Because of the preceding corollary, the self-decomposability property of a ch.f. f, necessarily of the form e^{ψ}, is as follows: for every $c \in (0, 1)$ the difference $\psi_c(u) = \psi(u) - \psi(cu)$ defines a ψ-function (a log of an i.d. ch.f.).

Upon replacing x by $c^{-1}x$, we can write

$$(1)\quad \psi(cu) = iu\left\{c\alpha + (1-c^2)\int \frac{x}{1+x^2}\,d\Psi(c^{-1}x)\right\} + \int\left(e^{iux} - 1 - \frac{iux}{1+x^2}\right)\frac{1+c^{-2}x^2}{c^{-2}x^2}\,d\Psi(c^{-1}x).$$

Thus

$$\psi_c(u) = iu\alpha_c + \int\left(e^{iux} - 1 - \frac{iux}{1+x^2}\right)\frac{1+x^2}{x^2}\,d\Psi_c(x),$$

where α_c is a finite number and

$$(2)\quad d\Psi_c(x) = d\Psi(x) - \frac{1+c^{-2}x^2}{c^{-2}(1+x^2)}\,d\Psi(c^{-1}x),\quad \Psi_c(-\infty) = 0.$$

Since Ψ_c is a difference of two Ψ-functions, its variation on R is bounded. It follows readily that ψ_c is a ψ-function if, and only if, Ψ_c is nondecreasing on R. Since

$$(3)\qquad \Psi_c(+0) - \Psi_c(-0) = (1 - c^2)\{\Psi(+0) - \Psi(-0)\} \geqq 0,$$

the self-decomposability property becomes $d\Psi_c(x) \geqq 0$ for every $c \in (0, 1)$ and $x \neq 0$ or, equivalently, on account of (2), for every $c \in (0, 1)$ and arbitrary $x' < x''$, $x'x'' > 0$,

$$(4)\quad \int_{x'}^{x''} \frac{1 + y^2}{y^2}\, d\Psi_c(y) = \int_{x'}^{x''} \frac{1 + y^2}{y^2}\, d\Psi(y) - \int_{x'}^{x''} \frac{1 + c^{-2}y^2}{c^{-2}y^2}\, d\Psi(c^{-1}y) \geqq 0.$$

It remains to show that this last inequality implies and is implied by the one asserted in the theorem.

If

$$J(x) = \int_{+\infty}^{e^x} \frac{1 + y^2}{y^2}\, d\Psi(y), \quad x \in R,$$

then, by setting in (4) $x' = e^{x-h}$, $x'' = e^x$, $c = e^{-h}$, we obtain

$$J(x) - J(x - h) \geqq J(x + h) - J(x) \quad \text{or} \quad J(x) \geqq \frac{J(x + h) + J(x - h)}{2}.$$

Therefore, the nondecreasing finite function J on R is convex (from above) and, consequently, J is continuous and its left and right derivatives $J'(x)$ exist and do not increase on R. Since

$$\frac{J(x + h) - J(x)}{e^h - 1} = \frac{1 + e^{2(x+\theta h)}}{e^{x+2\theta h}} \cdot \frac{\Psi(e^{x+h}) - \Psi(e^x)}{e^{x+h} - e^x}, \quad 0 \leqq \theta \leqq 1,$$

it follows, letting $h \to 0$ and setting $e^x = y$, that the left and right derivatives $\Psi'(y)$ exist and that $\dfrac{1 + y^2}{y}\Psi'(y)$ do not increase on $(0, \infty)$. Similarly, introducing $J^-(x) = \displaystyle\int_{-\infty}^{-e^x} \frac{1 + y^2}{y^2}\, d\Psi(y)$, we find that the same is true on $(-\infty, 0)$. Thus (4) implies the asserted property of Ψ.

Conversely, if this asserted property is true then, for every $c \in (0, 1)$ and $x' < x''$, $x'x'' > 0$

$$\int_{x'}^{x''} \frac{1+y^2}{y^2}\, d\Psi(y) = \int_{x'}^{x''} \frac{1+y^2}{y}\, \Psi'(y)\, \frac{dy}{y}$$

$$\geqq \int_{x'}^{x''} \frac{1+c^{-2}y^2}{c^{-1}y}\, \Psi'(c^{-1}y)\, \frac{dy}{y}$$

$$= \int_{x'}^{x''} \frac{1+c^{-2}y^2}{c^{-2}y^2}\, d\Psi(c^{-1}y),$$

so that the inequality in (4) holds and the conclusion is reached.

REMARK. Since Poisson laws correspond to functions Ψ discontinuous at some $x \neq 0$, they do not belong to the family $\mathfrak{N}$. This explains the isolation in which they remained as long as only limit laws of normed sums were considered.

23.4 Identically distributed summands and stable laws. The first *family* $\mathfrak{N}_I$ of limit laws to be investigated by P. Lévy, was that of limit laws of normed sums $\dfrac{S_n}{b_n} - a_n$ of independent and identically distributed summands X_k with an arbitrary common ch.f. f_0. In other words, $\mathfrak{N}_I$ is defined as the family of laws whose ch.f.'s f are such that

$$g_n(u) = e^{-iua_n} f_0^{\,n}\left(\frac{u}{b_n}\right) \to f(u), \quad u \in R.$$

Clearly, the uan condition is satisfied, so that $\mathfrak{N}_I \subset \mathfrak{N}$. The self-decomposability concept and the criteria for $\mathfrak{N}$ are easily particularized for $\mathfrak{N}_I$, as follows; we exclude degenerate limit laws which, clearly, belong to $\mathfrak{N}_I$. Let a law and its ch.f. f be called *stable* if, for arbitrary $b > 0$, $b' > 0$, there exist finite numbers a and $b'' > 0$ such that

$$f(b''u) = e^{iua} f(bu) f(b'u), \quad u \in R.$$

Upon replacing $b''u$ by u and setting $c = \dfrac{b}{b''}$, $c' = \dfrac{b'}{b''}$, we obtain

$$f(u) = e^{iu\frac{a}{b''}} f(cu) f(c'u) = f(cu) f_c(u)$$

where

$$f_c(u) = e^{iu\frac{a}{b''}} f(c'u).$$

The self-decomposability criterion for $\mathfrak{N}$ becomes

A. STABILITY CRITERION. *A law belongs to $\mathfrak{N}_I$ if, and only if, it is stable.*

Proof. The "if" assertion follows from the fact that stability of f implies, taking $f_0 = f$, that the ch.f. of S_n is of the form $f^n(u) = e^{iua_n}f(b_n u)$ so that, norming S_n with these quantities a_n and b_n, we have $g_n = f$.

Conversely, leaving out—to simplify the writing—factors of the form e^{iua}, which does not restrict the generality, we have to prove that $f_0^n\left(\frac{u}{b_n}\right) \to f(u)$, $u \in R$, implies that to arbitrary $b > 0$, $b' > 0$, there corresponds $b'' > 0$ such that $f(b''u) = f(bu)f(b'u)$. Since $b_n \to \infty$ and $\frac{b_{n+1}}{b_n} \to 1$, we can assign to every integer n integers m and m' such that $\frac{b_m}{b_n} \to b$, $\frac{b_{m'}}{b_n} \to b'$. Then

$$f_0^{m+m'}\left(\frac{b_{m+m'}}{b_n} \cdot \frac{u}{b_{m+m'}}\right) = f_0^m\left(\frac{b_m}{b_n} \cdot \frac{u}{b_m}\right) f_0^{m'}\left(\frac{b_{m'}}{b_n} \cdot \frac{u}{b_{m'}}\right)$$

and the right-hand side converges to $f(bu)f(b'u)$, while, according to the convergence of types theorem, there exists $b'' > 0$ such that the left-hand side converges to $f(b''u)$. The conclusion is reached.

Thus, a stable law is self-decomposable and, moreover, f_c belongs to the positive type of f; in particular f is an i.d. ch.f.

The Ψ-criterion for $\mathfrak{N}$ is easily transformed and, furthermore, the stable ch.f.'s are obtained in terms of elementary functions of analysis, as follows.

B. *A function f is a stable ch.f. if, and only if, either*

(i) $$\log f(u) = i\alpha u - b|u|^\gamma \left\{1 + ic\frac{u}{|u|}\tan\frac{\pi}{2}\gamma\right\}$$

or

(ii) $$\log f(u) = i\alpha u - b|u| \left\{1 + ic\frac{u}{|u|}\cdot\frac{2}{\pi}\log|u|\right\}$$

with

$$\alpha \gtreqless 0, \quad b \geqq 0, \quad |c| \leqq 1, \quad \gamma \in (0, 1) \cup (1, 2].$$

We observe that $\gamma = 2$ gives the normal laws and that real stable ch.f.'s are of the form $e^{-b|u|^\gamma}$.

Proof. If the asserted forms of f are ch.f.'s, then they are clearly stable. Thus, we have to prove that these forms are ch.f.'s and that stable ones are of this form. The first assertion will follow if we can determine functions Ψ such that $\log f = (\alpha, \Psi)$.

Let $f = e^\psi$ be a stable ch.f., that is, for arbitrary $b > 0$ and $b' > 0$, there exist a and $b'' > 0$ such that

$$iua + \psi(bu) + \psi(b'u) = \psi(b''u).$$

1° We follow the pattern of Ψ-criterion's proof $\left(\text{with } c = \frac{b}{b'}\right)$. Upon replacing ψ by its representation in terms of α and Ψ, the foregoing requirement reduces to

$$\frac{1 + b^2x^2}{b^2}\,d\Psi(bx) + \frac{1 + b'^2x^2}{b'^2}\,d\Psi(b'x) = \frac{1 + b''^2x^2}{b''^2}\,d\Psi(b''x).$$

Upon introducing the functions J and J^- defined on R by

$$J(x) = \int_\infty^{e^x} \frac{1 + y^2}{y^2}\,d\Psi(y), \quad J^-(x) = \int_{-\infty}^{-e^x} \frac{1 + y^2}{y^2}\,d\Psi(y), \quad x \in R,$$

and setting $e^h = b$, $e^{h'} = b'$, $e^{h''} = b''$, this requirement becomes

(1) $$\{\Psi(+0) - \Psi(-0)\}(b^2 + b'^2 - b''^2) = 0$$

and

(2) $$J(x + h) + J(x + h') = J(x + h''),$$

$$J^-(x + h) + J^-(x + h') = J^-(x + h''), \quad x \in R,$$

where h, h' are arbitrary numbers and h'' is a function of h and h'.

Let $\Psi(+\infty) - \Psi(+0) > 0$ so that J does not vanish. If, in the foregoing relation in J, we set repeatedly $h' = h$, it follows that, for arbitrary positive integers n and sn,

$$nJ(x + h) = J(x + h''_n), \quad snJ(x + h) = J(x + h''_{sn}).$$

Therefore, to every rational $s > 0 (s' > 0)$ there corresponds a number $t(t')$, such that, for every x,

(3) $$sJ(x) = J(x + t).$$

Since J is continuous from the left and nondecreasing, with $J \leqq 0$,

$J(+\infty) = 0$, it follows that $t' \downarrow t$ as $s' \uparrow s$, so that J is continuous, (3) holds for irrational s as well, and

$$t \downarrow {}^{\infty}_{-\infty} \quad \text{as} \quad s \uparrow {}^{\infty}_{0}.$$

Since J does not vanish, we can assume—by changing the origin if necessary—that $J(0) \neq 0$. Then, setting $J_0 = J/J(0)$, it follows, by

$$sJ(0) = J(t), \quad s'J(0) = J(t'), \quad s'J(t) = J(t+t'),$$

that

$$J_0(t)J_0(t') = J_0(t+t'), \quad t, t' \in R.$$

The only nonvanishing continuous solution of this functional equation, with $J_0(\infty) = 0$, is proportional to $e^{-\gamma t}$ with $\gamma > 0$. Therefore, setting $y = e^t$ and going back to Ψ, the derivative $\Psi'(y)$ exists for $y > 0$ and

$$\frac{1+y^2}{y}\Psi'(y) = \beta' y^{-\gamma}, \quad \beta' \geqq 0,$$

taking into account the vanishing case. Since Ψ is of bounded variation on $(0, +\infty)$, it follows that $\int_\epsilon^0 y^{1-\gamma}\,dy$ is finite for $\epsilon > 0$ and, hence, $\gamma < 2$. Furthermore, replacing J in (2) by its above-found expression, we have

$$b^\gamma + b'^\gamma = b''^\gamma, \quad 0 < \gamma < 2.$$

Similarly, with J^-: for $y < 0$

$$\frac{1+y^2}{y}\Psi'(y) = -\beta|y|^{-\gamma'}, \quad \beta \geqq 0,$$

with $b^{\gamma'} + b'^{\gamma'} = b''^{\gamma'}$, hence $\gamma = \gamma'$ (set $b = b' = 1$).

Therefore, on account of (1), either $b^2 + b'^2 = b''^2$ so that J and J^- vanish and f is a normal ch.f., or $\Psi(+0) - \Psi(-0) = 0$ and, for $y \neq 0$, $\Psi'(y)$ is given by the foregoing relations.

2° According to what precedes, a stable ch.f. f is either normal or of the form

$$(1)\quad \log f(u) = iu\alpha + \beta\int_{-\infty}^{-0}\left(e^{iux} - 1 - \frac{iux}{1+x^2}\right)\frac{dx}{|x|^{1+\gamma}} + \beta'\int_0^{+\infty}\left(e^{iux} - 1 - \frac{iux}{1+x^2}\right)\frac{dx}{x^{1+\gamma}}.$$

If $0 < \gamma < 1$, then it is possible to take out of the bracket the term $-\dfrac{iux}{1+x^2}$ and, by modifying α, we obtain

$$(2)\quad \log f(u) = iu\alpha' + \beta\int_{-\infty}^{0}(e^{iux} - 1)\frac{dx}{|x|^{1+\gamma}} + \beta'\int_0^{\infty}(e^{iux} - 1)\frac{dx}{x^{1+\gamma}}.$$

Let $u > 0$. Setting $ux = v$ and integrating along the closed contour formed by the positive halves of the real and imaginary axes and a circumference centered at the origin of radius $r \to \infty$, it follows, by the Cauchy theorem, that

$$(3) \qquad \int_0^\infty (e^{iux} - 1)\frac{dx}{x^{1+\gamma}} = |u|^\gamma e^{-\frac{\pi}{2}\gamma i}\Gamma(-\gamma),$$

where

$$\Gamma(-\gamma) = \int_0^\infty (e^{-v} - 1)\frac{dv}{v^{1+\gamma}} < 0.$$

The first integral in (1) follows by taking the complex-conjugate of (3) and, for $u < 0$, $\log f(u)$ is obtained by taking the complex-conjugate of $\log f(|u|)$. Upon substituting in (2) and setting

$$b = -\Gamma(-\gamma)(\beta + \beta')\cos\frac{\pi}{2}\gamma, \quad c = \frac{\beta - \beta'}{\beta + \beta'},$$

so that $b \geqq 0$, $|c| \leqq 1$, we obtain the asserted form (i) of $\log f(u)$.

If $1 < \gamma < 2$, then we can take out of the bracket in (1) the term $-\dfrac{iux}{1 + x^2} + iux$, and (2) is replaced by

$$(4) \quad \log f(u) = iu\alpha'' + \beta\int_{-\infty}^0 (e^{iux} - 1 - iux)\frac{dx}{|x|^{1+\gamma}}$$

$$+ \beta'\int_0^\infty (e^{iux} - 1 - iux)\frac{dx}{|x|^{1+\gamma}}.$$

Proceeding as above we obtain the same form (i) of $\log f(u)$.

If $\gamma = 1$, the foregoing modifications of the third term in the bracket in (1) are no more possible. But, for $u > 0$,

$$\int_{+0}^{+\infty}\left(e^{iux} - 1 - \frac{iux}{1 + x^2}\right)\frac{dx}{x^2}$$

$$= \int_0^\infty \frac{\cos ux - 1}{x^2}dx + i\int_{+0}^\infty\left(\sin ux - \frac{ux}{1 + x^2}\right)\frac{dx}{x^2}$$

$$= -\frac{\pi}{2}u + iu\lim_{\epsilon\downarrow 0}\left\{\int_{\epsilon u}^{+\infty}\frac{\sin v}{v^2}dv - \int_\epsilon^\infty \frac{dv}{v(1 + v^2)}\right\}$$

$$= -\frac{\pi}{2}u - iu\lim_{\epsilon\downarrow 0}\int_\epsilon^{\epsilon u}\frac{\sin v}{v^2}dv + iu\lim_{\epsilon\downarrow 0}\int_\epsilon^\infty\left(\frac{\sin v}{v^2} - \frac{1}{v(1 + v^2)}\right)dv.$$

The limit of the second integral exists and is finite, and that of the first one is $\log u$. The asserted form (ii) of $\log f(u)$ readily follows, and the conclusion is reached.

COMPLEMENTS AND DETAILS

1. Prove Lindeberg's theorem without using Liapounov's bounded case theorem. Then deduce Liapounov's theorem.

For Lindeberg's theorem use the expansion

$$f_k\left(\frac{u}{s_n}\right) = 1 - \frac{\sigma_k^2}{2s_n^2}u^2 + \theta_k \frac{|u|^3}{s_n^2}\left(\epsilon\sigma_k^2 + \int_{|x|\geqq \epsilon s_n} x^2\,dF_k(x)\right).$$

To deduce Liapounov's theorem observe that

$$\frac{1}{s_n^2}\int_{|x|\geqq\epsilon s_n} x^2\,dF_k \leqq \frac{1}{\epsilon^\delta}\cdot\frac{E|X_k|^{2+\delta}}{s_n^{2+\delta}}.$$

2. Prove directly the sufficiency of Kolmogorov's conditions for degenerate convergence. Then deduce the condition in $(1+\delta)$.

3. Deduce the Kolmogorov and Lindeberg-Feller theorems from the degenerate and normal convergence criteria—where existence of moments is not assumed.

4. Deduce the bounded variances limit theorem from the Central Limit theorem.

5. Let $\sum_k X_{nk}$ be sums of independent uan summands centered at expectations with $\sum_k \sigma^2 X_{nk} = 1$ whatever be n. Then

$$\mathfrak{L}(\sum_k X_{nk}) \to \mathfrak{N}(0, 1) \Leftrightarrow \sum_k X_{nk}^2 \xrightarrow{P} 1.$$

(Observe that the last convergence is equivalent to $\sum \int_{|x|\geqq\epsilon} x^2\,dF_{nk} \longrightarrow 0$ whatever be $\epsilon > 0$.)

6. Let $\zeta(t+iu)$, $t>1$, be the Riemann function defined by

$$\zeta(t+iu) = \sum_n n^{-t-iu} = \prod_p (1 - p^{-t-iu})$$

where p varies over all primes. $f_t(u) = \zeta(t+iu)/\zeta(t)$ is an i.d. ch.f.

$(\log f_t(u) = \sum_p \sum_n p^{-nt}(e^{-inu\log p} - 1)/n.)$

7. An i.d. law may be composed of two non i.d. laws. In fact, there exists a non i.d. ch.f. f such that $|f|^2$ is i.d.: form the ch.f. f of X with $P[X=-1] = p(1-p)/(1+p)$, $P[X=k] = (1-p)(1+p^2)p^k/(1+p)$, $k = 0, 1, \cdots$, $0 < p < 1$.

(Put f in the form (α, Ψ); observe that Ψ so found does not satisfy the necessary requirements. Put $|f|^2$ in the form (α, Ψ).)

8. An i.d. law may be composed of an i.d. law and an indecomposable one: let $X = 0$ or 1 with pr.'s $2/3$ and $1/3$, respectively; the ch.f. f is indecomposable

$$\log f(u) = \log \frac{2+e^{iu}}{3} = \sum a_n(e^{inu}-1), \quad \sum |a_n| < \infty.$$

Set

$$\log f^+(u) = \sum{}^+ a_n(e^{inu}-1), \quad \log f^-(u) = -\sum{}^- a_n(e^{inu}-1)$$

where $\sum^+ (\sum^-)$ denotes summation over positive (negative) a_n. Then f^+ and f^- are i.d. and $f^+ = ff^-$.

Also an i.d. law may be the product of an i.d. law and two indecomposable ones: proceed as above but with f defined by $\dfrac{5+4\cos u}{9} = \left|\dfrac{2+e^{iu}}{3}\right|^2$.

9. *P. Lévy's form of i.d. laws.* The family of i.d. laws coincides with laws defined by

$$\log f(u) = i\alpha u - \frac{\beta^2 u^2}{2} + ⨍ \left(e^{iux} - 1 - \frac{iux}{1+x^2}\right) dL(x)$$

where L is defined on R, except at the origin, is nondecreasing on $(-\infty, -0)$ and on $(+0, +\infty)$, with $L(\mp\infty) = 0$ and $⨍_{-\tau}^{+\tau} x^2\, dL(x) < \infty$ for some $\tau > 0$; the barred integral sign means that the origin is excluded.

Also

$$\log f(u) = i\alpha(\tau)u - \frac{\beta^2}{2}u^2 + ⨍_{-\tau}^{+\tau} (e^{iux} - 1 - iux)\, dL(x)$$
$$+ \left(\int_{-\infty}^{-\tau} + \int_{+\tau}^{+\infty}\right)(e^{iux}-1)\, dL(x).$$

The relation between the representations (α, β^2, L) and (α, Ψ) is defined by

$$\beta^2 = \Psi(+0) - \Psi(-0), \quad dL(x) = \frac{1+x^2}{x^2} d\Psi(x), \quad x \neq 0;$$

also

$$\alpha(\tau) = \alpha + \int_{|x|<\tau} x\, d\Psi(x) - \int_{|x|\geqq\tau} \frac{1}{x}\, d\Psi(x).$$

Write the convergence theorems (of i.d. laws, the central limit theorem, the central convergence theorem) in terms of $\alpha(\tau)$, β^2, and L.

Express the self-decomposability criterion conditions in terms of L.

(Setting $\log|x| = y$, and $L(x) = M_j(y)$, $j = 1$ or 2 according as $x > 0$ or $x < 0$, it becomes: the functions $(-1)^j M_j(y)$ are convex.)

10. Let r.v.'s $X_{n,k}$ with d.f.'s $F_{n,k}$, $k = 1, \cdots, k_n \to \infty$, $n = 1, 2, \cdots$, be independent in k and uniformly asymptotically distributed in k, that is, there exist d.f.'s F_n such that $F_{n,k} - F_n \to 0$ uniformly in k. The nondecreasingly ranked numbers $X_{n,k}(\omega)$ into $X^*_{n,1}(\omega) \leqq \cdots \leqq X^*_{n,k_n}(\omega)$ determine "ranked" $X^*_{n,r}$ of "rank" r; the ${}^*X_{n,s} = X_{n,k_n+1-s}$ are of "end rank" s. Set

$$L_n = \sum_k F_{n,k}, \quad M_n = \sum_k (F_{n,k} - 1),$$
$$g_{n,r_n} = \left(r_n - \sum F_{n,k}\right)\Big/\sqrt{\sum_k F_{n,k}(1-F_{n,k})},$$
$$I_n = \sum_k I_{n,k}, \quad \bar{I}_n = (I_n - EI_n)/\sigma I_n, \quad I_{n,k}(x) = I_{[X_{n,k}<x]}.$$

Use throughout the fundamental relation

$$[X^*_{n,r} < x] = [I_n(x) \geqq r].$$

a) The $X^*_{n,r}$ are r.v.'s.
b) For fixed ranks r, the class of limit laws of ranked r.v.'s $X_{n,r}$ is that of laws $\mathfrak{L}(X^*_r)$ with d.f.'s $F_r^L = \int_0^L \frac{t^{r-1}}{(r-1)!} e^{-t}\,dt$, where the functions L on R are nondecreasing, nonnegative, and not necessarily finite.

These limit laws are laws of r.v.'s if and only if $L(-\infty) = 0$, $L(+\infty) = +\infty$. And

$$F^*_{n,r} \xrightarrow[c]{w} F_r^L \Leftrightarrow L_n \xrightarrow[c]{w} L.$$

c) For fixed endranks s, the class of limit laws of ranked r.v.'s $^*X_{ns}$ is that of laws $\mathfrak{L}(^*X_s)$ with d.f.'s $^MF_s = \int_{-M}^{+\infty} \frac{t^{s-1}}{(s-1)!} e^{-t}\,dt$ where the functions M on R are nondecreasing, nonpositive, and not necessarily finite.

These limit laws are laws of r.v.'s if and only if $M(-\infty) = -\infty$, $M(+\infty) = 0$. And

$$^*F_{ns} \xrightarrow[c]{w} {}^MF_s \Leftrightarrow M_n \xrightarrow[c]{w} M.$$

d) For variable ranks $r_n \to \infty$ with $s_n = k_n + 1 - r_n \to \infty$, the class of limit laws of ranked r.v.'s X^*_{n,r_n} is that of laws with d.f.'s $F^g = \frac{1}{\sqrt{2\pi}} \int_g^\infty e^{-t^2/2}\,dt$, where the functions g on R are nonincreasing, and not necessarily finite.

These limit laws are those of r.v.'s if and only if $g(-\infty) = +\infty$, $g(+\infty) = -\infty$. And

$$F^*_{n,r_n} \xrightarrow[c]{w} F^g \Leftrightarrow g_{n,r_n} \xrightarrow[c]{w} g.$$

e) What if the X_{nk} are uniformly asymptotically negligible? What if, moreover, $\mathfrak{L}(\sum_k X_{nk}) \to \mathfrak{L}(X)$?
f) What about joint limit laws of ranked r.v.'s?

11. Let $\mathfrak{L}(X_n - a_n) \to (\alpha, \beta^2, L)$ where $X_n = \sum_k X_{nk}$ are sums of uan independent r.v.'s.

(a) The sequence $\mathfrak{L}(\max_k |X_{nk}|)$ converges. Find the limit law $\mathfrak{L}(X)$. Why can necessary and sufficient conditions for normality of the limit law of the sequence $\mathfrak{L}(X_n - a_n)$ be expressed in terms of $\mathfrak{L}(X)$? Are there other i.d. laws for which this is possible? (For n sufficiently large and $x > 0$

$$\log P[\max_k |X_{nk}| < x] = -(1 + o(1)) \sum_k P[|X_{nk}| \geqq x].)$$

(b) Let $a_{nk} = \int_{|x|<\tau} x\,dF_{nk}$, $\tau > 0$ finite, $\bar{F}_{nk}(x) = F_{nk}(x + a_{nk})$ and let F'_{nk} be the d.f. of $X'_{nk} = |X_{nk} - a_{nk}|^r$ for a fixed $r > 1$.

If $\mathfrak{L}(\sum_k X_{nk} - a_n) \to (\alpha, \beta^2, L)$, then there exist constants a'_n such that $\mathfrak{L}(\sum_k X'_{nk} - a'_n) \to (\alpha', 0, L')$ with $L'(x) = 0$ or $L(x^{1/r}) - L(-x^{1/r})$ according as $x < 0$ or $x > 0$. (If $g \geqq 0$ is even, then, for every $c > 0$,

$$\int_0^c g\,dF'_{nk} = \int_{|x|<c^{1/r}} g(|x|^r)\,d\bar{F}_{nk}, \quad \int_c^\infty g\,dF'_{nk} = \int_{|x|>c^{1/r}} g(|x|^r)\,d\bar{F}_{nk}.$$

Take $g = 1$ and $g(x) = x^2$. Observe that

$$0 \leqq \int_0^{\epsilon} x^2\, dF'_{nk} - \left(\int_0^{\epsilon} x\, dF'_{nk}\right)^2 \leqq \epsilon^{2(r-1)/r} \int_{|x|<\epsilon^{1/r}} x^2\, d\bar{F}_{nk}.)$$

(c) $\mathcal{L}(\sum_k X_{nk} - a_n) \to \mathfrak{N}(0, \beta^2)$ if, and only if, $\sum_k X'^2_{nk} \xrightarrow{P} \beta^2$. What about limit Poisson laws?

In what follows and unless otherwise stated, degenerate laws are excluded; f, with or without affixes, is a ch.f.; and, without restricting the generality, the type of f is the family of all ch.f.'s defined by $f(cu)$ for some $c > 0$.

12. f is decomposable by every f^n, $n = 2, 3, \cdots$, if, and only if, f is degenerate.

13. f is decomposable and every component belongs to its type with $f(u) = \sum f(c_j u)$, $\sum c_j^2 \geqq 1$, if and only if f is normal.

14. If for an $r > 0$ and $\neq 1$, f^r belongs to the type of f, then f is i.d. If there are two such values r' and r'' of r and $\log r'/\log r''$ is irrational, then f is stable.

15. If $f_n \to f, f'_n \to f'$ and $f_n = f'_n f''_n$ for every n, then f' is a component of f.

16. f is c-decomposable if $f(u) = f(cu)f_c(u)$ for some fixed c necessarily between 0 and 1. L_c is the family of all c-decomposable laws, L_0 is the family of all laws, and L_1 is that of self-decomposable ones.

(a) $L_0 \supset L_c \supset L_1$, and if $\log c/\log c'$ is rational, then $L_c = L_{c'}$. Every L_c is closed under compositions and passages to the limit.

(b) $f \in L_c$ if, and only if, it is limit of a sequence of ch.f.'s of normed sums S_n/b_n of independent r.v.'s with $b_n/b_{n+1} \to c$.

(c) $f \in L_c$ if, and only if, it is ch.f. of $X(c) = \sum_{k=0}^{\infty} \xi_k c^k$ where the law of the series converges and the ξ_k are independent and identically distributed. Then the series converges a.s., and $f_{\xi_k} = f_c$. If ξ_k is bounded, then f is not i.d.

(d) $g(x)$ is said to be γ-convex ($\gamma > 0$ fixed) if every polygonal line inscribed in its graph with vertices projecting at distance γ on the x-axis is convex.

If ξ_k is i.d., so is $X(c)$. f i.d. with Lévy's function L belongs to L_c and f_c is i.d. only if $(-1)^j M_j$ are γ-convex for $\gamma = |\log c|$ where M_j are defined as in 9. Is the converse true?

(e) If $E\xi_k = 0$, $\sigma^2 \xi_k = 1$, then, for $c, c' \in (-1, +1)$, the covariance $EX(c)X(c') = 1/(1 - cc')$, and the random function $X(c)$ on $(-1, +1)$ exists in q.m. and is continuous and indefinitely differentiable in q.m.

17. Attraction. A d.f. F is attracted by a law $\mathcal{L}$ if $\mathcal{L}$ is a limit law of suitably normed sums of independent r.v.'s with common d.f. F.

F is attracted by a normal law if, and only if,

$$\int_{|x|\geqq c} dF = o(c^{-2}\int_{|x|<c} x^2\, dF) \quad \text{for } c \to \infty.$$

F is attracted by a stable law with exponent $\gamma < 2$, if, and only if, as $x \to \infty$,

$$\frac{F(-x)}{1 - F(x)} \to c, \quad \frac{1 - F(x) + F(-x)}{1 - F(\beta x) + F(-\beta x)} \to \beta^\gamma$$

for every $\beta > 0$.

If F is attracted by a stable law with $\gamma \leqq 2$, then $\int |x|^r\, dF < \infty$ for $0 \leqq r < \gamma$.

Part Four

DEPENDENCE

For about two centuries probability theory has been concerned almost exclusively with independence. Yet, very particular forms of dependence appear already in the theory of games of chance. But a first general type of dependence—chains—was introduced only at the beginning of this century by Markov. Another type of dependence—stationarity—appears in ergodic theory, and a related type—second order stationarity—is then introduced in probability theory by Khintchine (1932). Centering at conditional expectations by P. Lévy (1935) gives rise to a new type of dependence—martingales.

At the very core of the study of dependence lies the concept of conditioning—with respect to a function—put in an abstract and rigorous form by Kolmogorov. In this part, the concept of conditioning is introduced in a more general form—with respect to a σ-field—and, as much as possible, the properties of various types of dependence are related to more general results, with emphasis given to the methods.

Chapter VII

CONDITIONING

§ 24. CONCEPT OF CONDITIONING

The concept of "conditioning" can be expressed in terms of sub σ-fields of events. Conditional probabilities of events and conditional expectations of r.v.'s "given a σ-field $\mathcal{B}$," to be introduced and investigated in this chapter, are $\mathcal{B}$-measurable functions defined up to an equivalence. If $\mathcal{B}$ is determined by a countable partition of the sure event, then these functions are elementary. In this "elementary case," a constructive approach with a definite intuitive appeal is possible and there are no technical difficulties. In the general case, there is no suitable and rigorous constructive approach, and a descriptive one, requiring more powerful tools, especially the Radon-Nikodym theorem, has to be used.

The R.-N. theorem was obtained in its abstract form in 1930 and the concept of conditional probabilities and of conditional expectations of integrable r.v.'s "given" a measurable function, finite or not, numerical or not, was then put on a rigorous basis by Kolmogorov in 1933.

24.1 Elementary case. Investigation of the elementary case will give us an insight into the ideas involved in the intuitive notion of conditioning and will lead "naturally" to the notions and problems which appear in the general case.

The notion of conditional probability of an event A "given an event B" corresponds to that of frequencies of A in the repeated trials where B occurs; it is one of the oldest probability notions. For every event A, the relation

$$PB \cdot P_B A = PAB$$

defines the *conditional probability* (*c.pr.*) $P_B A$ *of* A *given* B as the ratio PAB/PB, provided B is a nonnull event; if B is null, so is AB, and the foregoing relation leaves P_BA undetermined. In what follows, we assume that, unless otherwise stated, B is nonnull.

The function P_B on the σ-field $\mathcal{A}$ of events, whose values are P_BA, $A \in \mathcal{A}$, is called *conditional pr. given B*. The defining relation shows at once that since P on $\mathcal{A}$ is normed, nonnegative, and σ-additive, so is P_B on $\mathcal{A}$:

$$P_B\Omega = 1, \quad P_B \geqq 0, \quad P_B \sum A_j = \sum P_BA_j.$$

Thus, the conditioning expressed by "given B" means that the initial pr. space $(\Omega, \mathcal{A}, P)$ is replaced by the pr. space $(\Omega, \mathcal{A}, P_B)$. The expectation, if it exists, of a r.v. X on this new pr. space is called *conditional expectation* (*c.exp.*) *given B* and is denoted by E_BX; in symbols

$$E_BX = \int X\,dP_B.$$

Since $P_B = 0$ on $\{AB^c, A \in \mathcal{A}\}$, the right-hand side reduces to $\int_B X\,dP_B$ and, since $P_B = \frac{1}{PB}P$ on $\{AB, A \in \mathcal{A}\}$, it becomes $\frac{1}{PB}\int_B X\,dP$. Therefore, the c.exp. of X given B can be defined directly by

$$PBE_BX = \int_B X\,dP$$

and is determined if B is a nonnull event. In particular,

$$PBE_BI_A = \int_B I_A\,dP = PAB$$

so that the c.pr. P_BA can be defined, thereafter, by

$$P_BA = E_BI_A.$$

Thus, if E_B is the *c.exp. given B*, with values E_BX on the family $\mathcal{E}_B$ of all r.v.'s X whose integral on B exists, the c.pr. P_B becomes the restriction of E_B to the family $I_{\mathcal{A}}$ of indicators of events. Furthermore, properties of P_B become particular cases of the immediate properties of E_B below.

If $X \geqq 0$ then $E_BX \geqq 0$, and if c is a constant then $E_Bc = c$. If the X_j are nonnegative, or if the X_j are integrable and their consecutive sums are uniformly bounded by an integrable r.v., then $E_B \sum X_j = \sum E_BX_j$.

C.exp.'s (hence c.pr.'s) acquire their full meaning when reinterpreted as values of functions, as follows. The number E_BX is no longer assigned to B but to every point of B, and similarly for $E_{B^c}X$, so that we have a two-valued function on Ω, with values $E_B X$ for $\omega \in B$ and $E_{B^c}X$ for $\omega \in B^c$. More generally, let $\{B_j\}$ be a countable partition of Ω and let $\mathcal{B}$ be the minimal σ-field over this partition. Let $\mathcal{E}$ be the family of all r.v.'s X whose expectation EX exists, so that their indefinite integrals, hence c.exp.'s given any nonnull event, exist. Consider the elementary functions

$$E^{\mathcal{B}}X = \sum (E_{B_j}X)I_{B_j}, \quad X \in \mathcal{E}.$$

If some B_j are null, then the corresponding values $E_{B_j}X$ are undetermined, so that $E^{\mathcal{B}}X$ is undetermined on the null event which is the sum of null B_j. Such a possibility, together with the definition of $E_{B_j}X$, leads to the following

CONSTRUCTIVE DEFINITION. *The elementary function $E^{\mathcal{B}}X$ defined up to an equivalence by*

$$(1) \qquad E^{\mathcal{B}}X = \sum \left(\frac{1}{PB_j}\int_{B_j} X\,dP\right) I_{B_j}, \quad X \in \mathcal{E},$$

is the c.exp. of X given $\mathcal{B}$.

Upon particularizing to indicators, the $\mathcal{B}$-measurable function $P^{\mathcal{B}}A$, defined up to an equivalence by setting

$$P^{\mathcal{B}}A = E^{\mathcal{B}}I_A, \quad A \in \mathcal{A},$$

will be the *c.pr. of A given* $\mathcal{B}$; the contraction of $E^{\mathcal{B}}$ on $I_{\mathcal{A}}$, to be denoted by $P^{\mathcal{B}}$, will be the *c.pr. given* $\mathcal{B}$, and its values are the $\mathcal{B}$-measurable functions $P^{\mathcal{B}}A$, $A \in \mathcal{A}$, defined up to an equivalence.

We say "given (the σ-field) $\mathcal{B}$" and not "given (the partition) $\{B_j\}$," because $E^{\mathcal{B}}X$ determines the c.exp. of X given an arbitrary nonnull event $B \in \mathcal{B}$. In fact, if $\sum'$ denotes the summation over some subclass of $\{B_j\}$, then every event $B \in \mathcal{B}$ is of the form $\sum' B_j$, and we have

$$PBE_BX = \int_{\Sigma' B_j} X\,dP = \sum{}' \int_{B_j} X\,dP = \sum{}' PB_jE_{B_j}X.$$

This relation can also be written as follows: If $P_{\mathcal{B}}$ is the restriction of P to $\mathcal{B}$, defined by

$$P_{\mathcal{B}}B = PB, \quad B \in \mathcal{B},$$

then the right-hand side becomes $\int_B (E^{\mathcal{B}}X)\,dP_{\mathcal{B}}$ while the left-hand side is $\int_B X\,dP$. This leads to the following

DESCRIPTIVE DEFINITION. *The c.exp.* $E^{\mathcal{B}}X$ *of* $X \in \mathcal{E}$ *given* $\mathcal{B}$ *is any* $\mathcal{B}$*-measurable function whose indefinite integral with respect to* $P_{\mathcal{B}}$ *is the restriction to* $\mathcal{B}$ *of the indefinite integral of* X *with respect to* P. *Since the indefinite integral with respect to* $P_{\mathcal{B}}$ *of a* $\mathcal{B}$*-measurable function determines this function up to an equivalence, this definition means precisely that, for every* $X \in \mathcal{E}$, $E^{\mathcal{B}}X$ *is defined by*

$$\int_B (E^{\mathcal{B}}X)\, dP_{\mathcal{B}} = \int_B X\, dP, \quad B \in \mathcal{B}.$$

To conclude the discussion of the elementary case, we revert to the initial approach, first defining c.pr.'s and, then, defining c.exp.'s as integrals. According to what precedes, we define $P^{\mathcal{B}}$ on $\mathcal{A}$ either by

$$\int_B (P^{\mathcal{B}}A)\, dP_{\mathcal{B}} = PAB, \quad A \in \mathcal{A}, \quad B \in \mathcal{B}$$

or, equivalently, by

$$P^{\mathcal{B}}A = \sum \frac{PAB_j}{PB_j} I_{B_j}, \quad A \in \mathcal{A},$$

up to an equivalence.

Let B_0 be the null sum of all null B_j and, for every A, *select* $P^{\mathcal{B}}A$ within its equivalence class by taking its values $P_\omega{}^{\mathcal{B}}A$ at $\omega \in B_j$ to be PAB_j/PB_j if B_j is nonnull and PA if B_j is null ($\subset B_0$). Then, for every $\omega \in \Omega$, the function $P_\omega{}^{\mathcal{B}}$ on $\mathcal{A}$, with values $P_\omega{}^{\mathcal{B}}A$, is a probability and we can form integrals with respect to it. Let $X \in \mathcal{E}$ and set

$$E_\omega{}^{\mathcal{B}}X = \int X\, dP_\omega{}^{\mathcal{B}}, \quad \omega \in \Omega.$$

Since, for every $\omega \in B_j$ not contained in the null event B_0, we have

$$\int X\, dP_\omega{}^{\mathcal{B}} = \frac{1}{PB_j} \int_{B_j} X\, dP,$$

it follows that the function on Ω with values $E_\omega{}^{\mathcal{B}}X$ belongs to the equivalence class of $E^{\mathcal{B}}X$. Thus, we can define $E^{\mathcal{B}}X$ to be $P_{\mathcal{B}}$-equivalent to the integral $\int X\, dP^{\mathcal{B}}$ where $P^{\mathcal{B}}$, hence the integral, are functions of $\omega \in \Omega$; in symbols

$$E^{\mathcal{B}}X = \int X\, dP^{\mathcal{B}} \text{ a.s.}$$

24.2 General case. The constructive approach fails at the very start as soon as the "given" σ-fields are not generated by countable partitions. However, the descriptive approach remains possible, thanks to the Radon-Nikodym theorem.

Let $(\Omega, \mathcal{A}, P)$ denote, as usual, the pr. space. Let $\mathcal{B}$, with or without affixes, denote a σ-field contained in $\mathcal{A}$, and let $P_{\mathcal{B}}$ denote the restriction of P to $\mathcal{B}$. Finally, let $\mathcal{E}$ be the family of all $\mathcal{A}$-measurable functions whose integral (hence indefinite integral) exists.

DEFINITION. *The c.exp.* $E^{\mathcal{B}}X$ of $X \in \mathcal{E}$ *given* $\mathcal{B}$ *is a* $\mathcal{B}$*-measurable function, defined up to a* $P_{\mathcal{B}}$*-equivalence by*

(1) $$\int_B (E^{\mathcal{B}}X)\, dP_{\mathcal{B}} = \int_B X\, dP, \quad B \in \mathcal{B}.$$

It follows at once that

1° $E(E^{\mathcal{B}}X) = EX.$

2° *If* $\mathcal{B} = \mathcal{A}$ *or* X *is* $\mathcal{B}$*-measurable, then* $E^{\mathcal{B}}X = X$ *a.s.*

3° $E^{\mathcal{B}}X = E^{\mathcal{B}}X^+ - E^{\mathcal{B}}X^-$ *a.s.*

The definition is justified: the indefinite integral φ of X being σ-additive and P-continuous, its restriction $\varphi_{\mathcal{B}}$ to $\mathcal{B}$ is σ-additive and $P_{\mathcal{B}}$-continuous, the extended Radon-Nikodym theorem applies, and the $\mathcal{B}$-measurable function $E^{\mathcal{B}}X$ defined by (1) exists and is defined up to a $P_{\mathcal{B}}$-equivalence.

If $\varphi_{\mathcal{B}}$ is σ-finite, then the Radon-Nikodym theorem applies, so that, moreover, $E^{\mathcal{B}}X$ is finite except on an arbitrary null event belonging to $\mathcal{B}$. If X is a r.v., then φ is σ-finite, but this does not imply that $\varphi_{\mathcal{B}}$ is σ-finite: take $\varphi(\Omega) = \infty$ and $\mathcal{B} = \{\emptyset, \Omega\}$. However, such a possibility is excluded in the case of integrable r.v.'s for, then, φ and hence $\varphi_{\mathcal{B}}$ are bounded.

We observe that as soon as it is understood that $E^{\mathcal{B}}X$ is, by definition, a $\mathcal{B}$-measurable function, we can replace $P_{\mathcal{B}}$ by P, and properties of $E^{\mathcal{B}}X$ valid, except on a $P_{\mathcal{B}}$-null event, may and will be said "a.s."

The function $E^{\mathcal{B}}$ on $\mathcal{E}$ to the space of $\mathcal{B}$-measurable functions (more precisely, on the space of equivalence classes of $\mathcal{A}$-measurable functions possessing an integral to the space of equivalence classes of $\mathcal{B}$-measurable functions) will be called *c.exp. given* $\mathcal{B}$. $E^{\mathcal{B}}$ can also be considered as a function on $\Omega \times \mathcal{E}$ to $\bar{R} = [-\infty, +\infty]$ with values $E_\omega{}^{\mathcal{B}}X$ for $\omega \in \Omega$, $X \in \mathcal{E}$, the value for all ω belonging to an arbitrary $P_{\mathcal{B}}$-null event being arbitrary.

The restriction of $E^{\mathcal{B}}$ to the family $I_{\mathcal{A}}$ of indicators of events is called *c.pr. given* $\mathcal{B}$ and is denoted by $P^{\mathcal{B}}$; in other words, $P^{\mathcal{B}}$ is a function on $\mathcal{A}$ whose values are $\mathcal{B}$-measurable functions $P^{\mathcal{B}}A$ defined up to an equivalence by $P^{\mathcal{B}}A = E^{\mathcal{B}}I_A$ or, directly, by

$$\int_B (P^{\mathcal{B}}A)\, dP_{\mathcal{B}} = PAB, \quad B \in \mathcal{B}.$$

Extension. It is "natural" to require of the definition of c.exp.'s that $E^{\mathcal{A}}X = X$ a.s., whatever be the measurable function X. Yet, the foregoing definition does not apply to those X whose integrals do not exist. However, it is possible to extend the definition so as to achieve the foregoing requirement, as follows: Write $X = X^+ - X^-$ where, as usual, X^+ and X^- are the positive and negative parts of X, respectively; $E^{\mathcal{B}}X^+$ and $E^{\mathcal{B}}X^-$ always exist but may be infinite. *Define* $E^{\mathcal{B}}X$ by $E^{\mathcal{B}}X = E^{\mathcal{B}}X^+ - E^{\mathcal{B}}X^-$ so that $E^{\mathcal{B}}X$ exists on the set on which the difference is not of the form $+\infty - \infty$ up to a $P_{\mathcal{B}}$-null event. This generalized c.exp. exists a.s. if the event $[X \geq 0]$ and hence $[X < 0]$ belong to $\mathcal{B}$ (and, in particular, if $\mathcal{B} = \mathcal{A}$), for then $E^{\mathcal{B}}X^+ \times E^{\mathcal{B}}X^- = 0$ a.s. If $\mathcal{B} = \mathcal{A}$, then $E^{\mathcal{A}}X^+ - E^{\mathcal{A}}X^- = X^+ - X^- = X$ a.s., whatever be the r.v. X.

24.3 Conditional expectation given a function. We connect now the foregoing definition with the usual definition of c.exp., but we do not assume, as usually done, that the c.exp.'s are restricted to those of integrable r.v.'s.

Let Y be a function on $(\Omega, \mathcal{A}, P)$ to a measurable space $(\Omega', \mathcal{A}')$ and let $\mathcal{B}_Y \subset \mathcal{A}$ and $\mathcal{B}'_Y \subset \mathcal{A}'$ be the σ-fields *induced* by Y on Ω and Ω' respectively: $\mathcal{B}'_Y$ is the σ-field of all sets of $\mathcal{A}'$ whose inverse images under Y are events ($\in \mathcal{A}$) and $\mathcal{B}_Y$ is the σ-field of these events. Let P_Y and P'_Y be the probabilities *induced* by Y on $\mathcal{B}_Y$ and $\mathcal{B}'_Y$, respectively, defined by

$$P_Y B = PB \quad B \in \mathcal{B}_Y; \quad P'_Y B' = PB, \quad B' \in \mathcal{B}'_Y, \quad B = Y^{-1}(B').$$

(If Y is measurable, then $\mathcal{B}'_Y = \mathcal{A}'$. If no $\mathcal{A}'$ is given, then we take $\mathcal{A}' = S(\Omega')$.)

If $\mathcal{B} = \mathcal{B}_Y$ in the definitions of the preceding subsection, then we replace every $\mathcal{B}_Y$ by Y. Thus, we write $E^Y X$ instead of $E^{\mathcal{B}_Y}X$, and call it *c.exp. of X given Y.* The reason for this terminology is that, as we shall show now, $E^Y X$ is a function of the function Y. We require the following proposition.

a. *For every numerical measurable function g on $(\Omega', \mathcal{B}'_Y, P'_Y)$*

$$\int_{B'} g\, dP'_Y = \int_B g(Y)\, dP_Y, \quad B' \in \mathcal{B}'_Y, \quad B = Y^{-1}(B'),$$

in the sense that, if one of these integrals exists, so does the other, and both are equal.

Proof. The asserted equality is true if $g = I_{A'}$ is the indicator of a set $A' \in \mathcal{B}'_Y$, for, setting $A = Y^{-1}(A')$, we have $g(Y) = I_A$ and, hence,

$$\int_{B'} I_{A'}\, dP'_Y = P'_Y(A'B') = P_Y(AB) = \int_B I_A\, dP_Y.$$

Being true for indicators, the equality is true for simple functions g and, by the monotone convergence theorem, for nonnegative measurable g. The assertion follows upon decomposing a measurable g into its positive and negative parts.

We are now in a position to prove the above-stated property.

A. *The c.exp. of $X \in \mathcal{E}$ given Y is a function of the function Y.*

Proof. If φ is the indefinite integral of X and φ' on $\mathcal{B}'_Y$ is defined by

$$\varphi'(B') = \varphi(B), \quad B' \in \mathcal{B}'_Y, \quad B = Y^{-1}(B'),$$

then φ' is σ-additive and P'_Y-continuous, the extended Radon-Nikodym theorem applies to φ' and P'_Y and defines a measurable function g on $(\Omega', \mathcal{B}'_Y)$ by

$$\int_{B'} g\, dP'_Y = \varphi'(B') = \int_B X\, dP.$$

Since

$$\int_B X\, dP = \int_B (E^Y X)\, dP,$$

it follows, upon applying **a**, that

$$\int_B g(Y)\, dP_Y = \int_{B'} g\, dP'_Y = \int_B (E^Y X)\, dP,$$

so that the indefinite integrals of the $\mathcal{B}_Y$-measurable functions $g(Y)$ and $E^Y X$ are the same, and the assertion is proved.

As defined, $E^Y X$ is a Y-measurable function *on the original space* $(\Omega, \mathcal{A}, P)$. However, the usual interpretation of the c.exp. of X given Y is that it is the function g on Ω' defined by

$$\int_{B'} g\, dP'_Y = \int_B X\, dP, \quad B' \in \mathcal{B}'_Y, \quad B = Y^{-1}(B').$$

We prefer to consider c.exp.'s as functions on the original pr. space. Yet, on account of the foregoing theorem, both interpretations are possible: either $E^Y X$ is considered as a *function of the function* Y with values

$E_\omega{}^Y X$ for $\omega \in \Omega$, or it is considered as a *function of* Y with values $E_y{}^Y X$ for $Y = y$, defined up to a P_Y or P'_Y-equivalence, respectively.

Notation. The following symbols are and will be used according to convenience:

$E^Y X$ or $E(X \mid Y)$ or $E(Y; X)$, $E_y{}^Y X$ or $E(X \mid y)$ or $E(y; X)$;
$P^Y A$ or $P(A \mid Y)$ or $P(Y; A)$, $P_y{}^Y A$ or $P(A \mid y)$ or $P(y; A)$;
and similarly with y replaced by ω and/or Y replaced by $\mathcal{B}$.

***24.4 Relative conditional expectations and sufficient σ-fields.** The Radon-Nikodym theorem applies to σ-finite and μ-continuous signed measures on $\mathcal{A}$ with σ-finite measures μ on $\mathcal{A}$. Therefore, the concept of c.exp. continues to apply if, in what precedes, P is replaced by any such measure μ. But, then, we have to specify that the c.exp.'s are taken *with respect to* μ—they are *relative c.exp.'s*. To simplify, we limit ourselves to finite and P-continuous measures μ. (Yet, we shall see in the next volume that, led by physics, we may have to replace pr.'s by σ-finite measures and, thus, use fully the foregoing conditioning.)

Given the pr. space $(\Omega, \mathcal{A}, P)$, the measures μ are indefinite integrals of nonnegative r.v.'s Z and we say that the relative c.exp.'s are taken *with respect to* Z. *In what follows, the r.v.'s* Z, *with or without affixes, are nonnegative and integrable, and the* μ, *with the same affixes if any, are their indefinite integrals.*

If a r.v. X possesses an integral with respect to μ, then the c.exp. of X given $\mathcal{B}$ *with respect to* Z is a $\mathcal{B}$-measurable function, defined up to a $\mu_{\mathcal{B}}$-equivalence by

$$\int_B (E_Z{}^{\mathcal{B}} X)\, d\mu_{\mathcal{B}} = \int_B X\, d\mu, \quad B \in \mathcal{B}.$$

Since

$$\mu A = \int_A Z\, dP, \quad A \in \mathcal{A},$$

it follows that

$$\mu_{\mathcal{B}} B = \int_B Z\, dP = \int_B (E^{\mathcal{B}} Z)\, dP_{\mathcal{B}}, \quad B \in \mathcal{B},$$

and this definition is equivalent to

$$\int_B (E^{\mathcal{B}} Z)(E_Z{}^{\mathcal{B}} X)\, dP_{\mathcal{B}} = \int_B ZX\, dP = \int_B (E^{\mathcal{B}} ZX)\, dP_{\mathcal{B}}, \quad B \in \mathcal{B},$$

which, in its turn, is equivalent to

(1) $$E^{\mathcal{B}} Z \cdot E_Z{}^{\mathcal{B}} X = E^{\mathcal{B}} ZX \text{ a.s.}$$

$E_Z{}^{\mathcal{B}}X$ is defined up to a $\mu_{\mathcal{B}}$-equivalence and $P_{\mathcal{B}}A = 0$ entails $\mu_{\mathcal{B}}A = 0$; furthermore,

$$\mu_{\mathcal{B}}[E^{\mathcal{B}}Z = 0] = \int_{[E^{\mathcal{B}}Z=0]} (E^{\mathcal{B}}Z)\, dP_{\mathcal{B}} = 0.$$

Therefore, up to a $\mu_{\mathcal{B}}$-equivalence, $E_Z{}^{\mathcal{B}}X$ is given by

$$(1') \qquad E_Z{}^{\mathcal{B}}X = E^{\mathcal{B}}ZX/E^{\mathcal{B}}Z,$$

so that

a. *Relative c.exp.'s are reducible, up to an equivalence, to ratios of ordinary c.exp.'s.*

It may happen that c.exp.'s given $\mathcal{B}$ relative to the r.v.'s of a family $\{Z_t\}$ collapse together: there exists a r.v. Z such that, for every t,

$$(2) \qquad E_{Z_t}{}^{\mathcal{B}}X = E_Z{}^{\mathcal{B}}X$$

in the sense that, whenever the left-hand side exists, so does the right-hand side, and both are equal. But these sides are determined up to $(\mu_t)_{\mathcal{B}}$- and $(\mu_{\mathcal{B}})$-equivalences, respectively. Thus, equality might be interpreted in the sense that the $\mu_{\mathcal{B}}$-equivalence class of $E_Z{}^{\mathcal{B}}X$ belongs to every $(\mu_t)_{\mathcal{B}}$-equivalence class of $E_{Z_t}X$'s. This is certainly true as soon as the equality holds for an element of each class, provided every μ_t is μ-continuous. Then, moreover, whenever $E_{Z_t}{}^{\mathcal{B}}X$ exists so does $E_Z{}^{\mathcal{B}}X$. Finally, we are led to the following definition.

Let X be "admissible" for the family $\{Z_t\}$ if its integrals with respect to every u_t exist. A sub σ-field $\mathcal{B}$ of events is *sufficient* (with Z) for the family $\{Z_t\}$ if there exists a Z such that every μ_t is μ-continuous and, for every admissible X, (2) holds up to a $(\mu_t)_{\mathcal{B}}$-equivalence. This concept of sufficient sub σ-fields is slightly more general than the usual concept of "sufficient statistics" which plays a considerable role in statistics. Clearly every σ-field P-equivalent to a sufficient $\mathcal{B}$ is sufficient. Thus, in what follows, we assume that every sufficient σ-field is defined up to a P-equivalence.

The basic result (originating with Neyman and put in its final form by Halmos and Savage—in terms of sufficient statistics) is as follows:

A. FACTORIZATION THEOREM. *The sub σ-field $\mathcal{B}$ of events is sufficient for the family $\{Z_t\}$ if, and only if, there exists a Z such that every $Z_t = g_tZ$ a.s. and every g_t is $\mathcal{B}$-measurable; then every $g_t = E^{\mathcal{B}}Z_t/E^{\mathcal{B}}Z$ up to a $\mu_{\mathcal{B}}$-equivalence.*

We require two properties of c.exp.'s: 1° $E^{\mathcal{B}}$ and $\mathcal{B}$-measurable functions commute (25.2, **3**); 2° if, for Y integrable or nonnegative, and for every indicator I_A of events, $E^{\mathcal{B}}YI_A = E^{\mathcal{B}}Y'I_A$ a.s., then $Y = Y'$ a.s. since, for every $A \in \mathcal{A}$,

$$\int_A Y\,dP = \int YI_A\,dP = \int (E^{\mathcal{B}}YI_A)\,dP_{\mathcal{B}} = \int (E^{\mathcal{B}}Y'I_A)\,dP_{\mathcal{B}}$$

$$= \int Y'I_A\,dP = \int_A Y'\,dP.$$

Proof. $Z_t = g_tZ$ a.s. entails μ-continuity of μ_t and, by (1),

$$E^{\mathcal{B}}Z_t \cdot E_{Z_t}{}^{\mathcal{B}}X = E^{\mathcal{B}}Z_tX = g_tE^{\mathcal{B}}ZX = g_tE^{\mathcal{B}}Z \cdot E_Z{}^{\mathcal{B}}X = E^{\mathcal{B}}Z_t \cdot E_Z{}^{\mathcal{B}}X \text{ a.s.}$$

The sets $[E^{\mathcal{B}}Z_t = 0]$ being $(\mu_t)_{\mathcal{B}}$-null, $E_{Z_t}{}^{\mathcal{B}}X = E_Z{}^{\mathcal{B}}X$ up to $(\mu_t)_{\mathcal{B}}$-equivalence. The set $[E^{\mathcal{B}}Z = 0]$ being $\mu_{\mathcal{B}}$-null, it follows from

$$E^{\mathcal{B}}Z_t = g_tE^{\mathcal{B}}Z \text{ a.s.}$$

that $g_t = E^{\mathcal{B}}Z_t/E^{\mathcal{B}}Z$ up to a $\mu_{\mathcal{B}}$-equivalence.

Conversely, if, for all indicators X, every $E_{Z_t}{}^{\mathcal{B}}X = E_Z{}^{\mathcal{B}}X$ up to a $(\mu_t)_{\mathcal{B}}$-equivalence, then, by (1),

$$E^{\mathcal{B}}Z \cdot E^{\mathcal{B}}Z_tX = E^{\mathcal{B}}Z_tE^{\mathcal{B}}ZE_Z{}^{\mathcal{B}}X = E^{\mathcal{B}}Z_tE^{\mathcal{B}}ZX \text{ a.s.}$$

or

$$E^{\mathcal{B}}(Z_tXE^{\mathcal{B}}Z) = E^{\mathcal{B}}(ZXE^{\mathcal{B}}Z_t) \text{ a.s.,}$$

so that

$$Z_tE^{\mathcal{B}}Z = ZE^{\mathcal{B}}Z_t \text{ a.s.}$$

and, hence, on $B = [E^{\mathcal{B}}Z > 0]$,

$$Z_t = \frac{E^{\mathcal{B}}Z_t}{E^{\mathcal{B}}Z}Z \text{ a.s.} \tag{3}$$

Since μ_t is μ-continuous, from

$$\mu B^c = \int_{B^c} Z\,dP = \int_{B^c} (E^{\mathcal{B}}Z)\,dP = 0,$$

so that $Z = 0$ on B^c except for P-null subsets, it follows that $\mu_tB^c = 0$; hence $Z_t = 0$ on B^c except for P-null subsets. Thus (3) is trivially true on B^c. This completes the proof.

Underlying the concept of sufficient σ-fields with Z is the fact that every μ_t is supposed to be μ-continuous. This alone implies that every

$Z_t = g_t Z$ a.s. where the g_t are measurable. Thus, the whole σ-field $\mathcal{A}$ of events is trivially sufficient with any such Z and, in particular, with $Z = 1$. And every sub σ-field $\mathcal{B}$ of events such that all the g_t are $\mathcal{B}$-measurable is sufficient with such a Z; in particular, the sub σ-field $\mathcal{B}$ induced by the family $\{g_t\}$ is the least fine sufficient with Z. The question arises whether there exists some Z, say Z_0, such that the least fine sufficient σ-field with Z_0 is the least fine of all possible sufficient σ-fields for the family $\{Z_t\}$—the *minimal sufficient* σ-field for the family $\{Z_t\}$. The answer is in the affirmative, as follows:

According to Chapter II: Complements and Details 23, there exists a Z_0 such that

(i) $\mu_0 A = 0 \Leftrightarrow$ every $\mu_t A = 0$;

or, equivalently,

(i′) up to P-null subsets, $Z_0 = 0$ on $A \Leftrightarrow$ every $Z_t = 0$ on A.

Since, on account of (i′), every $E_Z{}^{\mathcal{B}}X$ common to all the equivalence classes $E_{Z_t}{}^{\mathcal{B}}X$ belongs also to the equivalence class of $E_{Z_0}{}^{\mathcal{B}}X$, it follows that every sufficient σ-field $\mathcal{B}$ with Z is also sufficient with Z_0. Therefore, the least fine sufficient σ-field with Z_0 is the minimal one. On account of (i′), the corresponding factorization—every $Z_t = g_t Z_0$ a.s.—is such that $Z_0 = 0$ a.s. $\Rightarrow$ every $Z_t = 0$ a.s. Thus:

B. MINIMALITY CRITERION. *Write every Z_t in the form $Z_t = g_t Z_0$ a.s., with Z_0 such that every $Z_t = 0$ a.s. $\Rightarrow$ $Z_0 = 0$ a.s.; this is always possible. Then the minimal sufficient σ-field for the family $\{Z_t\}$ is the one induced by the family $\{g_t\}$.*

§ 25. PROPERTIES OF CONDITIONING

To avoid constant repetitions, it will be assumed in this and the following section that the integrals of all functions which figure under the integration and c.exp. signs exist. We recall that an a.s. relation between $\mathcal{B}$-measurable functions is a $P_{\mathcal{B}}$-equivalence.

25.1 Expectation properties. Loosely speaking, *c.exp.'s have a.s. all properties of expectations.*

Let x_k, c, and c' be numbers.

1. *If $X = c$ a.s. then $E^{\mathcal{B}}X = c$ a.s., and if $X \geqq Y$, a.s. then $E^{\mathcal{B}}X \geqq E^{\mathcal{B}}Y$ a.s.*

$E^{\mathcal{B}}$ is an a.s. linear operation: $E^{\mathcal{B}}(cX + c'X') = cE^{\mathcal{B}}X + c'E^{\mathcal{B}}X'$ *a.s. In particular,*

$$P^{\mathcal{B}}\Omega = 1 \text{ a.s.}, \quad P^{\mathcal{B}}\emptyset = 0 \text{ a.s.}, \quad P^{\mathcal{B}}A \geqq 0 \text{ a.s.}$$

and

$$E^{\mathcal{B}}\left(\sum_{k=1}^{n} x_k I_{A_k}\right) = \sum_{k=1}^{n} x_k P^{\mathcal{B}} A_k \text{ a.s.}$$

These properties follow at once from the definition of c.exp.'s and properties of integrals.

Conditional inequalities. *Upon replacing E by $E^{\mathcal{B}}$, the c_r-, Minkowski and Hölder inequalities, as well as their consequences and the inequalities for convex functions, remain valid, almost surely.*

For, on account of **1**, their proofs remain valid up to a $P_{\mathcal{B}}$-equivalence (for Hölder's inequality use also 25.2, **3**).

2. Convergence in the rth mean. *If* $X_n \xrightarrow{r} X$, *then* $E^{\mathcal{B}}X_n \xrightarrow{r} E^{\mathcal{B}}X$ *for* $r \geqq 1$.

Monotone convergence. *If* $0 \leqq X_n \uparrow X$ *a.s., then* $0 \leqq E^{\mathcal{B}}X_n \uparrow E^{\mathcal{B}}X$ *a.s. In particular,* $P^{\mathcal{B}} \sum_{k=1}^{\infty} A_k = \sum P^{\mathcal{B}} A_k$ *a.s.*

Fatou-Lebesgue convergence. *Let Y and Z be integrable. If* $Y \leqq X_n$ *a.s. or* $X_n \leqq Z$ *a.s., then* $E^{\mathcal{B}} \liminf X_n \leqq \liminf E^{\mathcal{B}} X_n$ *a.s., resp.,* $\limsup E^{\mathcal{B}} X_n \leqq E^{\mathcal{B}} \limsup X_n$ *a.s.*

In particular, if $Y \leqq X_n \uparrow X$ *a.s., or* $Y \leqq X_n \leqq Z$ *a.s. and* $X_n \xrightarrow{\text{a.s.}} X$, *then* $E^{\mathcal{B}}X_n \xrightarrow{\text{a.s.}} E^{\mathcal{B}}X$.

The first assertion follows by

$$E|E^{\mathcal{B}}X_n - E^{\mathcal{B}}X|^r = E|E^{\mathcal{B}}(X_n - X)|^r$$
$$\leqq E(E^{\mathcal{B}}|X_n - X|^r) = E|X_n - X|^r \to 0.$$

As for the monotone convergence assertion, since $X_{n+1} \geqq X_n$ a.s. implies $E^{\mathcal{B}}X_{n+1} \geqq E^{\mathcal{B}}X_n$ a.s., it follows that $E^{\mathcal{B}}X_n \uparrow X'$ a.s. where X' is $\mathcal{B}$-measurable. Therefore, the monotone convergence criterion applies to both sequences X_n and $E^{\mathcal{B}}X_n$, for every $B \in \mathcal{B}$,

$$\int_B X'\,dP \uparrow \int_B (E^{\mathcal{B}}X_n)\,dP_{\mathcal{B}} = \int_B X_n\,dP \uparrow \int_B X\,dP = \int_B (E^{\mathcal{B}}X)\,dP_{\mathcal{B}},$$

and the assertion follows. Upon taking $X_n = \sum_{k=1}^{n} I_{A_k}$ so that $E^{\mathcal{B}}X_n = \sum_{k=1}^{n} P^{\mathcal{B}}A_k$ a.s., the particular case is proved.

The Fatou-Lebesgue assertion follows from that of monotone convergence as in the nonconditional case.

25.2 Smoothing properties. Loosely speaking, *the operation* $E^{\mathcal{B}}$ *is a* $\mathcal{B}$*-smoothing.*

1. *On every nonnull atom* $B \in \mathcal{B}$, $E^{\mathcal{B}}X$ *is constant and its value* $E_B X$ *is the average of values of* X *on* B *with respect to* P.

By definition, B is a nonnull atom of $\mathcal{B}$ if $PB > 0$ and B contains no other sets belonging to $\mathcal{B}$ than itself and the empty set.

Proof. The first assertion follows from the fact that $E^{\mathcal{B}}X$ is a $\mathcal{B}$-measurable function defined up to a $P_{\mathcal{B}}$-equivalence and a $\mathcal{B}$-measurable function is constant on atoms of $\mathcal{B}$. Therefore, on every atom B of $\mathcal{B}$,

$$E_B X \cdot PB = \int_B (E^{\mathcal{B}}X)\, dP_{\mathcal{B}} = \int_B X\, dP$$

and, for $PB > 0$,

$$E_B X = \frac{1}{PB}\int_B X\, dP.$$

This proves the second assertion and completes the proof.

Thus, $E^{\mathcal{B}}X$ is a $\mathcal{B}$-smoothed X, in the sense that on atoms of $\mathcal{B}$ which are not atoms of $\mathcal{A}$, $E^{\mathcal{B}}X$ is an "averaged X" and, on the whole, has "fewer values" than X. In particular, if $\mathcal{B}$ is the minimal σ-field over a countable partition $\{B_j\}$ of Ω, so that the B_j are atoms of $\mathcal{B}$, then, as is to be expected,

$$E^{\mathcal{B}}X = \sum (E_{B_j}X) I_{B_j} \text{ a.s.};$$

the right-hand side is a.s. defined since the $E_{B_j}X$ are determined except for null B_j whose countable sum is necessarily null. For the "least fine" or "smallest" of all possible σ-fields $\mathcal{B} \subset \mathcal{A}$, that is, for $\mathcal{B}_0 = \{\emptyset, \Omega\}$, we obtain $E^{\mathcal{B}_0}X = EX$ a.s. The same conclusion holds for every $\mathcal{B}$ independent of the σ-field $\mathcal{B}_X$ of events induced by X:

2. *If* $\mathcal{B}$ *and* $\mathcal{B}_X$ *are independent, then* $E^{\mathcal{B}}X = EX$ *a.s.*

For, X and I_B being independent for every $B \in \mathcal{B}$,

$$\int_B (E^{\mathcal{B}}X)\, dP_{\mathcal{B}} = \int_B X\, dP = E(XI_B) = EX \cdot PB = \int_B (EX)\, dP_{\mathcal{B}}.$$

In particular, since $E^Y X$ denotes $E^{\mathcal{B}_Y}X$ and independence of X and Y means independence of $\mathcal{B}_X$ and $\mathcal{B}_Y$, we have

If X *and* Y *are independent, then* $E^Y X = EX$ *a.s.,* $E^X Y = EY$ *a.s.*

The operation $E^{\mathcal{B}}$ transforms $\mathcal{A}$-measurable functions (whose integrals exist) into $\mathcal{B}$-measurable functions (whose integrals exist); in fact, it transforms classes of P-equivalence into classes of $P_{\mathcal{B}}$-equivalence. In particular, as is to be expected, the operation $E^{\mathcal{B}}$ does not modify classes of $P_{\mathcal{B}}$-equivalence, in the sense that, if X is $\mathcal{B}$-measurable, then $E^{\mathcal{B}}X = X$ a.s.; since, then, for every $B \in \mathcal{B}$,

$$\int_B (E^{\mathcal{B}}X)\,dP_{\mathcal{B}} = \int_B X\,dP = \int_B X\,dP_{\mathcal{B}}.$$

More generally, $E^{\mathcal{B}}$ and $\mathcal{B}$-measurable factors commute, as follows:

3. *If X is $\mathcal{B}$-measurable, then $E^{\mathcal{B}}XY = XE^{\mathcal{B}}Y$ a.s.*

The assertion holds for $X = I_{B'}$ where $B' \in \mathcal{B}$, since, for every $B \in \mathcal{B}$,

$$\int_B (E^{\mathcal{B}}I_{B'}Y)\,dP_{\mathcal{B}} = \int_B I_{B'}Y\,dP = \int_{BB'} (E^{\mathcal{B}}Y)\,dP_{\mathcal{B}} = \int_B (I_{B'}E^{\mathcal{B}}Y)\,dP_{\mathcal{B}}.$$

Therefore, it holds for simple functions X_n:

$$E^{\mathcal{B}}X_nY = X_nE^{\mathcal{B}}Y \text{ a.s.}$$

and, by the monotone convergence theorem for c.exp.'s, it holds for nonnegative functions—take $0 \leq X_n \uparrow X$ and let $n \to \infty$ in the foregoing relation. The assertion follows.

4. *If $\mathcal{B} \subset \mathcal{B}'$, then*

$$E^{\mathcal{B}}(E^{\mathcal{B}'}X) = E^{\mathcal{B}}X = E^{\mathcal{B}'}(E^{\mathcal{B}}X) \text{ a.s.}$$

Since $\mathcal{B} \subset \mathcal{B}'$ implies that $P_{\mathcal{B}}$ is restriction of $P_{\mathcal{B}'}$ to $\mathcal{B}$, we have, for every $B \in \mathcal{B}$,

$$\int_B (E^{\mathcal{B}}(E^{\mathcal{B}'}X))\,dP_{\mathcal{B}} = \int_B (E^{\mathcal{B}'}X)\,dP_{\mathcal{B}'} = \int_B X\,dP = \int_B (E^{\mathcal{B}}X)\,dP_{\mathcal{B}},$$

and the left-hand side equality is proved.

Since $\mathcal{B} \subset \mathcal{B}'$ implies that a $\mathcal{B}$-measurable function is $\mathcal{B}'$-measurable, the right-hand equality follows either from **3** or directly from

$$\int_B (E^{\mathcal{B}'}(E^{\mathcal{B}}X))\,dP_{\mathcal{B}'} = \int_B (E^{\mathcal{B}}X)\,dP = \int_B (E^{\mathcal{B}}X)\,dP_{\mathcal{B}}, \quad B \in \mathcal{B}.$$

Thus, the smoothing $E^{\mathcal{B}}$ can be performed in steps and remains a.s. invariant under "finer" smoothings.

Together, **3** and **4** yield the

A. Basic smoothing property. *If* $\mathcal{B} \subset \mathcal{B}'$ *and* X' *is* $\mathcal{B}'$*-measurable, then*

$$E^{\mathcal{B}}XX' = E^{\mathcal{B}}(X'E^{\mathcal{B}'}X).$$

In particular, denoting by $E^{X',Y}$ the c.exp. given the σ-field $\mathcal{B}_{X',Y}$ of events induced by the couple (X', Y), we have

$$E^{Y}XX' = E^{Y}(X'E^{X',Y}X) \text{ a.s.}$$

***25.3 Concepts of conditional independence and of chains.** Under conditioning, the concept of independence extends as follows:

We say that $\mathcal{B}_1$ and $\mathcal{B}_2$ are *conditionally independent* (*c.ind.*) given $\mathcal{B}$ if, for every $B_1 \in \mathcal{B}_1$ and $B_2 \in \mathcal{B}_2$,

$$P^{\mathcal{B}}B_1B_2 = P^{\mathcal{B}}B_1 \cdot P^{\mathcal{B}}B_2 \text{ a.s.}$$

If $\mathcal{B} = \mathcal{A}$, then this relation becomes $I_{B_1B_2} = I_{B_1}I_{B_2}$ a.s., so that two sub σ-fields of events are always c.ind. given the σ-field $\mathcal{A}$ of events, and the concept of c.ind. given $\mathcal{A}$ is trivial.

If $\mathcal{B} = \mathcal{B}_0 = \{\emptyset, \Omega\}$, then this relation becomes $PB_1B_2 = PB_1 \cdot PB_2$ a.s., so that independence is c.ind. given $\mathcal{B}_0$, the "smallest" of all sub σ-fields of events.

In what follows, we drop the parentheses and commas in writing compound σ-fields.

A. $\mathcal{B}_1$ *and* $\mathcal{B}_2$ *are c.ind. given* $\mathcal{B}$ *if, and only if, for every* $B_2 \in \mathcal{B}_2$,

$$P^{\mathcal{B}\mathcal{B}_1}B_2 = P^{\mathcal{B}}B_2 \text{ a.s.;}$$

the subscripts 1 *and* 2 *can be interchanged.*

Proof. Let $B_1 \in \mathcal{B}_1$ and $B_2 \in \mathcal{B}_2$ be arbitrary. We have to prove that

$$E^{\mathcal{B}}I_{B_1}I_{B_2} = E^{\mathcal{B}}I_{B_1} \cdot E^{\mathcal{B}}I_{B_2} \text{ a.s.} \tag{1}$$

is equivalent to

$$E^{\mathcal{B}\mathcal{B}_1}I_{B_2} = E^{\mathcal{B}}I_{B_2} \text{ a.s.} \tag{2}$$

Since, on account of smoothing properties (25.2),

$$E^{\mathcal{B}}I_{B_1}I_{B_2} = E^{\mathcal{B}}(I_{B_1}E^{\mathcal{B}\mathcal{B}_1}I_{B_2}) \text{ a.s.}$$

and

$$E^{\mathcal{B}}I_{B_1} \cdot E^{\mathcal{B}}I_{B_2} = E^{\mathcal{B}}(I_{B_1}E^{\mathcal{B}}I_{B_2}) \text{ a.s.,}$$

it suffices to prove that (2) is equivalent to

$$(3) \qquad E^{\mathcal{B}}(I_{B_1}E^{\mathcal{B}\mathcal{B}_1}I_{B_2}) = E^{\mathcal{B}}(I_{B_1}E^{\mathcal{B}}I_{B_2}) \text{ a.s.}$$

Upon multiplying both sides in (2) by I_{B_1} and performing the operation $E^{\mathcal{B}}$, (3) follows.

Conversely, (3) implies that, for every $B \in \mathcal{B}$,

$$\int_B (I_{B_1}E^{\mathcal{B}\mathcal{B}_1}I_{B_2})\, dP = \int_B (I_{B_1}E^{\mathcal{B}}I_{B_2})\, dP$$

or, both c.exp.'s being $\mathcal{B}\mathcal{B}_1$-measurable,

$$\int_{BB_1} (E^{\mathcal{B}\mathcal{B}_1}I_{B_2})\, dP_{\mathcal{B}\mathcal{B}_1} = \int_{BB_1} (E^{\mathcal{B}}I_{B_2})\, dP_{\mathcal{B}\mathcal{B}_1}.$$

Since bounded indefinite integrals coinciding on the class of all sets BB_1 coincide on the σ-field $\mathcal{B}\mathcal{B}_1$, it follows that the integrands $E^{\mathcal{B}\mathcal{B}_1}I_{B_2}$ and $E^{\mathcal{B}}I_{B_2}$ are $P_{\mathcal{B}\mathcal{B}_1}$-equivalent, and the proof is complete.

Upon following literally the pattern used for the investigation of the concept of independence, the concept of c.ind. extends to arbitrary families of σ-fields and hence of r.v.'s, random vectors and random functions, and the investigations of the case of independence (Part III) can be transposed to the case of c.ind.

Furthermore, the concept of c.ind. leads to another generalization of that of independence, as follows: Let $\mathcal{B}_n$ be a sequence of sub σ-fields of events. The $\mathcal{B}_n$ are said to form a *chain*, or to be *chained* (or *chain-dependent* or *Markov-dependent*) if, for arbitrary integers m and n, the σ-fields $\mathcal{B}_1, \cdots, \mathcal{B}_{n-1}$, and $\mathcal{B}_{n+1}, \cdots, \mathcal{B}_{n+m}$ are c.ind. given $\mathcal{B}_n$. In symbols, the $\mathcal{B}_n$ are chained, if, for every m, n, $B_k \in \mathcal{B}_k$,

$$(1) \quad P^{\mathcal{B}_n}B_1 \cdots B_{n-1}B_{n+1} \cdots B_{n+m} = P^{\mathcal{B}_n}B_1 \cdots B_{n-1}P^{\mathcal{B}_n}B_{n+1} \cdots B_{n+m} \text{ a.s.}$$

or, equivalently, on account of **A**,

$$(2) \qquad P^{\mathcal{B}_1\mathcal{B}_2\cdots\mathcal{B}_n}B_{n+1} \cdots B_{n+m} = P^{\mathcal{B}_n}B_{n+1} \cdots B_{n+m} \text{ a.s.}$$

or

$$(3) \qquad P^{\mathcal{B}_n\mathcal{B}_{n+1}\cdots\mathcal{B}_{n+m}}B_1 \cdots B_{n-1} = P^{\mathcal{B}_n}B_1 \cdots B_{n-1} \text{ a.s.}$$

If $n = 1, 2, \cdots$, is interpreted as the "time," we can say, loosely speaking, that the $\mathcal{B}_n$ form a chain if the "past" and the "future" are a.s. independent when the "present" is given, or, equivalently, the "future" ("past") is a.s. independent of the given "past" ("future"), when the "present" is given. We shall use mostly the defining property (2).

Let $k_1 < \cdots < k_{n-1} < k_n < k_{n+1} < \cdots < k_{n+m}$ be arbitrary integers and apply the operation $E^{\mathcal{B}_{k_1}\cdots\mathcal{B}_{k_n}}$ to both sides of (2) where n and $n+m$ are replaced by k_n and k_{n+m} respectively. It follows, by 24.2, and upon replacing by Ω all events whose subscripts are different from $k_{n+1}, \cdots, k_{n+m}$, that we have the seemingly more general property

$$P^{\mathcal{B}_{k_1}\cdots\mathcal{B}_{k_n}} B_{k_{n+1}} \cdots B_{k_{n+m}} = P^{\mathcal{B}_{k_n}} B_{k_{n+1}} \cdots B_{k_{n+m}} \text{ a.s.}$$

Loosely speaking, whatever be the "future" it depends a.s. only upon the last given "past."

As usual, if $\mathcal{B}_n = \mathcal{B}_{X_n}$ are σ-fields of events induced by r.v.'s (random vectors, random functions) X_n, we replace above $\mathcal{B}_n$ by X_n and speak about the *chain* of r.v.'s (random vectors, random functions) X_n.

§ 26. REGULAR PR. FUNCTIONS

26.1 Regularity and integration. Since c.exp.'s behave at first sight as integrals with respect to c.pr.'s, the question arises whether c.exp.'s can be so defined. More precisely, according to 25.1,

1° Properties of functions $P^{\mathcal{B}}A$ are almost surely those of pr. values:

$$P^{\mathcal{B}}\Omega = 1 \text{ a.s.}, \quad P^{\mathcal{B}}A \geqq 0 \text{ a.s.}, \quad P^{\mathcal{B}} \sum A_j = \sum P^{\mathcal{B}} A_j \text{ a.s.}$$

2° Properties of functions $E^{\mathcal{B}}X$ are almost surely those defining integrals with respect to $P^{\mathcal{B}}$:

$$E^{\mathcal{B}} \sum_{k=1}^{n} x_k I_{A_k} = \sum_{k=1}^{n} x_k P^{\mathcal{B}} A_k \text{ a.s.},$$

$$0 \leqq X_n \uparrow X \text{ implies } E^{\mathcal{B}} X_n \uparrow E^{\mathcal{B}} X \text{ a.s.},$$

$$E^{\mathcal{B}} X = E^{\mathcal{B}} X^+ - E^{\mathcal{B}} X^- \text{ a.s.}$$

Yet, to speak about integrals with respect to the $P_\omega^{\mathcal{B}}$, we have to know that the $P_\omega^{\mathcal{B}}$ are pr.'s for every $\omega \in \Omega$ or, the c.exp.'s being defined up to an equivalence, that at the least the $P_\omega^{\mathcal{B}}$ are pr.'s except for ω belonging to some null event. Thus, we have to assume that $P^{\mathcal{B}}$ is "regular."

A c.pr. $P^{\mathcal{B}}$ is said to be *regular* if, for every $A \in \mathcal{A}$, it is possible to select $P^{\mathcal{B}}A$ within its class of equivalence in such a manner that the $P_\omega^{\mathcal{B}}$ are pr.'s on $\mathcal{A}$ except for points ω belonging to a $P_{\mathcal{B}}$-null event N. A regular pr.f. $P^{\mathcal{B}}$ can be said to be defined up to an equivalence, in the sense that if all the functions $P^{\mathcal{B}}A$ are modified arbitrarily on an arbitrary but fixed $P_{\mathcal{B}}$-null event, the new c.pr. is still regular. In particular, a regular c.pr. $P^{\mathcal{B}}$ can be selected within its equivalence class so that $P_\omega^{\mathcal{B}}$ is a pr. on $\mathcal{A}$ for *every* $\omega \in \Omega$. For example, for every ω belonging to the exceptional $P_{\mathcal{B}}$-null event N set $P_\omega^{\mathcal{B}} = P_N$ where P_N is a pr. on $\mathcal{A}$. Unless otherwise stated, regular c.pr.'s will be so selected. In other words,

a regular c.pr. $P^{\mathcal{B}}$, with values $P^{\mathcal{B}}(\omega; A)$, will be a function on $\Omega \times \mathcal{A}$ with the following properties:

(i) *$P^{\mathcal{B}}(\omega; A)$ is $\mathcal{B}$-measurable in $\omega(\in\Omega)$ for every fixed A and is a pr. in $A(\in\mathcal{A})$ for every fixed ω.*
(ii) *For every $A \in \mathcal{A}$ and $B \in \mathcal{B}$,*

$$\int_B (P^{\mathcal{B}}A)\, dP_{\mathcal{B}} = PAB.$$

In the case of regular c.pr.'s the answer to the question stated at the beginning of this section is, as might be expected, in the affirmative.

A. INTEGRATION THEOREM. *If $P^{\mathcal{B}}$ is a regular c.pr., then*

$$E^{\mathcal{B}}X = \int X\, dP^{\mathcal{B}} \text{ a.s.}$$

Proof. Since all $P_\omega^{\mathcal{B}}$ are pr.'s on $\mathcal{A}$, we can write

$$P_\omega^{\mathcal{B}}A = \int I_A\, dP_\omega^{\mathcal{B}}, \quad \omega \in \Omega,$$

that is,

$$E^{\mathcal{B}}I_A = P^{\mathcal{B}}A = \int I_A\, dP^{\mathcal{B}}.$$

It follows, on account of relations 2°, that

$$E^{\mathcal{B}}\sum_{k=1}^{n} x_k I_{A_k} = \int\left(\sum_{k=1}^{n} x_k I_{A_k}\right) dP^{\mathcal{B}} \text{ a.s.};$$

$0 \leq X_n \uparrow X$, where the X_n are simple functions, implies that $E^{\mathcal{B}}X = \lim E^{\mathcal{B}}X_n = \lim\int X_n\, dP^{\mathcal{B}} = \int X\, dP^{\mathcal{B}}$ a.s.;

$$E^{\mathcal{B}}X = E^{\mathcal{B}}X^+ - E^{\mathcal{B}}X^- = \int X^+\, dP^{\mathcal{B}} - \int X^-\, dP^{\mathcal{B}} = \int X\, dP^{\mathcal{B}} \text{ a.s.};$$

and the assertion is proved.

The basic smoothing property becomes

B. BASIC INTEGRATION PROPERTY. *If* $\mathcal{B} \subset \mathcal{B}' \subset \mathcal{A}$ *and* $P^{\mathcal{B}}$, $P^{\mathcal{B}'}$ *are regular, then, for* $\mathcal{A}$*-measurable functions* X *and* $\mathcal{B}'$*-measurable functions* X',

$$\int XX'\,dP^{\mathcal{B}} = \int X'\,dP^{\mathcal{B}} \int X\,dP^{\mathcal{B}'} \quad a.s.$$

The iterated integrations are to be read, as usual, from right to left. The foregoing relation can be written explicitly as follows: except for ω belonging to a $P_{\mathcal{B}}$-null event

$$\int P^{\mathcal{B}}(\omega; d\omega')X(\omega')X'(\omega') = \int P^{\mathcal{B}}(\omega; d\omega'')X'(\omega'') \int P^{\mathcal{B}'}(\omega''; d\omega')X(\omega').$$

***26.2 Decomposition of regular c.pr.'s given separable σ-fields.** The "elementary" case investigated in 24.1 corresponds to a given σ-field $\mathcal{B}$ generated by a countable disjoint class of events. It can then be assumed, without restricting the generality, that the class is a partition of the form $\sum B_j + B_0$, where every $PB_j > 0$ and B_0 is null but not necessarily empty. The corresponding "elementary" c.pr.'s can be written as

(1) $$P^{\mathcal{B}} = \sum (P_{B_j})I_{B_j} + (P_{B_0})I_{B_0},$$

where every P_{B_j} is a pr. on $\mathcal{A}$ defined by

(2) $$P_{B_j}A = \frac{PAB_j}{PB_j}, \quad A \in \mathcal{A},$$

so that $P_{B_j}B_j = 1$, and P_{B_0} is an arbitrary pr. on $\mathcal{A}$ which disappears when B_0 is empty. Thus, an "elementary" c.pr. is regular and can be said to be "decomposed" into a countable set of pr.'s. We intend to show that regular c.pr.'s given separable σ-fields can be decomposed in an analogous manner.

A σ-field $\mathcal{B}$ is *separable* if it is generated by (is minimal over) a countable class of sets.

a. *If a σ-field* $\mathcal{B}$ *is separable, then every set* $B \in \mathcal{B}$ *is a sum of atoms* B_t *of* $\mathcal{B}$ *such that* $\sum_{t \in T} B_t = \Omega$ *with* $T \subset R$.

Proof. Let B_j be the generators of $\mathcal{B}$ and let B_t be the nonempty distinct sets of the form $\bigcap \bar{B}_j$ where $\bar{B}_j = B_j$ or B_j^c. Since the set of j's is countable, the power of the set T of t's is at most that of the continuum, so that T can be supposed to lie in R. Since the B_t are disjoint and any $\omega \in \Omega$ belongs to one of them, $\sum_{t \in T} B_t = \Omega$. Since $\mathcal{B}$ is the σ-field generated by the B_j, every B_t belongs to $\mathcal{B}$, and by construction contains no other sets belonging to $\mathcal{B}$ than itself and the empty set and is not empty; further, every $B \in \mathcal{B}$ is a sum of B_t's. The assertion is proved.

The functions $P^{\mathcal{B}}A$, being $\mathcal{B}$-measurable, reduce to constants on atoms of $\mathcal{B}$. In fact, they reduce to constants on possibly larger events, namely, on atoms of the σ-field $\mathcal{B}_P \subset \mathcal{B}$ induced by these functions for A varying over $\mathcal{A}$. The σ-field $\mathcal{B}_P$ is generated by events of the form $[P^{\mathcal{B}}A \in S]$ where S are arbitrary Borel sets in R; and it suffices to take events of the form $[P^{\mathcal{B}}A < r]$ where the r are positive rationals. The atoms of $\mathcal{B}_P$ will be called *$P^{\mathcal{B}}$-atoms* and every event contained in a $P^{\mathcal{B}}$-atom will be called *$P^{\mathcal{B}}$-indecomposable;* for example, atoms of $\mathcal{B}$ are $P^{\mathcal{B}}$-indecomposable.

A. DECOMPOSITION THEOREM. *If $P^{\mathcal{B}}$ is a regular c.pr. and $\mathcal{B}$ contains a separable σ-field $\mathcal{B}'$ whose atoms are $P^{\mathcal{B}}$-indecomposable, then there exists a partition $\Omega = \sum_{t \in T} B_t + N$ with $T \subset R$ and $P_{\mathcal{B}}N = 0$ such that, except on $N \times \mathcal{A}$,*

$$P^{\mathcal{B}} = \sum_{t \in T} (P_{B_t}) I_{B_t},$$

where the P_{B_t} are pr.'s on $\mathcal{A}$ and $P_{B_t}B_t = 1$.

Proof. Let the countable class $\{B_j\}$ generate $\mathcal{B}' \subset \mathcal{B}$. The field generated by the B_j is a countable class and, hence, it may be assumed that $\{B_j\}$ is a field.

Let $\sum_{t \in T'} B_t = \Omega$ be the partition into atoms $B_t \in \mathcal{B}'$, as constructed in **a**. Since, by assumption, these atoms are $P^{\mathcal{B}}$-indecomposable, the functions $P^{\mathcal{B}}A (A \in \mathcal{A})$ reduce to constants $P_{B_t}A$ on B_t. Since $P^{\mathcal{B}}$ is regular, the P_{B_t} are pr.'s on $\mathcal{A}$. It remains to show that, upon lumping together some atoms B_t into a $P_{\mathcal{B}}$-null event, $P_{B_t}B_t = 1$ for the remaining ones.

For all $B \in \mathcal{B}$, the indicators I_B being $\mathcal{B}$-measurable coincide with their c.exp. $P^{\mathcal{B}}B$ given $\mathcal{B}$ except on a $P_{\mathcal{B}}$-null event. Let N_j be the $P_{\mathcal{B}}$-null event on which $P^{\mathcal{B}}B_j \neq I_{B_j}$. Since the functions $P^{\mathcal{B}}B_j$ do not vary on the atoms B_t, N_j is the sum of some B_t and the union $N = \bigcup N_j$ of all those exceptional atoms is $P_{\mathcal{B}}$-null. Fix ω belonging to a remaining atom B_t. Since $P_{B_t}B' (= P_\omega{}^{\mathcal{B}}B')$ and $I_{B'}(\omega)$ are values of pr.'s on $\mathcal{B}'$ and coincide on the generating field $\{B_j\}$, it follows that $P_{B_t}B' = I_{B'}(\omega)$

for every $B' \in \mathcal{B}'$; in particular, $P_{B_t}B_t = I_{B_t}(\omega) = 1$, and the proof is concluded.

COROLLARY 1. *If $P^{\mathcal{B}}$ is a regular c.pr., and one of the σ-fields $\mathcal{A}$ or $\mathcal{B}$ or $\mathcal{B}_P$ is separable, then the foregoing decomposition holds.*

Proof. Since atoms of $\mathcal{B}$ and of $\mathcal{B}_P(\subset\mathcal{B})$ are $P^{\mathcal{B}}$-indecomposable, we have only to prove the assertion when $\mathcal{A}$ is separable. Thus, let $\{A_j\}$ be the countable class which generates $\mathcal{A}$; the class can be assumed to be a field. The countable class of events of the form $[P^{\mathcal{B}}A_j < r]$, r rational, generates a separable σ-field $\mathcal{B}' \subset \mathcal{B}_P \subset \mathcal{B}$. It suffices to show that its atoms B_t are $P^{\mathcal{B}}$-indecomposable.

The functions $P^{\mathcal{B}}A_j$ reduce to constants $P_{B_t}A_j$ on atoms B_t of $\mathcal{B}'$ and, for $\omega \in B_t$, the pr.'s $P_\omega^{\mathcal{B}}$ and P_{B_t} on $\mathcal{A}$ coincide on the field $\{A_j\}$. Therefore, they coincide on $\mathcal{A}$ and, hence, for every $A \in \mathcal{A}$, the functions $P^{\mathcal{B}}A$ reduce to constants $P_{B_t}A$ on atoms B_t. The proof is terminated.

COROLLARY 2. *Under conditions of the decomposition theorem*

$$E^{\mathcal{B}}X = \sum_{t \in T} (E_{B_t}X)I_{B_t} \ a.s.,$$

where

$$E_{B_t}X = \int X\, dP_{B_t}, \quad t \in T.$$

Apply the integration theorem 26.1**A**.

In the elementary case, relation (2) can be written

$$P_{B_j}A = \int_A p_{B_j}(\omega')\, dP(\omega'), \quad A \in \mathcal{A}$$

with

$$p_{B_j}(\omega') = \frac{1}{PB_j} I_{B_j}(\omega'), \quad \omega' \in \Omega.$$

Therefore the decomposition (1) becomes, for $\omega \notin B_0$,

$$P^{\mathcal{B}}(\omega; A) = \int_A p^{\mathcal{B}}(\omega, \omega')\, dP$$

with

$$p^{\mathcal{B}}(\omega, \omega') = \sum \frac{1}{PB_j} I_{B_j}(\omega) I_{B_j}(\omega'), \quad \omega \notin B_0, \quad \omega' \in \Omega,$$

and, taking $P_{B_0} = P$, the integral relation holds for all $\omega \in \Omega$, provided we add $I_{B_0}(\omega)$ to $p^{\mathcal{B}}(\omega, \omega')$.

In general, let μ be a σ-finite measure on $\mathcal{A}$. We say that a regular c.pr. $P^{\mathcal{B}}$ is *μ-continuous* if there exists an $\mathcal{A}$-measurable function $p_\mu^{\mathcal{B}}(\omega, \omega')$ in ω' such that, for every $\omega \in \Omega$ and $A \in \mathcal{A}$,

$$P^{\mathcal{B}}(\omega; A) = \int_A p_\mu^{\mathcal{B}}(\omega, \omega')\, d\mu(\omega').$$

The function $p_\mu^{\mathcal{B}}$ will be called the *conditional pr. density* given $\mathcal{B}$ with respect to μ. It can and will be assumed to be nonnegative and finite. Furthermore, μ can and will be assumed to be a pr. on $\mathcal{A}$. If μ is finite, it suffices to set

$$\mu' = \mu/\mu\Omega, \quad p_{\mu'}^{\mathcal{B}} = p_\mu^{\mathcal{B}} \cdot \mu\Omega.$$

If μ is strictly σ-finite and $\sum A_n = \Omega$ is a partition such that every $\mu A_n < \infty$, it suffices to set

$$\mu' A = \sum \mu A A_n / 2^n \mu A_n, \quad A \in \mathcal{A},$$

and

$$p_{\mu'}^{\mathcal{B}}(\omega, \omega') = p_\mu^{\mathcal{B}}(\omega, \omega') \cdot 2^n \mu A_n, \quad \omega \in \Omega, \quad \omega' \in A_n.$$

COROLLARY 3. *Under conditions of the decomposition theorem, if $P^{\mathcal{B}}$ is μ-continuous, then the decomposition is countable; more precisely, the decomposition is*

$$\Omega = \sum B_j + N, \quad \mu B_j > 0, \quad PN = 0.$$

Proof. Since μ is a pr. on $\mathcal{A}$ and hence on $\mathcal{B}$, there exists only a countable class $\{B_j\}$ of non μ-null atoms B_t. On the other hand, if B_t is one of the μ-null atoms then, for any $\omega \in B_t$,

$$1 = P_{B_t} B_t = \int_{B_t} p_\mu^{\mathcal{B}}(\omega, \omega')\, d\mu(\omega') = 0$$

so that B_t must be empty.

§ 27. CONDITIONAL DISTRIBUTIONS

27.1 Definitions and restricted integration. A regular c.pr. $P^{\mathcal{B}}$ restricted to a sub σ-field of events still has the regularity properties: it is $\mathcal{B}$-measurable and it is a pr. on the sub σ-field to which it is restricted. However, the converse is not necessarily true. Thus, in the search for regular c.pr.'s, it will be convenient to begin by investigating the weaker "restricted regularity." In fact, it will prove useful to extend this concept to functions of a point in a measurable space $(\Omega_1, \mathcal{A}_1)$ with points ω_1 and measurable sets A_1 and of a measurable set in a measurable space $(\Omega_2, \mathfrak{A}_2)$ with points ω_2 and measurable sets A_2.

We shall not hesitate to proceed to the usual abuse of language, that is, according to convenience and the possible degree of confusion, we shall speak of "the function $h(\omega_1, A_2)$" instead of "the function h on $\Omega_1 \times \mathfrak{A}_2$." We say that the function $h(\omega_1, A_2)$ is an *$\mathfrak{A}_1$-measurable pr.* if it is $\mathfrak{A}_1$-measurable in ω_1 for every fixed A_2 and is a pr. in A_2 for every fixed ω_1. Observe that, whenever there exists a pr. P_1 on $\mathfrak{A}_1$, then the function

$$P_{12}(A_1 \times A_2) = \int_{A_1} P_1(d\omega_1)h(\omega_1, A_2)$$

determines, by the extension theorem (for measures) a pr. on the product-measurable space $(\Omega_1 \times \Omega_2, \mathfrak{A}_1 \times \mathfrak{A}_2)$.

Let X be a family of r.v.'s on the pr. space $(\Omega, \mathfrak{A}, P)$. Let $\mathfrak{A}_X$ be the σ-field of events induced by X, that is, the σ-field of the inverse images $[X \in S]$ of Borel sets S in the range space $\mathfrak{X}$ of X. If a c.pr. $P^{\mathfrak{B}}(\omega, A)$, where A varies only over $\mathfrak{A}_X$, is a pr. on $\mathfrak{A}_X$ for every fixed $\omega \in \Omega$, we say that it is a *conditional distribution* (*c.d.*) of X given $\mathfrak{B}$. Clearly

A function $P^{\mathfrak{B}}(\omega, A)$, where A varies over $\mathfrak{A}_X$, is a c.d. of X given $\mathfrak{B}$ if, and only if,

(CD_1) *$P^{\mathfrak{B}}(\omega, A)$ is a $\mathfrak{B}$-measurable pr.*

(CD_2) $\displaystyle\int_B P(d\omega)P^{\mathfrak{B}}(\omega, A) = PAB$

for every $A \in \mathfrak{A}_X$ and every $B \in \mathfrak{B}$.

To c.pr.'s $P^{\mathfrak{B}}(\omega, A)$ restricted to $\mathfrak{A}_X$, we make correspond $\mathfrak{B}$-functions $Q^{\mathfrak{B}}(\omega, S)$ such that, for every fixed Borel set S in $\mathfrak{X}$,

(C) $Q^{\mathfrak{B}}(\omega, S) = P^{\mathfrak{B}}(\omega, [X \in S])$ *up to a $P_{\mathfrak{B}}$-equivalence.*

If a function $Q^{\mathfrak{B}}(\omega, S)$ in (C) is a pr. on the Borel sets S, we say that it is a *mixed c.d.* of X given $\mathfrak{B}$. Clearly, if there exists a c.d. of X given $\mathfrak{B}$, then there exists a mixed c.d. of X given $\mathfrak{B}$ but the converse is not necessarily true.

The importance of c.d.'s and mixed c.d.'s of X is due to the fact that they still have the integration property of regular c.pr.'s, provided the integrand depends only upon X.

A. RESTRICTED INTEGRATION THEOREM. *Let g be a Borel function on the range space $\mathfrak{X}$ of a family X of r.v.'s, such that $Eg(X)$ exists.*

If there exists a c.d. (mixed c.d.) of X given $\mathfrak{B}$, then, except for points ω in a $P_{\mathfrak{B}}$-null set,

$$E^{\mathfrak{B}}(\omega, g(X)) = \int_{\Omega} P^{\mathfrak{B}}(\omega, d\omega')g(X(\omega'))(= \int_{\mathfrak{X}} Q^{\mathfrak{B}}(\omega, dx)g(x)).$$

Proof. By definition of a c.d. (mixed c.d.), the asserted equality holds for indicators $g = I_S$. It follows, as usual, that it holds for simple, then for nonnegative, Borel functions, and the theorem follows.

If $Q^{\mathfrak{B}}(\omega, S)$ is a mixed c.d. of a random vector $X = (X_1, \cdots, X_n)$, we set

$$F^{\mathfrak{B}}(\omega, x) = Q^{\mathfrak{B}}(\omega, (-\infty, x)), \quad x = (x_1, \cdots, x_n),$$

and call this function a *conditional distribution function* (*c.d.f.*) of X given $\mathfrak{B}$; it is $\mathfrak{B}$-measurable in ω and a d.f. in x. Thus, we can form its Fourier-Stieltjes transform

$$f^{\mathfrak{B}}(\omega, u) = \int e^{iux}\, dF^{\mathfrak{B}}(\omega, x), \quad u = (u_1, \cdots, u_n)$$

where $ux = u_1x_1 + \cdots + u_nx_n$, and shall call this function a *conditional characteristic function* (*c.ch.f.*) of X given $\mathfrak{B}$; it is $\mathfrak{B}$-measurable in ω and a ch.f. in u.

COROLLARY. *To a c.ch.f. $f^{\mathfrak{B}}(\omega, u)$ of a random vector X given $\mathfrak{B}$, there correspond c.exp.'s $E^{\mathfrak{B}}e^{iuX}$ such that, for every ω and every u,*

$$E^{\mathfrak{B}}(\omega, e^{iuX}) = f^{\mathfrak{B}}(\omega, u).$$

For, we can select the c.exp.'s such that, according to the theorem, the equality holds for every rational point u, and then use the continuity property of ch.f.'s in passing to the limit along rational points.

27.2 Existence. The problem of existence of regular c.pr.'s has been investigated principally by Doob who begins by solving the problem of c.d.'s as follows.

a. EXISTENCE LEMMA. *If there exists a c.d. of a family X of r.v.'s given $\mathfrak{B}$, then there exists a mixed c.d. of X given $\mathfrak{B}$. The converse is true when the range of X is a Borel set.*

We recall that the range of X is the set of values $X(\omega)$ as ω varies over Ω.

Proof. We use repeatedly the correspondence relation (C). The direct assertion follows at once by setting, for every $\omega \in \Omega$ and every Borel set S in the range space of X,

$$Q^{\mathfrak{B}}(\omega, S) = P^{\mathfrak{B}}(\omega, [X \in S]).$$

In general, the converse is not true because a set $A \in \mathcal{A}_X$ may be inverse image of different Borel sets, say, S and S'. However, when the range of X is itself a Borel set S_X, then

$$Q^{\mathcal{B}}(\omega, S_X^c) = P^{\mathcal{B}}(\omega, [X \in S_X^c]) = P^{\mathcal{B}}(\omega, \emptyset) = 0$$

except for points ω of some $P_{\mathcal{B}}$-null set N. Therefore, when there exists a mixed c.d. $Q^{\mathcal{B}}(\omega, S)$, that is, a $\mathcal{B}$-measurable pr., then, SS'^c and S^cS' being in S_X^c, we have

$$Q^{\mathcal{B}}(\omega, S) = Q^{\mathcal{B}}(\omega, S') = Q^{\mathcal{B}}(\omega, SS'), \quad \omega \notin N.$$

It follows that, in (C), we can select a $\mathcal{B}$-measurable pr. by setting, for every $A \in \mathcal{A}_X$,

$$P^{\mathcal{B}}(\omega, A) = Q^{\mathcal{B}}(\omega, S), \quad \omega \notin N$$

$$P^{\mathcal{B}}(\omega, A) = Q^{\mathcal{B}}(\omega_0, S), \quad \omega \in N, \quad \omega_0 \notin N,$$

where S is any image of A. This function is an asserted c.d., and the proof is complete.

A. C.D.'S EXISTENCE THEOREM. *There always exists a mixed c.d. of a countable family X of r.v.'s given $\mathcal{B}$. If the range of X is a Borel set, then there exists a c.d. of X given $\mathcal{B}$.*

Proof. On account of the existence lemma, it suffices to prove that there exists a mixed c.pr. We show first that a c.d.f. exists; the proof is based upon the fact that the countable set of rational points $r = (r_1, \cdots, r_n)$ of an n-dimensional euclidean space is dense in it.

Let x, x' and r, r' denote points and rational points, respectively, of the range space of a random vector $X = (X_1, \cdots, X_n)$. Let $P(\omega, A)$ be a c.pr. given $\mathcal{B}$, and, for every r, set

$$F^{\mathcal{B}}(\omega, r) = P^{\mathcal{B}}(\omega, [X < r]), \quad \omega \in \Omega;$$

the right-hand sides are selected arbitrarily within their $P_{\mathcal{B}}$-equivalence classes and kept fixed. Let N, with or without affixes, denote $P_{\mathcal{B}}$-null sets. On account of a.s. properties of c.pr.'s, we have

$$F^{\mathcal{B}}(\omega, -\infty) = 0, \quad F^{\mathcal{B}}(\omega, +\infty) = 1, \quad \omega \notin N_0$$

$$\Delta_{r'-r} F^{\mathcal{B}}(\omega, r) \geq 0, \ 1, \quad r < r', \quad \omega \notin N_{rr'}$$

$$F^{\mathcal{B}}(\omega, r) \uparrow F^{\mathcal{B}}(\omega, r') \quad \text{as} \quad r \uparrow r', \quad \omega \notin N_{r'}.$$

The countable union

$$N = N_0 \cup \bigcup_{r<r'} N_{rr'} \cup \bigcup_{r'} N_{r'}$$

of $P_{\mathcal{B}}$-null sets is $P_{\mathcal{B}}$-null. For every x, set

$$F^{\mathcal{B}}(\omega, x) = \lim_{r \uparrow x} F^{\mathcal{B}}(\omega, r), \quad \omega \notin N,$$

$$F^{\mathcal{B}}(\omega, x) = F^{\mathcal{B}}(\omega_0, x), \quad \omega \in N, \quad \omega_0 \notin N.$$

For every $\omega \in \Omega$, the function so defined is a d.f. and, by the correspondence theorem (for d.f.'s), the relation

$$Q^{\mathcal{B}}(\omega, (-\infty, x)) = F^{\mathcal{B}}(\omega, x)$$

determines a pr. $Q^{\mathcal{B}}(\omega, S)$ in Borel sets S.

This function is an asserted mixed c.d., provided we prove that this function is $\mathcal{B}$-measurable in ω and that, for every S,

$$Q^{\mathcal{B}}(\omega, S) = P^{\mathcal{B}}(\omega, [X \in S])$$

up to a $P_{\mathcal{B}}$-equivalence. By construction, the assertion is true for every $S = (-\infty, r)$. Hence, on account of the a.s. properties of c.pr.'s, it is true on the field of all finite sums of intervals S and, by monotone passages to the limit, it is still true on the minimal monotone field over this field, that is, on the σ-field of all Borel sets S.

Now, let $X = (X_1, X_2, \cdots)$ be a countable family of r.v.'s. Once the $F^{\mathcal{B}}(\omega; r_1, \cdots, r_n)$ are selected, we can select the $F^{\mathcal{B}}(\omega; r_1, \cdots, r_n, r_{n+1})$ within the defining $P_{\mathcal{B}}$-equivalence classes so that, for every $\omega \in \Omega$,

$$F^{\mathcal{B}}(\omega; r_1, \cdots, r_n, r_{n+1}) \to F^{\mathcal{B}}(\omega; r_1, \cdots, r_n) \quad \text{as} \quad r_{n+1} \to \infty.$$

Then, for every $\omega \in \Omega$, the foregoing construction yields consistent d.f.'s, hence consistent pr.'s, and, proceeding step by step with $n = 1, 2, \cdots$, we obtain a consistent family of pr.'s which, by the consistency theorem (for measures), determines a mixed c.d. $Q^{\mathcal{B}}(\omega, S)$ on the σ-field of all Borel sets S in the range space of X. The theorem is proved.

Sample pr. spaces. As long as we are concerned only with a given family of r.v.'s, we can always take for pr. space, the sample pr. space of the family. To simplify the statements and the notations, we consider a countable family $X = (X_1, X_2, \cdots)$ of r.v.'s (or random vectors or random sequences). Set $R_{k_1 \cdots k_m} = \prod_{j=1}^{m} R_{k_j}$, denote by $S_{k_1 \cdots k_m}$ and

$\mathcal{A}_{k_1\cdots k_m}$ the Borel sets and their σ-field in this real space, and let $P_{k_1\cdots k_m}$ be the distribution of $X_{k_1\cdots k_m} = (X_{k_1}, \cdots, X_{k_m})$ defined by

$$P_{k_1\cdots k_m}(S_{k_1\cdots k_m}) = P[X_{k_1,\cdots,k_m} \in S_{k_1\cdots k_m}].$$

If the same affixes occur inside and outside a bracket, we shall omit either the inside or the outside ones, according to our convenience.

The sample pr. space of X consists of the space $R_1 \times R_2 \times \cdots$, the σ-field of Borel sets in this space, and the distribution of X on this σ-field. We take it for our pr. space $(\Omega, \mathcal{A}, P)$. Then

$$X(x_1, x_2, \cdots) = (x_1, x_2, \cdots)$$

and the range of any X_n coincides with its range space R_n. Therefore, the existence theorem applies and, for every σ-field $\mathcal{B} \subset \mathcal{A}$, there exists a c.d. of any subfamily of X given $\mathcal{B} \subset \mathcal{A}$; in fact, there exists a c.d. of the countable family X given $\mathcal{B}$, that is, a regular c.pr. $P^{\mathcal{B}}$. Thus

B. REGULARITY THEOREM. *C.pr.'s in sample pr. spaces of countable families of r.v.'s can be regularized and c.d.'s of their subfamilies always exist.*

In the remainder of this section we take for pr. space of $X = (X_1, X_2, \cdots)$ its sample pr. space and can and will assume that the c.pr.'s given a measurable function Y on Ω are expressed as functions of Y and are regularized. By applying repeatedly the restricted integration theorem, we obtain

b. *If g is a Borel function on $R_{1\cdots n}$, such that $Eg(X_1, \cdots, X_n)$ exists, then*

$$Eg(X_1, \cdots, X_n)$$
$$= \int g\, dP_{1\cdots n}$$
$$= \int P(dx_1) \int P(x_1; dx_2) \cdots \int P(x_1, \cdots, x_{n-1}; dx_n) g(x_1, \cdots, x_n)$$

and, except for a P_1-null set of points x_1,

$$E^{\mathcal{B}}(x_1; g(X_1, \cdots, X_n))$$
$$= \int P(x_1; dx_{2\cdots n}) g(x_1, \cdots, x_n)$$
$$= \int P(x_1; dx_2) \cdots \int P(x_1, \cdots, x_{n-1}; dx_n) g(x_1, \cdots, x_n).$$

The c.d.'s defining properties separate as follows:

c. *Property*

(CD$_1$) $P(x_1, S_2)$ *is an* $\mathfrak{A}_1$*-measurable pr. on* $\mathfrak{A}_2$,

characterizes c.d.'s of X_2 *given* X_1, *and property*

(CD$_2$) $P(S_1 \times S_2) = \int_{S_1} P(dx_1)P(x_1, S_2)$

relates the distributions of X_1 *and* (X_1, X_2).

Applications. 1° The law of the countable family $(X_1, X_2, \cdots)$ is defined by the distribution of this family which determines and is determined by a consistent family of distributions PS_1, PS_{12}, $\cdots$. Because of the consistency requirement, this family of distributions is superabundant. Conditioning permits us to determine the law by means of a nonsuperabundant family of measurable pr.'s (that is, with no required relations among members). For, by applying repeatedly the above propositions, we find that

The law of the countable family $X_1, X_2, \cdots$ *determines and is determined by a family* $P(S_1)$, $P(x_1; S_2)$, $P(x_1, x_2; S_3)$, $\cdots$ *of c.d.f.'s.*

Clearly, we can replace c.d.'s by c.d.f.'s or by c.ch.f.'s.

2° Let $X_1, X_2, \cdots$ be r.v.'s on their sample space $(\Omega, \mathfrak{A}, P)$ with joint d.f.'s $F_{k_1\cdots k_m}$ and c.d.f.'s $F_{k_1\cdots k_m}{}^{\mathfrak{B}}$ of $(X_{k_1}, \cdots, X_{k_m})$. We can define conditional independence of the X's given $\mathfrak{B}$ by the property

$$F_{k_1\cdots k_m}{}^{\mathfrak{B}} = F_{k_1}{}^{\mathfrak{B}} \cdots F_{k_m}{}^{\mathfrak{B}}$$

for arbitrary finite subsets $k_1, \cdots, k_m$ of subscripts. Then

$$F_{k_1\cdots k_m} = E(F_{k_1}{}^{\mathfrak{B}} \cdots F_{k_m}{}^{\mathfrak{B}})$$

where the expectation is obtained by integrating with respect to $P_{\mathfrak{B}}$.

Conversely, any family $X_1, X_2, \cdots$ is trivially conditionally independent—given $\mathfrak{A}$; we exclude this trivial case.

If the r.v.'s are conditionally independent with common c.d.f. $F^{\mathfrak{B}}$, then

$$F_{k_1\cdots k_m}(x_1, \cdots, x_m) = E(F^{\mathfrak{B}}(x_1) \cdots F^{\mathfrak{B}}(x_m)).$$

Thus, the joint d.f.'s of any m of the r.v.'s do not depend upon their subscripts but only upon their number m. If the joint d.f.'s have this property, that is, for every finite subset $k_1, \cdots, k_m$

$$G_m = F_{k_1\cdots k_m},$$

we say that the r.v.'s are *exchangeable.* This concept was introduced by de Finetti and his basic result, in terms of conditional independence and in a somewhat more precise form, is as follows.

The concept of exchangeability is equivalent to that of conditional independence with common c.d.f.

The second concept implying the first one, it suffices to prove the converse. Thus, let $G_m = F_{k_1\cdots k_m}$ and set, for every $x \in R$,

$$\xi_n(x) = \frac{1}{n}\sum_{j=1}^{n} I_{[X_{k_j}<x]}.$$

Since, as $m, n \to \infty$,

$$E(\xi_m(x) - \xi_n(x))^2 = \frac{|m-n|}{mn}(G_1(x) - G_2(x, x)) \to 0,$$

it follows that there exists a r.v. $\xi(x)$ such that $E(\xi_n(x) - \xi(x))^2 \to 0$, and hence $\xi_n(x) \xrightarrow{P} \xi(x)$. Since the $\xi_n(x)$ are bounded by 1, it follows, by the dominated convergence theorem and a.s. invariance under finite permutations of X's of $B \in \mathcal{B} = \mathcal{B}(\xi(x), x \in R)$, that

$$E(\xi(x_1)\cdots\xi(x_m)I_B) \leftarrow E(\xi_n(x_1)\cdots\xi_n(x_m)I_B) \\ \to P([X_1 < x_1,\cdots, X_m < x_m]I_B).$$

Thus $P^{\mathcal{B}}[X_1 < x_1,\cdots, X_m < x_m] = \xi(x_1)\cdots\xi(x_m)$ a.s. Finally, since the function $\xi_n(\omega, x)$ is a d.f. in x, it follows that the function $\xi(\omega, x)$ has a.s. the properties of a d.f. in x and therefore, in the preceding relation, $\xi(x)$ can be replaced by a c.d.f. (use, for example, the same method as in the proof of the c.d.'s existence theorem).

27.3 Chains; the elementary case. In the case of random vectors the definition of chain is as follows:

A sequence X_n of random vectors is a *chain* if, for every integer n, a c. distribution of X_{n+1} given $X_1, \cdots, X_n$ can (and will) be so selected that it coincides with a c. distribution of X_{n+1} given X_n; in symbols $P_{X_{n+1}}^{X_1,\cdots,X_n} = P_{X_{n+1}}^{X_n}$ or, equivalently, the c. distribution $P(x_1, \cdots, x_n; S_{n+1})$ is independent of the *a priori* arguments $x_1, \cdots, x_{n-1}$. (On account of 27.2b, this definition entails chain-dependence as defined in 25.3; apply the second relation with n replaced by $m+1$, $\alpha_1 = (1, \cdots, n)$, $\alpha_2 = n+1, \cdots, \alpha_{m+1} = n+m$, and $g = I_{S_{n+1}\times\cdots\times S_{n+m}}$.)

Usually, the chained random vectors have a common range-space; to fix the ideas, consider a chain X_n of r.v.'s. The terminology used is phenomenological. The chain is a "system" X whose "state" at "time" n is X_n and has for values points $x \in R$—the "possible" states. The c. distribution $P_{X_{n+1}}^{X_n}$ is the *one-step transition pr.* at "time" n. By the classical abuse of language, it is represented by the same symbol as its values. This symbol will be $P^{n,n+1}(x; S)$ and is read "pr. of passage from x at time n into S at time $n + 1$." The very language used contains implicitly the assumption of chain-dependence.

If $P^{n,n+1}(x; S) = P(x; S)$ is independent of n, then the chain is said to be *constant* (in time); $P(x; S)$ is called the *transition pr. function* (f.) of the chain and read "pr. of passage from x into S in one step." From a phenomenological point of view, a constant chain represents a "random system" whose "law of evolution" does not vary in time. Let $P_n(S)$ denote the distribution of X_n. Since, for every n,

$$P_{n+1}(S) = \int P_n(dx)P(x; S),$$

it follows, by induction, that, for every pair m, n of integers,

$$P_{m+n}(S) = \int P_m(dx)P^n(x; S)$$

where

$$P^n(x; S) = \int P(x; dx_1) \int P(x_1; dx_2) \cdots \int P(x_{n-2}; dx_{n-1})P(x_{n-1}; S).$$

Clearly, this relation implies and is implied by the relation

$$P^{n+p}(x; S) = \int P^n(x; dx')P^p(x'; S); \quad n, p = 1, 2, \cdots.$$

$P^n(x; S)$ is called the *n-step transition pr.* and read "pr. of passage from x into S in n-steps." Upon applying 27.2**b**, it is easily seen that the n-step transition pr. is a c. distribution of X_{m+n} given $X_m (m = 1, 2, \cdots)$.

Upon applying 27.2**a** and 27.2**A**, we can summarize the basic properties of constant chains, as follows:

A. *A function $P(x; S)$, of points $x \in R$ and Borel sets $S \subset R$, is the transition pr.f. of a constant chain of r.v.'s if, and only if, it is a Borel function in x for every fixed S and a pr. in S for every fixed x.*

The law of a constant chain of r.v.'s X_n, with distributions $P_n(S)$, is determined by the initial distribution $P_1(S)$ and the transition pr.f. $P(x; S)$.

For every pair m, n of integers,

$$P_{m+n}(S) = \int P_m(dx)P^n(x; S),$$

where the function $P^n(x; S)$ is determined by the relation

$$P^{n+p}(x; S) = \int P^n(x; dx')P^p(x'; S); \quad n, p = 1, 2, \cdots.$$

Let $P(x; S)$ be a transition pr.f. If there exists an initial distribution $P_1(S)$ such that, for every n, the consecutive distributions $P_n(S)$ coincide with the initial one, the transition pr.f. is said to *possess an invariant distribution* $P_1(S)$ and $P_1(S)$ is said to be *invariant* under the transition pr.f. $P(x; S)$; the chain whose law is determined by the invariant distribution $P_1(S)$ and by $P(x; S)$ is said to be *stationary*. In symbols, $P_1(S)$ is invariant under $P(x; S)$ if, for every n,

$$P_1(S) = \int P_1(dx)P^n(x; S).$$

Since, for every n,

$$P_{n+1}(S) = \int P_n(dx)P(x; S),$$

it suffices to require that this relation be valid for $n = 1$. It easily follows from 27.2**b** that, if the chain X_n is stationary, then, for every n, the distribution of $(X_m, \cdots, X_{m+n})$ is independent of m.

A transition pr.f. $P(x, S)$ is *elementary* if there exists a countable partition $\sum S_k = \Omega$ such that $P(x, S_k) = P_{jk}$ for all $x \in S_j$; thus it reduces to a *transition pr. matrix* and the only values of initial distributions which matter are of the form $P_j = P_1(S_j)$. We set $P_j^n = P_n(S_j)$ and $P_{jk}^n = P^n(x, S_k)$ for $x \in S_j$ and the basic properties of constant chains become

$$P_{jk}^n \geqq 0, \quad \sum_k P_{jk}^n = 1, \quad P_{jk}^{m+n} = \sum_h P_{jh}^m P_{hk}^n,$$

$$P_j^n \geqq 0, \quad \sum_j P_j^n = 1, \quad P_j^{m+n} = \sum_h P_{jh}^m P_{hj}^n.$$

EXPONENTIAL CONVERGENCE. The basic limit problem for constant chains is that of the asymptotic behavior of n-step transition pr.f.'s $P^n(x, S)$. A particularly simple yet a cornerstone case, which in essence goes back to Markov, is the *exponential convergence case:* there exists a set function $\bar{P}(S)$ and positive constants a, b such that, for n sufficiently large, $|P^n(x, S) - \bar{P}(S)| \leqq ae^{-bn}$ whatever be x and S. This implies at once that $\bar{P}(S)$ is a pr.
In what follows we use repeatedly the fact that differences $\varphi(S)$ of two pr.'s vanish for $S = R$ so that $2\varphi(S)$ and $2|\varphi(S)|$ attain the same supremum $\text{Var}\,\varphi = \int |\varphi(dy)|$ at a positive Hahn decomposition set of $\varphi(S)$ to be denoted by H, with or without affixes.

a. INVARIANCE LEMMA. *$\bar{P}(S)$ is invariant under transition pr.f.'s and $|P_{n+1}(S) - \bar{P}(S)| \leqq ae^{-bn}$ whatever be $P_1(S)$.*

For

$$\left|\int\left\{P^n(x, dy) - \bar{P}(dy)\right\}P^m(y, S)\right| \leqq \int\left|P^n(x, dy) - \bar{P}(dy)\right| \leqq 2ae^{-bn}$$

implies that

$$\bar{P}(S) \leftarrow P^{n+m}(S) = \int\{P^n(x, dy)P^m(y, S) \rightarrow \int\bar{P}(dy)P^m(S),$$

and

$$|P_{n+1}(S) - \bar{P}(S)| \leq \int P_1(dx)\,|P^n(x, S) - \bar{P}(S)| \leqq ae^{-bn}.$$

We introduce now a "measure" of chain dependence which originated with Markov. Let $\Delta_{n,n+m} = \sup_{x,y}\sup_S\{P^n(x, S) - P^{n+m}(y, S)\}$. The (generalized) *Markov measure* is $\Delta_n = \Delta_{n,n}$. Clearly $0 \leqq \Delta_n \leqq 1$, and in the independence case $\Delta_n = 0$ (since x, y disappear) while in the deterministic case $\Delta_n = 1$ (since $P^n(x, S) = I(x, S)$).

b. BASIC INEQUALITIES:

$$\Delta_{n,n+m} \leqq \Delta_n \quad \textit{and} \quad \Delta_{n+m} \leqq \Delta_n\Delta_m.$$

For

$$|P^n(x, S) - P^{n+m}(y, S)| \leqq \int P^m(y, dz)\,|P^n(x, S) - P^n(z, S)| \leqq \Delta_n$$

and if $\varphi_n(S) = P^n(x, S) - P^n(y, S)$, then

$$\begin{aligned}|\varphi_{n+m}(S)| = \left|\int_{H_n+H_n^c}\varphi_n(dz)P^m(z, S)\right| &\leqq \varphi_n(H_n)\sup_z P^m(z, S)\\ &\quad + \varphi_n(H_n^c)\inf_z P^m(z, S)\\ &= \varphi_n(H)\{\sup P^m(z, S) - \inf P^m(z, S)\} \leqq \Delta_n\Delta_m.\end{aligned}$$

B. EXPONENTIAL CONVERGENCE CRITERION. *Exponential convergence holds if, and only if, $\Delta_h < 1$ for some integer h.*

Proof. If exponential convergence holds, then

$$\begin{aligned}|P^n(x, S) - P^n(y, S)| &\leqq |P^n(x, S) - \bar{P}(S)|\\ &\quad + |P^n(y, S) - \bar{P}(S)| \leqq 2ae^{-bn}\end{aligned}$$

Conversely, if $\Delta_h < 1$ then, by **b**, as $m, n \rightarrow \infty$,

$$|P^n(x, S) - P^{n+m}(y, S)| \leqq \Delta_n \leqq \Delta_h^{[n/h]} \rightarrow 0,$$

hence $\lim P^n(x, S) = \bar{P}(x, S)$ exists and this limit is a function $\bar{P}(S)$ of S only (set $m = 0$), so that $|P^n(x, S) - \bar{P}(S)| \leqq \Delta_h{}^{[n/h]}$ (let $m \to \infty$).

Let μ be a σ-finite measure. By the Lebesgue decomposition theorem

$$P^h(x, S) = \int p^n(x, y)\mu(dy) + P_s^n(x, S)$$

where $p^n(x, y) \geqq 0$ and $P_s^n(x, S)$ is μ-singular.

Markov case (generalized). *If* $\inf_x p^h(x, y) \geqq \delta > 0$ *for all* y *in some* μ*-positive set* S, *then exponential convergence holds.*

For, if H is a Hahn set for the difference of pr.'s in Δ_h, then

$$\begin{aligned} P^h(x, S') - P^h(y, S') &\leqq 1 - \{P^h(x, H^c) + P^h(y, H)\} \\ &\leqq 1 - \{P^h(x, H^cS) + P^h(y, HS)\} \\ &\leqq 1 - \left\{\int_{H^cS} p^h(x, y)\mu(dy) + \int_{HS} p^h(y, z)\mu(dz)\right\} \\ &\leqq 1 - \delta\mu(S) < 1. \end{aligned}$$

c. *Let* $X_1, X_2, \cdots$ *be a sequence of chained r.v.'s in the exponential convergence case and let* Y *be a r.v. bounded by* c *defined on* $X_{n+m}, X_{n+m+1}, \cdots$. *If* $\bar{E}$ *refers to* $\bar{P}$, *then*

$$|\bar{E}Y - E(Y|X_n)| \leqq 2ace^{-bm}.$$

For,

$$\begin{aligned} &|\bar{E}Y - E(Y|X_n = x)| \\ &\quad = \left|\int E(Y|X_{n+m} = y)\{\bar{P}(dy) - P^m(x, dy)\}\right| \leqq 2ace^{-bm}. \end{aligned}$$

This sequence behaves asymptotically as in the case of independence, as follows:

C. Exponential convergence theorem. *In the exponential convergence case with chained r.v.'s* $X_1, X_2, \cdots$, *whatever be* $P_1(S)$,

(i) $\{g(X_1) + \cdots + g(X_n)\}/n \xrightarrow{\text{a.s.}} \int g(x)\bar{P}(dx)$

for every Borel function g *for which the integral exists*

(ii) *the limit laws of normed sums* $\dfrac{g(X_1) + \cdots + g(X_n)}{b_n} - a_n, b_n \to \infty$, *where* g *is a finite Borel function, are stable and independent of* $P_1(S)$.

Proof:

1° If $P_1 = \bar{P}$ then, by **a**, the sequence $g(X_1), g(X_2), \cdots$ is stationary and also is indecomposable since, for every invariant set C, $I_C(x) = P(x, C) = P^n(x, C) \to \bar{P}(C)$, so that assertion (i) follows by the stationarity theorem. Since the limit on and the indicator of the convergence set of the averages are tail functions on $X_1, X_2, \cdots$, it follows from **a** that (i) holds for any P_1.

2° To prove (ii) we can take $a_n = 0$ on account of the convergence of types theorem. Thus, let $\mathcal{L}(S_n/b_n) \to \mathcal{L}(X)$ with ch.f. f, where $S_n = \sum_{k=1}^{n} g(X_k)$. Then, by the same theorem, upon excluding the trivial case of degenerate limit laws, $b_n/b_{n+1} \to 1$ so that, given positive constants, c, c', there exists a sequence $m = m(n)$ such that $b_m/b_n \to c'/c$.

Let $P_1 = \bar{P}$. Then, by **a**, in

(1) $cS_n/b_n + c(S_{n+p} - S_n)/b_n + c(S_{m+n+p} - S_{n+p})/b_n = cS_{m+n+p}/b_n,$

the law of the middle term is $\mathcal{L}(S_p/b_n) \to \mathcal{L}(0)$ for every fixed p. Thus we can and do select $p = p(n) \uparrow \infty$ such that, for these p, $\mathcal{L}((S_{n+p} - S_n)/b_n) \to \mathcal{L}(0)$ and hence, in passing to limit laws, we neglect the corresponding term in (1) while the "distance" p between S_n and $S_{m+n+p} - S_{n+p}$ increases indefinitely. But, by **a** and **c**,

$$\begin{aligned} E(\exp\{iuc(S_{m+n+p} - S_{n+p})/b_n\} \mid S_n) \\ &= E(\exp\{iuc(S_{m+1+p} - S_{1+p})/b_n\} \mid X_1) \\ &= E(\exp\{iuc(S_{m+1+p} - S_{1+p})/b_n\} + o(1) \\ &= E(\exp\{iuc(b_m/b_n)(S_m/b_m)\}) \to f(c'u), \end{aligned}$$

so that the ch.f. $E(\exp\{iucS_n/b_n\}E(\exp\{iuc(S_{m+n+p} - S_{n+p})/b_n\} \mid S_n))$ of the sum of the extreme terms in the left side of (1) converges to $f(cu)f(c'u)$. It follows, by the convergence of types theorem, that there exists a constant c'' such that the ch.f. of cS_{m+n+p}/b_n converges to $f(c''u) = f(cu)f(c'u)$ so that the limit law is stable.

Let $P_1 \neq \bar{P}$. For every fixed k, we can replace S_n by $S_{n+k} - S_n$ in $\lim_n |\bar{E}(\exp\{iuS_n/b_n\}) - E(\exp\{iuS_n/b_n\} \mid X_1)|$ so that, by **c**, this expression is bounded by $2ae^{-bk} \to 0$ as $k \to \infty$. Therefore, the limit ch.f. given X, reduces to the limit ch.f. under $\bar{P}$, so does its expectation, and the proof is terminated

COMPLEMENTS AND DETAILS

1. Let $\mathcal{B}$ be the σ-field in $\Omega = [0, 1]$ of Borel sets B, with or without affixes, and let λ be the Lebesgue measure on $\mathcal{B}$. Let $C \subset \Omega$ be a set of outer Lebesgue measure 1 and inner Lebesgue measure 0. Take for pr. space $(\Omega, \mathcal{A}, P)$, where $\mathcal{A}$ is the σ-field of all sets of the form $A = B_1C + B_2C^c$ and $PA = \frac{1}{2}\lambda B_1 + \frac{1}{2}\lambda B_2$. Then $PB = \lambda B$, $PC = \frac{1}{2}$, and there is no regular c.pr. given $\mathcal{B}$.

Chapter VIII

FROM INDEPENDENCE TO DEPENDENCE

The problems in and the methods developed for the independence case can be transposed to the general case. This permits us to enlarge the domains of validity of the results obtained in the independence case and also to realize the range of the methods.

In the last section of this chapter appears a different method—of indefinite expectations—which leads to more general results for a.s. convergence and is used extensively in the next chapter.

§ 28. CENTRAL ASYMPTOTIC PROBLEM

The Central Limit Problem is concerned with convergence of sequences $\mathfrak{L}(X_n)$ of laws of sums $X_n = \sum_{k=1}^{k_n} X_{nk}$ of r.v.'s. In order to investigate this problem in the case of dependent summands, we have to extend it to a *Central Asymptotic Problem* concerned with the comparison of the asymptotic behaviors (as $n \to \infty$) of $\mathfrak{L}(X_n)$ and of suitably chosen laws $\mathfrak{L}(Y_n)$. In fact, already in the case of independent summands, the investigation of the Central Limit Problem was based upon the comparison of laws of sums with suitably chosen infinitely decomposable laws.

The tools we shall require are, naturally enough, extensions of those used in the Central Limit Problem for independent summands. We write

$$H = F - G, \quad h = f - g, \quad \hat{h} = \hat{f} - \hat{g},$$

with the same affixes (if any) throughout, for differences of d.f.'s F, G and corresponding ch.f.'s f, g and integral ch.f.'s $\hat{f}$, $\hat{g}$.

28.1 Comparison of laws. In what follows we state properties which either result at once from those of d.f.'s and ch.f.'s or are obtained by means of identical arguments.

I. *Every function $H = F - G$ is bounded by* 1 (*in absolute value*), *is continuous from the left, has a countable discontinuity set, and*

$$H(x+0) = F(x+0) - G(x+0),$$

$$\int dH = \operatorname{Var} F - \operatorname{Var} G, \quad \operatorname{Var} H = \int |\, dH \,| \leqq 2.$$

We write $H_n \xrightarrow{w} H$ (up to additive constants) when $\int g dH_n \to \int g dH$ for all $g \in C_0$—the family of continuous functions on R vanishing at infinity. Note that in the case of d.f.'s, by the weak convergence criterion, this convergence is their weak convergence.

The weak compactness theorem is valid for functions H: every sequence H_n is weakly compact.

We write $H_n \xrightarrow{c} H$ when $H_n \xrightarrow{w} H$ and $\int dH_n \to \int dH$.

The Helly-Bray theorem is not valid for functions H.
Its proof breaks down the moment we use convergence of variations, since $\int dH_n \to \int dH$ does not entail $\int |\, dH_n \,| \to \int |\, dH \,|$.

II. *The functions h and $\hat{h}$ are defined by*

$$h(u) = \int e^{iux}\, dH(x), \quad \hat{h}(u) = \int_0^u h(v)\, dv = \int \frac{e^{iux} - 1}{ix}\, dH(x).$$

h on R is continuous and bounded by 2 *but the relation* $|\, h \,| \leqq h(0)$ *is not valid.*

The inversion formula is valid:

$$H(a) - H(b) = \lim_{U \to \infty} \frac{1}{2\pi} \int_{-U}^{+U} \frac{e^{-iua} - e^{-iub}}{-iu}\, h(u)\, du.$$

The weak convergence criterion is valid: $H_n \xrightarrow{w} H$ up to additive constants if, and only if, $\hat{h}_n \to \hat{h}$.

The continuity theorem is valid: if $h_n \to k$ continuous at $u = 0$, then $H_n \xrightarrow{c} H$ up to additive constants and $h = k$.

However, *the converse is not valid*, for the proof given for d.f.'s breaks down when the Helly-Bray theorem—which is no longer valid—is to be applied; for example, if $h_n(u) = e^{-inu} - e^{+inu}$, then the sequence h_n does not converge on R, while $H_n(x) = 1$ for $|x| < n$ and $=0$ for $|x| > n$ so that $H_n \xrightarrow{c} 1$.

III. Expansion of h. *If S is a Borel set and $0 < \delta \leqq 1$ then, provided the integrals exist,*

$$|h(u)| \leqq \left|\int dH\right| + \sum_{j=l+1}^{m} \frac{|u|^j}{j!}\left|\int_S x^j\, dH\right| + c_{m\delta}|u|^{m+\delta}\int_S |x|^{m+\delta}\, dH$$

$$+ \sum_{j\leqq l} \frac{|u|^j}{j!}\left|\int_{S^c} x^j\, dH\right| + c_{l\gamma}|u|^{l+\gamma}\int_{S^c} |x|^{l+\gamma}\, dH$$

where, if $l \geqq 1$, then $0 < \gamma \leqq 1$, and, if $l = 0$, then $0 \leqq \gamma \leqq 1$; the c's depend only on their subscripts.

If the right side is infinite, the inequality is trivially true. If the right side is finite, it follows from

$$h(u) = \int e^{iux}\, dH(x) = \int dH + \int_S (e^{iux} - 1)\, dH + \int_{S^c} (e^{iux} - 1)\, dH;$$

use limited expansions of e^{iux} of order $l + \gamma$ and $m + \delta$; and for $l + \gamma = 0$ use $|e^{iux} - 1| \leqq 2$.

We can now proceed to the comparison of sequences of laws. Two sequences $\mathfrak{L}(X_n)$ and $\mathfrak{L}(Y_n)$ are said to be *weakly equivalent*, and we write $\mathfrak{L}(X_n) \overset{w}{\sim} \mathfrak{L}(Y_n)$, if the two sequences have the same weak limit laws for same subsequences of subscripts; in other words, if $\mathfrak{L}(X_{n'}) \xrightarrow{w} \mathfrak{L}$, then $\mathfrak{L}(Y_{n'}) \xrightarrow{w} \mathfrak{L}$ and conversely. We observe that $\mathfrak{L}(X_{n'}) \xrightarrow{w} \mathfrak{L}$ means that $F_{n'} \xrightarrow{w} F$ up to additive constants. We define complete equivalence $\mathfrak{L}(X_n) \overset{c}{\sim} \mathfrak{L}(Y_n)$ by replacing in what precedes "weakly" by "completely."

In what follows we use repeatedly properties I and II without further comment.

A. Weak equivalence criterion. $\mathfrak{L}(X_n) \overset{w}{\sim} \mathfrak{L}(Y_n)$ *if, and only if, $F_n - G_n \xrightarrow{w} 0$ up to additive constants or $\hat{f}_n - \hat{g}_n \to 0$.*

Proof. It suffices to consider $F_n - G_n$; the assertion with $\hat{f}_n - \hat{g}_n$ follows.

Let $F_n - G_n \xrightarrow{w} 0$ up to additive constants. The weakly compact sequence F_n contains subsequences $F_{n'} \xrightarrow{w} F$ to which correspond subsequences $G_{n'} = F_{n'} - (F_{n'} - G_{n'}) \overset{w}{\rightarrow} F$ up to additive constants. It follows that $\mathfrak{L}(X_n) \overset{w}{\sim} \mathfrak{L}(Y_n)$.

Conversely, let $\mathfrak{L}(X_n) \overset{w}{\sim} \mathfrak{L}(Y_n)$. The weakly compact sequence $F_n - G_n$ contains subsequences $F_{n'} - G_{n'} \overset{w}{\rightarrow}$ some H and the weakly compact sequence $F_{n'}$ contains subsequences $F_{n''} \overset{w}{\rightarrow}$ some F. By hypothesis $G_{n''} \overset{w}{\rightarrow} F$ up to additive constants and, hence, $F_{n''} - G_{n''} \overset{w}{\rightarrow} H = F - F = 0$ up to additive constants. It follows that the weakly compact sequence $F_n - G_n \overset{w}{\rightarrow} 0$ up to additive constants. The proof is concluded.

B. Complete equivalence criterion. *Let the sequences $\mathfrak{L}(X_n)$ or $\mathfrak{L}(Y_n)$ be completely compact. Then $\mathfrak{L}(X_n) \overset{c}{\sim} \mathfrak{L}(Y_n)$ if, and only if, $F_n - G_n \overset{c}{\rightarrow} 0$ up to additive constants or $f_n - g_n \rightarrow 0$.*

Proof. Since $f_n - g_n \rightarrow 0$ implies that $F_n - G_n \overset{c}{\rightarrow} 0$ up to additive constants, it suffices to prove the "if" assertion with $F_n - G_n$ and the "only if" assertion with $f_n - g_n$.

If $F_n - G_n \overset{c}{\rightarrow} 0$ up to additive constants, then to every completely convergent subsequence $F_{n'} \overset{c}{\rightarrow}$ some F there corresponds the subsequence $G_{n'} \overset{c}{\rightarrow} F$ up to additive constants, and conversely. It follows that $\mathfrak{L}(X_n) \overset{c}{\sim} \mathfrak{L}(Y_n)$.

If $\mathfrak{L}(X_n) \overset{c}{\sim} \mathfrak{L}(Y_n)$, then one of the sequences f_n or g_n being completely compact in the sense of convergence to continuous functions, the same is true of both sequences and, hence, of the sequence $f_n - g_n$. If $f_{n'} - g_{n'} \rightarrow h$, then the sequence $f_{n'}$ contains a subsequence $f_{n''} \rightarrow$ some f and, by hypothesis, $g_{n''} \rightarrow f$. Therefore, $f_{n''} - g_{n''} \rightarrow h = 0$—unique limit element of the completely compact sequence $f_n - g_n$. It follows that $f_n - g_n \rightarrow 0$, and the proof is concluded.

Remark. In the proof of the "if" assertion we made use of the complete compactness of $\mathfrak{L}(X_n)$ only to assert that $F_{n'} \overset{c}{\rightarrow}$ some F. Let us make the natural convention that, when neither of the sequences $\mathfrak{L}(X_n)$ and $\mathfrak{L}(Y_n)$ has a complete limit element, then $\mathfrak{L}(X_n) \overset{c}{\sim} \mathfrak{L}(Y_n)$. Thus, $F_n - G_n \overset{c}{\rightarrow} 0$ up to additive constants implies that if the sequence $\mathfrak{L}(X_n)$ has no complete limit element the same is true of the sequence $\mathfrak{L}(Y_n)$, and conversely. In other words, with the foregoing convention the assumption of complete compactness is unnecessary for the "if" assertion:

If $F_n - G_n \xrightarrow{c} 0$ *up to additive constants, or* $f_n - g_n \to 0$, *then* $\mathcal{L}(X_n) \overset{c}{\sim} \mathcal{L}(Y_n)$.

We shall frequently center the X_n at some suitably chosen conditional expectations ξ_n. We observe that

COROLLARY. *If* $\xi_n \xrightarrow{P} 0$, *then* $\mathcal{L}(X_n - \xi_n) \overset{c}{\sim} \mathcal{L}(X_n)$.

This follows from the law-equivalence lemma.

28.2 Comparison of summands. Let

$$X_n = \sum_k X_{nk}, \quad Y_n = \sum_k Y_{nk}, \quad k = 1, \cdots, k_n.$$

$$Z_{nk} = X_{n0} + \cdots + X_{n,k-1} + Y_{n,k+1} + \cdots + Y_{n,k_n+1},$$
$$X_{n0} = Y_{n,k_n+1} = 0.$$

To X and Y with or without affixes there correspond their d.f.'s F and G and ch.f.'s f and g with same affixes if any; *primes will denote conditioning by* Z_{nk}, *unless otherwise stated*; for example,

$$F'_{nk} = P[X_{nk} < x \mid Z_{nk}], \quad f'_{nk}(u) = E'e^{iuX_{nk}} = E(e^{iuX_{nk}} \mid Z_{nk}).$$

For every fixed value of Z_{nk}, the selected conditional d.f.'s and ch.f.'s have all the properties of d.f.'s and ch.f.'s, and all properties of differences $H = F - G$, $h = f - g$ given in the preceding subsection are valid for the conditional differences $H' = F' - G'$, $h' = f' - g'$.

We intend to compare the sequences $\mathcal{L}(X_n)$ and $\mathcal{L}(Y_n)$ through the summands X_{nk} and Y_{nk}. (Let us observe that it is frequently convenient to compare suitably selected partial sums, each partial sum to be considered as a single summand.) We are at liberty to introduce any suitable dependence between the sets $\{X_{nk}\}$ and $\{Y_{nk}\}$, provided the laws of each of these sets are not modified and, in fact, provided the sequences $\mathcal{L}(X_n)$ and $\mathcal{L}(Y_n)$ remain the same.

A. COMPARISON THEOREM. $\mathcal{L}(X_n) \overset{c}{\sim} \mathcal{L}(Y_n)$

if

$$\sum_k E|f'_{nk} - g'_{nk}| \to 0$$

or if, S being a Borel set fixed or not (depending on n and/or k or not),

(i) $$\sum_k E\left|\int x^j(dF'_{nk} - dG'_{nk})\right| \to 0, \quad j \leq l,$$

(ii) $$\sum_k E\int_{S^c} |x|^{l+\gamma} |dF'_{nk} - dG'_{nk}| \to 0,$$

(iii) $$\sum_k E\left|\int_S x^j(dF'_{nk} - dG'_{nk})\right| \to 0, \quad j = l+1, \cdots, m,$$

(iv) $$\sum_k E\int_S |x|^{m+\delta}|dF'_{nk} - dG'_{nk}| \to 0, \quad 0 < \delta \leqq 1 \text{ fixed.}$$

If $l = 0$ *condition* (i) *disappears and* $0 \leqq \gamma \leqq 1$; *if* $l \geqq 1$, *then* $0 < \gamma \leqq 1$.

Proof. The first assertion follows by the complete convergence criterion from the inequality

$$|f_n - g_n| \leqq \sum_k E|f'_{nk} - g'_{nk}|$$

given by

$$\left|E\left(e^{iu\sum_k X_{nk}} - e^{iu\sum_k Y_{nk}}\right)\right| = \left|E\sum_k (e^{iuX_{nk}} - e^{iuY_{nk}})e^{iuZ_{nk}}\right|$$
$$\leqq \sum_k E\left|E'(e^{iuX_{nk}} - e^{iuY_{nk}})\right|.$$

The second assertion follows then from the expansion 28.1 III, and the theorem is proved.

It is important to observe that the theorem, and hence all which follows, remain valid with "finer" conditioning than by Z_{nk}. In other words, we can condition by any collection of X_{nk}'s and Y_{nk}'s of which Z_{nk} is a function. In particular, we can condition by the random vectors $Z'_{nk} = (X_{n0}, \cdots, X_{n,k-1}, Y_{n,k+1}, \cdots, Y_{n,k_n+1})$ or $Z''_{nk} = (X_{n0} + \cdots + X_{n,k-1}, Y_{n,k+1} + \cdots + Y_{n,k_n+1})$.

FIRST APPROACH. To $X_n = \sum_k X_{nk}$ we make correspond $Y_n = X^*_n = \sum_k X^*_{nk}$ where the summands X^*_{nk} are independent, and independent of the X_{nk}, and $\mathfrak{L}(X^*_{nk}) = \mathfrak{L}(X_{nk})$. Loosely speaking, $\mathfrak{L}(X^*_n)$ is obtained from $\mathfrak{L}(X_n)$ by suppressing the dependence between summands. If $\mathfrak{L}(X_n) \overset{c}{\sim} \mathfrak{L}(X^*_n)$, we say that the summands X_{nk} are *asymptotically independent.* The foregoing equivalence and comparison theorems yield conditions for asymptotic independence upon replacing g' by f and G' by F. We can thus transform the results of the investigation of the Central Limit Problem in the case of independence. Furthermore, we use the conditioning by the vector Z''_{nk}. It is easily seen that, because of the independence assumption, it reduces to conditioning by $X_{n0} + \cdots + X_{n,k-1}$. As an example, let us give a first extension of Liapounov's theorem.

Let $X_{n1}, X_{n2}, \cdots$ be r.v.'s centered at their expectations. The conditioning is by $X_{n0} + \cdots + X_{n,k-1}$.

B. UNDER LIAPOUNOV'S CONDITION

$$\sum_k E|X_{nk}|^{2+\delta} \to 0,$$

if

$$\sum_k E|E'X_{nk}| \to 0 \quad \textit{and} \quad \sum_k E|E'X_{nk}^2 - EX_{nk}^2| \to 0,$$

then $\mathfrak{L}(\sum_k X_{nk}/s_n) \to \mathfrak{N}(0, 1)$, *where* $s_n^2 = \sum_k EX_{nk}^2$.

It suffices to apply the comparison theorem with $l = 0$, $m = 2$, and $S = R$ to X_{nk} and $Y_{nk} = X^*_{nk}$, use the inequality

$$E\int |x|^{2+\delta} |dF'_{nk} - dF_{nk}| \leqq EE'|X_{nk}|^{2+\delta} + E|X_{nk}|^{2+\delta}$$
$$= 2E|X_{nk}|^{2+\delta},$$

and apply Liapounov's theorem to the X^*_{nk}.

SECOND APPROACH. We can obtain *directly* results which even in the case of independence are more general than those we obtained for the Central Limit Problem (since they pertain to the more general Central Asymptotic Problem):

For every fixed n, the comparison summands Y_{nk} are selected so that

—*the Y_{nk} are independent and $\mathfrak{L}(Y_n)$ belongs to the family of limit laws we seek to obtain*

—*the sets $\{Y_{nk}\}$ and $\{X_{nk}\}$ are independent.*

As an example, let us prove a Lindeberg type of normal convergence. Let the X_{nk} be centered at their conditional expectations so that $EX_{nk} = EE'X_{nk} = 0$, and set $\sigma_{nk}^2 = EX_{nk}^2$, $\sigma'^2_{nk} = E'X_{nk}^2$.

C. UNDER LINDEBERG'S CONDITION:

(i) $$\sum_k \int_{|x| \geqq \epsilon} x^2\, dF_{nk} \to 0 \quad \textit{for every} \quad \epsilon > 0, \quad \textit{and}$$

(ii) $$\sigma_n^2 = \sum_k \sigma_{nk}^2 \leqq \sigma^2 < \infty \textit{ for every } n,$$

if

(iii) $$\sum_k E|\sigma'^2_{nk} - \sigma_{nk}^2| \to 0,$$

then $\mathfrak{L}(X_n) \overset{c}{\sim} \mathfrak{N}(0, \sigma_n^2)$.

Proof. Since, by (i), as $n \to \infty$ and then $\epsilon \to 0$,

$$\max_k \sigma_{nk}^2 \leqq \epsilon^2 + \sum_k \int_{|x| \geqq \epsilon} x^2 \, dF_{nk} \to 0,$$

it follows, by (ii), that

$$(1) \qquad \sum \sigma_{nk}^3 \leqq \sigma^2 \max_k \sigma_{nk} \to 0.$$

Take the summands Y_{nk} to be mutually independent and normal $\mathfrak{N}(0, \sigma_{nk}^2)$, and take $S = (-\epsilon, +\epsilon)$, $l = \gamma = 1$, $m = 2$, $\delta = 1$. The comparison conditions for $j = 1$ and 2 are fulfilled, the first because of the centering and the second because of (iii) and because the condition for $l + \gamma$ is fulfilled by (i) and (1) since

$$\sum_k E \int_{|x| \geqq \epsilon} x^2 |\, dF'_{nk} - dG'_{nk} \,| \leqq \sum_k \int_{|x| \geqq \epsilon} x^2 \, dF_{nk} + c\epsilon^{-1} \sum_k \sigma_{nk}^3 \to 0.$$

Finally, the condition for $m + \delta$ is fulfilled by (ii) and (1), since

$$\sum_k E \int_{|x| < \epsilon} |\, x \,|^{2+\delta} |\, dF'_{nk} - dG_{nk} \,| \leqq 2\epsilon^\delta \sigma^2$$

and $\epsilon > 0$ is arbitrarily small. The theorem is proved.

The reader may proceed in a similar fashion and obtain or extend other results of the case of independence.

***28.3 Weighted prob. laws.** The second approach outlined in the preceding subsection yields the same prob. laws as in the case of independence. However, as we shall see, under similar but less restrictive conditions, disappearance of independence brings forth not the same prob. laws but their "weighted averages." The conditional law of a r.v. X given a sub σ-field $\mathcal{B}$ of events is defined by the conditional d.f. $F^{\mathcal{B}}$ or the conditional ch.f. $f^{\mathcal{B}}$. The d.f. F and the ch.f. f are then given by

$$F = EF^{\mathcal{B}}, \quad f = Ef^{\mathcal{B}}.$$

If the conditioning σ-field $\mathcal{B}$ is induced by some measurable function V not necessarily finite, nor even necessarily numerical, then, denoting by W the pr. distribution of V, we write

$$F = \int F^v \, dW(v), \quad f = \int f^v \, dW(v).$$

We say that W is the *weight function* of the *parameter* V and F (or f) represents the *weighted law* over the family F^v (or f^v) of laws.

Examples

1° A weighted law over the family of degenerate laws is of the form

$$f_W(u) = \int e^{iu\alpha}\, dW(\alpha)$$

where W on R is a d.f. In other words, if w is the ch.f. corresponding to W, we simply have $f_W = w$. It follows that, if the only "weighted" parameter is the shift-parameter, that is, the family of laws consists of $f_\alpha(u) = e^{iu\alpha}f(u)$, $\alpha \in R$, then

$$f_W(u) = \int e^{iu\alpha}f(u)\, dW(\alpha) = w(u)f(u),$$

and the "weighting" over the family reduces to the composition of a law with that represented by f. In other words, the weighting of the shift-parameter alone reduces to the composition of two laws and presents no new interest.

2° The limit laws which emerged in the development of pr. theory are the normal, Poisson, and, more generally, the infinitely decomposable laws. The corresponding weighted laws are

weighted normal:

$$f_W(u) = \int \exp\left[iu\alpha - \frac{\sigma^2u^2}{2}\right] dW(\alpha, \sigma^2)$$

weighted Poisson:

$$f_W(u) = \int_0^\infty \exp\,[\lambda(e^{iu} - 1)]\, dW(\lambda)$$

weighted infinitely decomposable:

$$f_W(u) = \int \exp\left[iu\alpha + \int g(x, u)\, d\Psi(x)\right] dW(\alpha, \Psi)$$

where $g(x, u) = \left(e^{iux} - 1 - \dfrac{iux}{1 + x^2}\right)\dfrac{1 + x^2}{x^2}$ and the functions Ψ on R are nondecreasing, continuous from the left, and of bounded variation.

If W degenerates at some element (α, σ^2) or (λ) or (α, Ψ), then we get back the corresponding nonweighted laws. A systematic investigation, with restrictions on α, say α constant (since any law is a weighted degenerate), would be of great interest. We say only a few words about the weighted symmetric stable laws.

A *weighted symmetric stable* is defined by

$$f_W(u) = \int_0^\infty \exp\,[-c|\,u\,|^\gamma/2]\,dW(c), \quad 0 < \gamma \leqq 2.$$

It is a Laplace-Stieltjes transform in $|\,u\,|^\gamma$. Hence, on account of the known properties of such transforms,

a. *There is a one-to-one correspondence between a weighted symmetric stable and the weight d.f. defined up to additive constants. In particular, a weighted symmetric stable reduces to a symmetric stable if, and only if, the weight function is a degenerate d.f. of a r.v.*

Furthermore, if $W_n \xrightarrow{w} W$ up to additive constants, then, by the extended Helly-Bray lemma and the fact that we can set $W_n(x) = 0$ for $x < 0$,

$$f_{W_n}(u) = \int_0^\infty \exp\,[-c|\,u\,|^\gamma/2]\,dW_n(c)$$
$$\to \int_0^\infty \exp\,[-c|\,u\,|^\gamma/2]\,dW(c) = f_W(u).$$

Conversely, let $f_{W_n} \to g$. By the weak compactness theorem, there is a d.f. W and a subsequence $W_{n'} \xrightarrow{w} W$, so that, by what precedes, $g(u) = \int_0^\infty \exp\,[-c|\,u\,|^\gamma/2]\,dW(c)$. But by **a**, g determines W up to additive constants. Hence $W_n \xrightarrow{w} W$ up to additive constants and $g = f_W$. Thus

b. *The limit elements of a sequence of weighted symmetric stable are weighted symmetric stable with same exponent.*

Weighted stable laws appear in the case of sequences of exchangeable r.v.'s since, by 27.2, 2°, they are conditionally independent given a sub σ-field and 23.4 applies under this conditioning. In a different guise, weighted laws appear in the third approach where

The conditional laws $\mathfrak{L}'(Y_{nk})$ will be of the limit type obtained under similar conditions in the independence case.

We use the following notation:

$$\alpha'_{nk}(\epsilon) = \int_{|x|<\epsilon} x\,dF'_{nk}, \quad \alpha'_n(\epsilon) = \sum_k \alpha'_{nk}(\epsilon),$$

$$\sigma'^2_{nk}(\epsilon) = \int_{|x|<\epsilon} x^2\,dF'_{nk} - \left(\int_{|x|<\epsilon} x\,dF'_{nk}\right)^2, \quad \sigma'^2_n(\epsilon) = \sum_k \sigma'^2_{nk}(\epsilon);$$

we drop the primes if F' is replaced by F, and drop ϵ if $\epsilon = +\infty$.

Before we attack the extension of the more general i.d. case, let us give, as an example, the extension of the historically important Liapounov's theorem.

A. *Let the X_{nk} be centered at their conditional expectations. If*

$$\sum_k E|X_{nk}|^{2+\delta} \to 0$$

for a $\delta > 0$, then $\mathcal{L}(X_n) \overset{c}{\sim} \mathcal{L}(\sum_k Y_{nk})$ with $\mathcal{L}'(Y_{nk}) = \mathfrak{N}(0, \sigma'^2_{nk})$.

Proof. Take the Y_{nk} to be conditionally normal $\mathfrak{N}(0, \sigma'^2_{nk})$, so that the law of Y_{nk} is the weighted symmetric normal $E\mathfrak{N}(0, \sigma'^2_{nk})$, and apply the comparison theorem with $S = R$, $l = 0$, $m = 2$, and $\delta \leqq 1$. The comparison conditions for $j = 1, 2$ are fulfilled, since the corresponding sums vanish, the first because of the centering and the second because of $E'Y_{nk}^2 = E'X_{nk}^2$. The condition with $m + \delta$ is fulfilled, since

$$E|Y_{nk}|^{2+\delta} = EE'|Y_{nk}|^{2+\delta} = cE\sigma'^{2+\delta}_{nk} \leqq cEE'|X_{nk}|^{2+\delta}$$
$$= cE|X_{nk}|^{2+\delta}$$

and hence

$$\sum_k E\int |x|^{2+\delta}|dF'_{nk} - dG'_{nk}| \leqq (1 + c)\sum_k E|X_{nk}|^{2+\delta} \to 0.$$

The theorem is proved.

REMARK. If we add the hypothesis that the d.f. W_n of σ'^2_n converges weakly to W, then the sequence $\mathcal{L}(X_n)$ converges to a weighted symmetric normal. This limit law is that of a r.v. if, and only if, $W_n \overset{c}{\to} W$, and then it is normal if, and only if, W degenerates. Similar considerations apply to what follows.

We pass now to the limit weighted i.d. laws. We require the notion of *conditional uniform asymptotic negligibility*, for short *uan'*, defined by

$$\max_k P'[|X_{nk}| \geqq \eta] \overset{P}{\to} 0 \quad \text{for every} \quad \eta > 0.$$

In the case of independent summands, the uan' condition reduces to the uan condition and in the general case implies it, since, by the dominated convergence theorem,

$$\max_k P[|X_{nk}| \geqq \eta] \leqq E(\max_k P'[|X_{nk}| \geqq \eta]) \to 0.$$

c. *Under uan′ condition, for every* $\epsilon > 0$

(i) $$\max_k \int_{|x|<\epsilon} |x|^s\, dF'_{nk} \xrightarrow{P} 0 \quad \text{for} \quad s > 0;$$

(ii) $$\max_k \int_{|x|<\epsilon} |x - \alpha'_{nk}(\epsilon)|^s\, dF'_{nk} \xrightarrow{P} 0 \quad \text{for} \quad s \geqq 1;$$

(iii) $$\max_k \gamma'_{nk}(u) = \max_k \int (e^{iu(x-\alpha'_{nk}(\epsilon))} - 1)\, dF'_{nk} \xrightarrow{P} 0.$$

Proof. Let $0 < \eta < \epsilon$. Since

$$\int_{|x|<\epsilon} |x|^s\, dF'_{nk} \leqq \eta^s + \epsilon^s P'[|X_{nk}| \geqq \eta]$$

assertion (i) follows by taking $\max_k$ and letting $n \to \infty$ and then $\eta \to 0$. Assertion (ii) follows, on account of (i), from

$$\int_{|x|<\epsilon} |x - \alpha'_{nk}(\epsilon)|^s\, dF'_{nk} \leqq 2^{s-1} \int_{|x|<\epsilon} |x|^s\, dF'_{nk} + 2^{s-1} |\alpha'_{nk}(\epsilon)|^s$$

and

$$|\alpha'_{nk}(\epsilon)|^s \leqq \epsilon^{s-1} |\alpha'_{nk}(\epsilon)|.$$

Finally, assertion (iii) follows, on account of (ii), from

$$|\gamma'_{nk}(u)| \leqq 2 \int_{|x|\geqq\epsilon} dF'_{nk} + |u| \int_{|x|<\epsilon} |x - \alpha'_{nk}(\epsilon)|\, dF'_{nk}.$$

Let

$$\log g'_{nk}(u) = iu\alpha'_{nk}(\epsilon) + \gamma'_{nk}(u).$$

d. $\sum_k E|f'_{nk}(u) - g'_{nk}(u)| \leqq c \sum_k E|\gamma'_{nk}(u)|^2.$

This follows (upon dropping the subscripts n, k) by $|\gamma'| \leqq 2$, hence $e^{|\gamma'|} \leqq e^2$, from

$$|f'(u) - g'(u)| = |e^{-iu\alpha'(\epsilon)} f'(u) - e^{\gamma'(u)}| = |1 + \gamma'(u) - e^{\gamma'(u)}|$$
$$\leqq \tfrac{1}{2} |\gamma'(u)|^2 e^{|\gamma'(u)|} \leqq c |\gamma'(u)|^2.$$

B. *Under uan′, if for every* n

(i) $$\sum_k \int_{|x|\geqq\epsilon} dF_{nk}(x) \leqq c' < \infty$$

and

(ii) $$\sum_k E\sigma'^2_{nk}(\epsilon) \leqq c'' < \infty,$$

then $\mathcal{L}(\sum_k X_{nk}) \overset{c}{\sim} \mathcal{L}(\sum_k Y_{nk})$ *where the summands* Y_{nk} *are conditionally i.d. with ch.f.'s* g'_{nk}.

Proof. On account of (i), condition (ii) is equivalent to

$$\text{(iii)} \qquad \sum_k E\int_{|x|<\epsilon} (x - a'_{nk}(\epsilon))^2\, dF'_{nk}(x) \leqq c''' < \infty,$$

since

$$\sum_k E\left| \int_{|x|<\epsilon} (x - a'_{nk}(\epsilon))^2\, dF'_{nk}(x) - \sigma'^2_{nk}(\epsilon) \right|$$
$$= \sum_k Ea'^2_{nk}(\epsilon) \int_{|x|\geqq\epsilon} dF'_{nk}(x) \leqq \epsilon^2 c'.$$

Therefore, upon substituting on $(-\epsilon, +\epsilon)$ the limited expansion of order 2 of the integrand in $\gamma'_{nk}(u)$,

$$\sum_k E|\gamma'_{nk}(u)| \leqq (2 + \epsilon|u|) \sum_k \int_{|x|\geqq\epsilon} dF_{nk}$$
$$+ \frac{u^2}{2} \sum_k E\int_{|x|<\epsilon} (x - a'_{nk}(\epsilon)^2\, dF'_{nk}$$
$$\leqq (2 + \epsilon|u|)c' + \frac{u^2}{2} c''$$

so that, by **c**,

$$\sum_k E|\gamma'_{nk}(u)|^2 \leqq \max_k |\gamma'_{nk}(u)| \sum_k E|\gamma'_{nk}(u)| \xrightarrow{P} 0.$$

But the left-hand side sum is a number and hence converges to 0. Thus, by **d**,

$$\sum_k E|f'_{nk}(u) - g'_{nk}(u)| \to 0,$$

the comparison theorem in terms of ch.f.'s applies and, hence, $\mathcal{L}(\sum_k X_{nk}) \overset{c}{\sim} \mathcal{L}(\sum_k Y_{nk})$, where the summands Y_{nk} are conditionally i.d. with ch.f. g'_{nk} and mutually independent. The theorem follows.

REMARK. In the case of independence it can be shown that, under uan condition, (i) and (ii) hold when the sequences $\mathcal{L}(X_n)$ or $\mathcal{L}(Y_n)$ are completely compact so that, then, $\mathcal{L}(X_n) \overset{c}{\sim} \mathcal{L}(Y_n)$. This extends the Central Limit theorem. The proof is left to the reader.

RANDOM VECTORS. The extension to random vectors X_{nk} can be obtained as usual either by reinterpreting the symbols used or by making correspond to the random vectors X_{nk} the r.v.'s vX_{nk}—scalar products of X_{nk} and of an undetermined sure vector v.

RANDOM NUMBER OF R.V.'S. Let the number ν_n of summands in the nth sum $\sum_{k=1}^{\nu_n} X_{nk}$ be a r.v. Set

$$p_n(r) = P[\nu_n = r], \quad P_n(s) = \sum_{r \geq s} p_n(r)$$

and denote by E_r the conditional expectation given $\nu_n = r$. Assume that the expressions below exist and are finite—they certainly do if $E\nu_n < \infty$. Then

$$\begin{aligned} f_n(u) - g_n(u) &= E(\exp [iu \sum_k X_{nk}] - \exp [iu \sum_k Y_{nk}]) \\ &= \sum_{r=1}^{\infty} p_n(r) E_r(\exp [iu \sum_{k=1}^{r} X_{nk}] - \exp [iu \sum_{k=1}^{r} Y_{nk}]). \end{aligned}$$

But, when all the expectations are conditioned by $\nu_n = r$, then the comparison theorem applies. Hence,

$$E_r(\exp [iu \sum_1^r X_{nk}] - \exp [iu \sum_1^r Y_{nk}]) \leq \sum_{k=1}^{r} E_r|f'_{nk}(u) - g'_{nk}(u)|$$

and

$$\begin{aligned} |f_n - g_n| &\leq \sum_{r=1}^{\infty} p_n(r) \{ \sum_{k=1}^{r} E_r|f'_{nk} - g'_{nk}| \} \\ &= \sum_{k=1}^{\infty} \{ \sum_{r \geq k} p_n(r) E_r \} |f'_{nk} - g'_{nk}|. \end{aligned}$$

Write E_{nk} for the operator $\sum_{r \geq k} p_n(r) E_r$. The relation becomes

$$|f_n - g_n| \leq \sum_{k=1}^{\infty} E_{nk}|f'_{nk} - g'_{nk}|,$$

and, hence,

C. *When the number of summands is random, the results obtained by using the comparison theorem remain valid provided* $\sum_k E$ *is replaced by* $\sum_r p_n(r) \sum_{k=1}^{r} E_r$ *or by* $\sum_k E_{nk}$.

If ν_n is independent of the X_{nk} and Y_{nk}, then $E_r = E$ and hence $E_{nk} = P_n(k)E$, and it suffices to multiply F'_{nk} and G'_{nk} by $P_n(k)$. If, moreover, ν_n degenerates at k_n, then $P_n(k) = 1$ or 0, according as $k \leq k_n$ or $k > k_n$, and, as is to be expected, we fall back on sure number k_n of summands.

§ 29. CENTERINGS, MARTINGALES, AND A.S. CONVERGENCE

29.1 Centerings. Conditions for a.s. convergence (and a.s. stability) of sums of independent r.v.'s were obtained by means of centerings at expectations or at medians. The methods continue to apply to the general case, provided the centering quantities are conditioned and, thus, become themselves r.v.'s. Furthermore, as has to be expected, the conditions so obtained will be sufficient but no more necessary. Since the proofs run parallel to those in the case of independence, we shall be content with essentials and shall leave the complete transcription of Chapter V to the reader.

Centering at conditional medians. We say that a r.v. $\mu^{\mathcal{B}}X$ is a *conditional median* of X given $\mathcal{B}$, where $\mathcal{B}$ is a sub σ-field of events, if

$$P^{\mathcal{B}}[X - \mu^{\mathcal{B}}X \geqq 0] \geqq \tfrac{1}{2} \leqq P^{\mathcal{B}}[X - \mu^{\mathcal{B}}X \leqq 0] \text{ a.s.}$$

When independence is not assumed, the proof of inequality 17.1C breaks down at the point where PA_kB_k is replaced by PA_kPB_k. Yet, if we observe that $PA_kB_k = E\{I_{A_k}P(B_k \mid S_1, \cdots, S_k)\}$ and replace medians $\mu(S_k - S_n)$ by conditional medians $\mu(S_k - S_n \mid S_1, \cdots, S_k)$, then the proof remains valid. Thus

A. EXTENDED P. LÉVY INEQUALITY. *If the sums S_k are centered at conditional medians $\mu(S_k - S_n \mid S_1, \cdots, S_k)$, then*

$$P[\max_{k \leqq n} | S_k | \geqq \epsilon] \leqq 2P[| S_n | \geqq \epsilon].$$

The propositions in 17.2 which result from P. Lévy's inequality continue to hold with similar modifications. Let us state the most important one.

B. CONVERGENCE THEOREM. *If the sequence of sums $S_n \xrightarrow{P} S$, then there exists a sequence ξ_n of conditional medians of suitably selected partial sums such that $\xi_n \xrightarrow{P} 0$ and $S_n - \xi_n \xrightarrow{\text{a.s.}} S$.*

REMARK. Propositions much more similar to those of the case of independence are obtainable by means of centerings at conditional expectations and, as we shall see in the next subsection, such centerings provide an important dependence model—of "martingales," which is a "natural" generalization of that of consecutive sums of independent r.v.'s centered at expectations. Yet the power of the centerings at medians accompanying symmetrizations in the case of independence leads one to think that it would be of interest to investigate in detail the dependence model that such centerings provide.

Centering at conditional expectations. We suppose that the r.v.'s X_n are integrable so that c.exp.'s $\xi_n = E(X_n \mid X_1, \cdots, X_{n-1})$ exist and are finite; for $n = 1$ the conditioning disappears and $\xi_1 = EX_1$ a.s. We have $E\xi_n = EX_n$ and

$$E(\xi_n \mid X_1, \cdots, X_{n-1}) = \xi_n \text{ a.s.}$$

Therefore, for $m < n$,

$$E\{(X_m - \xi_m)(X_n - \xi_n) \mid X_1, \cdots, X_{n-1}\}$$
$$= (X_m - \xi_m)E(X_n - \xi_n \mid X_1, \cdots, X_{n-1}) = 0 \text{ a.s.,}$$

so that

$$E(X_m - \xi_m)(X_n - \xi_n) = 0$$

and, hence,

$$E\left\{\sum_{k=1}^{n}(X_k - \xi_k)\right\}^2 = \sum_{k=1}^{n} E(X_k - \xi_k)^2.$$

We say that the r.v.'s X_n of a sequence are *centered at c.exp.'s given the predecessors*, if $\xi_n = 0$ a.s. (Such centerings were first systematically used by P. Lévy.) Thus

a. EXTENDED BIENAYMÉ EQUALITY. *If the r.v.'s X_n of a sequence are centered at c.exp.'s given the predecessors, then they are centered at exp.'s and*

$$\sigma^2 S_n = \sum_{k=1}^{n} \sigma^2 X_k.$$

In fact, more is true. If $\xi_n = 0$ a.s. and $A_{n-1} \in \mathcal{B}(X_1, \cdots, X_{n-1})$ is an event defined in terms of $X_1, \cdots, X_{n-1}$, then

$$E(S_{n-1}I_{A_{n-1}} X_n \mid X_1, \cdots, X_{n-1}) = S_{n-1}I_{A_{n-1}} E(X_n \mid X_1, \cdots, X_{n-1})$$
$$= 0 \text{ a.s.}$$

and, hence,

$$E(S_{n-1}I_{A_{n-1}} X_n) = 0.$$

Because of this orthogonality property, the proof of the right-hand side of Kolmogorov's inequality remains valid word for word. Thus

C. EXTENDED KOLMOGOROV INEQUALITY. *If the r.v.'s X_k, $k = 1, \cdots n$, are centered at c.exp.'s given the predecessors, then*

$$P[\max_{k \leq n} |S_k| \geq \epsilon] \leq \frac{1}{\epsilon^2} \sum_{k=1}^{n} \sigma^2 X_k.$$

The propositions in 16.3 which result from Kolmogorov's inequality and those in 16.4 hold with similar modifications. Let us state the most important ones.

D. CONVERGENCE THEOREM. *If the series $\sum \sigma^2 X_n$ converges and the series $\sum \xi_n$ converges a.s., then the series $\sum X_n$ converges a.s.*

More generally, if for some positive constant c the series $\sum P[|X_n| \geqq c]$ and $\sum E\{X_n^c - E(X_n^c \mid X_1, \cdots, X_{n-1})\}^2$ converge and the series $\sum E(X_n^c \mid X_1, \cdots, X_{n-1})$ converges a.s., then the series $\sum X_n$ converges a.s.

E. STABILITY THEOREM. *If $\sum \frac{\sigma^2 X_n}{b_n^2} < \infty$ with $b_n \uparrow \infty$, then*

$$\frac{1}{b_n}\sum_{k=1}^{n}\{X_k - E(X_k \mid X_1, \cdots, X_{k-1})\} \xrightarrow{\text{a.s.}} 0.$$

Let X be a r.v. and let x vary on $[0, +\infty)$. If $E|X|^r < \infty$, $r < 2$, $P[|X_n| \geqq x] \leqq P[|X| \geqq x]$ or $P\{|X_n| \geqq x \mid X_1, \cdots, X_{n-1}\} \leqq P\{|X| \geqq x \mid X_1, \cdots, X_{n-1}\}$ a.s., according as $r \neq 1$ or $r = 1$, then

$$\frac{1}{n^{1/r}}\sum_{k=1}^{n}(X_k - \eta_k) \xrightarrow{\text{a.s.}} 0$$

with $\eta_k = 0$ or $E(X_k \mid X_1, \cdots, X_{k-1})$ according as $0 < r < 1$ or $1 \leqq r < 2$.

Let the r.v.'s X_n of a sequence be centered at c.exp.'s given the predecessors. Since $S_1 = X_1, \cdots, S_n = X_1 + \cdots + X_n$ determine and are determined by $X_1, \cdots, X_n$, it follows that

$$\begin{aligned} E(S_n \mid S_1, \cdots, S_{n-1}) &= E(S_{n-1} + X_n \mid S_1, \cdots, S_{n-1}) \\ &= S_{n-1} + E(X_n \mid X_1, \cdots, X_{n-1}) = S_{n-1} \text{ a.s.} \end{aligned}$$

This property of the sequence S_n is called a "martingale" property. Conversely, if a sequence S_n has the martingale property, then setting $X_n = S_n - S_{n-1} (S_0 = 0)$, we have

$$\begin{aligned} E(X_n \mid X_1, \cdots, X_{n-1}) &= E(S_n - S_{n-1} \mid S_1, \cdots, S_{n-1}) \\ &= S_{n-1} - S_{n-1} = 0 \text{ a.s.} \end{aligned}$$

Thus, the martingale property characterizes consecutive sums of r.v.'s centered at c.exp.'s given the predecessors. Since we are interested in a.s. properties of such sums, it is "natural" to investigate them directly without writing them as sums.

29.2 Martingales: generalities. A possible interpretation of a "fair game" is as follows: Let X_t represent the debt or fortune of a gambler at time s. The game is fair if the gambler's expected fortune at time t, given the past up to the time $s < t$, equals his fortune at time s. To this interpretation corresponds the concept of martingale. It has been introduced and investigated in the form of consecutive sums by P. Lévy, then studied by Ville and systematically explored by Doob—to whom most of the results are due—and, finally, extended to "advantageous games" or submartingales by Doob and, in a different formulation, by Andersen and Jessen.

In this section we assume that, unless otherwise stated, *the expectations of the r.v.'s under consideration exist*, and denote by

$$\mathcal{B}_n = \mathcal{B}(X_1, \cdots, X_n), \quad \mathcal{C}_n = \mathcal{B}(X_n, X_{n+1}, \cdots),$$

$$\mathcal{B} = \mathcal{C}_1 = \mathcal{B}(X_1, X_2, \cdots), \quad \mathcal{C} = \bigcap \mathcal{C}_n,$$

the sub σ-fields of events induced by the families of r.v.'s $(X_1, \cdots, X_n)$, $(X_n, X_{n+1}, \cdots)$, $(X_1, X_2, \cdots)$ and the tail of the sequence $\{X_n\}$, respectively.

DEFINITIONS. Let $\{X_t, t \in T\}$ be a family of r.v.'s on a set T ordered by the relation "$<$," and let $\mathcal{B}_t = \mathcal{B}\{X_{t'}, t' < t\}$ be the sub σ-field of events induced by the subfamily of all the $X_{t'}$ with $t' < t$.

The family is said to be a *martingale* if, for every pair $s < t$,

$$X_s = E^{\mathcal{B}_s} X_t \text{ a.s., equivalently, } \int_{B_s} X_s = \int_{B_s} X_t, \quad B_s \in \mathcal{B}_s.$$

The martingale is said to be *closed* on the left or on the right according as it has a first or a last member (it may have neither or both).

If in the foregoing definitions "$=$" is replaced by "$\leqq$," the family is said to be a *submartingale* (or *semi-martingale*). If the inequality sign is reversed, it is a *supermartingale*. Changing the X_t into $-X_t$ interchanges "sub" and "super."

Note that the X_t being r.v.'s, the above c.exp.'s are a.s. finite for martingales while their negative parts are a.s. finite for submartingales.

We intend to investigate submartingales $\{X_n, n = 1, 2, \cdots\}$. The subscripts are ordered either by the relation "$\leqq$" and then we have a *submartingale sequence* $X_1, X_2, \cdots$ (closed on the left by X_1), or by the relation "$\geqq$" and then we have a *submartingale reversed sequence* $\cdots X_2, X_1$ (closed on the right by X_1). Because of the basic smoothing property of c.exp.'s the foregoing definitions reduce as follows. The r.v.'s X_n, $n = 1, 2, \cdots$ form

a martingale sequence, if $X_n = E(X_{n+1} \mid X_1, \cdots, X_n)$ a.s.,
a closed (by X) martingale sequence, if $X_n = E(X_{n+1} \mid X_1, \cdots, X_n)$, $X_n = E(X \mid X_1, \cdots, X_n)$ a.s.,
a martingale reversed sequence, if $X_{n+1} = E(X_n \mid X_{n+1}, X_{n+2}, \cdots)$ a.s.
a closed (by X) martingale reversed sequence, if $X_{n+1} = E(X_n \mid X_{n+1}, X_{n+2}, \cdots, X)$ a.s., $X = E(X_n \mid X)$ a.s.

For example, in the first case, $\mathcal{B}_m \subset \mathcal{B}_{m+1} \subset \cdots \subset \mathcal{B}_{n-1}$ for $m < n$ and, by the basic smoothing property, we have

$$E^{\mathcal{B}_m} X_n = E^{\mathcal{B}_m} E^{\mathcal{B}_{m+1}} \cdots E^{\mathcal{B}_{n-1}} X_n = X_m \text{ a.s.;}$$

similarly in the other cases.

If, in the foregoing relations, "=" is replaced by "≦," martingales become *submartingales.s.*

Examples

1° Let $X_n = \sum_{k=1}^{n} Y_k$, $n = 1, 2, \cdots$. If the r.v.'s Y_k are independent with $EY_k = 0$, or dependent with $E(Y_k \mid Y_1, \cdots, Y_{k-1}) = 0$ a.s. for $k > 1$, then, according to 29.1, the X_n form a martingale sequence.

2° Let $\mathcal{A}_1, \mathcal{A}_2, \cdots$ be a sequence of sub σ-fields of events and let every X_n be $\mathcal{A}_n$-measurable.

If $\mathcal{A}_1 \subset \mathcal{A}_2 \subset \cdots$ and every $X_n = E^{\mathcal{A}_n} X_{n+1}$ a.s., then the X_n form a martingale sequence, since $\mathcal{B}_n \subset \mathcal{A}_n$ and, by the smoothing property,

$$E^{\mathcal{B}_n} X_{n+1} = E^{\mathcal{B}_n} E^{\mathcal{A}_n} X_{n+1} = X_n \text{ a.s.}$$

Similarly, if $\mathcal{A}_1 \subset \mathcal{A}_2 \subset \cdots$ and every $X_n = E^{\mathcal{A}_n} X$, then the X_n form a martingale sequence closed on the right by X. For example, for any r.v. X and random sequence Y_n, the sequence $E(X \mid Y_1, \cdots, Y_n)$ is such a martingale.

Similarly, if $\mathcal{A}_1 \supset \mathcal{A}_2 \supset \cdots$ and every $X_n = E^{\mathcal{A}_n} X$ a.s., then the X_n form a martingale reversed sequence. For example, for any r.v. X and random sequence Y_n, the reversed sequence $\cdots E(X \mid Y_n, Y_{n+1}, \cdots) \cdots E(X \mid Y_2, Y_3, \cdots), E(X \mid Y_1, Y_2, \cdots)$ is a martingale.

Decomposition of submartingales. To simplify, we assume that the r.v.'s below are integrable, and leave to the reader the discussion of the case when their expectations exist but are not necessarily finite.

1° Let $X_1, X_2, \cdots$ be a sequence of r.v.'s and set $X''_1 = 0$

$$X_n = X'_n + X''_n, \quad X''_n = \sum_{k=2}^{n} \{E(X_k \mid X_1, \cdots, X_{k-1}) - X_{k-1}\}.$$

It follows that

$$E(X'_{n+1} \mid X_1, \cdots, X_n) = X'_n \text{ a.s.},$$

and hence

$$E(X'_{n+1} \mid X'_1, \cdots, X'_n, = X'_n \text{ a.s.}$$

Thus the sequence $X'_1, X'_2, \cdots$ is a martingale. In particular, if the sequence $X_1, X_2, \cdots$ is a submartingale, then every summand in X''_n is a.s. nonnegative. Therefore, a submartingale sequence X_n is decomposable into a martingale sequence X'_n and an a.s. nonnegative and nondecreasing sequence X''_n; more precisely,

$$X_n = X'_n + X''_n, \quad E(X'_{n+1} \mid X_1, \cdots, X_n) = X'_n \text{ a.s.}, \quad 0 \leqq X''_n \uparrow \text{ a.s.}$$

and, hence,

$$EX_n = EX'_n + EX''_n, \quad E|X'_n| \leqq E|X_n| + EX''_n, \quad 0 \leqq EX''_n \uparrow.$$

Let $\sup E|X_n| < \infty$. Then, it follows that X'_n and X''_n are integrable and $\sup E|X'_n| < \infty$, $\sup EX''_n < \infty$. Thus $0 \leqq X''_n \uparrow X''$ a.s. finite, and the study of the convergence of the submartingale sequence reduces to that of the martingale sequence X'_n with $\sup E|X'_n| < \infty$. Moreover, the limits, if any, differ by an integrable r.v.

2° Similarly, let the reversed sequence $\cdots X_2, X_1$ be a submartingale with EX_1 finite and set

$$X_n = X'_n + X''_n, \quad X''_n = \sum_{k=n}^{\infty} \{E(X_k \mid X_{k+1}, X_{k+2}, \cdots) - X_{k+1}\}.$$

The summands of the infinite sum are a.s. nonnegative, so that a.s. $0 \leqq X''_n \downarrow$ with $EX''_1 = EX_1 - \lim EX_n$ (the limit exists since, clearly, $EX_1 \geqq EX_2 \geqq \cdots$). Let $\lim EX_n > -\infty$ so that EX''_1 is finite. Then X''_1 is a.s. finite, $0 \leqq X''_n \downarrow 0$ a.s., and the study of the convergence of the submartingale reversed sequence reduces to that of the martingale reversed sequence $\cdots X'_2, X'_1$ with $E|X'_1| < \infty$; moreover, the limit, if any, is the same.

The interpretation of a martingale as a sequence of fortunes of a gambler raises the question whether in the long run ($n \to \infty$) his fortune was or becomes stabilized, that is, whether there is convergence—in some sense. To answer the question we require a few inequalities.

a. *Let g be convex and continuous on R with $g(+\infty) = +\infty$. If EX exists and $E^{\mathfrak{B}}X > -\infty$ a.s., then $g(E^{\mathfrak{B}}X) \leqq E^{\mathfrak{B}}g(X)$ a.s.*

For, if $E^{\mathfrak{B}}X < \infty$ a.s. the conditional convexity inequality applies; otherwise apply it to $X_n = XI_{[X<n]} + nI_{[X \geqq n]}$ and let $n \to \infty$.

A. Submartingale inequalities. *Let the r.v.'s X_j form a countable submartingale. Then*

(i) *the r.v.'s X_j^+ form a submartingale; and if the $X_j \geqq 0$ a.s. or the X_j form a martingale, then, for every $r \geqq 1$, the $|X_j|^r$ form a submartingale.*

(ii) *if the r.v. Y closes on the right the submartingale, then, for every $c > 0$,*

$$cP[\sup X_j > c] \leqq \int_{[\sup X_j > c]} Y;$$

and if the $X_j \geqq 0$ a.s. or the X_j form a martingale then, for every $r \geqq 1$,

$$c^r P[\sup |X_j| > c] \leqq \int_{[\sup |X_j| > c]} |Y|^r.$$

Proof. (i) follows from **a** by taking, respectively, $g(x) = x^+$, or $g(x) = 0$ for $x < 0$ and $g(x) = x^r$ for $x \geqq 0$, or $g(x) = |x|^r$.

To prove (ii), set $A_j = [X_j > c$, the predecessors $\leqq c]$, so that $B = [\sup X_j > c] = \sum A_j$ and, since Y closes the submartingale,

$$\int_B Y = \sum \int_{A_j} Y = \sum \int_{A_j} E(Y \mid X_j \text{ and the predecessors})$$

$$\geqq \sum \int_{A_j} X_j \geqq \sum c\, PA_j = cPB,$$

so that, letting $n \to \infty$, the first inequality is proved and the second follows on account of (i).

29.3 Martingales: convergence and closure. The limit properties of submartingales are summarized in the convergence theorem below. The proof is based on an ingenious inequality due to Doob.

Let x_k, $k = 1, \cdots, n$, be finite numbers. The number h of *crossings* from the left of the interval $[a, b]$ is the number of times that, starting with x_1 and proceeding to x_n, we pass from the left of the interval to its right. More precisely, let

$$x_{k_1} \leqq a, \quad x_{k_2} \geqq b, \quad x_{k_3} \leqq a, \quad x_{k_4} \geqq b, \cdots$$

where k_1 is the first subscript k, if any, such that $x_{k_1} \leqq a$, then k_2 is the first subscript $k > k_1$, if any, such that $x_{k_2} \geqq b$, and so on. If k_{j_0} is the last subscript so obtained, set $k_j = n + 1$ for $j_0 < j \leqq n$; if there is none, then $k_1 = \cdots = k_n = n + 1$. Thus, to every $k > k_1$, if any, there corresponds an integer j determined by the values of $x_1, \cdots, x_{k-1}$ and such that $k_j < k \leqq k_{j+1}$. For $k > 1$, if $k \leqq k_1$ set $i_k = 0$, and if $k > k_1$ set $i_k = 0$ or 1 according as the corresponding j is odd or even. When $k_2 \leqq n$, the number of crossings is the largest integer h such that $k_{2h} \leqq n$; when $k_2 > n$, the number of crossings $h = 0$.

Let $h > 0$. If $k_{2h+1} \leqq n$, then

$$\sum_{k=2}^{n} i_k(x_k - x_{k-1}) = (x_{k_3} - x_{k_2}) + \cdots + (x_{k_{2h+1}} - x_{k_{2h}}) \leqq (a - b)h.$$

If $k_{2h+1} > n$, then

$$\sum_{k=2}^{n} i_k(x_k - x_{k-1}) = (x_{k_3} - x_{k_2}) + \cdots + (x_{k_{2h-1}} - x_{k_{2h-2}}) + (x_n - x_{k_{2h}})$$
$$\leqq (a - b)h + (x_n - a).$$

Let $h = 0$. Then the left-hand sum is null, and the first inequality is trivially true.

Thus, in either case

$$\sum_{k=2}^{n} i_k(x_k - x_{k-1}) \leqq (a - b)h + (x_n - a)^+.$$

Now, to r.v.'s $X_1, \cdots, X_n$ we make correspond a r.v. H_n and r.v.'s $I_k (k > 1)$ determined by $X_1, \cdots, X_{k-1}$. We define them by $H_n(\omega) = h$, $I_k(\omega) = i_k$ for $x_1 = X_1(\omega), \cdots, x_n = X_n(\omega)$, $\omega \subset \Omega$, where h and i_k are the numbers introduced above. The inequality established above becomes

$$\sum_{k=2}^{n} I_k(X_k - X_{k-1}) \leqq (a - b)H_n + (X_n - a)^+$$

and, by taking expectations assumed finite, we have

$$\sum_{k=2}^{n} \int_{[I_k=1]} (X_k - X_{k-1}) \leqq (a - b)EH_n + E(X_n - a)^+.$$

If $X_1, \cdots, X_n$ is an integrable submartingale, then every left-hand integral is nonnegative and, hence,

$$(b - a)EH_n \leqq E(X_n - a)^+ = \sup_{k \leq n} E(X_k - a)^+.$$

If $E(X_n - a)^+ = \infty$, the inequality is trivially true. If $E(X_n - a)^+ < \infty$, note that H_n is also the number of crossings of $[0, b - a]$ by the integrable submartingale $(X_1 - a)^+, \cdots, (X_n - a)^+$. It follows that the inequality is always true.

Similarly, but proceeding from x_n to x_1 instead of from x_1 to x_n, if $X_n, \cdots, X_1$ is a submartingale, then

$$(b-a)EH_n \leqq E(X_1-a)^+ = \sup_{k\leqq n} E(X_k-a)^+.$$

To summarize (see also 36.2)

a. *If $X_1, \cdots, X_n$ or $X_n, \cdots, X_1$ is a submartingale, then*

$$(b-a)EH_n \leqq \sup_{k\leqq n} E(X_k-a)^+.$$

We are now in a position to prove the basic

A. SUBMARTINGALES CONVERGENCE THEOREM. *Let the r.v.'s X_n form a submartingale sequence or reversed sequence.*

(i) *If* $\sup EX_n^+ < \infty$, *then* $X_n \xrightarrow{\text{a.s.}} X < \infty$ *with* $EX \leqq \sup EX_n^+$ *and* $E|X| \leqq \sup E|X_n|$.

(ii) $X_n \xrightarrow{r} X$ *where* $r \geqq 1$ *if, and only if, the* $|X_n|^r$ *are uniformly integrable, and then* $X_n \xrightarrow{\text{a.s.}} X$.

Proof. 1° Since

$$[X_n \not\to] = \bigcup_{a,b} A_{a,b} \quad \text{with} \quad A_{a,b} = [\liminf X_n < a < b < \limsup X_n],$$

where a, b vary over the denumerable set of all rationals, the divergence set is null if, and only if, every set $A_{a,b}$ is null.

We apply the foregoing lemma to an arbitrary set $A_{a,b}$. Since $H_n \uparrow H = \infty$ on $A_{a,b}$, this set is null, provided $P[H = \infty] = 0$; it will be so, whether the submartingale is a sequence or a reversed sequence, provided

$$EH = \sup EH_n \leqq \sup E(X_n - a)^+/(b-a) < \infty.$$

Therefore, $\sup EX_n^+ < \infty$ and hence $\sup E(X_n - a)^+ < \infty$, for every $a \in R$, imply that $X_n \xrightarrow{\text{a.s.}}$ some X finite or not, and, by the Fatou-Lebesgue theorem, $E|X| \leqq \sup E|X_n|$.

It follows that $X_n^+ \xrightarrow{\text{a.s.}} X^+$ and, by the same theorem, $EX^+ \leqq \sup EX_n^+ < \infty$. Thus, X^+ is integrable hence a.s. finite. Therefore, upon modifying if necessary X^+ hence X on a null set, we can take X^+ to be finite so that $X \leqq X^+ < \infty$. Also, EX exists and

$$EX \leqq EX^+ \leqq \sup EX_n^+.$$

The first assertion is proved.

2° If $X_n \xrightarrow{r} X$ for some $r \geqq 1$, then, by the L_r-convergence theorem, the $|X_n|^r$ are uniformly integrable. Conversely, let the $|X_n|^r$ be uniformly integrable for some $r \geqq 1$. Then the $E|X_n|^r$, *a fortiori* the $E|X_n| \leqq E^{1/r}|X_n|^r$, are uniformly bounded. Therefore, by (i), $X_n \xrightarrow{\text{a.s.}} X$ and, by the L_r-convergence theorem, $X_n \xrightarrow{r} X$. The second assertion is proved.

The foregoing convergence theorem yields

B. SUBMARTINGALES CLOSURE THEOREM. *Let* $r \geqq 1$.

(i) *Let* $\{X_n\}$ *be a martingale or a nonnegative submartingale sequence or reversed sequence. If* $Y \in L_r$ *closes it on the right, then* $X_n \underset{\text{a.s.}}{\xrightarrow{r}} X$. *If* $\sup E|X_n|^r < \infty$ *with* $r > 1$, *then such a* Y *exists.*

(ii) *Let* $\{X_n\}$ *be a (sub)martingale sequence or reversed sequence. If* $X_n \xrightarrow{r} X$, *then* $X_n \xrightarrow{\text{a.s.}} X$ *and* X *closes on the right, respectively, on the left the (sub)martingale; in fact,* X *is the nearest of the closing r.v.'s.*

Proof. 1° Let $Y \in L_r$ close $\{X_n\}$. Set $B_n = [|X_n| > c]$ so that $B = [\sup|X_n| > c = \bigcup B_n$, and use 29.2**A**. Since $c^r PB \leqq \int_B |Y|^r$ $\leqq E|Y|^r < \infty$ and $\int_{B_n} |X_n|^r \leqq \int_B |Y|^r$, it follows that, as $c \to \infty$, $PB \to 0$, hence $\int_B |Y|^r \to 0$, and the $|X_n|^r$ are uniformly integrable.

Thus **A** applies, and $X_n \underset{\text{a.s.}}{\xrightarrow{r}} X$.

If $\sup E|X_n|^r < \infty$ with $r > 1$, then, by 9.4**C**, Cor. 2, and by **A**, $X_n \xrightarrow{1} X$. Since $E|X|^r \leqq \sup E|X_n|^r$ and, by (ii), X closes $\{X_n\}$, (i) is proved, provided we prove (ii);

2° Let the assumptions of (ii) hold. Then $X_n \xrightarrow{r} X$ implies that $X_n \xrightarrow{1} X$ and also, by **A**, that $X_n \xrightarrow{\text{a.s.}} X$. Thus we can pass to the limit under the integration sign, as follows:

In the submartingale sequence case, we have, for every $B_n \in \mathcal{B}_n$,

$$\int_{B_n} X_n \leqq \int_{B_n} X_{n+m},$$

and, by letting $m \to \infty$, we obtain

$$\int_{B_n} X_n \leqq \int_{B_n} X = \int_{B_n} E^{\mathcal{B}_n} X.$$

Therefore, $X_n \leqq E^{\mathcal{B}_n} X$ a.s., that is, the submartingale sequence is closed on the right by X. If an integrable Y also closes the sequence on the right, that is, for every $B_n \in \mathcal{B}_n$,

$$\int_{B_n} X_{n+m} \leqq \int_{B_n} Y,$$

then, by letting $m \to \infty$, we obtain

$$\int_{B_n} X \leqq \int_{B_n} Y.$$

Therefore, on every $\mathcal{B}_n$ and hence on $\bigcup \mathcal{B}_n$, the indefinite integral of $Y - X$ (finite since also X is integrable) is a finite measure and, by the extension theorem, determines a finite measure on $\mathcal{B}$. Thus, the indefinite integral of $Y - X$ on $\mathcal{B}$ is nonnegative, that is, for every $B \in \mathcal{B}$,

$$\int_B X \leqq \int_B Y = \int_B E^{\mathcal{B}} Y.$$

Since X is equivalent to a $\mathcal{B}$-measurable function, it follows that $X \leqq E^{\mathcal{B}} Y$ a.s. and hence the submartingale $X_1, X_2, \cdots, X$ is closed on the right by Y; that is, X is the "nearest" of the integrable closing r.v.'s.

Similarly, in the case of a submartingale reversed sequence, for every $C \in \mathcal{B}(X)$, as $m \to \infty$,

$$\int_C X \leftarrow \int_C X_{n+m} \leqq \int_C X_n = \int_C E(X_n \mid X),$$

so that X is a closing r.v. on the left and if Y is another closing r.v., then for $C \in \mathcal{B}(Y)$

$$\int_C Y \leqq \int_C X_n \to \int_C X = \int_C E(X \mid Y)$$

so that Y closes on the left the submartingale $X, \cdots, X_2, X_1$. Finally, for martingales all foregoing inequalities become equalities. The proof is terminated.

Various cases. Let us put together the properties of the various types of martingales and submartingales which are contained in 29.2**A** and 29.3**A** and **B**. In what follows $r \geqq 1$.

I MARTINGALE SEQUENCE $X_1, X_2, \cdots$

Inequalities:

$$EX_1 = EX_2 = \cdots; \quad EX_1^+ \leqq EX_2^+ \leqq \cdots; \quad E|X_1|^r \leqq E|X_2|^r \leqq \cdots.$$

Convergence. *If* $\lim EX_n^+ < \infty$ *or* $\lim EX_n^- < \infty$, *then* $X_n \xrightarrow{\text{a.s.}} X$.

Closure. *The martingale is closed on the right by a r.v.* $Y \in L_r$ *if, and only if, the* $|X_n|^r$ *are uniformly integrable; then* $X_n \underset{r}{\xrightarrow{\text{a.s.}}} X$ *and* X *is the nearest of the closing r.v.'s. In particular, the martingale is closed by a r.v.* $\in L_r$ *when* $\lim E|X_n|^r < \infty$ *with* $r > 1$.

II SUBMARTINGALE SEQUENCE $X_1, X_2, \cdots$

Inequalities:

$$EX_1 \leqq EX_2 \leqq \cdots; \quad EX_1^+ \leqq EX_2^+ \leqq \cdots;$$
$$X_n \geqq 0 \text{ a.s.} \Rightarrow EX_1^r \leqq EX_2^r \leqq \cdots.$$

Convergence. *If* $\lim EX_n^+ < \infty$, *then* $X_n \xrightarrow{\text{a.s.}} X < \infty$. *If*

$$\sup E|X_n| < \infty,$$

in particular if either every $X_n \leqq 0$ a.s. *or every* $X_n \geqq 0$ a.s. *and* $\lim |EX_n| < \infty$, *then* $X_n \xrightarrow{\text{a.s.}} X$ *finite.*

Closure. *If the* $|X_n|^r$ *are uniformly integrable, then* $X_n \xrightarrow{\text{a.s.}} X \in L_r$ *and* X *is the nearest of closing r.v.'s. If every* $X_n \geqq 0$ a.s., *then the* X_n^r *are uniformly integrable, if, and only if, there is a closing on the right r.v.* $Y \in L_r$, *and there is one when* $\lim EX_n^r < \infty$ *with* $r > 1$.

III MARTINGALE REVERSED SEQUENCE $\cdots, X_2, X_1$

Inequalities:

$$\cdots = EX_2 = EX_1; \quad \cdots \leqq EX_2^+ \leqq EX_1^+; \quad \cdots \leqq E|X_2|^r \leqq E|X_1|^r.$$

Convergence. *If* $EX_1^+ < \infty$ *or* $EX_1^- < \infty$, *then* $X_n \xrightarrow{\text{a.s.}} X < \infty$ *or* $> -\infty$, *respectively.*

Closure. *If* $E|X_1|^r < \infty$, *then* $X_n \underset{r}{\xrightarrow{\text{a.s.}}} X \in L_r$ *and* X *is the nearest of the closing r.v.'s.*

IV SUBMARTINGALE REVERSED SEQUENCE $\cdots, X_2, X_1$.

Inequalities:

$\cdots \leqq EX_2 \leqq EX_1$; $\quad \cdots \leqq EX_2^+ \leqq EX_1^+$;

every $X_n \geqq 0$ a.s. $\Rightarrow \cdots \leqq EX_2^r \leqq EX_1^r$.

Convergence. *If* $EX_1^+ < \infty$, *then* $X_n \xrightarrow{\text{a.s.}} X < \infty$.

Closure. *If the* $|X_n|^r$ *are uniformly integrable, then* $X_n \underset{r}{\xrightarrow{\text{a.s.}}} X \in L_r$ *and X is the nearest of the closing r.v.'s. In particular,* $X_n \underset{1}{\xrightarrow{\text{a.s.}}} X$ *if and only if* $\sup E|X_n| < \infty$, *equivalently*, $E|X_1| < \infty$, $\lim EX_n > -\infty$ (see 36.1c).

REMARK. By using the decomposition of submartingales given in 29.2, we can deduce their properties from those of martingales:

1° Let $X_1, X_2, \cdots$ be a submartingale sequence with $\sup E|X_n| < \infty$. Then $X_n = X'_n + X''_n$ where $X'_1, X'_2, \cdots$ is a martingale sequence with $\sup E|X'_n| < \infty$, and $0 \leqq X''_n \uparrow X''$ finite a.s. Therefore, $X'_n \xrightarrow{\text{a.s.}} X'$ finite and $X_n \xrightarrow{\text{a.s.}} X = X' + X''$ finite.

2° Let $\cdots X_2, X_1$ be a submartingale reversed sequence with $E|X_1| < \infty$ and $\lim EX_n > -\infty$. Then $X_n = X'_n + X''_n$ where $\cdots X'_2, X'_1$ is a martingale reversed sequence with $E|X'_1| < \infty$ and $0 \leqq X''_n \downarrow 0$ a.s., $EX''_1 < \infty$. Therefore $X'_n \underset{1}{\xrightarrow{\text{a.s.}}} X'$ and $X_n \underset{1}{\xrightarrow{\text{a.s.}}} X'$.

29.4 Applications. We use now the properties of martingales in order to extend various properties obtained in the case of independence. In general, we shall revert to P. Lévy's form of martingale sequences $X_n = \sum_{k=1}^{n} Y_k$ with $E(Y_{k+1} \mid Y_1, \cdots, Y_k) = 0$ a.s.; then $\mathfrak{B}_n = \mathfrak{B}(X_1, \cdots, X_n) = \mathfrak{B}(Y_1, \cdots, Y_n)$, and we set $\mathfrak{B}_0 = \{\emptyset, \Omega\}$.

We shall have use for a truncation of subscripts, first introduced by P. Lévy and which transforms martingales into martingales. Let ν be an integer-valued measurable function, finite or not, and such that the events $[\nu > n]$ are defined on the first n terms of a sequence $Y_1, Y_2, \cdots$ of r.v.'s, that is, $[\nu > n] \in \mathfrak{B}_n$. We set $Y'_n = Y_n I_{[\nu \geqq n]}$, so that $\mathfrak{B}(Y'_1, \cdots, Y'_n) \subset \mathfrak{B}_n$ and $E(Y'_{n+1} \mid Y_1, \cdots, Y_n) = I_{[\nu \geqq n+1]} E(Y_{n+1} \mid Y_1, \cdots, Y_n)$. Thus, if every $E(Y_{n+1} \mid Y_1, \cdots, Y_n) = 0$ a.s., then $E(Y'_{n+1} \mid Y_1, \cdots, Y_n) = 0$ a.s., and the martingale sequence $X_n = \sum_{k=1}^{n} Y_k$ is transformed by the above "ν-truncation" into the martingale sequence $X'_n = \sum_{k=1}^{n} Y'_k$. Observe that, the $\mathfrak{B}_n$ being closed under countable operations, we have

$$[\nu > n] \in \mathcal{B}_n \Leftrightarrow [\nu \leqq n] \in \mathcal{B}_n \Leftrightarrow [\nu = n] \in \mathcal{B}_n, \quad n = 1, 2, \cdots.$$

I. ZERO-ONE LAWS. The zero-one laws of the case of independence extend as follows:

Let $Y, Y_1, Y_2, \cdots$ be r.v.'s and apply 29.3**A**. The sequence $Z_n = E(Y \mid Y_1, \cdots, Y_n)$ is a martingale closed on the right by Y whose expectation is assumed to exist, say, $EY^+ < \infty$. Since every $EZ_n^+ \leqq EY^+$ and hence $\sup EZ_n^+ < \infty$, it follows that $Z_n \xrightarrow{\text{a.s.}} Z < \infty$. If $E|Y|^r < \infty$ for some $r \geqq 1$, then $Z_n \underset{r}{\xrightarrow{\text{a.s.}}} Z = E(Y \mid Y_1, Y_2, \cdots)$. If, moreover, Y is defined on the Y_n, then $Z_n \underset{r}{\xrightarrow{\text{a.s.}}} Y$. To summarize

A. *If $EY < \infty$, then $E(Y \mid Y_1, \cdots, Y_n) \xrightarrow{\text{a.s.}} Z < \infty$. If $E|Y|^r < \infty$ for some $r \geqq 1$, then $E(Y \mid Y_1, \cdots, Y_n) \underset{r}{\xrightarrow{\text{a.s.}}} E(Y \mid Y_1, Y_2, \cdots)$, which reduces a.s. to Y when Y is defined on the Y_n.*

We specialize now these properties. If $Y = I_B$ where the event B is defined on the Y_n, whence B is a "property" of the sequence, then $P(B \mid Y_1, \cdots, Y_n) \xrightarrow{\text{a.s.}} I_B$. (P. Lévy.)
In more intuitive terms

The sequence $P(B \mid Y_1, \cdots, Y_n)$ of c.pr.'s of a property B of the sequence $Y_1, Y_2, \cdots$, given the first n terms of the sequence, converges a.s. to 1 or to 0 according as the sequence has or has not this property.

In particular, if $P(B \mid Y_1, \cdots, Y_n) = PB$ a.s. for every value of n (or for a sequence of values of n), then $PB = I_B$ a.s. (Kolmogorov.)
In more intuitive terms

a. *If the c.pr. of a property of a sequence of r.v.'s, given any finite number of its terms, degenerates into a constant, then, a.s., the property is either sure or impossible.*

Also Borel's zero-one law extends as follows: Let $B_1, B_2, \cdots$ be a sequence of events and set $\mathcal{B}_n = \mathcal{B}(I_{B_1}, \cdots, I_{B_n})$. The two events $[\sum I_{B_n} < \infty]$ and $[\sum P^{\mathcal{B}_{n-1}} B_n < \infty]$ are equivalent (P. Lévy). In more intuitive terms

b. *The number of occurrences of the events B_n is a.s. finite or infinite according as the series of their c.pr.'s $\sum P^{\mathcal{B}_{n-1}} B_n$ is a.s. finite or infinite.*

Proof. The sequence $X_n = \sum_{k=1}^{n} Y_k$ where $Y_k = I_{B_k} - P^{\mathcal{B}_{k-1}} B_k$ (hence $|Y_k| \leqq 1$ a.s.) is a martingale. Let $a > 0$ be a finite number and define

ν by $[\nu = n] = [\sup_{k<n} X_k \leqq a,\ X_n > a]$, $[\nu = \infty] = [\sup X_n \leqq a]$. The ν-truncated sequence $X'_n = \sum_{k=1}^{n} Y'_k$ is a martingale bounded above by $a + 1$, and hence $X'_n \xrightarrow{\text{a.s.}} X'$ finite.

Since $X_n = X'_n$ on $[\sup X_n < a]$ and $a > 0$ is an arbitrary finite number, it follows that $X_n \to X$ finite on $[\sup X_n < \infty]$ except for a null event. Since changing the X_n into $-X_n$ preserves the martingale property, it follows that $X_n \to X$ finite on $[\inf X_n > -\infty]$ except for a null event.

But $0 \leqq \sum_{k=1}^{n} I_{B_k} \uparrow$ and $0 \leqq \sum_{k=1}^{n} P^{\mathfrak{B}_{n-1}} B_k \uparrow$ a.s. so that both sequences have a.s. a limit, finite or not. If one of them is finite and the other is infinite, then $\sup X_n(\omega) = +\infty$ or $\inf X_n(\omega) = -\infty$. Thus both limits are a.s. simultaneously either finite or infinite. The assertion is proved.

II. CONVERGENCE OF SERIES. Let $X_n = \sum_{k=1}^{n} Y_k$ form a martingale sequence. According to 29.2**A**

$$P[\sup_{k \leqq n} |X_k| \geqq c] \leqq E|X_n|^r/c^r, \quad r \geqq 1.$$

For $r = 2$, the summands are orthogonal, $EX_n^2 = \sum_{k=1}^{n} EY_k^2$, the inequality reduces to the extended Kolmogorov inequality and yields the results of 29.1. The martingales convergence properties yield more, but the assumptions to be made are not easily expressed in terms of the summands. However, the ν-truncation method yields a direct extension of the convergence property established in the case of independent and uniformly bounded summands, as follows (P. Lévy):

If the summands of the martingale sequence $X_n = \sum_{k=1}^{n} Y_k$ are uniformly bounded ($|Y_k| \leqq c < \infty$), then $X_n \xrightarrow{\text{a.s.}} X$ finite if, and only if, the series $\sum E^{\mathfrak{B}_{n-1}} Y_n^2$ is a.s. finite.

Proof. Let $a > 0$ be a finite number and define ν by $[\nu = n] = [\sup_{k<n} |X_k| \leqq a,\ |X_n| > a]$, $[\nu = \infty] = [\sup |X_n| \leqq a]$. The sequence $X'_n = \sum_{k=1}^{n} Y'_k$ of ν-truncated summands is a martingale bounded by $a + c < \infty$, and hence $X'_n \xrightarrow{r} X'$ for every $r \geqq 1$. In particular, for $r = 2$, we have $EX'^2 = \sum EY'^2_n < \infty$, so that the series $\sum E^{\mathfrak{B}_{n-1}} Y'^2_n$, whose expectation EX'^2 is finite, is a.s. finite. Since $E^{\mathcal{B}_{n-1}}Y'^2_n = E^{\mathcal{B}_{n-1}}Y_n^2$ on $[\nu \geqq n]$ and vanishes on $[\nu < n]$, it follows that the series $\sum E^{\mathcal{B}_{n-1}}Y_n^2$ is a.s. finite on $[\nu = \infty] = [\sup |X_n| \leqq a]$. Since a is arbitrary, this series is a.s. finite on $[X_n \to X \text{ finite}]$.

Conversely, define ν by

$$[\nu = n] = \left[\sum_{k=1}^{n-1} E^{\mathcal{B}_{k-1}}Y_k^2 \leqq a, \quad \sum_{k=1}^{n} E^{\mathcal{B}_{k-1}}Y_k^2 > a\right],$$

$$[\nu = \infty] = [\sum E^{\mathcal{B}_{n-1}}Y_n^2 \leqq a].$$

The corresponding sequence of ν-truncated sums is a martingale bounded by a. Upon taking the expectations, it follows that

$$E^2|X'_n| \leqq EX'^2_n \leqq \sum EY'^2_n \leqq a + c^2,$$

so that, by 29.3**A**, $X'_n \xrightarrow{\text{a.s.}} X'$ finite; and, as above, $X_n \xrightarrow{\text{a.s.}} X$ finite on $[\sum P^{\mathcal{B}_{n-1}}Y_n^2 < \infty]$. The proof is terminated.

III. STRONG LAWS OF LARGE NUMBERS. Let $X_n = \sum_{k=1}^{n} Y_k$ where the Y_k are "conditionally exchangeable" with respect to addition (implied by ordinary exchangeability), that is, for every n and $k \leqq n$,

$$E(Y_k \mid X_n, X_{n+1}, \cdots) = E(Y_1 \mid X_n, X_{n+1}, \cdots) \text{ a.s.}$$

According to 29.3**B**, if $E|X_1|^r < \infty$ for some $r \geqq 1$, then

$$E(Y_1 \mid X_n, X_{n+1}, \cdots) \underset{r}{\xrightarrow{\text{a.s.}}} E^{\mathcal{C}}Y_1.$$

Therefore

$$\frac{X_n}{n} = E\left(\frac{X_n}{n} \middle| X_n, X_{n+1}, \cdots\right) = \frac{1}{n}\sum_{k=1}^{n} E(Y_k \mid X_n, X_{n+1}, \cdots)$$

$$= E(Y_1 \mid X_n, X_{n+1}, \cdots) \underset{r}{\xrightarrow{\text{a.s.}}} E^{\mathcal{C}}Y_1.$$

To summarize

If $X_n = \sum_{k=1}^{n} Y_k$ *where the* Y_k *are exchangeable and* $E|Y_1|^r < \infty$ *for some* $r \geqq 1$, *then* $\dfrac{X_n}{n} \underset{r}{\xrightarrow{\text{a.s.}}} E^{\mathcal{C}}Y_1$.

IV. INDEPENDENCE. The foregoing results, in fact all the results of this chapter, were obtained under the guidance but not by the use of similar results in the case of independence. Thus we have new proofs of the latter under the supplementary assumptions of independence.

First consider results in I. Proposition **a** reduces to the zero-one law for tail events on sequences Y_n of independent r.v.'s; since any tail event $B \in \mathcal{C}_n = \mathcal{B}(Y_{n+1}, Y_{n+2}, \cdots)$ whatever be n, and $\mathcal{B}_n$ and $\mathcal{C}_n$ are independent σ-fields, it follows that $P^{\mathcal{B}_n}B = PB$ a.s. whatever be n. Similarly, proposition **b** reduces to the Borel zero-one criterion, since then $P^{\mathcal{B}_{n-1}}B_n = PB_n$ a.s.

Now, let $X_n = \sum_{k=1}^{n} Y_k$ where the summands are independent r.v.'s centered at expectations. Then, in II, the inequality extends Kolmogorov's inequality for $r = 2$ to any $r \geqq 1$, while the proposition proved there reduces to the fact that, when the summands are uniformly bounded, the series $\sum Y_n$ converges a.s. if, and only if, it converges in q.m. ($r = 2$). As for III, it yields Kolmogorov's strong law of large numbers, since then the tail σ-field is $\{\emptyset, \Omega\}$ a.s. and the limit is a tail function.

To summarize, as had to be expected, the results in I, II, and III provide the basic convergence properties of the case of independence, but nothing new. However, if we do not limit ourselves to results expressed in terms of summands, we get more (Marcinkiewicz).

A. *Let* $X_n = \sum_{k=1}^{n} Y_k$ *be consecutive sums of independent r.v.'s centered at expectations, and let* $r \geqq 1$. *Then* $X_n \xrightarrow{\text{a.s.}} X \in L_r$ *if, and only if,* $X_n \xrightarrow{r} X$.

Proof. If $X_n \xrightarrow{r} X$, then, by 29.3**B**, $X_n \xrightarrow{\text{a.s.}} X \in L_r$. Conversely, let $X_n \xrightarrow{\text{a.s.}} X \in L_r$. Since $r \geqq 1$, the r.v. X is integrable. Since, for every p, $X_{n+p} - X_n$ is independent of $X_1, \cdots, X_n$, it follows that $X - X_n$ is independent of $X_1, \cdots, X_n$, so that

$$E^{\mathcal{B}_n}X = E^{\mathcal{B}_n}(X - X_n) + E^{\mathcal{B}_n}X_n = E(X - X_n) + X_n = EX + X_n \text{ a.s.}$$

But $X_n \xrightarrow{\text{a.s.}} X$, while, by I**A**, $E^{\mathcal{B}_n}X \xrightarrow{\text{a.s.}} X$. Therefore, $EX = 0$ so that $E^{\mathcal{B}_n}X = X_n$ a.s. and the martingale sequence X_n is closed by $X \in L_r$. Thus, theorem 29.3**B** applies and, hence, $X_n \xrightarrow{r} X$. The proof is terminated.

***29.5 Indefinite expectations and a.s. convergence.** Convergence properties of martingales and, hence, all the applications of the preceding subsection are but particular cases of a convergence theorem that we establish now.

We consider sequences X_n of r.v.'s whose indefinite expectations φ_n defined by

$$\varphi_n(A) = \int_A X_n, \quad A \in \mathcal{A},$$

exist. We recall that $\mathcal{B} = \mathcal{B}(X_1, X_2, \cdots)$ is the minimal σ-field over the field $\mathcal{B}_0 = \bigcup \mathcal{B}(X_1, \cdots, X_n)$, and $\mathcal{C} = \bigcap \mathcal{B}(X_n, X_{n+1}, \cdots)$ is the tail σ-field of the sequence X_n. We introduce the following hypothesis.

(H): *There exists a set function φ on $\mathcal{C}$ such that, as $n \to \infty$ and then $m \to \infty$,*

$$\sum_{k=m}^{n} \varphi_k(B_{mk}C) \to \varphi(C'C)$$

whatever be the disjoint in k events $B_{mk} \in \mathcal{B}\{X_m, \cdots, X_k\}$ such that $\sum_{k=m}^{n} B_{mk} \to C'$ and whatever be the tail events C and C'.

If the foregoing events B_{mk} are replaced by disjoint in k events $B_{kn} \in \mathcal{B}\{X_k, \cdots, X_n\}$, the so modified hypothesis will be denoted by (H′).

a. BASIC INEQUALITIES. *Under* (H) *or* (H′),

$$\varphi(\underline{C}_aC) \leqq aP(\underline{C}_aC) \quad \textit{and} \quad \varphi(\overline{C}_bC) \geqq bP(\overline{C}_bC)$$

whatever be the tail event C and whatever be the finite numbers a, b in the tail events

$$\underline{C}_a = [\liminf X_n \leqq a], \quad \overline{C}_b = [\limsup X_n \geqq b].$$

Proof. Let $a < a_m \downarrow a$ as $m \to \infty$ and set

$$B_{mm} = [X_m < a_m], \quad B_{mk} = [X_m \geqq a_m, \cdots, X_{k-1} \geqq a_m, X_k < a_m],$$

so that, as $n \to \infty$ and then $m \to \infty$,

$$\sum_{k=m}^{n} B_{mk} = [\inf_{m \leqq k \leqq n} X_k < a_m] \to \underline{C}_a.$$

Since, for every event C,

$$\sum_{k=m}^{n} \varphi_k(B_{mk}C) = \sum_{k=m}^{n} \int_{B_{mk}C} X_k \leqq a_m P\left(\sum_{k=m}^{n} B_{mk}C\right),$$

it follows, upon letting $n \to \infty$ and then $m \to \infty$, that (H) entails

$$\varphi(\underline{C}_a C) \leqq aP(C_a C).$$

The same inequality is entailed by (H') upon setting $B_{kn} = [X_n \geqq a_m, \cdots, X_{k+1} \geqq a_m, X_k < a_m]$, and the first asserted inequality is proved. The second one follows by changing the X_n into $-X_n$ and a into $-b$. The proof is complete.

A. Basic convergence theorem. *Under* (H) *or* (H'), $X_n \xrightarrow{\text{a.s.}} X$ *a.s. finite above or below according as* φ *is bounded above or below. If, moreover,* φ *on* $\mathcal{C}$ *is* σ*-additive and* σ*-finite, then* $X_n \xrightarrow{\text{a.s.}} X = \dfrac{d\varphi}{dP_{\mathcal{C}}}$ *a.s. finite.*

$\dfrac{d\varphi}{dP_{\mathcal{C}}}$ denotes the $\mathcal{C}$-measurable function whose indefinite integral is the $P_{\mathcal{C}}$-continuous part φ_c of φ, and $P_{\mathcal{C}}$ is the restriction of P to $\mathcal{C}$.

Proof. Since the set $D = [\liminf X_n \neq \limsup X_n]$ of divergence of the sequence X_n can be written as a denumerable union $D = \bigcup C_{ab}$ where

$$C_{ab} = [\liminf X_n \leqq a < b \leqq \limsup X_n]$$

and a, $b (a < b)$ vary over all rationals, it suffices to prove that every event C_{ab} is null. But, by taking $C = C_{ab}$ in the basic inequality so that

$$\underline{C}_a C_{ab} = \overline{C}_b C_{ab} = C_{ab},$$

it follows that

$$bPC_{ab} \leqq \varphi(C_{ab}) \leqq aPC_{ab},$$

and, since $b > a$, we have $PC_{ab} = 0$. Thus, if $X = \lim X_n$ on D^c and we set, say, $X = 0$ on D, then $X_n \xrightarrow{\text{a.s.}} X$ where X is a $\mathcal{C}$-measurable function—not necessarily finite. If $\varphi \leqq c < \infty$, then, taking $C = \Omega$ in the second basic inequality, we have, for every finite $b > 0$,

$$P[X = +\infty] = P[\limsup X_n = +\infty] \leqq P\overline{C}_b \leqq c/b \to 0 \quad \text{as} \quad b \to \infty,$$

so that $X < +\infty$ a.s. Similarly when φ is bounded below, and the first assertion is proved.

Let now φ be σ-additive and σ-finite. By taking, if necessary, a denumerable partition of Ω into events of φ-finite measure, it suffices as usual to prove the last assertion for φ on $\mathcal{C}$ σ-additive and finite, hence bounded. Then X is a.s. finite, and we can take X to be finite by including the null event $[X = \pm\infty]$ in the null event D and setting, say, $X = 0$ on D.

Let

$$C_{nk} = \left[\frac{k}{2^n} \leqq X < \frac{k+1}{2^n}\right], \quad k = 0, \pm 1, \pm 2, \cdots$$

and set

$$X'_n = \sum_{k=-\infty}^{+\infty} \frac{k}{2^n} I_{C_{nk}}$$

so that

$$X'_n \leqq X < X'_n + \frac{1}{2^n} \quad \text{on} \quad D^c.$$

On the other hand, the basic inequalities with C replaced by $CC_{nk}D^c$ and a,b replaced by $\frac{k+1}{2^n}, \frac{k}{2^n}$, respectively, become

$$\frac{k}{2^n} P(CC_{nk}D^c) \leqq \varphi(CC_{nk}D^c) \leqq \frac{k+1}{2^n} P(CC_{nk}D^c)$$

so that, by summing over k, we obtain

$$\int_{CD^c} X'_n \leqq \varphi(CD^c) \leqq \int_{CD^c} X'_n + \frac{1}{2^n}.$$

It follows that

$$\varphi(CD^c) - \frac{1}{2^n} \leqq \int_{CD^c} X \leqq \varphi(CD^c) + \frac{1}{2^n}$$

and, letting $n \to \infty$,

$$\int_C X = \int_{CD^c} X = \varphi(CD^c) = \varphi_c(C), \quad C \in \mathcal{C}.$$

Since X is $\mathcal{C}$-measurable, the last assertion is proved.

Variant. It may happen that φ is defined on the σ-field $\mathcal{B}$ of the whole sequence X_n and at the same time (H) or (H′) continues to hold when arbitrary $C \in \mathcal{C}$ are replaced by arbitrary $B \in \mathcal{B}$. The so modified hypotheses will be denoted by (H_0) and (H'_0), respectively. The same proofs continue to apply with B's instead of C's and we obtain

a₀. *Under* (H_0) *or* (H'_0),

$$\varphi(\underline{C}_a B) \leqq aP(\underline{C}_a B), \quad \varphi(\overline{C}_b B) \geqq bP(\overline{C}_b B)$$

whatever be $B \in \mathcal{B}$ and the finite numbers a, b.

A₀. *Under* (H_0) *or* (H'_0), $X_n \xrightarrow{\text{a.s.}} X$ *a.s. finite above or below according as φ is bounded above or below. If, moreover, φ on $\mathcal{B}$ is σ-additive and σ-finite, then* $X_n \xrightarrow{\text{a.s.}} X = \frac{d\varphi}{dP_{\mathcal{B}}}$ *a.s. finite.*

COROLLARY 1. *Let φ on $\mathcal{B}_0$ be a σ-finite signed measure, so that it determines its σ-additive and σ-finite extension φ on $\mathcal{B}$.*

If (H) *or* (H′) *hold when $C \in \mathcal{C}$ are replaced by $B_0 \in \mathcal{B}_0$, then $X_n \xrightarrow{\text{a.s.}} X = \dfrac{d\varphi}{dP_{\mathcal{B}}}$ a.s. finite.*

Proof. The basic inequalities hold with $C \in \mathcal{C}$ replaced by $B_0 \in \mathcal{B}_0$. Therefore, by continuity of P and φ, they hold with $B \in \mathcal{B}$ and the above variant applies.

COROLLARY 2. *Let φ on $\mathcal{B}_0$ be a σ-finite signed measure. If, given any $\epsilon > 0$, for k sufficiently large*

$$|\varphi_k(B) - \varphi(B)| \leqq \epsilon PB$$

whatever be $B \in \mathcal{B}(X_k, X_{k+1}, \cdots)(B \in \mathcal{B}(X_1, \cdots, X_k))$, then $X_n \xrightarrow{\text{a.s.}} X = \dfrac{d\varphi}{dP_{\mathcal{C}}}\left(\dfrac{d\varphi}{dP_{\mathcal{B}}}\right)$.

Proof. In the first case, hypothesis (H′) holds, since φ extends to $\mathcal{B}$ and, for m sufficiently large,

$$\left|\sum_{k=m}^{n} \varphi_k(B_{kn}C) - \varphi\left(\sum_{k=m}^{n} B_{kn}C\right)\right| \leqq \epsilon \sum_{k=m}^{n} PB_{kn}C \leqq \epsilon \to 0$$

as $n \to \infty$, then $m \to \infty$, and then $\epsilon \to 0$. Theorem **A** applies, and the assertion is proved. Similarly, in the second case, hypothesis (H_0) holds, and theorem $\mathbf{A}_0$ applies.

APPLICATION TO MARTINGALES. 1° Let the r.v.'s X_n form a martingale sequence or a martingale reversed sequence, closed by a r.v. Y whose expectation exists, that is, $X_n = E(Y \mid X_1, \cdots, X_n)$ a.s. or $X_n = E(Y \mid X_n, X_{n+1}, \cdots)$ a.s. Then Corollary 2 applies with φ indefinite integral of Y; in fact, in each case $E_B X_n = E_B Y = \varphi(B)/PB$. Thus $X_n \xrightarrow{\text{a.s.}} X = E^{\mathcal{C}}Y$ or $E^{\mathcal{B}}Y$, respectively.

2° Let the r.v.'s X_n with $\sup E|X_n| \leqq c < \infty$ form a martingale sequence. Take for pr. space the "sample pr. space," that is, the range space of the sequence together with its Borel field $\mathcal{A}_\infty$ and the pr. distribution P_∞ of the sequence. Then $\mathcal{B}(X_1, \cdots, X_n)$ is the σ-field $\mathcal{A}_n$ of all Borel cylinders whose bases are Borel sets in the range space of $(X_1, \cdots, X_n)$. Since the X_n form a martingale sequence, we have $\varphi_n(A_n) = \varphi_{n+1}(A_n) = \cdots$ for every $A_n \in \mathcal{A}_n$ and, hence, $\varphi_n \to \varphi$ on $\mathcal{A}_0$—the field of all Borel cylinders in the range space. We apply the extension 4.3**A**, 2°. Since the indefinite expectations $\bar{\varphi}_n$ of $|X_n|$ are bounded by c, and form a nondecreasing sequence, it follows that $\lim \bar{\varphi}_n$ exists and is bounded and σ-additive on $\mathcal{A}_0$. *A fortiori*, φ is bounded and σ-additive on $\mathcal{A}_0$. Thus, Corollary 2 applies and $X_n \xrightarrow{\text{a.s.}} X = \dfrac{d\varphi}{dP_{\mathcal{B}}}$.

We seek now necessary *and* sufficient conditions for a.s. convergence of sequences X_n of r.v.'s.

B. Dominated convergence criterion. *Let* $|X_n| \leqq Y$ *integrable. Then* $X_n \xrightarrow{\text{a.s.}} X$ *(necessarily integrable) if, and only if,* (H) *or* (H′) *or* (H_0) *or* (H'_0) *holds.*

Thus, if $|X_n| \leqq Y$ integrable and $X_n \xrightarrow{\text{a.s.}} X$, then all the hypotheses H are equivalent.

Proof. The "if" assertion is contained in **A** and $\mathbf{A}_0$. Conversely, let $X_n \xrightarrow{\text{a.s.}} X$ with indefinite expectation φ so that, for every $\epsilon > 0$,

$$PB_\epsilon = P[\sup_{m \leqq k \leqq n} |X_k - X| \geqq \epsilon] \to 0 \quad \text{as} \quad m \to \infty.$$

Whatever be the disjoint events $B_k \in \mathcal{B}$ varying or not with m and n and whatever be $B \in \mathcal{B}$ such that $\sum_{k=m}^{n} B_k \to B$, upon summing over $k = m, \cdots, n$ the relations

$$\varphi_k(B_k) - \varphi(B_k) = \int_{B_k} (X_k - X) = \int_{B_k B_\epsilon^c} (X_k - X) + \int_{B_k B_\epsilon} (X_k - X),$$

it follows from $|X_k| \leqq Y$ integrable that

$$\left|\sum_{k=m}^{n} \varphi_k(B_k) - \varphi(B)\right| \leqq \epsilon PB_\epsilon^c + 2\int_{B_\epsilon} Y + \left|\varphi\left(\sum_{k=m}^{n} B_k\right) - \varphi(B)\right|.$$

Letting $n \to \infty$, then $m \to \infty$, and then $\epsilon \to 0$, the "only if" assertion follows.

C. Convergence criterion. *A sequence X_n of r.v.'s converges a.s. to a r.v. if, and only if, for every $\epsilon > 0$ there exist events B_ϵ with $PB_\epsilon > 1 - \epsilon$ on which $|X_n| \leqq Y_\epsilon$ integrable and* (H) *or* (H′) *or* (H_0) *or* (H'_0) *holds.*

Proof. If for $\epsilon_m \downarrow 0$ as $m \to \infty$, we have $|X_n| \leqq Y_{\epsilon_m}$ integrable and, say, (H) holds on $B_m = B_{\epsilon_m}$; then by **B**, the sequence X_n converges to a r.v. on $B_m - N_m$ where N_m are null events; hence it converges to a r.v. on $\bigcup B_m - \bigcup N_m$. Since, for every integer m',

$$P \bigcup B_m \geqq PB_{m'} \geqq 1 - \epsilon_{m'},$$

it follows that $P \bigcup B_m = 1$ and the "if" assertion is proved.

Conversely, let $X_n \xrightarrow{\text{a.s.}} X$ r.v. and apply Egorov's theorem which asserts that for every $\epsilon > 0$ there exists an event B with $PB < \frac{\epsilon}{2}$ such that $X_n \to X$ uniformly on B^c. Let n and $c > 0$ be sufficiently large so that, on the one hand, $|X_n| < |X| + 1$ on B^c and, on the other hand,

$$PC = P[|X| > c] < \frac{\epsilon}{2}.$$

Then $|X_n| < c + 1$ on B^cC^c and, by **B**, (H) holds. Since

$$1 - PB^cC^c = P(B \cup C) \leqq PB + PC < \epsilon,$$

the "only if" assertion is proved.

COMPLEMENTS AND DETAILS

1. Let $S_n = \sum_{k=1}^{n} X_k$ where $E(X_n \mid X_1, \cdots, X_{n-1}) = 0$ a.s. $(X_0 = 0)$ and $E(X_n^2 \mid X_1, \cdots, X_{n-1}) = \sigma^2 X_n$ a.s. Find conditions for degenerate and for normal convergence of suitably normed sums.

2. Take one by one the results relative to the degenerate, the normal, and the Poisson convergence obtained in the case of independent summands, and transpose them to the case of dependent summands by using successively the various approaches given and illustrated in the text.

3. Let $S_{\nu_n} = \sum_{k=1}^{\nu_n} X_k$, $X_0 = 0$. The summands are independent with common ch.f. f and finite $a = EX_n$, $\sigma^2 = \sigma^2 X_n$. The ν_n are integer-valued r.v.'s independent of all the summands with $p_{nk} = P[\nu_n = k]$, $k = 0, 1, \cdots$, and with finite $\alpha_n = E\nu_n$, $\beta_n^2 = \sigma^2\nu_n$. Set $\sigma_n^2 = \sigma^2 S_{\nu_n}$ and let g_n be the ch.f. of $(S_{\nu_n} - E_{\nu_n})/\sigma_n$. Then

$$g_n(u) = \sum_{k=0}^{\infty} p_{nk} e^{-i\alpha_n u/\sigma_n} f_k \cdots + a^2\beta_n^2.$$

Let $\sigma_n^2 \to \infty$, $\beta_n^2 = O(\sigma_n^2)$. Then $g_n(u)$ is of the form

$$g_n(u) = h_n(cu) e^{-\frac{u^2}{2}(1 - c_n^2)} + o(1), \quad c_n = a_n\beta_n/\sigma_n.$$

If also $a^2\beta_n^2 = O(\alpha_n)$, then $g_n(u) \to e^{-u^2/2}$.

If also $\mathfrak{L}(\nu_n) \sim \mathfrak{N}(\alpha_n, \beta_n^2)$, then $\mathfrak{L}(S_{\nu_n}) \sim \mathfrak{N}(a\alpha_n, \sigma_n^2)$.

What happens when $\mathcal{L}(\nu_n)$ is a Poisson law $\mathcal{P}(n)$, or when $\mathcal{L}(\nu_n)$ is a binomial law with

$$p_{nk} = \frac{n!}{k!(n-k)!} p^k q^{n-k}?$$

4. Search for conditions under which the various limit laws obtained in the case of independent summands remain the same when the numbers of summands are r.v.'s ν_n independent of the summands. What happens when $\nu_n \xrightarrow{\text{a.s.}} \infty$?

5. By following the indications given in the text transpose to the case of dependent summands as many as possible of the a.s. convergence and a.s. stability theorems obtained in the case of independence.

6. Let Y be an integrable r.v. defined on a sequence X_n of r.v.'s. Take for pr. space the sample space of the sequence: $\Omega = \prod R_n$, $\mathcal{A}$-Borel field in Ω, P—pr. distribution of the sequence.

If the X_n are independent, with pr. distributions P_n, then

$$E(Y \mid X_1, \cdots, X_n) = \int Y \, dP_{n+1} \, dP_{n+2} \cdots \xrightarrow{\text{a.s.}} Y$$

$$E(Y \mid X_n, X_{n+1}, \cdots) = \int Y \, dP_1 \cdots dP_{n-1} \xrightarrow{\text{a.s.}} EY.$$

What becomes of the integrals when the X_n are dependent?

7. A net is a sequence of countable partitions $\Omega = \sum_j A_{nj}$ into events such that every partition is finer than the preceding one. Every partition determines a σ-field $\mathcal{B}_n$, and $\mathcal{B}_n \uparrow$. Let φ on $\bigcup \mathcal{B}_n$ be bounded and let it be σ-additive on every $\mathcal{B}_n$. Set $X_n = \sum_j x_{nj} I_{A_{nj}}$ with $x_{nj} = \varphi(A_{nj})/P(A_{nj})$ and throw out the null union of all null events A_{nj}.

The sequence X_n is a martingale and its a.s. limit, if it exists, is called the derivative of φ with respect to P given the net.

If φ is σ-additive on $\bigcup \mathcal{B}_n$, then it extends to φ on $\mathcal{B}$ bounded and σ-additive, and $X_n \xrightarrow{\text{a.s.}} X = \dfrac{d\varphi}{dP_{\mathcal{B}}}$. The X_n are uniformly integrable if, and only if, φ is P-continuous and then $X_n \underset{1}{\xrightarrow{\text{a.s.}}} X$.

Particular case. Let $\Omega = [0, 1]$, P is the Lebesgue measure, φ is determined by a function H on Ω of bounded variation. Consider a net of partitions into intervals such that the length of the largest converges to 0. Then $\mathcal{B}$ is the Borel field in Ω and $X_n \xrightarrow{\text{a.s.}} H'$-derivative of H known to exist a.s.

8. If g on R is a Lebesgue integrable function of period 1 and $g_n(x) = \dfrac{1}{2^n} \sum_{k=0}^{2^n-1} g\left(x + \dfrac{k}{2^n}\right)$, then $g \xrightarrow{\text{a.e.}} \int_0^1 g(x)\,dx$. (Let $\Omega = R$, $\mathcal{A}$ = σ-field of Lebesgue sets A of period 1, PA = Lebesgue measure of a period of A. Let $\mathcal{B}_n$ be a similar σ-field of sets of period $1/2^n$. The σ-field $\bigcap \mathcal{B}_n$ consists only of sets of pr. 0 or 1. Use martingale reversed sequences convergence theorem.)

9. Let φ on $\mathcal{A}$ be bounded and σ-additive. Then $X_n = \frac{d\varphi}{dP_{\mathcal{B}_n}} \xrightarrow{\text{a.s.}} X = \frac{d\varphi}{dP_{\mathcal{B}}}$ or $\frac{d\varphi}{dP_{\mathcal{C}}}$ according as $\mathcal{B}_n \uparrow$ or $\mathcal{B}_n \downarrow$. (Either start with $\varphi \geqq 0$ and observe that the sequence $-X_n$ is a submartingale; or first extend the basic convergence theorem 29.4**A** to $X_n = \frac{d\varphi_n}{dP}$ where $\varphi_n = \varphi_n{}^c + \varphi_n{}^s$ and X_n a.s. finite with $(\varphi_n{}^s)^{\pm}(A) = \varphi_n(A[X = \pm\infty])$.)

10. Let $m, n \to \infty$ and let $r \geqq 1$. Let $\mathcal{B}_m$, $\mathcal{B}$ be sub-σ-fields of events with $\mathcal{B}_m \uparrow \mathcal{B}$ or $\mathcal{B}_m \downarrow \mathcal{B}$

(a) $X_n \xrightarrow{r} X \in L_r \Rightarrow E(X_n \mid \mathcal{B}_m) \xrightarrow{r} E(X \mid \mathcal{B})$.

For, by the c_r-inequality and martingales convergence, $E|\, E(X_n \mid \mathcal{B}_m) - E(X \mid \mathcal{B})\,|^r \leqq c_r E|\, E(X_n - X \mid \mathcal{B}_m)\,|^r + c_r E|\, E(X \mid \mathcal{B}_m) - E(X \mid \mathcal{B})\,|^r \leqq c_r E|\, X_n - X\,|^r + c_r E|\, E(X \mid \mathcal{B}_m) - E(X \mid \mathcal{B})\,|^r \to 0$.

(b) $0 \leqq X_n \uparrow X \in L_r \Rightarrow E(X_n \mid \mathcal{B}_m) \underset{r}{\xrightarrow{\text{a.s.}}} E(X \mid \mathcal{B})$.

Use (a) and, by martingale convergence and conditional monotone convergence,

$$\inf_{j \geqq m, k \geqq n} E(X_k \mid \mathcal{B}_j) \geqq \inf_{j \geqq m} E(X_n \mid \mathcal{B}_j) \Rightarrow \liminf_{m,n} E(X_n \mid \mathcal{B}_m)$$
$$\geqq \lim_n \lim_m E(X_n \mid \mathcal{B}_m) = E(X \mid \mathcal{B}) \text{a.s.}$$

$$\sup_{j \geqq m, k \geqq n} E(X_k \mid \mathcal{B}_j) \leqq \sup_{j \geqq m} E(X \mid \mathcal{B}_j) \Rightarrow \limsup_{m,n} E(X_n \mid \mathcal{B}_m)$$
$$\leqq E(X \mid \mathcal{B}_m) = E(X \mid \mathcal{B}) \text{ a.s.}$$

(c) $\inf_n X_n \in L_r, \liminf_n X_n \in L_r \Rightarrow E(\liminf_n X_n \mid \mathcal{B}) \leqq \liminf_{m,n} E(X_n \mid \mathcal{B}_m)$ a.s.

$\sup_n X_n \in L_r, \limsup_n X_n \in L_r \Rightarrow \limsup_{m,n} E(X_n \mid \mathcal{B}_m) \leqq E(\limsup_n X_n \mid \mathcal{B})$ a.s.

Use (b) as in the Fatou-Lebesgue Theorem.

(d) $\sup |\, X_n \,| \in L_r, X_n \xrightarrow{\text{a.s.}} X \Rightarrow E(X_n \mid \mathcal{B}_m) \underset{r}{\xrightarrow{\text{a.s.}}} E(X \mid \mathcal{B})$. Use (a) and (c). What if $X_n \xrightarrow{P} X$?

Chapter IX

ERGODIC THEOREMS

Ergodic theory has a phenomenological origin which, on account of the Liouville theorem, leads to the study of measure-preserving transformations. The "classical period" (1930–1944) is concerned with one-to-one measure-preserving transformations T_1 of a measure space onto itself (in what follows we limit ourselves to a pr. space $(\Omega, \mathcal{A}, P)$ and leave out results with which we are not concerned). Let X, Y be r.v.'s and define X_n, Y_n, by $X_n(\omega) = X(T_1^{n-1}\omega)$, $Y_n(\omega) = Y(T_1^{n-1}\omega)$.

The first two basic results are

von Neumann's: if $X \in L_2$, then $\frac{1}{n}\sum_{k=1}^{n} X_k$ converges in q.m.

Birkhoff's: if $X \in L_1$, then $\frac{1}{n}\sum_{k=1}^{n} X_k$ converges a.s.

Then Khintchine gets rid of the supplementary but unnecessary assumptions made by Birkhoff and, at the same time, simplifies very considerably his proof; Hopf extends this theory to ratios of sequences $\sum_{k=1}^{n} X_k / \sum_{k=1}^{n} Y_k$ and proceeds to a systematic investigation of ergodic properties; Yosida and Kakutani extend von Neumann's theorem to transformations on Banach spaces and apply them to Markov chains.

The "modern period" (1944–) is characterized by several weakenings of the ergodic setup. Hurewicz and Halmos abandon, at least partly, the initial setup and start with set functions from which r.v.'s are derived, whereas Dunford and Miller abandon definitely the measure-preserving property and obtain necessary and sufficient conditions for convergence in the first mean. Then, F. Riesz gives very simple proofs of the Birkhoff theorem and of the sufficiency part of the Dunford-Miller theorem, and, at the same time, he abandons the one-to-one assumption about transformations. Doob applies Birkhoff's theorem to stationary chains. Finally, Hartman and Ryll-Nardzewski, employing methods developed by Y. Dowker in investigating "potentially invariant" measures, obtain necessary and sufficient conditions for a.s. convergence on L_1 to L_1.

In pr. theory we are interested not in the behavior of averages of sequences $\{X_n\}$ due to *any* r.v. $X \in L_1$ but in that of individual sequences. Stripped of all considerations due to their phenomenological origin, *ergodic theorems* assert conditions, in terms of point or, more generally, set transformations, under which averages $\frac{1}{n}\sum_{k=1}^{n} X_k$ of r.v.'s converge in some sense and, then, assert conditions under which the limits are degenerate. In the case of independence, such theorems reduce to limit theorems as expounded at length in Part III. Thus, Bernoulli's law of large numbers is to be thought of as the first ergodic theorem with convergence in pr. and Tchebichev's proof as an improvement of the conclusion—with convergence in q.m. Above all, Borel's strong law of large numbers is to be thought of as the first "best" ergodic theorem—with a.s. convergence.

What characterizes the *ergodic method*—as compared with those of Part III—is that the conditions for convergence are to be in terms of iterated translations, in a sense to be made precise in this chapter. We shall establish a basic ergodic inequality and deduce from it ergodic theorems which contain the mentioned results (sometimes improved or completed). The reader will recognize the proofs (and hence the statements) which remain valid when P is replaced by a σ-finite or an arbitrary measure μ.

§ 30. TRANSLATION OF SEQUENCES; BASIC ERGODIC THEOREM AND STATIONARITY

***30.1 Phenomenological origin.** Let $q = (q_1, \cdots, q_N)$, $p = (p_1, \cdots, p_N)$ be the generalized coordinates and momenta of a "conservative" mechanical system with N degrees of freedom. The equations of motion of the system are

$$\frac{dq_i}{dt} = \frac{\partial H}{\partial p_i}, \quad \frac{dp_i}{dt} = -\frac{\partial H}{\partial q_i}, \quad i = 1, \cdots, N,$$

where $H = H(q, p)$ is the Hamiltonian of the system, independent of time t. The "states" of the system are represented by points $\omega = (q, p)$ of the $2N$-dimensional real space Ω—the "phase space" of the system.

Under the equations of motion, every point $\omega \in \Omega$ which represents the "initial state"—at time $t = 0$—moves along a trajectory and this motion describes the evolution of the system; the state at time t is the point ω_t. Thus the phase space can be envisioned as a fluid in motion within itself. The celebrated Liouville's theorem asserts that in this motion the Lebesgue measure λ of Lebesgue sets A remains invariant. More precisely, in a unit of time the phase space undergoes a one-to-one transformation $\omega \leftrightarrow T_1\omega$ such that $\lambda T_1 A = \lambda A$. To simplify, we consider only discrete values of time $0, 1, 2, \cdots$. In fact, the conservative systems have constant energy E and the possible trajectories lie on the surface $H = E$ whence the term "ergodic" from *ergos* meaning energy; then the invariant measure μ is defined by $d\mu = d\sigma/\text{grad}\, H$ where $d\sigma$ is the differential area of an element of the surface and grad H is taken at a point of the element.

The comparison between theory and reality is made by measuring "observables" or "phase-functions"—λ-measurable functions of the state. The ideal would be to observe the consecutive states (their components are phase functions). Yet, in statistical mechanics, insurmountable difficulties of experimental as well as computational nature arise. In the systems under investigation the number of constituents or "particles" is *extremely large* and so is the number of degrees of freedom, as well as the number of microscopic phenomena in a unit of time of the observer (such as collisions of particles between themselves and with the walls of the container). Thus from the microscopic point of view, the time required by the observer to measure an observable X is extremely large, and its observed values are to be compared not to its instantaneous theoretical values but to the theoretical time-averages

$$\overline{X^n}(\omega) = \frac{1}{n}\{X(\omega) + X(T_1\omega) + \cdots + X(T_1^{n-1}\omega)\}$$

for n extremely large. However, computation of time-averages requires knowledge of consecutive states $T_1^n\omega$ given the initial state ω. Yet the *exact* knowledge of the initial state is experimentally unattainable. Even if it were attainable, then theoretical knowledge of succeeding states would require integration of an extremely large number of equations of motion—which is practically impossible. Thus, some other way of evaluating theoretical time-averages is to be found or postulated. Physicists were led to replace the exact initial state by the set of all possible states compatible with the precision of the experimental data and to postulate equality of time-averages with phase-averages over the multiplicity described by the trajectories of the initial states compatible with the data; this is the "ergodic hypothesis."

From the physicist's point of view, the justification of the ergodic hypothesis lies in its practical success. For the mathematician, ergodic theory was at first an attempt to *justify* theoretically the ergodic hypothesis. But bitter experience imposed supplementary hypotheses. Ergodic theorems assert conditions under which sequences of time-averages converge in some sense and *then* conditions under which the limits are independent of the initial state and reduce to phase averages. From the experimental point of view, comparison between theory and observation will be best when measurements of the observables yield approximate values of the time-averages, that is, when the convergence will be an everywhere or at least an a.e. convergence.

30.2 Basic ergodic inequality. Let the pr. space $(\Omega, \mathcal{A}, P)$ be fixed. On a family of one or several sequences of r.v.'s, say $\{X_n\}$, $\{Y_n\}$, we define Borel functions of the family, say ξ. These functions are measurable and, more precisely, are $\mathcal{B}$-measurable where $\mathcal{B}$ is the sub σ-field of events induced by the family. The *translate by* $k - 1$ of ξ is the function ξ_k (so that $\xi_1 = \xi$) obtained by adding $k - 1 (k = 1 \text{ or } 2 \text{ or } \cdots)$ to the subscripts of all those r.v.'s of the family which figure in the definition of ξ. Thus ξ_k is defined on the family $\{X_{n+k-1}\}$, $\{Y_{n+k-1}\}$—the translate by $k - 1$ of the given family—exactly as ξ is defined on the original one. We say that ξ is *invariant* (under translations) if it coincides with all its translates. Translations of indicators of events of $\mathcal{B}$ define translations of the events; in other words, the above definitions and notation apply to events $B \in \mathcal{B}$—replace ξ by B and ξ_k by B_k. Because of the definition, translations preserve countable set operations on events belonging to $\mathcal{B}$. It follows that the class of all invariant events is closed under countable set operations, hence is a sub σ-field $\mathcal{C}$ of events—the *invariant σ-field* defined on the family, and invariance of measurable functions defined on the family means $\mathcal{C}$-measurability.

The concept of translation can be interpreted by means of the range space of the family: the Borel space of points $(x_1, y_1, x_2, y_2, \cdots)$. For example, the translate $[X_k < Y_k]$ of $[X_1 < Y_1]$ is obtained as follows. The event $[X_1 < Y_1]$ is the inverse image (under the family) of the Borel set $[x_1 < y_1]$ in the range space. Then the translate $[X_k < Y_k]$ is, by definition, the inverse image of the Borel set $[x_k < y_k]$. In fact, to avoid any ambiguity, in what follows it suffices to think of the pr. space as being the sample pr. space of the double sequence.

Convention and notation. In this chapter we shall reserve subscripts to indicate translations and otherwise use superscripts—not to be confused with power indices. Denote by $\mathcal{B}$ the σ-field of events induced by the family $\{X_n\}$, $\{Y_n\}$, by $\mathcal{C}$ the σ-field of invariant events, and set

$$X^n = \sum_{k=1}^{n} X_k, \quad Y^n = \sum_{k=1}^{n} Y_k.$$

To avoid undefined ratios $\frac{X^n}{Y^n}$ and to make the limits independent of an arbitrary but finite number of the summands, we assume once and for all that

$$Y_n > 0, \quad Y^n \uparrow \infty.$$

Then, from

$$\frac{X_2 + \cdots + X_{n+1}}{Y_2 + \cdots + Y_{n+1}} = \frac{X^{n+1} - X_1}{Y^{n+1} - Y_1} = \frac{X^{n+1}}{Y^{n+1}} \cdot \frac{Y^{n+1}}{Y^{n+1} - Y_1} - \frac{X_1}{Y^{n+1} - Y_1}$$

it follows that $\liminf \frac{X^n}{Y^n}$ and $\limsup \frac{X^n}{Y^n}$ are invariant and, hence, so are the sets of convergence and of divergence of the sequence $\frac{X^n}{Y^n}$:

$$C = \left[\liminf \frac{X^n}{Y^n} = \limsup \frac{X^n}{Y^n}\right], \quad D = \left[\liminf \frac{X^n}{Y^n} \neq \limsup \frac{X^n}{Y^n}\right]$$

as well as the events

$$\underline{C}_a = \left[\liminf \frac{X^n}{Y^n} < a\right], \quad \overline{C}_b = \left[\limsup \frac{X^n}{Y^n} > b\right].$$

So far, the defined concepts do not contain the pr. P; they are expressed only in terms of the measurable space $(\Omega, \mathcal{A})$ and of measurable functions on this space. However, when the r.v.'s represent points of L_r-spaces, the transformations have to be equivalence-preserving. It suffices for the translates of null events to be null, *to replace the assumptions on the Y_n's by $Y_n > 0$ a.s. and $Y_n \uparrow \infty$ a.s., and at the same time to replace invariance by a.s. invariance—invariance when a null event is neglected.*

We establish now the basic inequality from which we deduce the basic ergodic theorem. But first we require an elementary lemma. Let $a_1, a_2, \cdots, a_{n+m}$ be finite numbers. We say that a_k is *m-positive* if at least one of the sums $a_k + a_{k+1}, \cdots$ containing no more than m summands is positive. In symbols, a_k is m-positive if $\sup (a_k + \cdots + a_l) > 0$ for $k \leq l \leq \min(k + m - 1, n + m)$.

a. F. RIESZ'S LEMMA. *If there exist m-positive terms, then their sum is positive.*

Proof. Let a_k be the first m-positive term and let $a_k + \cdots + a_l$ be the *shortest* positive sum starting with a_k. If one of the terms a_h of this sum is not m-positive, then $a_h + \cdots + a_l \leqq 0$ so that $a_k + \cdots + a_{h-1} > 0$ and the sum is not the shortest positive one. Thus, all its terms are m-positive. Hence, the successive m-positive terms form disjoint stretches of positive sums. The assertion follows.

A. BASIC ERGODIC INEQUALITY. *If* $B^m = \left[\sup_{j \leqq m} \frac{X^j}{Y^j} > b\right]$ *and* $Z^n > 0$, *then for every integer n and every invariant event C*

$$\sum_{k=1}^{n} \int_{B_k{}^m C} \left(\frac{X_k}{Z^n} - b\frac{Y_k}{Z^n}\right) + \sum_{k=n+1}^{n+m} \int_C \left(\frac{X_k}{Z^n} - b\frac{Y_k}{Z^n}\right)^+ \geqq 0,$$

provided the sum exists.

Proof. Let $k = 1, 2, \cdots, n + m$, $k \leqq l \leqq \min(k + m - 1, n + m)$, and let

$$B^{mk} = [X_k - bY_k \text{ is } m\text{-positive}] = \left[\sup_l \frac{X_k + \cdots + X_l}{Y_k + \cdots + Y_l} > b\right].$$

If $k \leqq n$, then l varies from k to $k + m - 1$ and, hence, $B^{mk} = B_k{}^m$ where $B_k{}^m$ is the translate by $k - 1$ of B^m. Since by Riesz's lemma

$$\sum_{k=1}^{n+m} (X_k - bY_k) I_{B^{mk}} \geqq 0$$

a fortiori

$$\sum_{k=1}^{n} (X_k - bY_k) I_{B_k{}^m C} + \sum_{k=n+1}^{n+m} (X_k - bY_k)^+ I_C \geqq 0,$$

the asserted inequality follows upon dividing by Z^n and integrating.

*Let A with or without affixes denote events defined on the X_n's and Y_n's.

B. BASIC ERGODIC THEOREM. *If*

(i) $$\sum_{k=1}^{n} \int_{A_k} \frac{X_k}{Y^n} \to 0 \quad \textit{and}$$

$$\sum_{k=1}^{n} \int_{A_k} \frac{Y_k}{Y^n} \to 0, \quad \textit{as} \quad n \to \infty \quad \textit{and then} \quad A \downarrow \emptyset,$$

(ii) $$\int \frac{|X_{n+k}|}{Y^n} \to 0 \quad \textit{and}$$

$$\int \frac{Y_{n+k}}{Y^n} \to 0, \quad \textit{for every fixed } k \textit{ as} \quad n \to \infty,$$

then the sequence $\frac{X^n}{Y^n}$ *converges a.s. Assumption* (i) *implies that, from some n on, the sum therein exists.*

Proof. We have to prove that the invariant event

$$D = \left[\liminf \frac{X^n}{Y^n} \neq \limsup \frac{X^n}{Y^n}\right]$$

is null. Since $D = \bigcup_{a,b} C_{ab}$ is the denumerable union of the invariant events

$$C_{ab} = \underline{C}_a \overline{C}_b = \left[\liminf \frac{X^n}{Y^n} < a < b < \limsup \frac{X^n}{Y^n}\right]$$

where $a < b$ vary over the set of all rationals, it suffices to prove that every event C_{ab} is null.

We set in the basic inequality $Z^n = Y^n$ and

$$C = C_{ab} = B^m C_{ab} + A^m.$$

Because of the invariance of C_{ab}, translation by $k - 1$ yields

$$C_{ab} = B_k{}^m C_{ab} + A_k{}^m,$$

and the basic inequality becomes

$$\int_{C_{ab}} \left(\frac{X^n}{Y^n} - b\right) - \sum_{k=1}^{n} \int_{A_k{}^m} \left(\frac{X_k}{Y^n} - b\frac{Y_k}{Y^n}\right)$$
$$+ \sum_{k=n+1}^{n+m} \int_{C_{ab}} \left(\frac{X_k}{Y^n} - b\frac{Y_k}{Y^n}\right)^+ \geqq 0.$$

Since by definition of B^m and C_{ab} we have $B^m C_{ab} \uparrow C_{ab}$ and hence $A^m \downarrow \emptyset$ as $m \to \infty$, it follows because of (i) and (ii) that, upon letting $n \to \infty$ and then $m \to \infty$, the foregoing inequality becomes

$$\liminf \int_{C_{ab}} \left(\frac{X^n}{Y^n} - b\right) \geqq 0.$$

By changing the X_k into $-X_k$, b into $-a$, a into $-b$, this inequality becomes

$$\liminf \int_{C_{ab}} \left(a - \frac{X^n}{Y^n}\right) \geqq 0.$$

Therefore, by adding up the two last inequalities, we have

$$(a - b)PC_{ab} \geqq 0, \quad a < b,$$

so that $PC_{ab} = 0$ and the proof is concluded.

Let Z^n be positive r.v.'s such that for every invariant event C defined on the X_n and Y_n

$$\text{(C)} \qquad \liminf \int_C \frac{Y^n}{Z^n} = 0 \Rightarrow PC = 0;$$

this is certainly true if $Z^n = Y^n$. It follows that, if throughout the preceding proof we divide by Z^n instead of by Y^n, this proof remains valid. In other words,

B′. *The basic ergodic theorem remains valid if in the assumptions therein Y^n are replaced by Z^n obeying condition* (C).

30.3 Stationarity. To the concept of invariance correspond weaker ones of invariance in terms of integrals and in terms of pr.'s.

Let $\{X_n\}$, $\{Y_n\}$, be the family of r.v.'s on which the translations are defined. Let the events A, B, C, with or without affixes, be defined on this family, that is, belong to the σ-field $\mathcal{B}$ induced by the family. Let ξ, with or without affixes, be a r.v. defined on the family. We say that ξ is *integral invariant* or that the sequence of translates $\xi_1, \xi_2, \cdots$ is *integral stationary* if the integrals of the ξ_k exist and if, for every A and every k,

$$\int_{A_k} \xi_k = \int_{A_1} \xi_1.$$

We say that ξ is *P-invariant* or that the sequence $\xi_1, \xi_2, \cdots$ is *stationary* if, for every A defined on it,

$$PA_k = PA_1.$$

In particular, the family itself is (*integral*) *stationary*, if the sequence $X_1, X_2, \cdots Y_1, Y_2, \cdots$ is (*integral*) *stationary*. It follows from the definitions that the sequences of translates of all r.v.'s ξ defined on a stationary family are stationary. Furthermore

a. STATIONARITY LEMMA. *The sequences of translates of r.v.'s, defined on a stationary family, and whose expectations exist, are integral stationary.*

Proof. We have to prove that, for every A and every k,

$$\int_{A_k} \xi_k = \int_{A_1} \xi_1,$$

provided the right-hand side exists. As usual, it suffices to prove it for indicators $\xi_1 = I_{B_1}$. Since set operations are preserved under translations, it follows, by stationarity, that

$$\int_{A_k} I_{B_k} = PA_kB_k = P(A_1B_1)_k = PAB = \int_{A_1} I_{B_1},$$

and the lemma follows.

In the integral stationarity case, the basic ergodic inequality takes very simple forms.

b. STATIONARITY INEQUALITIES. *Let the family $\{X_n\}$, $\{Y_n\}$ be integral stationary and let X_1 or Y_1 be integrable. Then for every invariant C*

$$\int_{C\underline{C}_a} (aY_1 - X_1) \geqq 0 \quad \text{and} \quad \int_{C\overline{C}_b} (X_1 - bY_1) \geqq 0$$

where, for X_1 and Y_1 integrable on C, $\underline{C}_a$ and $\overline{C}_b$ ($< a$ and $> b$) can be replaced by $\underline{B}_a = [\inf X^n/Y^n < a]$ and $\overline{B}_b = [\sup X^n/Y^n > b]$ ($\leqq a$ and $\geqq b$).

Proof. On account of integral stationarity and of integrability of X_1 or Y_1 hence of all the X_k or Y_k, the integrals below exist and

$$\int_{A_k} (X_k - bY_k) = \int_{A_1} (X_1 - bY_1).$$

It follows from the hypotheses that the basic ergodic inequality, with $Z^n = n$, can be written

$$\int_{B^mC} (X_1 - bY_1) + \frac{m}{n} \int_C (X_1 - bY_1)^+ \geqq 0$$

where, as $m \to \infty$,

$$B^m = \left[\sup_{j \leqq m} \frac{X^j}{Y^j} > b\right] \uparrow \overline{B}_b.$$

Therefore

$$\int_{C\overline{C}_b} (X_1 - bY_1) \geqq 0$$

since, either $\int_{C\overline{C}_b} (X_1 - bY_1)^+ = \infty$ and this inequality is trivially true, or $\int_{C\overline{C}_b} (X_1 - bY_1)^+ < \infty$ and this inequality is obtained upon replacing C by $C\overline{C}_b$ and letting $n \to \infty$ then $m \to \infty$. By changing b into $-a$ and $X_1, X_2, \cdots$ into $-X_1, -X_2, \cdots$, the other asserted inequality follows; similarly for the remaining assertions.

When the family is integral stationary and X_1 and Y_1 are integrable, then 30.2B′, with $Z^n = n$, yields $\frac{X^n}{Y^n} \xrightarrow{\text{a.s.}} U$ invariant. However, by using directly the basic ergodic inequality in its foregoing forms, the assumptions can be weakened and the result be made more precise, as follows.

A. Integral stationarity theorem. *Let the family* $\{X_n\}$, $\{Y_n\}$ *be integral stationary, and let* X_1 *or* Y_1 *be integrable. Then*

$$\frac{X_1 + \cdots + X_n}{Y_1 + \cdots Y_n} \xrightarrow{\text{a.s.}} U \quad \textit{invariant.}$$

If X_1 *is integrable, then* U *is a.s. finite.*
If Y_1 *is integrable, then* $U = E^{\mathfrak{C}}X_1/E^{\mathfrak{C}}Y_1$ *a.s.*

Proof. Upon replacing C by

$$C_{ab} = \left[\liminf \frac{X^n}{Y^n} < a < b < \limsup \frac{X^n}{Y^n}\right],$$

the stationarity inequalities become

$$\int_{C_{ab}} (X_1 - bY_1) \geqq 0, \quad \int_{C_{ab}} (aY_1 - X_1) \geqq 0,$$

so that

$$\int_{C_{ab}} (a - b)Y_1 \geqq 0.$$

Since $a < b$ and $Y_1 > 0$ a.s., it follows that $PC_{ab} = 0$. Since

$$D = \left[\liminf \frac{X^n}{Y^n} \neq \limsup \frac{X^n}{Y^n}\right] = \bigcup_{a,b} C_{ab},$$

where a and $b > a$ vary over all rationals, is a countable union of null sets, $PD = 0$ and the first assertion follows. From now on, we throw out of Ω the null set D; this does not modify the values of the integrals below.

According to the stationarity inequalities, if $\int X_1^+ < \infty$, then, as $b \to \infty$,

$$\int_{[U=+\infty]} Y_1 \leqq \int_{\overline{C}_b} Y_1 \leqq \frac{1}{b}\int_{\overline{C}_b} X_1^+ \to 0,$$

and, since $Y_1 > 0$ a.s., $P[U = +\infty] = 0$; similarly, if $\int X_1^- < \infty$ then $P[U = -\infty] = 0$. The second assertion follows.

Let Y_1 be integrable and set

$$C^m = [(m-1)\epsilon \leqq U < m\epsilon], \quad \epsilon > 0,$$

so that

$$\sum_{m=-\infty}^{+\infty} C^m = [|U| < \infty].$$

The stationarity inequalities with a, b, C replaced by $m\epsilon$, $(m-1)\epsilon$, CC^m, respectively, yield

$$(m-1)\epsilon\int_{CC^m} Y_1 \leqq \int_{CC^m} X_1 \leqq m\epsilon\int_{CC^m} Y_1,$$

while, by definition of C^m,

$$(m-1)\epsilon\int_{CC^m} Y_1 \leqq \int_{CC^m} UY_1 \leqq m\epsilon\int_{CC^m} Y_1.$$

If U is finite, then, by summing over $m = 0, \pm 1, \pm 2, \cdots$ and taking into account that $\int_C X_1$ exists, and $\int_C Y_1 < \infty$, we find that

$$\int_C X_1 - \epsilon\int_C Y_1 \leqq \int_C UY_1 \leqq \int_C X_1 + \epsilon\int_C Y_1$$

and, by letting $\epsilon \to 0$, it follows that

$$\int_C UY_1 = \int_C X_1, \quad C \in \mathcal{C}.$$

Since U is $\mathcal{C}$-measurable, we can write

$$UE^{\mathcal{C}}Y_1 = E^{\mathcal{C}}(UY_1) = E^{\mathcal{C}}X_1 \text{ a.s.},$$

and, since $Y_1 > 0$ a.s. implies that $E^{\mathfrak{C}} Y_1 > 0$ a.s., we have $U = E^{\mathfrak{C}} X_1 / E^{\mathfrak{C}} Y_1$ a.s.

If U is not finite, the above equality of integrals continues to hold provided C is replaced by $C[|U| < \infty]$, so that $U = E^{\mathfrak{C}} X_1 / E^{\mathfrak{C}} Y_1$ on $[|U| < \infty]$ outside a null subset. But, by the stationarity inequalities with C replaced by $C[U = +\infty]$, we have

$$\int_{C[U=+\infty]} X_1 \geqq b \int_{C[U=+\infty]} Y_1,$$

so that, by letting $b \to \infty$, if $PC[U = +\infty] > 0$, then

$$\int_{C[U=+\infty]} E^{\mathfrak{C}} X_1 = +\infty,$$

and hence $E^{\mathfrak{C}} X_1 = +\infty$ on $[U = +\infty]$ outside a null subset. Since $E^{\mathfrak{C}} Y_1 < \infty$ a.s., it follows that

$$U = +\infty = E^{\mathfrak{C}} X_1 / E^{\mathfrak{C}} Y_1 \quad \text{on} \quad [U = +\infty]$$

outside a null subset; similarly, for $[U = -\infty]$. Thus, $U = E^{\mathfrak{C}} X_1 / E^{\mathfrak{C}} Y_1$ a.s. whether U is finite or not, and the last assertion is proved.

From now on, we consider only families consisting of one sequence $\{X_n\}$ so that the events and translations are defined in terms of the $\{X_n\}$. We use the fact that if EX_1 exists, then, by the stationarity lemma, stationarity of $\{X_n\}$ is equivalent to integral stationarity of the family $\{X_1, X_2, \cdots\}$ and $\{1, 1, \cdots\}$.

B. STATIONARITY THEOREM. *Let the family $\{X_n\}$ be stationary. If EX_1 exists, then*

$$\frac{X_1 + \cdots + X_n}{n} \xrightarrow{\text{a.s.}} E^{\mathfrak{C}} X_1.$$

If $E|X_1|^r < \infty$ for an $r \geqq 1$, then

$$\frac{X_1 + \cdots + X_n}{n} \underset{\text{a.s.}}{\xrightarrow{r}} E^{\mathfrak{C}} X_1.$$

Proof. The first assertion follows from the integral stationarity theorem with $Y_n = 1$. The second assertion will follow from the first one on account of the L_r-convergence theorem, if we prove that the rth powers of $|\overline{X^n}| = \left| \frac{1}{n} (X_1 + \cdots + X_n) \right|$ are uniformly integrable. But, by the stationarity lemma and integrability of $|X_1|^r$, for every k,

$$\int_{[|X_k|\geqq c]} |X_k|^r = \int_{[|X_1|\geqq c]} |X_1|^r < \epsilon,$$

where $\epsilon > 0$ is arbitrarily small and $c \geqq c_\epsilon$ sufficiently large. Therefore, for PA sufficiently small and for every k,

$$\int_A |X_k|^r = \int_{A[|X_k|\geqq c]} |X_k|^r + \int_{A[|X_k|<c]} |X_k|^r \leqq \epsilon + c^r PA < 2\epsilon$$

and hence, by Minkowski's inequality, $\int_A |\overline{X^n}|^r < 2\epsilon$ whatever be n. By the same lemma and inequality $E|\overline{X^n}|^r \leqq E|X_1|^r$. The assertion is proved.

COROLLARY. *If $\{X_n\}$ is stationary, then*

$$\frac{X_1 + \cdots + X_n}{n} \to E^{\mathfrak{C}}X_1^+ - E^{\mathfrak{C}}X_1^-$$

on the set on which the right-hand side generalized c.exp of X_1 exists, outside a null subset.

Apply the stationarity theorem to $\frac{1}{n}(X_1^+ + \cdots + X_n^+)$ and to $\frac{1}{n}(X_1^- + \cdots + X_n^-)$.

REMARK 1. In the case of integrable X_1, the equality $\overline{X} = \text{a.s.} \lim \overline{X^n} = E^{\mathfrak{C}}X_1$ a.s., results also from convergence in the first mean, implied, according to the above theorem, by $E|X_1| < \infty$. For then we can pass to the limit under the integration sign, so that

$$\int_C X_1 = \int_C \overline{X^n} \to \int_C \overline{X}, \quad C \in \mathfrak{C}.$$

REMARK 2. It is useful to observe that convergence in the rth mean follows from (1) bounded sequences which converge a.s. (or even only in pr.) converge in the rth mean, and (2) bounded functions are dense in L_r. The first property is immediate. The second property is exploited by setting

$$X_1 = \xi_1 + \eta_1, \quad \xi_1 = X_1 I_{[|X_1|<c]}, \quad \eta_1 = X_1 I_{[|X_1|\geqq c]},$$

so that $E|X_1|^r < \infty$ implies that as $m, n \to \infty$ then $c \to \infty$,

$$\|\overline{X^m} - \overline{X^n}\| \leqq \|\overline{\xi^m} - \overline{\xi^n}\| + \|\overline{\eta^m}\| + \|\overline{\eta^n}\|$$
$$\leqq \|\overline{\xi^m} - \overline{\xi^n}\| + 2\|\eta_1\| \to 0.$$

30.4 Applications; ergodic hypothesis and independence. We say that a sequence X_n of r.v.'s is indecomposable if all invariant functions defined on it degenerate into constants or, equivalently, if its invariant σ-field consists of $\emptyset$ and Ω only, up to an equivalence. We say that the *ergodic hypothesis* is true for the sequence X_n if $\frac{1}{n}\sum_{k=1}^{n}\xi_k \xrightarrow{\text{a.s.}} E\xi_1$ whatever be the r.v. $\xi \in L$ defined on the sequence. Since for an indecomposable sequence $E^{\mathcal{C}}\xi = E\xi$ a.s., we have, on account of the stationarity theorem,

If the sequence X_n is stationary, then the ergodic hypothesis is true if, and only if, the sequence X_n is indecomposable.

Let the r.v.'s X_n be independent. Then the sequence X_n is stationary if, and only if, the X_n are identically distributed. On the other hand, by the zero-one law, its tail σ-field reduces to $\emptyset$ and Ω up to an equivalence and, the invariant events being tail-events, the sequence is indecomposable. Thus

The ergodic hypothesis is true for sequences of independent and identically distributed r.v.'s.

In particular, if $E|X_1| < \infty$, then $\frac{1}{n}\sum_{k=1}^{n} X_k \xrightarrow{\text{a.s.}} EX_1$ finite (and, moreover, converges in the first mean). Conversely, if $\frac{1}{n}\sum_{k=1}^{n} X_k \xrightarrow{\text{a.s.}} c$ finite, then $X_n/n \xrightarrow{\text{a.s.}} 0$ and, by Borel's zero-one law,

$$E|X_1| \leqq 1 + \sum P[|X_n| \geqq n] < \infty.$$

Thus, we have Kolmogorov's strong law of large numbers. In fact, it can be made more precise, as follows:

Let the r.v.'s X_n be independent and identically distributed. If EX_1 exists, then $\frac{1}{n}\sum_{k=1}^{n} X_k \xrightarrow{\text{a.s.}} EX_1$. If $\frac{1}{n}\sum_{k=1}^{n} X_k \xrightarrow{\text{a.s.}} \overline{X}$ necessarily degenerate at some constant c, then c finite implies that EX_1 exists and is finite, while $c = +\infty$ $(-\infty)$ implies that EX_1^+ $(EX_1^-) = +\infty$.

First observe that degeneracy follows by the zero-one law.

The convergence assertion follows from the stationarity theorem.

If $\frac{1}{n}\sum_{k=1}^{n} X_k \xrightarrow{\text{a.s.}} c$ finite, then, according to the converse above, c finite implies that EX_1 exists and is finite. If $\frac{1}{n}\sum_{k=1}^{n} X_k \xrightarrow{\text{a.s.}} c = +\infty$, then

$$EX_1^+ \leftarrow \frac{1}{n}\sum_{k=1}^{n} X_k^+ = \frac{1}{n}\sum_{k=1}^{n} X_k + \frac{1}{n}\sum_{k=1}^{n} X_k^- \xrightarrow{\text{a.s.}} +\infty + EX_1^-$$

and, hence, $EX_1^+ = +\infty$. Similarly, if $c = -\infty$, then $EX_1^- = \infty$. The proposition is proved.

Because the ergodic hypothesis cannot be true for decomposable stationary sequences, the brutal answer to the ergodic problem is in the negative. However, its wreck can be salvaged in various ways. One way is to assert that it is true "in general" with a suitable definition of this term—the best is a category definition. Another approach can be stated as follows: Observe that when the sequence X_n is stationary, then, for all events B defined on it, $\frac{1}{n}\sum_{k=1}^{n} I_{B_k} \xrightarrow{\text{a.s.}} P^e B$. If the decomposition theorem 26.2**A** applies, then

$$P_{\mathcal{L}} = \sum_{t \in T} P_{B_t} \cdot I_{B_t}, \quad P_{B_t} B_t = 1,$$

provided a null event N is thrown out of Ω. Then the ergodic hypothesis is true, provided P is replaced by any of the P_{B_t} into which P is decomposed, or Ω is replaced by any of the B_t. This is the *ergodic decomposition.*

***30.5 Applications; stationary chains.** Let X_n be a sequence of r.v.'s (or more generally random vectors) with same range space R and let P_n on the Borel field $\mathcal{B}$ in R be the distribution of X_n. We recall that when the sequence X_n is a constant chain its law is described by means of the initial distribution P_1 of X_1 and the (one-step) transition probability (tr.pr.)

$$P(x, S) = P[X_m = x, X_{m+1} \in S], \quad x \in R, \quad S \in \mathcal{B}, \quad m = 1, 2, \cdots.$$

We can and do select the tr.pr. to be regular; in other words, $P(x, S)$ is a pr. in S for every fixed x and is a Borel function in x for every fixed S. Then the same is true of its iterates

$$P^n(x, S) = P[X_{m+n} \in S \mid X_m = x]$$

and

$$P^{m+n}(x, S) = \int P^m(x, dy) P^n(y, S)$$

$$P_n S = \int P_1(dx) P^{n-1}(x, S).$$

The chain is stationary if, and only if, the initial distribution is invariant under the tr.pr., that is, under translations: $P_n = \bar{P}$, $n = 1, 2, \cdots$.

From now on, we assume that the X_n form a stationary chain ($P_n = \bar{P}$, $n = 1, 2, \cdots$) with regular tr.pr. $P(x, S)$ and invariant initial distribution $\bar{P}$.

Since the stationary chain is described in terms of the common distribution $\bar{P}$ of single r.v.'s X_n and the common conditional pr.'s $P^n(x, S)$ of events defined on single r.v.'s X_{m+n} given X_m, the limit properties of the chain are essentially related to those of single r.v.'s. All the more so since the invariant events defined on the chain are a.s. defined on any single r.v., say X_1. For, every invariant event C is defined on X_n, $X_{n+1}, \cdots$ whatever be n and, because of stationarity, chain dependence, and martingale convergence theorem,

$$P^{X_1}C = P^{X_n}C = P^{X_1\cdots X_n}C \to I_C \text{ a.s.}$$

Thus, the inverse image of the sub σ-field $\mathcal{C}$ of Borel sets S such that $P(x, S) = I_S(x) = 1$ or 0 according as $x \in S$ or $x \notin S$ is equivalent to the σ-field of invariant events defined on the stationary chain X_n; by abuse of language, we shall call $\mathcal{C}$ the σ-field of *invariant Borel sets*. We are ready to investigate the asymptotic behavior of the tr.pr.'s of our stationary chain and follow Doob.

A. Invariant tr.pr. theorem. *There exists a tr.pr. $\bar{P}(x, S)$ such that*
(i) *For every S and every $x \notin N_S$ with $\bar{P}N_S = 0$*

$$\frac{1}{n}\sum_{k=1}^{n} P^k(x, S) \to \bar{P}(x, S)$$

and

$$\bar{P}(x, S) = \int \bar{P}(x, dy)P(y, S) = \int P(x, dy)\bar{P}(y, S) = \int \bar{P}(x, dy)\bar{P}(y, S).$$

(ii) *For every S and every invariant S'*

$$\bar{P}SS' = \int_{S'} \bar{P}(dx)\bar{P}(x, S).$$

The equalities in (i) say that, except for $x \in N_S$, the tr.pr.'s $\bar{P}(x, S)$ and $P(x, S)$ are invariant under one another and that $\bar{P}(x, S)$ is idempotent. Property (ii) says that $\bar{P}(x, S)$ is a Radon-Nikodym derivative of $\bar{P}$ given $\mathcal{C}$.

Proof. Since the sequence $I_{[X_n \in S]}$ of translations of the indicator $I_{[X_1 \in S]}$ is defined on the stationary sequence X_n, it follows that

$$\frac{1}{n}\sum_{k=1}^{n} I_{[X_k \in S]} \xrightarrow{\text{a.s.}} \bar{I}(S) \quad \text{invariant.}$$

Therefore, upon taking the conditional exp.'s given X_1 and applying the conditional dominated convergence theorem, we have

$$\frac{1}{n}\sum_{k=1}^{n} P^{X_1}[X_k \in S] \xrightarrow{\text{a.s.}} E^{X_1}\bar{I}(S).$$

Thus, denoting the limit by $\bar{P}(x, S)$, we have

$$\frac{1}{n}\sum_{k=1}^{n} P^k(x, S) \rightarrow \bar{P}(x, S)$$

except for $x \in N_S$ such that $\bar{P}_1 N_S = P[X_1 \in N_S] = 0$.

On account of stationarity, if S' is an invariant Borel set, then

$$\begin{aligned}\int_{S'} \bar{P}(dx)\left\{\frac{1}{n}\sum_{k=1}^{n} P^k(x, S)\right\} &= \frac{1}{n}\sum_{k=1}^{n} P[X_k \in S, X_1 \in S'] \\ &= \frac{1}{n}\sum_{k=1}^{n} P[X_k \in S, X_k \in S'] \\ &= P[X_1 \in SS'] = \bar{P}SS'.\end{aligned}$$

Therefore, upon letting $n \rightarrow \infty$ and using the dominated convergence theorem, we obtain

$$\int_{S'} \bar{P}(dx)\bar{P}(x, S) = \bar{P}SS'.$$

Since $\bar{P}(x, S)$ is a conditional pr. of the distribution $\bar{P}$ given the sub σ-field $\mathcal{C}$ of Borel sets in R, we can and do regularize it. Then $\bar{P}(x, S)$ is $\mathcal{C}$-measurable in x for every fixed S and the equalities in (i) follow from the fact that the indefinite integrals on $\mathcal{C}$ of their terms coincide. This concludes the proof.

The exceptional $\bar{P}$-null sets N_S of starting points x vary in general with the entrance sets S. The question arises how to recognize points which do not belong to the exceptional sets and to find conditions under which these sets do not vary with the entrance sets. We denote by $P_s^n(x, S)$ the $\bar{P}$-singular part of $P^n(x, S)$ and by S_n a $\bar{P}$-null set such that $P^n(x, S_n) = P_s^n(x, R)$; S_n depends upon n and x.

a. SINGULAR TR.PR. LEMMA. *For every fixed $x \in R$, the sequence $P_s^n(x, R)$ is nonincreasing and hence converges.*

For, if $m < n$ and S_0 is the $\bar{P}$-null set of points y for which $P(X_m = y; X_n \in S_n) > 0$, then

$$P_s^n(x, R) = P^n(x, S_n)$$

$$= \int P^m(x, dy)P(X_m = y; X_n \in S_n) \leqq P^m(x, S_0)$$

$$\leqq P_s^m(x, R).$$

We set $\bar{P}_s(x, R) = \lim P_s^n(x, R) = \lim P^n(x, S_n)$ and call it *singular tr.pr.*

B. VANISHING SINGULAR TR.PR. THEOREM. *For every x such that $\bar{P}_s(x, R) = 0$ and for every S,*

$$\frac{1}{n}\sum_{k=1}^{n} P^k(x, S) \rightarrow \bar{P}(x, S)$$

where $\bar{P}(x, S)$ is $\bar{P}$-continuous and

$$\bar{P}(x, S) = \int \bar{P}(x, dy)P(y, S).$$

If $\bar{P}_s(x, R) = 0$ except for $x \in N_s$ with $\bar{P}N_s = 0$, then moreover for every $x \notin N_s$ and every S, $\bar{P}(x, S)$ is idempotent:

$$\bar{P}(x, S) = \int \bar{P}(x, dy)\bar{P}(y, S).$$

Proof. Fix $x \in [\bar{P}_s(x, R) = 0]$. For $m < n$ and arbitrary S, we have

$$\frac{1}{n}\sum_{k=1}^{n} P^k(x, S) = \frac{1}{n}\sum_{k=1}^{m} P^k(x, S) + \int P^m(x, dy)\left\{\frac{1}{n}\sum_{k=1}^{n-m} P^k(y, S)\right\}.$$

According to theorem **A**, as $n \rightarrow \infty$ the integrand converges to $\bar{P}(y, S)$ except on a $\bar{P}$-null set N_S.

Since the integration measure is $\bar{P}$-continuous for events in S_m^c, we can apply the dominated convergence theorem for the integral taken over S_m^c. As for the integral taken over S_m, it is bounded by $P^m(x, S_m) = P_s^m(x, R) \rightarrow 0$ as $m \rightarrow \infty$. Therefore, by letting $n \rightarrow \infty$ and then

$m \to \infty$, the limit assertion follows. If S_0 is a $\bar{P}$-null set, then

$$\bar{P}(x, S_0) = \lim \frac{1}{n} \sum_{k=1}^{n} P^k(x, S_0) \leq \lim \frac{1}{n} \sum_{k=1}^{n} P_s{}^k(x, R) = 0$$

so that $\bar{P}(x, S)$ is $\bar{P}$-continuous. Moreover, by letting $n \to \infty$ in

$$\frac{1}{n} \sum_{k=m+1}^{m+n} P^k(x, S) = \int \left\{ \frac{1}{n} \sum_{k=1}^{n} P^k(x, dy) P^m(y, S) \right\},$$

we obtain

$$\bar{P}(x, S) = \int \bar{P}(x, dy) P^m(y, S);$$

hence

$$\bar{P}(x, S) = \int \bar{P}(x, dy) \left\{ \frac{1}{n} \sum_{k=1}^{n} P^k(y, S) \right\}.$$

Now, assume that the set of points x such that what precedes does not hold is a $\bar{P}$-null set. Because of the $\bar{P}$-continuity of $\bar{P}(x, S)$, this exceptional set is also null in the integration measure and we can apply the dominated convergence theorem. This yields idempotency of $\bar{P}(x, S)$ and concludes the proof.

COROLLARY. *If $\bar{P}_s(x, R) = 0$ for every $x \in R$, then*

$$\bar{P}(x, S) = \int P^m(x, dy) \bar{P}(y, S)$$

This follows from the limit assertion in the foregoing theorem upon letting $n \to \infty$ in

$$\frac{1}{n} \sum_{k=1+m}^{n+m} P^k(x, S) = \int P^m(x, dy) \left\{ \frac{1}{n} \sum_{k=1}^{n} P^k(y, S) \right\}.$$

C. DECOMPOSITION THEOREM. *There exists a partition*

$$R = \sum_{t \in T} S_t + N, \quad T \subset R, \quad \bar{P}N = 0$$

such that $P(x, S_t) = 1$ for every $x \in S_t$, and pr.'s $\bar{P}_t$ such that

$$\bar{P}(x, S) = \sum_{t \in T} I_{S_t}(x) \bar{P}_t S, \quad x \notin N, \quad \bar{P}_t B_t = 1$$

and the $\bar{P}_t$ are invariant:

$$\bar{P}_t S = \int \bar{P}_t(dx) P(x, S).$$

In fact, every pr. $\bar{P}'$ *of the form*

$$\bar{P}'S = \int \bar{P}_t S \mu(dt)$$

where μ *is a pr. on a* σ*-field in* T *is invariant, and if an invariant pr. is* $\bar{P}$*-continuous, then the converse is true.*

Proof. Since the Borel field $\mathfrak{B}$ is generated by a denumerable field $\{S(n)\}$ of Borel sets (say, the field of finite sums of intervals with rational extremities) and the conditional pr. $\bar{P}(x, S)$ given the σ-field of invariant Borel sets is regularized, the decomposition theorem 26.2**A** applies. This yields the asserted decomposition with invariant atoms S_t, and the asserted properties of $P(x, S_t)$ and $\bar{P}(x, S)$. As for the invariance of the $\bar{P}_t$, since, by theorem **A**, for every fixed S,

$$\bar{P}(x, S) = \int \bar{P}(x, dy)P(y, S), \quad x \notin N_S, \quad \bar{P}N_S = 0,$$

it follows that

$$\bar{P}_t S = \int \bar{P}_t(dx)P(x, S)$$

except for indices $t \in T_S \subset T$ corresponding to sets S_t of total $\bar{P}$-pr. zero. We add such sets for $S = S(1), S(2), \cdots$ to the P-null set N of the partition so that the last equality holds for all remaining t's and every $S(n)$. Since the $S(n)$ form a field, it follows as usual that the equality holds for all Borel sets S. This proves the invariance of the pr.'s P_t, and by integrating the invariance relation with respect to the pr. μ in t we find that the pr. $\bar{P}'$ is invariant:

$$\bar{P}'S = \int \bar{P}'(dx)P(x, S).$$

Conversely, if $\bar{P}'$ is an invariant pr., then for every n

$$\bar{P}'S = \int \bar{P}'(dx) \left\{ \frac{1}{n} \sum_{k=1}^{n} P^k(x, S) \right\}.$$

By theorem **A**, the integrand converges to $\bar{P}(x, S)$ except for x belonging to a $\bar{P}$-null set. Thus, if $\bar{P}'$ is $\bar{P}$-null, then the exceptional set is $\bar{P}'$-null, the dominated convergence theorem applies and

$$\bar{P}'S = \int \bar{P}'(dx)\bar{P}(x, S).$$

Upon using the decomposition of $\bar{P}(x, S)$, the converse assertion is proved, and the proof is concluded.

COROLLARY. *Every component chain* $\{\bar{P}_t, P(x, S)\}$ *is stationary and indecomposable.*

*§ 31. ERGODIC THEOREMS AND L_r-SPACES

Let $(\Omega, \mathfrak{A}, P)$ be our pr. space. In the usual ergodic theory the primary datum is a one-to-one transformation T_1 on Ω to Ω. In fact, the basic underlying concept is that of the inverse transformation T_1^{-1} operating on sets and not on points: $T_1^{-1}A = [T_1^{-1}\omega;\ \omega \in A]$. T_1^{-1} preserves all sets operations and is said to be *measurable* if it transforms measurable sets into measurable sets. But this is precisely what the translations along, say, a sequence $X_1, X_2, \cdots$ of r.v.'s do; they transform events defined on the sequence into events defined on the same sequence, that is, they transform the sub σ-field $\mathfrak{B}$ of events induced by the sequence into itself. Once the translates of these events, hence of their indicators, are determined, they determine the translates of simple and then measurable functions defined on the sequence and, in particular, determine the sequence itself, given its first term. Thus the primary datum becomes that of translations of events; it is more general than that of point transformations. In the sequel, the σ-field of events to be translated is the whole σ-field $\mathfrak{A}$ of events, but it may as well be any fixed sub σ-field $\mathfrak{B}$, whether induced by a random function or not.

31.1 Translations and their extensions. We say that a single-valued transformation T on the σ-field $\mathfrak{A}$ of events into itself is a *translation* (by 1) if it preserves (commutes with) all countable operations on events and preserves Ω and $\emptyset$: for any $A, A_j \in \mathfrak{A}$

$$T(A^c) = (TA)^c,\quad T\bigcap A_j = \bigcap TA_j,\quad T\bigcup A_j = \bigcup TA_j,$$

$$T\sum A_j = \sum TA_j,\quad T\Omega = \Omega,\quad T\emptyset = \emptyset.$$

In fact, it suffices that T preserve complementations and countable intersections (unions); for preservation of countable unions (intersections) follows by the de Morgan rules, while that of Ω, hence of $\emptyset$, and consequently of disjunctions and countable sums, follows by

$$T\Omega = T(A \cup A^c) = TA \cup TA^c = TA \cup (TA)^c = \Omega.$$

Thus, T translates the σ-field $\mathfrak{A}$ into a σ-field $T\mathfrak{A}$, then $T\mathfrak{A}$ into $T(T\mathfrak{A})$

$= T^2\mathfrak{A}$, and so on. The translation T^k (by $k = 1, 2, \cdots$) is the kth iterate of T and $T^0 = I$ is the identity transformation: $IA = A, A \in \mathfrak{A}$.

Let ξ on Ω be a measurable function. The *translate* by $k - 1$ of ξ is the measurable function $\xi_k = T^{k-1}\xi$ (so that $\xi_1 = \xi$) determined by

(1) $$[\xi_k \in S] = T^{k-1}[\xi \in S]$$

for all Borel sets $S \subset \overline{R}$; for, then, ξ_k assigns to any $\omega \in \Omega$ a value $x \in \overline{R}$ by the correspondence

$$\xi_k(\omega) = x \Leftrightarrow \omega \in T^{k-1}[\xi = x];$$

in other words, any atom $[\xi_k = x]$ of the sub σ-field of events induced by ξ_k is the translate by $k - 1$ of the atom $[\xi = x]$ of the sub σ-field of events induced by ξ. Conversely, letting r vary over the rationals, $T^{k-1}[\xi = y] = \bigcap_{r>y>s} T^{k-1}[s \leq \xi < r]$ implies that

$$[\xi_k \leq x] = \bigcap_{r>s} T^{k-1}[\xi < r] = T^{k-1}[\xi \leq x].$$

Let $\mathfrak{M}$ be the family of measurable functions on $(\Omega, \mathfrak{A})$, to be denoted by ξ with or without affixes. Let $I_{\mathfrak{A}}$ be the subfamily of indicators of events. Translation T on $\mathfrak{A}$ can also be considered as defined on $I_{\mathfrak{A}}$ (to $I_{\mathfrak{A}}$), and relation (1) extends it to a transformation T on $\mathfrak{M}$ (to $\mathfrak{M}$) which we continue to call a translation. We intend to show that this extension is linear: $T(a\xi + a'\xi') = aT\xi + a'T\xi'$ and continuous: $T(\lim \xi^{(n)}) = \lim T\xi^{(n)}$. More precisely

A. EXTENSION THEOREM. *Relation* (1) *extends the translation* T *on* $I_{\mathfrak{A}}$ *to a linear and continuous transformation* T *on* $\mathfrak{M}$, *with* $T1 = 1$ *and* $TI_{AB} = TI_A \cdot TI_B$. *Conversely, the restriction of such a transformation to* $I_{\mathfrak{A}}$ *is a translation on* $I_{\mathfrak{A}}$.

Proof. The converse assertion is immediate. As for the direct assertion, it is obvious that $T1 = 1$; linearity follows from the relations below where $a > 0$ and r varies over the set of all rationals:

$$[T(-\xi) < x] = T[-\xi < x] = T[\xi > -x] = [T\xi > -x] = [-T\xi < x],$$

$$[T(a\xi) < x] = T[a\xi < x] = T[\xi < x/a] = [T\xi < x/a] = [aT\xi < x],$$

$$[T(\xi + \xi') < x] = T[\xi + \xi' < x] = T \bigcup_r [\xi < r][\xi' < x - r]$$

$$= \bigcup_r [T\xi < r][T\xi' < x - r] = [T\xi + T\xi' < x];$$

continuity follows from the relations

$$[T \sup \xi^{(n)} > x] = T \bigcup [\xi^{(n)} > x] = \bigcup [T\xi^{(n)} > x] = [\sup T\xi^{(n)} > x]$$

which mean that T commutes with $\sup_n$, hence with $\inf_n \xi^{(n)} = -\sup_n -(\xi^{(n)})$ and, consequently, with $\lim \sup_n = \inf \sup_n$ and $\lim \inf_n = \sup \inf_n$.

COROLLARY 1. *Given a translation T on $I_{\mathfrak{A}}$, the translate (by 1) of a simple function $\sum_{k=1}^{n} x_k I_{A_k}$ is $\sum_{k=1}^{n} x_k I_{TA_k}$, and the translate of any ξ is the limit of the translates of any sequence of simple functions which converges to ξ.*

COROLLARY 2. *A translation T on $I_{\mathfrak{A}}$ has a unique extension to translation T on $\mathfrak{M}$: a linear continuous transformation with $T1 = 1$ and $TI_{AB} = TI_A \cdot TI_B$.*

This follows from Corollary 1.

31.2 A.s. ergodic theorem. From now on, T denotes a fixed translation, and we set

$$\overline{T^n} = \frac{1}{n} \sum_{k=1}^{n} T^{k-1},$$

so that

$$\overline{T^n}\xi = \overline{\xi^n} = (\xi_1 + \cdots + \xi_n)/n.$$

We denote by $\mathfrak{C}$ the sub σ-field of events invariant under T, to be called *invariant events: $TC = C$, $C \in \mathfrak{C}$*, set

$$\overline{P^n}A = E\overline{T^n}I_A = \frac{1}{n} \sum_{k=1}^{n} PT^{k-1}A, \quad A \in \mathfrak{A},$$

and observe that $\overline{P^n}$ so defined on $\mathfrak{A}$ is a pr. coinciding with P on $\mathfrak{C}$.

According to the definitions, the translates of events defined on the sequence of translates $X_n = T^{n-1}X$ of a r.v. X coincide with the translates along this sequence as defined in the preceding section. Thus, all propositions therein apply to such sequences. Yet, the primary datum being now the translation and not the sequence, the outlook changes and new problems arise:

Find conditions to be imposed upon T under which the sequences $\overline{T^n}X$ converge in some sense for (as large as possible) families of r.v.'s. Furthermore, find families for which these conditions on T are not only sufficient but also necessary for various types of convergence.

We begin by observing that, upon setting $Y_n = 1$ in the basic ergodic theorem, it yields

a. ERGODIC LEMMA. *Let*

$$\limsup \overline{P^n}A \to 0 \quad as \quad A \downarrow \emptyset.$$

Then $\overline{T^n}X \xrightarrow{\text{a.s.}} \overline{T}X$ *invariant, for every r.v.* X *such that*

$$\frac{1}{n}\int |T^{n-1}X| \to 0 \quad and \quad \frac{1}{n}\sum_{k=1}^{n}\int_{T^{k-1}A} T^{k-1}X \to 0,$$

as $n \to \infty$ *and then* $A \downarrow \emptyset$.

In particular, $\overline{T^n}X \xrightarrow{\text{a.s.}} \overline{T}X$ *bounded, for every bounded r.v.* X.

The particular case follows from $|X| \leq c < \infty$ and hence $|T^kX| \leq c$, so that the foregoing integral is bounded by c/n while the sum of integrals is bounded by $c\overline{P^n}A$.

b. INVARIANT PR. LEMMA. *The three properties below are equivalent:*

(i) $\limsup \overline{P^n}A \to 0$ *as* $A \downarrow \emptyset$.

(ii) $\lim \overline{P^n} = \overline{P}$ *exists and is a pr. on* $\mathfrak{A}$.

(iii) *There exists on* $\mathfrak{A}$ *a pr.* $\overline{P}$ *invariant under* T *and coinciding with* P *on the* σ*-field* $\mathfrak{C}$ *of invariant events (and then* $\lim \overline{P^n} = \overline{P}$*).*

We shall denote by $\overline{E}^{\mathfrak{C}}X$ the c.pr. of X given $\mathfrak{C}$ with respect to $\overline{P}$, defined by

$$\int_C \overline{E}^{\mathfrak{C}}X\,d\overline{P}_{\mathfrak{C}} = \int_C X\,d\overline{P}, \quad C \in \mathfrak{C},$$

which exists when $\int X\,d\overline{P}$ exists.

Proof. 1° Since a finite measure is continuous at $\emptyset$, (ii) implies (i). Conversely, if (i) holds, then, by the particular case of the ergodic lemma and the dominated convergence theorem,

$$\overline{P^n}A = \int \overline{T^n}I_A\,dP \to \int \overline{T}I_A\,dP = \overline{P}A, \quad A \in \mathfrak{A}.$$

Clearly, $\overline{P}$ so defined on $\mathfrak{A}$ is nonnegative and finitely additive, with $\overline{P}\Omega = 1$. Moreover, (i) becomes $\overline{P}A \to 0$ as $A \downarrow \emptyset$, so that $\overline{P}$ is also continuous at $\emptyset$. Thus, $\overline{P}$ is a pr. on $\mathfrak{A}$, and (i) implies (ii).

2° If (ii) holds, then, as $n \to \infty$,

$$|\bar{P}TA - \bar{P}A| \leftarrow |\overline{P^n}TA - \overline{P^n}A| = \frac{1}{n}|PT^nA - PA| \leqq \frac{1}{n} \to 0$$

so that $\bar{P}$ is invariant under T. Since for an invariant event C, $PC = \overline{P^n}C = \bar{P}C$, it follows that $\bar{P} = P$ on $\mathcal{C}$. Thus (ii) implies (iii).

Conversely, if there exists on $\mathcal{A}$ an invariant pr. $\bar{P}$ which coincides with P on $\mathcal{C}$, then, by the stationarity theorem, $\overline{T^n}I_A \to \bar{E}^{\mathcal{C}}I_A$ outside an invariant $\bar{P}$-null and hence P-null, event and, by the dominated convergence theorem and $P_{\mathcal{C}} = \bar{P}_{\mathcal{C}}$,

$$\overline{P^n}A = \int \overline{T^n}I_A\,dP \to \int \bar{E}^{\mathcal{C}}I_A\,dP_{\mathcal{C}} = \int \bar{E}^{\mathcal{C}}I_A\,d\bar{P}_{\mathcal{C}} = \bar{P}A.$$

Thus (iii) implies (ii), and the proof is terminated.

A. A.S. ERGODIC THEOREM. *Let*

$$\limsup \overline{P^n}A \to 0 \quad as \quad A \downarrow \emptyset.$$

Then $\overline{P^n} \to \bar{P}$ invariant pr. on $\mathcal{A}$ with $\bar{P} = P$ on $\mathcal{C}$ and

(i) *For every nonnegative r.v. X*

$$\overline{T^n}X \xrightarrow{\text{a.s.}} \bar{E}^{\mathcal{C}}X$$

while for every r.v. X

$$\overline{T^n}X \to \bar{E}^{\mathcal{C}}X^+ - \bar{E}^{\mathcal{C}}X^-$$

on the invariant set on which the right-hand side generalized c.exp. exists, outside an invariant null subset.

(ii) *If $\int X\,d\bar{P}$ exists or if the sequences $\overline{T^n}X^{\pm}$ converge in pr. (a fortiori, if they converge in the* r*th mean), then $\bar{E}^{\mathcal{C}}X$ exists and in the second case is finite, and*

$$\overline{T^n}X \xrightarrow{\text{a.s.}} \bar{E}^{\mathcal{C}}X.$$

Proof. The first assertion follows by the invariant pr. lemma. Since all sequences $\overline{T^n}X$ are $\bar{P}$-stationary, assertion (i) follows by the stationarity theorem and the fact that an invariant $\bar{P}$-null set is P-null. The first case of assertion (ii) is immediate and the second case follows from (i) by the fact that the limits of sequences of r.v.'s which converge in pr. are r.v.'s; hence $\bar{E}^{\mathcal{C}}X = \bar{E}^{\mathcal{C}}X^+ - \bar{E}^{\mathcal{C}}X^-$ a.s. exists, outside an invariant null set.

So far, whenever $\bar{P} = \lim \overline{P^n}$ appeared, we either proved or assumed that $\bar{P}$ is a pr. Since, by Complements and Details, 19, of Chapter I, the limit of the sequence of pr.'s $\overline{P^n}$ is a pr., the fact that $\bar{P} = \lim \overline{P^n}$ *implies* that $\bar{P}$ is a pr. Thus, in this subsection, we can *drop the assumption* that $\lim \overline{P^n}$ is a pr. Then, the a.s. ergodic theorem yields at once the following

B. A.S. ERGODIC CRITERION. *The sequences $\overline{T^n}X$ converge a.s. for every nonnegative r.v. X if, and only if, the sequences $\overline{P^n}A$ converge for every event A.*

The ergodic hypothesis corresponds to T being P-indecomposable, that is, the σ-field $\mathcal{C}$ of invariant sets reducing a.s. to $\emptyset$ and Ω or, equivalently, the invariant functions degenerating into constants (finite or not).

C. INDECOMPOSABILITY THEOREM. *The following properties are equivalent.*

(i) $\overline{T^n}X \xrightarrow{\text{a.s.}} \bar{T}X$ *degenerate for every* $X \geqq 0$.

(ii) T *is P-indecomposable and* $\overline{P^n} \rightarrow \bar{P}$.

(iii) $\frac{1}{n}\sum_{k=1}^{n} I_{T^{k-1}A} \xrightarrow{\text{a.s.}} \bar{P}A$ *for every* $A \in \mathcal{a}$.

(iv) $\frac{1}{n}\sum_{k=1}^{n} P(T^{k-1}A)B \longrightarrow \bar{P}A \cdot PB$ *for every pair* $A, B \in \mathcal{a}$.

Proof. (i) $\Leftrightarrow$ (ii) by the a.s. ergodic criterion.

(ii) $\Rightarrow$ (iii) by the a.s. ergodic theorem and the fact that $\bar{P}^{\mathcal{C}}A$ degenerates into $\bar{P}A$.

(iii) $\Rightarrow$ (iv) by integrating over B with respect to P and using the dominated convergence theorem.

(iv) $\Rightarrow$ (ii) by setting $B = \Omega$ so that $\overline{P^n} \rightarrow \bar{P}$ and hence $\bar{P} = P$ on $\mathcal{C}$; then, setting $A = B = C \in \mathcal{C}$ so that $PC = \bar{P}C \cdot PC = (PC)^2$, we have $PC = 0$ or 1.

The proof is complete.

31.3 Ergodic theorems on spaces L_r. We can now attack the problem of convergence a.s. or in pr. or in the rth mean of sequences $\overline{T^n}X$ to a limit $\bar{T}X \in L_r$ *for every* $X \in L_r$, $r \geqq 1$. Since a point $X \in L_r$ is an arbitrary element of a class of equivalence and not a specific r.v., the transformation $X \rightarrow \bar{T}X$ must not split classes of equivalence, that is, is to be a mapping on L_r to L_r. This is accomplished if we assume, and *we shall do so in the sequel*, that the translation T is *null-preserving*, that is, if $PA = 0$, then $PTA = 0$. For, then $X = X'$ a.s.

implies that $\overline{T^n}X = \overline{T^n}X'$ a.s.; hence $\overline{T}X = \overline{T}X'$ a.s.

We recall that $\| X \|_r = \left(\int | X |^r\right)^{\frac{1}{r}}$ is the *norm* of $X \in L_r$ $(r \geqq 1)$, and a mapping M on L_r to L_r is

linear if $M(aX + a'X') = aMX + a'MX'$ a.s., $a, a' \in R$,
nonnegative if $X \geqq 0$ a.s. $\Rightarrow MX \geqq 0$ a.s.,
bounded (or normed) if $\| MX \|_r \leqq c\| X \|_r$, where c is a finite constant independent of $X \in L_r$; the smallest of these constants is the *norm* $\| M \|_r$ of M, defined by $\| M \|_r = \sup \| MX \|_r / \| X \|_r$ for all $X \neq 0$ a.s.

We drop the subscript r when $r = 1$. Also, denoting by "c" some type of convergence, we write $M_n \xrightarrow{c} M$ on L' when $M_n X \xrightarrow{c} MX$ for every $X \in L' \subset L_r$, and drop "on L'" when $L' = L_r$.

We require three properties of linear mappings, of which the second and the third extend at once to arbitrary Banach spaces (Banach-Steinhaus) and the first extends to partially ordered Banach spaces under a supplementary assumption on the norms.

a. LINEAR MAPPINGS LEMMA. *Let M, M_n be linear mappings on L_r to L_r, $r \geqq 1$.*

(i) *If M is nonnegative, then it is bounded.*

(ii) *If the M_n are bounded and* lim sup $\| M_n X \|_r < \infty$ *for every $X \in L_r$, then they are uniformly bounded.*

(iii) *If the M_n are uniformly bounded and $M_n \xrightarrow{r} M$ on the subspace L_∞ of all bounded r.v.'s, then $M_n \xrightarrow{r} M$ on L_r.*

Proof. 1° Because of the linearity of M, to prove (i) it suffices to show that a nonnegative M is bounded on the subspace of all a.s. nonnegative $X \in L_r$. If M is not bounded, then there exists a sequence $X_n \geqq 0$ a.s. such that $\| X_n \|_r = 1$, while $\| MX_n \|_r > n^2$. Thus

$$\sum \| X_n \|_r / n^2 < \infty; \quad \text{hence} \quad X = \sum X_n / n^2 \in L_r;$$

while, by the elementary inequality $(a + b)^r \geqq a^r + b^r$, $a, b \geqq 0$, $r \geqq 1$, as $n \to \infty$

$$\int (MX)^r \geqq \int \left\{M\left(\sum_{k=1}^{n} X_k/k^2\right)\right\}^r \geqq \sum_{k=1}^{n} \int (MX_k/k^2)^r > n \to \infty.$$

Thus $\| MX \|_r = \infty$, the assertion follows *ab contrario*, and (i) is proved.

2° Let the linear mappings M_n be bounded and let $\limsup \| M_n X \|_r < \infty$ for every $X \in L_r$. To prove that all $\| M_n \| \leqq c < \infty$, it suffices to show that all $\| M_n X' \|_r \leqq c' < \infty$ for all points X' belonging to some sphere $S = [X' : | X' - X_0 | < s]$. For then, if $\| X \| < s$, we have $\| M_n X \|_r = \| M_n(X + X_0) - M_n X_0 \|_r \leqq 2c'$; hence for any $X \in L_r$

$$\| M_n X \|_r = \left\| \frac{\| X \|_r}{s} M_n \left(\frac{sX}{\| X \|_r} \right) \right\|_r \leqq \frac{2c'}{s} \| X \|_r.$$

But $L_r = \bigcup_{m=1}^{\infty} L_r^m$, where L_r^m is the closed set of those points X for which all $\| M_n X \|_r \leqq m$. Since the space L_r is complete, it follows, by Baire's category theorem, that at least one of the L_r^m is of second category and, consequently, contains a sphere S we were looking for. Assertion (ii) is proved.

3° Let all $\| M_n \|_r \leqq c < \infty$ and let $M_n \xrightarrow{r} M$ on L_∞. For every $X \in L_r$ set $X = X' + X''$, where $X' = X I_{[| X | < k]}$ and $X'' = X I_{[| X | \geqq k]}$. Then, by linearity of the M_n and completeness of L_r, as $m, n \to \infty$, then $k \to \infty$,

$$\begin{aligned} \| M_m X - M_n X \|_r &\leqq \| M_m X' - M_n X' \|_r + \| M_m X'' \|_r + \| M_n X'' \|_r \\ &\leqq \| M_m X' - M_n X' \|_r + 2c \| X'' \|_r \to 0. \end{aligned}$$

Assertion (iii) follows.

In what follows, c denotes some finite positive constant *independent of n.*

A. L_r-ERGODIC THEOREM. *The following implications hold on L_r to L_r with $r \geqq 1$:*

$$\begin{array}{ccccc} \overline{T^n} \xrightarrow{r} \overline{T} & \Rightarrow & \overline{T^n} \xrightarrow{P} \overline{T} \Leftrightarrow \overline{T^n} \xrightarrow{a.s.} \overline{T} \Leftrightarrow \limsup \| \overline{T^n} \|_r < \infty \text{ on } L_\infty \\ \Updownarrow & & \Downarrow \\ \| \overline{T^n} \|_r \leqq c \Rightarrow \overline{P^n} \leqq c P^{\frac{1}{r}} & \Rightarrow & \overline{P^n} \to \overline{P} \leqq c P^{\frac{1}{r}} \Rightarrow \overline{T^n} \underset{r}{\xrightarrow{a.s.}} \overline{T} \text{ on } L_\infty. \end{array}$$

Proof. 1° The implications

$$\overline{T^n} \xrightarrow{r} \overline{T} \Rightarrow \overline{T^n} \xrightarrow{P} \overline{T} \Leftarrow \overline{T^n} \underset{r}{\xrightarrow{a.s.}} \overline{T}$$

require no proof.

$\overline{T^n} \xrightarrow{P} \overline{T} \Rightarrow \overline{P^n} \to \overline{P} \leqq c P^{\frac{1}{r}} \Rightarrow \overline{T^n} \underset{r}{\xrightarrow{a.s.}} \overline{T}$ on L_∞. For, $\overline{T}$ being obviously a linear and nonnegative mapping on L_r to L_r, the linear mappings lemma (i) applies, so that $\| \overline{T} \|_r \leqq c$ and, by the dominated convergence theorem,

$$\overline{P^n}A = \int \overline{T^n} I_A \, dP \to \int \overline{T} I_A \, dP$$

$$\leqq \left(\int (\overline{T} I_A)^r \, dP \right)^{\frac{1}{r}} \leqq c \left(\int (I_A)^r \, dP \right)^{\frac{1}{r}} = cP^{\frac{1}{r}}A.$$

Therefore, $\limsup \overline{P^n}A \to 0$ as $A \downarrow \emptyset$ and, by the ergodic lemma and the same theorem, $\overline{T^n} \underset{r}{\xrightarrow{\text{a.s.}}} \overline{T}$ on L_∞.

$\overline{T^n} \xrightarrow{\text{P}} \overline{T} \Rightarrow \overline{T^n} \xrightarrow{\text{a.s.}} \overline{T}$ with $\overline{T} = \overline{E}^{\mathfrak{C}}$. For, then $\overline{P^n} \to \overline{P} \leqq cP^{\frac{1}{r}}$ while $\overline{T^n}X^{\pm} \xrightarrow{\text{P}} \overline{T}X^{\pm}$ for every $X \in L_r$ and hence, by the a.s. ergodic theorem, $\overline{T^n} \xrightarrow{\text{a.s.}} \overline{T}$ with $\overline{T} = \overline{E}^{\mathfrak{C}}$.

2° $\overline{T^n} \xrightarrow{\text{a.s.}} \overline{T} \Rightarrow \lim \| \overline{T^n} \|_r \leqq c$ on L_∞. For then $\overline{T^n} \xrightarrow{\text{r}} \overline{T}$ on L_∞ and hence $\| \overline{T^n} \|_r \to \| \overline{T} \|_r$ on L_∞; and we can take $c = \| \overline{T} \|_r$.

$\limsup \| \overline{T^n} \|_r \leqq c$ on $L_\infty \Rightarrow \overline{T^n} \xrightarrow{\text{a.s.}} \overline{T}$. For then, on the one hand, $\overline{P} = \limsup \overline{P^n} \leqq cP^{\frac{1}{r}}$ and hence, by 1°, $\overline{T^n} \xrightarrow{\text{r}} \overline{T}$ on L_∞, lim sup become lim, and $\| \overline{T^n} \|_r \to \| \overline{T} \|_r$ on L_∞; on the other hand, by the a.s. ergodic theorem, $\overline{T^n}X \xrightarrow{\text{a.s.}} \overline{E}^{\mathfrak{C}}X$ for every nonnegative r.v. X. Thus, to prove the assertion, it suffices to show that, for $0 \leqq X \in L_r$, we have $\overline{E}^{\mathfrak{C}}X \in L_r$. But, setting $X_m = XI_{[|X|<m]}$ so that $0 \leqq X_m \uparrow X$ as $m \to \infty$ and $X_m \in L_\infty$, we have, by what precedes, $\int (\overline{E}^{\mathfrak{C}}X_m)^r \, dP \leqq c^r \int (X_m)^r \, dP$. Letting $m \to \infty$, it follows, by the monotone and conditional monotone convergence theorems, that

$$\int (\overline{E}^{\mathfrak{C}}X)^r \, dP \leqq c^r \int (X)^r \, dP < \infty.$$

Thus, $\overline{E}^{\mathfrak{C}}X \in L_r$, and the proof is complete.

3° $\overline{T^n} \xrightarrow{\text{r}} \overline{T} \Rightarrow \| \overline{T^n} \|_r \leqq c$. For, then, clearly for every $X \in L_r$, $TX, T^2X, \cdots \in L_r$ and hence $\overline{T^n}X \in L_r$. Thus every $\overline{T^n}$ is a mapping on L_r to L_r and, these mappings being obviously linear and nonnegative, the linear mappings lemma (i) applies. Therefore, every $\overline{T^n}$ is bounded, while $\limsup \| \overline{T^n}X \|_r = \| \overline{T}X \|_r < \infty$ and hence, by the linear mappings lemma (ii), all $\| \overline{T^n} \|_r \leqq c$.

$\|\overline{T^n}\|_r \leqq c \Rightarrow \overline{P^n} \leqq cP^{\frac{1}{r}} \Rightarrow \overline{P^n} \to \overline{P} \leqq cP^{\frac{1}{r}}$. For, then

$$\overline{P^n}I_A = \int \overline{T^n}I_A\,dP \leqq \left(\int (\overline{T^n}I_A)^r\,dP\right)^{\frac{1}{r}} \leqq c\left(\int (I_A)^r\,dP\right)^{\frac{1}{r}} = cP^{\frac{1}{r}}A,$$

and hence

$$\limsup \overline{P^n}A \leqq cP^{\frac{1}{r}}A \to 0 \quad \text{as} \quad A \downarrow \emptyset.$$

Thus, by the invariant pr. lemma, $\overline{P^n} \to \overline{P} \leqq cP^{\frac{1}{r}}$.

$\|\overline{T^n}\|_r \leqq c \Rightarrow \overline{T^n} \xrightarrow{r} \overline{T}$. For, then $\overline{T^n} \xrightarrow{\text{a.s.}} \overline{T}$ on L_∞ and, hence, by the dominated convergence theorem, $\overline{T^n} \xrightarrow{r} \overline{T}$ on L_∞, the linear mappings lemma (iii) applies, and $\overline{T^n} \xrightarrow{r} \overline{T}$. The theorem is proved.

B. L_r-ERGODIC CRITERIA. *The following equivalences hold on L_r, $r \geqq 1$.*

A.s. ergodic criterion:

$$\overline{T^n} \xrightarrow{P} \overline{T} \Leftrightarrow \overline{T^n} \xrightarrow{\text{a.s.}} \overline{T} \Leftrightarrow \limsup \|\overline{T^n}\|_r \leqq c \textit{ on } L_\infty.$$

Mean ergodic criterion:

$$\overline{T^n} \xrightarrow{r} \overline{T} \Leftrightarrow \sup \|\overline{T^n}\|_r \leqq c.$$

And for $r = 1$:

$$\overline{T^n} \xrightarrow{\text{a.s.}} \overline{T} \Leftrightarrow \overline{P^n} \to \overline{P} \leqq cP$$

$$\overline{T^n} \xrightarrow{1} \overline{T} \Leftrightarrow \sup \overline{P^n} \leqq cP$$

Proof. According to the L_r-ergodic theorem, it suffices to prove that

$$\overline{P^n} \to \overline{P} \leqq cP \Rightarrow \overline{T^n} \xrightarrow{\text{a.s.}} \overline{T} \quad \text{and} \quad \overline{P^n} \leqq cP \Rightarrow \|\overline{T^n}\| \leqq c.$$

The first implication follows by the a.s. ergodic theorem from the inequality

$$\int |X|\,d\overline{P} \leqq c\int |X|\,dP < \infty$$

which holds by hypothesis for all indicators $X = I_A$; hence, as usual, it holds for simple r.v. and then for any r.v. $X \in L$. Similarly, the second assertion follows by the L_r-ergodic theorem from the inequality

$$\int |\overline{T^n}X|\,dP \leqq c\int |X|\,dP$$

which holds by hypothesis for all indicators and hence for all r.v.'s $X \in L$.

REMARK. In this subsection, the translation T was assumed to be null-preserving, that is, $PA = 0$ implies that $PTA = PT^2A = \cdots = 0$. Thus, every $\overline{P^n}$ is P-continuous and, if $\overline{P^n} \to \bar{P}$ pr. on $\mathfrak{A}$, then $\bar{P}$ is P-continuous. In other words, we can select nonnegative integrable r.v.'s $\overline{p^n}$ and $\bar{p}$ (with $E\overline{p^n} = E\bar{p} = 1$) such that

$$\overline{P^n}A = \int_A \overline{p^n}\,dP, \quad \bar{P}A = \int_A \bar{p}\,dP, \quad A \in \mathfrak{A}.$$

The reader is invited to play around with these r.v.'s and the results of this subsection. For example, the following relations hold:

$$E^{\mathfrak{C}}\overline{p^n} = E^{\mathfrak{C}}\bar{p} = 1 \text{ a.s.}, \quad \bar{T}X = \bar{E}^{\mathfrak{C}}X = E^{\mathfrak{C}}(\bar{p}X) \text{ a.s.};$$

$$\overline{P^n} \to \bar{P} \text{ pr.} \Leftarrow \overline{p^n} \xrightarrow{1} \bar{p} \Leftarrow \overline{p^n} \xrightarrow{P} \bar{p};$$

$$\overline{P^n} \leqq cP \Leftrightarrow \overline{p^n} \leqq c \text{ a.s.}, \quad \bar{P} \leqq cP \Leftrightarrow \bar{p} \leqq c \text{ a.s.}$$

*§ 32. ERGODIC THEOREMS ON BANACH SPACES

The implication $\|\overline{T^n}\|_r \leqq c \Rightarrow \|\overline{T^n}X - \bar{T}X\|_r \to 0$ for every $X \in L_r$ remains meaningful for transformations T on a Banach space B of points X to itself, provided the norms are interpreted as norms in B. The question arises whether a similar ergodic implication can be obtained for Banach spaces and this *without reference to an underlying pr. space*. It is to be expected that some supplementary condition will be required; for sequences of bounded or uniformly bounded transformations have compactness properties in L_r that they do not have in general Banach spaces. We shall follow Yosida and Kakutani.

32.1 Norms ergodic theorem. Let T be a linear transformation on a Banach space B to itself, and set $\overline{T^n} = \frac{1}{n}\sum_{k=1}^{n} T^{k-1}$. Let X, with or without affixes, denote a point in B and let c be some finite positive constant independent of n.

We say that a sequence X_n converges *strongly* to X and write $X_n \xrightarrow{s} X$, if the convergence is in norm, that is, if $\|X_n - X\| \to 0$. If there exists a transformation $\bar{T}$ on B to B such that $\overline{T^n}X \xrightarrow{s} \bar{T}X$ for every $X \in B$, then we write $\overline{T^n} \xrightarrow{s} \bar{T}$. A sequence X_n converges *weakly* to X if $f(X_n) \to f(X)$ for all bounded linear functionals f on B. The term "weakly" as opposed to "strongly" is justified by the fact that $X_n \xrightarrow{s} X$ implies $X_n \xrightarrow{w} X$, for

$$|f(X_n) - f(X)| = |f(X_n - X)| \leqq \|f\|\cdot\|X_n - X\|.$$

The *null-transformation*, to be denoted by 0, maps every $X \in B$ onto the null element $\theta \in B$.

A subset B' is *range* of a transformation T' on B to B if T' maps B onto B', and we write $B' = T'B$.

B' is *linear* if it is closed under all (finite) linear combinations of its elements; observe that if T' is linear, then $T'B$ is a linear subspace.

B' is *weakly* (*strongly*) *closed* if it is closed under weak (strong) passages to the limit; observe that a strongly closed linear subspace is a Banach space. We denote the strong closure of B' by $\overline{B'}$. In fact, the strong closure of B' is also its weak closure. For, if $X_n \in B'$ and $X_n \xrightarrow{w} X$, then $X \notin \overline{B'}$ implies that the distance d of X to $\overline{B'}$ is positive. But, by Corollary 2 of the Hahn-Banach theorem, there exists a functional f such that $0 = f(X_n) \to f(X) = d > 0$, and we reach a contradiction.

B' is *weakly* (*strongly*) *compact* if every sequence in B' contains a weakly (strongly) convergent subsequence; observe that strong compactness implies weak compactness.

a. CONVERGENCE LEMMA. *Let all* $\|T^n\| \leqq c$. *Then*

$$(T - I)\overline{T^n} = \overline{T^n}(T - I) \xrightarrow{s} 0$$

and

$$\overline{T^n} \xrightarrow{s} 0 \quad on \quad \overline{(T - I)B}.$$

Proof. Since for every $X \in B$

$$\|(T - I)\overline{T^n}X\| = \|\overline{T^n}(T - I)X\| = \frac{1}{n}\|T^nX - X\| \to 0,$$

the first assertion is true. Since, given $\epsilon > 0$, for every $X \in \overline{(T - I)B}$ there exists an $X' \in (T - I)B$ such that $\|X - X'\| < \epsilon$ and there exists an $X'' \in B$ such that $X' = (T - I)X''$, it follows that, as $n \to \infty$ and then $\epsilon \to 0$,

$$\begin{aligned}\|\overline{T^n}X\| &\leqq \|\overline{T^n}X'\| + \|\overline{T^n}(X - X')\|\\ &\leqq \|\overline{T^n}(T - I)X''\| + c\epsilon \to 0,\end{aligned}$$

and the second assertion is proved.

b. NORMS ERGODIC LEMMA. *Let all* $\|T^n\| \leqq c$. *Then every weakly compact sequence* $\overline{T^n}X$ *converges strongly to a point invariant under* T.

Proof. If a subsequence $\overline{T^{n'}}X \xrightarrow{w} \overline{X}$ as $n' \to \infty$, then $(T - I)\overline{T^{n'}}X \xrightarrow{w} (T - I)\overline{X}$; hence, by the first part of the convergence lemma, $(T - I)\overline{X} = \theta$, that is, $T\overline{X} = \overline{X}$. Therefore, $\overline{T^n}\overline{X} = \overline{X}$ whatever be n and, setting $X = \overline{X} + (X - \overline{X})$, it remains to be proved that $\overline{T^n}(X - \overline{X}) \xrightarrow{s} 0$. Because of the second part of the convergence lemma, it suffices to prove that $X - \overline{X} \in \overline{(T - I)B}$. But

$$(\overline{T^n} - I)X = (T - I)\left(\frac{n-1}{n}I + \frac{n-2}{n}T + \cdots + \frac{1}{n}T^{n-2}\right)X$$

so that all $(\overline{T^n} - I)X \in (T - I)B$, while a subsequence $(\overline{T^{n'}} - I)X \xrightarrow{w} \overline{X} - X$ as $n' \to \infty$. It follows that $\overline{X} - X \in \overline{(T - I)B}$, and the proof is concluded.

A. NORMS ERGODIC THEOREM. *Let all* $\| T^n \| \leqq c$. *If all sequences* $\overline{T^n}X$ *are weakly compact, then* $\overline{T^n} \xrightarrow{s} \overline{T}$ *linear and*

$$\| \overline{T} \| \leqq c, \quad \overline{T}T = T\overline{T} = \overline{T}\overline{T} = \overline{T}.$$

Proof. According to the norms ergodic lemma, every sequence $\overline{T^n}X$ converges strongly so that the passage to the limit is a transformation $\overline{T}$ on B to B, obviously linear and of norm bounded by c. Since, on account of the convergence lemma,

$$0 \xleftarrow{s} \overline{T^n}(T - I) \xrightarrow{s} \overline{T}(T - I) = (T - I)\overline{T},$$

it follows that

$$\overline{T} = \overline{T}T = T\overline{T} = \overline{T^n}\overline{T} \xrightarrow{s} \overline{T}\overline{T},$$

and the proof is concluded.

The set B_λ of all points X such that $TX = \lambda X$ is said to be the *proper subspace* of T corresponding to the *proper value* λ of T (λ real or complex), provided this set does not consist of the null-point θ only. Thus, $T - \lambda I = 0$ on $B_\lambda \neq \{\theta\}$, and λ is not a proper value of T if, and only if, $TX = \lambda X$ implies that $X = \theta$. Since, in this section, T is bounded and linear, every B_λ is, clearly, a strongly closed linear subspace of B.

COROLLARY. *In the norms ergodic theorem,* $\overline{T} \neq 0$ *if and only if* $\lambda = 1$ *is a proper value of* T, *and then* $\overline{T}B = B_1$.

This follows by the implications

$$TX = X \Rightarrow \overline{T^n}X = X \Rightarrow \overline{T}X = X$$

and

$$\overline{T}X = X \Rightarrow TX = T\overline{T}X = \overline{T}X = X.$$

B. Extended norms ergodic theorem. *Let all* $\| T^n \| \leqq c$ *and let all sequences* $\overline{T^n}_\lambda X$ *be weakly compact with* $T_\lambda = T/\lambda$, *where* λ *with or without affixes have modulus* 1. *Then*

(i) $\overline{T_\lambda{}^n} \xrightarrow{s} \overline{T}_\lambda$ *linear,*

$$\| \overline{T}_\lambda \| \leqq c, \quad \overline{T}_\lambda T_\lambda = T_\lambda \overline{T}_\lambda = \overline{T}_\lambda \overline{T}_\lambda = \overline{T}_\lambda, \quad \overline{T}_\lambda \overline{T}_{\lambda'} = 0 \quad \text{for} \quad \lambda \neq \lambda',$$

$\overline{T}_\lambda \neq 0 \Leftrightarrow \lambda$ *is a proper value of* T, *and then* $B_\lambda = \overline{T}_\lambda B$.

(ii) *If* $T' = T - \sum_{j=1}^{m} \lambda_j \overline{T}_{\lambda_j}$ *with all* $\overline{T}_{\lambda_j} \neq 0$ *and all* λ_j *distinct, then*

$$T^n = T'^n + \sum_{j=1}^{m} \lambda_j{}^n \overline{T}_{\lambda_j}, \quad \overline{T}_{\lambda_j} T' = T' \overline{T}_{\lambda_j} = 0, \quad TT' = T'T = T'T',$$

λ *is a proper value of* $T' \Leftrightarrow \lambda$ *is a proper value of* T *and all* $\lambda_j \neq \lambda$.

Proof. The norms ergodic theorem and its corollary, with T replaced by $T_\lambda = T/\lambda$, yield directly properties (i) except for $\overline{T}_\lambda \overline{T}_{\lambda'} = 0$, $\lambda \neq \lambda'$. The latter follows from

$$\overline{T}_\lambda \overline{T}_{\lambda'} \xleftarrow{s} \overline{T^n_\lambda} \overline{T}_{\lambda'} = \frac{1}{n} \left\{ \sum_{k=1}^{n} (\lambda'/\lambda)^{k-1} \right\} \overline{T}_{\lambda'} \xrightarrow{s} 0.$$

Properties (ii) follow from (i) by elementary computations; only the last one deserves a proof. Let all $\overline{T}_{\lambda_j} \neq 0$, the λ_j be distinct, and $X \neq \theta$. If $TX = \lambda X$ and all $\lambda_j \neq \lambda$, then, by (i), $\overline{T}_{\lambda_j} X = \overline{T}_{\lambda_j} T_\lambda X = 0$, and forming $T'X$ it follows that $T'X = TX = \lambda X$. Conversely, if $T'X = \lambda X$, then

$$TX = TT'X/\lambda = T'TX/\lambda = T'T'X/\lambda = \lambda X;$$

moreover, if $\lambda_j = \lambda$, then by (i) and the second relation in (ii),

$$\theta \neq X = \overline{T^n_{\lambda_j}} X \xrightarrow{s} \overline{T}_{\lambda_j} X = \overline{T}_{\lambda_j} T'X/\lambda = 0,$$

and the converse follows *ab contrario*.

Remark. The condition (i) $\sup \| T^n \| < \infty$ implies that (i′) $\sup \| \overline{T^n} \| < \infty$ and (i″) $\| T^n \|/n \to 0$. In fact, the foregoing proofs and hence the propositions remain valid if (i) is replaced by (i′) and (i″). Furthermore, if all sequences $\overline{T^n}X$ are weakly compact, then it is easily proved that $\sup \| \overline{T^n} \| < \infty$ (use the Banach-Steinhaus parts of the linear mappings lemma extended to Banach spaces). Thus, in the norms ergodic theorem, $\sup \| T^n \| < \infty$ can be replaced by $\| T^n \|/n \to 0$. This weakening is due to Dunford.

As for the converse, let us only mention that, if $\overline{T^n} \to \overline{T}$, then, by the same lemma, $\sup \| \overline{T^n} \| < \infty$. Also, if the Banach space is reflexive, that is, can be identified with its second adjoint (for example, if it is a space L_r with $r > 1$), then the condition $\sup \| \overline{T^n} \| < \infty$ implies that every sequence $\overline{T^n} X$ is weakly compact.

32.2 Uniform norms ergodic theorems. We recall that in this section all mappings are bounded linear ones on B to B.

The weak compactness assumption in the norms ergodic theorem is fulfilled, under the condition that all $\| T^n \| \leqq c$, if T is *weakly* (*strongly*) *compact*, that is, maps the sphere of points X with $\| X \| \leqq 1$ onto a weakly (strongly) compact set. In fact, as we shall see, it suffices that T be *quasi-weakly* (*strongly*) *compact*, that is, that there be an integer h such that $T^h = V + W$, where V is weakly (strongly) compact and $\| W \| \leqq w < 1$. And it is to be expected that in the quasi-strongly compact case the norms ergodic theorem can be made more precise; it will become the uniform norms ergodic theorem toward which this subsection is directed. This theorem was first given by Krylov and Bogoliubov for Markov chains.

Observe that, if M_1 and M_2 are weakly (strongly) compact and M is bounded, then $M_1 + M_2$ as well as MM_1 and M_1M are weakly (strongly) compact. Therefore, when T is quasi-weakly (strongly) compact, then for every integer k

(Q) $$T^{hk} = V_k + W^k, \quad \| W^k \| \leqq w^k, \quad w < 1,$$

where $V_k = (V + W)^k - W^k$, being a finite sum of terms with at least one weakly (strongly) compact factor V, is itself weakly (strongly) compact.

A. *Let all* $\| T^n \| \leqq c$ *and let* T *be quasi-weakly compact. Then the conclusions of the extended norms ergodic theorem are valid.*

Proof. It suffices to prove that all sequences $\overline{T^n} X$ are weakly compact. We use repeatedly the relation (Q), denote by f an arbitrary bounded linear functional on B, and set $Y_{m,p} = p\overline{T^p}X/m$. Thus

$$\overline{T^m} X = \frac{p}{m} \overline{T^p} X + \frac{m-p}{m} T^p \overline{T^{m-p}} X, \quad m > p,$$

will become

$$\overline{T^m} X = Y_{m,hk} + V_k Y_{m,m-hk} + W^k Y_{m,m-hk}, \quad m > hk.$$

Since V_k is weakly compact, there exists a subsequence $Y_n = V_k Y_{m_n, m_n - hk}$

such that $f(Y_n) \to f(\overline{Y}_k)$, where $\overline{Y}_k$ is some point which may depend upon k. Since

$$|f(Y_{m_n, hk})| \leqq \frac{chk}{m_n} \|f\| \cdot \|X\| \to 0$$

and

$$|f(W^k Y_{m_n, m_n - hk})| \leqq cw^k \|f\| \cdot \|X\|,$$

it follows that

$$\limsup |f(Y_n) - f(\overline{Y}_k)| \leqq cw^k \|f\| \cdot \|X\|.$$

By selecting successive subsequences for $k = 1, 2, \cdots$ and applying the diagonal procedure, we may assume that this inequality holds for all k whatever be f. Since, on account of a corollary of the Hahn-Banach theorem, there exists an f such that $|f(\overline{Y}_k) - f(\overline{Y}_l)| = \|\overline{Y}_k - \overline{Y}_l\|$, it follows that, as $k, l \to \infty$,

$$\|\overline{Y}_k - \overline{Y}_l\| \leqq c(w^k + w^l) \|f\| \cdot \|X\| \to 0$$

so that $\overline{Y}_k \xrightarrow{s} \overline{Y} \in B$ and, whatever be f,

$$\limsup |f(Y_n) - f(Y)| \leqq cw^k \|f\| \cdot \|X\| + |f(Y_k - Y)| \to 0.$$

Thus $f(Y_n) \to f(Y)$, and the assertion is proved.

To pass to the quasi-strongly compact case, we require the following properties.

a. *If B' is a strict Banach subspace of B, then for every $\epsilon \in (0, 1)$ there exists an $X \in B$ such that*

$$\|X\| = 1, \quad \|X - X'\| \geqq 1 - \epsilon \quad \textit{for all} \quad X' \in B'.$$

Proof. There exists a point $Y \notin B'$ such that $d = \inf_{Y' \in B'} \|Y - Y'\| > 0$ and hence, given $\epsilon' = d\epsilon/1 - \epsilon$, there exists a point $Y_0 \in B'$ such that $d \leqq \|Y - Y_0\| < d + \epsilon'$. Therefore, if $X = (Y - Y_0)/\|Y - Y_0\|$ and $X' \in B'$, then $Y' = Y_0 + \|Y - Y_0\| X' \in B'$ and

$$\|X\| = 1,$$

$$\|X - X'\| = \|Y - Y'\| / \|Y - Y_0\| \geqq d/(d + \epsilon') \doteq 1 - \epsilon.$$

A set S of linearly independent points of B *generates* (*or spans*) a Banach subspace, to be denoted by $B(S)$, if all points of the subspace are linear combinations, or strong limits thereof, of the points of S; when S is a finite set, $B(S)$ is said to be *finite-dimensional.*

b. *If $X_1, X_2, \cdots$ are linearly independent, then there exist points $Y_n \in B(X_1, \cdots, X_n)$ such that $\| Y_n \| = 1$ and $\| X_m - Y_n \| \geqq \frac{1}{2}$ for every $X \in B(X_1, \cdots, X_m)$ and $m < n$.*

It suffices to apply the foregoing lemma with $\epsilon = \frac{1}{2}$ to the strictly increasing sequence of Banach spaces $B(X_1, \cdots, X_n) \subset B$, starting with $Y_1 = X_1/\| X_1 \|$.

c. *Let T be quasi-strongly compact. Then:*

(i) *No sequence of distinct proper values of T may converge to a limit λ with $|\lambda| \geqq 1$, so that the number of distinct proper values of T of modulus 1 is finite and they are isolated proper values.*

(ii) *The proper subspaces B_λ of T with $|\lambda| \geqq 1$ are finite-dimensional.*

Proof. Because of (Q) we can assume, without loss of generality, that $T^h = V + W$, where V is strongly compact and $\| W \| < \frac{1}{4}$.

1° Suppose there is a sequence $\lambda_n \to \lambda$ of proper values λ_n, and $|\lambda| \geqq 1$. Then there is a sequence $X_n \neq \theta$ such that $TX_n = \lambda_n X_n$. If the X_n are not linearly independent, then there exists a smallest integer n such that $X_1, \cdots, X_n$ are linearly independent and $X_{n+1} = \sum_{k=1}^{n} c_k X_k$ with at least one $c_k \neq 0$. But then

$$\sum_{k=1}^{n} c_k(\lambda_{n+1} - \lambda_k)X_k = T(X_{n+1} - \sum_{k=1}^{n} c_k X_k) = 0,$$

so that $X_1, \cdots, X_n$ are linearly dependent and we reach a contradiction. Thus, the X_n are linearly independent and, by **B**, there exist $Y_n = \sum_{k=1}^{n} c_k X_k$ such that $\| Y_n \| = 1$ and $\| X - Y_n \| \geqq \frac{1}{2}$ for every $X \in B(X_1, \cdots, X_m)$ and $m < n$. Let $Z_n = Y_n/\lambda_n{}^h$ and observe that $Y_n - T^h Z_n = \sum_{k=1}^{n-1} d_k X_k$ so that $\| T^h(Z_m - Z_n) \| \geqq \frac{1}{2}$. Since the sequence Z_n is uniformly bounded, there exists a subsequence $Z'_n = Z_{k_n}$ such that $V(Z'_m - Z'_n) \to 0$ as $m \to \infty$, and hence

$$\| T^h(Z'_m - Z'_n) \| \leqq \| V(Z'_m - Z'_n) \| \\ + \| W \| (|\lambda_{k_m}{}^{-h}| + |\lambda_{k_n}{}^{-h}|) \to 2 \| W \| |\lambda^{-h}| \geqq 2 \| W \|.$$

It follows that $2 \| W \| \geqq \frac{1}{2}$. Since $\| W \| < \frac{1}{4}$, we reach a contradiction and (i) follows.

2° If the proper space B_λ with $|\lambda| \geqq 1$ is not finite-dimensional, then it contains a sequence X_n of linearly independent points, the argument in 1° applies with $\lambda_n = \lambda$ for every n, and (ii) follows.

d. *Let all* $\|T\|'^n \leqq c' < \infty$ *and let* T' *be quasi-strongly compact. If* T' *has no proper value of modulus* 1, *then there exists two constants* c_1 *and* $c_2 \in (0, 1)$ *such that for all* n

$$\| T'^n \| \leqq c_1(1 - c_2)^n.$$

We give only an indication of the proof. To begin with, T' has no proper values λ with $|\lambda| > 1$. Otherwise, $T'X = \lambda X$ for an $X \neq \theta$ hence $\|T'^n X\| = |\lambda^n| \cdot \|X\| \to \infty$ as $n \to \infty$, and this contradicts $\|T'^n X\| \leqq c' \|X\|$, $c' < \infty$. Since T' has no proper value of modulus 1, it follows, by the preceding lemma, that it has no proper value λ with $|\lambda| \geqq 1 - \gamma$, for some $\gamma > 0$. Now, it is easily seen that it suffices to consider the case $T' = V + W$ where V is strongly compact and $\|W\| \leqq w < 1$. It can then be proved that $T' - \lambda I$ has an inverse for $|\lambda| \geqq \max(1 - \gamma, \frac{1}{2}w) = 1 - c_2$, so that the series $I + \sum T'_\lambda{}^n$ converges ($T'_\lambda = T'/\lambda$). Therefore there exists a c_1 such that $\|T'^n\| \leqq c_1(1 - c_2)^n$.

B. Uniform ergodic theorem. *Let all* $\|T^n\| \leqq c$ *and let* T *be quasi-strongly compact. Then:*

(i) *The conclusions of the extended norms ergodic theorem hold.*

(ii) T *can have only a finite number* m *of proper values* λ_j *of modulus* 1 (*and has no proper values of modulus greater than* 1).

For every λ *of modulus* 1 *there exists a finite constant* c' *such that for all* n

$$\| \overline{T_\lambda^n} - \overline{T}_\lambda \| \leqq c'/n,$$

$\overline{T}_\lambda \neq 0$ *if, and only if, some* $\lambda_j = \lambda$, *and every* $\overline{T}_{\lambda_j}$ *maps* B *onto the finite-dimensional proper subspace* B_{λ_j} *of* T.

(iii) *If* $T' = T - \sum_{j=1}^{m} \lambda_j \overline{T}_{\lambda_j}$, *then* T' *is quasi-strongly compact and the* $\|T'^n\|$ *are uniformly bounded, in fact there exist constants* c_1 *and* $c_2 \in (0, 1)$ *such that for all* n

$$\| T'^n \| \leqq c_1(1 - c_2)^n.$$

Proof. Since a quasi-strongly compact T is quasi-weakly compact, (i) follows by theorem **A**. Together with lemmas **c** and **d**, (i) implies (ii) and (iii) by elementary computations. For (iii), observe that

$$\| T'^n \| \leqq \| T^n \| + \sum_{j=1}^{m} \| \overline{T}_{\lambda_j} \| \leqq (m+1)c = c'$$

while, if $T^h = V + W$ and $V' = V - \sum_{j=1}^{m} \lambda_j{}^h \overline{T}_{\lambda_j}$, then $\| T'^h - V' \| = \| T^h - V \|$. For (ii), observe that

$$\overline{T^n_\lambda} - \overline{T}_\lambda = \overline{T'^n}_\lambda + \sum_{j=1}^{m} \left(\frac{1}{n} \sum_{k=1}^{n} (\lambda_j/\lambda)^k \overline{T}_{\lambda_j} \right) - \overline{T}_\lambda,$$

where $\| \overline{T'^n} \| \to 0$ faster than any fixed power of $1/n$, $\overline{T}_\lambda = 0$ or $\overline{T}_{\lambda_j}$ according as all $\lambda_j \neq \lambda$ or a $\lambda_j = \lambda$, and $\frac{1}{n} \sum_{k=1}^{n} (\lambda_j/\lambda)^k = O(1/n)$ or 1 according as $\lambda_j \neq \lambda$ or $\lambda_j = \lambda$. One may also (take $\lambda = 1$ to simplify the writing) observe that

$$\overline{T^n} - \overline{T} = (I - \overline{T})\overline{T^n}, \quad (T^n - I)/n = (T - I)(I - \overline{T})\overline{T^n},$$

so that

$$\| (T - I)(\overline{T^n} - \overline{T}) \| = \| (T^n - I)/n \| \leqq (c+1)/n;$$

then examine boundedness of $(T - I)^{-1}$ on $B_1{}^c$.

32.3 Application to constant chains. Let $P^n(x, S)$ be an n-step transition pr.f. from a point $x \in R$ into a Borel set $S \subset R$. It is a Borel function in x for every fixed S, a pr. in S for every fixed x, and

$$P^{m+n}(x, S) = \int P^m(x, dy) P^n(y, S), \quad m, n = 1, 2, \cdots.$$

It is easily checked that:

1. The space of all complex valued σ-additive functions φ of bounded variation on the Borel field in R is a Banach space B when the norm $\| \varphi \|$ of φ is defined by $\| \varphi \| = \text{Var}\, \varphi$.

2. The transformation T with *kernel* $P(x, S)$, defined by

$$\varphi \to T\varphi = \int \varphi(dx) P(x, S)$$

is a linear transformation on B to B of norm 1; its nth iterate T^n is given by

$$\varphi \to T^n \varphi = \int \varphi(dx) P^n(x, S)$$

and has norm 1 and kernel $P^n(x, S)$.

We say that the kernel $P(x, S)$ and the chain are *quasi-strongly compact* if T is quasi-strongly compact. The uniform ergodic theorem, where, obviously, we can replace $\overline{T^n} = \frac{1}{n}\sum_{k=1}^{n} T^{k-1}$ by $\frac{1}{n}\sum_{k=1}^{n} T^k$ without changing the conclusions, yields at once, on account of uniformity in all σ-additive functions of bounded variation, the following result. We set

$$\overline{P^n}_\lambda(x, S) = \frac{1}{n}\sum_{k=1}^{n} P^k(x, S)/\lambda^k, \quad \bar{P}_j(x, S) = \bar{P}_{\lambda_j}(x, S),$$

and omit the subscript if it equals 1.

Let the chain be quasi-strongly compact. Then:

(i) *There exists only a finite number m of proper values λ_j of modulus 1 with corresponding finite-dimensional proper subspaces B_{λ_j} of the transformation.*

(ii) *For every λ of modulus* 1, *there exists a constant c' such that for all n*

$$\sup_{x,S} |\overline{P^n}_\lambda(x, S) - \bar{P}_\lambda(x, S)| \leqq c'/n$$

and

$$\int P^n(x, dy)\bar{P}_\lambda(x, S) = \int \bar{P}_\lambda(x, dy)P^n(y, S) = \lambda^n \bar{P}_\lambda(x, S)$$

$$\int \bar{P}_\lambda(x, dy)\bar{P}_{\lambda'}(y, S) = \delta_{\lambda\lambda'}\bar{P}_\lambda(x, S).$$

(iii) $\bar{P}_\lambda(x, S) \neq 0$ *if, and only if, some* $\lambda_j = \lambda$,

$$P^n(x, S) = \sum_{j=1}^{m} \lambda_j^n \bar{P}_j(x, S) + P'^n(x, S)$$

with

$$\int \bar{P}_j(x, dy)P'^n(y, S) = \int P'^n(x, dy)\bar{P}_j(y, S)$$

and constants c_1 and $c_2 \in (0, 1)$ such that for all n

$$\sup_{x,S} |P'^n(x, S)| \leqq c_1(1 - c_2)^n.$$

Let us examine the case $\lambda = 1$. On account of (ii) $\lambda = 1$ *is a proper value of the transformation, and $\bar{P}(x, S)$ is a transition pr.f. which, for every fixed x, is an invariant element of the Banach space, with*

$$\sup_{x,S} |\overline{P^n}(x, S) - \bar{P}(x, S)| < c'/n.$$

It suffices to observe that from the last relation it follows that $\bar{P}(x, S)$ is a Borel function in x for every fixed S and a pr. in S for every fixed x. Thus, it is a nonvanishing solution ($\bar{P}(x, R) = 1$) of the proper value 1 equation $\varphi(S) = \int \varphi(dy)P(y, S)$.

Observe that $\varphi^+ = (-\varphi)^-$ whatever be $\varphi \in B$, let $\varphi, \varphi' \in B$, and set $\varphi \wedge \varphi' = \varphi - (\varphi - \varphi')^+ = \varphi' - (\varphi' - \varphi)^+$. Nonnegative φ and φ' are *orthogonal* if $\varphi \wedge \varphi' = 0$ or, equivalently, if there exist two disjoint sets S, S' such that $\varphi(S) = \varphi(\Omega)$ and $\varphi'(S') = \varphi'(\Omega)$. Clearly, φ^+ and φ^- are orthogonal and so are $\varphi - \varphi \wedge \varphi'$ and $\varphi' - \varphi \wedge \varphi'$.

There exists a finite number l of mutually orthogonal invariant pr.'s $\bar{P}_i$ such that a pr. $\bar{P}$ is invariant if, and only if,

$$\bar{P} = \sum_{i=1}^{l} p_i\bar{P}_i, \quad p_i \geqq 0, \quad \sum_{i=1}^{l} p_i = 1.$$

Consider the family of all mutually orthogonal pr.'s $\bar{P}_i$ which are invariant, that is, are solutions of the proper value 1 equation. Since mutual orthogonality implies linear independence and the proper space B_1 is finite dimensional, the number of such solutions is finite. The "if" assertion follows and it suffices to prove that if a pr. $\bar{P}$ is invariant, then it can be written in the asserted form; to simplify the writing, we drop the bars over the P's.

First, we show that $Q_i = P \wedge P_i = p_iP_i$. We exclude the trivial case $Q_i = 0$ and set $P'_i = Q_i/\| Q_i \|$. If the pr. P'_i does not coincide with P_i, then $Q'_i = (P_i - P'_i)^+$ and $Q''_i = (P_i - P'_i)$ do not vanish. Since $Q'_i/\| Q'_i \|$, $Q''_i/\| Q''_i \|$, and the P_j with $j \neq i$, form a family of $l + 1$ mutually orthogonal invariant pr.'s, we reach a contradiction. Hence, $Q_i = p_iP_i$ where, necessarily, $0 \leqq p_i \leqq 1$.

It remains to show that $Q = P - \sum p_iP_i = 0$. Either $p_i = 1$ and then $P \wedge P_i = P_i$, hence $P = P_i$; or $p_i < 1$ and then

$$(1 - p_i)((P - Q_i) \wedge P_i) \leqq (P - Q_i) \wedge (1 - p_i)P_i$$

$$= (P - Q_i) \wedge (P_i - Q_i) = 0,$$

hence $(P - Q_i) \wedge P_i = 0$. It follows that every $Q \wedge P_i = 0$, so that, if $Q \neq 0$, then $Q/\| Q \|$ and the P_i form a family of $m + 1$ mutually orthogonal invariant pr.'s. We reach a contradiction and the assertion follows.

There exists a finite measurable partition $\sum_{i=1}^{l} S_i = R$ *such that*

$$\bar{P}_i S_i = 1,\quad \bar{P}(x, S) = \sum_{i=1}^{l} I_{S_i}(x)\bar{P}_i S,\quad P(x, S_i) = 1 \quad \textit{for} \quad x \in S_i;$$

and for all n

$$\sup_{x \in S_i, S} |\overline{P^n}(x, S) - \bar{P}_i S| < c'/n.$$

These relations imply that, if the system starts at any point of S_i, then it remains a.s. in S_i and the uniform limit $\bar{P}_i$ is independent of the starting point.

The proof goes as follows: Since $\bar{P}(x, S)$ is an invariant pr. for every fixed x, we have

$$\bar{P}(x, S) = \sum p_i(x)\bar{P}_i S,\quad p_i(x) \geqq 0,\quad \sum p_i(x) = 1.$$

Since the $\bar{P}_i$ are linearly independent and $\bar{P}(x, S)$ is invariant under $P(x, S)$ and under itself, it follows that

$$p_i(x) = \int P(x, dy)p_i(y),\quad \int \bar{P}_i(dx)p_j(x) = \delta_{ij}.$$

By setting $S_i = [x; p_i(x) = 1]$ and $S'_n = \left[x; \frac{1}{n+1} \leqq 1 - p_i(x) < \frac{1}{n}\right]$, the last relation yields

$$1 = \int \bar{P}_i(dx)p_j(x) \leqq \bar{P}_i S_i + \sum \bar{P}_i S'_n = \bar{P}_i R = 1.$$

But the equality holds only if every $\bar{P}_i S'_n = 0$. Hence $\bar{P}_i S_i = 1$, and the other assertions follow.

Let us mention a few classical cases in which $P(x, S)$ is quasi-strongly compact and invite the reader to check it.

1. Finite constant chains with transition pr. matrix P_{jk}, $j, k = 1, \cdots, l$.

2. Take $P(x, S) = \int_S p(x, y)\mu(dy)$, where μ is a pr. on Borel sets S and $p(x, y)$ is a bounded nonnegative Borel function on $R \times R$; the second iterate is strongly compact.

3. Take $P^h(x, S) \leqq cP^h(y, S)$ for some integer h and all x, y, S.

4. Take $P^h(x, S) \leqq c\mu S$, where μS is a pr., for some integer h and all x, S.

5. Take $P^h(x, S) \leqq 1 - \epsilon$ for $\mu S \leqq \epsilon$, where μS is a pr. and $\epsilon > 0$, for some integer h and all x, S. This is the Doblin condition and encompasses the preceding ones. Form

$$P^h(x, S) = \int_S p_0(x, y)\mu(dy) + P_s^h(x, N_x S)$$

where $p_0(x, y)$ can be selected to be nonnegative and Borel measurable on $R \times R$, and N_x is the μ-null set on which the μ-singular part of $P^h(x, S)$ does not vanish. Then set $p(x, y) = \min(p_0(x, y), 1/\epsilon)$ and observe that

$$P^h(x, S) = \int_S p(x, y)\mu(dy) + Q(x, S)$$

with $0 \leqq Q(x, S) \leqq 1 - \epsilon$ for all x and S. Thus $T^h = T' + Q$ where T' has for kernel $\int_S p(x, y)\mu(dy)$ and Q has for kernel $Q(x, S)$, hence $\|Q\| \leqq 1 - \epsilon$. In the expansion of $T^{hk} = (T' + Q)^k$, all terms containing T' at least twice are strongly compact and there are $k + 1$ other terms of norm $\leqq (1 - \epsilon)^{k-1}$. Thus, for k sufficiently large

$$\|T^{hk} - V_k\| \leqq (k + 1)(1 - \epsilon)^{k-1} < 1$$

and every V_k is strongly compact. Thus, Doblin's condition implies quasi-strong compactness and the foregoing results apply.

COMPLEMENTS AND DETAILS

In what follows T operates on $\mathfrak{A}$ to $\mathfrak{A}$ and t operates on Ω to Ω.

1. Let T be null-preserving. To every P-invariant event A there corresponds an invariant event A' such that $AA'^c + A^cA'$ is null. Example: $\limsup T^nA$.

2. Let $\Omega = [0, 1)$, P = Lebesgue measure, g integrable of period 1.

$$g_n(x) = \frac{1}{n}\sum_{k=0}^{n-1} g(x + kc) \xrightarrow{\text{a.e.}} \int_0^1 g\,dx, \quad c \text{ irrational fixed;}$$

$$g_n(x) = \frac{1}{n}\sum_{k=0}^{n-1} g(2^k x) \xrightarrow{\text{a.e.}} \int_0^1 g\,dx.$$

(In the first case introduce $t(x) = x + c$ modulo 1; in the second case introduce $t(x) = 2x$ modulo 1. Show that the transformations preserve the measure and are indecomposable.)

3. Construct examples of non P-preserving T such that for every $X \in L$, $\overline{T^nX} \xrightarrow[1]{\text{a.s.}} \overline{T}X \in L$. For instance, take suitable T such that $T^2 = I$ on $\Omega = [0, 1)$ with P = Lebesgue measure; in particular, take suitable linear transformations of $[0, c)$ and $[c, 1)$ into $[0, 1)$.

4. Let $Y_1, Y_2, \cdots$ be a stationary sequence of r.v.'s. There exists a stationary sequence $\cdots X_{-1}, X_0, X_1, \cdots$ such that the laws of $(X_1, X_2, \cdots)$ and of $(Y_1, Y_2, \cdots)$ are the same. (Take, for every finite subfamily of X's,

$$\mathfrak{L}(X_{k_1}, \cdots, X_{k_m}) = \mathfrak{L}(Y_{k_1+h}, \cdots, Y_{k_m+h})$$

where h is so large that the subscripts of Y's are positive, and apply the consistency theorem.) Since all pr. properties of the sequence $Y_1, Y_2, \cdots$ are the same as those of $X_1, X_2, \cdots$, we can replace in what follows every X_k with $k > 0$ by Y_k.

If $E|X_1|^r < \infty$ for an $r \geqq 1$, then

$$E(X_1 \mid X_0, \cdots, X_{-n+1}) \xrightarrow[\text{a.s.}]{r} E(X_1 \mid X_0, X_{-1}, \cdots) = U.$$

(Observe that the left-hand side r.v.'s form a martingale sequence.)

If $E|X_1|^r < \infty$ for an $r \geqq 1$, then stationarity of $\cdots X_{-1}, X_0, X_1, \cdots$ entails

$$\mathcal{L}\{E(X_{n+1} \mid X_1, \cdots, X_n)\} \to \mathcal{L}(U)$$

$$U^{(n)} = \frac{1}{n}\sum_{k=1}^{n} E(X_{k+1} \mid X_1, \cdots, X_k) \xrightarrow{r} U.$$

(By stationarity $\mathcal{L}\{E(X_{n+1} \mid X_1, \cdots, X_n)\} = \mathcal{L}\{E(X_1 \mid X_0, \cdots, X_{-n+1})\}$. For every $Z \in L_r$, set $\| Z \| = \left(\int | Z |^r\right)^{1/r}$ and observe that

$$\| U^{(n)} - U \| \leqq \frac{1}{n}\sum_{k=1}^{n} \| E(X_{k+1} \mid X_1, \cdots, X_k) - U_{k+1} \| + \left\| \frac{1}{n}\sum_{k=1}^{n} (U_{k+1} - U) \right\|$$

where the U_{k+1} are the translates of U by k. By the stationarity assumption, the sequence $U_1, U_2, \cdots$ is stationary and, since $E|U|^r < E|X_1|^r < \infty$, the second right-hand side term converges to 0. The first one reduces to

$$\frac{1}{n}\sum_{k=1}^{n} \| E(X_1 \mid X_{-1}, \cdots, X_{-k+1}) - U \|,$$

every term of the sum converges to 0 as $k \to \infty$, and so does their arithmetic mean.)

5. Let Ω be a compact metric space, $\mathcal{A}$ the minimal σ-field over the class of open sets, $PA > 0$ for every open set A. Let $T = t^{-1}$ be P-preserving, indecomposable, and the t^n be equicontinuous.

If g on Ω is continuous and integrable, then $\overline{g^n} \longrightarrow \int g$. The same holds if, for every $\epsilon > 0$, there exist continuous bounds $g' \leqq g \leqq g''$ with $\int | g' - g'' | < \epsilon$.

Application. If g on R of period 1 is Riemann integrable, then

$$g_n(x) = \frac{1}{n}\sum_{k=0}^{n-1} g(x + kc) \longrightarrow \int_0^1 g\,dx, \quad c \text{ irrational fixed.}$$

(Replace, in *2*, Ω by a circumference of length 1.)

6. lim sup $\overline{P^n} \leqq cP$ does not imply $\overline{P^n} \leqq cP$ for every n. What are the consequences of this statement? A counter example: Let $\Omega = \{\omega_n\}$, $\Omega = \sum A_n$ where A_n consist of 2^{n+1} points. Denoting by $1, 2, \cdots, 2^{n+1}$ the points of a fixed A_n, set for every $A_n\mu\{k\} = 2^{k-1}$ or 2^{2n-k} or 1 according as $1 \leqq k \leqq n$ or $n < k \leqq 2n$ or $2n < k \leqq 2^{n+1}$, and set $t(k) = 2^{n+1}$ or $k - 1$ according as $k = 1$ or $1 < k \leqq 2^{n+1}$. Finally, set $PA = c\sum \frac{1}{3^n}\mu(AA_n)$.

7. Let Ω be the set of irrationals in (0, 1) and let P = Lebesgue measure. Let $t(x) = 1/x$ modulo 1. Show that with the usual notation for continued fraction

$$t(x) = t\left(\frac{1\,|}{|\,c_1(x)} + \frac{1\,|}{|\,c_2(x)} + \frac{1\,|}{|\,c_3(x)} + \cdots\right) = \frac{1\,|}{|\,c_2(x)} + \frac{1\,|}{|\,c_3(x)} + \cdots$$

(a) The transformation $T = t^{-1}$ is P-indecomposable.

(Let A be invariant with $P = PA < 1$. To prove that $P = 0$, take a fixed x_0 set $c_n(x_0) = c_n$, and denote by a/b and a'/b' the $(2n-1)$th and $2n$th approximations of x_0, so that

$$y = \frac{1|}{|c_1} + \cdots + \frac{1\,|}{|\,c_{2n-1}} + \frac{1\quad|}{|\,x + c_{2n}} = \frac{ax + a'}{bx + b'}.$$

Then $t^{2n+1}(y) = x$; hence $I_A(x) = I_A(y)$. Set $\alpha = \dfrac{a}{b}$ and $\beta = \dfrac{a + a'}{b + b'}$. Then

$$\frac{P(A(0, 1))}{\beta - \alpha} = b'(b + b')\int_0^1 I_A(x)\frac{dx}{(bx + b')^2} < 1.$$

As x_0 varies over Ω and n over all the integers, the intervals (α, β) form a Vitali covering of Ω and, by Lebesgue's density theorem, $P = 0$.)

(b) Let $\mu A = \dfrac{1}{\log 2}\displaystyle\int_A \frac{dx}{1 + x}$. Then $T = t^{-1}$ is μ-preserving. (This follows from $\mu(t^{-1}A) = \mu A$ for all $A = (0, x)$ by $\displaystyle\int_0^x \frac{dx'}{1 + x'} = \sum \int_{1/(n+x)}^{1/n} \frac{dx'}{1 + x'}$.)

(c) The null sets and the integrable functions g are the same for μ and P, the a.s.-ergodic theorem holds for $T = t^{-1}$ and both measures, and

$$\frac{1}{n}\sum_{k=0}^{n-1} g(t^k x) \xrightarrow{\text{a.s.}} \frac{1}{\log 2}\int_0^1 \frac{g(x)}{1 + x}\,dx.$$

Applications. 1° For almost all $x \in \Omega$ and every integer p, the frequency of p in $\{c_n(x)\}$ is $\dfrac{1}{\log 2}\log\dfrac{(p + 1)^2}{p(p + 2)}$.

(Take $g = I_A$ where A is the set of all x such that $c_1(x) = p$.)

2° $$\sqrt[n]{c_1(x)c_2(x)\cdots c_n(x)} \xrightarrow{\text{a.s.}} \prod\left(1 + \frac{1}{n^2 + 2n}\right)^{\log n/\log 2}.$$

(Take $g(x) = \log c_1(x)$.)

3° $$\frac{1}{n}\sum_{k=1}^{n} c_k(x) \xrightarrow{\text{a.s.}} +\infty.$$

(Take $g(x) = c_1(x)$.)

Chapter X

SECOND ORDER PROPERTIES

We consider sequences, and more generally random functions, formed by r.v.'s whose second moments and hence mixed second moments are finite.

Their second order properties are those which can be expressed in terms of these moments. Up to equivalences, the r.v.'s in question can be interpreted as points in a Hilbert space, and such spaces are a "natural" generalization of euclidean spaces for which all the classical tools were developed. Thus, it is to be expected that the study of second order properties—to which this chapter is devoted—will only require analytical tools similar to the familiar ones. In fact, except for a few concepts and properties, this chapter is practically independent of the preceding ones.

§ 33. ORTHOGONALITY

We examine in this section the elementary properties of orthogonal r.v.'s. The concepts become almost obvious if geometric intuition is used; we expect the reader to do it. The only difficulties consist in the justification of the intuitive conclusions in the nonfinite case, and in the use of pr. concepts such as a.s. convergence which have no direct geometric equivalent.

Let $(\Omega, \mathcal{A}, P)$ be a fixed pr. space. Let $X, Y, \cdots$, with or without affixes, be *second order r.v.'s*, in general complex-valued:

$$E|X|^2 < \infty, \quad E|Y|^2 < \infty, \cdots,$$

so that by Schwarz's inequality their mixed second moments $EX\bar{Y}$ exist and are finite. The "bar" means "complex-conjugate."

The space L_2 of equivalence classes of such r.v.'s is a Hilbert space: equivalent r.v.'s represent the same *point*, and $EX\bar{Y}$ defines the *scalar product* of the points represented by X and Y. Scalar products determine the *norms* $\|X\| = (E|X|^2)^{1/2} = (EX\bar{X})^{1/2}$, and norms determine *distances* $\|X - Y\|$. Convergence "in norm" is convergence "in quadratic mean" $X_n \xrightarrow{\text{q.m.}} X$: $\|X_n - X\| \to 0 \Leftrightarrow E|X_n - X|^2 \to 0$. The space L_2 is *complete* in the sense that $X_n \xrightarrow{\text{q.m.}} X \Leftrightarrow X_m - X_n \xrightarrow{\text{q.m.}} 0$, $m, n, \to \infty$.

33.1 Orthogonal r.v.'s; convergence and stability. X and Y are *orthogonal*, and we write $X \perp Y$, if $EX\bar{Y} = 0$. In particular, $X \perp X$ if, and only if, $E|X|^2 = 0$, that is, $X = 0$ a.s.; in fact, $X = 0$ a.s. is orthogonal to every Y. Since independent r.v.'s in L_2 are orthogonal when centered at expectations, we assume, for sake of analogy, that *all our r.v.'s are centered at expectations, unless otherwise stated.* Since our r.v.'s have finite second hence first moments, they can always be so centered and the assumption made does not restrict the generality. Then $E|X|^2$ is the *variance* of X and $EX\bar{Y}$ is the *covariance* of X and Y.

From $X \perp Y$ it follows, upon expanding, that $E|X + Y|^2 = E|X|^2 + E|Y|^2$. More generally, if $X_1, X_2, \cdots$ are orthogonal r.v.'s, then (Pythagorean relation)

$$E\left|\sum_{k=1}^{n} X_k\right|^2 = \sum_{k=1}^{n} E|X_k|^2, \quad n = 1, 2, \cdots,$$

and, as $n \to \infty$,

$$E\left|\sum_{k=1}^{n} X_k\right|^2 \to \sum_{k=1}^{\infty} E|X_k|^2.$$

Since, by the mutual convergence in q.m. criterion, the sequence of sums $\sum_{k=1}^{n} X_k$ converges in q.m. if, and only if, $E\left|\sum_{k=m}^{n} X_k\right|^2 = \sum_{k=m}^{n} E|X_k|^2 \to 0$ as $m, n \to \infty$, we have

A. CONVERGENCE AND STABILITY IN Q.M. *Let the r.v.'s X_n be orthogonal.*

(i) *The series $\sum X_n$ converges in q.m. if, and only if, $\sum E|X_n|^2 < \infty$; and then $E|\sum X_n|^2 = \sum E|X_n|^2$ (Pythagorean relation).*

(ii) *If* $\sum \dfrac{E|X_n|^2}{b_n^2} < \infty$, $b_n \uparrow \infty$, *then* $\dfrac{1}{b_n} \sum_{k=1}^{n} X_k \xrightarrow{\text{q.m.}} 0$.

The second assertion follows from the first by Kronecker's lemma.

The foregoing properties correspond to those of the case of independence if independence and convergence a.s. are replaced by orthogonality and convergence in q.m., respectively. In fact, not much more is needed in the case of orthogonality to obtain an inequality (Rademacher) similar to that of Kolmogorov and, consequently, a.s. convergence and stability theorems well-known in theory of orthogonal functions.

a. *If* $S_n = \sum_{k=1}^{n} X_k$ *are consecutive sums of orthogonal r.v.'s, then*

$$E(\max_{h \leqq n} | S_h |)^2 \leqq (\log 4n/\log 2)^2 \sum_{k=1}^{n} E| X_k |^2.$$

Proof. For $n = 1$ the inequality is trivial. For $n > 1$, let m be the integer such that $2^{m-1} < n \leqq 2^m$, set $X'_k = X_k$ or 0 according as $k \leqq n$ or $n < k \leqq 2^m$, and assign X'_k to the point of abscissa k. Divide the interval $(0, 2^m]$ into intervals $(0, 2^{m-1}]$ and $(2^{m-1}, 2^m]$, each of these two intervals into two halves, and so on; the elements of the jth partition are of length 2^j and $j = 0, 1, \cdots, m$. Every interval $(0, h]$ is the sum of at most m disjoint intervals each of which belongs to a different partition; in other words, we have the dyadic representation of h in geometric terms. We can write $S_h = \sum_{j=0}^{m} Y_{jh}$ where any Y_{jh} is sum of the r.v.'s belonging to the interval of length 2^j which may or may not figure in the representation of h, so that some Y_{jh} may vanish. It follows, by the elementary Schwarz inequality

$$| \sum_{j=0}^{m} a_j |^2 \leqq (m+1) \sum_{j=0}^{m} | a_j |^2,$$

that, whatever be $h \leqq n$,

$$| S_h |^2 \leqq (m+1) \sum_{j=0}^{m} | Y_{jh} |^2 \leqq (m+1)T,$$

where T is the sum of all r.v.'s $| Y_{jh} |^2$ as j and h vary. But the expectation of the sum of all those r.v.'s $| Y_{jh} |^2$ which belong to the jth partition is $\sum_{k=1}^{n} E| X_k |^2$, so that $ET = (m+1) \sum_{k=1}^{n} E| X_k |^2$. Therefore

$$E(\max_{h \leqq n} | S_h |^2) \leqq (m+1)^2 \sum_{k=1}^{n} E| X_k |^2$$

and, since

$$(m+1)^2 \leqq \left(\frac{\log n}{\log 2} + 2\right)^2 = \left(\frac{\log 4n}{\log 2}\right)^2,$$

the asserted inequality is proved.

b. *If* $S_n = \sum_{k=1}^{n} X_k$ *are consecutive sums of orthogonal r.v.'s and* $\sum b_n E|X_n|^2 < \infty$, $b_n \uparrow \infty$, *then* $S_n \xrightarrow{\text{q.m.}} S$ *and there exists a subsequence* $S_{n_k} \xrightarrow{\text{a.s.}} S$ *such that, for every integer* k, b_{n_k} *be the first* $b_n \geq k$.

Proof. The hypothesis implies that $\sum E|X_n|^2 < \infty$ so that **A** applies and $S_n \xrightarrow{\text{q.m.}} S$.

Let

$$r_n = E|S - S_n|^2 = \sum_{j=n+1}^{\infty} E|X_j|^2$$

so that

$$\sum_{k=1}^{\infty} E|S - S_{n_k}|^2 = \sum_{k=1}^{\infty} r_{n_k} = \sum_{k=1}^{\infty} k(r_{n_k} - r_{n_{k+1}}) + \lim_{k \to \infty} k r_{n_{k+1}}$$

$$\leqq \sum_{n=1}^{\infty} b_n E|X_n|^2 < \infty$$

and, hence, $S_{n_k} \xrightarrow{\text{a.s.}} S$ on account of the Borel-Cantelli lemma and the Tchebichev inequality.

B. A.S. CONVERGENCE AND STABILITY. *Let the r.v.'s* X_n *be orthogonal.*

(i) *If* $\sum \log^2 n E|X_n|^2 < \infty$, *then the series* $\sum X_n$ *converges in q.m. and a.s.*

(ii) *If* $\sum \left(\frac{\log n}{b_n}\right)^2 E|X_n|^2 < \infty$, $b_n \uparrow \infty$, *then* $\frac{1}{b_n} \sum_{k=1}^{n} X_k \xrightarrow{\text{a.s.}} 0$.

Proof. Let $S_n = \sum_{k=1}^{n} X_k$. Under hypothesis (i), $S_n \xrightarrow{\text{q.m.}} S$ according to **A**, and lemma **b** with $b_n = \log n/\log 2$ yields $S_{2^k} \xrightarrow{\text{a.s.}} S$; thus (i) will follow if we prove that $T_k = \max_{2^k \leqq n < 2^{k+1}} |S_n - S_{2^k}| \xrightarrow{\text{a.s.}} 0$. But lemma **a**, with $n = 2^{k+1} - 2^k = 2^k$, yields, by elementary computations,

$$\sum_{k=1}^{\infty} E|T_k|^2 \leqq (3/\log 2)^2 \sum_{n=1}^{\infty} \log^2 n E|X_n|^2 < \infty,$$

and the assertion follows by the Borel-Cantelli lemma and the Tchebichev inequality. This proves (i), and (ii) follows by Kronecker's lemma.

Corollary. *If the series* $\sum X_n$ *of orthogonal r.v.'s converges in q.m., then* $\sum_{k=1}^{n} X_k = o(\log n)$ *a.s.*

This follows by **A**(i) from **B**(ii) with $b_n = \log n$.

33.2 Elementary orthogonal decomposition. Let $\{X_j\}$ be a countable family of (second order) r.v.'s. We intend to show that the X_j's are linear combinations of orthogonal r.v.'s.

Without restricting the generality, we can exclude those X_j's which are degenerate at 0 or are linear combinations of other r.v.'s of the family. Let S be an arbitrary finite subset of indices j and let c_j be complex numbers. If $\sum_{j \in S} c_j X_j = 0$ a.s., then

$$E(\sum_{j \in S} c_j X_j)\bar{X}_h = \sum_{j \in S} c_j E X_j \bar{X}_h = 0, \quad h \in S.$$

Conversely, this set of relations implies readily that $E|\sum_{j \in S} c_j X_j|^2 = 0$ and hence $\sum_{j \in S} c_j X_j = 0$ a.s. On the other hand, if there exist some nonvanishing c_j such that the foregoing set of relations holds, then the determinant $\| EX_j\bar{X}_h \| = 0$, $j, h \in S$. Thus, what we have excluded is the possibility for such determinants to vanish. It follows, by elementary computations, that the r.v.'s $Y_1 = X_1$ and, for $n > 1$,

$$Y_n = \begin{vmatrix} X_1 & X_2 & \cdots & X_n \\ EX_1\bar{X}_1 & EX_2\bar{X}_1 & \cdots & EX_n\bar{X}_1 \\ \cdot & \cdot & \cdots & \cdot \\ EX_1\bar{X}_{n-1} & EX_2\bar{X}_{n-1} & \cdots & EX_n\bar{X}_{n-1} \end{vmatrix}$$

are nondegenerate. Furthermore, they are linear combinations of the X_n and are easily verified to be mutually orthogonal. Upon setting $\xi_j = Y_j/(E|Y_j|^2)^{1/2}$, we have $E\xi_j\bar{\xi}_h = \delta_{jh}(= 1$ or 0 according as $j = h$ or $j \neq h)$ and

$$X_j = \sum_{h=1}^{j} c_{jh}\xi_h, \quad c_{jh} = EX_j\bar{\xi}_h.$$

In general, r.v.'s ξ_j, ξ_h such that $E\xi_j\bar{\xi}_h = \delta_{jh}$ are said to be *orthonormal*. Given a finite or denumerable set of orthonormal r.v.'s $\xi_1, \xi_2, \cdots$, we can set

$$X = X'_n + \sum_{k=1}^{n} c_k\xi_k, \quad c_k = EX\bar{\xi}_k.$$

Then $EX'_n\bar{\xi}_k = 0$ for $k \leqq n$ and, by the Pythagorean relation,

$$E|X|^2 = E|X'_n|^2 + \sum_{k=1}^{n} |c_k|^2 \geqq \sum_{k=1}^{n} |c_k|^2.$$

It follows, by letting $n \to \infty$, that, given an orthonormal sequence ξ_n,

$$\sum |c_n|^2 \leqq E|X|^2 < \infty$$

so that, by **A**, the series $\sum c_n\xi_n$ converges in q.m. and hence $X'_n \xrightarrow{\text{q.m.}} X'$ orthogonal to every ξ_n. Thus, we obtain the orthogonal decomposition

$$X = X' + \sum c_n\xi_n \text{ a.s.}, \quad c_n = EX\bar{\xi}_n,$$

with

$$E|X|^2 = E|X'|^2 + \sum |c_n|^2.$$

The r.v. X', being orthogonal to all ξ_n, is orthogonal to all linear combinations of the ξ_n and, hence, to their limits in q.m. This leads to the introduction of linear subspaces; we recall that here all r.v.'s under consideration are of second order.

A *linear subspace* $\mathfrak{L}$ is a family of r.v.'s closed under formation of all a.s. linear combinations of its elements. If, also, $\mathfrak{L}$ is closed under passages to the limit in q.m., then it is a *closed linear subspace*. Given a family $\{X_t\}$ of r.v.'s, the linear space $\mathfrak{L}_0\{X_t\}$ of all their linear combinations and the closed linear subspace $\mathfrak{L}\{X_t\}$, the closure of $\mathfrak{L}_0\{X_t\}$ under passages to the limit in q.m., are said to be *generated* by the r.v.'s X_t. If we keep only those X_t which are linearly independent, we obtain a *base* of $\mathfrak{L}_0\{X_t\}$ and of $\mathfrak{L}\{X_t\}$.

A r.v. X is *orthogonal* to $\mathfrak{L}$, and we write $X \perp \mathfrak{L}$, if $X \perp Y$ whatever be $Y \in \mathfrak{L}$. For example, in what precedes, an arbitrary r.v. X was decomposed into $X' \perp \mathfrak{L}\{\xi_n\}$ and $\sum c_n\xi_n \in \mathfrak{L}\{\xi_n\}$. This orthogonal decomposition with respect to $\mathfrak{L}\{\xi_n\}$ can be generalized for arbitrary closed linear subspaces, as follows:

a. *Let $\mathfrak{L} \subset L_2$ be a closed linear subspace. To every $X \in L_2$ there corresponds an $X_0 \in \mathfrak{L}$ such that $E|X - X_0|^2 = \inf_{Y \in \mathfrak{L}} E|X - Y|^2 = \alpha$, and then $X - X_0 \perp \mathfrak{L}$.*

Proof. Let $\{X_n\} \subset \mathfrak{L}$ be such that $E|X - X_n|^2 \to \alpha$. Since

$$E|X_m - X_n|^2 = 2E|X_m - X|^2 + 2E|X - X_n|^2 - 4E\left|\frac{X_m + X_n}{2} - X\right|^2$$

and $(X_m + X_n)/2 \in \mathfrak{L}$, it follows by letting $m, n \to \infty$ that

$$0 \leqq E|X_m - X_n|^2 \leqq 2E|X_m - X|^2 + 2E|X - X_n|^2 - 4\alpha \to 4\alpha - 4\alpha = 0.$$

Therefore, by the mutual convergence criterion and the closure of $\mathfrak{L}$ under passages to the limit in q.m., $X_n \xrightarrow{\text{q.m.}} X_0 \in \mathfrak{L}$, and

$$E|X - X_0|^2 = \lim E|X - X_n|^2 = \alpha.$$

Since $X_0 + cY \in \mathfrak{L}$ whatever be $Y \in \mathfrak{L}$ and the complex number c, we have $E|X - (X_0 + cY)|^2 \geqq \alpha$. Thus, by taking $c = bE(X - X_0)\overline{Y}$ with $0 < b < 2/E|Y|^2$, we have

$$0 \leqq E|X - (X_0 + cY)|^2 - E|X - X_0|^2 = b(bE|Y|^2 - 2)|E(X - X_0)\overline{Y}|^2 \leqq 0.$$

Hence, $E(X - X_0)\overline{Y} = 0$ and the proof is terminated.

A. PROJECTION THEOREM. *Let $\mathfrak{L}$ be a closed linear subspace. For every X there exists an a.s. unique orthogonal decomposition*

$$X = X' + X'', \quad X' \perp \mathfrak{L}, \quad X'' \in \mathfrak{L}.$$

Proof. There exists such a decomposition: according to **a** it suffices to set $X'' = X_0$. The decomposition is a.s. unique, since if $X = X'_1 + X''_1$ is another such a decomposition, then $X' - X'_1 = X''_1 - X''$ and $X' - X'_1 \perp X''_1 - X''$, so that $X' = X'_1$ a.s. and $X'' = X''_1$ a.s.

COROLLARY. *If $\{\xi_t\}$, $E\xi_t\bar{\xi}_{t'} = \delta_{t,t'}$, is an orthonormal base of the closed linear subspace $\mathfrak{L}$, then, for every X, there exists an a.s. unique orthogonal decomposition*

$$X = X' + \sum c_{t_j}\xi_{t_j}, \quad X' \perp \mathfrak{L}.$$

Proof. It suffices to prove that if $X'' \in \mathfrak{L}$, then there exists a countable subset of indices j such that $X'' = \sum c_{t_j}\xi_{t_j}$. Set $c_t = EX''\bar{\xi}_t$ so that, whatever be the summation set of t's,

$$0 \leqq E|X'' - \sum c_t\xi_t|^2 = E|X''|^2 - \sum |c_t|^2,$$

and hence

$$\sum |c_t|^2 \leqq E|X''|^2 < \infty.$$

Thus, there can be only a countable number of nonvanishing c_t, and the assertion is proved.

33.3 Projection, conditioning, and normality. If $X = X' + X''$ where $X' \perp \mathcal{L}$ and $X'' \in \mathcal{L}$, we say that X' is the *perpendicular* from X to $\mathcal{L}$ and X'' is the *projection* of X on $\mathcal{L}$. According to the projection theorem, the perpendicular X' and the projection X'' exist and are a.s. unique whatever be X and the closed linear subspace $\mathcal{L}$. We denote the projection by $E_2(X \mid \mathcal{L})$ and if $\mathcal{L} = \mathcal{L}\{X_t\}$ we also write it $E_2(X \mid \{X_t\})$. The reasons for this notation similar to that of conditional expectations are many. The operation of projection plays the role of a "second order" conditioning, for, as is readily verified, it is an a.s. linear operation and has the smoothing property of the operation of conditioning if σ-fields are interpreted as closed linear subspaces.

Furthermore, if $\mathcal{L}(\mathcal{B})$ is the space of all $\mathcal{B}$-measurable square integrable r.v.'s then $E(X \mid \mathcal{B}) = E_2(X \mid \mathcal{L}(\mathcal{B}))$ a.s., since, for $B \in \mathcal{B}$, $E\{E(X \mid \mathcal{B})I_B\} = EXI_B = E\{E_2(X \mid \mathcal{L}(\mathcal{B}))I_B\}$. Also, this similarity is related to normality. To begin with, observe that if $X = Y + iZ$, where $Y = \mathcal{R}X$ and $Z = \mathcal{J}X$ with same affixes as X if any, then

$$EX\bar{X}' = EYY' + EZZ' - i(EYZ' - EZY')$$

$$EXX' = EYY' - EZZ' + i(EYZ' + EZY').$$

It follows that $Y \perp X'$ and $Z \perp X'$ if, and only if, $EX\bar{X}' = 0$ *and* $EXX' = 0$ (if X or X' is real-valued, then $EX\bar{X}' = 0$ implies that $EXX' = 0$).

Let the r.v.'s X, X' be jointly normal, that is, let the r.v.'s Y, Z, Y', Z' be jointly normal. If X and X' are orthogonal and real-valued, hence $EXX' = EX\bar{X}' = 0$, then they are independent, since

$$\log Ee^{i(uX+u'X')} = -\frac{u^2}{2}EX^2 - \frac{u'^2}{2}EX'^2 = \log Ee^{iuX} + \log Ee^{iu'X'}.$$

Similarly, if X and X' are orthogonal and complex-valued and $EXX' = 0$, then they are independent, that is, the pairs $\{Y, Z\}$ and $\{Y', Z'\}$ are independent.

We shall say that a family of r.v.'s $X_t = Y_t + iZ_t$, where t with or without affixes varies on some set T, is *strongly normal*, if it is normal (that is, all finite subfamilies of the r.v.'s Y_t, Z_t, are normal) and if either all the X_t are real-valued or all the $EX_tX_{t'} = 0$.

A. *Within a strongly normal family, orthogonality is equivalent to independence and projection is equivalent to conditioning.*

Proof. The first assertion follows from the foregoing discussion. As for the second assertion, let X, X_t, $t \in T$, form a strongly normal family. Since $X' = X - E_2(X \mid \{X_t\}) = X - \sum_j c_{t_j} X_{t_j}$ is orthogonal to and strongly jointly normal with every X_t, it follows, by the first assertion, that X' is independent of every X_t, hence $E(X' \mid \{X_t\}) = EX' = 0$ a.s. Therefore

$$E(X \mid \{X_t\}) = E(\sum_j c_{t_j} X_{t_j} \mid \{X_t\}) = \sum c_{t_j} X_{t_j} = E_2(X \mid \{X_t\}) \text{ a.s.,}$$

and the proof is concluded.

We shall see in the next section that, to any family of second order r.v.'s, we can make correspond a strongly normal family with same second order moments. Thus, a projection can always be considered as a conditioning within suitably selected normal families. Furthermore, to every concept in terms of conditional expectations corresponds a second order concept in terms of projections which, according to what precedes, coincides with the specialization within suitable normality. Let us give two examples.

The concept of a martingale $\{X_n\}$ becomes

$$E_2(X_n \mid X_1, \cdots, X_{n-1}) = X_{n-1} \text{ a.s.}$$

or, setting $Y_n = X_n - X_{n-1}$,

$$E_2(Y_n \mid Y_1, \cdots, Y_{n-1}) = 0 \text{ a.s.,}$$

that is, $Y_n \perp Y_k$ for $k < n$. Thus, a "second order martingale" is simply a sequence of consecutive sums of orthogonal r.v.'s and the results of 33.1 apply. As is to be expected, the a.s. properties of martingales become properties in q.m. of second order martingales. For example,

a. *If the sequence X_n is a second order martingale, then $E|X_n|^2 \uparrow$; if, moreover,* $\lim E|X_n|^2 < \infty$, *then* $X_n \xrightarrow{\text{q.m.}} X$ *and X closes this martingale.*

Proof. It suffices to write $X_n = \sum_{k=1}^{n} Y_k$ where the Y_k are orthogonal and hence $E|X_n|^2 = \sum_{k=1}^{n} E|Y_k|^2 \uparrow$. If, moreover, $\lim E|X_n|^2 = \sum_{k=1}^{\infty} E|Y_k|^2 < \infty$, then by **A**, $X_n \xrightarrow{\text{q.m.}} X$. Furthermore, $X_n - X_m \perp X_k$

for $k \leqq m \leqq n$ and letting $n \to \infty$, it follows that $X - X_m \perp X_k$ for $k \leqq m$, hence $E_2(X \mid X_1, \cdots, X_m) = X_m$ a.s. This proves the last assertion.

The concept of a chain $\{X_n\}$ yields

$$E(X_n \mid X_1, \cdots, X_m) = E(X_n \mid X_m) \text{ a.s.}$$

for $m < n$, $n = 2, 3, \cdots$. A "second order chain" is defined by replacing E by E_2. (In the strongly normal case, both concepts coincide.)

Set $r_{mn} = E(X_n\bar{X}_m)/E|X_m|^2$ or 0 according as $E|X_m|^2 > 0$ or $= 0$; then $E_2(X_n \mid X_m) = r_{mn}X_m$.

b. *A sequence* $\{X_n\}$ *is a second order chain if, and only if,*

$$r_{mp} = r_{mn}r_{np}, \quad m < n < p.$$

Proof. The relation is trivially true if $E|X_m|^2 = 0$. Otherwise, if

$$E_2(X_p \mid X_1, \cdots, X_n) = E_2(X_p \mid X_n) = r_{np}X_n \text{ a.s.,}$$

then $X_p - r_{np}X_n \perp X_m$ for $m < n$ hence

$$r_{mp}E|X_m|^2 = E(X_p\bar{X}_m) = r_{np}E(X_n\bar{X}_m) = r_{mn}r_{np}E|X_m|^2$$

and the asserted relation holds. Conversely, if this relation is true, then, for every $m < n$,

$$r_{np}E(X_n\bar{X}_m) = E(X_p\bar{X}_m);$$

hence $X_p - r_{np}X_n \perp X_m$, that is,

$$r_{np}X_n = E_2(X_p \mid X_n) = E_2(X_p \mid X_1, \cdots, X_n) \text{ a.s.}$$

The proposition is proved.

§ 34. SECOND ORDER RANDOM FUNCTIONS

Second order stationary random functions were introduced by Khintchine (1934) who gave the harmonic decomposition of their covariances. Slutsky (1937) obtained a first harmonic decomposition of such random functions. Kolmogorov (1941) proceeded to a detailed study of second order stationary random sequences by means of Hilbert space methods. Cramér (1941) extended Khintchine's results to the vector case and (1942) obtained a decomposition theorem in functional spaces, essentially equivalent to the harmonic decomposition of second order stationary random functions; this decomposition is also an immediate consequence of Stone's theorem (1930) on groups of unitary operators in Hilbert spaces. All this research is limited to second order stationarity.

The author formulated a calculus of general second order random functions and gave (1945–46) the results of this section; they contain as special cases the second order stationarity properties (the foregoing decomposition was stated there explicitly for the first time).

34.1 Covariances. We indulge in the usual abuse of notation by using the same symbol for a function and for its value. The argument t, with or without affixes, will vary over a fixed set T. The only requirement is that the operations performed on the elements of T be meaningful; this will always be so if $T = R = (-\infty, +\infty)$ and, to make his life easier, the reader may assume that $T = R$, unless otherwise stated.

A random function $X(t)$ on T is the family of r.v.'s $\{X(t), t \in T\}$; in general, the r.v.'s will be complex-valued. According to the essential feature of pr. theory, pr. properties are those which are described by the consistent set of laws of all finite subfamilies of the r.v.'s $X(t)$. Conversely, according to the consistency theorem, every such consistent set of laws is the law of some random function.

A *second order random function* $X(t)$ on T is a family of second order r.v.'s: $E|X(t)|^2 < \infty$ whatever be $t \in T$. Without restricting the generality, we can and do assume that the second order random functions under consideration are centered at expectations, unless otherwise stated. Then the second moments $E|X(t)|^2$ are variances and the function defined on $T \times T$ by

$$\Gamma_X(t, t') = EX(t)\bar{X}(t')$$

is, by definition, the *covariance* of the random function $X(t)$ on T. According to the Schwarz inequality, this covariance exists and is finite. Conversely, if $\Gamma_X(t, t')$ on $T \times T$ exists and is finite, then $E|X(t)|^2 = \Gamma_X(t, t) < \infty$, $t \in T$. Thus, second order random functions can be defined as those having covariances. Their *second order properties* are those which can be defined or determined by means of covariances. It is to be expected that to a covariance corresponds more than one random function. For example, the covariances of the random functions $X(t)$ and $Y(t) = \eta X(t)$ on T, where η is a r.v. independent of all $X(t)$, $t \in T$, with $E|\eta|^2 = 1$, coincide. In fact, this example shows that our convention which consists in centering second order random functions at their expectations is immaterial. If we take $E\eta = 0$, then the function $EX(t)\bar{X}(t')$ where the random function $X(t)$ is not centered at its expectation is still a covariance of the random function $Y(t)$ centered at its expectation:

$$EY(t) = E\eta X(t) = E\eta EX(t) = 0,$$

$$EY(t)\overline{Y}(t') = E|\eta|^2 X(t)\overline{X}(t') = E|\eta|^2 EX(t)\overline{X}(t') = EX(t)\overline{X}(t').$$

Since we consider only first and second (mixed or not) moments, it is natural to try to make correspond to a covariance a random function whose law is determined by these moments only. This will be done in the next theorem. We require the following definition.

A function $\Gamma(t, t')$ on $T \times T$ is of *nonnegative-definite type* if, for every finite subset $T_n \subset T$ and every function $h(t)$ on T_n

$$\sum_{t,t' \in T_n} \Gamma(t, t')h(t)\bar{h}(t') \geqq 0.$$

Then it is *hermitian*, that is, $\Gamma(t, t') = \bar{\Gamma}(t', t)$. For, with $T_1 = \{t\}$ and $h(t) = 1$, we have $\Gamma(t, t) \geqq 0$; then, with $T_2 = \{t, t'\}$, the expression $\Gamma(t, t')h(t)\bar{h}(t') + \Gamma(t', t)h(t')\bar{h}(t)$ is real, and hermiticity follows by taking $h(t) = 1$, $h(t') = 1, i$. The reason for the terminology is that a nonnegative-definite type function $\Gamma(t, t') = f(t - t')$ which depends only upon the difference $u = t - t'$ of its arguments reduces to a nonnegative-definite function $f(u)$. We require the following lemma.

a. *Let j, k vary over* $1, \cdots, n$ *and let the u_k vary over* $(-\infty, +\infty)$. *If* $Q(u) = \sum m_{jk}u_ju_k \geqq 0$, $m_{jk} \in (-\infty, +\infty)$, *then there exist jointly normal real-valued r.v.'s X_k (some of which may be degenerate at* 0) *such that* $m_{jk} = EX_jX_k$.

Proof. According to the classical properties of quadratic forms, the assumption means that

$$Q(u) = R(v) = \sum \sigma_k^2 v_k^2, \quad \sigma_k^2 \geqq 0,$$

where the v's are linear combinations of the u's. But $e^{-R(v)/2} = \prod e^{-\sigma_k^2 v_k^2/2}$ is the joint ch.f. of independent normal r.v.'s Y_k (centered at expectations) with $\sigma_k^2 = EY_k^2$. Therefore, by going back to the u's,

$$R(v) = E(\sum Y_k v_k)^2 = E(\sum X_k u_k)^2 = \sum m_{jk}u_ju_k,$$

where the X's are linear combinations of the Y's hence are jointly normal, and $EX_jX_k = m_{jk}$. The assertion is proved.

A. COVARIANCE CRITERION AND NORMALITY. *A function $\Gamma(t, t')$ on $T \times T$ is a covariance if, and only if, it is of nonnegative-definite type.*

And every covariance is also the covariance of a strongly normal random function which can be selected to be real-valued when the covariance is real-valued.

Proof. Let $\Gamma(t, t') = EX(t)\bar{X}(t')$, $t, t' \in T$. Then

$$\sum_{t,t' \in T_n} \Gamma(t, t')h(t)\bar{h}(t') = E \sum_{t,t' \in T_n} X(t)\bar{X}(t')h(t)\bar{h}(t')$$
$$= E\left|\sum_{t \in T_n} X(t)h(t)\right|^2 \geqq 0$$

and the "only if" assertion is proved.

Conversely, let $\Gamma(t, t')$ on $T \times T$ be of nonnegative-definite type. The "if" assertion means that there exists a random function $X(t)$ on T such that $EX(t)\bar{X}(t') = \Gamma(t, t')$ on $T \times T$. By hypothesis,

$$Q(u, v) = \sum_{t,t' \in T_n} \tfrac{1}{2}\Gamma(t, t')h(t)\bar{h}(t') \geqq 0$$

or, setting $h(t) = u_t - iv_t$,

$$Q(u, v) = \sum_{t,t' \in T_n} \tfrac{1}{2}\{\mathcal{R}\Gamma(t, t')(u_t u_{t'} + v_t v_{t'}) - \mathcal{J}\Gamma(t, t')(u_t v_{t'} - u_{t'} v_t)\} \geqq 0$$

whatever be u_t, v_t, $t \in T_n$. Therefore, by **a**, the function $f(u, v) = e^{-\frac{1}{2}Q(u,v)}$ is a normal ch.f. of $2n$ r.v.'s (centered at expectations) $Y(t)$ and $Z(t)$ corresponding to u_t and v_t, respectively, with

$$EY(t)Y(t') = EZ(t)Z(t') = \tfrac{1}{2}\mathcal{R}\Gamma(t, t'), \quad EY(t)Z(t') = -\tfrac{1}{2}\mathcal{J}\Gamma(t, t').$$

It follows by setting $X(t) = Y(t) + iZ(t)$ that $EX(t)X(t') = 0$ and $EX(t)\bar{X}(t') = \Gamma(t, t')$, $t, t' \in T_n$.

The normal laws of finite subfamilies of r.v.'s $X(t)$ so defined for every $T_n \subset T$ are consistent, since the law for $T_m \subset T_n$ coincides with the marginal law on T_m obtained by setting $u_t = v_t = 0$ for $t \in T_n - T_m$. Thus is defined the law of a normal random function $X(t)$ on T with covariance $\Gamma(t, t')$ on $T \times T$. If the covariance is real-valued, we can simply set $EX(t)X(t') = \Gamma(t, t')$ and the ch.f.'s $e^{-Q(u,0)}$ determine the law of a real-valued normal random function $X(t)$. The proof is concluded.

COROLLARY. *The real part of a covariance $\Gamma(t, t')$ is a covariance, while the imaginary part is not a covariance except when it vanishes.*

Proof. The first assertion follows from the above proof by $\mathcal{R}\Gamma(t, t') = E\{\sqrt{2}\,Y(t)\cdot\sqrt{2}\,Y(t')\}$. The second assertion follows from the fact that $\mathcal{J}\Gamma(t, t) = 0$ so that, if $\mathcal{J}\Gamma(t, t') = EX(t)\bar{X}(t')$ on $T \times T$, then $E|X(t)|^2 = 0$ on T, hence $X(t) = 0$ a.s. and $\mathcal{J}\Gamma(t, t') = 0$.

We examine a few operations which preserve covariances (see also the following subsections), and leave to the reader the specialization of what follows to stationary covariances, that is, to nonnegative-definite functions.

B. CLOSURE THEOREM. *The class of covariances is closed under additions, multiplications, and passages to the limit.*

Proof. Let $\Gamma_1(t, t')$ and $\Gamma_2(t, t')$ be two covariances and let $X_1(t)$ and $X_2(t)$ be random functions whose covariances are $\Gamma_1(t, t')$ and $\Gamma_2(t, t')$, respectively. If we select these random functions to be orthogonal, that is, $EX_1(t)\bar{X}_2(t') = 0$ for $t, t' \in T$, then

$$E\{X_1(t) + X_2(t)\}\{\bar{X}_1(t') + \bar{X}_2(t')\} = EX_1(t)\bar{X}_1(t') + EX_2(t)\bar{X}_2(t')$$
$$= \Gamma_1(t, t') + \Gamma_2(t, t').$$

If we select them to be independent, then

$$E\{X_1(t)X_2(t)\}\{\bar{X}_1(t')\bar{X}_2(t')\} = EX_1(t)\bar{X}_1(t')\cdot EX_2(t)\bar{X}_2(t')$$
$$= \Gamma_1(t, t')\Gamma_2(t, t').$$

Such selections are possible, for it suffices to take the random functions $X_1(t)$ and $X_2(t)$ to be normal on pr. spaces $(\Omega_1, \mathcal{A}_1, P_1)$ and $(\Omega_2, \mathcal{A}_2, P_2)$, respectively, and then form the product pr. space. Thus, the first two assertions are proved.

Finally, let $\Gamma_s(t, t')$ be covariances for $s \in S$ arbitrary with s_0 a limit point of S (not necessarily in S) and let $\Gamma_s(t, t') \to \Gamma(t, t')$ on $T \times T$ as $s \to s_0$.

Since passage to the limit and finite summation on $T_n \times T_n$ can be interchanged, it follows by **A** that

$$\sum_{T_n}\sum_{T_n}\Gamma(t, t')h(t)\bar{h}(t') = \lim \sum_{T_n}\sum_{T_n}\Gamma_s(t, t')h(t)\bar{h}(t') \geqq 0$$

and by the same theorem $\Gamma(t, t')$ is a covariance.

Applications. 1° If $\Gamma(t, t')$ is a covariance, so is its real part and hence so is the real part $(\mathcal{R}\Gamma(t, t'))^2 - (\mathfrak{I}\Gamma(t, t'))^2$ of $\Gamma^2(t, t')$; similarly for higher powers.

2° Every nonnegative number being a covariance, so is every polynomial in covariances with positive coefficients, and so is every limit of such polynomials. For example, $1/(1 - tt')$ is covariance of a random function $\sum_0^\infty t^n\xi_n$ analytic in $(-1, +1)$, where the ξ_n are orthonormal. This random function was encountered in studying new classes of limit laws and is at the origin of the investigations of this section.

3° If $\Gamma_s(t, t')$ is continuous in $s \in R$ and $F(s)$ on R is nondecreasing, then $\int \Gamma_s(t, t')\, dF(s)$ is a covariance, provided it exists and is finite.

4° Let Δ_h and Δ'_h be difference operators of step h operating on t and $t'(\in R)$, respectively. If $\Gamma(t, t')$ is the covariance of the random function $X(t)$, then, by computing the covariance of $\Delta_h{}^n X(t)$ where $X(t)$ has for covariance $\Gamma(t, t')$, we find that $\Delta_h{}^n \Delta'_h{}^n \Gamma(t, t')$ is a covariance and the variance $\Delta_h{}^n \Delta'_h{}^n \Gamma(t, t) \geqq 0$. It follows that if $\dfrac{\partial^{2n}}{\partial t^n \partial t'^n} \Gamma(t, t')$ exists and is finite, then it is a covariance; we shall see that it is the covariance of the nth derivative in q.m. of $X(t)$.

34.2 Calculus in q.m.; continuity and differentiation. Let s, s' vary over some set S and let s_0, s'_0 be limit points of S; they do not necessarily belong to S (for example, $S = (-\infty, +\infty)$ and $s_0 = +\infty$).

a. *If* $X_s \xrightarrow{\text{q.m.}} X$ *as* $s \to s_0$ *and* $X'_{s'} \xrightarrow{\text{q.m.}} X'$ *as* $s' \to s'_0$, *then* $EX_s\overline{X}_{s'} \to EX\overline{X}'$.

Proof. This follows from

$$E(X_s\overline{X}'_{s'} - X\overline{X}') = E(X_s - X)(\overline{X}'_{s'} - \overline{X}') + E(X_s - X)\overline{X}' + EX(\overline{X}'_{s'} - \overline{X}'),$$

since, as $s \to s_0$ and $s' \to s'_0$,

$$| E(X_s - X)(\overline{X}'_{s'} - \overline{X}') |^2 \leqq E| X_s - X |^2 \cdot E |X'_{s'} - X'|^2 \to 0,$$

and similarly for the two other r.h.s. terms.

A. CONVERGENCE IN Q.M. CRITERION. *Second order random functions* $X_s(t)$ *on* T *converge in q.m. as* $s \to s_0$ *to some random function* $X(t)$ *on* T *(necessarily of second order) if, and only if, the functions* $EX_s(t)\overline{X}_{s'}(t)$ *converge to a finite function on* T, *as* s, $s' \to s_0$ *in whatever way* s *and* s' *converge to* s_0. *Then* $\Gamma_{X_s}(t, t') \to \Gamma_X(t, t')$ *on* $T \times T$.

Proof. The "if" assertion follows, by the mutual convergence in q.m. criterion, from

$$\begin{aligned} E| X_s(t) - X_{s'}(t) |^2 &= E| X_s(t) |^2 - EX_s(t)\overline{X}_{s'}(t) - EX_{s'}(t)\overline{X}_s(t) \\ &\quad + E| X_{s'}(t) |^2 \to \Gamma(t, t) - 2\Gamma(t, t) + \Gamma(t, t) \\ &= 0, \quad s, s' \to s_0. \end{aligned}$$

The "only if" and the last assertions follow from the foregoing lemma upon replacing X_s by $X_s(t)$ and $X'_{s'}$ by $X_{s'}(t')$.

A second order random function $X(t)$ on T is *continuous in q.m.* at $t \in T$ if

$$X(t+h) \xrightarrow{\text{q.m.}} X(t) \quad \text{as} \quad h \to 0, \quad t+h \in T.$$

B. CONTINUITY IN Q.M. CRITERION. *$X(t)$ is continuous in q.m. at $t \in T$ if, and only if, $\Gamma_X(t, t')$ is continuous at (t, t).*

Proof. This follows by the convergence in q.m. criterion from

$$\lim_{h,h' \to 0} EX(t+h)\bar{X}(t+h') = \lim_{h,h' \to 0} \Gamma_X(t+h, t+h'), \quad t+h, t+h' \in T.$$

COROLLARY. *If a covariance $\Gamma(t, t')$ on $T \times T$ is continuous at every diagonal point $(t, t) \in T \times T$, then it is continuous on $T \times T$.*

It suffices to observe that if $\Gamma(t, t')$ is the covariance of $X(t)$, then

$$X(t+h) \xrightarrow{\text{q.m.}} X(t), \quad X(t'+h') \xrightarrow{\text{q.m.}} X(t'), \quad h, h' \to 0,$$

imply by **a** that

$$EX(t+h)\bar{X}(t'+h') \to EX(t)\bar{X}(t').$$

A second order random function $X(t)$ on T has a *derivative in q.m.* $\frac{dX(t)}{dt}$ (or $X'(t)$) at $t \in T$ if

$$\frac{X(t+h) - X(t)}{h} \xrightarrow{\text{q.m.}} X'(t), \quad h \to 0, \quad t+h \in T.$$

C. DIFFERENTIATION IN Q.M. CRITERION. *$X(t)$ has a derivative in q.m. at $t \in T$ if, and only if, the second generalized derivative of $\Gamma_X(t, t')$ exists and is finite at (t, t).*

This follows by the convergence in q.m. criterion from

$$\lim_{h,h' \to 0} E\left\{\frac{X(t+h) - X(t)}{h} \cdot \frac{\bar{X}(t+h') - \bar{X}(t)}{h'}\right\} = \lim_{h,h' \to 0} \frac{1}{hh'} \Delta_h \Delta'_{h'} \Gamma(t, t).$$

COROLLARY 1. *If the second generalized derivative of a covariance $\Gamma(t, t')$ on $T \times T$ exists and is finite at every diagonal point $(t, t) \in T \times T$,* *then the derivatives* $\frac{\partial}{\partial t}\Gamma(t, t')$, $\frac{\partial}{\partial t'}\Gamma(t, t')$, $\frac{\partial^2}{\partial t \partial t'}\Gamma(t, t')$ *exist and are finite on* $T \times T$.

It suffices to observe that, if $\Gamma(t, t')$ is the covariance of a random function $X(t)$, then $X'(t)$ exists, and since "E" and "lim q.m." can be interchanged, it follows by **a** that

$$EX'(t)\overline{X}(t') = \lim_{h \to 0} \frac{\Gamma(t+h, t') - \Gamma(t, t')}{h} = \frac{\partial}{\partial t}\Gamma(t, t').$$

Similarly for $\frac{\partial}{\partial t'}\Gamma(t, t')$, and also for

$$EX'(t)\overline{X}'(t') = \lim_{h' \to 0} \frac{1}{h'}\left\{\frac{\partial}{\partial t}\Gamma(t, t'+h') - \frac{\partial}{\partial t}\Gamma(t, t')\right\} = \frac{\partial^2}{\partial t \partial t'}\Gamma(t, t').$$

COROLLARY 2. *If* $X'(t)$ *on* T *exists, then* $\Gamma_{X'}(t, t') = \frac{\partial^2}{\partial t \partial t'}\Gamma_X(t, t')$ *on* $T \times T$.

This property extends at once to

$$EX^{(n)}(t)\overline{X}^{(n')}(t') = \frac{\partial^{n+n'}}{\partial t^n \, \partial t'^{n'}}\Gamma(t, t').$$

Assume that $\Gamma(t, t')$ is indefinitely differentiable on $T \times T$, select for origin a fixed value of the argument, set

$$X_n(t) = X(0) + \frac{t}{1}X'(0) + \cdots + \frac{t^n}{n!}X^{(n)}(0),$$

and form $E|X(t) - X_n(t)|^2$. Elementary computations yield

COROLLARY 3. *A second order random function* $X(t)$ *on* T *is analytic in q.m. if, and only if,* $\Gamma_X(t, t')$ *is analytic at every diagonal point* $(t, t) \in T \times T$; *and then* $\Gamma_X(t, t')$ *is analytic on* $T \times T$.

34.3 Calculus in q.m.; integration. The investigation of integrals in q.m. follows the foregoing pattern but is somewhat more involved.

Let $X(t)$ and $Y(t)$ on T be second order random functions with covariances $\Gamma_X(t, t')$ and $\Gamma_Y(t, t')$. Contrary to our convention, we do not assume that they are centered at expectations. The reason is that we shall have cases in which either one or the other of these random functions degenerates into a nonvanishing sure function, while if they were centered at expectations the sure function would have to vanish.

The *Riemann-Stieltjes integrals in q.m.* are defined as follows: Let

$$D_I: a = t_1 < t_2 < \cdots < t_{n+1} = b$$

be a finite set of consecutive points defining a partition of the finite interval $I = [a, b)$ and set $|D_I| = \max_{k \leq n} (t_{k+1} - t_k)$. If $X_{D_I}(t) = \sum_{k=1}^{n} X_k I_{[t_k, t_{k+1})}(t)$ is a random step-function, we set

$$\int_I X_{D_I}(t)\, dY(t) = \sum_{k=1}^{n} X_k\{Y(t_{k+1}) - Y(t_k)\}.$$

In the general case of a second order random function $X(t)$, we set $X_k = X(t'_k)$, $t_k \leq t'_k < t_{k+1}$,

$$\int_I X(t)\, dY(t) = \lim \text{ q.m.} \int_I X_{D_I}(t)\, dY(t),$$

$$\int X(t)\, dY(t) = \lim_{\substack{a \to -\infty \\ b \to +\infty}} \text{q.m.} \int_I X(t)\, dY(t),$$

(and similarly if only $a \to -\infty$ or $b \to +\infty$), provided the second limit exists and the first limit exists for some sequence of partitions D_I and is independent of the choices of the corresponding t'_k; they are necessarily defined up to an equivalence.

It is important to bear in mind that the preceding definition depends not upon the random function $Y(t)$ but upon its increments $\Delta Y(t)$; in other words, the random functions $Y(t)$ are to be considered as defined up to an additive r.v. Similarly, it is not the covariance $\Gamma_Y(t, t')$ of $Y(t)$ which matters below but its increments $\Delta\Delta'\Gamma_Y(t, t') = E\Delta Y(t)\Delta\overline{Y}(t')$.

A. Integration in q.m. criterion. *Let the second order random functions $X(t)$, with or without affixes, be independent of the second order increment function $\Delta Y(t')$ on an interval $I \times I$ finite or not. Then*

$$\int_I X(t)\, dY(t) \textit{ exists if, and only if,} \int_I\int_I \Gamma_X(t, t')\, dd'\Gamma_Y(t, t') \textit{ exists}$$

and, if the integrals in q.m. which figure below exist, then

$$E\left\{\int_I X^s(t)\, dY(t) \int_{I'} \overline{X}^{s'}(t')\, d\overline{Y}(t')\right\} = \int_I\int_{I'} E\{X^s(t)\overline{X}^{s'}(t')\}\, dd'\Gamma_Y(t, t').$$

The double integrals are usual Riemann-Stieltjes integrals.

Proof. The first assertion follows, upon starting with finite intervals, from the convergence in q.m. criterion applied to

$$E\left\{\int X_{D_I}(t)\,dY(t)\int \overline{X}_{D'_I}(t')\,d\overline{Y}(t')\right\};$$

similarly for the second assertion upon replacing X_{D_I} by $X_{D_I}{}^s$ and $\overline{X}_{D'_I}$ by $\overline{X}_{D'_{I'}}{}^{s'}$.

REMARK. The independence condition is certainly fulfilled when the random functions $X(t)$ and $Y(t)$ are independent or when $X(t)$ or $\Delta Y(t)$ degenerate into sure functions. On the other hand, it is used only to assert that for $t, t' \in I$

$$EX(t)\overline{X}(t')\Delta Y(t)\Delta'\overline{Y}(t') = EX(t)\overline{X}(t')\cdot E\Delta Y(t)\Delta'\overline{Y}(t'),$$

and, hence, it can be replaced by this less restrictive condition. Finally, it can be suppressed altogether, provided the elements of double integrals are replaced by, say, $dd'E\{X(t)\overline{X}(t')Y(t)\overline{Y}(t')\}$.

COROLLARY 1. *Formal properties of Riemann-Stieltjes integrals such as finite additivity hold a.s. for corresponding integrals in q.m.*

The corollary follows by elementary computations.

Let $D: a = t_1 < t_2 < \cdots < t_{n+1} = b$ and $D': a = t'_1 < t'_2 < \cdots < t'_{n'+1} = b$ be finite sets of points defining partitions of the finite interval $I = [a, b)$. Let $\Delta Y(t) = Y(t_{k+1}) - Y(t_k)$ and $\Delta' Y(t') = Y(t'_{k'+1}) - Y(t'_{k'})$ denote the corresponding increments of $Y(t)$ for $t = t_k \in D$ and $t' = t'_{k'} \in D'$. We say that the function $\Gamma_Y(t, t')$ is of *bounded variation* on $I \times I$ if there exists a constant c_I such that

$$\sum_{t\in D}\sum_{t'\in D'} |E\,\Delta Y(t)\Delta'\overline{Y}(t')| = \sum_{t\in D}\sum_{t'\in D'} |\Delta\Delta'\Gamma_Y(t, t')| \leqq c_I < \infty$$

whatever be D and D'. We say that $\Gamma_Y(t, t')$ is of *bounded variation* on the infinite interval $I \times I$ if there exists a constant c such that $c_{I'} \leqq c < \infty$ whatever be $I' \subset I$ and then we write, for short, $\int |dd'\Gamma_Y(t, t')| < \infty$. Clearly

COROLLARY 2. *If on $I \times I$ the random function $X(t)$ is continuous in q.m. and independent of the increment random function $\Delta Y(t')$, and if the covariance $\Gamma_X(t, t')$ is bounded while the covariance $\Gamma_Y(t, t')$ is of bounded variation, then $\int_I X(t)\,dY(t)$ exists.*

Let us observe that, under the bounded variation condition, the covariance $\Gamma_Y(t, t')$ can be assumed to have the property $\lim_{t,t' \to -\infty} \Gamma_Y(t, t') = 0$; it can also be normalized, that is, replaced by

$$\hat{\Gamma}_Y(t, t') = \tfrac{1}{4}\{\Gamma_Y(t+0, t'+0) + \Gamma_Y(t-0, t'+0) + \Gamma_Y(t+0, t'-0) + \Gamma_Y(t-0, t'-0)\}$$

where the limits exist and are finite. Furthermore, because of the boundedness and continuity condition on $\Gamma_X(t, t')$, the integral $\int_I\int_I \Gamma_X(t, t')\, dd'\Gamma_Y(t, t')$ remains the same if Γ_Y is replaced by $\hat{\Gamma}_Y$.

34.4 Fourier-Stieltjes transforms in q.m. A covariance $\Gamma(t, t')$ is said to be *harmonizable* if there exists a covariance $\gamma(s, s')$ of bounded variation on $R \times R$ such that

$$(\mathrm{H}_\Gamma) \qquad \Gamma(t, t') = \iint e^{i(ts - t's')}\, dd'\gamma(s, s').$$

A second order random function $X(t)$ is said to be *harmonizable* if there exists a second order random function $\xi(s)$ with a covariance $\gamma(s, s')$ of bounded variation on $R \times R$ such that

$$(\mathrm{H}_X) \qquad X(t) = \int e^{its}\, d\xi(s) \text{ a.s.}$$

We intend to show that harmonizability of a random function implies that of its covariance, and conversely. The direct assertion follows at once from the integration in q.m. criterion, and the problem lies in the proof of the converse assertion. We shall use repeatedly the convergence and integration in q.m. criteria, without further comment.

To begin with, let us observe that the bounded variation condition on $\gamma(s, s')$ implies that harmonizable random functions are continuous in q.m. and harmonizable covariances are continuous and bounded. Moreover, $\gamma(s \pm 0, s' \pm 0)$ and $\xi(s \pm 0)$, $\xi(\mp\infty)$ exist.

We denote by $\hat{\xi}(s)$ and $\Delta_0\xi(s)$ the "normalized" $\xi(s)$ and the "jump" of $\xi(s)$ at s, defined by

$$\hat{\xi}(s) = \tfrac{1}{2}\{\xi(s+0) + \xi(s-0)\}, \quad \Delta_0\xi(s) = \xi(s+0) - \xi(s-0).$$

Similarly, we denote by $\hat{\gamma}(s, s')$ and $\Delta_0\Delta'_0\gamma(s, s')$ the "normalized" $\gamma(s, s')$ and the "jump" of $\gamma(s, s')$ at (s, s'), defined by

$$\hat{\gamma}(s, s') = \tfrac{1}{4}\{\gamma(s+0, s'+0) + \gamma(s+0, s'-0) + \gamma(s-0, s'+0) + \gamma(s-0, s'-0)\}$$

$$\Delta_0\Delta'_0\gamma(s, s') = \gamma(s+0, s'+0) - \gamma(s+0, s'-0) - \gamma(s-0, s'+0) + \gamma(s-0, s'-0).$$

It follows that

$$E\hat{\xi}(s)\bar{\hat{\xi}}(s') = \hat{\gamma}(s, s'), \quad E\Delta_0\hat{\xi}(s)\Delta'_0\hat{\xi}(s') = \Delta_0\Delta'_0\hat{\gamma}(s, s').$$

Let Δ_h and $\Delta'_{h'}$ be difference operators of step h and h' acting on s and s', respectively. Let $a_\tau(u) = \sin \tau u/\tau u$ and let

$$b_\tau(v, h) = \frac{1}{\pi}\int_{\tau(v-h)}^{\tau v} \frac{\sin u}{u}\, du.$$

We use repeatedly the fact that $b_\tau(v, h) \to 0$, 1, or $\frac{1}{2}$, according as $v(v-h) > 0$, <0, or $=0$.

a. Inversion of covariances. *If a covariance $\Gamma(t, t')$ is harmonizable, then, as $\tau, \tau' \to \infty$,*

$$\frac{1}{4\tau\tau'}\int_{-\tau}^{+\tau}\int_{-\tau'}^{+\tau'} e^{-i(st-s't')}\Gamma(t, t')\, dt\, dt' \to \Delta_0\Delta'_0\gamma(s, s')$$

and

$$\frac{1}{4\pi^2}\int_{-\tau}^{+\tau}\int_{-\tau'}^{+\tau'} \frac{1}{tt'}\Delta_h\Delta'_{h'}e^{-i(st-s't')}\Gamma(t, t')\, dt\, dt' \to \Delta_h\Delta'_{h'}\hat{\gamma}(s, s').$$

Proof. (H_Γ) entails, by elementary transformations, that the integrals are, respectively,

$$\iint a_\tau(u-s)a_{\tau'}(u'-s')\, dd'\gamma(u, u'),$$

$$\iint b_\tau(u-s, h)b_{\tau'}(u'-s', h')\, dd'\gamma(u, u'),$$

and the assertions follow by letting $\tau, \tau' \to \infty$.

b. Inversions of random functions. *If $X(t)$ is a random function with a harmonizable covariance $\Gamma(t, t')$, then there exist random functions*

$\Delta_0\xi(s)$ with covariance $\Delta_0\Delta'_0\gamma(s, s')$ and $\hat{\xi}(s)$ with covariance $\hat{\gamma}(s, s')$ such that, as $\tau \to \infty$,

$$\frac{1}{2\tau}\int_{-\tau}^{+\tau} e^{-ist}X(t)\,dt \xrightarrow{\text{q.m.}} \Delta_0\xi(s)$$

$$\frac{1}{2\pi}\int_{-\tau}^{+\tau} -\frac{1}{it}\Delta_h e^{-ist}X(t)\,dt \xrightarrow{\text{q.m.}} \Delta_h\hat{\xi}(s).$$

If the random function $X(t)$ is harmonizable with respect to $\xi(s)$, then $\Delta_0\xi(s)$ is the jump function of $\xi(s)$ and $\hat{\xi}(s)$ is the normalized $\xi(s)$.

Proof. The first assertions follow readily from the foregoing lemma. The remaining ones can be deduced from the first ones or follow directly from (H_X), upon observing that, by elementary transformations, the foregoing integrals become, respectively,

$$\int a_\tau(u - s)\,d\xi(u), \quad \int b_\tau(u - s, h)\,d\xi(u).$$

A. HARMONIZABILITY THEOREM. *A random function is harmonizable if, and only if, its covariance is harmonizable.*

Proof. From

$$X(t) = \int e^{its}\,d\xi(s), \quad \iint |\,dd'\gamma(s, s')\,| < \infty$$

where $\gamma(s, s')$ is the covariance of $\xi(s)$, it follows at once that

$$EX(t)\bar{X}(t') = \iint e^{i(ts - t's')}\,dd'\gamma(s, s').$$

Conversely, let $X(t)$ have for covariance

$$\Gamma(t, t') = \iint e^{i(ts - t's')}\,dd'\gamma(s, s') \quad \textit{with} \quad \iint |\,dd'\gamma(s, s')\,| < \infty.$$

Since the integrand is continuous and bounded and $\gamma(s, s')$ is of bounded variation, we can assume without restricting the generality that $\gamma(s, s')$ is normalized. According to the foregoing lemma, there exists a random function $\xi(s)$ whose covariance is $\gamma(s, s')$ such that

$$\frac{1}{2\pi}\int_{-\tau}^{+\tau} -\frac{1}{it}\Delta_h e^{-ist}X(t)\,dt \xrightarrow{\text{q.m.}} \Delta_h\xi(s), \quad \tau \to \infty.$$

Upon applying the second parts of the foregoing lemmas, it follows by elementary computations that $Y(t) = \int e^{its}\,d\xi(s)$ exists and $E|\,X(t)\,|^2$

$= EX(t)\bar{Y}(t) = E|\,Y(t)\,|^2$, so that $E|\,X(t) - Y(t)\,|^2 = 0$, and the proof is concluded.

Particular cases

1° If s varies over a countable set only, then the integrals with respect to $\xi(s)$ and $\gamma(s, s')$ reduce to countable sums, and what precedes continues to hold. In fact, the proofs reduce to those of the first parts of **a** and **b**.

2° If t varies over a countable set only, then the integrals with respect to dt, $dt\,dt'$ reduce to countable sums and what precedes continues to hold. In fact, the same proofs apply, provided $\Gamma(t, t')$ and $X(t)$ are first extended by letting t, t' vary over R in (H_Γ) and (H_X).

Remark. The following analytical problem is of interest: characterize harmonizable covariances $\Gamma(t, t')$, that is, harmonizable functions of nonnegative-definite type. The answer ought to reduce to Bochner's theorem in the particular case of a continuous covariance $\Gamma(t, t') = f(t - t')$. The necessary condition is that $\Gamma(t, t')$ be continuous and bounded. Is this condition sufficient? If not, what supplementary conditions—which ought to disappear in the preceding particular case—are required?

34.5 Orthogonal decompositions. Among various decompositions of second order random functions, the orthogonal ones play a prominent role. The physical reason is that orthogonal components can be isolated experimentally by means of suitable "filters." The mathematical reason is that orthogonal decompositions correspond to the introduction of a general form of cartesian frames of reference which allow the use of a general form of Pythagorean relation. We saw in the preceding subsection that in the case of a random function defined on a countable set of values of the argument such a decomposition is always possible and the frame of reference can be obtained by linear combinations of the random values of the function. We intend to proceed to more general orthogonal decompositions of the same character. First let us give two countable decompositions. The one below follows from 34.2**B**, Corollary 3 by elementary computations.

A. Orthogonal expansion theorem. *The expansion*

$$X(t) = X(0) + \frac{t}{1!}X'(0) + \frac{t^2}{2!}X''(0) + \cdots, \quad t \in I,$$

of a second order analytic random function $X(t)$ is an orthogonal decom-

position if, and only if, its (analytic) covariance $\Gamma(t, t')$ *is a function of the product* tt' *of its arguments in* $I \times I$.

In fact, every random function $X(t)$ continuous in q.m. on a closed interval I has a countable orthogonal decomposition. We shall use Mercer's theorem which states that if a nonnegative-definite type function $\Gamma(t, t')$ is continuous on $I \times I$, then

$$\Gamma(t, t') = \sum |\lambda_n|^2 \psi_n(t)\bar{\psi}_n(t'),$$

where the series converges absolutely and uniformly on $I \times I$, and the continuous functions $\psi_n(t)$ are "proper functions" of $\Gamma(t, t')$ corresponding to "proper values" $|\lambda_n|^2$:

$$\int \Gamma(t, t')\psi_n(t')\, dt' = |\lambda_n|^2 \psi_n(t).$$

Proper functions which correspond to (necessarily finitely) multiple proper values are written with distinct indices, and all proper functions are orthonormalized on I:

$$\int \psi_m(t)\bar{\psi}_n(t)\, dt = \delta_{mn}.$$

B. Proper orthogonal decomposition theorem. *A random function* $X(t)$ *continuous in q.m. on a closed interval* I *has on* I *an orthogonal decomposition*

$$X(t) = \sum \lambda_n \xi_n \psi_n(t)$$

with

$$E\xi_m\bar{\xi}_n = \delta_{mn}, \quad \int \psi_m(t)\bar{\psi}_n(t)\, dt = \delta_{mn},$$

if, and only if, the $|\lambda_n|^2$ *are the proper values and the* $\psi_n(t)$ *are the orthonormalized proper functions of its covariance. Then the series converges in q.m. uniformly on* I.

Proof. Let t, t' vary over I, and

$$X_n(t) = \sum_{k=1}^{n} \lambda_n \xi_n \psi_n(t).$$

If $X(t)$ has the asserted decomposition, then

$$EX(t)\bar{X}(t') = \lim EX_n(t)\bar{X}_n(t') = \sum |\lambda_n|^2 \psi_n(t)\bar{\psi}_n(t'),$$

and the "only if" assertion follows. Conversely, let the $\psi_n(t)$ be the orthonormalized proper functions of the covariance $\Gamma(t, t')$ of $X(t)$, and form the integrals

$$\lambda_n \xi_n = \int X(t)\bar{\psi}_n(t)\, dt.$$

These integrals exist, since $X(t)$ (in q.m.) and $\psi_n(t)$ are continuous on the closed interval I, and

$$E\xi_m\bar{\xi}_n = \delta_{mn}, \quad EX(t)\bar{\xi}_n = \lambda_n\psi_n(t).$$

It follows from Mercer's theorem that $E|X(t) - X_n(t)|^2 \to 0$ uniformly on I, and the "if" assertion is proved.

In physics, the most important orthogonal decomposition is the harmonic one, for, loosely speaking, it yields "amplitudes," and hence "energies," corresponding to the various parts of the "spectrum" of the random function, and we seek it now. But, first, we have to introduce random functions which correspond to sums of orthogonal r.v.'s. It will be convenient to denote the increment of a function, say $\xi(t)$, on an interval $[a, b)$ by $\xi[a, b) = \xi(b) - \xi(a)$. Increment functions are characterized by their additivity:

$$\xi[a, b) + \xi[b, c) = \xi[a, c)$$

and determine the point functions $\xi(t)$ up to additive quantities. While what follows is valid for more general ordered sets T, we shall assume, to simplify the language, that $T \subset R$.

A second order random function $\xi(t)$ has *orthogonal increments* if, for disjoint intervals $[a, b)$, $[a', b')$,

$$E\xi[a, b)\bar{\xi}[a', b') = 0.$$

Then

$$E|\xi[a, b) \pm \xi[a', b')|^2 = E|\xi[a, b)|^2 + E|\xi[a', b')|^2,$$

and it follows, by setting, for some fixed a,

$$E|\xi[a, t)|^2 = F(t), \quad E|\xi[t, a)|^2 = -F(t),$$

that

$$E|\xi[t, t')|^2 = F[t, t');$$

for short,

$$E|d\xi(t)|^2 = dF(t).$$

If $\gamma(t, t')$ is the covariance of $\xi(s)$, this relation becomes

$$\gamma(t', t') - \gamma(t', t) - \gamma(t, t') + \gamma(t, t) = F(t') - F(t), \quad t' \geqq t;$$

for short,

$$dd'\gamma(t, t') = dF(t), \quad t' \geqq t.$$

In words, the increment of the foregoing covariance over a two-dimensional rectangle with sides parallel to the axes of t and of t' reduces to its increment over the square whose diagonal is the part of the line $t = t'$ belonging to the rectangle (draw the figure).

Observe that the second moments of the increments of $\xi(t)$ are bounded (by some fixed finite number) if, and only if, the finite nondecreasing function $F(t)$ is bounded on T. Whether bounded or not, this nondecreasing function extends to a nondecreasing function (finite or not) on $R = (-\infty, +\infty)$ and, in fact, on $\bar{R} = [-\infty, +\infty]$. This leads to

C. EXTENSION THEOREM. *If the random function $\xi(t)$ on T has orthogonal increments with bounded second moments, then, preserving this property, $\xi(t)$ can be extended by continuity to the closure of T and can be extended on $\bar{R}$.*

Proof. Let τ be a limit point of T from the left (it may or may not belong to T and may be $+\infty$). Since $F(t)$ is nondecreasing and bounded on T, $F(\tau - 0) = \lim_{t \uparrow \tau} F(t)$ exists and is finite, and as $t, t' \uparrow \tau (t > t')$

$$E|\xi(t) - \xi(t')|^2 = F(t) - F(t') \to 0. \tag{1}$$

It follows that the r.v. $\xi(\tau - 0) = \lim_{t \uparrow \tau} \text{q.m. } \xi(t)$ exists, and for $t \geqq \tau$

$$E|\xi(t) - \xi(\tau - 0)|^2 = \lim_{t' \uparrow \tau} E|\xi(t) - \xi(t')|^2 = F(t) - F(\tau - 0).$$

Similarly if τ is a limit point of T from the right. If τ is a limit point from both sides, then, by letting $t \downarrow \tau$ and $t' \uparrow \tau$ in (1), we also have

$$E|\xi(\tau + 0) - \xi(\tau - 0)|^2 = F(\tau + 0) - F(\tau - 0).$$

Now, if $\tau \notin T$ is a limit point from the left set $F(\tau) = F(\tau - 0)$, $\xi(\tau) = \xi(\tau - 0)$, and if it is a limit point from the right but not from the left set $F(\tau) = F(\tau + 0)$, $\xi(\tau) = \xi(\tau + 0)$. This provides the asserted extension on the closure $\bar{T}$ of T. For it suffices to write that the increments are orthogonal on T and let one or more end points approach points $\tau \in \bar{T} - T$; in particular $\xi(\tau + 0) - \xi(\tau - 0)$ is orthogonal to increments on intervals disjoint from $\{\tau\}$.

Finally, $\bar{R} - \bar{T}$ is a countable sum of intervals I with at least one end point τ of I belonging to $\bar{T}$. By setting $F(t) = F(\tau)$, $\xi(t) = \xi(\tau)$ on the I's, we obtain the asserted extension on $\bar{R}$.

When the boundedness condition is not satisfied, then the same proof shows that the asserted extension exists on the smallest interval containing T, provided the end points which do not belong to T are excluded.

Corollary 1. *Under the hypotheses of the extension theorem, the random function $\xi(t)$ is decomposable into two parts with mutually orthogonal increments:*

$$\xi(t) = \xi_d(t) + \xi_c(t),$$

where $\xi_d(t)$ is a sum of orthogonal r.v.'s (converging in q.m. when denumerable) and $\xi_c(t)$ is continuous in q.m., and

$$E|\, d\xi_d(t)\,|^2 = dF_d(t), \quad E|\, d\xi_c(t)\,|^2 = dF_c(t),$$

where $F_d(t)$ and $F_c(t)$ are the purely discontinuous and the continuous parts of $F(t)$, respectively.

Extend on $\bar{R}$, let $\{t_j\}$ be the (countable) discontinuity set of F, and set

$$\xi_d(t) = \sum_{t_j < t} (\xi(t_j + 0) - \xi(t_j - 0)), \quad \xi_c(t) = \xi(t) - \xi_d(t).$$

The assertions follow by elementary computations.

We say that a random function $\xi(t)$ on T with covariance $\gamma(t, t')$ is *orthogonal to its increments*, if for $t < t'$

$$E\xi(t)(\bar{\xi}(t') - \bar{\xi}(t)) = 0, \quad \text{that is,} \quad E\xi(t)\bar{\xi}(t') = E|\, \xi(t)\,|^2.$$

Clearly, every such random function $\xi(t)$ has orthogonal increments and the definition is equivalent to

$$\gamma(t, t') = F(t) \text{ nondecreasing}, \quad t \leqq t';$$

if, moreover, $F(t)$ is bounded on T, then the foregoing extension on $\bar{R}$ preserves this relation. Conversely

Corollary 2. *If a random function $\xi(t)$ on T has orthogonal increments with bounded second moments, then it can be made orthogonal to its increments by a suitable change of origin of its random values.*

Extend $\xi(t)$ on $\bar{R}$ and subtract $\xi(-\infty)$.

Thus, in the integrals below with respect to random functions $\xi(t)$ which have orthogonal increments with bounded second moments, we can assume that the $\xi(t)$ are orthogonal to their increments.

Consider now the integration in q.m. criterion and assume that the random function $Y(t)$ therein is a function $\xi(t)$ with orthogonal increments. Then, clearly, the double integrals therein reduce to simple integrals with respect to the nondecreasing function $F(t)$ defined above. In particular, the harmonizability theorem becomes

a. *Let* $E|d\xi(s)|^2 = dF(s)$. *A second order random function* $X(t)$ *is of the form*

$$X(t) = \int e^{its}\, d\xi(s)$$

if, and only if, its covariance $\Gamma(t, t')$ *is of the form*

$$\Gamma(t, t') = \int e^{i(t-t')s}\, dF(s).$$

For such covariances, the question raised at the end of the preceding subsection has a very simple answer (Khintchine, Kolmogorov).

b. *A covariance* $\Gamma(t, t')$ *is of the form*

$$\Gamma(t, t') = \int e^{i(t-t')s}\, dF(s), \quad \text{Var } F < \infty$$

where t, t' *vary over* R *or over* $\{\cdots -1, 0, +1, \cdots\}$ *if, and only if, it depends only upon the difference* $t - t'$ *of its arguments, and when* t, t' *vary over* R *it is, moreover, continuous.*

Proof. The "only if" assertion is obvious. The "if" assertion follows from the fact that a covariance is of nonnegative-definite type. For, the definition of this type reduces to that of nonnegative-definite functions when $\Gamma(t, t') = f(t - t')$, and then Herglotz's and Bochner's theorems apply.

A covariance which depends only upon the difference of its arguments is said to be *stationary*. A random function with a stationary covariance is said to be *second order stationary*. This concept is closely related to the usual concept of stationarity (of distributions); for, if a random function is stationary in the usual sense and is of second order, then it is *a fortiori* second order stationary, and in the strongly normal case the converse is true. A stationary random function need not be second order stationary since it may not be of second order.

Together, the preceding two lemmas and the inversion lemma for random functions become

D. HARMONIC ORTHOGONAL DECOMPOSITION THEOREM. *A second order random function $X(t)$ on $T = R$ or $\{\cdots -1, 0, +1, \cdots\}$ has a harmonic orthogonal decomposition*

$$X(t) = \int e^{its}\, d\xi(s), \quad t \in T,$$

where $E|\, d\xi(s)\,|^2 = dF(s)$, and $F(s)$ is of bounded variation on T

if, and only if,

$X(t)$ is second order stationary and in the case $T = R$ also continuous in q.m. at one point t.

Then, as $\tau \to \infty$,

$$\frac{1}{2\tau}\int_{-\tau}^{+\tau} e^{-ist}X(t)\, dt \xrightarrow{\text{q.m.}} \Delta_0\xi(s),$$

$$\frac{1}{2\tau}\int_{-\tau}^{+\tau} -\frac{1}{it}\Delta_h e^{-ist}X(t)\, dt \xrightarrow{\text{q.m.}} \Delta_h\hat{\xi}(s),$$

where the integrals reduce to sums when $T = \{\cdots -1, 0, 1, \cdots\}$.

COROLLARY 1. *A second order stationary random function on $\{\cdots -1, 0, +1, \cdots\}$ extends on R to a second order stationary random function continuous in q.m.*

COROLLARY 2. *A second order stationary random function $X(t)$ continuous in q.m. is decomposable into two orthogonal and second order stationary parts $X(t) = X_d(t) + X_c(t)$, with*

$$X_d(t) = \int e^{its}\, d\xi_d(s), \quad X_c(t) = \int e^{its}\, d\xi_c(t)$$

where the first integral reduces to a countable sum converging in q.m.

What precedes applies to second order r.f.'s of the form $X(t) = \{X_u(t),\ u \in U\}$, $t \in T$, that is, whose random values $X(t)$ are second order random functions of t: $E|\, X_u(t)\,|^2 < \infty$ whatever be t and u. It suffices to consider the argument $(u, t) \in U \times T$. However, the arguments u and t may play a nonsymmetric role. For example, the above random function is *second order stationary in t* if

$$EX_u(t)\overline{X}_{u'}(t') = f_{u,u'}(t - t'), \quad t, t' \in R, \quad u, u' \in U.$$

Clearly, this property is equivalent to the following one: the random function $Y_{U_n}(t) = \sum_{u \in U_n} c_u X_u(t)$, $t \in T$, are second order stationary whatever be the complex numbers c_u and whatever be the finite subset $U_n \subset U$. Upon applying the harmonic orthogonal decomposition theorem to the random functions $Y_{U_n}(t)$, we obtain by elementary computations

Corollary 3. *A random function $X(t) = \{X_u(t), u \in U\}$ continuous in q.m. on R is second order stationary if, and only if,*

$$EX_u(t)\overline{X}_{u'}(t') = \int e^{i(t-t')s}\, dF_{u,u'}(s), \quad u, u' \in U$$

where the functions $F_{u,u'}(s)$ are of bounded variation on R and the functions $\Delta_h F_{u,u'}(s)$ are of nonnegative-definite type in u, u' for any s and $h > 0$

or if, and only if,

$$X_u(t) = \int e^{its}\, d\xi_u(s), \quad u \in U$$

where the second order random functions $\xi_u(s)$ have mutually orthogonal increments with

$$E\xi_u(s)\overline{\xi}_{u'}(s') = F_{u,u'}(s) \quad \text{for} \quad s \leqq s'.$$

If U contains only two elements, the first assertion reduces to a result of Khintchine. If U contains only a finite number of elements, then it reduces to a result of Cramér.

Another nonsymmetric role of the two arguments u and t appears in considering them as the real and imaginary parts of a complex argument $z = u + it$. The definition of second order stationarity becomes: the random function $X(z)$ is *second order stationary* if its covariance

$$\Gamma_X(z, z') = f(z + \overline{z}') = f(u + u' + i(t - t'))$$

depends only upon $z + \overline{z}'$. Then, for every fixed $u + u'$, the covariance is stationary in t, t', the harmonic orthogonal decomposition theorem applies, and it is not difficult to obtain

Corollary 4. *A random function $X(z)$, $z = u + it$, continuous in q.m. in the complex-plane strip S: $a < u < b$, is second order stationary if, and only if, for $z, z' \in S$,*

$$\Gamma_X(z, z') = \int e^{(z+\overline{z}')s}\, dF(s), \quad \text{Var}\, F < \infty,$$

or, if and only if, for $z \in S$,

$$X(z) = \int e^{izs}\, d\xi(s),$$

where the random function $\xi(s)$ is orthogonal to its increments with $\Gamma_\xi(s, s') = F(s)$ for $s < s'$.

Remark. By replacing the argument t by a complex argument z in the preceding extension, Corollary 3 extends to random functions of z. It suffices to replace therein t by z and t' by $\bar{z}'$.

34.6 Normality and almost-sure properties. The class of normal r.v.'s is closed under linear combinations and passages to the limit in q.m. Since differentiation and integration in q.m. (with one of the two functions being a sure function) are obtained by such operations, it follows that the stability of normal laws (so far considered only for sums of independent r.v.'s) extends to the calculus in q.m.:

A. Normal stability theorem. *Normality is preserved under differentiations and integrations in q.m.*

In fact, the foregoing remarks apply word for word to second order random functions which obey infinitely decomposable laws—necessarily with finite variances (that is, all the linear combinations of the random values obey such laws). But more is true when the second order random functions are normal: many of the properties in q.m. established in this section become then a.s. properties. We saw that to every covariance there corresponds a strongly normal random function and, for the random values of such functions, orthogonality becomes independence. If a series of these orthogonal values converges in q.m., then, the summands being independent, the convergence is almost sure. Therefore, upon applying the foregoing theorem to the three orthogonal decomposition theorems of the preceding subsection, we obtain

Corollary 1. *If the covariance $\Gamma_X(t, t')$ is an analytic function of tt', then the random function $X(t)$ is normal if, and only if, its derivatives in q.m. are normal. If, moreover, $X(t)$ is real-valued, then its derivatives are independent and the Taylor expansion of $X(t)$ converges a.s.*

Corollary 2. *If the covariance $\Gamma_X(t, t')$ is continuous on a closed interval I, then the random function $X(t)$ is normal if, and only if, the r.v.'s $\lambda_n \xi_n = \int_I X(t)\bar{\psi}_n(t)\,dt$ are normal.*

If, moreover, $X(t)$ is real-valued, then the ξ_n are independent and the proper decomposition of $X(t)$ converges a.s.

Corollary 3. *If the continuous covariance $\Gamma_X(t, t')$ is stationary, then*

(i) *The random function $X(t)$ is normal if, and only if, the random function $\xi(s)$ which figures in its harmonic decomposition $X(t) = \int e^{its}\,d\xi(s)$*

is normal. If $\xi(t)$ *is strongly normal, then its increments are independent and* $\int_a^b \xrightarrow{\text{a.s.}} \int$ *as* $a \to -\infty$, $b \to +\infty$.

(ii) *If* $\xi(t) = \xi_d(t) + \xi_c(t)$ *is the decomposition of* $\xi(t)$ *with independent increments into its purely discontinuous and its continuous in q.m. parts, then* $X(t) = X_d(t) + X_c(t)$ *is decomposed into two independent parts:* $X_d(t) = \int e^{its}\, d\xi_d(s)$ *which is an a.s. convergent series of independent r.v.'s, and* $X_c(t) = \int e^{its}\, d\xi_c(s)$ *which obeys an infinitely decomposable law with finite variances.*

Only the very last assertion deserves proof. It is due to the fact that $E|\, d\xi(s)\,|^2 = dF_c(s)$, where the function $F_c(s)$ on R is nondecreasing, bounded, and continuous. Thus, R can be subdivided into intervals I on which the increments of $F_c(s)$ are bounded by $\epsilon > 0$ arbitrarily small. It follows that $X_c(t)$ is a sum of an arbitrarily large number of independent r.v.'s $\int_I e^{its}\, d\xi_c(s)$ whose variances are uniformly bounded by an arbitrarily small number, and this implies the assertion for every r.v. $X_c(t)$. Since the same is true of every linear combination of these r.v.'s, the random function $X_c(t)$ obeys an infinitely decomposable law.

Observe that the "if and only if" assertions remain valid when "normal" is replaced by "infinitely decomposable."

34.7 A.s. stability. If the second order random function $X(t)$ with covariance $\Gamma(t, t')$ is not normal, we have to impose supplementary restrictions upon the covariance in order to transform properties in q.m. into a.s. properties. We shall content ourselves with conditions for a.s. stability corresponding to the strong law of large numbers. If the random function is defined on the set of all integers, we look for $Y(n) = \frac{1}{n}\sum_{k=1}^{n} X(k) \xrightarrow{\text{a.s.}} 0$, and if it is defined on $T = [0, +\infty)$, we look for $Y(\tau) = \frac{1}{\tau}\int_0^\tau X(t)\, dt \xrightarrow{\text{a.s.}} 0$ where we assume $X(t)$ continuous in q.m. on T.

In both cases, the conditions we shall find and the proof are the same, except that in the second case the integrals are to be replaced by sums. Therefore, we shall give the proof in the slightly more involved case of $T = [0, +\infty)$.

First, we reduce the random function $Y(\tau)$ to a sequence of r.v.'s. We consider a sequence m^a, $m = 1, 2, \cdots$, and the symbols a, c, c'

are finite positive constants while b is a nonnegative constant. For $m^a \leqq \tau < (m+1)^a$, we can set

$$\frac{\tau}{m^a} Y(\tau) = Y(m^a) + Z(m^a, \tau), \quad Z(m^a, \tau) = \frac{1}{m^a}\int_{m^a}^{\tau} X(t)\, dt.$$

Since

$$U(m^a) = \sup_{m^a \leqq \tau < (m+1)^a} | Z(m^a, \tau) | \leqq \frac{1}{m^a}\int_{m^a}^{(m+1)^a} | X(t) | \, dt,$$

we have

$$E| U(m^a) |^2 \leqq \frac{1}{m^{2a}}\int_{m^a}^{(m+1)^a}\int_{m^a}^{(m+1)^a} E| X(t)\overline{X}(t') | \, dt\, dt'$$

$$\leqq \left(\frac{1}{m^a}\int_{m^a}^{(m+1)^a} \sqrt{\Gamma(t, t)}\, dt\right)^2.$$

We are led to assume that $\Gamma(t, t) \leqq ct^{2b}$ so that

$$\sum_{m=1}^{\infty} E| U(m^a) |^2 \leqq \sum_{m=1}^{\infty} \frac{c'}{m^{2a}} | (m+1)^{a(b+1)} - m^{a(b+1)} |^2 \sim \sum_{m=1}^{\infty} \frac{1}{m^{2(1-ab)}}$$

and the series converges when $ab < 1/2$. Then by Tchebichev's inequality and the Borel-Cantelli lemma $U(m^a) \xrightarrow{\text{a.s.}} 0$.

Similarly, by taking the sequence $q^m = (1+\epsilon)^m$ with $\epsilon > 0$, we find that

$$U(q^m) \leqq \frac{1}{q^m}\int_{q^m}^{q^{m+1}} | X(t) | \, dt \text{ a.s.}$$

and are led to assume that $| X(t) | \leqq c$, so that

$$U(q^m) \leqq c\frac{q^{m+1} - q^m}{q^m} = c\epsilon \text{ a.s.}$$

and, hence, $U(q^m)$ is arbitrarily small for $\epsilon > 0$ sufficiently small. Thus

a. *If, for t sufficiently large,*

(i) $\Gamma(t, t) \leqq ct^b$ *and* $ab < 1/2$, *then* $Y(\tau) - Y(m^a) \xrightarrow{\text{a.s.}} 0$ *as* $\tau \to \infty$.

(ii) $| X(t) | \leqq c$, *then* $Y(\tau) - Y(q^m)$ *becomes a.s. arbitrarily small for* $q - 1 > 0$ *sufficiently small.*

It remains to insure the a.s. convergence of the sequences $Y(m^a)$ or $Y(q^m)$ to zero. This is obtained as follows:

A. A.S. STABILITY THEOREM. *Let the random function* $X(t)$ *on* $T = [0, +\infty)$ *with covariance* $\Gamma(t, t')$ *be continuous in q.m., and let* c, c', γ *be finite positive constants.*

If, for large $\tau > 0$

$$\text{(i)} \qquad \Gamma(t, t) \leqq c \quad \text{and} \quad \frac{1}{\tau^2}\int_0^\tau\int_0^\tau \Gamma(t, t')\, dt\, dt' \leqq \frac{c'}{\tau^\gamma}$$

or

$$\text{(ii)} \qquad |X(t)| \leqq c \quad \text{and} \quad \int_0^\infty \frac{d\tau}{\tau^3}\int_0^\tau\int_0^\tau |\Gamma(t, t')|\, dt\, dt' \leqq c,$$

then, as $\tau \to \infty$

$$\text{(iii)} \qquad \frac{1}{\tau}\int_0^\tau X(t)\, dt \xrightarrow{\text{a.s.}} 0.$$

The same is true if $X(t)$ *is defined on the set of all integers, the integrals being replaced by the corresponding sums.*

Proof. The first condition in (i) permits to apply the first part of the preceding lemma with $b = 0$ and arbitrary $a > 0$. The second condition in (i) yields

$$\text{(1)} \qquad \sum_p E|Y(p)|^2 = \sum_p \frac{1}{p^2}\int_0^p\int_0^p \Gamma(t, t')\, dt\, dt' < \infty$$

for $p = m^a$ with $a > 1/\gamma$. The first assertion follows. Similarly, the first condition in (ii) permits to apply the second part of the preceding lemma, and the second condition in (ii) yields by elementary computations (1) with $p = q^m$ however small be $q - 1 > 0$. The second assertion follows.

If the random function $X(t)$ is second order stationary either on $T = \{\cdots -1, 0, +1, \cdots\}$ or continuous in q.m. on $T = R$, then the first condition of (i) holds for the random function $e^{-ist}X(t)$ whose covariance is $\int e^{-i(t-t')(s-s')}\, dF(s')$ with the function $F(s')$ of bounded variation. Upon replacing $\Gamma(t, t')$ by this expression in the second condition of (i), we obtain

COROLLARY. *Let the random function* $X(t)$ *be second order stationary and continuous in q.m. on* $T = R$. *If there exist two finite positive constants* c *and* γ *such that, for large* $\tau > 0$

$$\int \frac{\sin^2 \frac{\tau}{2}(s - s')}{\frac{\tau^2}{4}(s - s')^2}\, dF(s') \leqq \frac{c}{\tau^\gamma},$$

then as $\tau \to \infty$,

$$\frac{1}{\tau}\int_0^\tau e^{-ist}X(t)\, dt \xrightarrow{\text{a.s.}} 0.$$

The same is true if $X(t)$ *is defined on* $T = \{\cdots -1, 0, +1, \cdots\}$, τ *being replaced by* n *and the last integral being replaced by the corresponding sum.*

The reader is invited to compare the last assertion with the second part of the stationarity theorem (with $r = 2$).

It is easily seen that the conclusion holds a.e. (in Lebesgue measure) in s. In fact, if the symmetric derivative

$$F'(s_0) = \lim_{h \to 0} \frac{F(s_0 + h) - F(s_0 - h)}{2h}$$

exists and is finite (and this is true a.e., since the d.f. F has a.e. a finite derivative), then the integral which figures in the hypothesis is $O(F'(s_0)/\tau)$, and hence the conclusion holds with $s = s_0$. It suffices to take s_0 for origin of values of s and use the relation

$$\lim_{T \to \infty} \frac{1}{\pi}\int \frac{\sin^2 Ts}{Ts^2}\, dF(s) = F'(0),$$

which can be proved as follows: Write the integral in the form

$$\int_0^\infty \frac{\sin^2 Ts}{Ts^2}\, d\{F(s) - F(-s)\},$$

split it into $\int_0^a + \int_a^\infty$, select $a > 0$ for a given $\epsilon > 0$ so as to have $\left|\frac{F(s) - F(-s)}{2s} - F'(0)\right| < \epsilon$ on $(0, a)$, integrate by parts the integral $\int_0^a$, and let $T \to \infty$ and then $\epsilon \to 0$.

COMPLEMENTS AND DETAILS

In what follows $\Gamma(t, t')$ denotes the covariance of a second order random function $X(t)$ with $EX(t) = 0$ on $T \subset R$, unless otherwise stated.

1. $\Gamma(t, t')$ on $R \times R$ is said to be a triangular covariance of $\Gamma(t, t') = F_1(t)\bar{F}_2(t')$, $t \leqq t'$. Such a product, with $F_2 \neq 0$, is a covariance if, and only if, F_1/F_2 is nonnegative and nondecreasing on R. Construct a few random functions with triangular covariances. What about random functions orthogonal to their increments?

2. Let $\tilde{M}(t, t')$ denote the complex-conjugate of the function $M(t, t')$ on $T \times T$.

If $\int_T\int_T M(t, \tau)\Gamma(\tau, \tau')\tilde{M}(\tau', t')\,d\tau\,d\tau'$ exists and is finite, then it is covariance of the random function $\int_T M(t, \tau)X(\tau)\,d\tau$. What about $\sum M(t, t_j)X(t_j)$?

Application. The iterated $\Gamma^{(n)}(t, t')$ is defined by

$$\Gamma^{(m+n)}(t, t') = \int_T \Gamma^{(m)}(t, \tau)\Gamma^{(n)}(\tau, t')\,d\tau$$

assumed to exist and be finite. Every iterate of a covariance is a covariance. (This follows for $\Gamma^{(2n+1)}(t, t')$ from what precedes. For $\Gamma^{(2n+2)}(t, t')$ begin by verifying directly that $\Gamma^{(2)}(t, t')$ is a covariance.)

3. Let H be a family of functions h on T forming an euclidean or, more generally, a Hilbert space; denote—by overabusing the notations—the scalar product by $(h(t), h'(t))$. A function $\Gamma(t, t')$ on $T \times T$ is said to be a *reproducing kernel* of H, if $\Gamma(t, t') \in H$ for every fixed t' and $\Gamma(t, t')$ reproduces every $h \in H$, that is, $h(t') = (h(t), \Gamma(t, t'))$.

$\Gamma(t, t')$ is a reproducing kernel of some family H if, and only if, it is covariance of a random function $X(t)$ on T. If the Y's are limits in q.m. of all possible linear combinations of random values of $X(t)$, then there exist Y's such that $h(t) = EY\bar{X}(t)$, $(h_1(t), h_2(t)) = EY_1Y_2$.

Examples

1° Let $T = \{t_1, t_2, \cdots\}$. Consider the space of all sequences $\{h(t_1), h(t_2), \cdots\}$ with $\sum | h(t_n) |^2 < \infty$, $(h_1(t), h_2(t)) = \sum h_1(t_n)h_2(t_n)$. Then $\Gamma(t_m, t_n) = \delta_{m,n}$, $X(t)$ on T are second order random functions with orthonormal random values.

2° Let $T = R$ and F on R be a nondecreasing function of bounded variation. Consider all functions of the form $h(t) = \int e^{-itx}g(x)\,dF(x)$ with $\int | g(x) |^2\,dF(x) < \infty$. Then $\Gamma(t, t') = \int e^{i(t-t')x}\,dF(x)$, $X(t)$ on T are second order stationary and continuous in q.m.

4. Let $\{t_1, t_2, \cdots,\} \subset T$, set

$$D\begin{pmatrix} t, t_1, \cdots, t_n \\ t', t_1, \cdots, t_n \end{pmatrix} = \begin{vmatrix} \Gamma(t, t') & \Gamma(t, t_1) & \cdots & \Gamma(t, t_n) \\ \Gamma(t_1, t') & \Gamma(t_1, t_1) & \cdots & \Gamma(t_1, t_n) \\ \cdots & \cdots & \cdots & \cdots \\ \Gamma(t_n, t') & \Gamma(t_n, t_1) & \cdots & \Gamma(t_n, t_n) \end{vmatrix}$$

and denote by $D(t, t_1, \cdots, t_n)$ the square root of this determinant when $t' = t$.

Assume that the $X(t_n)$ are nondegenerate and are linearly independent. Then $D^2(t_1, \cdots, t_n) > 0$.

(a) The foregoing determinant is a covariance, say, of the random function $D(t_1, \cdots, t_n)Y_n(t)$ where

$$Y_n(t) = \frac{1}{D^2(t_1, \cdots, t_n)} \begin{vmatrix} X(t) & \Gamma(t, t_1) & \cdots & \Gamma(t, t_n) \\ X(t_1) & \Gamma(t_1, t_1) & \cdots & \Gamma(t_1, t_n) \\ \cdot & \cdot & \cdot & \cdot \\ X(t_n) & \Gamma(t_n, t_1) & \cdots & \Gamma(t_n, t_n) \end{vmatrix}.$$

Set $Y_0(t) = X(t)$, $Y_n(t) = X(t) - \sum_{k=1}^{n} c_{nk}(t)X(t_k)$, $n = 1, 2, \cdots$. Then $Y_n(t)$ are determined either by the condition that they be orthogonal to the $X(t_k)$ or by the condition that its variance be the smallest possible. We have

$$D^2(t) \geqq D^2(t, t_1)/D^2(t_1) \geqq \cdots \geqq D^2(t, t_1, \cdots, t_n)/D^2(t_1, \cdots, t_n) \geqq \cdots$$

and

$$D^2(t_1)D^2(t_2) \cdots D^2(t_n) \geqq D^2(t_1, \cdots, t_n).$$

(b) Set $\xi_k = Y_{k-1}(t_k)/\sqrt{E|Y_{k-1}(t_k)|^2}$. Then $E\xi_k\bar{\xi}_l = \delta_{k,l}$ and

$$Y_n(t) = X(t) - \sum_{k=1}^{n} a_k(t)\xi_k, \quad a_k(t) = EX(t)\bar{\xi}_k.$$

The sequence $Y_n(t) \xrightarrow{\text{q.m.}} Y(t)$ such that

$$Y(t_n) = 0,\ EY(t)\bar{X}(t_n) = 0,\ EY(t)\bar{Y}(t') = \lim D\begin{pmatrix} t, t_1, \cdots, t_n \\ t', t_1, \cdots, t_n \end{pmatrix}/D^2(t_1, \cdots, t_n)$$

and

$$X(t) = X_\infty(t) + Y(t)$$

with

$$EX_\infty(t)\bar{Y}(t') = 0, \quad X_\infty(t) = \sum_{n=0}^{\infty} a_n(t)\xi_n.$$

Let $t_0 = t$, $t_n = U^n t$ where U is a transformation on T to T, so that $T_\infty = \{t_n\}$ moves with t. Let $\eta(t_n) = Y(t_n)/\sqrt{E|Y(t_n)|^2}$, $n = 0, 1, \cdots$, so that the $\eta(t_n)$ form an orthonormal system moving with t.

The random function $X(t)$ is decomposable into $X(t) = X_1(t) + X_2(t)$ where the random functions $X_1(t)$ and $X_2(t)$ are orthogonal on $\{U^n t\}$,

$$X_1(t) = \sum_{n=0}^{\infty} a_n(t)X_1(t_n), \quad X_2(t) = \sum_{n=0}^{\infty} b_n(t)\eta(t_n)$$

with

$$\frac{D(t, t_1, \cdots)}{D(t_1, t_2, \cdots)}\eta(t) = X(t) - \sum_{k=1}^{\infty} a_k(t)\xi_k, \quad E\xi_k\bar{\xi}_l = \delta_{k,l}.$$

(c) What are the properties of ch.f.'s and second order stationary random functions which follow from (a) and (b).

5. To every continuous stationary covariance there corresponds a random function of the form $X(t) = \alpha e^{\beta t}$ where α and β are independent r.v.'s. Find its harmonic decomposition.

6. Let $\xi(t)$ on $[0, +\infty)$ be a random function with orthogonal increments and $E|\xi(t)|^2 = t$. Investigate the random functions $X(t) = e^{-t}\xi(e^{2t})$, $X(t) = \xi(t+1) - \xi(t)$, $X(t) = \int g(t+\tau)\, d\xi(\tau)$ with $\int |g(t)|^2\, dt = 1$,

$$X(t) = \int_{-\infty}^{t} e^{\tau - t}(t-\tau)^n\, d\xi(t), \quad X(t) = \int_0^\infty e^{-\tau - \frac{1}{\tau}}\, d\xi(t-\tau).$$

7. Let the random function $X(t)$ be defined for $t = n = 0, 1, 2, \cdots$. $X(t)$ is second order stationary and a second order chain if, and only if, $\Gamma(m, n) = f(m-n) = f(0)e^{(a+ib)(m-n)}$ with $a \leqq 0$. What can be said about the harmonic decomposition of such random functions and of their covariances?

Extend to $X(t)$ on $[0, +\infty)$.

8. It is assumed in what follows that the random functions under consideration are of second order and that the integrals and derivatives are in q.m. and exist. In every case the reader shall write the conditions for existence in terms of covariances and assume or prove that they are fulfilled.

Let the random functions $\xi(s)$ and $\eta(s)$ on R be independent. Let $X(t) = \int e^{its}\, d\xi(s)$ and $Z(t) = \int e^{its}\eta(s)\, d\xi(s)$. We say that $Z(t)$ is obtained from $X(t)$ by a linear operation with gain $\eta(s)$. Usually, it is assumed that $\eta(s)$ is degenerate (into a sure function) and that $X(t)$ is second order stationary and continuous in q.m., that is, $E|d\xi(s)|^2 = dF(s)$, $\text{Var}\, F < \infty$.

(a) Express Γ_Z in terms of Γ_ξ and Γ_η; find its form in the usual case.

Interpret $\dfrac{dX(t)}{dt}$ as a linear operation; what is the corresponding gain?

$\{X(t), Z(t)\}$ is second order stationary and continuous in q.m., if, and only if, so is $X(t)$.

(b) Express $\Delta_\theta(t) = X(t+\theta) - Z(t)$ in terms of $\xi(s)$ and $\eta(s)$. We say that $\Delta_\theta(t)$ is the error function when $X(t+\theta)$ is replaced by $Z(t)$. Express Γ_θ in terms of Γ_ξ and Γ_η. In the usual case $\Gamma_\theta(t, t) = \int |e^{i\theta s} - \eta(s)|^2\, dF(s)$ and is minimized for $\eta(s) = \eta_0(s)$ such that for all t, $\int e^{its}\eta_0(s)\, dF(s) = \int e^{its}e^{i\theta s}\, dF(s)$, provided such an $\eta_0(s)$ exists; the difference for $\eta_0(s)$ and any $\eta(s)$ is given by $\int |\eta_0(s) - \eta(s)|^2\, dF(s)$. The linear prediction problem is that of existence and determination of $\eta_0(s)$.

(c) If $\eta(s) = \int e^{-ist}\, dY(t)$, with same affixes for $\eta(s)$ and $Y(t)$ if any, then $Z(t) = \int X(t-\tau)\, dY(\tau)$ is equivalent to $Z(t) = \int e^{its}\eta(s)\, d\xi(s)$. We say that the convolution is a filtering of $X(t)$ by $Y(t)$ with gain $\eta(s)$. What are $X(t)$ and $Y(t)$ in the usual case?

Filtering by $Y_1(t) + Y_2(t)$ is a filtering with gain $\eta_1(s) + \eta_2(s)$. Filtering by $Y_1(t)$ and then by $Y_2(t)$ is a filtering with gain $\eta_1(s)\eta_2(s)$; find the corresponding $Y(t)$.

(d) A convolution defined by $Z(t) = \int X(t-\tau)\, dY(\tau)$ (without reference to $\eta(s)$ which may not exist) will be called averaging of $Y(t)$ by $X(t)$. Usually, this terminology is used when $X(t)$ is degenerate and $E|dY(t)|^2 = dt$.

Let $X(t)$ be degenerate, let $E|\,dY(t)\,|^2 = dG(t)$ and denote by μ the σ-finite measure on the Borel field in R determined by the finite function G on R. Then, up to a constant factor Γ_Z is a ch.f. with a μ-continuous d.f. and we can write $Z(t) = \int e^{its}g(s)\,dY(s)$; find $g(s)$ in the usual case. Conversely, if Γ_Z has the foregoing properties, then $Z(t)$ is an averaging of a $Y(t)$ with the foregoing property by a degenerate $X(t)$.

Part Five

ELEMENTS OF RANDOM ANALYSIS

As soon as random functions on more general sets than sets of integers appear, random analysis comes into its own. It is concerned with analytical properties and, in particular, with local ones such as continuity. The most important types of random functions isolated so far are the decomposable, martingale, and Markov ones. Foundations of random analysis and analysis of decomposable and martingale types are due primarily to P. Lévy and to Doob. Analysis of the Markov type was founded primarily by Kolmogorov and by Feller.

Investigations of random functions rely very heavily upon the particular case of random sequences. But, by their very nature, they are on a higher level of mathematical sophistication. The less involved portions may be covered first: 35.1, 35.2, 36.1, 37.3 and 39.1, 39.2.

Chapter XI

FOUNDATIONS; MARTINGALES AND DECOMPOSABILITY

§ 35. FOUNDATIONS

Random functions were defined as families $X_T = (X_t, t \in T)$ of r.v.'s on some pr. space $(\Omega, \mathfrak{A}, P)$. According to convenience, the random values at t are denoted by X_t or $X(t)$, the values of X_t at ω are denoted by $X_t(\omega)$ or $X(\omega, t)$ and, unless otherwise stated, ω and s, t, u with or without affixes, are elements of Ω and of T, respectively.

Since a function is a mapping of a space—its domain, to a space—its range space, the above definition is to be completed by specifying the domain and the range space of the random function. We are at liberty to select them according to the argument ω, t or (ω, t). Then, in order to proceed to the analytical study of random functions or *random analysis*, concepts such as extrema, continuity, measurability, are to be introduced. Thus σ-fields and/or topologies (that is, concepts of limit) are to be selected in the domains and/or range spaces. There was no such problem for random sequences: The domain was the set of natural integers $n = 1, 2, \cdots$ with limit as $n \to \infty$. The range space was the space of r.v.'s on some pr. space with limits in pr., a.s., in the rth mean. Analytical questions, such as continuity in the argument or measurability of the limits, did not arise except for the existence of limits as $n \to \infty$. However, for more general families of r.v.'s these questions are to be given a precise meaning before proceeding to the study of analytical properties of various types of r.f.'s.

In the literature, the terms "random function," "random process" and "stochastic process" are treated as synonymous. In fact, "process" means sometimes a family of r.v.'s and sometimes a class of such families, and it is deemed preferable to separate these meanings. A *random function* (*r.f.*) will be a family $X_T = (X_t, t \in T)$ of r.v.'s. A (*random* or *stochastic*) *process* will be a class of r.f.'s with a common conditional law (given the "initial" or "boundary" or "lateral" conditions). In intuitive terms, consider the argument t belonging to, say, $T = [0, \infty)$ as "time" and the conditional law of a r.f. X_T given, say, X_{T_0}, $T_0 \subset T$, as its "law of evolution." Then a process $(X_T \mid X_{T_0})$ is the class of all r.f.'s on our pr. space with the same law of evolution $\mathfrak{L}(X_T \mid X_{T_0})$, and to every choice of X_{T_0} there corresponds a r.f. of the process. "Markov" processes, whose analytical properties are investigated in the next chapter, are of this type with $T_0 = \{0\}$ and those which lend themselves to a detailed analysis are regular:

We say that $(X_T \mid X_0)$ is a *regular process* if there exists a regular c.pr. P^{X_0} of events defined on the process. There are two ways of viewing a regular process. The first one corresponds to underlying laws only: the process is a family of r.f.'s X_T with laws determined by the family of laws $\mathfrak{L}(X_T \mid X_0 = x)$, $x \in \mathfrak{X}$, and the choice of the initial law $\mathfrak{L}(X_0)$ on $\mathcal{S}$. The second one corresponds to underlying functions—the process is a family X_T of measurable functions X_t, $t \in T$, on a measurable space $(\Omega, \mathfrak{A})$ to a measurable space $(\mathfrak{X}, \mathcal{S})$ and a family $(P^x, x \in \mathfrak{X})$ of pr.'s on $\mathfrak{B}(X_T)$; to every choice of the initial distribution P_0 there corresponds a r.f. X_T on the pr. space $(\Omega, \mathfrak{B}, P)$ with $\mathfrak{B} = \mathfrak{B}(X_T)$ and P on $\mathfrak{B}$ defined by $PB = \int P_0(dx)P^xB$, $B \in \mathfrak{B}$. We shall choose the point of view according to convenience. But first we have to install the apparatus of random analysis.

35.1 Generalities. We examine possible complete definitions of r.f.'s considered as mappings. We take them on some unspecified but fixed pr. space $(\Omega, \mathfrak{A}, P)$.

Analogy with random sequences yields the following

$\mathfrak{R}$-DEFINITION. *A r.f. X_T is a function on a set T to the space $\mathfrak{R}$ of r.v.'s on the pr. space.*

In general, T is some set in the Euclidean line $R = (-\infty, +\infty)$ or the compactified Euclidean line $\bar{R} = [-\infty, +\infty]$. Unless otherwise stated, it will be so, and we shall denote by $\bar{T}$ the closure (or adherence) of T in the corresponding topology.

The values X_t at t are r.v.'s, that is, finite measurable functions on the pr. space to the Borel line. We have encountered more general random elements such as complex-valued r.v.'s or random vectors. Their common feature is that they take their values in (linear complete) separable metric spaces $\mathfrak{X}$ and the Borel sets in these spaces are topological Borel sets—the elements of the σ-fields $\mathcal{S}$ generated by the topology (the class of open sets). What follows, either is directly applicable to such random elements or can be transposed without difficulty. Therefore, we shall denote the *state space*—the measurable space of values of each r.v. by $(\mathfrak{X}, \mathcal{S})$ but, to fix the ideas, $\mathfrak{X}$ *will be a Borel set in* R *or* $\bar{R}$, unless otherwise stated.

In the range space $\mathfrak{R}$ we have at our disposal various types of limit based upon probability. However, in order that such limits, when they exist, be unique (that is, the corresponding topology be separated) we are led, as in the case of random sequences, to identify equivalent r.v.'s and, more generally, equivalent measurable functions: X and $\tilde{X}$ are equivalent when $X = \tilde{X}$ on N^c; N, *with or without affixes, will denote a null event.* In turn, this identification already led us to a slight extension of the concept of r.v., to be considered either as a representative element only of a class of equivalence or as an a.s. defined, a.s. finite, real-valued measurable function on the pr. space.

In the case of limits in pr. and in the rth mean, we know that the spaces of equivalence classes are linear complete metric spaces with the corresponding distances defined by $d(X, Y) = E\{|X - Y|/(1 + |X - Y|)\}$ and $d(X, Y) = E|X - Y|^r$ or $E^{1/r}|X - Y|^r$ according as $r < 1$ or $r \geqq 1$; for $r \geqq 1$ the spaces L_r are Banach spaces with norm $\|X\|_r = E^{1/r}|X|^r$. Classes of equivalence are partially ordered by the relation $X \leqq Y$ a.s. or, equivalently, $g(X) \leqq g(Y)$ a.s. where g is some real-valued strictly increasing finite Borel function, otherwise arbitrary; in particular, we can select a bounded and continuous function g, say, $g =$ Arctan. Thus, whenever we are concerned with order relations or existence and uniqueness of limits along T, we may assume without loss of generality that our r.f. X_T is bounded, that is, all its values X_t are uniformly bounded by some finite constant.

At first sight, identification of equivalent r.v.'s leads to identification of equivalent r.f.'s: We say that X_T and $\tilde{X}_T$ are *equivalent r.f.'s*, and write, $X_T \sim \tilde{X}_T$, if $X_t = \tilde{X}_t$ a.s., that is, $X_t = \tilde{X}_t$ on N_t^c, for every $t \in T$; thus, the equivalence class of a r.f. X_T is characterized by the common law of all the elements of the class. When, moreover, the set $N = \bigcup_{t \in T} N_t$ is null, that is, $X_t = \tilde{X}_t$ on N^c for all $t \in T$, we say that X_T and $\tilde{X}_T$ are *a.s. equal r.f.'s*. When the set T is not countable, the foregoing union set may be nonnull or even nonmeasurable, so that equivalence of r.f.'s does not imply their a.s. equality and a.s. equality classes are subclasses of equivalence classes of r.f.'s. Identification of equivalent r.f.'s looks natural since the pr. space is to be but a frame of reference, that is, the pr. properties of r.f.'s are to be describable in terms of their laws. Yet, while no difficulties arise in the case of random sequences, in the general case we are faced with analytical properties of r.f.'s which vary within the equivalence classes and thus are not describable in terms of the common laws alone. For example, let $T = [0, 1]$, take for pr. space the Lebesgue interval $[0, 1]$, and consider r.f.'s X_T defined as follows: $X_t(\omega) = 0$ except at ω_t where $X_t(\omega_t) = 1$; the ω_t can be varied in any way from one r.f. to another without altering their equivalence. Consider $Y = \sup_{t \in T} X_t$, which takes at most two values 0 and 1, and set $A = [Y = 1] = (\omega_t, t \in T)$. We can select the ω_t so that A be any set in Ω. For instance, $A = \emptyset$ and $Y \equiv 0$ for the r.f. with every $\omega_t \in \emptyset$, $A = \Omega$ and $Y = 1$ for the r.f. with every $\omega_t = t$, and Y is not measurable for a r.f. with ω_t ranging over a nonmeasurable set. Thus we are forced to back down and to limit ourselves to subclasses of equivalence classes of r.f.'s, their choice to be based upon the requirement that limits along T be measurable, hence the possibility of expressing their analytical properties in terms of the common laws. Luckily such a choice is always possible within any equivalence class and the next subsection is devoted to a specific choice—that of "separable" r.f.'s, due to Doob.

Instead of emphasizing the argument "t" we may emphasize the argument "ω" and thus consider r.f.'s X_T as functions on Ω to the space $\mathfrak{X}_T = \prod_{t \in T} \mathfrak{X}_t$ where the $\mathfrak{X}_t$ are replicas of the range space $\mathfrak{X}$ of the X_t. The values $X_T(\omega)$ at ω of a r.f. X_T will be called *sample functions* or *trajectories* or *paths* of the r.f.; they are elements of the space $\mathfrak{X}_T$, that is, functions on T with values $X_t(\omega) \in \mathfrak{X}_t$. However, we have to insure that the sections $X_t = (X_t(\omega), \omega \in \Omega)$ at every t be r.v.'s. This necessitates the introduction of the Borel field $\mathcal{S}_T$ in $\mathfrak{X}_T$ generated by the class of Borel cylinders $C(S_{T_n})$ whose bases S_{T_n} are Borel sets in finite-dimensional subspaces $\mathfrak{X}_{T_n} = \prod_{t \in T_n} \mathfrak{X}_t$; it suffices to take finite product bases $S_{T_n} = \prod_{t \in T_n} S_t$ of Borel sets S_t in $\mathfrak{X}_t$. But the class of cylinders with countably-dimensional Borel bases is contained in $\mathcal{S}_T$, contains the generating class of cylinders, and is itself a σ-field, hence

The Borel field $\mathcal{S}_T$ in $\mathfrak{X}_T$ coincides with the class of cylinders with countably-dimensional Borel bases.

The measurable space $(\mathfrak{X}_T, \mathcal{S}_T)$ will be called the *sample space* of X_T.

SAMPLE DEFINITION. *A r.f. X_T is a measurable function on a pr. space $(\Omega, \mathcal{A}, P)$ to a sample space $(\mathfrak{X}_T, \mathcal{S}_T)$; in symbols $X_T^{-1}(\mathcal{S}_T) \subset \mathcal{A}$.*

If a r.f. X_T is so defined then, for every Borel set $S_t \subset \mathfrak{X}_t$, $[X_t \in S_t] = X_T^{-1}(C(S_t)) \in \mathcal{A}$ so that the X_t are r.v.'s. Conversely, if according to the $\mathcal{R}$-definition the X_t are r.v.'s then, for every finite product $S_{T_n} = \prod_{t \in T_n} S_t$ of Borel sets S_t in $\mathfrak{X}_t$, $X_T^{-1}(C(S_{T_n})) = \bigcap_{t \in T_n} X_T^{-1}(C(S_t)) \in \mathcal{A}$ so that X_T is a r.f. according to the sample definition. Thus these definitions are equivalent.

The $\mathcal{R}$-definition leads to analytical properties of r.f.'s X_T in terms of limits in $\mathcal{R}$, say, X_T is *continuous in pr. or a.s. or in the rth mean at t* according as $X_{t'} \xrightarrow{P} X_t$ or $X_{t'} \xrightarrow{\text{a.s.}} X_t$ or $X_{t'} \xrightarrow{r} X_t$ as $t' \to t$; we drop "at t" when the property holds for all t. The sample definition leads to analytical properties of r.f.'s X_T in terms of those of their sample functions $X_T(\omega)$, say, X_T is *sample continuous or sample measurable or sample integrable at* ω if the property is true for $X_T(\omega)$; we replace "at ω" by "on A" when the property holds for all $\omega \in A$, drop it when $A = \Omega$, and say that the sample property is *almost sure* (*a.s.*), or holds for *almost all* sample functions, when $A = N^c$.

In general, analytical properties of a r.f. relative to the sample space are "finer" than the corresponding properties relative to the space $\mathcal{R}$ of r.v.'s. Let us consider a.s. continuity properties: *a.s. sample continuity of X_T implies a.s. continuity of X_T* but the converse is not necessarily true. For, as $t' \to t$, in the first case $X_{t'} \to X_t$ on N^c where the null event N is independent of t, while in the second case $X_{t'} \to X_t$ on N_t^c where the null events N_t may depend upon t and $\bigcup_{t \in T} N_t$ may not be a null event or may even not be an event.

It may be convenient to describe a.s. continuity and (a.s.) sample continuity in negative terms. When $X_{t'}$ does not converge a.s. to X_t as $t' \to t$, we say that t is a *fixed* discontinuity point of X_T. A discontinuity point $t = t(\omega)$ of $X_T(\omega)$ which is not a fixed discontinuity point of X_T will be called a *moving* (with ω) discontinuity point of X_T. Thus a.s. continuity of X_T means no fixed discontinuity points while (a.s.) sample continuity of X_T means neither fixed nor (outside a null event) moving discontinuity points.

An advantage of the sample definition is that it introduces directly the σ-field $\mathcal{B}(X_T) = X_T^{-1}(\mathcal{S}_T)$ of events induced by the r.f. X_T—the union σ-field of the σ-fields $\mathcal{B}(X_t)$ induced by the r.v.'s X_t. As long as we are concerned with the properties of X_T alone, the only events and (not necessarily finite) r.v.'s we have to consider are those *defined on* X_T, that is, events belonging to $\mathcal{B}(X_T)$ and $\mathcal{B}(X_T)$-measurable r.v.'s. Since the Borel sets in the sample space are cylinders with countably-dimensional Borel bases S_{T_c} and $X_T^{-1}(C(S_{T_c})) = X_{T_c}^{-1}(S_{T_c})$, events defined on X_T are defined on countable sections X_{T_c} of X_T (those sections varying in general with the events). More generally

a. Countability lemma. *A r.v. ξ is defined on X_T if and only if it is a Borel function g of X_T, in fact, of a countable section X_{T_c} of X_T. In particular, if $\xi = E(Y \mid X_T)$ a.s. then $\xi = E(Y \mid X_{T_c})$ a.s.*

Proof. If $\xi = g(X_T)$ then $\xi^{-1} = X_T^{-1}g^{-1}$ implies that $\mathcal{B}(\xi) \subset \mathcal{B}(X_T)$ and the "if" assertion is proved.

Conversely, let $\mathcal{B}(\xi) \subset \mathcal{B}(X_T)$ so that the events $A_{nk} = \left[\frac{k}{2^n} \leqq \xi < \frac{k+1}{2^n}\right]$ for k finite and $A_{nk} = [\xi = -\infty]$ or $[\xi = +\infty]$ for $k = -\infty$ or $+\infty$, belong to $\mathcal{B}(X_T)$. By the sample definition there exist Borel sets $S_{nk} \in \mathcal{S}_T$ such that $A_{nk} = X_T^{-1}(S_{nk})$ and, since the events A_{nk} are disjoint in k, we also have $A_{nk} = X_T^{-1}(S'_{nk})$ where the Borel sets $S'_{nk} = S_{nk}(\bigcup_{j \neq k} S_{nj})^c$ are disjoint in k. The functions $g_n = \sum_k \frac{k}{2^n} I_{S'_{nk}}$ are Borel functions on $\mathfrak{X}_T$ and $g_n(X_T) \to \xi$ on the range $S' = X_T(\Omega)$ of X_T. Thus, $g = \lim g_n$ exists on some Borel set $S \supset S'$ and setting, say, $g = 0$ on S^c, we obtain a Borel function g on $\mathfrak{X}_T$ with $\xi = g(X_T)$. This proves the "only if" assertion.

The "countable section" assertion follows from what precedes since the events A_{nk} are defined on countable sections of X_T; or it suffices to observe that the events $[\xi < r]$ where r varies over the rationals generate $\mathcal{B}(\xi)$ and, every $[\xi < r]$ being defined on a countable section X_{T_r}, the r.v. ξ is defined on the countable section X_{T_c} where $T_c = \bigcup_r T_r$. In particular, if $\xi = E(Y \mid X_T)$ a.s. then $\xi = E(Y \mid X_{T_c})$ a.s. upon conditioning by X_{T_c}. The proof is terminated.

We may emphasize the argument (ω, t) and consider r.f.'s as functions on $\Omega \times T$ to $\mathfrak{X}$ with values $X(\omega, t)$. But then we have to insure that sections $X_t = (X(\omega, t), \omega \in \Omega)$ at every t be r.v.'s, and this brings us back to the previous definitions. Yet, the present interpretation leads to an important type of r.f. as follows. Let $\mathfrak{J}$ be some σ-field in T.

Measurable r.f. definition. *An $(\mathcal{A} \times \mathfrak{J})$-measurable r.f. is a measurable function on $(\Omega \times T, \mathcal{A} \times \mathfrak{J})$ to the state space $(\mathfrak{X}, \mathcal{S})$.*

Since every section at t of an $(\mathfrak{A} \times \mathfrak{I})$-measurable function X_T is $\mathfrak{A}$-measurable, it follows that the sections X_t are $\mathfrak{A}$-measurable; similarly for $\mathfrak{I}$-measurability of sections $X_T(\omega)$ at ω:

$(\mathfrak{A} \times \mathfrak{I})$-measurable r.f.'s are r.f.'s and their sample functions are $\mathfrak{I}$-measurable.

A measurable X_T is a *Borel* r.f. if T is a Borel set and $\mathfrak{I}$ is the σ-field of Borel sets in T. Then we introduce on $\mathfrak{I}$ the Lebesgue measure λ (that is, its restriction to Borel measure on $\mathfrak{I}$). If X_T coincides with a Borel r.f. outside a $(P \times \lambda)$-null set, we say that X_T is an *a.e. Borel* r.f. We emphasize that measurability of r.f.'s is relative to the product σ-field $\mathfrak{A} \times \mathfrak{I}$ and not, as usual, to the completion of this σ-field with respect to the product measure $P \times \mu$ where μ is a measure on $\mathfrak{I}$. The importance of measurable r.f.'s is due to the fact that, as we shall see later, under some continuity conditions a r.f. is equivalent to a measurable one, and to the following immediate consequences of measurability and of Fubini's theorem.

b. Measurability lemma. *If X_T is an $(\mathfrak{A} \times \mathfrak{I})$-measurable r.f. then the sample functions are $\mathfrak{I}$-measurable and, for $X_T \geqq 0$ or $\int_T E|X_t|\,d\mu(t) < \infty$ where μ is a σ-finite measure on $\mathfrak{I}$,*

$$E\left(\int_T X_t\,d\mu(t)\right) = \int_T (EX_t)\,d\mu(t).$$

If X_T is a Borel r.f. then, for every r.v. τ with range in T, the function $X_\tau = (X(\omega, \tau(\omega)), \omega \in \Omega)$ is a r.v.

Application. Let X_T be a Borel or, more generally, an a.e. Borel stationary r.f. with $T = [0, \infty)$; stationarity means as usual that $\mathfrak{L}(X_{t_1}, \cdots, X_{t_n}) = \mathfrak{L}(X_{t_1+h}, \cdots, X_{t_n+h})$ for all finite subsets $(t_1, \cdots, t_n) \subset T$ and all $h > 0$. It follows at once that if a r.v. ξ is defined on X_T and ξ_h is its translate by h, then ξ and ξ_h have the same distribution. Let $E|X_0|^r$ be finite for some $r \geqq 1$ so that the r.v.'s $Y_n = \int_{n-1}^n X_s\,ds$ and $Z_n = \int_{n-1}^n |X_s|\,ds$ exist and $EY_n = EX_0$, $EZ_n = E|X_0|$, $E|Y_n|^r \leqq EZ_n^r \leqq E|X_0|^r$ are finite. Since the sequences are stationary, it follows that $(Y_1 + \cdots + Y_n)/n \xrightarrow[r]{\text{a.s.}} \bar{Y}$ and $(Z_1 + \cdots + Z_n)/n \xrightarrow[r]{\text{a.s.}} \bar{Z}$ hence $Z_n/n \xrightarrow[r]{\text{a.s.}} 0$. Therefore, if $n = n_t$ is the largest integer contained in t then, as $t \to \infty$,

$$U_t = \frac{1}{t}\int_0^t X_s\,ds = \frac{n}{t}\left(\frac{1}{n}\int_0^n X_s\,ds + \frac{1}{n}\int_n^t X_s\,ds\right) \xrightarrow[r]{\text{a.s.}} \bar{Y},$$

since $n/t \to 1$, the first divided integral in parentheses is $(Y_1 + \cdots + Y_n)/n$ and the second one is bounded by Z_{n+1}/n. Finally, if $\mathfrak{C}$ is the σ-field of invariant (under translations) events on X_T then, for every $C \in \mathfrak{C}$,

$$\int_C E^{\mathfrak{C}} X_0\,dP = \int_C X_0\,dP = \int_C X_s\,dP = \int_C U_t\,dP \to \int_C \bar{Y}\,dP$$

so that $\bar{Y} = E^{\mathfrak{C}} X_0$ a.s. Thus

A. R.F.'S STATIONARITY THEOREM. *If $X_{[0,\infty)}$ is a stationary a.s. Borel r.f. with $E|X_0|^r < \infty$ for an $r \geqq 1$, then*

$$\frac{1}{t}\int_0^t X_s\,ds \xrightarrow[r]{\text{a.s.}} E^{\mathfrak{C}} X_0$$

where $\mathfrak{C}$ is the σ-field of invariant events on X_T.

35.2 Separability. Let $X_T = (X_t, t \in T)$ be a r.f. with domain T. Let $\bar{T}$ be the closure of T and set $I_n = \left(t - \frac{1}{n}, t + \frac{1}{n}\right)$. By definition, for every $t \in \bar{T}$,

$$\underline{X}_t = \liminf_{t' \to t} X_{t'} = \sup_n \inf_{t' \in I_n T} X_{t'}$$

$$\bar{X}_t = \limsup_{t' \to t} X_{t'} = \inf_n \sup_{t' \in I_n T} X_{t'}.$$

Thus, to every r.f. X_T on T there correspond two limit functions $\underline{X}_T$ and $\bar{X}_T$ on $\bar{T}$, respectively, lower and upper semi-continuous. Since the limits are taken using nondeleted neighborhoods,

$$\underline{X}_t \leqq X_t \leqq \bar{X}_t, \quad t \in T,$$

and $\underline{X}_t = X_t = \bar{X}_t$ at every isolated point $t \in T$. Since the X_t are a.s. finite, it follows that $\underline{X}_t < +\infty$ a.s., $\bar{X}_t > -\infty$ a.s. Also

If $\lim_{t' \to t} X_{t'}$ exists at $t \in T$, then it coincides with X_t.

The basic difficulty consists in that the functions $\underline{X}_t$ and $\bar{X}_t$ of $\omega \in \Omega$ may not be measurable, for then analytical properties of the r.f. X_T would not be expressible in probability terms. Since these functions of

ω are formed by means of sequences of extrema of the X_t on sets of the form IT where the I are open intervals, we are led to require that all such extrema be measurable. This requirement is automatically fulfilled when T is countable and would be fulfilled were it possible to replace T by some fixed countable subset S of T in the formation of these extrema. If there is such a set S, we say that the r.f. X_T is *separable* and is *separated by S—a separating set* of X_T. In fact

A. Separability criteria. *The following properties are equivalent and define separability of a r.f. X_T: There exists a countable subset $S = \{s_j\}$ of T such that*

—for every open interval I whose intersection with T is not empty

$$(S_1)\quad \inf_{s_j \in IS} X_{s_j} = \inf_{t \in IT} X_t, \quad \sup_{t \in IT} X_t = \sup_{s_j \in IS} X_{s_j}$$

$$(S_2)\quad \inf_{s_j \in IS} X_{s_j} \leqq \inf_{t \in IT} X_t, \quad \sup_{t \in IT} X_t \leqq \sup_{s_j \in IS} X_{s_j}$$

$$(S_3)\quad \inf_{s_j \in IS} X_{s_j} \leqq X_t \leqq \sup_{s_j \in IS} X_{s_j}, \quad t \in IT$$

—for every $t \in \overline{T}$

$$(S'_1)\quad \liminf_{s_j \to t} X_{s_j} = \liminf_{t' \to t} X_{t'}, \quad \limsup_{t' \to t} X_{t'} = \limsup_{s_j \to t} X_{s_j}$$

$$(S'_2)\quad \liminf_{s_j \to t} X_{s_j} \leqq \liminf_{t' \to t} X_{t'}, \quad \limsup_{t' \to t} X_{t'} \leqq \limsup_{s_j \to t} X_{s_j}$$

$$(S'_3)\quad \liminf_{s_j \to t} X_{s_j} \leqq X_t \leqq \limsup_{s_j \to t} X_{s_j}, \quad t \in T.$$

Proof. The three nonprimed properties are equivalent: For, $(S_1) \Rightarrow (S_2)$ while $(S_2) \Rightarrow (S_1)$ since for $S \subset T$ the reverse inequalities are always true, and $(S_2) \Rightarrow (S_3)$ while $(S_3) \Rightarrow (S_2)$ upon taking extrema over $t \in IT$.

The three primed properties are equivalent: For, $(S'_1) \Rightarrow (S'_2)$ while $(S'_2) \Rightarrow (S'_1)$ since for $S \subset T$ the reverse inequalities are always true, and $(S'_2) \Rightarrow (S'_3)$ while $(S'_3) \Rightarrow (S'_2)$ upon replacing t by t' in (S'_3) letting $t' \to t \in I\overline{T}$ and using the semi-continuity of inferior and superior limits.

It follows that the primed and unprimed properties are equivalent: For, $(S_2) \Rightarrow (S'_2)$ upon replacing I by $I_n = \left(t - \frac{1}{n}, t + \frac{1}{n}\right)$ and letting $n \to \infty$, and $(S'_3) \Rightarrow (S_3)$ since $I_n \subset I$ from some n on for any fixed

$t \in IT$ so that the extreme terms of (S_3) are then farther away from X_t than those of (S'_3). The proof is terminated.

The use of separable r.f.'s is justified in the sense that every r.f. is equivalent to a separable one. In fact, more is true, as follows.

Separability criterion (S_1) means that for every open interval I and every closed interval C

$$\text{(S)} \qquad [X_t \in C, t \in IT] = [X_{s_j} \in C, s_j \in IS](\in \mathcal{A}),$$

that is, *separability as defined is separability for closed intervals*. This interpretation leads to the general concept of separability for sets of a given class. In particular, if the above equality holds for all closed sets C, the r.f. is *separable for closed sets;* it is then *a fortiori* separable (for closed intervals). Yet, given any r.f. X_T, there exists an equivalent r.f. separable for closed sets, because of the following

a. SEPARABILITY LEMMA. *Given a r.f. X_T and a Borel set C in the state space, there exists a countable subset $S = \{s_j\}$ of T such that for all $t \in T$ the intersections $A_t = [X_{s_j} \in C, s_j \in S][X_t \notin C]$ are null events.*
In fact, given a class $\mathcal{C}$ of countable intersections of a countable class $\{C_k\}$ of Borel sets, there exists a countable subset S of T such that the intersections A_t are subsets of null events N_t which do not vary with $C \in \mathcal{C}$.

Proof. Set $A_{nt} = [X_{s_k} \in C, k \leq n][X_t \notin C]$ and $B_n = A_{ns_{n+1}}$. The PA_{nt} and $p_n = \sup_{t \in T} PA_{nt}$ form nonincreasing sequences and the B_n being disjoint $\sum PB_n \leq 1$ hence $PB_n \to 0$. Select $s_1 \in T$ arbitrarily and, for every n such that $p_n > 0$, select s_{n+1} so that $PB_n \geq \left(1 - \frac{1}{n}\right) p_n$. If $p_n = 0$ the asserted set is $(s_1, \cdots, s_n)$, and if no p_n vanishes the asserted set is $(s_1, s_2, \cdots)$ since

$$PA_t \leq PA_{nt} \leq p_n \leq PB_n \bigg/ \left(1 - \frac{1}{n}\right) \to 0.$$

If $C \in \mathcal{C}$ is intersection of some sets $C_{k'}$ of a countable class $\{C_k\}$, denote by A_{tk} and S_k the sets A_t and S relative to C_k. Then the set $S = \bigcup_k S_k$ is a countable subset of T and the event $N_t = \bigcup_k A_{tk}$ is null. Since $[X_t \notin C] \subset \bigcup_{k'} [X_t \notin C_{k'}]$ and $[X_{s_j} \in C, s_j \in S][X_t \notin C_{k'}] \subset A_{tk'} \subset N_t$, the event $[X_{s_j} \in C, s_j \in S][X_t \notin C]$ is a subset of N_t, and the proof is terminated.

B. SEPARABILITY EXISTENCE THEOREM. *Given a r.f. X_T, there exists a separable for closed sets r.f. $\tilde{X}_T$ equivalent to X_T and defined on it.*

Proof. We seek a r.f. $\tilde{X}_T$ equivalent to X_T with $\mathcal{B}(\tilde{X}_T) \subset \mathcal{B}(X_T)$, and a countable subset S of T such that for every open interval I (with $IT \neq \emptyset$) and every closed set C

$$[\tilde{X}_{s_j} \in C, s_j \in IS] = [\tilde{X}_t \in C, t \in IT].$$

Since the left side event contains the right side one, it suffices that it be contained in the right side one. Since every open interval I is a countable union of open intervals I_r with rational or infinite endpoints, it suffices to realize this property simultaneously for all I_r.

The class of closed sets is contained in the class $\mathcal{C}$ of countable intersections C of finite unions C_k of open or closed intervals with rational or infinite endpoints, so that we can apply the separability lemma with T replaced by I_rT, S by I_rS, and N_t by N_{tr}. Then the set $S = \bigcup_r S_r$ is a countable subset of T, the event $N_t = \bigcup_r N_{tr}$ is null and is the same for all C and all I_r and, for $t \in I_rT$,

$$[X_{s_j} \in C, s_j \in I_rS][X_t \notin C] \subset N_t.$$

Given I_r, let $C_r(\omega)$ be the closure of the set $\{X_{s_j}(\omega), s_j \in I_rS\}$ in the closure $\bar{\mathfrak{X}} \subset \bar{R}$ of the state space. The set $C_t(\omega) = \bigcap_{I_r \ni t} C_r(\omega)$ is non-empty and closed, and $X_t(\omega) \in C_t(\omega)$ for $\omega \notin N_t$. We set $\tilde{X}_t(\omega) = X_t(\omega)$ for $t \in S$, $\omega \in \Omega$ and for $t \notin S$, $\omega \notin N_t$ and we set, say, $\tilde{X}_t(\omega) = \liminf_{s_j \to t} X_{s_j}(\omega)$ for $t \notin S$, $\omega \in N_t$. Thus $\tilde{X}_T$ is equivalent to X_T with $\mathcal{B}(\tilde{X}_T) \subset \mathcal{B}(X_T)$ and $\tilde{X}_t(\omega) \in C_t(\omega)$ for every ω and t. Given $C \in \mathcal{C}$, if ω is such that $\tilde{X}_{s_j}(\omega) \in C$ for all $s_j \in I_rS$, then $C_t(\omega) \subset C$ for all $t \in I_rS$ and, by definition of $\tilde{X}_T$, $\tilde{X}_t(\omega) \in C$ for all $t \in I_rT$. Therefore the set $\{\tilde{X}_{s_j}(\omega) \in C, s_j \in I_rS\}$ is contained in the set $\{\tilde{X}_t(\omega) \in C, t \in I_rT\}$, hence coincides with it. The proof is terminated.

Separability implications. Separability implies properties of separating sets and of one-sided limits, and it will be convenient to weaken slightly this concept, as follows.

1° A.S. SEPARABILITY. The equivalent definitions of separability were assumed to hold on Ω. Yet, the proof of their equivalence remains valid when Ω is replaced by any fixed event. If, in particular, they hold outside some fixed null event, we say that the r.f. X_T is *a.s. separable* and continue to say that S *separates it*. When X_T is a.s. separable so is every r.f. a.s. equal to it, and there always exists a separable r.f. a.s. equal to X_T—it suffices to change X_T to a constant on the exceptional null event. Thus

In the study of a.s. properties of a given r.f., a.s. separability may be replaced by separability without loss of generality.

Note that in **A**

If the separability criteria in terms of open intervals I hold outside null events $N(I)$ which may vary with I, then the r.f. is a.s. separable.

For then, they hold for all I outside the fixed null event $N = \bigcup_r N(I_r)$ where the I_r are all open intervals with rational or infinite endpoints.

2° SEPARATING SETS. Since every $t \in \overline{T}$ has to be a limit along any separating set of a separable r.f. X_T, such sets are to be dense in T and, in particular, contain the set of isolated points of T. On account of (S_2), the union of a separating set with any countable subset of T is also a separating set. Therefore,

A separable r.f. X_T is separated by a sufficiently large countable subset of T dense in it and by any larger countable subset of T.

Thus, whenever convenient, we may include within a separating set a suitable countable set, say, that of all rationals or of all dyadic numbers belonging to T. Note that

If the separability criteria in terms of limits hold outside null events N_t which may depend upon t and the r.f. X_T is known to be a.s. separable, then the set S therein separates X_T.

For, (S'_3) outside N_t implies (S_3) outside N_t, while a.s. separability implies (S_2) outside a fixed null event N' with a separating set $S' = \{s'_k\}$ (in lieu of S). Therefore, outside the fixed null event $N = N' \cup (\bigcup_k N_{s'_k})$,

$$\inf_{s_j \in IS} X_{s_j} \leq \inf_{s'_k \in IS'} X_{s'_k} \leq \inf_{t \in IT} X_t,$$

and similarly for the suprema.

3° ONE-SIDED LIMITS. Separability allows us to replace two-sided limits along the domain of a r.f. by two-sided limits along a suitable countable set. The question arises whether the same is true of one-sided limits which are defined like the two-sided ones but with $I'_n = \left(t - \frac{1}{n}, t\right)$

for left limits and $I''_n = \left(t, t + \frac{1}{n}\right)$ for right limits, in lieu of $I_n = \left(t - \frac{1}{n}, t + \frac{1}{n}\right)$. The answer is as follows.

Let X_T be a r.f. separated by $S = \{s_j\}$ and let t be a left limit point of T. There exists in S a sequence $t_n \uparrow t$ such that, outside a null event N_t,

$$\underline{X}_{t-0} = \liminf_{t' \to t-0} X_{t'} = \liminf_{t_n \uparrow t} X_{t_n}, \quad \overline{X}_{t-0} = \limsup_{t' \to t-0} X_{t'} = \limsup_{t_n \uparrow t} X_{t_n}.$$

In particular, if $\lim_{t_n \uparrow t} X_{t_n}$ exists a.s. for every sequence $t_n \uparrow t$, then $X_{t-0} = \lim_{t' \to t-0} X_{t'}$ exists a.s. even when the exceptional null events vary with the sequence.

Similarly for right limit points of T.

It will suffice to prove the general assertion for, say, a left limit point t. We may assume that the r.f. is bounded (replacing, if necessary, X_T by Arctan X_T), so that the limits are finite. Because of separability there exist finite subsets $S_{nm} = (s_{nk}, k \leq m)$ of S in $I'_n T$ such that

$$P[Y_{nm} - Y_n > 1/n] < 1/n, \quad P[Z_n - Z_{nm} > 1/n] < 1/n$$

where

$$Y_n = \inf_{t' \in I'_n T} X_{t'} \leq Y_{nm} = \inf_k X_{s_{nk}},$$

$$Z_{nm} = \sup_k X_{s_{nk}} \leq Z_n = \sup_{t' \in I'_n T} X_{t'}.$$

The elements of all these subsets can be reordered into a sequence $t_n \uparrow t$, for they are all less than t and for every n only a finite number of them are less than $t - \frac{1}{n}$. Since all $S_{nm} \subset I'_n S$, it follows that

$$Y_n \leq Y'_n = \inf_{t_k \in I'_n S} X_{t_k} \leq Y_{nm}$$

so that $P[Y'_n - Y_n > 1/n] < 1/n$ and

$$0 \xleftarrow{P} Y'_n - Y_n \to \liminf_{t_n \uparrow t} X_{t_n} - \underline{X}_{t-0}.$$

Thus, $\underline{X}_{t-0} = \liminf_{t_n \uparrow t} X_{t_n}$ outside a null event N'_t and similarly $\overline{X}_{t-0} = \limsup_{t_n \uparrow t} X_{t_n}$ outside a null event N''_t, so that both equalities hold outside $N_t = N'_t \cup N''_t$ and the general assertion is proved.

If the one-sided limits $g_{t\pm 0}$ of a numerical function exist and are finite but unequal, the function is said to have a *simple* (or *first kind*) *discontinuity* at t. If, moreover, g_t lies between the one-sided limits (inclusively), the discontinuity is said to be a *jump*.

If a r.f. X_T is separable then the simple discontinuities of almost all its sample functions are jumps except perhaps at fixed discontinuity points, and if X_T is separable for closed sets then at these jumps the sample functions are left or right continuous.

Let $S = \{s_j\}$ separate X_T and use separability relation (S) for closed intervals and for closed sets with all open intervals $I \ni t$ where $t \notin S$ is a simple discontinuity point of $X_T(\omega)$. Upon taking the closed interval with endpoints $X_{t\pm 0}(\omega)$, (S) implies that $X_t(\omega)$ belongs to the interval so that t is a jump point. Upon observing that these endpoints are the common limit points of all sets $(X_{s_j}(\omega), s_j \in IS)$, separability for closed sets implies that $X_t(\omega)$ coincides with one of these common limit points. Thus the assertions are true for all sample functions except perhaps at the s_j. At those s_j which are not fixed discontinuity points the sample functions $X_T(\omega)$ are continuous for ω's outside null events N_{s_j}, hence outside their null union. The assertions are proved.

Random analysis is based upon the fact that limits along T for a r.f. X_T separated by S are limits along S—necessarily measurable. The immediate problem is that of finding separating sets. There is no difficulty whenever the r.f. is continuous in pr.:

C. CONTINUITY SEPARATION THEOREM. *Let X_T be an a.s separable r.f. continuous in pr. or a.s. Then every countable subset dense in T separates X_T. Let $I = [a, b]$ be intervals with endpoints in T and let $S_n = (s_{nk}, k \leqq k_n)$ form sequences of finite subsets of IT becoming dense in it: $\sup_{t\in IT}\inf_k | t - s_{nk} | \to 0$. Then*

$$\inf_k X_{s_{nk}} \xrightarrow{\mathrm{P}} \inf_{t\in IT} X_t, \quad \sup_k X_{s_{nk}} \xrightarrow{\mathrm{P}} \sup_{t\in IT} X_t$$

and if X_T is continuous a.s., then the convergence is a.s., while if $S_1 \subset S_2 \subset \cdots$ then the convergence is a.s. monotone.

Proof. Let $S = \{s_j\}$ be a subset of T dense in it. Convergence in pr. along S is implied by convergence a.s. and implies convergence a.s. of a subsequence. Thus, if $s_j \to t$ then there exists a subsequence $s_{j'} \to t$ such that $X_{s_{j'}} \to X_t$ outside a null event N_t so that on N_t^c

$$\liminf_{s_j\to t} X_{s_j} \leqq X_t \leqq \limsup_{s_j\to t} X_{s_j}.$$

The separation assertion results then from separability implications 2° for separating sets and implies the pointwise convergence assertion.

It remains to prove the convergence assertions. It will suffice to do so for, say, the infima. We can assume X_{IT} bounded. Let $S = \{s_j\}$ with $a, b \in S$ denote now a separating set of X_{IT} and set

$$Y = \inf_{t \in IT} X_t, \quad Y_n = \inf_k X_{s_{nk}}, \quad Z_m = \inf_{j \leq m} X_{s_j}.$$

Since $Z_m \downarrow Y$ a.s. as $m \to \infty$, it follows that for every $\epsilon > 0$, as $n \to \infty$ then $m \to \infty$,

$$P[Y_n - Y \geqq 2\epsilon] \leqq P[Y_n - Z_m \geqq \epsilon] + P[Z_m - Y \geqq \epsilon] \to 0,$$

so that $Y_n \xrightarrow{P} Y$. If X_T is continuous a.s. hence limsup $Y_n \leqq X_{s_j}$ outside null events N_{s_j} then, outside their null union, limsup $Y_n \leqq Y$ while $Y_n \geqq Y$, so that $Y_n \xrightarrow{\text{a.s.}} Y$. The proof is terminated.

In fact, less than continuity in pr. is required provided a suitable countable subset is excluded, because of

D. CONTINUITY EXTENSION THEOREM. *Let X_T be an a.s. separable r.f. such that limits in pr. (in the rth mean) X_{t-0} or X_{t+0} exist for every $t \in T' \subset \overline{T}$. Then there exists a countable subset T_c of T' such that for every $t \in T' - T_c$, $X_{t-0} = X_{t+0}$ and $= X_t$ for $t \in T$, outside a null event N_t. In particular, if $T' = T$ then X_T is continuous in pr. (in the rth mean) on $T - T_c$, and if $T' = \overline{T}$ then X_T extends on $\overline{T} - T_c$ to a r.f. continuous in pr. (in the rth mean) and determined up to an equivalence.*

Proof. It suffices to prove the general assertion with convergence in pr. The space of equivalence classes of r.v.'s on our pr. space, with distance $d(X, Y) = E(|X - Y|/1 + |X - Y|)$ is a complete metric space in which convergence in distance is equivalent to convergence in pr. The family of r.f.'s equivalent to X_T is a function ξ_T on T to this space, and the general assertion reduces to a classical one about functions ξ_T which take their values in a complete metric space, proved as follows.

At least one of the one-sided limits $\xi_{t\pm0}$ has no meaning at isolated or one-sided limit points of $T(\subset R)$. But their set is countable and thus may be included in T' provided it is also included in the asserted exceptional countable set T_c. Thus it suffices to consider two-sided limit points of T belonging to T'. But the set T'_n of such points and at which the oscillation of ξ_T is at least $1/n$ is countable for, by hypothesis, if $t \in T'_n$ then at least one of the one-sided limits, say, ξ_{t-0} exists, so that t is right endpoint of an interval containing no other points of T'_n. Thus we include the countable set T'_n in T_c. The remaining set $T' - T_c$ is that of two-sided limit points at which the oscillation of ξ_T vanishes, and the general assertion follows. The proof is terminated.

Continuous numerical functions on a Borel set in R are Borel functions. The question arises whether similar properties hold for r.f.'s X_T with some continuity on T based upon pr. The answer is in the affirmative in the following sense.

Let $\mathfrak{J}$ be the σ-field of Borel sets in a Borel set T and λ the Lebesgue measure on $\mathfrak{J}$.

E. MEASURABILITY THEOREM. *Let X_T be a r.f. with a Borel set T. If X_T is a.s. continuous and separable, then it is an a.e. Borel r.f. and the discontinuity sets of almost all its sample functions are λ-null. If X_T is continuous in pr., then there exists an equivalent a.e. Borel r.f. separable for closed sets.*

Proof. 1° Let $T_k^{(n)} = \left[\frac{k-1}{2^n}, \frac{k}{2^n}\right) T$, $k = 0, \pm 1, \pm 2, \cdots$, and form the infimum $\underline{X}_k^{(n)}$ and the supremum $\overline{X}_k^{(n)}$ over each nonempty $T_k^{(n)}$. Set $\underline{X}_T^{(n)} = \sum_k \underline{X}_k^{(n)} I_{T_k^{(n)}}$, $\overline{X}_T^{(n)} = \sum_k \overline{X}_k^{(n)} I_{T_k^{(n)}}$ and note that

$$\underline{X}_T \uparrow \underline{X}_T^{(n)} \leqq X_T \leqq \overline{X}_T^{(n)} \downarrow \overline{X}_T.$$

Since $\underline{X}_T^{(n)}$ and $\overline{X}_T^{(n)}$ are Borel r.f.'s so are their limits $\underline{X}_T$ and $\overline{X}_T$, hence the set $L^c = [(\omega, t): \underline{X}_t(\omega) = \overline{X}_t(\omega)]$ is $\mathfrak{A} \times \mathfrak{J}$-measurable and, by the preceding inequality $\underline{X}_t(\omega) = X_t(\omega) = \overline{X}_t(\omega)$ for $(\omega, t) \in L^c$.

If X_T is a.s. continuous then $\underline{X}_t(\omega) = X_t(\omega) = \overline{X}_t(\omega)$ for $\omega \notin N_t$, $t \in T$, hence the ($\mathfrak{A}$-measurable) sections $L_t \subset N_t$ are null events. It follows that L is $(P \times \lambda)$-null, hence almost all its sections L_ω are λ-null. Thus X_T is an a.e. Borel r.f. and the discontinuity sets L_ω of almost all its sample functions $X_T(\omega)$ are λ-null. The first assertion is proved.

2° It suffices to prove the second assertion for X_T separated for closed sets by $S = \{s_j\}$ (on account of the separability theorem and the fact that equivalence preserves continuity in pr.) with X_T and T bounded (replace, if necessary, X_T by Arctan X_T and transform T into a bounded set, say, by $x' = -2 - \frac{1}{x}$ for $x < -1$, $x' = x$ for $-1 \leqq x \leqq 1$, $x' = 2 - \frac{1}{x}$ for $x > 1$). Reorder the n first s_j into $s_1^{(n)} < \cdots < s_n^{(n)}$ and set $s_0^{(n)} = -\infty$, $T_k^{(n)} = [s_{k-1}^{(n)}, s_k^{(n)})T$.

The $Y_T^{(n)} = \sum_k X_{s_k^{(n)}} I_{T_k^{(n)}}$ are Borel r.f.'s and, by continuity in pr., $Y_t^{(n)} \xrightarrow{P} X_t$ hence $Y_t^{(m)} - Y_t^{(n)} \xrightarrow{P} 0$ as $m, n \to \infty$. Since T and the sequence $Y_T^{(n)}$ are bounded it follows, by the Fubini theorem, that

$$\int_{\Omega\times T} | Y_t^{(m)}(\omega) - Y_t^{(n)}(\omega) | \, d(P\lambda)(\omega, t) = \int_T E| Y_t^{(m)} - Y_t^{(n)} | \, d\lambda(t) \to 0.$$

A fortiori, $Y_T^{(n)} \xrightarrow{P\times\lambda} Y_T$ where Y_T is a Borel r.f. and there is a subsequence $Y_T^{(n')} \to Y_T$ outside a $(P \times \lambda)$-null set M. It follows that $Y_t^{(n')} \xrightarrow{\text{a.s.}} Y_t$ for all t outside a λ-null set T_0 into which we can and do include the countable separating set S. But, by hypothesis, $Y_t^{(n')} \xrightarrow{P} X_t$ and therefore $X_t = Y_t$ outside a null event N_t for every $t \in T - T_0$. Thus, the a.e. Borel r.f. $\tilde{X}_T$ with values $Y_t(\omega)$ for $(\omega, t) \in M^c \cap (\Omega \times (T - T_0))$ and values $X_t(\omega)$ elsewhere is equivalent to X_T. Since we included S in T_0 so that $\tilde{X}_S = X_S$ and outside the section M_t every Y_t is limit of X_{s_j} while on these sections $\tilde{X}_t = X_t$, it follows that the set S which separates X_T for closed sets does the same for $\tilde{X}_T$. The proof is terminated.

VARIANTS. 1° The theorem remains valid when continuity in pr. or a.s. holds outside a λ-null Borel subset of T: it suffices to increase the exceptional $(P \times \lambda)$-null set.

2° The theorem remains valid when continuity in pr. is one-sided. The continuity extension theorem reduces it to 1°.

3° The theorem and its foregoing variants remain valid when "Borel set T" is replaced by "Lebesgue set T": proceed as for 1°.

4° The Borel r.f.'s of the theorem are a.e. limits of sample continuous, in fact, of sample polygonal r.f.'s: We may replace the approximating simple measurable r.f.'s by polygonal ones with vertices $\underline{X}_k^{(n)}$, $\overline{X}_k^{(n)}$, $X_{s_k}^{(n)}$.

35.3 Sample continuity. The presence in r.f.'s of two arguments ω and t and the requirements of uniqueness then of measurability in ω of limits in t led us to equivalence classes of r.f.'s then to a partial retreat to their never empty subclasses of separable r.f.'s. The *a priori* weakest type of limit in t of r.v.'s X_t, based upon pr. and yielding r.v.'s, is the limit in pr. Random analysis is concerned primarily with separable r.f.'s X_T such that one-sided limits in pr. exist on T and hence, by the continuity extension lemma, with r.f.'s continuous in pr. outside countable subsets of T. When investigating specific types of r.f.'s, our first problem will thus be that of existence of one-sided limits in pr. or a.s.

But the next problem and, in fact, the essential problem of random analysis is that of the behavior of sample functions and especially that of the a.s. sample behavior, that is, behavior common to almost all sample functions. In particular, it seeks conditions for some kind of continuity: a.s. sample continuity, or a.s. sample continuity except for simple discontinuities (that is, except for jumps—because of separability) or except for nonsimple discontinuities. In this subsection we are concerned with some general conditions for such kinds of behavior when the types of r.f.'s are not specified.

In general, pointwise continuity or continuity based upon pr. is required or reduced to continuity on compact domains—when it becomes uniform. For any type "c" of convergence, X_T is "c"-*uniformly continuous* if $X_{t'} - X_t \xrightarrow{c} 0$ uniformly in $t \in T$, as $t' \to t$. For ordinary pointwise convergence hence for sample functions, the classical fact that continuity is uniform on compact domains is due to the Heine-Borel characterization of compact sets and to the equivalence of convergence and mutual convergence—which is also true of convergences based upon pr. Thus

a. Uniform continuity lemma. *For separable r.f.'s on compact domains, continuity of a sample function and continuities based upon pr. are uniform.*

In what follows T will be compact so that there will be no distinction between continuity and uniform continuity. To simplify the writing we shall take $T = [a, b]$ and, whenever convenient, replace it by $[0, 1]$—which does not restrict the generality. The reader is invited to rewrite the results for any compact T and examine their validity for noncompact T.

Let $\alpha(I) = \sup\limits_{t',t'' \in IT} | X_{t'} - X_{t''} |$ be the oscillation of X_T on an interval I and let $\beta(I) = \inf\limits_{t \in IT} \{\alpha(I \cap (-\infty, t)) + \alpha(I \cap (t, +\infty))\}$ be the "left-right" oscillation of X_T on I. Given $t \in T$, let $I_1 \supset I_2 \supset \cdots$ be a nonincreasing sequence of intervals converging to $\{t\}$ and to which t is interior. Then $\alpha(I_n)$ converges to the oscillation α_t of X_T at t and every sample function $X_T(\omega)$ for which $\alpha_t(\omega) = 0$ is continuous at t, and conversely. Similarly $\beta(I_n)$ converges to the "left-right" oscillation of X_T at t and every sample function $X_T(\omega)$ for which $\beta_t(\omega) = 0$ has no nonsimple discontinuity at t, and conversely. Finally, if $X_{t'_n}(\omega) - X_{t''_n}(\omega) \to 0$ for some $t'_n \uparrow t$ and $t''_n \downarrow t$ then the sample function $X_T(\omega)$ cannot have a simple discontinuity at t.

A. SAMPLE CONTINUITY THEOREM. *Let a r.f. X_T, $T = [a, b]$, be separable for closed sets, let $a, b \in S_n = (s_{n1}, \cdots, s_{nk_n}) \uparrow S$ dense in T, and let $I_{nk} = [s_{nk}, s_{n,k+1}]$.*

If $P[\max_k \alpha(I_{nk}) \geqq \epsilon] \rightarrow 0$ for every $\epsilon > 0$, then X_T is a.s. sample continuous.

If $P[\max_k \beta(I_{nk}) \geqq \epsilon] \rightarrow 0$ for every $\epsilon > 0$, then X_T is a.s. sample continuous except perhaps for simple discontinuities.

If $P[\max_k |X_{s_{n,k}} - X_{s_{n,k+1}}| \geqq \epsilon] \rightarrow 0$ for every $\epsilon > 0$, then X_T is a.s. sample continuous except perhaps for nonsimple discontinuities.

The $\beta(I)$ are assumed measurable. Otherwise, in their definition replace T by S, or in the corresponding condition replace P by its outer extension.

Proof. If $p_n = P[\max_k \alpha(I_{nk}) \geqq \epsilon] \rightarrow 0$ then there exists a subsequence n' such that $\sum p_{n'} < \infty$. It follows, by the Borel-Cantelli lemma that there exists a finite integer-valued r.v. ν_ϵ such that $\max_k \alpha(I_{n'k}) \leqq \epsilon$ outside a fixed null event N_ϵ for all $n' \geqq \nu_\epsilon$. But, given $t \in [a, b]$, there exists a subsequence $I_{n'k}$ converging nonincreasingly to $\{t\}$ so that $\alpha_t \leqq \epsilon$ and $\epsilon > 0$ being arbitrary, $\alpha_t = 0$. For, if $t \notin S$ then it is interior to the $I_{n'k}$ and if $t \in S$ then for n' sufficiently large it suffices to replace ϵ by 2ϵ and $I_{n'k}$ by its union with the other interval of which t is the endpoint, unless $t = a$ or b—in which case the asserted continuity is automatically one-sided.

Similarly for the other assertions, and the proof is terminated.

PARTICULAR CASES. 1° Let the suprema be taken over all intervals $[t, t + h]$ in $[a, b]$ and $o_\epsilon(h)/h \rightarrow 0$ as $h \rightarrow 0$ for every $\epsilon > 0$.

If $\sup P[\alpha[t, t + h] \geqq \epsilon] = o_\epsilon(h)$ *or* $\sup P[\beta[t, t + h] \geqq \epsilon] = o_\epsilon(h)$, *then the r.f. is a.s. sample continuous or a.s. sample continuous except perhaps for simple discontinuities.*

If $\sup P[|X_{t+h} - X_t| \geqq \epsilon] = o_\epsilon(h)$ *then the r.f. is a.s. sample continuous, except perhaps for nonsimple discontinuities.*

Replace $[a, b]$ by $[0, 1]$, take for S the set of dyadic numbers kh_n, $h_n = 2^{-n}$, and note that

$$P[\max \alpha(I_{nk}) \geqq \epsilon] \leqq \sum_k P[\alpha(I_{nk}) \geqq \epsilon] \leqq o_\epsilon(h_n)/h_n,$$

and similarly for the other assertions. The first proposition is due to Dynkin and the second one to Dobrushin.

2° In the case of consecutive sums of r.v.'s, convergence in pr. entailed a.s. convergence under conditions which for $T = [a, b]$ take the form

(C) $p_{\epsilon,t}(h)P[\sup_{t' \in [t,t+h]} |X_{t'} - X_t| \geqq g(\epsilon)] \leqq P[|X_{t+h} - X_t| \geqq \epsilon]$ for every $[t, t+h] \subset T$ and every $\epsilon > 0$, with $g(\epsilon) \to 0$ as $\epsilon \to 0$.

Inequalities of this type were obtained in the case of sums $X_1, X_2, \cdots$ by a procedure which, similarly, yields the following property:

If $P(|X_n - X_k| < \epsilon \mid X_1, \cdots, X_k) \geqq p_\epsilon$ *a.s.,* $k = 1, \cdots, n$, *then* $p_\epsilon P[\max_k |X_k| \geqq 2\epsilon] \leqq P[|X_n| \geqq \epsilon]$.

For, setting $A_k = [|X_k| \geqq 2\epsilon, \max_{j<k} |X_j| < 2\epsilon]$ (with $X_0 = 0$) and $B_k = [|X_n - X_k| < \epsilon]$ so that $\sum_k A_k = [\max |X_k| \geqq 2\epsilon]$ and $|X_n| \geqq \epsilon$ on the $A_k B_k$, we have

$$P[|X_n| \geqq \epsilon] \geqq \sum_k PA_kB_k = \sum_k E(I_{A_k}E(I_{B_k} \mid X_1, \cdots, X_k)) \geqq p_\epsilon \sum_k PA_k.$$

Upon setting $X_k = X_{t_k} - X_t$ it follows, by separability and the usual limiting procedure, that

If
(C′) $P(|X_{t+h} - X_{t_k}| < \epsilon \mid X_{t_1} - X_t, \cdots, X_{t_k} - X_t) \geqq p_{\epsilon,t}(h)$ *a.s.,* $k = 1, \cdots, n$, *for every finite subset* $t_1 < \cdots < t_n$ *in every* $(t, t+h) \subset T$, *then* (C) *holds with* $g(\epsilon) > 2\epsilon$.

Therefore, on account of 1° and

$$[\sup_{t',t'' \in [t,t+h]} |X_{t'} - X_{t''}| \geqq 2g(\epsilon)] \subset [\sup_{t' \in [t,t+h]} |X_{t'} - X_t| \geqq g(\epsilon)],$$

Under (C) *or* (C′) *with* $p_{\epsilon,t}(h) \geqq p_{\epsilon,t} > 0$ *for h sufficiently small, the r.f. is (left, right) a.s. continuous if and only if it is (left, right) continuous in pr., and if* $p_{\epsilon,t} \geqq p_\epsilon > 0$ *then* $\sup_t P\|X_{t+h} - X_t| \geqq \epsilon| = o_\epsilon(h)$ *entails a.s. sample continuity.*

In its turn, (C′) is implied by

(C″) $P(|X_{t+h} - X_{t_k}| < \epsilon \mid X_{t_1}, \cdots, X_{t_k}) \geqq p_{\epsilon,t}(h)$ a.s. for every finite subset $t_1 < \cdots < t_n$ in every $[t, t+h) \subset T$.

Then, taking $n = 1$ and $t_1 = t$, it follows that

$$P[|X_{t+h} - X_t| \geqq \epsilon] = EP(|X_{t+h} - X_t| \geqq \epsilon \mid X_t) \leqq 1 - p_{\epsilon,t}(h)$$

and hence

Under (C″) *with* $1 - p_{\epsilon,t}(h) \leqq o_\epsilon(h)$, *the r.f. is a.s. sample continuous.*

Conditions on moduli of continuity in pr. yield moduli of sample continuity, as follows:

Let $g(h)$ with $|h| \leqq h_0$ be an even, nondecreasing in $h > 0$ function such that $g(h) \to 0$ as $h \to 0$. Let $h_n = q^{-n}$, $q > 1$ integer.

B. SAMPLE CONTINUITY MODULI THEOREM. *Let* X_T, $T = [a, b]$, *be a separable r.f. such that, for every* $t, t + h \in T$,

$$P[|X_{t+h} - X_t| \geqq g(h)] \leqq q(h) \to 0 \text{ as } h \to 0.$$

(i) *If* $\sum q(h_n)/h_n < \infty$ *and* $\sum g(h_n) < \infty$, *then* X_T *is a.s. sample continuous.*

(ii) *If* $\sum q(jh_n)/h_n < \infty$ *for every integer* j *and* $\sum_{r=1}^{\infty} g(h_{n+r})/g(h_n) \leqq \alpha < \infty$ *for all* n, *then* X_T *is a.s. sample continuous and there exists an a.s. positive r.v.* H *and a finite constant* c *such that*

$$|X_{t+h} - X_t| < cg(h), \quad |h| < H.$$

If, moreover, $g(h_n)/g(jh_n)$ *is arbitrarily small for* n, j *sufficiently large, then for every* $\epsilon > 0$ *there exists an a.s. positive r.v.* H_ϵ *such that*

$$|X_{t+h} - X_t| < (1 + \epsilon)g(h), \quad |h| < H_\epsilon.$$

Proof. To simplify the writing we take $T = [0, 1]$. Since the hypotheses about $q(h)$ imply continuity in pr., we can and do separate the r.f. by the dense subset of all q-adic numbers kh_n, $k = 0, \cdots, 1/h_n$, $n = 1, 2, \cdots, h_n = q^{-n}$, $q > 1$ integer.

1° By the covering rule and the first hypothesis in (i)

$$\sum_n P[\max_k |X_{(k+1)h_n} - X_{kh_n}| \geqq g(h_n)]$$
$$\leqq \sum_n \sum_k P[|X_{(k+1)h_n} - X_{kh_n}| \geqq g(h_n)]$$
$$\leqq \sum_n q(h_n)/h_n < \infty$$

so that, by the Borel-Cantelli lemma, $\max_k |X_{(k+1)h_n} - X_{kh_n}| \geqq g(h_n)$ finitely often, that is, there exists an a.s. finite and integer-valued r.v. ν

such that for all k and $n \geqq \nu$

$$| X_{(k+1)h_n} - X_{kh_n} | < g(h_n).$$

Since every q-adic number in $I_{nk} = [kh_n, (k+1)h_n]$ is of the form $t = kh_n + \sum_{r=1}^{m} \theta_r h_{n+r}$, where $\theta_r = 0$ or 1 or $\cdots$ or $q-1$ and m is some integer, it follows, by repeated applications of the triangular inequality that for all k and $n \geqq \nu$

$$| X_t - X_{kh_n} | \leqq \sum_{r=1}^{m} \theta_r g(h_{n+r}) \leqq q \sum_{r=1}^{\infty} g(h_{n+r})$$

and, because of separability, the same is true for every $t \in I_{nk}$. By the second hypothesis in (i), $q \sum_{r=1}^{\infty} g(h_{n+r}) < \epsilon/2$ for any given $\epsilon > 0$ and $n \geqq n_\epsilon$ sufficiently large. Therefore, by applying the triangular inequality, for $n \geqq \max(n_\epsilon, \nu)$ hence $|h| < H_\epsilon$ a.s. positive r.v. and all $t \in T$, it follows that $|X_{t+h} - X_t| < \epsilon$ hence X_T is a.s. sample continuous.

2° The first two hypotheses in (ii) imply those in (i). Therefore, by the second of these hypotheses in (ii), there exists an a.s. positive r.v. ν_0 such that, for all kh_n, $t \in I_{nk}$ and $n \geqq \nu_0$,

$$| X_t - X_{kh_n} | < q \sum_{r=1}^{\infty} g(h_{n+r}) \leqq \alpha q g(h_n).$$

On the other hand, by the first hypothesis in (ii), for every fixed j,

$$\sum_n P[\max_k | X_{(k+j)h_n} - X_{kh_n}] \geqq g(jh_n) | \leqq \sum_n q(jh_n)/h_n < \infty$$

so that there exists a similar r.v. ν_j such that, for all k and $n \geqq \nu_j$

$$| X_{(k+j)h_n} - X_{kh_n} | < g(jh_n).$$

Let m be an integer to be selected later and let $\nu = \max(\nu_0, \cdots, \nu_{qm})$. Given t and $h > 0$ there exists an n such that $mh_n < h < qmh_n$, so that there exist k and j with $m \leqq j \leqq qm$ such that

$$(k-1)h_n < t \leqq kh_n < (k+j)h_n \leqq t + h < (k+j+1)h_n.$$

Since

$$| X_{t+h} - X_t | \leqq | X_{t+h} - X_{(k+j)h_n} | \\ + | X_{(k+j)h_n} - X_{kh_n} | + | X_{kh_n} - X_t |,$$

it follows that for $n \geqq \nu$ hence $h \leqq H$ a.s. positive r.v.

$$| X_{t+h} - X_t | < g(jh_n) + 2\alpha q g(h_n) \leqq (1 + 2\alpha q)g(h) = cg(h).$$

If, moreover, given $\epsilon > 0$, for $n(\geqq n_\epsilon)$ and m hence j sufficiently large, $g(h_n)/g(jh_n) < \epsilon/2\alpha q$ then, for $n \geqq \max(n_\epsilon, \nu)$, hence $h < H_\epsilon$ a.s. positive r.v.

$$| X_{t+h} - X_t | < (1 + \epsilon)g(jh_n) \leqq (1 + \epsilon)g(h),$$

and the proof is terminated.

Corollary. *Let X_T with $T = [a, b]$ be a separable r.f. such that for some $r(> 0)$ and all $t, t + h \in T$ with h sufficiently small*

$$E| X_{t+h} - X_t |^r \leqq \rho(h).$$

If $\rho(h) = c| h |^{1+s}$ with $s > 0$ or $\rho(h) = c| h |/| \log | h ||^{1+s}$ with $s > r$, then X_T is a.s. sample continuous and in the first case for every $0 < \alpha < s/r$ and $\epsilon > 0$ there exists an a.s. positive r.v. H_ϵ such that for all $t, t + h \in T$ and $| h | < H_\epsilon$

$$| X_{t+h} - X_t | < (1 + \epsilon)| h |^\alpha.$$

The assertions follow from the theorem and Markov inequality:

$$P[| X_{t+h} - X_t | \geqq g(h)] \leqq q(h) = \rho(h)/g^r(h),$$

upon taking in the first case $g(h) = | h |^\alpha$ so that $q(h) = c| h |^{1+s-\alpha r}$, and in the second case $g(h) = | \log | h ||^{-\beta}$ with $1 < \beta < s/r$ so that $q(h) = c | h |/| \log | h ||^{1+s-\beta r}$.

The fact that $\rho(h) = c| h |^{1+s}$, $s > 0$, implies a.s. sample continuity is due to Kolmogorov.

Application to second order calculus. Analytical properties of second order r.f.'s which can be described in terms of their covariances constitute the second order calculus. This was done in Section 34 for continuity, differentiability and integrability, all in q.m., and required no concepts introduced in this section. We now proceed to use them and go back to the usual abuse of notation: $X(t)$ represents a second order r.f.—complex-valued with t varying over an interval $[a, b]$, and $\Gamma(t, t') = EX(t)\overline{X}(t')$ represents its covariance; it is not assumed that $EX(t) = 0$. We recall that $\Delta_h X(t) = X(t + h) - X(t)$, $\Delta_h \Delta'_{h'} \Gamma(t, t') = E\Delta_h X(t)\Delta'_{h'}\overline{X}(t')$, $\text{Var}\,\Gamma = \iint | dd'\Gamma(t, t') |$ and that if $\Gamma(t, t')$ is of bounded variation, then the limits $\Gamma(t \pm 0, t' \pm 0)$ exist and differ from $\Gamma(t, t')$ on a countable set of (t, t') only, so that the one-sided limits in q.m. $X_{t\pm 0}$ exist and $X(t)$ is continuous in q.m. outside a countable set of values of t. This permits us to extend, with obvious modifications, Gavce's theorem which follows to the case of covariances of bounded variation. Since all relations below between r.v.'s will be a.s. relations, we drop "a.s.".

C. SECOND ORDER CALCULUS THEOREM. *Let $X(t), t \in [a, b]$, be a second order r.f. with a continuous covariance $\Gamma(t, t')$. Then, up to equivalences,*

(i) *The indefinite integral in q.m. $Y(t) = \int_a^t X(s)\, ds$ exists, $X(t)$ is its derivative in q.m., and if $Y_0(t)$ is a primitive in q.m. of $X(t)$ then $Y(t) = Y_0(t) - Y_0(a)$.*

The r.f. $X(t)$ continuous in q.m. is Borelian, sample integrable and sample square integrable, and its indefinite sample and q.m. integrals coincide.

(ii) *If $\Delta_h \Delta'_h \Gamma(t, t) \leqq ch^2$ for h sufficiently small, then $X(t)$ is sample continuous, is the sample derivative of its indefinite sample integral, and for every $\epsilon > 0$ and $0 < \alpha < \frac{1}{2}$ there exists an a.s. positive r.v. $H_{\epsilon,\alpha}$ such that $|X(t + h) - X(t)| < (1 + \epsilon)|h|^\alpha$ for $|h| < H_{\epsilon,\alpha}$.*

If the derivative $\dfrac{\partial^2}{\partial t\, \partial t'} \Gamma(t, t')$ exists and is finite then $X(t)$ is differentiable in q.m., and if $\Delta_h \Delta'_h \dfrac{\partial^2}{\partial t\, \partial t'} \Gamma(t, t') \leqq c'h^2$ for h sufficiently small then $X(t)$ is sample differentiable.

(iii) *If the derivative $\dfrac{\partial^{2n+2}}{\partial t^{n+1} \partial t'^{n+1}} \Gamma(t, t')$ exists and is finite then $X(t)$ is n times sample differentiable, and if $\Gamma(t, t')$ is infinitely differentiable, then $X(t)$ is infinitely sample differentiable.*

Proof. We use throughout the convergence in q.m. criterion without further comment.

1° Since $\Gamma(t, t')$ is continuous hence bounded on $[a, b] \times [a, b]$, the indefinite integral in q.m. $Y(t)$ exists and, as $h \to 0$,

$$E\left|\frac{1}{h}\int_t^{t+h} X(s)\, ds - X(t)\right|^2$$

$$= \frac{1}{h^2}\int_t^{t+h}\int_t^{t+h} \Gamma(s, s')\, ds\, ds' - \frac{1}{h}\int_t^{t+h} \Gamma(s, t)\, ds - \frac{1}{h}\int_t^{t+h} \Gamma(t, s')\, ds' + \Gamma(t, t) \to 0,$$

so that $X(t)$ is the derivative in q.m. of $Y(t)$. Therefore, if $X(t)$ is also the derivative in q.m. of $Y_0(t)$ then the derivative $\Delta'(t)$ in q.m. of $\Delta(t) =$

$Y(t) - (Y_0(t) - Y_0(a))$ is 0 while $\Delta(a) = 0$. It follows that $\Delta(t) = 0$ since $\frac{d}{dt} E|\Delta(t)|^2 = E\{\Delta(t)\bar{\Delta}'(t) + \Delta'(t)\bar{\Delta}(t)\} = 0$, and the first set of assertions in (i) is proved.

$X(t)$ is continuous in q.m. hence in pr. and consequently, by the measurability theorem, the r.f. $X(t)$ is equivalent to a separable a.e. Borel r.f. Boundedness of $E|X(t)|^2 = \Gamma(t, t)$ implying that of $E|X(t)| \leqq E^{1/2}|X(t)|^2$, the second set of assertions follows from the measurability lemma, except for the assertion of equivalence of $Y(t)$ and the sample integral $Z(t)$.

Since $Y_n(t) \xrightarrow{\text{q.m.}} Y(t)$ where

$$Y_n(t) = \sum_k X(t_{nk})(t_{nk} - t_{n,k-1}),$$

$$a = t_{n0} < \cdots < t_{nk_n} = t, \quad \max(t_{nk} - t_{n,k-1}) \to 0$$

the assertion will be proved if we show that

$$E|Y_n(t) - Z(t)|^2$$
$$= E|Y_n(t)|^2 - EY_n(t)\bar{Z}_n(t) - E\bar{Y}_n(t)Z(t) + E|Z(t)|^2 \to 0.$$

But

$$E|Y_n(t)|^2 \to \int_a^t \int_a^t \Gamma(s, s')\, ds\, ds'$$

while, by the measurability lemma,

$$E|Z(t)|^2 = E\int_a^t \int_a^t X(s)\bar{X}(s')\, ds\, ds' = \int_a^t \int_a^t \Gamma(s, s')\, ds\, ds',$$

and

$$EY_n(t)\bar{Z}(t) = \sum_k \left(\int_a^t \Gamma(t_{nk}, t)\, dt\right)(t_{nk} - t_{n,k-1}) \to \int_a^t \int_a^t \Gamma(t, t')\, dt\, dt'.$$

The assertion is proved and the proof of (i) is terminated.

2° Since, by the first hypothesis in (ii),

$$E|X(t+h) - X(t)|^2 = \Delta_h \Delta'_h \Gamma(t, t) \leqq ch^2,$$

the first and third assertions in (ii) result from the Corollary of the sample continuity moduli theorem with $r = 2$ and $s = 1$ applied to an equivalent separable r.f. The second one results from the fundamental theorem of ordinary calculus, since $Z(\omega, t) = \int_a^t X(\omega, s)\, ds$ where $X(\omega, s)$

is continuous in s. The fourth assertion is immediate. The fifth and last one results from

$$E|X'(t+h) - X'(t)|^2 = \Delta_h \Delta'_h \left[\frac{\partial^2}{\partial t\, \partial t'} \Gamma(t, t')\right]_{t'=t} \leqq c'h^2$$

upon applying what precedes with $X(t)$ replaced by its derivative in q.m. $X'(t)$, and assertions (iii) readily follow.

§ 36. MARTINGALES

We apply the foregoing concepts and generalities to a specific type of r.f.'s—the submartingales. While important in their own right, martingales play the role of a tool in the investigation of other types of r.f.'s and in the applications of pr. methods to various branches of analysis. Because of that and because the passage from sequences to functions is particularly neat in the case of martingales, we include them in this chapter. Martingale sequences and reversed sequences were studied in Section 29 and we shall have to complete this study and use the submartingales and convergence and closure theorems in 29.3. As usual, *all our r.f.'s are taken to be separable.*

36.1 Continuity. Let r, s, t with or without affixes denote the elements of $T \subset \bar{R}$, unless otherwise stated. We say that a r.f. X_T is a *martingale* if its expectation $EX_T = (EX_t, t \in T)$ exists and if, for every $s < t$ and every $B \in \mathfrak{B}(X_r, r \leqq s)$,

(M) $\displaystyle\int_B X_s = \int_B X_t$, equivalently, $X_s = E(X_t \mid X_r, r \leqq s)$ a.s.

In fact, it suffices to require that for every finite set $r_1 < \cdots < r_n < s < t$ and every $B \in \mathfrak{B}(X_{r_1}, \cdots, X_{r_n}, X_s)$

(M′) $\displaystyle\int_B X_s = \int_B X_t$, equivalently, $X_s = E(X_t \mid X_{r_1}, \cdots, X_{r_n}, X_s)$ a.s.

For, (M) implies (M′) since $\mathfrak{B}(X_{r_1}, \cdots, X_{r_n}, X_s) \subset \mathfrak{B}(X_r, r \leqq s)$ and (M′) implies (M) on account of the following property, already used in various guises, which we isolate now:

a. *Let a σ-field $\mathfrak{B}$ be generated by a field $\mathfrak{C}$ which consists of finite sums of events belonging to a class $\mathfrak{D}$. If φ and φ' are signed measures on $\mathfrak{B}$ σ-finite on $\mathfrak{C}$ then $\varphi \leqq \varphi'$ ($\varphi = \varphi'$) on $\mathfrak{D}$ implies the same relation on $\mathfrak{B}$, and if φ and φ' are σ-finite measures on $\mathfrak{D}$ then the same implication holds for their unique extensions to measures on $\mathfrak{B}$.*

Proof. The assertion about measures on $\mathfrak{D}$ reduces to the one about signed measures. The equality assertion reduces to the inequalities one upon applying that one to $\varphi \leqq \varphi'$ and to $\varphi' \leqq \varphi$. Since the class of events on which $\varphi \leqq \varphi'$ is closed under countable summations and contains $\mathfrak{D}$, it contains the field $\mathfrak{C}$ over $\mathfrak{D}$. Furthermore, upon taking a countable partition of Ω, we may assume that φ and φ' are finite on this class so that it is also closed under monotone passages to the limit hence contains the monotone field $\mathfrak{B}$ over the field $\mathfrak{C}$. The inequalities assertion follows, and the proof is terminated.

We say that a martingale X_T *is closed on the left* or *on the right* if it has a first element X_a or a last element X_b, that is, if T has a first element a or a last element b. More generally, we say that X_T *is a martingale closed on the left or on the right* by Y (with existing EY) if (Y, X_T) or (X_T, Y) is a martingale, equivalently, for $s < t$,

$$X_s = E(X_t \mid Y, X_r, r \leqq s) \text{ a.s. and } Y = E(X_t \mid Y) \text{ a.s.}$$

or

$$X_t = E(Y \mid X_s, s \leqq t) \text{ a.s.}$$

In other words, we may add virtual points, say, θ or θ' to T, to be the first or the last element of the increased domain T' and, setting $Y = X_\theta$ or $X_{\theta'}$, the r.f. $X_{T'}$ is a martingale closed on the left or on the right. If there are two closing r.v.'s Y and Z on the same side and if the ordered (Z, Y, X_T) or (X_T, Y, Z) is a martingale, then we say that Y is *nearer* than Z to X_T, and we say that Y is the *nearest* closing r.v. when there are no nearer ones. Trivially, if T has a first element a or a last element b then X_a or X_b are the nearest closing r.v.'s from the left or from the right.

We replace the term "martingale" by "*submartingale*" whenever in the relations which precede the sign "=" is replaced by "$\leqq$." Note that the X_t being r.v.'s hence a.s. finite, in a martingale we have $-\infty < E(X_t \mid X_r, r \leqq s) < +\infty$ a.s. while in a submartingale we have only $-\infty < E(X_t \mid X_r, r \leqq s)$ a.s.

Extensions. Let $\mathfrak{B}_s$ denote sub σ-fields of events nondecreasing with s and let $\mathfrak{B}_0$ be their union field and $\mathfrak{B}$ their union σ-field—the σ-field generated by $\mathfrak{B}_0$. Let $s < t$.

1° If $\mathfrak{B}_s = \mathfrak{B}(X_r, r \leqq s)$ then the martingales defining relation takes the form $X_s = E(X_t \mid \mathfrak{B}_s)$ a.s. Conversely, if we have such a relation then $\mathcal{B}_s \supset \mathcal{B}(X_r, r \leqq s)$ and conditioning both sides by $\mathcal{B}(X_r, r \leqq s)$ we have the martingales defining relation. We then speak of a *martingale* $(X_T, \mathcal{B}_T)$ (but it may be more restrictive—it suffices that some $\mathcal{B}_s$ be strictly larger than $\mathcal{B}(X_r, r \leqq s)$). Similarly for submartingales.

2° Since the martingales defining relation is in terms of the indefinite integrals φ_{X_t} of the X_t: $\varphi_{X_s} = \varphi_{X_t}$ on $\mathcal{B}(X_r, r \leqq s)$, we may say that the family $\varphi_{X_T} = (\varphi_{X_t}, t \in T)$ is a martingale or, upon taking into account 1°, we may say that the family $(\varphi_{X_t}, \mathcal{B}_t, t \in T)$ is a martingale. This leads to a somewhat more general concept of martingales: Let φ_t on $\mathcal{B}_t$ be signed measures where the $\mathcal{B}_t$ are sub σ-fields of a field $\mathcal{A}$ in a space Ω. We may say that the family $(\varphi_t, \mathcal{B}_t, t \in T)$ is a martingale if $\varphi_s = \varphi_t$ on $\mathcal{B}_s$. If a pr. P on $\mathcal{A}$ is given and the φ_t are σ-finite and $P_{\mathcal{B}_t}$-continuous hence have a.s. finite Radon-Nikodym derivatives X_t (defined up to equivalences), then the r.f. X_T is a martingale.

3° If Y closes a martingale X_T on the right, then φ_{X_t} is the restriction of φ_Y on $\mathcal{B}(X_s, s \leqq t)$ and this property suffices for X_T to be a martingale. This leads to saying that the family $(\varphi, \mathcal{B}_t, t \in T)$ is a martingale $(\varphi_t, \mathcal{B}_t, t \in T)$ closed on the right by the signed measure φ when φ_t is the restriction of φ on $\mathcal{B}_t$. If $(\varphi_t, \mathcal{B}_t, t \in T)$ is a martingale then the set function φ_0 on $\mathcal{B}_0$ defined by its restrictions φ_t on $\mathcal{B}_t$ is finitely additive but not necessarily σ-additive. In fact, σ-additivity of φ_0 and boundedness from above or from below yields existence of closing signed measures. For then, φ_0 has extensions to signed measures φ on $\mathcal{B}$ and $(\varphi, \mathcal{B}_t, t \in T)$ are the martingales $(\varphi_t, \mathcal{B}_t, t \in T)$ closed by φ. If, moreover, φ_0 is σ-finite then φ is unique and, in fact, is the nearest closing signed measure.

We recall that if g on R is a continuous convex function with $g(x) \to g(+\infty)$ as $x \to +\infty$ and if EX exists with $E^{\mathcal{B}}X > -\infty$ a.s., then $g(E^{\mathcal{B}}X) \leqq E^{\mathcal{B}}g(X)$ a.s.

b. SUBMARTINGALES INEQUALITIES. *Let $(X_T, \mathcal{B}_T)$ be a separable submartingale.*

(i) *$(X_T^+, \mathcal{B}_T)$ is a submartingale, and $2EX_t^+ \geqq EX_r + E|X_s|$ when $r < s < t$ and $EX_r > -\infty$.*

$(|X_T|^r, \mathcal{B}_T)$, $r \geqq 1$, is a submartingale when $X_T \geqq 0$ or is a martingale.

(ii) *If Y closes X_T on the right, then, for every finite constant c,*

$$cP[\sup_{t \in T} X_t > c] \leqq \int_{[\sup_{t \in T} X_t \geqq c]} Y$$

and if, moreover, Y is integrable, then $>c$ can be replaced by $\geqq c$ and

$$\alpha = \inf_{t\in T} EX_t \leqq \int_{[\inf_{t\in T} X_t \geqq c]} Y + cP[\inf_{t\in T} X_t < c].$$

Proof. 1° The assertions in (i), except for the asserted inequality, result from the recalled proposition upon taking $g(x) = x^+$ and $g(x) = |x|^r$. As for the inequality, since $EX_s \geqq EX_r > -\infty$, the identity $2EX_s^+ = EX_s + E|X_s|$ has content and, X_T^+ being a submartingale, $EX_t^+ \geqq EX_s^+$ so that

$$2EX_t^+ \geqq 2EX_s^+ \geqq EX_r + E|X_s|.$$

2° In order to prove the asserted inequalities (ii), it suffices to show that given a submartingale $X_1, X_2, \cdots$ closed on the right by Y, for every n,

$$cPA_n \leqq \int_{A_n} Y, \quad EX_1 \leqq \int_{B_n^c} Y + cPB_n$$

with $A_n = [\max_k X_k > c]$, $B_n = [\min_k X_k < c]$, $k = 1, \cdots, n$. For then, by an obvious limiting procedure they hold for t varying over a (countable) separating set S hence over T. Note that in the second asserted inequality $\alpha = \lim EX_t$ as $t \to \inf_{t\in T} t$ and $\alpha = EX_a$ if T has a first element a; the inequality is trivially true if $\alpha = -\infty$.

Let $C_1 = [X_1 > c]$ and $C_k = [X_k > c, \max_{j<k} X_j \leqq c]$ for $k > 1$, so that $A_n = \sum_k C_k$ and

$$\int_{A_n} Y = \sum_{k=1}^{n} \int_{C_k} Y \geqq \sum_{k=1}^{n} \int_{C_k} X_k \geqq c \sum_k PC_k = cPA_n.$$

If Y is integrable, replace c by c' and let $c' \downarrow c$.

Let $D_1 = [X_1 < c]$ and $D_k = [X_k < c, \min_{j<k} X_j \geqq c]$ for $k > 1$, so that $B_n = \sum_k D_k$ and in order to prove the second inequality above it suffices to show that

$$\text{(I)} \qquad \int_{D_1^c} X_1 \leqq \int_{B_n^c} X_n + cPB_nD_1^c.$$

For then, from $\int_{D_1} X_1 \leqq cPD_1$ and $D_1 = B_nD_1$, hence $B_n = D_1 + B_nD_1^c$, it would follow that

$$EX_1 = \int_{D_1^c} X_1 + \int_{D_1} X_1 \leqq \int_{B_n^c} X_n + cPB_nD_1^c + cPD_1 \leqq \int_{B_n^c} Y + cPB_n.$$

We proceed by induction. Inequality (I) is trivially true for $n = 1$. If it holds for n it holds for $n + 1$: For, $B_n^c = B_{n+1}^c + B_n^cD_{n+1}$ so that

$$\int_{B_n^c} X_n \leqq \int_{B_n^c} X_{n+1} = \int_{B_{n+1}^c} X_{n+1} + \int_{B_n^cD_{n+1}} X_{n+1}$$

$$\leqq \int_{B_{n+1}^c} X_{n+1} + cPB_n^cD_{n+1}$$

hence, on account of $B_nD_1^c + B_n^cD_{n+1} \subset B_{n+1}D_1^c$, we have

$$\int_{D_1^c} X_1 \leqq \int_{B_n^c} X_n + cPB_nD_1^c \leqq \int_{B_{n+1}^c} X_{n+1} + cPB_nD_1^c + cPB_n^cD_{n+1}$$

$$\leqq \int_{B_{n+1}^c} X_{n+1} + cPB_{n+1}D_1^c.$$

The proof is terminated.

Continuity conditions for separable submartingales will be obtained essentially by means of submartingales convergence and closure theorems. Thus these conditions will be in terms of integrability. We denote by I_{ab} the interval with endpoints $a = \inf_{t \in T} t$, $b = \sup_{t \in T} t$, these points being included only when they belong to T, and set $\overline{T}_{ab} = I_{ab}\overline{T}$.

A. SUBMARTINGALES A.S. CONTINUITY THEOREM. *Let X_T be a separable submartingale. Let $t \in \overline{T}$ be a left or right limit point of T.*

(i) *If* $\lim_{t' \uparrow t} EX_{t'}^+ < \infty$ *then* $X_{t'} \xrightarrow{\text{a.s.}} X_{t-0}$ *with* $EX_{t-0} \leqq EX_{t-0}^+ \leqq \lim_{t' \uparrow t} EX_{t'}^+ < \infty$, *and with* $EX_{t-0} \leqq EX_t$ *when* $t \in T$.

If $\lim_{t' \downarrow t} EX_{t'}^+ < \infty$ *then* $X_{t'} \xrightarrow{\text{a.s.}} X_{t+0}$ *with* $EX_{t+0} \leqq EX_{t+0}^+ = \lim_{t' \downarrow t} EX_{t'}^+ < +\infty$, *and with* $EX_t \leqq EX_{t+0}$ *when* $t \in T$.

(ii) *If X_T^+ is integrable, then X_T has a.s. left (right) limits at left (right) limit points of T belonging to $\overline{T}_{ab}$ and X_T is a.s. continuous on T except for a countable subset.*

Note that the foregoing limits of exp.'s exist because of the monotone character of the exp.'s of r.v.'s of submartingales.

Proof. 1° Let $t' \uparrow t \in \bar{T}$ and let $\lim_{t' \uparrow t} EX_{t'}^+ < \infty$. On account of the submartingale convergence theorem and of separability implications 3° for one-sided limits, X_{t-0} exists a.s. with $EX_{t-0} \leqq EX_{t-0}^+ \leqq \lim_{t' \uparrow t} EX_t^+$. Furthermore, the asserted inequality $EX_{t-0} \leqq EX_t$ when $t \in T$ is trivially true when $EX_t = +\infty$ and we prove it when $EX_t < \infty$ as follows.

Set $X^c = (X - c)^+ + c = c$ or X according as $X < c$ or $X \geqq c$ and take $c \leqq 0$; thus $c \leqq X^c \leqq X^+$ and $X^c \downarrow X$ as $c \downarrow -\infty$. It follows from **b** that X_T^c is a submartingale so that $\lim_{t' \uparrow t} EX_{t'}^c \leqq EX_t^c$ and, by the submartingale convergence theorem, $X_{t'}^c \xrightarrow{\text{a.s.}} X_{t-0}^c$. Since the $X_{t'}^c$ are bounded from below by c integrable and X_t^c is bounded from above by X_t^+ integrable, the Fatou-Lebesgue theorem applies,

$$EX_{t-0} \leqq EX_{t-0}^c \leqq \lim_{t' \uparrow t} EX_{t'}^c \leqq EX_t^c,$$

and, as $c \to -\infty$,

$$EX_{t-0} \leqq \limsup_{c \to -\infty} EX_t^c \leqq EX_t.$$

The left limit assertions are proved.

2° Let $t' \downarrow t \in \bar{T}$ and $\lim_{t' \downarrow t} EX_{t'}^+ < \infty$ so that, by the submartingales convergence theorem and separability implications 3°, X_{t+0} exists a.s. and $X_{t'}^{\pm} \xrightarrow{\text{a.s.}} X_{t+0}^{\pm}$. On the one hand, by the Fatou-Lebesgue theorem, $EX_{t+0}^- \leqq \lim_{t' \downarrow t} EX_{t'}^-$. On the other hand, there exists $t'' > t$ such that $EX_{t''}^+ < \infty$ so that the submartingale $(X_{t'}^+, t' \in (t, t'')T)$ is closed on the right by $X_{t''}^+$ integrable hence, by the submartingales closure theorem, the $X_{t'}^+$ are uniformly integrable and thus $EX_{t+0} \leqq EX_{t+0}^+ = \lim_{t' \downarrow t} EX_{t'}^+$. It follows that

$$EX_{t+0} = EX_{t+0}^+ - EX_{t+0}^- \geqq \lim_{t' \downarrow t} EX_{t'} \ (\geqq EX_t \text{ when } t \in T),$$

and the right limit assertions are proved.

3° Let X_T^+ be integrable. For every $t \in \bar{T}_{ab}$ there exists $t'' \in T$ such that $t \leqq t''$. Therefore, if t is left (right) limit point of T then $\lim_{t' \uparrow t} EX_{t'}^+ \ (\lim_{t' \downarrow t} EX_{t'}^+) \leqq EX_{t''}^+ < \infty$ and (i) applies. The a.s. continuity assertion follows from the continuity extension theorem, and the proof is terminated.

Continuity in the first mean properties result from the lemma below, the first part of which is based upon the following proposition.

If $X_n \geqq 0$, *then* $X_n \xrightarrow{1} X$ *if and only if* $X_n \xrightarrow{P} X$ *and finite* $EX_n \to EX$ *finite.*

For, the "only if" assertion is always true and the "if" assertion results from $0 \xleftarrow{P} (X - X_n)^+ \leqq X^+$ integrable and $\int (X - X_n)^+ - \int (X - X_n)^- \to 0$, so that $\int (X - X_n)^+ \to 0$ and $\int (X - X_n)^- \to 0$ hence $\int | X - X_n | \to 0$.

c. *If* $X_1, X_2, \cdots$, *is a submartingale sequence, then the* X_n *are uniformly integrable if and only if* $X_n \xrightarrow{P} X$ *closing r.v. and finite* $EX_n \to EX$ *finite, and then* $X_n \underset{\text{a.s.}}{\xrightarrow{1}} X$. *If* $\cdots X_2, X_1$ *is a submartingale reversed sequence, then the* X_n *are uniformly integrable if and only if* $\sup E| X_n | < \infty$, *and then* $X_n \underset{\text{a.s.}}{\xrightarrow{1}} X$ *closing r.v.*

Proof. We use throughout the submartingales convergence and closure theorems without further comment.

1° Let $X_1, X_2, \cdots$ be a submartingale sequence. If the X_n are uniformly integrable, then $X_n \underset{\text{a.s.}}{\xrightarrow{1}} X$ closing r.v., hence $X_n \xrightarrow{P} X$ and finite $EX_n \to EX$ finite. Conversely, if $X_n \xrightarrow{P} X$ closing *r.v.* and finite $EX_n \to EX$ finite, then X^+ integrable closes the nonnegative submartingale X_n^+ so that $X_n^+ \underset{\text{a.s.}}{\xrightarrow{1}} X^+$, hence the X_n^+ are uniformly integrable and finite $EX_n^+ \to EX^+$ finite. Therefore, $X_n^- \xrightarrow{P} X^-$ and finite $EX_n^- \to EX^-$ finite hence $X_n^- \xrightarrow{1} X^-$ integrable and the X_n^- are uniformly integrable. Thus, the X_n are uniformly integrable. The first assertion is proved.

2° Let $\cdots X_2, X_1$ be a submartingale reversed sequence. If the X_n are uniformly integrable hence $\sup E| X_n | < \infty$, then $X_n \underset{\text{a.s.}}{\xrightarrow{1}} X$ closing r.v. Conversely, if $\sup E| X_n | < \infty$ then, setting $A_n = [| X_n | \geqq c]$, for $n > m$,

$$\int_{A_n} | X_n | = 2\int_{A_n} X_n^+ + \int_{A_n^c} X_n - EX_n \leqq 2\int_{A_n} X_m^+ + \int_{A_n^c} X_m - EX_n$$

$$= \int_{A_n} | X_m | + EX_m - EX_n$$

and, noting that $EX_n \downarrow \alpha$ finite, it follows that given $\epsilon > 0$ we can fix m sufficiently large so that $\int_{A_n} |X_n| \leqq \int_{A_n} |X_m| + \epsilon$. But, by the Markov inequality, as $c \to \infty$, $PA_n \leqq E|X_n|/c \leqq \sup E|X_n|/c \to 0$ uniformly in n, so that for $n > m$ and c sufficiently large $\int_{A_n} |X_m| < \epsilon$ hence $\int_{A_n} |X_n| < 2\epsilon$. Uniform integrability of the X_n follows, and the proof is terminated.

B. Submartingales mean continuity theorem. *Let X_T be a separable submartingale. Let t be a left or right limit point of T and let $h > 0$ be sufficiently small.*

(i) *For $t' \in (t - h, t)T$:*

$X_{t'} \xrightarrow[\text{a.s.}]{1} X_{t-0} \Leftrightarrow X_{t'}$ *uniformly integrable* $\Leftrightarrow X_{t'} \xrightarrow{P} X_{t-0}$ *closing r.v. and finite* $EX_{t'} \to EX_{t-0}$ *finite.*

For $t' \in (t, t + h)T$:

$X_{t'} \xrightarrow[\text{a.s.}]{1} X_{t+0} \Leftrightarrow X_{t'}$ *uniformly integrable* $\Leftrightarrow \sup_{t'} E|X_{t'}| < \infty$.

(ii) *If X_T is integrable, then it has a.s. (a.s. and in the first mean) left (right) limits at left (right) limit points of T belonging to $\overline{T}_{ab}$ and X_T is a.s. and in the first mean continuous on T outside a countable subset and extends to a submartingale on I_{ab}.*

Proof. The assertions result from **A**, **b** and **c**, except for the extension assertion. But it suffices to set $X_t = X_{t+0}$ at every right limit point $t \in \overline{T}_{ab}$ of T which does not belong to T and, for every interval $[c, d]$ whose endpoints only belong to $\overline{T}_{ab}$, set $X_t = X_d$ for $c < t < d$, and also for $t = c$ unless X_c is already defined. The r.f. so defined on I_{ab} is a submartingale, and the proof is terminated.

C. Submartingales sample continuity theorem. *Let X_T be a separable submartingale. Then, for almost all sample functions, on every $[t_1, t_2]T$ with $t_1, t_2 \in T$,*

(i) *If $X_{t_2}^+$ is integrable, then the left (right) limits exist at left (right) limit points and the sample functions are continuous outside countable subsets.*

(ii) *If X_{t_1} and X_{t_2} are integrable, then these sample functions are bounded and their discontinuities are jumps except perhaps at the fixed discontinuity points.*

Proof. Reorder the first n points of a separating set in $[t_1, t_2]T$ into $s_{n1} < \cdots < s_{nn}$, let H_n be the number of crossings (from the left) of

$[r, s]$ by $X_{sn1}, \cdots, X_{snn}$. Set $A_{nm} = [H_n \geqq m]$ so that $A_{nm} \uparrow A_m = \bigcup_n A_{nm}$. Set $A_{rs} = \bigcap_m A_m$ and take the union $A = \bigcup_{r<s} A_{rs}$ over all rationals r, s. The limit $X_{t-0}(\omega)$ fails to exist if and only if there exist rationals r, s such that $\underline{X}_{t-0}(\omega) < r < s < \overline{X}_{t-0}(\omega)$, that is, only if $\omega \in A_{rs}$. Thus to prove that $X_{t-0}(\omega)$ exists for almost all ω it suffices to show that $PA = 0$. But by 29.3a and the Markov inequality, as $n \to \infty$ then $m \to \infty$,

$$PA_{rs} \leqq PA_m \leftarrow PA_{nm} \leqq EH_n/m \leqq (EX_{t_2}^+ + |r|)/m(s - r) \to 0$$

so that the countable union A of null events A_{rs} is null. Similarly for the $X_{t+0}(\omega)$, and assertion (i) follows.

If X_{t_1} and X_{t_2} are integrable then, upon letting $c \to \infty$ in the submartingales inequalities,

$$P[\sup_{t \in [t_1, t_2]T} X_t \geqq c] \leqq \frac{1}{c}\int |X_{t_2}| \to 0,$$

$$P[\inf_{t \in [t_1, t_2]T} X_t < -c] \leqq \frac{1}{c}(E|X_{t_2}| - EX_{t_1}) \to 0.$$

Thus, the extrema and hence the limits for almost all sample functions are finite, and (ii) follows by separability implications 3°. The proof is terminated.

36.2 Martingale times. Let $\mathfrak{B}_T = (\mathfrak{B}_t, t \in T)$, $T \subset \overline{R}$, be a nondecreasing family of sub-σ-fields of events: $\mathfrak{B}_s \subset \mathfrak{B}_t$ for $s < t$ and set $a = \inf \{t : t \in T\}$, and $b = \sup \{t : t \in T\}$. We say that a measurable function τ on Ω to $T \cup \{b\}$ is a (*random*) *time of* $\mathfrak{B}_T$ (or *of* X_T if every $\mathfrak{B}_t = \mathfrak{B}(X_r, r \leqq t)$) if $[\tau \leqq t] \in \mathfrak{B}_t$ for every $t \in T$; in particular any $t \in T$ is a "degenerate" time of any $\mathfrak{B}_T$. To times τ of $\mathfrak{B}_T$ are associated sub-σ-fields of events $\mathfrak{B}_\tau = \{B : B[\tau \leqq t] \in \mathfrak{B}_t, t \in T\}$. The following properties of times σ, τ of $\mathfrak{B}_T$ are immediate: (i) Every τ is $\mathfrak{B}_\tau$-measurable, (ii) If $\sigma \leqq \tau$ then $\mathfrak{B}_\sigma \subset \mathfrak{B}_\tau$. (iii) $\inf(\sigma, \tau)$ and $\sup(\sigma, \tau)$ are times of $\mathfrak{B}_T$. (iv) If $T = 1, 2, \cdots$ or $T = [0, \infty)$ then the $\tau + t$, $t \in T$ are times of $\mathfrak{B}_T$. Note that if above $\tau \leqq t$ is replaced by $\tau < t$, then $[\tau \leqq t] \in \mathfrak{B}_{t+}$, i.e., τ is a time of $\mathfrak{B}_{T+} = \{\mathfrak{B}_{t+}, t \in T\}$, where $\mathfrak{B}_{t+} = \bigcap_{u>t} \mathfrak{B}_u$. If $b \in T$, we take b_+ to be b.

Random times appear naturally in sample analysis and will play a basic role in the next chapter. Here we examine *submartingale times* which originated with P. Lévy and were made by Doob into a powerful tool in pr. theory and in Analysis. Under "natural" conditions, *replacing degenerate times* $s < t$ *by times* $\sigma \leqq \tau$ *of a submartingale* $(X_T, \mathfrak{B}_T)$, *i.e., of times of* $\mathfrak{B}_T$, preserves the sub martingale property.

The proofs of the propositions which follow are given for submartingales, the arguments being the same for martingales.

We start with the discrete case $(X_J, \mathcal{B}_J)$ with $J = \{1, \cdots, n\}$ then $J = \{1, 2, \cdots\}$. To times τ of $\mathcal{B}_J$ are associated functions $X_\tau = \sum X_j I_{[\tau=j]}$; clearly they are $\mathcal{B}_\tau$-measurable r.v.'s.

a. DISCRETE SUBMARTINGALE TIMES LEMMA. *Let $\sigma \leqq \tau \leq n$ be times of the submartingale $(X_k, \mathcal{B}_k)$, $k = 1, \cdots, n$.*
If EX_σ, EX_τ exist, then $(X_\sigma, \mathcal{B}_\sigma)$, $(X_\tau, \mathcal{B}_\tau)$ form a submartingale and $EX_1 \leqq EX_\sigma \leqq EX_\tau \leqq EX_n$.

If $EX_1 > -\infty$ or (and) $EX_n^+ < +\infty$ then all EX_τ exist and $E|X_\tau| \leqq 2EX_n^+ - EX_1 (< \infty)$ provided this difference has content.

Proof. If $B_k \in \mathcal{B}_k$ and $j > k$ then $B_k[\tau > j-1] \in \mathcal{B}_{j-1}$ hence

$$\int_{B_k[\tau>j-1]} X_{j-1} \leqq \int_{B_k[\tau>j-1]} X_j = \int_{B_k[\tau=j]} X_j + \int_{B_k[\tau>j]} X_j$$

and, summing over $j = k+1, \cdots, n$,

$$\int_{B_k[\tau>k]} X_k \leqq \sum_{j=k+1}^{n} \int_{B_k[\tau=j]} X_j = \int_{B_k[\tau>k]} X_\tau.$$

Since for $B_\sigma \in \mathcal{B}_\sigma$, we can take $B_k = B_\sigma[\sigma = k] (\in \mathcal{B}_k)$ and, for $j < k$, $[\sigma = k][\tau < k] = \emptyset$, it follows that

$$\int_{B_\sigma[\sigma=k]} X_\sigma = \int_{B_\sigma[\sigma=k]\,[\tau=k]} X_k + \int_{B_\sigma[\sigma=k]\,[\tau>k]} X_k \leqq \int_{B_\sigma[\sigma=k]} X_\tau$$

and, summing over $k = 1, \cdots, n$, $\int_{B_\sigma} X_\sigma \leqq \int_{B_\sigma} X_\tau$; in particular, $EX_\sigma \leqq EX_\tau$. Therefore, taking $1 = \sigma' \leqq \sigma \leqq \tau \leqq \tau' = n$, so that $EX_1 \leqq EX_\sigma \leqq EX_\tau \leqq EX_n$, and noting that in the martingale case the foregoing inequalities become equalities, the first assertions are proved.

If $EX_1 > -\infty$ or $EX_n^+ < +\infty$ then the $EX_k I_{[\tau=k]}$ are all $> -\infty$ or all $< +\infty$, so that $EX_\tau = \sum EX_k I_{[\tau=k]} > -\infty$ or $< +\infty$ exists. Since the $(X_k^+, \mathcal{B}_k)$ form a submartingale so that $EX_\tau^+ \leqq EX_n^+$, we have $E|X_\tau| = 2EX_\tau^+ - EX_\tau \leqq 2EX_n^+ - EX_1$.

APPLICATIONS. Limit properties of submartingales were derived from the submartingale inequalities in 29.2**A** and 36.1**b** and from the crossings inequality 29.3**a**. These inequalities result at once from **a**:

1°. *Basic submartingales inequalities.* Let c be a (finite) constant and $\sigma(\omega)(\tau(\omega))$ be the smallest k, if any, with $X_k(\omega) \geqq c (X_k(\omega) \leqq c)$ or be n if none. Set $A_n = [\sup_k X_k \geqq c]$, $B_n = [\inf_k X_k \leqq c]$ and note that

EX_σ always exists and EX_τ exists when $EX_1 > -\infty$ or $EX_n^+ < \infty$. Then

$$cPA_n \leqq \int_{A_n} X_\sigma \leqq \int_{A_n} X_n$$

and

$$EX_1 \leqq \int_{B_n} X_\tau + \int_{B_n^c} X_\tau \leqq cPB_n + \int_{B_n^c} X_n.$$

2°. *Crossings inequality.* The number H_n of crossings (from the left) of $[a, b]$ by the submartingale formed by the X_k is also that of crossings of $[0, b-a]$ by the submartingale formed by the $Y_k = (X_k - a)^+$. Let $\tau_1(\omega) = 1$, $\tau_2(\omega)$ be the smallest $k \geqq \tau_1(\omega)$ (if any) for which $Y_k(\omega) = 0$, $\tau_3(\omega)$ be the smallest $k \geqq \tau_2(\omega)$ (if any) for which $Y_k(\omega) \geqq b - a$ and so on, proceeding alternately to 0 and to values $\geqq b - a$, up to $\tau_n(\omega) = n$; from the first undefined $\tau_j(\omega)$ (if any) on, set $\tau_j(\omega) = \cdots = \tau_n(\omega) = n$. Then $Y_n = \sum_{j=2}^{n} (Y_{\tau_j} - Y_{\tau_{j-1}}) + Y_1$ where Y_{τ_j} form a submartingale. If $EY_n < \infty$ (hence all $EY_{\tau_j} < \infty$), the sum of the first H_n summands with odd j, being at least $(b-a)H_n$, contributes at least $(b-a)EH_n$ to the right-hand sum of expectations while the expectations of the remaining summands are nonnegative. Thus $(b-a)EH_n \leqq E(X_n - a)^+$; if $EY_n = \infty$ this crossing inequality is trivially true. (This modification of Doob's proof is due to Hunt.)

A. DISCRETE SUBMARTINGALE TIMES THEOREM. *Let $\sigma \leqq \tau < \infty$ be times of the (sub)martingale $(X_n, \mathcal{B}_n)$, $n = 1, 2, \cdots$.*

(i) *If EX_σ, EX_τ exist, then* ($\liminf \int_{[\tau>n]} X_n^+ = 0$. $\liminf \int_{[\tau>n]} |X_n| = 0$ *implies that $(X_\sigma, \mathcal{B}_\sigma)$, $(X_\tau, \mathcal{B}_\tau)$ form a (sub)martingale and* $EX_1 \,{}^{(}\!\leqq\!{}^{)}\, EX_\sigma \,{}^{(}\!\leqq\!{}^{)}\, EX_\tau$; (*if moreover* $\liminf \int_{[\tau>n]} X_n = 0$ *then* $EX_\tau \leqq \sup EX_n$).

If $EX_1 > -\infty$ and $\sup EX_n^+ < +\infty$ *then all $E|X_\tau|$ are uniformly bounded with* $E|X_\tau| \leqq 2 \sup EX_n^+ - EX_1 < \infty$.

(ii) *For all $\tau \leqq n$, and also for all $\tau \leqq \infty$ when the X_n are uniformly integrable, the conclusions hold and the X_τ are uniformly integrable.*

Proof. We use **a** without further comment.

1°. Let $\sigma_n = \inf(\sigma, n)$, $\tau_n = \inf(\tau, n)$, so that $\sigma_n \leqq \tau_n \leqq n$ are times of the submartingale $(X_1, \mathcal{B}_1), \cdots, (X_n, \mathcal{B}_n)$. Since, for $B_\sigma \in \mathcal{B}_\sigma$, $B_\sigma[\sigma \leqq n] \in \mathcal{B}_{\sigma_n}$ and $[\sigma \leqq n][\tau \leqq n] = [\tau \leqq n]$,

$$\int_{B_\sigma[\sigma \leqq n]} X_\sigma \leqq \int_{B_\sigma[\tau \leqq n]} X_\tau + \int_{B_\sigma[\sigma \leqq n]\,[\tau > n]} X_n$$

hence, by hypothesis, letting $n \to \infty$ along a suitable subsequence, $\int_{B_\sigma} X_\sigma \leqq \int_{B_\sigma} X_\tau$; and, taking $1 = \sigma' \leqq \sigma$, $EX_1 \leqq EX_\sigma \leqq EX_\tau$.

If moreover $\liminf \int_{[\tau>n]} X_n = 0$ then, as $n \to \infty$ along a suitable subsequence, the relation

$$EX_{\tau_n} = \int_{[\tau \leqq n]} X_\tau + \int_{[\tau>n]} X_n \leqq \sup EX_n$$

yields $EX_\tau \leqq \sup EX_n$. The first assertions follow.

If $EX_1 > -\infty$ and $\sup EX_n^+ < \infty$ then, by the Fatou-Lebesgue theorem, letting $n \to \infty$ in $E|X_{\tau_n}| \leqq 2 \sup EX_n^+ - EX_1 = \alpha < \infty$, it follows from $X_{\tau_n} \to X_\tau$ that $E|X_\tau| \leqq \alpha$.

2°. If the X_n are uniformly integrable then $\sup E|X_n| < \infty$ and, decomposing the submartingale (pp. 389–390), $X_n = X'_n + X''_n$ where the X'_n form a martingale and the X''_n form a nonnegative submartingale with $X''_n \uparrow X''$ integrable. By the closure theorem, $X_n \xrightarrow{1} X$, $X''_n \xrightarrow{1} X''$ so that $X'_n \xrightarrow{1} X'$. Therefore, $|X_n| \leqq |X'_n| + X''_n = Y_n \xrightarrow{1} Y = |X'| + X''$ and the Y_n form a submartingale closed by Y. Since $\int_{[|X_\tau|>c]} |X_\tau| \leqq \int_{[|X_\tau|>c]} Y$ with $P[|X_\tau| > c] \leqq EY/c \to 0$ as $c \to \infty$ uniformly in τ, the X_τ are uniformly integrable, and the last assertions follow. (For martingales, $X''_n = 0$ and $Y = |X|$).

The brutal transposition of the definition of X_τ in the discrete case to the general case $(X_T, \mathcal{B}_T)$ yields *formally* $X_\tau = \sum X_t I_{[\tau=t]}$. However, as written, X_τ may not be a r.v. and to avoid such a possibility we have to impose conditions on X_T and modify accordingly the terms of the above sum. In the most important case of X_T a.s. sample right continuous on an interval T, no change is required. In the more general case of an arbitrary T and almost all sample functions of X_T having right limits on the subset $T_+ \subset T$ of its right limit points, we set $X_\tau(\omega) = X_{t_j}(\omega)$ for $\tau(\omega) = t_j \in S_\tau$—the countable set of points $t_j \in T$ with $P[\tau = t_j] > 0$, and we set $X_\tau(\omega) = X_{t+0}(\omega)$ for $\tau(\omega) = t \in T_+ S_\tau^c$. Thus X_τ is undefined only on the exceptional null event of X_T and on the countable union of null events $[\tau = t]$ for the remaining $t \notin T_+$.

Then X_τ is a r.v., as follows. Let $r_{n0} < r_{n1} < \cdots < r_{nk_n}$ be points of finite subsets $T_n (\subset T_{n+1} \subset \cdots \subset T)$, containing the endpoints of T whenever they belong to T and the n first $t_j \in S_\tau$, and such that the distance of any $t \in [-n, +n]T$ to some r_{nk} is less than $1/n$. Set $S_{nk} = (r_{n,k-1}, r_{n,k}]$ and $g_n = \sum_{k=1}^{k_n} r_{nk} I_{S_{nk}}$, $\tau_n = g_n(\tau)$. We are in the discrete case with times $\tau_n \downarrow \tau$ and r.v.'s $X_{\tau_n} \xrightarrow{\text{a.s.}} X_\tau$; set $\mathcal{B}_{\tau+} = \bigcap_n \mathcal{B}_{\tau_n}$. Since the X_{τ_n} are $\mathcal{B}_{\tau_n}$-measurable, it follows that the X_τ are $\mathcal{B}_{\tau+}$-measurable and, to emphasize this fact, we shall also write $X_{\tau+}$ for X_τ. Note that $\mathcal{B}_{\tau+} = \{B : B[\tau < t] \in \mathcal{B}_t, t \in T\}$.

b. GENERAL SUBMARTINGALE TIMES LEMMA. *Let $\sigma \leqq \tau \leqq b$ be times of the separable submartingale $(X_T, \mathcal{B}_T)$ with last element X_b with almost all sample functions having right limits on T_+, and with* $\inf EX_t > -\infty$, $EX_b^+ < +\infty$.

(i) *Then $(X_{\sigma+}, \mathcal{B}_{\sigma+})$, $(X_{\tau+}, \mathcal{B}_{\tau+})$ form a submartingale with* $\inf EX_t \leqq EX_{\sigma+} \leqq EX_{\tau+} \leqq EX_b$, *and* $E|X_\tau| \leqq 2EX_b^+ - \inf EX_t$.

(ii) *In the martingale or nonnegative submartingale case, all X_τ are uniformly integrable.*

Proof. Since all $E|X_t| = 2EX_t^+ - EX_t \leqq 2EX_b^+ - \inf EX_t = \alpha < \infty$, hence $E|X_{\tau_n}| \leqq \alpha$, by the Fatou-Lebesgue theorem, $X_{\tau_n} \xrightarrow{\text{a.s.}} X_{\tau+}$ implies that $E|X_\tau| \leqq \alpha$. Since $\tau_n \downarrow \tau$ and the $(X_{\tau_n}, \mathcal{B}_{\tau_n})$ form a submartingale reversed sequence with $\sup E|X_{\tau_n}| < \infty$, 36.1**c** implies that $X_{\tau_n} \xrightarrow{1} X_{\tau+}$; similarly $X_{\sigma_n} \xrightarrow{1} X_{\sigma+}$ where $\sigma_n = g_n(\sigma) \downarrow \sigma$. But $\sigma \leqq \tau$ implies that $\sigma_n \leqq \tau_n$ hence, by **A**,

$$\int_{B_{\sigma+}} X_{\sigma_n} \leqq \int_{B_{\sigma+}} X_{\tau_n} \quad \text{for} \quad B_{\sigma+} \in \mathcal{B}_{\sigma+} = \bigcap_n \mathcal{B}_{\sigma_n}.$$

Thus, we can pass to the limit under the integration sign, $\int_{B_{\sigma+}} X_{\sigma+} \leqq \int_{B_{\sigma+}} X_{\tau+}$, $EX_{\sigma+} \leqq EX_{\tau+}$, and taking $\tau \leqq \tau' = b$, $EX_{\tau+} \leqq EX_b$.

If X_T has a first element X_a then, taking $a = \sigma' \leqq \sigma$, we have $\inf EX_t = EX_a \leqq EX_{\sigma+}$. Otherwise, $\sup E|X_t| < \infty$ and 36.1**b** imply that $X_t \xrightarrow{1} X$ as $t \downarrow a$, we can close X_T on the left by $X_a = X$, and the preceding inequality remains valid.

Finally, (ii) follows by the argument at the very end of proof of **A**, with Y replaced by $|X_b|$.

B. GENERAL SUBMARTINGALE TIMES THEOREM. *Let $\sigma \leqq \tau < b$ be times of the separable submartingale $(X_T, \mathcal{B}_T)$ with almost all sample functions having right limits on T_+ and with* $\inf EX_t > -\infty$, $\sup EX_t^+ < \infty$.

(i) *Then* $E|X_\tau| \leqq 2 \sup EX_t^+ - \inf EX_t$, *and* $(\liminf_{t\to b} \int_{[\tau>t]} X_t^+ = 0)$ $\liminf_{t\to b} \int_{[\tau>t]} |X_t| = 0$ *implies that $(X_{\sigma+}, \mathcal{B}_{\sigma+})$, $(X_{\tau+}, \mathcal{B}_{\tau+})$ form a submartingale and* $\inf EX_t (\leqq) EX_{\sigma+} (\leqq) = EX_{\tau+}$; (*if moreover* $\liminf_{t\to b} \int_{[\tau>t]} X_t = 0$ *then* $EX_\tau \leqq \sup EX_t$)

(ii) *For all $\tau \leqq t (\in T)$, and also for all $\tau \leqq b$, when the X_t are uniformly integrable, the conclusions hold and, in the martingale or nonnegative submartingale case, the X_τ are uniformly integrable.*

Proof. If $b \in T$, the theorem reduces to the preceding lemma. Also if the X_t are uniformly integrable: then $X_t \xrightarrow{1} X$ as $t \uparrow b$ and we can close on the right the submartingale by setting $X_b = X$.
In the general case of $b \notin T$, we proceed as in the proof of **A** but use **b** in lieu of **a**. Set $\sigma_t = \inf(\sigma, t)$, $\tau_t = \inf(\tau, t)$, $t \in T$, so that $\sigma_t \leqq \tau_t \leqq t$ are times of the submartingale $(X_s, \mathcal{B}_s)$ with $s \leqq t$. Since

$$\int_{B_{\sigma+}[\sigma \leqq t]} X_{\sigma+} \leqq \int_{B_{\sigma+}[\tau \leqq t]} X_{\tau+} + \int_{B_{\sigma+}[\sigma \leqq t][\tau > t]} X_t,$$

by hypothesis, letting $t \to b$ along a suitable sequence, $\int_{B_{\sigma+}} X_{\sigma+} \leqq \int_{B_{\sigma+}} X_{\tau+}$. In particular, closing on the left our submartingale at a and taking $a = \sigma' \leqq \sigma$, we obtain $\inf EX_t \leqq EX_{\sigma+} \leqq EX_{\tau+}$.
If moreover $\liminf \int_{[\tau > t]} X_t = 0$ then, as $t \to b$ along a suitable subsequence, the relation

$$EX_{\tau_t} = \int_{[\tau \leqq t]} X_{\tau+} + \int_{[\tau > t]} X_t \leqq \sup EX_t$$

yields $EX_{\tau+} \leqq \sup EX_t$.
Finally, by the Fatou-Lebesgue theorem, letting $t \to b$ in $E|X_{\tau_t}| \leqq 2 \sup EX_t - \inf EX_t = \alpha < \infty$, it follows from $X_{\tau_t} \to X_{\tau+}$ that $E|X_\tau| \leqq \alpha$. The theorem follows.

Under the various hypotheses made, a submartingale $(X_T, \mathcal{B}_T)$ is transformed into a submartingale $(X_{\tau_u+}, \mathcal{B}_{\tau_u+}, u \in U \in \overline{R})$ by a family $\{\tau_u, u \in U\}$ of its times with $\tau_u \leqq \tau_v$ for $u \leqq v$. We can transform it by means of one of its times τ only. It may be used as (and, in fact, very frequently is called) "stopping time:" For every $t \in T$, set $\tau_t = \inf(\tau, t)$; the new submartingale is obtained from the given one by "stopping it at time τ." We did it to pass from the above lemmas to the corresponding theorems.

§ 37. DECOMPOSABILITY

Decomposability sprang forth fully armed from the forehead of P. Lévy (1934). His analysis of "integrals with independent elements" or "r.f.'s with independent increments" or "additive r.f.'s" or "differential r.f.'s" or "P. Lévy r.f.'s" or, as we shall call them, "decomposable r.f.'s" was so complete that since then only improvements of detail have been added. Before his work there were only pioneering ones by de Finetti

(1929) and Kolmogorov (1932). The only decomposable r.f.'s known were the Poisson and the Brownian processes, both born from physical phenomena. (After a pioneering work by Bachelier (1900), the first rigorous study of the Brownian process was by Wiener (1923) who discovered its a.s. sample continuity.) Furthermore, the basic concepts and problems of random analysis appeared in and were born from the P. Lévy analysis of decomposability. Thus, decomposability is at the root of the concepts and problems of random analysis.

37.1 Generalities. A r.f. X_T is said to be *decomposable* if its increments $X_{st} = X_t - X_s$ for disjoint intervals $[s, t)$ are independent. A *decomposable process* on T is the family of decomposable r.f.'s (on some pr. space) with the same increments. But there is a one-to-one correspondence between the increment function on T—an additive r.f. X_{st} of intervals whose endpoints belong to T—and the family of r.f.'s on T defined by $X_t = X_a + X_{at}$ or $X_a - X_{ta}$ according as $t \geqq a$ or $t < a$, where $a \in T$ and X_a are selected arbitrarily. Therefore, we may consider a decomposable process on T as a decomposable r.f. on T defined up to an arbitrarily selected value X_a at an arbitrarily selected point $a \in T$. Thus, a decomposable process will be *represented* by one of its r.f.'s X_T either an unspecified one or selected according to convenience, but *always separable.*

The law of a decomposable process determined by the joint laws of all its finite sections is, in fact, determined by the individual laws of its increments. For, because of decomposability, setting $f_{st}(u) = E \exp \{iuX_{st}\}$, the joint law of, say, X_{ac} and X_{bd} with $a < b < c < d$ is given by

$$E \exp \{iu_1X_{ac} + iu_2X_{bd}\} = E \exp \{iu_1X_{ab} + i(u_1 + u_2)X_{bc} + iu_2X_{cd}\}$$
$$= f_{ab}(u_1)f_{bc}(u_1 + u_2)f_{cd}(u_2).$$

Also, if X_T is one of the r.f.'s of a decomposable process then, upon setting $f_s(u) = E \exp \{iuX_s\}$ and $v_k = u_k + \cdots + u_n$, we have $f_t = f_sf_{st}$ and, for $t_1 < \cdots t_n$,

$$E \exp \{iu_1X_{t_1} + \cdots + iu_nX_{t_n}\}$$
$$= E \exp \{iv_1X_{t_1} + iv_2X_{t_1t_2} + \cdots + iv_nX_{t_{n-1}t_n}\}$$
$$= f_{t_1}(v_1)f_{t_1t_2}(v_2 \cdots f_{t_{n-1}t_n}(v_n).$$

Therefore, if the individual laws of all increments are of a specific type, say, normal or Poisson or infinitely decomposable, we shall say that the process is of this type: *normal decomposable* or *Poisson decomposable* or *infinitely decomposable.* In fact, we shall find that deleting the fixed discontinuities the remaining part of any decomposable process is infinitely decomposable. It is by a deep sample analysis of this remaining part that P. Lévy discovered the general form of infinitely decomposable laws with ch.f.'s e^{ψ}, $\psi = (\alpha, \Psi)$. Since we have at our disposal this general form (§22), we shall proceed from it to the sample interpretation. In order to do so, it will prove convenient to write ψ in P. Lévy's form $\psi = (\alpha, \beta^2, L)$, explicitly

$$\psi(u) = i\alpha u - \frac{\beta^2}{2}u^2 + \int\!\!\!\!-_{-\infty}^{+\infty} \left(e^{iux} - 1 - \frac{iux}{1+x^2}\right) dL(x)$$

where the bar which crosses the integral excludes the origin from the domain of integration and where we can and do take $L(-\infty) = L(+\infty) = 0$. The correspondence between the two forms of ψ is given by

$$\beta^2 = \Psi(+0) - \Psi(-0), \quad dL(x) = \frac{1+x^2}{x^2}\, d\Psi(x) \text{ for } x \neq 0$$

and

$$\operatorname{Var} \Psi < \infty \Leftrightarrow \int\!\!\!\!-_{-1}^{+1} x^2\, dL(x) < \infty.$$

In order to reach the infinitely decomposable part of the process, we shall have to delete its degenerate discontinuities by "centering" it, then delete the "fixed discontinuities" part. This will require recalling (Sections 16 and 17) and adding to convergence properties of series of independent summands, as follows.

Let $X_1, X_2, \cdots$ be a sequence of independent r.v.'s with ch.f.'s $f_1, f_2, \cdots$. The series $\sum_n X_n$ is said to be *convergent* if it converges a.s. to a r.v., equivalently, if $\prod_{k=1}^{n} f_k \to f$ ch.f., or if $\prod f_n > 0$ on an argument set of positive Lebesgue measure. The series $\sum X_n$ is said to be *essentially convergent* if there exist centering constants c_n such that the series $\sum (X_n - c_n)$ is convergent, equivalently, if the series $\sum X_n^s$ of symmetrized summands is convergent, or if $\prod |f_n|^2 > 0$ on an argument set of positive Lebesgue measure. Clearly, if c_n are centering constants then c'_n are also centering constants if and only if the series $\sum (c_n - c'_n)$ converges (to a finite limit).

a. *If the series $\sum X_n$ is essentially convergent, then the constants $c_n = s_n - s_{n-1} (s_0 = 0)$, determined by the relation E Arctan $(S_n - s_n) = 0$ where $S_n = X_1 + \cdots + X_n$, are centering constants.*

Note that since Arctan is a bounded increasing and continuous function, the stated relation determines the (finite) constant s_n.

Proof. By hypothesis, there exists a sequence s'_n such that $S_n - s'_n \xrightarrow{\text{a.s.}} S$ r.v. Therefore, for every subsequence n' such that $s_{n'} - s'_{n'} \to s$ finite or not,

$$S_{n'} - s_{n'} = (S_{n'} - s'_{n'}) - (s_{n'} - s'_{n'}) \xrightarrow{\text{a.s.}} S - s$$

and, by the dominated convergence theorem,

$$E \operatorname{Arctan}(S - s) = \lim_{n'} E \operatorname{Arctan}(S_{n'} - s_{n'}) = 0.$$

Thus, the constant s is finite and independent of the subsequence n' so that the whole sequence $s_n - s'_n \to s$, and $S_n - s_n \xrightarrow{\text{a.s.}} S - s$ r.v. The assertions follow.

b. *If there exists a r.v. X such that the r.v.'s Y_n defined by $X_1 + \cdots + X_n + Y_n = X$ are independent of $X_1, \cdots, X_n$, then the series $\sum X_n$ is essentially convergent.*

Proof. Let $g_n = \prod_{k=1}^{n} f_k$ and let h_n and f be the ch.f.'s of Y_n and X. By hypothesis and because ch.f.'s are continuous and bounded by $f(0) = 1$, $|g_n|^2 \geqq |g_n|^2 |h_n|^2 = |f|^2 > 0$ on some neighborhood of the origin. Since $|g|^2 = \lim |g_n|^2 = \prod |f_n|^2$ exists, it follows that $|g|^2 > 0$ on this neighborhood. The assertion follows.

The series $\sum X_n$ is *unconditionally convergent* if it is convergent under all reorderings. Constants $\bar{c}_n$ such that the series $\sum (X_n - \bar{c}_n)$ is unconditionally convergent are said to be *unconditionally centering*, and constants $\bar{c}'_n$ are so if and only if the series $\sum (\bar{c}_n - \bar{c}'_n)$ is absolutely convergent, since for numerical series unconditional and absolute convergence are the same. For example, if the second moments EX_n^2 are finite and $\sum \sigma^2 X_n < \infty$ then the series $\sum (X_n - EX_n)$ converges in q.m. hence is convergent. In fact, it is unconditionally convergent since so is the series $\sum \sigma^2 X_n$ (and the EX_n are unconditionally centering constants), while the series $\sum X_n$ itself is unconditionally convergent if and only if $\sum |EX_n| < \infty$.

c. *If the series $\sum X_n$ is essentially convergent, then the constants $\bar{c}_n = \mu X_n + E(X_n - \mu X_n)^c$ (where μX_n is a median of X_n and c is some truncating constant) are unconditionally centering.*

Proof. According to the two-series criterion, essential convergence of the series $\sum X_n$ is equivalent to convergence of the two series $\sum P[|X_n - \mu X_n| \geqq c]$ and $\sum \sigma^2(X_n - \mu X_n)^c$, and then the series $\sum (X_n - \bar{c}_n)$ is convergent. Since the two series are convergent under all reorderings so is the series $\sum (X_n - \bar{c}_n)$, and the assertion is proved.

d. *The series $\sum X_n$ is unconditionally convergent if and only if the series $\sum |f_n - 1|$ converges on some argument set U of positive Lebesgue measure.*

Proof. If $\sum |f_n - 1| < \infty$ on some U hence converges on U under all reorderings, then so does $\prod f_n$ and the "if" assertion follows.

Conversely, if the series $\sum X_n$ is unconditionally convergent, then so is the series $\sum (X_n - \bar{c}_n)$ and $\sum |\bar{c}_n| < \infty$. Let $\bar{f}_n(u) = e^{-i\bar{c}_n u} f_n(u)$ so that, for $|u| \leqq b$,

$$|f_n(u) - 1| = |\bar{f}_n(u) - e^{-i\bar{c}_n u}| \leqq |\bar{f}_n(u) - 1| + b|\bar{c}_n|.$$

According to 22.2$\mathbf{B}_1$ and 22.2$\mathbf{B}_2$ where we take $\tau = c$ and center at a median μ—which does not change $|f|$, for $|u| \leqq b$ sufficiently small so that $\log |f_n(u)|$ exists and is finite, there exists a constant a such that

$$|\bar{f}_n(u) - 1| \leqq a \int_0^b |\log |f_n(v)|| \, dv.$$

But the series $\sum (X_n - \bar{c}_n)$ being convergent, we can take b sufficiently small so that $\prod |f_n(u)| > \frac{1}{2}$ for $|u| \leqq b$, and then

$$\sum |f_n(u) - 1| \leqq (a \log 2 + \sum |\bar{c}_n|) b < \infty.$$

The "only if" assertion is proved.

Let $\sum T_j$ be an arbitrary partition of the set of integers $(1, 2, \cdots)$. The series $\sum_j (\sum_{k \in T_j} X_k)$ is a "partitioned summation" of the series $\sum X_n$.

e. *If the series $\sum X_n$ is unconditionally convergent then it converges a.s. to a same limit r.v. under all reorderings and partitioned summations, and so does any of its subseries.*

Proof. If $\sum |f_n(u) - 1| < \infty$ then $\sum |f_{n'}(u) - 1| < \infty$ for any subsequence n' hence, by **d**, the subseries $\sum X_{n'}$ is unconditionally convergent.

The difference between the series $\sum X_n$ and the reordered series $\sum X'_n$ is defined on the independent r.v.'s X_n and is independent of $X_1, \cdots, X_n$ for any n. Therefore, by the zero-one law, this difference degenerates into a constant c. Since $\sum |f_n - 1|$ is absolutely convergent so that $\prod f_n = \prod f'_n$, it follows that $c = 0$. Similarly for partitioned summations because of the elementary inequality $\sum_j | \prod_{k \in T_j} f_k - 1 | \leqq \sum_j \sum_{k \in T_j} | f_k - 1 |$ valid for arbitrary complex numbers $f_k(u)$ bounded by one (proceed by induction).

37.2 Three parts decomposition. We separate decomposable processes into three parts: a numerical function, an interpolated series of fixed discontinuities and an a.s. continuous part. More precisely

A. THREE PARTS DECOMPOSITION THEOREM. *Every decomposable process is the sum*

$$X_T = x_T + X_T{}^d + X_T{}^c$$

of three independent parts (not necessarily all present):

(i) *A centering function x_T.*

(ii) *A fixed discontinuities part $X_T{}^d$, centered decomposable, with almost all sample functions continuous except at the countable fixed discontinuities set.*

(iii) *An a.s. continuous part $X_T{}^c$, centered decomposable, with almost all sample functions continuous except for countable sets of jumps.*

We proceed in steps and, first, we have to give a meaning to the term "centered."

Let a, b, be the extreme points of the closure $\bar{T}$ of T and let $\bar{T}_{ab}$ be this closure with a and/or b excluded unless they belong to T. We intend to show that we can delete from X_T its degenerate discontinuities, if any, by including them within a *centering function* x'_T so as to leave a *centered* decomposable process X'_T: a process such that for every $t \in \bar{T}_{ab}$ which is a left (right) limit point of T the a.s. limits $X_{t-0}(X_{t+0})$ exist and there are no degenerate discontinuities, that is, if $X_{t-0,t} = X_t - X_{t-0}(X_{t,t+0} = X_{t+0} - X_t)$ degenerates then it degenerates at zero. Note that

The fixed discontinuities set of a centered decomposable process is countable and $X_{t-0,t+0} = X_{t+0} - X_{t-0}$ degenerates at zero if and only if $X_{t-0,t}$ and $X_{t,t+0}$ degenerate at zero.

The first assertion results from the continuity extension theorem. The second assertion results from the fact that $X_{t-0,t+0} = X_{t-0,t} + X_{t,t+0}$ is the sum of two independent r.v.'s, since $f_{t-0,t+0} = f_{t-0,t} f_{t,t+0} = 1$ if and only if $f_{t-0,t}(u) = e^{-iuc}, f_{t,t+0}(u) = e^{+iuc}$ where $c = 0$ by definition of a centered process.

a. CENTERING LEMMA. *Let X_T be decomposable. There exist centering functions x'_T such that $X'_T = X_T - x'_T$ is centered decomposable. The* *fixed discontinuities set of X'_T is independent of the choice of the centering function and the set D_s of its points to the right of any $s \in T$ forms the discontinuity set of the function* $d_{st} = \int_0^1 |f_{st}(u)|\, du,\ t > s.$

Proof. Let x'_T be determined by the relation E Arctan $(X_T - x'_T) = 0$. If $s_n \uparrow t \in \overline{T}_{ab}$ then, from $t \leqq t' \in T$ and the equality $\sum_{k=1}^{n} (X_{s_{k+1}} - X_{s_k}) + (X_{t'} - X_{s_{n+1}}) = X_{t'} - X_{s_1}$ it follows, by 37.1**a** and **b**, that the sequence $X'_{s_{n+1}} = X'_{s_1} + \sum_{k=1}^{n} (X'_{s_{k+1}} - X'_{s_k})$ converges a.s. to a r.v. If $s'_n \to t - 0$ and $s''_n \to t - 0$ then both sequences can be reordered and combined into a sequence $s_n \uparrow t$ so that, because of separability, the one-sided a.s. limit r.v. X_{t-0} exists; similarly for X_{t+0}. Since E Arctan $X'_T = 0$ hence, by the dominated convergence theorem, E Arctan $X'_{t\pm 0} = 0$, it follows that all degenerate discontinuities and, in fact, all degenerate increments degenerate at zero. Thus the decomposable process X'_T is centered. Since any change of centering function changes any given discontinuity by a constant hence cannot reduce nondegenerate discontinuities to zero, it follows that the fixed discontinuities set of X'_T does not vary with the centering function. The first assertion is proved.

The relation $f_{s,t+h} = f_{st}f_{t,t+h}$ implies that $|f_{st}|$ hence d_{st} is nonincreasing in t. But, setting $f'_{st}(u) = E \exp \{iuX'_{st}\}$ we have $|f'_{t-0,t+0}(u)| = 1$ for $t \notin D_s$ and for all u, and $|f'_{t-0,t+0}(u)| < 1$ for $t \in D_s$ and an u-set of positive Lebesgue measure in [0, 1]. Therefore, from $|f'_{st}| = |f_{st}|$ it follows that d_{st} is discontinuous at t if and only if $t \in D_s$. The proof is terminated.

So far, T was an arbitrary set in $\overline{R}$. In fact, we can and whenever convenient we shall take as domain an interval in the interval I_{ab} whose endpoints a, b are those of $\overline{T}$ except that they are to be excluded from I_{ab} unless they belong to T. For

b. DECOMPOSABILITY EXTENSION LEMMA. *A decomposable process X_T can be extended on I_{ab} with preservation of decomposability, of centering, and of type provided the type is invariant under additions and passages to the limit.*

Proof. Let X'_T be the process after centering. Set $X_t = X'_{t+0}$ for right limit points $t \in I_{ab} - T$. The remaining points of $I_{ab} - T$ form intervals (c, d) or $[c, d)$ where the d are right limit points belonging or not to T. Set $X_t = X_d$ on such intervals. The process so extended has the asserted properties.

c. CENTERED SAMPLE LEMMA. *Almost all sample functions of a centered decomposable process X'_T are bounded on every set $[c, d]T$, $c, d \in T$, and are continuous except for countable sets of jumps outside the fixed discontinuities set.*

Proof. We can take T to be $[c, d]$, without restricting the generality.

1° Let μ_t be a median of $X'_d - X'_t$. If $s_n \uparrow t$ then every limit value of the sequence μ_{s_n} is a median of $X'_d - X'_{t-0}$, and similarly for $s_n \downarrow t$. Therefore, the function μ_t is bounded by some constant α. Since the decomposable process defined by $Y_t = X'_t - X'_c + \mu_t$ is such that $Y_d - Y_t$ has 0 for a median it follows by 17.1c that, for $C = s_0 < \cdots < s_n = d$,

$$P[\sup_{k \leqq n} | Y_{s_k} | \geqq \beta] \leqq 2P[| Y_d | \geqq \beta]$$

hence

$$P[\sup_{k \leqq n} | X'_{s_k} - X'_c | \geqq \alpha + \beta] \leqq 2P[| X'_d - X'_c | \geqq \beta]$$

and, by separability,

$$P[\sup_{t} | X'_t - X'_c | \geqq \alpha + \beta] \leqq 2P[|X'_d - X'_c | \geqq \beta],$$

which implies the boundedness assertion.

2° The second assertion will be proved if we show that one-sided limits exist for almost all sample functions. Let u vary over $[-\gamma, +\gamma]$ where γ is sufficiently small so that $f'_{cd}(u) \neq 0$. Since $f'_{ct}(u)f'_{td}(u) = f'_{cd}(u)$, the same is true of all $f'_{ct}(u)$. Then the $Z_t(u) = \exp\{iuX'_{ct}\}/f'_{ct}(u)$ are bounded (complex-valued) r.v.'s. Since a.s.

$$E(Z_t(u) \mid Z_r(u), r \leqq s) = \exp\{iX'_{cs}u\}f'_{st}(u)/f'_{cs}(u)f'_{st}(u) = Z_s(u),$$

the $Z_t(u)$ form a (complex-valued) martingale for every $u \in [-\gamma, +\gamma]$. But, the decomposable process X'_T being centered, the functions $f'_{ct}(u)$ of t have one-sided limits. Thus, to prove that almost all sample functions $X'_T(\omega)$ have one-sided limits, it suffices to show that the same is true of every martingale $Z_t(u)$. If these martingales are separable, this follows from 36.1**C**. However, the martingales may not be separable. But, X'_T being separable, it suffices to prove the assertion for its restriction X'_S on a separating set S and hence for the restrictions $Z_S(u)$ of the martingales on this countable set, and then what precedes applies. The proof is terminated.

For the next proposition, it will be convenient to think of T as an interval. Let $\{t_j\}$ be the fixed discontinuities in $[a, b]T$ of a centered decomposable process X'_T and let $U'_j = X'_{t_j-0,t_j}$, $V'_j = X'_{t_j,t_j+0}$; at most one of these one-sided jumps may degenerate and then it will be at zero. Let $U_j = U'_j - \bar{c}_j$ where $\bar{c}_j = \mu U'_j + E(U'_j - \mu U'_j)^c$ and let V_j be similarly defined in terms of V'_j. Since the one-sided jumps are all independent and, setting $\sum_{k=1}^{n} U'_k + Y_n = X'_b - X'_a$, lemma 37.1**b** applies, and the same is true for the V'_j, it follows, by 37.1**c** and **e**, that the series $\sum_{j\in I} U_j$ and $\sum_{j\in I} V_j$ are unconditionally convergent for every interval $I \subset T$ and converge to the same a.s. limit under all reorderings and partitioned summations. Thus, selecting arbitrarily $t_0 \in T$ and setting

$$X_t^d = \sum_{t_0\leqq t_j\leqq t} U_j + \sum_{t_0\leqq t_j<t} V_j \quad \text{or} \quad -\sum_{t<t_j<t_0} U_j - \sum_{t\leqq t_j<t_0} V_j$$

according as $t \geqq t_0$ or $t < t_0$, the X_t^d are a.s. determined and we can and do select them so that X_T^d be separable.

d. DECOMPOSITION LEMMA. *A centered decomposable process X'_T is the sum*

$$X'_T = x''_T + X_T^d + X_T^c$$

of three independent parts: a centering function x''_T, a centered decomposable process X_T^d corresponding to the fixed discontinuities of X'_T, and a centered decomposable process X_T^c with no fixed discontinuities.

Proof. If $s_n \uparrow t \notin \{t_j\}$, then a.s. $X_{s_{n+1}}^d = X_{s_1}^d + \sum_{k=1}^{n} (X_{s_{k+1}}^d - X_{s_k}^d)$, hence $X_{s_n}^d \xrightarrow{\text{a.s.}} X_t^d$. Similarly, if $s_n \uparrow t_j$, then $X_{s_n}^d \xrightarrow{\text{a.s.}} X_{t_j}^d - U_j$. In the same manner, if $s_n \downarrow t \notin \{t_j\}$, then $X_{s_n}^d \xrightarrow{\text{a.s.}} X_t^d$; and if $s_n \downarrow t_j$, then $X_{s_n}^d \xrightarrow{\text{a.s.}} X_{t_j}^d + V_j$. Thus $U_j = X_{t_j-0,t_j}^d$, $V_j = X_{t_j,t_j+0}^d$, $X_{t_j-0,t_j+0}^d = U_j + V_j$ nondegenerate, and X_T^d is centered with fixed discontinuities set $\{t_j\}$. If x''_T is a centering function of $X'_T - X_T^d$ it follows that the process $X_T^c = X'_T - X_T^d - x''_T$ is centered decomposable and has no fixed discontinuity points. The proof is terminated.

e. DISCONTINUOUS PART SAMPLE LEMMA. *Almost all sample functions of the centered decomposable process X_T^d are continuous outside the fixed discontinuities set $\{t_j\}$.*

Proof. We already know that for a centered decomposable process almost all sample functions are continuous except for countable sets of jumps outside the fixed discontinuities set $\{t_j\}$. Yet, this does not imply that they are continuous except at the t_j's.

To prove the assertion, we take $T = [a, b]$, denote by D the ω-set of sample functions with discontinuities outside $\{t_j\}$, and note that $\sup_t |X_t^d(\omega)| \geqq \epsilon$ for those sample functions $X_T^d(\omega)$ which have a discontinuity outside the t_j at which their oscillation is at least 2ϵ. Furthermore, if a finite number n of t_j's with the corresponding U_j and V_j are deleted from T and from X_T^d, the ω-set D_ϵ of these sample functions remains the same. Thus, to prove the assertion, we may assume this deletion made (including a and V_a if necessary).

Since the series $\sum U_j$ and $\sum V_j$ are unconditionally convergent, it follows from the three series criterion that

$$\sum_{j>n} \sigma_j^2 = \sum_{j>n} \sigma^2(U_j^c + V_j^c) \to 0$$

and, for $n \geqq n(\epsilon)$ sufficiently large,

$$\sum_{j>n} \{P[U_j \neq U_j^c] + P[V_j \neq V_j^c]\} < \epsilon, \quad \sum_{j>n} \{|EU_j^c)| + |EV_j^c|\} < \epsilon$$

Let Y_T be a process defined by means of the U_j^c, V_j^c as X_T^d was defined by means of the U_j, V_j, with same deletion. According to Kolmogorov's inequality, if $a = s_0 < \cdots < s_m$ then

$$P[\sup_{j \leqq m} |Y_{s_j} - EY_{s_j}| > \epsilon] \leqq \sigma^2 Y_{s_m}/\epsilon^2 \leqq \sum_{j>n} \sigma_j^2/\epsilon^2$$

hence

$$P[\sup_{j \leqq m} |X_{s_j}^d| > 2\epsilon] \leqq \epsilon + \sum_{j>n} \sigma_j^2/\epsilon^2$$

and, by separability,

$$P[\sup_t |X_t^d| > 2\epsilon] \leqq \epsilon + \sum_{j>n} \sigma_j^2/\epsilon^2.$$

Since the bracketed condition defines an event which contains the set of sample functions with $\omega \in D_\epsilon$ and we can let $n \to \infty$, it follows that $D_\epsilon \subset A_\epsilon$ with $PA_\epsilon \leqq \epsilon$. If $\epsilon_n = \epsilon/2^n$, then $D_{\epsilon_n} \uparrow D = \bigcup D_{\epsilon_n} \subset \bigcup A_{\epsilon_n} = B$ with $PB \leqq \sum \epsilon_n = \epsilon$. Thus the set D is contained in events of pr. at most equal to $\epsilon > 0$ arbitrarily small and hence is contained in a null event. The proof is terminated.

Upon gathering the foregoing lemmas and setting $x_T = x'_T + x''_T$, the three-parts decomposition follows.

37.3 Infinite decomposability; normal and Poisson cases. We proceed now to the analysis of the a.s. continuous decomposable parts. To simplify the writing, unless otherwise stated, the domain T will be an interval $[a, b]$ and *the processes X_T will be represented by their r.f.'s determined by the condition $X_a = 0$*, so that $X_{at} = X_t$ and $f_{at} = f_t$. We shall also use the following notation: partitions of $[a, t]$ are given by $a = s_{n0} < \cdots < s_{nn} = t$ with $\max_k (s_{nk} - s_{n,k-1}) \to 0$ and the increments $X_{nk} = X_{s_{nk}} - X_{s_{n,k-1}}$ have for d.f.'s F_{nk} and for ch.f.'s f_{nk}.

a. Continuity equivalence lemma. *For a decomposable process X_T, $T = [a, b]$, continuity in law, in pr., a.s., are equivalent and are uniform, and imply that the process is centered.*

Proof. Since continuity a.s. implies continuity in pr. which implies continuity in law, the equivalence assertion will be proved by showing that continuity in law, that is, continuity of the ch.f.'s f_t in t implies a.s. continuity. Since the fixed discontinuities set of the centered process $X'_T = X_T - x_T$ is the discontinuities set of the function $d_t = \int_0^1 |f_t(u)|\,du$ and f_t is continuous in t, it is empty. Therefore X'_T is continuous a.s. hence in law, that is, the function $e^{-iux_t}f_t(u)$ as well as the function $f_t(u)$ are continuous in t. Thus the centering function is continuous, hence $X_T = x_T + X'_T$ has no fixed discontinuities and is centered. The uniformity assertion follows from 35.3**a**, and the proof is terminated.

Because of the above lemma, we drop "a.s." in "a.s. continuous decomposable process" and write *c.d. process*, for short.

We say that $\Psi_t(x)$ is continuous in t if $\Psi_{t'}(x) \to \Psi_t(x)$ as $t' \to t$ for every continuity point x of the function Ψ_t, $t \in T$.

A. Continuous decomposability theorem. *Let X_T, $T = [a, b]$, be a separable c.d. process. Then*

(i) *X_T is infinitely decomposable and* $\log f_t = \psi_t = (\alpha_t, \beta_t^2, L_t) = (\alpha_t, \Psi_t)$ *with α_t continuous and $\Psi_t(x)$ continuous in t and nondecreasing in t and x.*

(ii) *Almost all sample functions of X_T are bounded and are continuous except for countable sets of jumps, and $|L_t(x)| = E\nu_t(x)$ is the expectation of the number $\nu_t(x)$ of jumps of these sample functions in $[a, t)$ of height less than $x < 0$ or at least equal to $x > 0$.*

Proof. 1° Since X_T is uniformly continuous in law, $f_{ss'} \to 1$ uniformly in s, s' as $s - s' \to 0$ so that, taking partitions of $[a, t]$, $X_t =$

$\sum_{k=1}^{n} X_{nk}$ is sum of uan independent r.v.'s. Therefore, by the central limit theorem, X_T is infinitely decomposable and $\log f_t = \psi_t = (\alpha_t, \beta_t^2, L_t) = (\alpha_t, \Psi_t)$. Since $\log f_t$ is continuous in t, it follows from the convergence theorem 22.1**D** that α_t, $\Psi_t(x)$ are continuous in t. Since $\log f_{st} = \psi_{st} = (\alpha_{st}, \Psi_{st})$ with $\Psi_{st}(x)$ nonnegative and nondecreasing in x, the function $\Psi_t(x)$ is nondecreasing in t and x.

2° Let $x < 0$ be a continuity point of $L_t(x)$, take partitions of $[a, t]$ and set $\nu_t^{(n)}(x) = \sum_k I_{[X_{nk} < x]}$. Consider the (almost all) sample functions which are bounded and continuous except for countable sets of jumps, according to the centered sample lemma **37.2c**.

Then $\nu_t^{(n)}(x) \xrightarrow{\text{a.s.}} \nu_t(x)$ while, by the central limit theorem,

$$E\nu_t^{(n)}(x) = \sum_{k=1}^{n} P[X_{nk} < x] = \sum_{k=1}^{n} F_{nk}(x) \to L_t(x)$$

and

$$\sigma^2(\nu_t^{(n)}(x)) = \sum_{k=1}^{n} F_{nk}(x) - \sum_{k=1}^{n} F_{nk}^2(x) \leqq E\nu_t^{(n)}(x),$$

so that the sequence of second moments

$$E(\nu_t^{(n)}(x))^2 \leqq E\nu_t^{(n)}(x) + (E\nu_t^{(n)}(x))^2 \to L_t(x) + L_t^2(x)$$

is bounded. Therefore $E\nu_t(x) = \lim E\nu_t^{(n)}(x) = L_t(x)$, and the nondecreasing and continuous from the left functions $E\nu_t(x)$ and $L_t(x)$ coincide for all continuity points $x < 0$ of $L_t(x)$ hence for all $x < 0$. Similarly for $x > 0$, and the proof is terminated.

Brownian and Poisson cases. Leaving out the trivial degenerate c.d. case which corresponds to vanishing $\Psi_t(x)$, the foregoing theorem yields properties of two extreme cases corresponding to $\Psi_t(x)$ with one fixed point of increase only: The *normal c.d. process* corresponds to the point of increase $x = 0$, that is, to vanishing $L_t(x)$, so that

$$\psi_t(u) = i\alpha_t u - \frac{\beta_t^2}{2} u^2$$

with α_t continuous and β_t^2 continuous and nondecreasing. The *Poisson c.d. process* corresponds to the point of increase $x = c \neq 0$ and, setting $\lambda_t = L_t(c + 0) - L_t(c - 0)$, $\gamma_t = \alpha_t - \frac{c}{1 + c^2} \lambda_t$, to

$$\psi_t(u) = i\gamma_t u - \lambda_t(e^{iuc} - 1)$$

with γ_t continuous and λ_t continuous and nondecreasing. The "reduced" forms obtained by centering and changing the instantaneous time-scale are called the *Brownian process*, $\psi_t(u) = -\frac{\sigma^2 t}{2} u^2$, $\sigma^2 > 0$, and the *Poisson process*, $\psi_t(u) = \lambda t(e^{iu} - 1), \lambda > 0$. We denote by $\mathfrak{L}(\alpha, \beta)$ a law with two values α and β only.

B. NORMAL AND POISSON C.D. CRITERIA. *Let X_T, $T = [a, b]$ be a c.d. process and take partitions of $[a, b]$.*

($\mathfrak{N}$) *X_T is a normal c.d. process if and only if almost all its sample functions are continuous, or X_{ab} is a normal r.v., or $\mathfrak{L}(\max_k |X_{nk}|) \to \mathfrak{L}(0)$.*

($\mathfrak{P}$) *X_T is a Poisson c.d. process if and only if almost all its sample functions are step-functions with jumps of constant height, or X_{ab} is a Poisson type r.v., or $\mathfrak{L}(\min_k X_{nk}) \to \mathfrak{L}(0)$ and $\mathfrak{L}(\max_k X_{nk}) \to \mathfrak{L}(0, c)$ where $c > 0$ or $\mathfrak{L}(\min_k X_{nk}) \to \mathfrak{L}(c, 0)$ and $\mathfrak{L}(\max_k X_{nk}) \to \mathfrak{L}(0)$ where $c < 0$, with X_{ab} lattice-valued.*

Proof. In ($\mathfrak{N}$) and in ($\mathfrak{P}$) the sample functions assertions follow from **A**(ii), the normality and Poisson type assertions follow from **A**(i) or from the composition and decomposition theorem 19.2**A**, and the extrema assertions follow from the extrema criterion 22.4**C**.

Brownian processes are born from and are used in physics to describe motions of particles such as molecules of a gas. However, by their very nature, they are first approximations only, for, while almost all sample functions are continuous, they are extremely irregular in nature and, in particular, almost none is differentiable. In fact

b. BROWNIAN SAMPLE CONTINUITY MODULI LEMMA. *Let X_T, $T = [a, b]$, be a Brownian process. Let $g(h) = \sqrt{2|h| \log 1/|h|}$. For almost all sample functions $X_T(\omega)$ when $c > 1$, but not when $c < 1$,*

$$|X_{t+h}(\omega) - X_t(\omega)| < cg(h), \quad |h| \leqq H(\omega) \text{ positive}, t, t + h \in T.$$

Proof. We can take $a = 0$, $b = 1$, $0 < h < 1$. Since $X_{t+h} - X_t$ is normal $\mathfrak{N}(0, h)$, we have

$$q(h) = P[|X_{t+h} - X_t| \geqq cg(h)] = \sqrt{2/\pi}\int_{c\sqrt{2 \log 1/h}}^{\infty} e^{-x^2/2}\, dx$$

$$\sim h^{c^2}/c\sqrt{\pi \log 1/h}.$$

It follows by elementary computations that when $c > 1$ conditions (ii) of **35.3B** are satisfied and the corresponding assertion is proved. When $c < 1$ then, the events $A_{nk} = [|X_{(k+1)h_n} - X_{kh_n}| \geqq cg(h_n)]$ being independent, $P\bigcup_k A_{nk} = 1 - (1 - q(h_n))^n \to 1$ so that at least one of the A_{nk} occurs with pr. as close to one as we wish for n sufficiently large, that is, for h_n sufficiently small, and the corresponding assertion follows.

Poisson processes are also born from and serve to describe various physical phenomena such as radioactive disintegrations. In fact, we can and do consider only the step sample functions (whose jumps correspond to disintegrations) taken to be rightcontinuous. Then

b′. POISSON SAMPLE JUMPS LEMMA. *Let X_T, $T = [0, \infty)$, be a Poisson process.*

(i) *The c. law of jumps in $I = [s, s + t)$ given that n occurred is that of n independent r.v.'s uniformly distributed in I.*

(ii) *The times $\tau_n(\tau_0 = 0)$ between the $(n - 1)$th and nth jumps are independent r.v.'s with $P[\tau_n > t] = e^{-\lambda t}$.*

Proof. The numbers of jumps Y and Z in $I' = [s', s' + t') \subset I$ and in $I - I'$ are independent Poisson r.v.'s with parameters $\lambda t'$ and $\lambda(t - t')$ and $Y + Z$ is also Poisson with parameter λt. Thus, by elementary computations,

$$P(Y = m \mid Y + Z = n) = P[Y = m, Z = n - m]/P[Y + Z = n]$$
$$= C_n^m(t'/t)^m(1 - t'/t)^{n-m},$$

and (i) is proved. Therefore, taking $t > t_1 + t_2$,

$$P(\tau_1 > t_1, \tau_2 > t_2 \mid X_t = n) = n\int_{t_1}^{t-t_2} d\left(\frac{s_1}{t}\right)\left(1 - \frac{s_1 + t_2}{t}\right)^{n-1}$$
$$= \left(1 - \frac{t_1 + t_2}{t}\right)^n = p^n$$

and $P[\tau_1 > t_1, \tau_2 > t_2] = e^{-\lambda t}(1 + \lambda tp + (\lambda tp)^2/2! + \cdots) = e^{-\lambda t_1}e^{-\lambda t_2}$; similar computations yield independence of any number of τ's, and (ii) is proved. Or, note that **41.1A** case 2° or **38.4A** corollary applies to X_T, set $\sigma_n = \tau_1 + \cdots + \tau_{n-1}$, and

$$P(\tau_n > t_n \mid \tau_{n-1}, \tau_{n-2}, \cdots) = P(X_{\sigma_n + t_n} - X_{\sigma_n} = 0 \mid \sigma_n)$$
$$= P[X_{t_n} - X_0 = 0] = e^{-\lambda t_n}.$$

Stationarity and law derivatives. The ch.f.'s of the increments $X_{t,t+h}$

of Brownian and Poisson processes are $f_{t,t+h} = f_h = e^{h\psi}$ with $\psi(u) = -\frac{\sigma^2}{2}u^2$ and $\psi(u) = \lambda(e^{iu} - 1)$. Thus, these processes are stationary and (equivalently) have stationary law derivatives, according to what follows.

The general concept of stationarity along an index set T is that of invariance under translations along T or, to use an intuitive terminology when $T = [0, \infty)$, of invariance under translations in time. Thus, a process X_T on $T = [0, \infty)$ *is stationary* if its law of evolution is stationary under translations in time. For a centered decomposable X_T it suffices that the individual laws of its increments $X_{t,t+h}$ be stationary: $f_{t,t+h} = f_h$ for all $h > 0$. For, it follows at once that the joint laws of its increments hence its law of evolution are stationary. But then, by decomposability, $f_{h+k} = f_h f_k$ for all $h, k > 0$ so that $f_h = f_{h/n}{}^n$ for n as large as we wish. Therefore $f_h = e^{\psi_h}$ is infinitely decomposable and $f_h \to 1$ as $h \to 0$. Thus the relation becomes $\psi_{h+k} = \psi_h + \psi_k$ with $\psi_h \to 0$ as $h \to 0$. It follows that a stationary centered decomposable is a c.d. process with $f_{t,t+h} = f_h = e^{h\psi}$, $\psi = (\alpha, \Psi)$.

We say that a ch.f. $\dot{f}_t$ represents the (*right*) (*left*) *law derivative* at t of a centered decomposable X_T if $(f_{t-h,t+k})^{[1/(h+k)]} \to \dot{f}_t$ as $h + k \to 0$ with $h, k \geqq 0, h + k > 0$ $(h \equiv 0)$ $(k \equiv 0)$. According to the central convergence theorem, when the limit ch.f. exists, then it is necessarily an i.d. ch.f. $\dot{f}_t = e^{\dot{\psi}_t}$, $\dot{\psi}_t = (\dot{\alpha}_t, \dot{\Psi}_t)$ and the process is (right) (left) law continuous; in particular, if X_T is stationary then $\dot{f}_t \leftarrow (e^{h\psi})^{1/h} = e^{\psi}$ so that the law derivative exists and is stationary. Thus, in searching for conditions of existence of law derivatives, we may assume that X_T is a c.d. process with $f_{st} = e^{\psi_{st}}$, $\psi_{st} = (\alpha_{st}, \Psi_{st})$. Then it follows from 22.1**D** that the law derivatives exists if and only if $\frac{1}{h+k}\psi_{t-h,t+k} \to \dot{\psi}_t$ or, equivalently, $\frac{1}{h+k}\Psi_{t-h,t+k} \xrightarrow{c} \dot{\Psi}_t$ and $\frac{1}{h+k}\alpha_{t-h,t+k} \to \dot{\alpha}_t$. In particular, if the law derivative exists and is stationary hence for every t and every u the derivative $\dot{\psi}(u)$ of $\psi_t(u)$ exists and is independent of t, it follows that the process is stationary.

We collect what precedes into

C. STATIONARY DECOMPOSABILITY CRITERION. *A centered decomposable X_T, $T = [0, \infty)$ is stationary if and only if the ch.f.'s of its increments are of the form $f_{t,t+h} = f_h = e^{h\psi}$, $\psi = (\alpha, \Psi)$ or, equivalently, its law derivative exists and is stationary.*

Integral decomposition. A look at the P. Lévy form of $\psi_t = (\alpha_t, \beta_t^2, L_t)$ makes one think of a c.d. process as a "sum" of a normal c.d. process and of the Poisson c.d. processes corresponding to all points of increase of the P. Lévy functions. In fact, P. Lévy showed that in a sense it was true and *then* obtained the general form for infinitely decomposable laws. Itō made this analysis precise. We shall give the P. Lévy-Itō result but proceeding *from* the general form of the infinitely decomposable laws—to be specific, from the continuous decomposability theorem to which it led—to the reconstruction of c.d. processes by means of their normal and Poisson "components," as follows.

Let X_T be a c.d. process, use the notation introduced in this subsection, and take partitions of $[a, t)$.

c. Number of jumps lemma. *The numbers $\nu_{st}[x, y)$ of jumps in $[s, t)$ of height in $[x, y)$, $xy > 0$, are Poisson r.v.'s with parameter $L_{st}[x, y)$ independent for disjoint time-intervals $[s, t)$ and independent for disjoint height-intervals $[x, y)$.*

Thus, the processes $(\nu_t(x),\ t \in T)$, $x \neq 0$, are Poisson c.d. processes and the processes $(\nu_t(x),\ x \in (-\infty, 0) \cup (0, +\infty))$, $t \in T$, are Poisson decomposable.

Proof. Independence of the $\nu_{st}[x, y)$ for disjoint time-intervals $[s, t)$ results from independence of the corresponding increments X_{st} on which they are defined $x < y < 0$,

For the $x, y \in C(L_t)$ the continuity set of L_t,

$$\varphi_t^{(n)}(u) = E \exp\{iu\nu_t^{(n)}[x, y)\} = \prod_k E \exp\{iuI_{[x \leqq X_{nk} < y]}\}$$
$$= \prod_k \{1 + (e^{iu} - 1)F_{nk}[x, y)\}$$

where $\max_k F_{nk}[x, y) \to 0$ and $\sum_k F_{nk}[x, y) \to L_t[x, y)$. Upon taking n sufficiently large and setting $\varphi_t(u) = E \exp\{iu\nu_t[x, y)\}$, it follows from $\nu_t^{(n)}[x, y) \xrightarrow{\text{a.s.}} \nu_t[x, y)$ that

$$\log \varphi_t(u) \leftarrow \log \varphi_t^{(n)}(u) = (1 + o(1))(e^{iu} - 1)\sum_k F_{nk}[x, y)$$
$$\to (e^{iu} - 1)L_t[x, y).$$

Therefore, upon letting $x' \uparrow x$ along $C(L_t)$ when $x \notin C(L_t)$, and similarly when $y \notin C(L_t)$, the result is valid for all $x, y < 0$, by continuity from the left of L_t; similarly for all $x, y > 0$. The Poisson r.v.'s assertion follows. In fact, it is an immediate consequence of the first criterion in **B**($\mathcal{P}$). We gave a direct proof because the one below generalizes it:

If $[x_j, y_j)$, $x_j y_j > 0$, are m disjoint height-intervals and all the x_j, $y_j \in C(L_t)$ then, as above, for n sufficiently large,

$$\log E \exp \{\sum_j iu_j \nu_t^{(n)}[x_j, y_j)\} = \sum_j \log \{1 + (e^{iu_j} - 1) \sum_k F_{nk}[x_j, y_j)\}$$
$$= (1 + o(1)) \sum_j (e^{iu_j} - 1) \sum_k F_{nk}[x_j, y_j)$$
$$\rightarrow \sum_j (e^{iu_j} - 1) L_t[x_j, y_j),$$

and the restriction to continuity endpoints is removed as above. The Poisson r.v.'s and independence assertions follow, and the proof is terminated.

d. INTEGRATION LEMMA. *The a.s. integrals*

$$I_t = \int_{-\infty}^{+\infty} \{x \, d\nu_t(x) - \frac{x}{1 + x^2} dL_t(x)\}$$

exist and are i.d. r.v.'s with

$$\log E \exp \{iuI_t\} = \int_{-\infty}^{+\infty} \left(e^{iux} - 1 - \frac{iux}{1 + x^2}\right) dL_t(x).$$

Proof. We drop the subscript t and consider only the almost all sample functions for which $\nu[a, b)$ has the properties stated in **A**(ii), so that we drop "a.s."

Let $\alpha < \beta < 0$. If the jumps of the step-function $\nu(\omega, [\alpha, \beta))$ occur at $x_1(\omega), \cdots, x_{n(\omega)}(\omega)$ then $I_\alpha^\beta(\omega) = \int_\alpha^\beta x \, d\nu(\omega, x) = \sum_{k=1}^{n(\omega)} x_k(\omega)$ defines a finite function I_α^β of ω. Set $x_{nk} = \alpha + k(\beta - \alpha)/2^n$, $\nu_{nk} = \nu[x_{nk}, x_{n,k+1})$, and $L_{nk} = E\nu_{nk} = L[x_{nk}, x_{n,k+1})$. Since $S_n = \sum_k x_{nk}\nu_{nk} \uparrow I_\alpha^\beta$, it follows that the integral I_α^β is measurable,

$$\log E \exp \{iuS_n\} \sim \sum_k (e^{iux_{nk}} - 1)L_{nk} \rightarrow \int_\alpha^\beta (e^{iux} - 1) \, dL(x)$$
$$= \log E \exp \{iuI_\alpha^\beta\},$$

and this integral is an i.d. r.v. We can define the integral $I_{-\infty}^\beta$ by $I_\alpha^\beta \uparrow I_{-\infty}^\beta$ as $\alpha \downarrow -\infty$ or directly by means of partitions of $[\alpha_n, \beta]$ where $\alpha_n \downarrow -\infty$. In either case, it follows that $I_{-\infty}^\beta$ is an i.d. r.v. with $\log E \exp \{iuI_{-\infty}^\beta\} = \int_{-\infty}^\beta (e^{iux} - 1) \, dL(x)$. Similarly for $\alpha, \beta > 0$ then $\beta \uparrow +\infty$.

We cannot do the same with, say, $I_{-\infty}{}^{\beta}$ as $\beta \uparrow 0$, since $\int_{-1}^{-0} (e^{iux} - 1)\, dL(x)$ may not exist. However, as $\beta \uparrow 0$, by the convergence in q.m. criterion, there is a limit in q.m.:

$$(1) \qquad J_{-1}{}^{\beta} = \int_{-1}^{\beta} x\, d\{\nu(x) - L(x)\} \xrightarrow{\text{q.m.}} \int_{-1}^{-0} x\, d\{\nu(x) - L(x)\},$$

since as $\beta, \beta' \uparrow 0$

$$\int_{-1}^{\beta}\int_{-1}^{\beta'} xx'\, dd'E\{\nu(x) - L(x)\}\{\nu(x') - L(x')\} \to \int_{-1}^{-0} x^2\, dL(x) < \infty.$$

But the left side of (1) is a martingale in $\beta \uparrow 0$ or may be considered as a sequence of consecutive sums of independent summands $\xi_n = \int_{\beta_n}^{\beta_{n+1}}$ with $\beta_n \uparrow 0$. Either way, the limit in q.m. is also an a.s. limit and

$$\log E \exp\{iuJ_{-1}{}^{\beta}\} = \int_{-1}^{\beta} (e^{iux} - 1 - iux)\, dL(x) \to$$

$$\int_{-1}^{-0} (e^{iux} - 1 - iux)\, dL(x)$$

$$= \log E \exp\{iuJ_{-1}{}^{-0}\};$$

similarly for $\int_{+0}^{+1} x\, d\{\nu(x) - L(x)\}$. Finally, upon adding to $\int_{-\infty}^{-1} x\, d\nu(x)$ the finite constant $-\int_{-\infty}^{-1} \frac{x}{1 + x^2}\, dL(x)$ and to $\int_{-1}^{-0} x\, d\{\nu(x) - L(x)\}$ the finite constant $-\int_{-1}^{-0} \frac{x^3}{1 + x^2}\, dL(x)$ so as to obtain under the integral signs the same expression $x\, d\nu(x) - \frac{x}{1 + x^2}\, dL(t)$, similarly for integrals over $(0, 1)$ and $[1, \infty)$, and adding the four integrals, the lemma follows.

We are now ready for the "integral decomposition" of c.d. processes. We go back to our partitions of $[a, t)$ and recall that the X_{nk} are uan independent (in k) r.v.'s with ch.f.'s $f_{nk} = e^{\psi_{nk}}$, $\psi_{nk} = (\alpha_{nk}, \Psi_{nk})$ and $\max_k |\psi_{nk}(u)| \to 0$. Let $\alpha_t \equiv 0$. Since $\psi_t = \sum_k \psi_{nk}$ where $\psi_t = (0, \Psi_t)$, it follows that there exist a finite $c(u)$ such that

$$\sum_k |f_{nk}(u) - 1| = \sum_k | e^{\psi_{nk}(u)} - 1 | = (1 + o(1)) \sum_k | \psi_{nk}(u) |$$

$$\leqq (1 + o(1)) \left\{ \sum_k \left| \int_{-\infty}^{+\infty} \left(e^{iuy} - 1 - \frac{iuy}{1+y^2} \right) \frac{1+y^2}{y^2} d\Psi_{nk}(y) \right| \right\}$$

$$\leqq (1 + o(1)) \{ c(u) \operatorname{Var} \Psi_t \}.$$

Thus,

$$\sum_k |f_{nk}(u) - 1|^2 \leqq \max_k |f_{nk}(u) - 1| \sum_k |f_{nk}(u) - 1| \to 0$$

and therefore

$$\sum_k (f_{nk}(u) - 1) = (1 + o(1)) \sum_k \psi_{nk}(u) \to \psi_t(u).$$

D. INTEGRAL DECOMPOSITION THEOREM. *Every c.d. process X_T, $T = [a, b]$, is sum of two independent processes*

$$X_T = \eta_T + \int_{-\infty}^{+\infty} \left\{ x \, d\nu_T(x) - \frac{x}{1+x^2} dL_T(x) \right\}$$

where η_T is a normal c.d. process, and ν_T has the properties stated in the number of jumps lemma; and conversely.

Proof. The converse is immediate and, by **A** and **c**, the direct assertion requires only the proof of the independence and normality parts. Since the integral process values I_t are defined on the process $(\nu_t(x), x \neq 0)$, it suffices to prove that η_t and $\nu_t(x)$ are independent for $x \in C(L_t)$. The proof is the same for $x < 0$ and $x > 0$. Let, say, $x < 0$. We can and do assume that $\alpha_t \equiv 0$, without restricting the generality.

If

$$I_{nt}(\epsilon) = \int_{|x| \geqq \epsilon} \left\{ x \, d\nu_t^{(n)}(x) - \frac{x}{1+x^2} dL_t(x) \right\}$$

then it is easily verified that $I_{nt}(\epsilon) \to I_t$ as $n \to \infty$ then $\epsilon \to 0$ so that, setting $\eta_t = X_t - I_t$, it follows that

$$Y_{nt}(\epsilon) = X_t - I_{nt}(\epsilon) \xrightarrow{\text{a.s.}} \eta_t$$

and

$$\varphi_{n,\epsilon}(u, v) = E \exp \{ iuY_{nt}(\epsilon) + iv\nu_t^{(n)}(x) \} \to \varphi(u, v)$$

$$= E \exp \{ iu\eta_t + iv\nu_t(x) \}.$$

We drop t, write ν_n in lieu of $\nu^{(n)}$, and can and do take $x < -\epsilon$, $\pm\epsilon \in C(L_t)$. Since

$$\int_{-\infty}^{-\epsilon-0} x\, dI_{[X_{nk}<x]} = X_{nk}I_{[X_{nk}<-\epsilon]}, \quad \int_{\epsilon+0}^{+\infty} x\, dI_{[X_{nk}\geqq x]} = -X_{nk}I_{[X_{nk}>\epsilon]},$$

we have

$$Y_n(\epsilon) = \sum_k X_{nk}I_{[|X_{nk}|<\epsilon]} + \int_{|y|\geqq\epsilon} \frac{y}{1+y^2}\, dL(y)$$

and

$$\varphi_{n,\epsilon}(u, v)$$

$$= \exp\left\{iu\int_{|y|\geqq\epsilon} \frac{y}{1+y^2}\, dL(y)\right\}\cdot\prod_k E\exp\{iuX_{nk}I_{[|X_{nk}|<\epsilon]} + ivI_{[X_{nk}<x]}\}.$$

Elementary computations yield for the factors in $\prod_k$ the expression

$$1 + (f_{nk}(u) - 1) - \int_{|y|\geqq\epsilon} (e^{iuy} - 1)\, dF_{nk}(y) + (e^{iv} - 1)F_{nk}(x).$$

Since

$$\max_k |f_{nk}(u) - 1| \to 0, \quad \max_k \int_{|y|\geqq\epsilon} dF_{nk}(y) \to 0, \quad \max_k F_{nk}(x) \to 0,$$

it follows that for n sufficiently large

$$\log\varphi_{n\epsilon}(u, v) = +iu\int_{|y|\geqq\epsilon} \frac{y}{1+y^2}\, dL(x)$$

$$+ (1 + o(1))\left\{\sum_k (f_{nk}(u) - 1) - \sum_k \int_{|y|\geqq\epsilon} (e^{iuy} - 1)\, dF_{nk} + (e^{iv} - 1)\sum_k F_{nk}(x)\right\}.$$

Therefore, as $n \to \infty$ then $\epsilon \to 0$,

$$\log\varphi(u, v) \leftarrow \log\varphi_{n\epsilon}(u, v) \to -\frac{\beta^2}{2}u^2 + (e^{iv} - 1)L(x).$$

The independence and normality assertions follow, and the proof is terminated.

COMPLEMENTS AND DETAILS

All r.f.'s are selected to be separable on some unspecified but fixed pr. space.

1. Let $X_T = (X_t, t \geqq 0)$ be separated for closed sets by S. Set $\underline{X}_t = \liminf_{t'\to t} X_{t'}$, $\underline{X}_{t\mp0} = \liminf_{t'\to t\mp0} X_{t'}$.

a) If $\underline{X}_t = X_t$ outside a null event N_t, then the r.f. $(\underline{X}_t, t \geqq 0)$ is equivalent to X_T, is separated for closed sets by S, and is a Borel r.f. with almost all sample functions lower semi-continuous. What if the limits are defined in terms of deleted neighborhoods?

b) If $\underline{X}_{t-0} = \underline{X}_{t+0} = X_t$ outside N_t, then the r.f. $(X_{t+0}, t \geqq 0)$ has the above asserted properties except that lower semi-continuity is replaced by right lower semi-continuity.

c) What if liminf is replaced by limsup?

2. If there exist constants $c, p, q \geqq 0, r > 0$ such that

$$E|X_{t_1} - X_{t_2}|^p |X_{t_2} - X_{t_3}|^q < c|t_1 - t_3|^{1+r}, \quad 0 \leqq t_1 < t_2 < t_3 \leqq 1,$$

then almost all sample functions of $(X_t, 0 \leqq t \leqq 1)$ are continuous except perhaps for simple discontinuities.

3. Let X_T, $T = [a, b]$, be an integrable martingale. If $\sup P[|X_{t+h} - X_t| > \epsilon] = o(h)$ for every $\epsilon > 0$, then almost all sample functions of X_T are continuous. What if integrability is not assumed? What if X_T is a sub-martingale?

4. P. Lévy's normal martingale theorem. An a.s. sample continuous square-integrable martingale X_T, $T = [a, b]$, $X_a = 0$, with $E(X_{st}^2 \mid X_r, r \leqq s) = EX_{st}^2$ a.s. is symmetric normal.

We can take X_T to be sample continuous and $T = [0, 1]$, without loss of generality. To show that X_T is symmetric normal, it will suffice to prove that X_1 is symmetric normal. For the same argument applies with obvious modifications to any fixed linear combination.

Given $\epsilon > 0$, let $\tau_n(\omega)$ be the first $t \leqq 1$ for which $\max |X_r(\omega) - X_s(\omega)| = \epsilon$ where max is taken over all $r \leqq t$, $s \leqq t$ with $|r - s| \leqq 1/n$; if there is no such t, let $\tau_n(\omega) = 1$. Thus, $0 < \tau_n \leqq 1$ and τ_n is a time of X_T since, for $0 < t < 1$, r, s rational with $0 \leqq r$, s, $\leqq t$, $|r - s| \leqq 1/n$ and integers m, $[\tau_n > t] = \bigcup_m \bigcap_{r,s}[|X_r - X_s| \leqq \epsilon - 1/m] \in \mathcal{B}(X_p, p \leqq t)$.

Therefore, by 36.2B applied to the integrable martingale X_T and to the integrable submartingale X_T^2, stopping at τ_n yields martingales $X_T^{\tau_n}$ with $EX_t^{\tau_n} = 0$ and $E(X_t^{\tau_n})^2 \leqq EX_1^2$. Partition $[0, 1]$ into n equal intervals and denote by Y_{nk} the increment of $X_t^{\tau_n}$ over the k-th subinterval. Thus, $X_1^{\tau_n} = \sum_k Y_{nk}$ with $|Y_{nk}| \leqq \epsilon$, $E(Y_{nk} \mid Y_{nj}, j < k) = 0$, $\sum_k EY_{nk}^2 \leqq EX_1^2$. Therefore, if $n = n(\epsilon) \to \infty$ as $\epsilon \to 0$, then $\sum_k E|Y_{nk}|^3 \leqq \epsilon EX_1^2 \to 0$ and, by 28.2B, $\mathcal{L}(X_1^{\tau_n}) \to \mathcal{L}$ symmetric normal. Since, given $\epsilon > 0$, $X_1^{\tau_n} \to X_1$ as $n \to \infty$, it follows that there exist $n = n(\epsilon) \to \infty$ as $\epsilon \to 0$ such that $\mathcal{L}(X_1^{\tau_n}) \to \mathcal{L}(X_1)$. Therefore $\mathcal{L}(X_1) = \mathcal{L}$. But the normal increments for disjoint intervals being orthogonal are independent, and the martingale is decomposable normal.

5. Second order integration. Let $\gamma(s, s') = E\xi(s)\bar{\xi}(s')$, $s, s' \in R$, be of bounded variation in every finite square. It determines a complex-valued σ-additive function $\gamma = \gamma_1 + i\gamma_2$ on bounded Borel sets S and $\int_S g\,d\gamma$ is defined by $\int_S g\,d\gamma_1 + i\int_S g\,d\gamma_2$ whenever this sum exists. Let $G(s)$, with or without affixes be second order r.f.'s "doubly orthogonal" to $\xi(s)$:

$$EG(s)\bar{G}'(s')\Delta_h\Delta'_{h'}\xi(s') = EG(s)\bar{G}'(s')\Delta_h\Delta'_{h'}\gamma(s, s)$$

(second order r.f.'s independent of the r.f. $\xi(s)$ are doubly orthogonal to it), and such that

$$(G, G') = \iint EG(s\bar{G}(s')\, dd'\gamma(s, s')$$

exist and are finite. Indicators of bounded Borel sets are G-functions.

Identify G and G' when they are "γ-equivalent": $(G - G', G - G') = 0$.

a) The (G, G') are inner products: $(aG + bG', G'') = a(G, G'') + b(G', G'')$, $(G, G') = (G', G)$, $\| G \|^2 = (G, G) \geq 0$ and $\| G \|$ is a norm—the "γ-norm" of G which yields a "γ-distance" and a "γ-convergence."

In what follows, consider only G-functions which are step r.f.'s ($\sum c_k I_k$, c_k r.v.'s, I_k indicators of disjoint intervals) or which are their γ-limits. Define $X = \int G\, d\gamma$ up to a P-equivalence, with same affixes if any for G and for X, so as to preserve linearity and inner products, as follows: Set $\int I_{[a,b)}\, d\xi(s) = \xi\,[a, b)$, $\int \sum c_{k'} I_{[a_k\, b_k)}(s)\, d\xi(s) = \sum c_k \xi[a_k, b_k)$, and if step r.f.'s $G_n \xrightarrow{\gamma} G$ set $X = \int G\, d\xi$ where $X = \lim$ q.m. X_n, $X_n = \int G_n\, d\xi$:

b) The definition is justified and the following properties hold up to P-equivalences in X's and up to γ-equivalences in G's: Integration is a linear operation which preserves inner products. It preserves passages to the limit: $X'_n \xrightarrow{\text{q.m.}} X' \Leftrightarrow G'_n \xrightarrow{\gamma} G$. There is a one-to-one correspondence between integrals and integrands: $E| X - X' |^2 = 0 \Leftrightarrow \| G - G' \| = 0$. $X(S) = \int_S G\, d\xi$ is σ-additive in Borel sets S. The "random measure" $\xi(S) = \int I_S\, d\xi$ is σ-additive in bounded Borel sets S.

6. *Second order integral decomposition.* Given a second order r.f. X_T with $EX_t\bar{X}'_{t'} = \iint g_t(s)\bar{g}_{t'}(s')\, dd'\gamma(s, s')$ where $\gamma(s, s')$ is of bounded variation on every finite square, there exists a second order r.f. $\xi(s)$ with $E\Delta_h\xi(s)\Delta'_{h'}\bar{\xi}(s') = \Delta_h\Delta'_{h'}\gamma(s, s')$ such that $X_t = \int g_t\, d\xi$.

Set $(X_t, X_{t'}) = (g_t, g_{t'})$, let c_k's be constants, $L(X)$ be the family of all linear combinations $\sum c_k X_{t_k}$, $L(g)$ be the family of all linear combinations $\sum c_k g_{t_k}$, and let $L_2(X)$ be the closure in q.m. of $L(X)$ and $L_\gamma(g)$ be the γ-closure of $L(g)$.

a) Extend the correspondence $X_t \leftrightarrow g_t$, $t \in T$, to $L(X)$ and $L_\gamma(g)$ preserving linearity and then extend it to $L_2(X)$ and $L_\gamma(g)$ preserving inner products. The correspondence between $L_2(X)$ and $L_\gamma(g)$ preserves linearity and inner products.

b) If all indicators $I_{[a,b)} \in L_\gamma(g)$, denote by $\xi[a, b)$ the corresponding elements of $L_2(X)$. Then $X_t = \int g_t\, d\xi$.

c) If some indicators $I_{[a,b)} \notin L_\gamma(g)$ introduce a set T' whose elements t are the missing indicators, extend g on $T + T'$ by taking g_t, $t \in T'$, such that $(g_t, g_{t'})$ exist and are finite for $t, t' \in T \cup T'$, extend X_T on $T \cup T'$ by taking X_t, $t \in T'$ so that $(X_t, X_{t'}) = (g_t, g_{t'})$, $t, t' \in T \cup T'$ and $X_{T'}$ independent of X_T, and apply *b*).

d) Give as particular cases the decompositions theorems in Section 34.

7. Let $E|\,d\xi(s)\,|^2 = dF(s)$ and let $g_t(s)$ be measurable with respect to the $dt\,dF(s)$-product measure with $|\,g_t(s)\,|^2\,dF(s) < \infty$ for almost all t.

a) The r.f. X_T with $X_t = \int g_t(s)\,d\xi(s)$ can be selected so as to be measurable: Start with $g_t(s) = \sum_k f_{k,t}h_k(s)$.

b) If $\int dF(s)\left(\int|\,g_t(s)\,|\,dt\right)^2 < \infty$, $\int dt\left(\int|\,g_t(s)\,|^2\,dF(s)\right)^{1/2} < \infty$, then the iterated integrals $\int d\xi(s)\left(\int g_t(s)\,dt\right)$ (where the bracketed integral is selected so as to be measurable), $\int dt\left(\int g_t(s)\,d\xi(s)\right)$ exist, are a.s. equal, and denoted by $\iint g_t(s)\,dt\,d\xi(s)$.

c) Let $g'(t)$ be the derivative of $g(t)$, $t \in [a, b]$, and $g_t(s) = g'(t)$ or 0 according as $t \leqq s$ or $t > s$. Then, for F continuous at a and b

$$\int_a^b\int_a^b g_t(s)\,dt\,d\xi(s) = (g(b) - g(a))\xi[a, b) - \int_a^b \xi[a, t)g'(t)\,dt.$$

What if $F(s)$ is not continuous at a and b?

8. Let $\xi(s)$ be a martingale with

$$E(|\,\xi[s, t)\,|^2, \xi_r, r \leqq s) = E|\,\xi[s, t)\,|^2 = F[s, t) \text{ a.s.}$$

a) The family of r.f.'s $G(s)$ with $\int E|\,G(s)\,|^2\,dF(s) < \infty$ contains those which are measurable with respect to the $ds\,dP$-measure with every r.v. $G(s)$ being $\mathfrak{B}(\xi_r, r \leqq s)$-measurable.

If $\xi(s) = \eta(s) - s$ where $\eta(s)$ is a Poisson r.f. with $\lambda = 1$, then $E\,d\xi(s) = 0$, $E(d\xi(s))^2 = dt$, and

$$\int_a^b \xi[a, s)\,d\xi(s) = \tfrac{1}{2}(\xi[a, b))^2 - \tfrac{1}{2}\xi[a, b).$$

If $\xi(s)$ is a Brownian r.f. with $\sigma^2 = 1$, then

$$\int_a^b \xi[a, s)\,d\xi(s) = \tfrac{1}{2}(\xi[a, b))^2 - \tfrac{1}{2}(b - a).$$

b) R.f.'s $X_T = (X_t, t \geqq a)$, $X_t = \int_a^t G(s)\,d\xi(s)$ are square-integrable martingales whose almost all sample functions have one-sided limits at all points and the fixed discontinuities are discontin ities of $F(s)$. If almost all sample functions of $\xi(s)$ are continuous so are those of X_T.

c) If X_T, $T = [a, b]$ is a square-integrable martingale whose almost all sample functions are continuous and $E(|\,X_{s,t}\,|^2|\,X_r, r \leqq s) = E\left(\int_s^t|\,G(s)\,|^2\,ds\,|\,X_r, r \leqq s\right)$ a.s., then there exists a Brownian r.f. $\xi(s)$, $a \leqq s \leqq b$, such that $X_{a,t} = \int_a^t G(s)\,d\xi(s)$ a.s.; however, the pr. space may have to be enlarged to a product of the given pr. space with an appropriate one unless $G(\omega, s)$ vanishes almost nowhere on (ω, t)-space. What about a converse?

9. Zero-one law. Let X_T, $T = [0, \infty)$, be decomposable. If ξ is a tail r.v. on the family of increments $X_t - X_s$, $s, t \geqq 0$, then ξ is degenerate.

10. Strong law of large numbers. Let X_T, $T = [0, \infty)$, be decomposable with stationary increments and $E(X_t - X_0) = 0$. Then $E(X_1 - X_0 \mid X_s - X_0, s \geqq t) = E(X_1 - X_0 \mid X_t - X_0) = (X_t - X_0)/t$ a.s. and $X_t/t \xrightarrow{\text{a.s.}} 0$. Extend as in Section 29.4III.

11. Let $X^{(n)} = (X_t^{(n)}, t \geqq 0)$, $n = 1, 2, \cdots$, be independent Poisson r.f.'s with parameter $\lambda > 0$. The pr. that all $X^{(n)}$ be constant in an interval of length h is $e^{-\lambda hn}$. Let $c_n > 0$, $\sum c_n < \infty$. The series $X_t = \sum c_n X_t^{(n)}$ is a.s. convergent. Are the following statements true? $X = (X_t, t \geqq 0)$ is a.s. continuous, strictly increasing, and its moving discontinuity sets form an everywhere dense set in $[0, \infty)$.

12. Let X_{st} be the number of occurrences in $[s, t)$ of *purely random events:* the X_{st} are a.s. finite for finite time-intervals and independent for disjoint ones with $EX_{st} = \lambda(t - s)$, the pr. of occurrence at any specified time and the pr. of two simultaneous occurrences are zero. The X_{st} form a Poisson process, and conversely.

13. Poisson distribution of particles. Let X_i, $i = \cdots -1, 0, +1, \cdots$ be r.v.'s and let ν_I be the number of $X_i \in I$; I, with or without affixes, are intervals of length $|I|$ and X_i and ν_I have same affixes if any. The X_i represent (positions of) particles and the ν_I are their numbers in I. The "distribution of particles" is the family of joint distributions of $\nu_{I_1}, \cdots, \nu_{I_n}$ for all $I_1, \cdots, I_n$, $n = 1, 2, \cdots$. The distribution is "Poissonian" if the ν_I are independent for disjoint intervals I and Poissonian with parameter $\lambda|I|$.

Let the particles $X_i(t)$, $t \geqq 0$, move "independently": for every fixed t, the increments $X_i[0, t)$ are mutually independent, identically distributed, and independent of all $X_i(0)$. Let $P_t[a, b] = P[X_i[0, t) \in [a, b]]$.

a) The Poisson distribution of particles is invariant in time (that is, if their distribution is Poissonian at time 0 then it is Poissonian with same λ at any time t).

b) If, as $t \to \infty$, for every $h > 0$,

$$\sum_{n=-\infty}^{+\infty} | P_t[nh, (n+1)h] - P_t[(n-1)h, nh] | \to 0$$

and $E|(\nu_I(0)/|I|) - \lambda| \to 0$ as $|I| \to \infty$, uniformly in all intervals I of length I, for some constant λ, then the distribution of particles converges to the Poisson distribution (the distribution "converges" to the distribution of particles $\{\overline{X}_i\}$ if the $\mathfrak{L}(\nu_{I_1}, \cdots, \nu_{I_n})$ converge to $\mathfrak{L}(\bar{\nu}_{I_1}, \cdots, \bar{\nu}_{I_n})$ for all $I_1, \cdots, I_n$, $n = 1, 2, \cdots$). What if λ is a r.v.? What if there is no convergence as $|I| \to \infty$?

14. Reflection principle of Desire Andre. Let X_T, $T = [0, \infty)$, be a Brownian r.f. Let $Z_n = Y_1 + \cdots + Y_n$ where the summands are independent and symmetrically distributed, and let $x, \epsilon > 0$.

a) From $P[\max_{k \leqq n} Z_k \geqq x] \leqq 2P[Z_n \geqq x]$, it follows that $P[\sup_{t \leqq a} (X_t - X_0) \geqq x] \leqq 2P[X_a - X_0 \geqq x]$, and X_T is a.s. sample continuous. From $P[Z_n \geqq x + 2\epsilon] - \sum_{k=1}^{n} P[Y_k \geqq \epsilon] \leqq \frac{1}{2}P[\max_{k \leqq n} Z_k \geqq x]$, it follows that $P[\sup_{t \leqq a} (X_t - X_0) \geqq x] = 2P[X_a - X_0]$.

b) Consider only continuous sample functions of X_T. Then

$$P[\max_{t \leqq a} (X_t - X_0) \geqq x, X_a - X_0 \geqq x] = P[X_a - X_0 \geqq x],$$

$$P[\max_{t \leqq a} (X_t - X_0) \geqq x, X_a - X_0 < x] = P[\max_{t \leqq a} (X_t - X_0 \geqq x, X_a - X_0 > x]$$
$$= P[X_a - X_0 > x],$$

and if τ is the first value of t for which $X_t - X_0 = x$, the changes of X_T after τ are independent of the changes before τ and are equally likely to be positive or negative: "the pr.'s are not changed upon reflecting X_T for $t > \tau$ in the line $y = x$."

15. If X_T, $T = [0, \infty)$, is Brownian then, at every fixed t, the sample functions have a.s. $+\infty$ for upper derivative and $-\infty$ for lower derivative: for every $x > 0$, as $h \to 0$,

$$P[\sup_{t \leqq t' \leqq t+h} (X_{t'} - X_t)/(t' - t) \geqq x] \geqq 2P[X_{t+h} - X_t \geqq hx] \to 1.$$

16. Let X_T, $T = [0, \infty)$, be Brownian with $\sigma^2 = 1$. Consider only the continuous sample functions, and set $\xi_t = \max_{t' \leqq t} X_{t'}$, $\eta_t = \min_{t' \leqq t} X_{t'}$. The r.v.'s $\xi_t, -\eta_t, | X_t |, \xi_t - X_t, X_t - \eta_t$ have the same d.f. $F(x) = \sqrt{2/\pi t}\int_0^x e^{-y^2/2t}\,dy$, $v > 0$. The joint pr. density of ξ_t and X_t is $\sqrt{2/\pi t}\,\dfrac{2x - y}{t}\,e^{-(2x-y)^2/2t}$, $x > 0$, $v < x$. The joint pr. density of ξ_t and $\xi_t - X_t$ is $\sqrt{2/\pi t}\,\dfrac{x + z}{t}\,e^{-(x+z)^2/2t}$, $x > 0, z > 0$.

17. Let X_T, $T = [0, \infty)$, be Brownian with $\sigma^2 = t$. Consider only the continuous sample functions. $\xi_t = \max_{t' \leqq t} X_{t'}$ is nonnegative and nondecreasing in t, and converges to $+\infty$ as $t \to +\infty$.

a) Let $\tau_x = \tau_{x-0}$ be the inverse function of ξ_t: $x = \xi_t$ corresponds to $t = \tau_x$—the smallest zero of $X_t = x = 0$, and $[\tau_x \leqq t] = [\xi_t \geqq x]$, $P[\tau_x < t] = \dfrac{x}{\sqrt{2\pi}}\displaystyle\int_0^t e^{-x^2/2v}v^{-3/2}\,dv$.

b) The r.f. $(\tau_x, x > 0)$ is increasing and decomposable with $\psi_x(u) = x(-1 + i \text{ sign } u)\sqrt{| u |} = \dfrac{x}{\sqrt{2\pi}}\displaystyle\int_0^\infty (e^{iuy} - 1)y^{-3/2}\,dy$. The number of jumps of τ_x in $(0, 1)$ of height greater than h is Poissonian with expectation $\sqrt{2/\pi h}$.

c) Given $| X_{t_0} | = x$, the pr. that X_t have at least one zero in (t_0, t_1) is $(x/\sqrt{2\pi})\displaystyle\int_0^{t_1-t_0} e^{-x^2/2v}v^{-3/2}\,dv$. The pr. that X_T have at least one zero in (t_0, t_1) is $(2/\pi)$ Arccos $\sqrt{t_0/t_1}$. If ζ'_t and ζ''_t are the zeros of X_t immediately preceding and following a given t, then $P[\zeta'_t < t'] = (2/\pi)$ Arcsin $\sqrt{t'/t}, 0 < t' < t, P[\zeta''_t < t'']$ $= (2/\pi)$ Arccos $\sqrt{t/t''}$, $t'' > t$, and $P[\zeta'_t < t', \zeta''_t > t''] = (2/\pi)$ Arcsin $\sqrt{t'/t''}$, $0 < t' < t < t''$.

18. Let X_T, $T = [0, \infty)$, be Brownian with $\sigma^2 = 1$.

a) Divide $[0, t]$ into $2^n = h_n^{-1}$ equal intervals and set $S_n = \sum_{k=0}^{2^n} (X_{(k+1)h_nt} -$

$X_{kh_nt})^2$. Then $E(S_n - t)^2 = 2h_nt^2 \to 0$, $P[|S_n - t| \geqq 2^{-n/2}nt] \leqq 2/n^2$ and, by Borel-Cantelli, $S_n \xrightarrow{\text{a.s.}} t$.

b) More generally, let $(t_1, t_2, \cdots)$ be dense in $[0, t)$, reorder the n first t_k into $t_{n1} < \cdots < t_{nn}$, and let $t_{n0} = 0$, $t_{n,n+1} = t$. Set $S_n = \sum_{k=0}^{n} (X_{t_{n,k+1}} - X_{t_{n,k}})^2$, $S'_n = \sum_{k=2}^{n-1} (X_{t_{n,k+1}} - X_{t_{n,k}})^2$. Then $S_n - S'_n \xrightarrow[\text{q.m.}]{\text{a.s.}} 0$ and $E(S_n - t)^2 = 2\sum_{k=0}^{n} (t_{n,k+1} - t_{n,k})^2 \to 0$. The S_n form a reversed martingale sequence, $S_n \xrightarrow[\text{q.m.}]{\text{a.s.}} t$, and the same is true of the S'_n.

c) Almost no sample function of X_T is of bounded variation in any $[0, t]$. What about their "lengths" in any $[0, t]$?

19. Law of the iterated logarithm. If X_T, $T = [0, \infty)$, is Brownian with $\sigma^2 = 1$, then $\limsup_{t\to\infty} X_t/\sqrt{2t \log_2 t} = 1$ a.s., that is, given $c > 1$, there exists an a.s. finite r.v. τ such that $X_t < c\sqrt{2t \log_2 t}$ for $t > \tau$ and, given $c < 1$, $X_t > c\sqrt{2t \log_2 t}$ infinitely often:

a) Let $t_n = q^n$, $q > 1$, and $X_n = \max_{t \leqq t_n} X_t$. Then, for some constant a and n sufficiently large,

$$p_n = P[X_n \geqq c\sqrt{2t_{n-1} \log_2 t_{n-1}}\,] < an^{-c^2/q}/\sqrt{\log n}.$$

Let $c > 1$ and take $q^2 < c$. Then $\sum p_n < \infty$, by Borel-Cantelli there exists an a.s. positive ν such that $X_n < c\sqrt{2t_{n-1} \log_2 t_{n-1}}$ for $n > \nu$ and $X_t < c\sqrt{2t \log_2 t}$ for $t > \tau = t_\nu$.

b) The $X_n = X_{t_n} - X_{t_{n-1}}$ are independent and

$$p'_n = P[X_n > c'\sqrt{(q-1)/q}\,\sqrt{2t_n \log_2 t_n}\,] \sim a'(n \log q)^{-c'^2}/\sqrt{\log n}.$$

Then, for $c' \leqq 1$, $\sum p'_n = \infty$ and, by Borel, $X_n > c'\sqrt{(q-1)/q}\,\sqrt{2t_n \log_2 t_n}$ infinitely often. For sufficiently large $n > \nu$

$$|X_{t_{n-1}}| < 2\sqrt{2(t_n/q) \log_2 t_n},\ X_{t_n} \geqq X_n - |X_{t_{n-1}}| \geqq b\sqrt{2t_n \log_2 t_n}.$$

where, for $c < 1$, $b = c'\sqrt{(q-1)/q} - 2/q$, $c' > c$ then q can be selected so that $b > c$.

c) What if X_t is replaced by $-X_t$, by $|X_t|$?

Chapter XII

MARKOV PROCESSES

Markov dependence was introduced by Markov (1906) as a natural extension of independence such that the asymptotic properties of sums of r.v.'s, say the law of large numbers and the normal convergence, may be expected to continue to hold under reasonable restrictions. As the restrictions were gradually removed and the setup expanded, new types of behavior and problems appeared. Independently, similar ones appeared in physics (Chapman, Fokker, Planck, etc.).

In his fundamental paper (1931) Kolmogorov introduces, rigorously, Markov dependence in the continuous case and shows that the transition pr.'s satisfy certain differential or integro-differential equations under various restrictions (primarily of the Lindeberg type for normal convergence—on the conditional moments—or of the continuity type $P[X_{t+h} = X_t] \to 1$ as $h \to 0$). The *leit-motif* is the search for local characteristics. Feller explores this new field of research in two series of basic papers essentially of a purely analytical character. First (1936, 1940) he pursues Kolmogorov's approach through local characteristics of transition pr.'s, investigating conditions under which there do exist transition pr.'s with various local characteristics. Then (1952 on) he uses and expands the semi-group theory (created by Hille and Yosida and immediately applied by them to Markov evolution in time) and introduces and analyzes "Feller" or "stable Markov" processes; his work is at the root of novel developments in Markov processes.

Meanwhile Doblin (1938–39) proceeds to a direct sample analysis under a uniform continuity condition on transition pr.'s which leads to step sample functions; Doob (1945), upon removing the uniformity restriction, discovers sample discontinuities more complicated than jumps; P. Lévy (1951) flushes the then monstrous sample possibilities into the open; and Kinney (1953) investigates sample continuity properties.

The purely analytical and the sample analysis gradually merge. Fortet early (1943) examines sample continuity in connection with the Kolmogorov-Feller approach. Later (1955) in connection with the Hille-Yosida-Feller semi-group approach, Neveu combines the two lines of attack. At the same time, Dynkin begins a series of basic papers in which he establishes and extends Feller's results and pursues an investigation of various Markov evolutions in time by means of an intimate blend of sample and semi-group analysis; for this purpose Dynkin and Yushkevich isolate and analyze the concept of strong Markov dependence, first mentioned by Doob (1945), and which is essential for the sample analysis of Markov processes. Hunt (1957) in a basic series of papers connects potential theory with Markov processes.

Let us only mention another line of attack by difference, differential, and integral stochastic equations in the r.f.'s themselves, now in the process of growth, thanks to the works of Bernstein, P. Lévy, Doob, Maruyama, and especially Itō (1951).

§ 38. MARKOV DEPENDENCE

38.1 Markov property. In order to describe and analyze Markov dependence in intuitive yet precise terms and also to simplify the writing, we recall and add some terminology and notation.

Let r, s, t with or without affixes denote the elements of a linearly ordered index set T. To fix the ideas, we take T to be a set of reals with the usual ordering, say, the set of all reals or of all integers or of all nonnegative reals or of all nonnegative integers; in fact, we shall end up by taking $T = [0, \infty)$. It is convenient to give the indices a phenomenological meaning: they will represent moments of "time." Thus, an ordered triplet $U \prec V \prec W$ of disjoint index subsets (that is, every element of U precedes every element of V, which precedes every element of W) represents a "past" U, a "present" V, and a "future" W.

R.v.'s $X, Y, \cdots$ with or without affixes are measurable functions on a pr. space $(\Omega, \mathcal{A}, P)$ to a measurable space $(\mathcal{X}, \mathcal{S})$ or *state space*. In general, the r.v.'s take their values in a topological space $\mathcal{X}$, and $\mathcal{S}$ is the σ-field of topological Borel sets, that is, the σ-field generated by the class of open sets. Most of the considerations in this chapter remain valid for more general state spaces and especially for locally compact separable metric state spaces and, thus, are frequently couched in general terms. However, to fix the ideas we assume that unless otherwise stated our r.v.'s are numerical: the state space is a Borel set of the (extended) real line together with the σ-field of its Borel sets. In general, the extension to the N-dimensional real spaces will be trivial. But the reader is invited to transpose what follows to more general state spaces.

$\mathcal{B}$ with or without affixes will denote sub σ-fields of events B with same affixes if any, unless otherwise stated. We denote by $\mathcal{B}_{T'}$ or by $(\mathcal{B}_t, t \in T')$ the union (or compound) σ-field of the $\mathcal{B}_t$, $t \in T'$, that is, the smallest σ-field containing all of them. Note that the finite sums of finite intersections $B_{t_1} \cap \cdots \cap B_{t_n}$ of their events over all finite subsets of T' form a field $\mathcal{C}$ which generates $\mathcal{B}_{T'}$—the smallest monotone field containing $\mathcal{C}$. In connection with this construction we shall use the properties of c.pr.'s and c.exp.'s without further comment. $\mathcal{B}_{T'}$-measurable functions whose expectation exists will be denoted by $Y_{T'}$, unless otherwise stated.

Finally, in accordance with 25.3**A**, we introduce the following equivalent terminologies for conditional independence

$\mathcal{B}_1$ *and* $\mathcal{B}_3$ *are c.ind. given* $\mathcal{B}_2$: $P(B_1B_3 \mid \mathcal{B}_2) = P(B_1 \mid \mathcal{B}_2)P(B_3 \mid \mathcal{B}_2)$ *a.s.*

$\mathcal{B}_1$ *is c.ind. of* $\mathcal{B}_3$ *given* $\mathcal{B}_2$: $P(B_1 \mid \mathcal{B}_{23}) = P(B_1 \mid \mathcal{B}_2)$ *a.s.*

$\mathcal{B}_3$ *is c.ind. of* $\mathcal{B}_1$ *given* $\mathcal{B}_2$: $P(B_3 \mid \mathcal{B}_{12}) = P(B_3 \mid \mathcal{B}_2)$ *a.s.*

A family $\mathcal{B}_t$, $t \in T$, of σ-fields of events is said to be *Markovian* or a *Markov family* if the following equivalent forms of *Markov property* hold:

For any past, present, and future

(M_{future}) *The future of the family is c.ind. of its past given its present:*

$$P(B_{\text{future}} \mid \mathcal{B}_{\text{present+past}}) = P(B_{\text{future}} \mid \mathcal{B}_{\text{present}}) \text{ a.s.}$$

(M_{past}) *The past of the family is c.ind. of its future given its present:*

$$P(B_{\text{past}} \mid \mathcal{B}_{\text{present+future}}) = P(B_{\text{past}} \mid \mathcal{B}_{\text{present}}) \text{ a.s.}$$

$(M_{\text{past, future}})$ *The past and the future of the family are c.ind. given its present:*

$$P(B_{\text{past}}B_{\text{future}} \mid \mathcal{B}_{\text{present}}) = P(B_{\text{past}} \mid \mathcal{B}_{\text{present}})P(B_{\text{future}} \mid \mathcal{B}_{\text{present}}) \text{ a.s.}$$

It will be convenient to have at our disposal seemingly weaker and seemingly stronger yet equivalent forms of Markov property. The lemma below permits the passage from events to functions.

a. *Let* $\mathcal{B}_1, \mathcal{B}_2, \mathcal{B}_3$ *be σ-fields of events and let* $\mathcal{B}_3$ *be generated by a field* $\mathcal{C}$ *of finite sums of events D belonging to a class* $\mathfrak{D}$.

If $P(D \mid \mathcal{B}_{12}) = P(D \mid \mathcal{B}_2)$ *a.s. then* $P(B_3 \mid \mathcal{B}_{12}) = P(B_3 \mid \mathcal{B}_2)$ *a.s. and, more generally,* $E(Y_{23} \mid \mathcal{B}_{12}) = E(Y_{23} \mid \mathcal{B}_2)$ *a.s.*

Proof. The class of events B such that $P(B \mid \mathcal{B}_{12}) = P(B \mid \mathcal{B}_2)$ a.s. is closed under countable summations and contains $\mathfrak{D}$ hence contains $\mathcal{C}$. It is also closed under monotone passages to the limit hence it contains the monotone field $\mathcal{B}_3$ generated by the field $\mathcal{C}$. This proves the particular assertion.

Since the family of functions Y (whose exp.'s exist) such that $E(Y \mid \mathcal{B}_{12}) = E(Y \mid \mathcal{B}_2)$ a.s. is closed under multiplications by the indicators I_{B_2} and we just proved that it contains the I_{B_3}, it follows that it contains the $I_{B_2B_3} = I_{B_2}I_{B_3}$. Therefore, by the particular assertion, it contains the $I_{B_{23}}$. But this family is also closed under linear combinations whose exp.'s exist and under passages to the limit by nondecreasing sequences of its nonnegative elements. Thus it contains the simple $\mathcal{B}_{23}$-measurable functions then the nonnegative $\mathcal{B}_{23}$-measurable functions and finally all the $\mathcal{B}_{23}$-measurable functions whose exp.'s exist. The proof is terminated.

The next lemma permits the extension of the futures.

b. *Let* $\mathcal{B}_1, \mathcal{B}_2, \mathcal{B}_3, \mathcal{B}_4$ *be σ-fields of events. If* (i) $P(B_3 \mid \mathcal{B}_{12}) = P(B_3 \mid \mathcal{B}_2)$ *a.s. and* (ii) $P(B_4 \mid \mathcal{B}_{123}) = P(B_4 \mid \mathcal{B}_3)$ *a.s., then* $P(B_{34} \mid \mathcal{B}_{12}) = P(B_{34} \mid \mathcal{B}_2)$ *a.s. and, more generally,* $E(Y_{34} \mid \mathcal{B}_{12}) = E(Y_{34} \mid \mathcal{B}_2)$ *a.s.*

Proof. According to **a**, (i) yields

(1) $$E(Y_{23} \mid \mathcal{B}_{12}) = E(Y_{23} \mid \mathcal{B}_2) \text{ a.s.}$$

while, upon multiplying by I_{B_3} and conditioning by $\mathcal{B}_{23}$, (ii) yields

(2) $$E(I_{B_3}I_{B_4} \mid \mathcal{B}_{23}) = E(I_{B_3B_4} \mid \mathcal{B}_3) = E(I_{B_3B_4} \mid \mathcal{B}_{123}) \text{ a.s.}$$

But we always have

(3) $$E(I_{B_3B_4} \mid \mathcal{B}_{12}) = E\{E(I_{B_3B_4} \mid \mathcal{B}_{123}) \mid \mathcal{B}_{12}\} \text{ a.s.}$$

where, by (2), the right side reduces to $E\{E(I_{B_3B_4} \mid \mathcal{B}_{23}) \mid \mathcal{B}_{12}\}$ a.s. and the c.exp. within the last expression is $\mathcal{B}_{23}$-measurable. Therefore, by (1), the right side of (3) reduces to

$$E\{E(I_{B_3B_4} \mid \mathcal{B}_{23}) \mid \mathcal{B}_2\} = E(I_{B_3B_4} \mid \mathcal{B}_2) \text{ a.s.}$$

and (3) becomes

$$P(B_3B_4 \mid \mathcal{B}_{12}) = P(B_3B_4 \mid \mathcal{B}_2) \text{ a.s.}$$

The lemma follows on account of **a**, and the proof is terminated.

We are primarily interested in the Markovian evolution and thus explore the future as time increases. Therefore, we shall use primarily the corresponding form (M_{future}) of the Markov property and be content with stating its equivalent formulations. Besides, the equivalent formulations of the other forms will then follow at once on account of 25.3**A**.

A. MARKOV EQUIVALENCE THEOREM. *The following equivalent relations characterize the Markov property:*

(M) $P(B_{\text{future}} \mid \mathcal{B}_{\text{present+past}}) = P(B_{\text{future}} \mid \mathcal{B}_{\text{present}})$ *a.s. or*

(M′) $E(Y_{\text{present+future}} \mid \mathcal{B}_{\text{present+past}}) = E(Y_{\text{present+future}} \mid \mathcal{B}_{\text{present}})$ *a.s.*

for

(i) *any past, present, and future or*

(ii) *any instant present s, the whole past* $[r\colon r < s]$, *and the whole future* $[t\colon t > s]$ *or*

(iii) *any finite past, instant present, and instant future.*

Note that it suffices that

(M″) *the c.pr.'s of future events given the present and the past depend only upon the present:*

$$P(B_{\text{future}} \mid \mathcal{B}_{\text{present+past}}) \text{ is } \mathcal{B}_{\text{present}}\text{-measurable.}$$

For, this property is implied by (M) and, upon conditioning it by $\mathcal{B}_{\text{present}}$, it yields (M).

Proof. Since (M) ⇒ (M′) on account of **a** and (M′) ⇒ (M) as a particular case, it suffices to consider (M). Since (i) ⇒ (ii) ⇒ (iii), it remains only to prove that (iii) ⇒ (i). Thus, let $U \prec V \prec W$ and

$$\text{(M)} \qquad P(B_W \mid \mathcal{B}_{U+V}) = P(B_W \mid \mathcal{B}_V) \text{ a.s.}$$

for every finite U, singleton V, and singleton W. Since, by **b**, we can increase the future point by point, (M) extends to all finite subsets W_n of any given future, whence, by **a**, to W.

Since for a singleton V and for all finite subsets U_n of any given past U

$$\int_B P(B_W \mid \mathcal{B}_{U+V}) = \int_B P(B_W \mid \mathcal{B}_{U_n+V}) = \int_B P(B_W \mid \mathcal{B}_V)$$

for every $B \in \mathcal{B}_{U_n+V}$ and since the $\mathcal{B}_{U_n+V}$ generate $\mathcal{B}_{U+V}$, it follows that the equality between the finite measures on every $\mathcal{B}_{U_n+V}$ defined by the extreme terms extends on $\mathcal{B}_{U+V}$. Therefore, the integrands coincide a.s. for any given U.

Finally, (M) extends to any given present V, as follows. If V has a last element t, then take $U + V - \{t\}$ as the past, $\{t\}$ as the present, and W as the future, so that

$$P(B_W \mid \mathcal{B}_{U+V}) = P(B_W \mid \mathcal{B}_t) \text{ a.s.}$$

It follows, upon conditioning by $\mathcal{B}_V$ that

$$P(B_W \mid \mathcal{B}_V) = P(B_W \mid \mathcal{B}_t) = P(B_W \mid \mathcal{B}_{U+V}) \text{ a.s.}$$

If V has no last element, decompose V into $V' + V''$ where $V' < V''$ and V' has a last element. Then take U as the past, V' as the present, and $V'' + W$ as the future. According to what precedes

$$P(B_{V''+W} \mid \mathcal{B}_{V'+U}) = P(B_{V''+W} \mid \mathcal{B}_{V'}) \text{ a.s.}$$

and, interchanging the past U and the future $V'' + W$, it follows that

$$P(B_U \mid \mathcal{B}_{V+W}) = P(B_U \mid \mathcal{B}_{V'}) \text{ a.s.,}$$

Finally, conditioning by $\mathcal{B}_V$ so that

$$P(B_U \mid \mathcal{B}_V) = P(B_U \mid \mathcal{B}_{V'}) = P(B_U \mid \mathcal{B}_{V+W}) \text{ a.s.}$$

and interchanging the past U and the future W so that

$$P(B_W \mid \mathcal{B}_V) = P(B_W \mid \mathcal{B}_{U+V}) \text{ a.s.,}$$

the extension of (M) to any given past is proved and the proof is terminated.

The time set T consists most frequently of all $t \geqq t_0$ with $t_0 = 0$ and the Markov evolution is analyzed from instant presents s to the futures $t \geqq s$. Thus, it is primarily the form (M_{future}) which is used. However, difficulties arise at once: The c.pr.'s in (M_{future}) may have no regular versions, so that we cannot treat them as measures; and even if they have, we may not be able to select the regular versions so as to transform all the a.s. equalities into strict equalities, and then we are faced with too many exceptional null events since T is uncountable. If none of these difficulties arises, so that by a suitable choice of c.pr.'s the equalities (M_{future}) become strict equalities and we have no exceptional null events, then we say that the Markov property is *regular* and further analysis of a regular Markov evolution becomes possible. If, moreover, the evolution is independent of the instant present, that is, the c.pr.'s are invariant under translations in time, then we say that the Markov property is *stationary* (or *homogeneous in time*). We shall be mostly concerned with the stationary Markov property.

38.2 Regular Markov processes. A r.f. $X_T = (X_t, t \in T)$ is said to be *Markovian* or a *Markov r.f.* if the family of σ-fields $\mathcal{B}_t = \mathcal{B}(X_t) = X_t^{-1}(\mathcal{S})$ of events induced by the X_t is Markovian. A process is said to be *Markovian* or a *Markov process* if it consists of Markov r.f.'s with same induced σ-fields. Since $\mathcal{B}(X_t) = \mathcal{B}(Y_t)$ means that there exists an invertible Borel g_t such that $Y_t = g_t(X_t)$ and $X_t = g_t^{-1}(Y_t)$, we may consider a Markov process as a Markov r.f. defined up to an invertible scale function g_T.

In this connection, it is sometimes convenient to modify the definitions so as not to require invertibility. A Markov r.f. becomes a family $(X_t, \mathcal{B}_t, t \in T)$ where the X_t are $\mathcal{B}_t$-measurable and the σ-fields $\mathcal{B}_t$ form a Markov family so that, *a fortiori*, $(X_t, t \in T)$ is a Markov r.f. according to our definition. Then, a Markov process becomes the family of all r.f.'s X_T with $\mathcal{B}_t$-measurable X_t's, where the given $\mathcal{B}_t$ form a Markov family.

For r.f.'s X_T, the Markov property (Mii), equivalently, (M'ii) which emphasizes the Markov evolution from any instant present becomes

$$P(B \mid X_r, r \leq s) = P(B \mid X_s) \text{ a.s., equivalently,}$$
$$E(Y \mid X_r, r \leq s) = E(Y \mid X_s) \text{ a.s.}$$

where B and Y are defined on $(X_t, t \geq s)$. In order to examine regularity possibilities, we take its most particular case. First, we take an instant past $\{r\}$ so that, upon conditioning by (X_r, X_s), we have

$$P(B \mid X_r, X_s) = P(B \mid X_s) \text{ a.s.}$$

On account of the general smoothing property $P(B \mid X_r) = E\{P(B \mid X_r, X_s) \mid X_r\}$, it yields the relation

$$P(B \mid X_r) = E\{P(B \mid X_s) \mid X_r\} \text{ a.s.}$$

Next, we take an instant future $\{t\}$ so that the relation becomes

$$P(X_t \in S \mid X_r) = E\{P(X_t \in S \mid X_s) \mid X_r\}, \quad r < s < t, \quad S \in \mathcal{S},$$

outside a null event of the form $[X_r \in S_0]$, where the Borel set S_0 may vary with r, s, t, and with S.

If there are regular versions of the c.pr.'s which figure in the last relation, then its right side can be written as an integral. If moreover the regular versions may be selected so as to make disappear the exceptional sets S_0 for all r, s, t and all S, then we have regularity at least in the particular case of instant present, past, and future. Thus, the requirement that it be possible to transform this relation into strict equality is a minimal one for regularity. We then give to the relation the name of (Chapman-)Kolmogorov equation and to the c.pr.'s therein the name of transition pr.'s. More precisely, a function $P_{st}(x, S)$ defined for all pairs $(s, t) \in T \times T$ with $s < t$ and for all $x \in \mathfrak{X}$, $S \in \mathcal{S}$, is a *transition pr.* (*tr.pr.*) if it is $\mathcal{S}$-measurable in x and a pr. in S and if the *Kolmogorov equation* holds: for all r, s, t with $r < s < t$, all x, and all S,

$$\text{(K)} \qquad P_{rt}(x, S) = \int P_{rs}(x, dy)P_{st}(y, S).$$

We add the *convention:* $P_{ss}(x, S) = I(x, S)(= 1$ or 0 according as $x \in S$ or $x \notin S)$ so that the tr.pr.'s are now defined for all pairs (s, t) with $s \leqq t$, and then the Kolmogorov equation clearly holds with $r \leqq s \leqq t$.

The Kolmogorov equation can be given a seemingly stronger form which parallels the passage of (M) to (M′), that is, from events to functions. Let G be the space of all bounded Borel functions on $\mathfrak{X}$ (so that all integrals below exist and are finite). The relation

$$T_{uv}g(x) = \int P_{uv}(x, dy)g(y), \quad u \leqq v,$$

defines a function $T_{uv}g$ which clearly also belongs to G. In particular, for $g(x) = I(x, S)$ we have $T_{uv}g(x) = P_{uv}(x, S)$, and $T_{uu} = I$—the identity operator on G. Thus, it defines a *tr. operator* T_{uv} on G (to G) which is an extension of the transformation of indicators $I(x, S)$ into tr.pr.'s $P_{uv}(x, S)$. Furthermore, according to the Kolmogorov equation, for $r \leqq s \leqq t$,

$$\begin{aligned}(T_{rt}g)(x) &= \int P_{rt}(x, dz)g(z) = \int\left\{\int P_{rs}(x, dy)P_{st}(y, dz)\right\}g(z)\\ &= \int P_{rs}(x, dy)\left\{\int P_{st}(y, dz)g(z)\right\} = \{T_{rs}(T_{st}g)\}(x).\end{aligned}$$

Therefore, the family of transformations T_{uv} on G has the so-called *generalized semi-group property:*

$$\text{(K}'\text{)} \qquad T_{rt} = T_{rs}T_{st}, \quad r \leqq s \leqq t.$$

Conversely, upon applying (K') to indicators $g(x) = I(x, S)$, Kolmogorov equation follows. Thus

a. *The Kolmogorov equation for tr.pr.'s and the generalized semi-group property for corresponding tr. operators are equivalent.*

What precedes leads to the following definition: A r.f. X_T is a *regular Markov r.f.* if there exists a tr.pr. $P_{st}(x, S)$ such that for all pairs (s, t) with $s \leqq t$

$$P(X_t \in S \mid X_r, r \leqq s) = P_{st}(X_s, S) \text{ a.s.}$$

The r.f. is Markovian, since conditioning by X_s yields

$$P(X_t \in S \mid X_s) = P_{st}(X_s, S) = P(X_t \in S \mid X_r, r \leqq s) \text{ a.s.};$$

by omitting "a.s.", we can and do select regular versions of its c.pr.'s so as to have a regular Markov property. Thus, the definition of a regular Markov r.f. X_T and the choice of regular versions of c.pr.'s coalesce into *the strict equality*

$$\text{(MR)} \qquad P(X_t \in S \mid X_r, r \leqq s) = P(X_t \in S \mid X_s)$$

with the identity

$$P(X_t \in S \mid X_s) \equiv P_{st}(X_s, S),$$

where the values $P_{st}(x, S)$ of the right side are those of a tr.pr.

Note that for $t = s$, the identity reduces to, hence justifies, our convention for tr.pr.'s.

From now on and unless otherwise stated the time set T has a first element t_0, $P_{st}(x, S)$ *is a tr.pr. and the "initial distribution"* P_{t_0} *is a pr. on* $\mathcal{S}$.

Let $(\mathfrak{X}_T, \mathcal{S}_T) = \prod_{t \in T} (\mathfrak{X}_t, \mathcal{S}_t)$ be the sample space of a regular Markov r.f. On account of (MR) the tr.pr. $P_{st}(x, S)$ determines its conditional distributions $P_T{}^x$ given $X_{t_0} = x$. For, according to the Tulcea theorem 8.3**A** and to 27.2b with $g(x_1, \cdots, x_n) = I_{S_{t_1}}(x_1) \times \cdots \times I_{S_{t_n}}(x_n)$, $t_1 < \cdots < t_n$, $S_t \in \mathcal{S}_t$, $P_T{}^x$ is determined by

$$\text{(CD)} \quad P_T{}^x C(S_{t_1} \times \cdots \times S_{t_n}) = \int_{S_{t_1} \cdots S_{t_n}} P_{t_0 t_1}(x, dx_1) \cdots P_{t_{n-1} t_n}(x_{n-1}, dx_n).$$

It follows that, for the law of X_0 given by an initial distribution P_{t_0}, the law of X_T is determined by its distribution defined by

$$\text{(D)} \qquad P_T S_T = \int_{\mathfrak{X}} P_{t_0}(dx) P_T{}^x S_T, \quad S_T \in \mathcal{S}_T.$$

Thus, the c.pr. P^x given $X_{t_0} = x$ on $\mathcal{B}(X_T)$—the σ-field of events $B = [X_T \in S_T]$, is given by

$$\text{(CP)} \qquad P^x[X_T \in S_T] = P_T{}^x S_T, \quad S_T \in \mathcal{S}_T.$$

On account of the integration theorem 26.1**A**, the c.exp. E^x given $X_{t_0} = x$ on the $\mathcal{B}(X_T)$-measurable functions ξ (defined on X_T) whose expectations exist, is given by

$$\text{(CE)} \qquad E^x\xi = \int P^x(d\omega)\xi(\omega),$$

and, for an initial pr. P_{t_0}, hence for an initial distribution $P_{t_0}(S) = P[X_{t_0} \in S]$,

$$\text{(P)}: \quad PB = \int P_{t_0}(dx)P^xB, \quad \text{(E)}: \quad E\xi = \int P_{t_0}(dx)E^x\xi.$$

What precedes leads us to the following definition: A *regular Markov process* is a family of regular Markov r.f.'s with common tr.pr. and arbitrary initial distributions. The tr.pr. $P_{st}(x, S)$ determines by means of (CD), (CP), and (CE) the common conditional distributions $P_T{}^x$, c.pr.'s P^x, and c.exp.'s E^x, given $X_{t_0} = x$. The choice of the initial distribution P_{t_0} determines the law of the corresponding r.f. of the process.

Since the regular Markov property (MR) is in terms of c.pr.'s only, we may also define a regular Markov process as follows. Let X_T be a family of measurable functions X_t on a measurable space $(\Omega, \mathcal{A})$ to a measurable space $(\mathfrak{X}, \mathcal{S})$ (with all singletons $\{x\} \in \mathcal{S}$) and let $(P^x, x \in \mathfrak{X})$ be a family of pr.'s on the induced σ-field $\mathcal{B} = \mathcal{B}(X_T)$ with

$$P^x[X_{t_0} = x] = P^x(X_{t_0}{}^{-1}\{x\}) = 1.$$

If for every $x \in \mathfrak{X}$ there exists a tr.pr. $P_{st}(x, S)$ such that (CD) holds for $P_T{}^x$ defined by (CP), then we may say that X_T is a *regular Markov process*. For every choice of an initial distribution P_{t_0} on $\mathcal{S}$, X_T becomes a regular Markov r.f. on the pr. space $(\Omega, \mathcal{B}, P)$ where P on $\mathcal{B}$ is determined by (P).

The two definitions correspond to the two ways of viewing regular processes and, as long as we are concerned with one regular Markov process with a given tr.pr., we may use either of these definitions according to convenience. However, this raises a basic question: whether given an arbitrary tr.pr. there always exists a corresponding Markov process. The answer is in the affirmative, as follows.

A. REGULAR MARKOV EXISTENCE THEOREM. *To any tr.pr. there corresponds a regular Markov process with a determined law of evolution.*

To any tr.pr. and any initial distribution there corresponds a regular Markov r.f. with a determined law.

Proof. The assertions relative to r.f.'s follow from those relative to processes on account of the necessary condition (P). The assertion relative to the law of evolution follows from the existence assertion on account of the necessary condition (CD). It remains to prove the existence assertion for processes. The necessary conditions (CD) and (CP) show the way: Take $(\Omega, \mathfrak{A}) = (\mathfrak{X}_T, \mathcal{S}_T)$ with points $x_T = (x_t, t \in T)$. Define $X_T = (X_t, t \in T)$ by $X_t(x_T) = x_t$ and using the given tr.pr. $P_{st}(x, S)$, by means of (CD) and (CP), construct the family of pr.'s $P^x = P_T{}^x, x \in \mathfrak{X}$. The process so defined is a regular Markov process with the given tr.pr., provided we can show that for every finite index set $t_1 < \cdots < t_n < s < t$ and every $S_t \in \mathcal{S}_t$ it is possible to select versions of c.pr.'s such that

$$P(X_t \in S_t \mid X_{t_1}, \cdots, X_{t_n}, X_s) = P_{st}(X_s, S_t)$$

or, equivalently,

$$P^x C(S_{t_1} \times \cdots \times S_{t_n} \times S_s \times S_t) = \int_{C(S_{t_1} \times \cdots \times S_s)} P^x(dx_T) P_{st}(x_s, S_t).$$

But by the construction of P^x, both sides reduce to

$$\int_{S_{t_1} \times \cdots \times S_{t_n} \times S_s \times S_t} P_{t_0 t_1}(x, dx_{t_1}) \cdots P_{t_n s}(x_{t_n}, dx_s) P_{st}(x_s, dx_t),$$

and the proof is terminated.

There is a one-to-one correspondence between tr.pr.'s $P_{st}(x, S)$ and *tr.d.f.'s* $F_{st}{}^x$ defined by $F_{st}{}^x(y) = P_{st}(x, (-\infty, y))$ (when $\mathfrak{X} = R$). The Kolmogorov equation in terms of tr.d.f.'s is

$$F_{rt}{}^x(z) = \int F_{rs}{}^x(dy) F_{st}{}^y(z).$$

Tr.d.f.'s $F_{st}{}^x$ are d.f.'s with a parameter x in which they are Borelian, and there is a one-to-one correspondence between tr.d.f.'s $F_{st}{}^x$ and *tr.ch.f.'s* $f_{st}{}^x$ defined by $f_{st}{}^x(u) = \int e^{iuy} F_{st}{}^x(dy)$. However, in the study of local characteristics under some continuity condition we are primarily interested in the Markov behavior in the neighborhood of given states x at times s. This leads to the centering at x of the tr.d.f. $F_{st}{}^x$ and of the tr.ch.f. f_{st} hence to the introduction of

$$\bar{F}_{st}{}^x(y) = P(X_t - X_s < y \mid X_s = x) = F_{st}{}^x(x + y), \quad \bar{f}_{st}{}^x(u) = e^{-iux} f_{st}{}^x(u).$$

A trivial example of regular Markov processes and of tr.d.f.'s is provided by processes $(X_t, t \geqq 0)$ with independent r.v.'s; then $F_{st}^x = F_t$ and Kolmogorov's equation reduces to an identity. An important and suggestive example of regular Markov processes and of tr.d.f.'s is provided by decomposable processes with increments X_{st}; then $\bar{F}_{st}^x = F_{st}$ and Kolmogorov's equation becomes the composition relation $F_{rt} = F_{rs} * F_{st}$. Concepts and problems relative to decomposable processes suggest similar ones for regular Markov processes. In particular, the concept of law derivative will lead us to the historically important local characterizations of tr.d.f.'s, as follows.

We say that a ch.f. $\dot{f}_t^x$ represents the *tr. law (right) (left) derivative* given x at t of a regular Markov process X_T if $(\bar{f}_{t-h,t+k}^x)^{[1/(h+k)]} \to \dot{f}_t^x$ as $h + k \to 0$ with $h, k \geqq 0$ $(h \equiv 0)$ $(k \equiv 0)$, $h + k > 0$. According to the central convergence theorem, if the limit ch.f. exists then it is necessarily an i.d. ch.f. $\dot{f}_t^x = e^{\psi_t^x}$, $\psi_t^x = (\alpha_t^x, \Psi_t^x)$, and the process is tr. law (right) (left) continuous given x at t: $F_{t-h,t+k}^x(y) \to 0$ or 1 according as $y < x$ or $y > x$. While in the decomposable case we could use the convergence theorem for i.d. laws here we need the more general central convergence criterion but for identically distributed summands, that is, with $F_{nk} = F_n$:

$$f_n^n \to f = e^{\psi}, \psi = (\alpha, \Psi) \textit{ if and only if } \Psi_n \xrightarrow{c} \Psi \textit{ and } \int d\Psi_n(x)/x \to \alpha$$
$$\textit{with } \Psi_n(x) = n\int_{-\infty}^{x} \frac{y^2}{1+y^2}\,dF_n.$$

It suffices to note that because of $F_{nk} \equiv F_n$, condition (iii) of the criterion yields $n\left(\int_{|x|<\epsilon} x\,dF_n\right)^2 \leq \left|\int_{|x|<\epsilon} x\,dF_n\right| \times \left|\int_{|x|<\epsilon} n\,dF_n\right| \to 0$ so that in its condition (ii) the left side can be replaced by $n\int_{|x|<\epsilon} \frac{x^2}{1+x^2}\,dF_n$, and the assertion follows by elementary computations.

Upon applying this particular form of the central convergence criterion to tr. law derivatives, we have

b. TR. LAW DERIVATIVE EXISTENCE CRITERION. *The tr. law (right) (left) derivative* $e^{\psi_t^x}$, $\psi_t^x = (\alpha_t^x, \Psi_t^x)$ *exists if and only if*

$$\Psi_{t-h,t+k}^x \xrightarrow{c} \Psi_t^x \textit{ and } \int d\Psi_{t-h,t+k}^x(y)/y \to \alpha_t^x$$
$$\textit{as } h + k \to 0 \ (h \equiv 0)\ (k \equiv 0)$$

where

$$\Psi_{t-h,t+k}^x(x') = \frac{1}{h+k}\int_{-\infty}^{x'} \frac{y^2}{1+y^2}\,dF_{t-h,t+k}^x$$

Next we note that, as for $g(x) = e^{iux}$, $x \in R$,

If a function g on R is bounded and twice differentiable then the function h_g on $R \times R$, given by

$$h_g(x, y) = \left(g(x+y) - g(x) - \frac{y}{1+y^2} g'(x)\right) \frac{1+y^2}{y^2}$$

and defined by continuity at $y = 0$ to be $h_g(x, 0) = g''(x)/2$, is bounded and continuous in y for every fixed x.

The two foregoing remarks yield

c. *If $f_n^n \to f = e^{\psi}$, $\psi = (\alpha, \Psi)$, then for every bounded and twice differentiable function g on R and every fixed $x \in R$*

$$n\left\{\int g(x+y)\,dF_n(y) - g(x)\right\} \to \alpha g'(x) + \int h_g(x, y)\,d\Psi(y).$$

It suffices to apply the Helly-Bray theorem to the left side expressed as

$$n\int (g(x+y) - g(x))\,dF_n(y) = g'(x)\int d\Psi_n(y)/y + \int h_g(x, y)\,d\Psi_n(y).$$

B. TR. LAW DERIVATIVES THEOREM. *Let $e^{\psi_s^x}$, $\psi_s^x = (\alpha_s^x, \Psi_s^x)$, $t \geqq 0$ represent the (right) (left) tr. law derivatives corresponding to a tr.pr. $P_{st}(x, S)$, and let g on R be a bounded twice differentiable function. Then as $h + k \to 0$ $(h \equiv 0)$ $(k \equiv 0)$*

$$\frac{1}{h+k}\left\{\int P_{s-h,s+k}(x, dy)g(y) - g(x)\right\} \to \alpha_s^x g'(x) + \int h_g(x, y)\,d\Psi_s^x(y).$$

In particular, for left tr. law derivatives, if $P_{st}(x, S)$ is twice differentiable in x then its left derivative $\frac{\partial^-}{\partial s} P_{st}(x, S) = \lim_{h\to 0} \frac{1}{h}\left\{P_{s-h,t}(x, S) - P_{st}(x, S)\right\}$ exists and

$$\frac{\partial^-}{\partial s} P_{st}(x, S) = \alpha_s^x \frac{\partial}{\partial x} P_{st}(x, S)$$

$$+ \int\left\{P_{st}(x+y, S) - P_{st}(x, S) - \frac{y}{1+y^2}\frac{\partial}{\partial x} P_{st}(x, S)\right\} \frac{1+y^2}{y^2}\,d\Psi_s^x(y).$$

Proof. The general assertion results from **c**, and the particular one follows upon setting $k = 0$, $g(x) = P_{st}(x, S)$, and using Kolmogorov's equation

$$P_{s-h,t}(x, S) = \int P_{s-h,s}(x, dy)P_{st}(y, S).$$

Upon introducing the P. Lévy form $\psi_s^x = (\alpha_s^x, (\beta_s^x)^2, L_s^x)$, the foregoing integro-differential equation is explicited into

$$\frac{\partial^-}{\partial s} P_{st}(x, S) = \alpha_s^x \frac{\partial}{\partial x} P_{st}(x, S) + \tfrac{1}{2}(\beta_s^x)^2 \frac{\partial^2}{\partial x^2} P_{st}(x, S)$$

$$+ \int \left\{ P_{st}(x + y, S) - P_{st}(x, S) - \frac{y}{1 + y^2} \frac{\partial}{\partial x} P_{st}(x, S) \right\} dL_s^x(y).$$

Kolmogorov's "continuous case" corresponds to vanishing P. Lévy functions L_s^x hence to normal tr. law derivatives $\psi_s^x(u) = i\alpha_s^x(u) - (\beta_s^x)^2 \frac{u^2}{2}$. Feller's "purely discontinuous" and "mixing" cases correspond to vanishing functions α_s^x, $(\beta_s^x)^2$ and to finite $L_s^x(\pm 0)$, respectively. In the first purely analytical approach the study of Markov processes was centered about the questions of existence, unicity, and tr.pr. properties of solutions of these equations. Itō, to whom the foregoing theorem is due, answers these questions by solving stochastic integral equations under somewhat stringent restrictions. As we shall see later, the semigroup approach leads to answers under weaker restrictions.

38.3 Stationarity. To discuss stationarity, that is, invariance under translations in time, it is convenient to take *once and for all* $T = [0, \infty)$, $u, v, r, s, t \geqq 0$ and to use the terminology and notation of Section **30** for translates. Let $\xi = g(X_T)$ denote Borel functions g of X_T and let B denote events defined on X_T; the ξ and the I_B are $\mathcal{B}(X_T)$-measurable functions. The translate X_{s+T} of $X_T = (X_t, t \geqq 0)$ by s is the family $(X_{s+t}, t \geqq 0)$ of the translates X_{s+t} of the X_t by s. The *translate* ξ_s of ξ by s is defined by $\xi_s = g(X_{s+T})$ and the *translate* B_s of B by s is defined by $I_{B_s} = (I_B)_s$; the ξ_s and the I_{B_s} are $\mathcal{B}(X_T)$-measurable, in fact, $\mathcal{B}(X_{s+T})$-measurable functions. To avoid ambiguities, it suffices to take for pr. space the sample space of X_T (see 30.2).

A tr.pr. $P_{u,v}(x, S)$ is *stationary* if it is invariant under translations in time: $P_{s+u,s+v}(x, S) = P_{u,v}(x, S)$ for all u, v, x, S. Thus, a tr.pr. $P_{u,v}(x, S)$ is stationary if and only if its dependence upon the time arguments u, v reduces to dependence upon their differences $t = v - u$ only, so that a stationary tr.pr. may be denoted by $P_t(x, S)$. Its complete definition is then as follows: A *stationary tr.pr.* $P_t(x, S)$, $t \geqq 0$, $x \in \mathfrak{X}$, $S \in \mathcal{S}$, is measurable in x, a pr. in S, with $P_0(x, S) = I(x, S)$, and it satisfies the *stationary Kolmogorov equation*

$$(\mathrm{K_{st}}) \qquad P_{s+t}(x, S) = \int P_s(x, dy) P_t(y, S).$$

The corresponding generalized semi-group property becomes then the *semi-group property*

$$(\mathrm{K'_{st}}) \qquad T_{s+t} = T_s T_t$$

of *stationary tr. operators* or *Markov endomorphisms* T_t on the space G of bounded Borel functions g on $\mathfrak{X}$, defined by

$$(T_t g)(x) = \int P_t(x, dy) g(y), \quad T_0 = I,$$

and 38.2**a** becomes

a. *The stationary Kolmogorov equation for stationary tr.pr.'s and the semi-group property for corresponding stationary tr. operators (Markov endomorphisms) are equivalent.*

A regular Markov process on $T = [0, \infty)$ is *stationary* if its law of evolution is invariant under translations in time. On account of 38.2(MR) the process is stationary if and only if its tr.pr. is stationary.

Let the c.pr. P^ξ be defined by $P^\xi = P^x$ when $\xi \cdot = x$, where P^x is defined by 38.2(CD) and (CP); similarly for E^ξ. Thus, P^ξ (E^ξ) is the c.pr. (c.exp.) on a regular Markov process given the initial random value $X_0 = \xi$. The Markov equivalence and existence theorems, together with (CD) and (CP) and integral definitions of c.pr.'s and c.exp.'s, yield without any difficulty the following theorem where the functions under exp. signs are limited to those whose exp.'s exist.

A. MARKOV STATIONARITY THEOREM. *To every stationary tr.pr. $P_t(x, S)$ there corresponds a stationary regular Markov process X_T. The following equivalent relations characterize regular Markov stationarity:*

$$\text{(i)} \qquad P(X_{s+t} \in S \mid X_r, r \leqq s) = P_t(X_s, S)$$

or

$$P(X_{s+t_1} \in S_1, \cdots, X_{s+t_n} \in S_n \mid X_r, r \leqq s)$$
$$= P^{X_s}[X_{t_1} \in S_1, \cdots, X_{t_n} \in S_n]$$

for all s, all finite index sets and all Borel sets.

(ii) $$P(B_s \mid X_r, r \leqq s) = P(B_s \mid X_s) = P^{X_s}B$$

or

$$E(\xi_s \mid X_r, r \leqq s) = E(\xi_s \mid X_s) = E^{X_s}\xi$$

for all events B and measurable functions ξ defined on X_T and their translates B_s, ξ_s.

(iii) $$P^x AB_s = \int_A P^x(d\omega)P^{X_s(\omega)}B$$

or

$$E^x\eta\xi_s = E^x(\eta E^{X_s}\xi)$$

for all $x \in \mathfrak{X}$, all events A and measurable functions η defined on $(X_r, r \leqq s)$, all events B and measurable functions ξ and their translates B_s, ξ_s.

Note that, in particular, stationarity implies that

$$P(B_s \mid X_s = x) = P(B \mid X_0 = x), \quad E(\xi_s \mid X_s = x) = E(\xi \mid X_0 = x).$$

A r.f. X_T is *stationary* if its law is invariant under translations in time, while the r.f.'s of a *stationary process* have only a stationary law of evolution. In order that X_T be stationary it is necessary but, in general, not sufficient that the initial distribution be stationary: $P_s = P_0$ for all s, where P_s is the distribution of X_s. However, if the law of evolution is stationary then this condition is also sufficient, since then, for all events B on X_T and their translates B_s,

$$PB_s = \int P_s(dx)P(B_s \mid X_s = x) = \int P_0(dx)P(B \mid X_0 = x) = PB.$$

Corollary. *A Markov r.f. with a stationary tr.pr. is stationary if and only if the initial distribution is stationary.*

For stationary tr.pr.'s the study of tr. law derivatives becomes as follows: The *stationary tr.d.f.* $F_t^x(y)$ is defined by $F_t^x = P_t(x, (-\infty, y))$ (when $\mathfrak{X} = R$) and the corresponding stationary tr.d.f. f_t^x is its ch.f. Upon centering at x, we have

$$\bar{F}_t^x(y) = F_t^x(x + y), \quad \bar{f}_t^x(u) = e^{-iux}f_t^x(u).$$

The representation of tr. law derivatives, when they exist, reduces to the limit ch.f.'s $(\bar{f}_h^x)^{[1/h]}$ as $h \to 0$ (independent of t), necessarily of the form e^{ψ^x}, $\psi^x = (\alpha^x, \Psi^x)$. Note that if e^{ψ^x} exists, then $P_h(x, (x - \epsilon, x + \epsilon)^c) \to 0$ for every $\epsilon > 0$ as $h \to 0$. The corresponding existence criterion becomes

b. *The stationary tr. law derivative* $e^{\psi^x}, \psi^x = (\alpha^x, \Psi^x)$ *exists if and only if*

$$\Psi_h^x \xrightarrow{c} \Psi^x \quad and \quad \int d\Psi_h^x(y)/y \to \alpha^x \quad as \quad h \to 0,$$

$$\Psi_h^x(x') = \frac{1}{h}\int_{-\infty}^{x'} \frac{y^2}{1+y^2}\, dF_h^x(y).$$

The corresponding theorem becomes

B. *Let* $e^{\psi^x}, \psi^x = (\alpha^x, \Psi^x)$, *represent the tr. law derivative corresponding to a stationary tr.pr.* $P_t(x, S)$ *and let g on R be a bounded twice differentiable function. Then, as* $h \to 0$,

$$\frac{1}{h}\left\{\int P_h(x, dy)g(y) - g(x)\right\} \to \alpha^x g'(x) + \int h_g(x, y)\, d\Psi^x(y).$$

In particular, if $P_t(x, S)$ *is twice differentiable in x then its right derivative* $\frac{\partial^+}{\partial t} P_t(x, S) = \lim_{h\to 0} \frac{1}{h}\{P_{t+h}(x, S) - P_t(x, S)\}$ *exists and, upon introducing the P. Lévy form* $\psi^x = (\alpha^x, (\beta^x)^2, L^x)$,

$$\frac{\partial^+}{\partial t} P_t(x, S) = \alpha^x \frac{\partial}{\partial x} P_t(x, S) + \tfrac{1}{2}(\beta^x)^2 \frac{\partial^2}{\partial x^2} P_t(x, S)$$

$$+ \int\!\!\!\!\!- \left\{P_t(x + y, S) - P_t(x, S) - \frac{y}{1+y^2}\frac{\partial}{\partial x} P_t(x, S)\right\} dL^x(y).$$

The Kolmogorov continuous case (L^x vanishes) becomes then the Fokker-Planck original case.

From now on and unless otherwise stated, all our Markov processes will be regular and stationary on $T = [0, \infty)$ *so that, whenever convenient, we will drop "regular" and "stationary."*

38.4 Strong Markov property. The central problem of random analysis is that of the sample functions behavior. As soon as this problem arises, random times appear, say, the time of appearance of the first discontinuity of sample functions or the time when the r.f. takes a given value. In the Markov case, we might expect the Markov property to hold under conditionings with a random time τ as "present," since it holds when every one of its values is used as "present." In fact, for a long time this possibility was not even questioned. Yet, let X_T be a stationary regular Markov r.f. with tr.pr.

$$P_t(x, S) = \frac{1}{\sqrt{2\pi t}} \int_S e^{-(x-y)^2/2t}\, dy, \quad x \neq x_0, \quad P_t(x_0, S) = I(x_0, S).$$

Let τ be the time X_T first reaches x_0: $X_t(\omega) \neq x_0$ for $t < \tau(\omega)$ and $X_t(\omega) = x_0$ for $t = \tau(\omega)$. For almost every one of its sample functions, if we know that $X_T(\omega)$ is in a state $x \neq x_0$ at a time t_0, then this sample function is a Brownian continuous one, and, when it passes through x_0, it does not stop there; while if we know that it is in the state x_0 at time t_0, then it stays there forever. It follows that the Markov property of X_T is no more true with τ as "present." Thus, the Markov property is to be strengthened if we want to be able to investigate the sample functions behavior. In other words, we restrict ourselves to those Markov r.f.'s, always separable, for which, *formally*,

$$P(B_\tau \mid X_r, r \leqq \tau) = P^{X_\tau} B,$$

where $B \in \mathcal{B}(X_T)$ and B_τ is its translate by a random time τ—a "random present." However, the introduction of random "presents" raises immediate difficulties. Let, say, τ be the first time X_T takes a given value x_0. To begin with, $\tau(\omega)$ does not exist for those sample functions $X_T(\omega)$ which never take the value x_0. If τ exists it may not be measurable; even if it is measurable it may take infinite values $\tau(\omega) = \infty$, and $X_{\tau(\omega)}(\omega)$ does not exist unless the point at infinity is added to the time interval. If τ is measurable and finite, $X_\tau = (X_{\tau(\omega)}(\omega), \omega \in \Omega)$ may not be measurable; even if all the $X_{\tau+t}$ are measurable, the formal Markov property above has still to be given a meaning. Thus, at first we have to consider and eliminate these difficulties.

A *random time* τ will be a nonnegative measurable function, not necessarily finite but not a.s. infinite: $P\Omega^\tau > 0$, $\Omega^\tau = [\tau < \infty]$. For random times corresponding to sample properties, existence and measurability are to be proved. However, if a random time exists only outside some event, we take it to be infinite on the exceptional event. This convention is acceptable as long as we consider sample functions on the given time interval $[0, \infty)$ only. In other words, *we limit ourselves to the restricted pr. space* $(\Omega^\tau, \mathcal{A}^\tau, P^\tau)$ where $\mathcal{A}^\tau = (\Omega^\tau A, A \in \mathcal{A})$ and $P^\tau = P/P\Omega^\tau$; this is one reason for excluding the possibility of $P\Omega^\tau = 0$. Since we seek sample properties which, at best, are those of almost all sample functions, that is, are valid outside a null event, the exclusion of $P\Omega^\tau = 0$ is not a restriction on random times. Note that whenever relations between pr.'s are homogeneous in P^τ *we may and will replace* P^τ *by* P (upon multiplying throughout by a suitable power of $P\Omega^\tau$).

Translates by τ of events and, more generally, random variables defined on X_T are defined as in 30.2: The translate of $A_t = [X_t \in S]$ is $A_{\tau+t} = [X_{\tau+t} \in S]$ where $X_{\tau+t}(\omega) = X_{\tau(\omega)+t}(\omega)$, the translate of X_t is $X_{\tau+t}$ and, in general, the translate of $\xi = g(X_T)$ is $\xi_\tau = g(X_{\tau+T})$ where $X_{\tau+T} = (X_{\tau+t}, t \in T)$. Translates by a random time τ are considered on Ω^τ only, and *we drop "on* Ω^τ." If a random time τ is elementary $\tau = \sum t_j I_{A_j}$ (on Ω^τ), then, for any r.f. $X_T = (X_t, t \geqq 0)$, the translates $X_{\tau+t}$ by τ are r.v.'s $X_{\tau+t} = \sum X_{t_j+t} I_{A_j}$ so that the translate $X_{\tau+T} = (X_{\tau+t}, t \geqq 0)$ is a r.f. If τ is not elementary, then we have to assume or to prove that $X_{\tau+T}$ is a r.f. It is a r.f. when X_T is Borelian. In particular, X_T is Borelian when it is sample right continuous. For then, it is Borelian as limit of Borel r.f.'s $X_T{}^{(n)} = (X_t{}^{(n)}, t \geqq 0)$ where

$$X_t{}^{(n)} = \sum_{k=0}^{\infty} X_{\frac{k+1}{2^n}} I_{\left[\frac{k}{2^n} \leqq t < \frac{k+1}{2^n}\right]}.$$

To summarize

a. *The translates of arbitrary r.f.'s by elementary random times are r.f.'s and so are the translates by arbitrary random times of Borel r.f.'s, in particular, of sample right continuous r.f.'s.*

A random time τ will be a *time of* X_T if $[\tau \leqq t] \in \mathcal{B}(X_s, s \leqq t)$ for every t; in other words, if we know what happened up to time t inclusive, that is, if we know the sample values $X_s(\omega)$, $s \leqq t$, then we know whether $\tau(\omega) \leqq t$ or not. In particular, every "degenerate" time t is time of any r.f. and if τ is a time of X_T so is every $\tau + t$. In fact, since the inverse images under a time τ of X_T of Borel sets in $[0, t]$ belong to $\mathcal{B}(X_s, s \leqq t)$, we have

b. *If* τ *is a time of* X_T *so are the random times* $\tau + t$ *and, in fact, so are the random times* $g(\tau) \geqq \tau$, *where the functions* g *are Borelian.*

For, by hypothesis, $[g(\tau) \leqq t] = \tau^{-1}g^{-1}[0, t] \subset \mathcal{B}(X_s, s \leqq t)$.

A trivial example of a time of X_T is any elementary time $\tau = \sum t_j I_{A_j}$, $t_1 < t_2 < \cdots$, where every $A_j \in \mathcal{B}(X_r, r \leqq t_j)$.

A nontrivial and important example is as follows. Let U be an open state set and let $\tau_U(\omega)$ be the infimum of all t such that the distance of the sets $(X_s(\omega), s \leqq t)$ and U^c be zero; if $\tau_U(\omega)$ does not exist we take it to be infinite. If X_T is sample right continuous, then τ_U is the time X_T

first *hits* U^c, and if X_T is sample continuous, then τ_U is the time X_T first *reaches* U^c.

c. *If X_T is sample continuous (from the right), then the time X_T first reaches (hits) U^c is a time of X_T.*

For, setting $S_n = [x: d(x, U^c) < 1/n]$ and letting $r < t$ vary over the rationals,

$$[\tau_U \leqq t] = [X_t \in U^c] \cup (\bigcap_n \bigcup_{r<t} [X_r \in S_n]).$$

If τ is a time of X_T, the events $A \in \mathcal{Q}^\tau$ such that $A[\tau \leqq t] \in \mathcal{B}(X_s, s \leqq t)$ for all t form the σ-field $\mathcal{B}(X_s, s \leqq \tau)$—in Ω^τ—of *events defined on* $(X_s, s \leqq \tau)$ or *on X_T up to time τ* (inclusive). Since the definitions of times τ of X_T and of events defined on X_T up to time τ are in terms of events on X_T, it follows that in the Markov case they pertain to all the r.f.'s of the Markov process simultaneously, that is, to the process as a whole. The same is true for what follows and, therein, a "stationary Markov X_T" will mean the process X_T as well as any r.f. belonging to it.

A time τ of a stationary Markov X_T with stationary Borel tr.pr. $P_t(x, S)$ is *Markovian* or *a Markov time of X_T* if all $X_{\tau+t}$ are r.v.'s and, given X_τ, the Markov evolution starts anew:

$$P(X_{\tau+t} \in S \mid X_r, r \leqq \tau) = P(X_{\tau+t} \in S \mid X_\tau)$$

with the same Markov law of evolution:

$$\text{(TM)} \qquad P(X_{\tau+t} \in S \mid X_{\tau+r}, r \leqq s) = P_{t-s}(X_{\tau+s}, S), \quad s \leqq t;$$

it may and will happen that X_τ is to be replaced by $X_{\tau+0}$ for these relations and hence for the following ones but, to simplify, we shall still speak of a Markov time τ; as usual, we write $|\cdot$ in lieu of $|\, \mathcal{B}(\cdot)$.

Upon setting $s = 0$ in the second relation, the first becomes equivalent to the relation

$$\text{(SM)} \qquad P(X_{\tau+t} \in S \mid X_r, r \leqq \tau) = P_t(X_\tau, S).$$

Thus Markov times τ are also characterized by (SM) and (TM). Furthermore, since our Markov evolution is stationary in terms of its degenerate Markov times t, it is natural to require that the same be true in terms of its random Markov times τ: we say that τ is a *stationary Markov time of X_T* if it is and remains a Markov time under translations in time, that is, if all $\tau + s$, $s \geqq 0$, are Markov times of X_T. Thus, stationary Markov times τ are characterized by (SM) and (TM) with τ replaced by any

$\tau + s$ or, equivalently, by

$$(\mathrm{SM}_{st})\quad P(X_{\tau+t} \in S \mid X_r, r \leqq \tau + s) = P_{t-s}(X_{\tau+s}, S), \quad s \leqq t.$$

For, (SM) with $\tau + s$ in lieu of τ yields (SM_{st}) with $t + s$ in lieu of t, while (SM_{st}), with $s + t$ and $s + s'$ in lieu of t and s, conditioned by $(X_{\tau+s+r}, r \leqq s')$, yields (TM) with $\tau + s$ and s' in lieu of τ and s..

A very useful example is that of elementary times:

d. *Elementary times of an arbitrary stationary Markov X_T are stationary Markov times of X_T.*

Proof. Let $\tau = \sum t_j I_{A_j}$ (on Ω^τ), $t_1 < t_2 < \cdots$, $A_j \in \mathcal{B}(X_r, r \leqq t_j)$, be an elementary time of X_T. The integral form of (SM_{st}) is

$$P^x A[X_{\tau+t} \in S] = \int_A P^x(d\omega) P_{t-s}(X_{\tau(\omega)+s}(\omega), S), \quad A \in \mathcal{B}(X_r, r \leqq \tau + s).$$

We have to prove that this relation holds for the elementary time τ, namely that

$$\sum_j P^x AA_j[X_{t_j+t} \in S] = \sum_j \int_{AA_j} P^x(d\omega) P_{t-s}(X_{t_j+s}(\omega), S).$$

Since $AA_j = A[\tau \leqq t_j] \in \mathcal{B}(X_r, r \leqq t_j + s)$, the ordinary Markov property of X_T applies, that is,

$$P(X_{t_j+t} \in S \mid X_r, r \leqq t_j + s) = P_{t-s}(X_{t_j+s}, S),$$

and its integral form is the equality between the terms with same j of both sums. Therefore, these sums are equal and the assertion is proved.

Note that it would have sufficed to prove that elementary times of X_T are Markovian since their translates are also elementary times, hence Markovian.

If all the times of a stationary Borel Markov X_T with a Borel tr.pr. are Markovian, we say that X_T is (*stationary*) *strongly Markovian* or has the *strong Markov property*. If τ is a time of X_T so are all $\tau + s$, hence all the times of a stationary strongly Markovian X_T are stationary Markov times of X_T. Therefore, (SM) holds for all times of X_T or (SM_{st}) holds for all times of X_T if and only if X_T is strongly Markovian.

If τ' varies over all the times of X_T so does $\tau = \tau' I_A + \infty I_{A^c}$ with A varying over all the events defined on $(X_r, r \leqq \tau')$. For, τ is a Borel function of τ' at least equal to τ' and $\tau = \tau'$ when $A = \Omega^{\tau'}$. The integral form of (SM) with τ' in lieu of τ is

$$P^x A[X_{\tau'+t} \in S] = \int_A P^x(d\omega) P_t(X_{\tau'(\omega)}(\omega), S)$$

and, in terms of τ, becomes

$$P^x[X_{\tau+t} \in S] = \int_{\Omega^\tau} P^x(d\omega) P_t(X_{\tau(\omega)}(\omega), S).$$

Set

$$P_\tau(x, S) \equiv P^x[X_\tau \in S]$$

by analogy with the identity $P_t(x, S) = P^x[X_t \in S]$. Then the above relation becomes: for every x

$$(\text{SK}_{\text{st}}) \qquad P_{\tau+t}(x, S) = \int P_\tau(x, dy) P_t(y, S).$$

This equality reduces to the stationary Kolmogorov equation for degenerate times and will be called *(stationary) strong Kolmogorov equation.* The same arguments as in 38.2 and 38.3 lead to the equivalent semigroup property

$$T_{\tau+t} = T_\tau T_t$$

of the Markov endomorphisms on G with T_τ defined as T_t by

$$(T_\tau g)(x) = \int P_\tau(x, dy) g(y), \quad g \in G.$$

In fact, the arguments in the preceding subsections remain valid when, therein, s is replaced by τ and, together with what precedes, yield

A. STRONG MARKOV EQUIVALENCE THEOREM. *Let $P_t(x, S)$ be a Borel stationary tr.pr., where x varies over $\mathfrak{X}$, S over $\mathcal{S}$, and let $s \leqq t$ vary over $[0, \infty)$. Let τ vary over all the times of Borel r.f. X_T.*

(i) *The following equivalent relations characterize the (stationary) strong Markov property of X_T with tr.pr. $P_t(x, S)$*

$$P(X_{\tau+t} \in S \mid X_r, r \leqq \tau) = P_t(X_\tau, S)$$

or

$$P^x AB_\tau = \int_A P^x(d\omega) P^{X_{\tau(\omega)}} B \text{ or } E^x \eta \xi_\tau = E^x(\eta E^{X_\tau} \xi)$$

for all events B and measurable functions ξ (whose exp.'s exist) on X_T, for all $x \in \mathfrak{X}$ and all events A and measurable η on $(X_r, r \leqq \tau)$.

(ii) *The following equivalent relations characterize the strong Markov property of a stationary Markov* X_T *with tr.pr.* $P_t(x, S)$

$$P_{\tau+t}(x, S) = \int P_\tau(x, dy)P_t(y, S) \quad or \quad T_{\tau+t} = T_\tau T_t.$$

An important example of strong Markov X_T is as follows (Dynkin and Ushkevitch).

COROLLARY. *Let* X_T *be stationary Markovian with Borel tr.pr. If* X_T *is sample right continuous and the Markov endomorphisms* T_t *transform bounded continuous functions into bounded continuous functions, then* X_T *is strongly Markovian.*

Proof. Let τ be a time of X_T and note that the elementary times

$$\tau_n = \sum_{k=1}^{\infty} \frac{k}{2^n} I_{\left[\frac{k-1}{2^n} \leqq \tau < \frac{k}{2^n}\right]}$$

converge to τ from the right (on Ω^τ). Let $s \geqq 0$ be arbitrary. Since X_T is sample right continuous, it follows by **a** that it is Borelian so that $X_{\tau+s}$ are r.v.'s, and $X_{\tau_n+s} \to X_{\tau+s}$. Therefore, if $g \in G$ is continuous then $g(X_{\tau_n+s}) \to g(X_{\tau+s})$ and

$$E^x g(X_{\tau_n+s}) \to E^x g(X_{\tau+s}) \quad \text{or} \quad T_{\tau_n+s}g \to T_{\tau+s}g.$$

Since, by **d**, elementary times τ_n of X_T are Markov times, hence, by **A**(ii), $T_{\tau_n+t}g = T_{\tau_n}T_tg$ and $T_tg \in C$, we have $T_{\tau+t}g = T_\tau T_tg$. Thus, the family of bounded functions g on $\mathfrak{X}$ for which this relation holds contains the continuous ones. It is closed under passages to the limit by bounded sequences. Therefore, by the Baire definition of Borel functions, it contains the family G of all bounded Borel functions on $\mathfrak{X}$ and, by **A**(ii), τ is a strong Markov time of X_T. The proof is terminated.

§ 39. TIME-CONTINUOUS TRANSITION PROBABILITIES

Let $P_t(x, S)$ be a stationary tr.pr., that is, a Borel function in x and a pr. in S with $P_0(x, S) = I(x, S)$, obeying the Kolmogorov equation:

$$P_{s+t}(x, S) = \int P_s(x, dy)P_t(y, S), \quad s, t \geq 0.$$

Unless otherwise stated, the time arguments r, s, t, *with or without affixes, vary over* $[0, \infty)$, x *varies over the state space* $\mathfrak{X}$, *and* S *varies over the* σ*-field of state sets* $\mathcal{S}$ *generated by the class of open sets in* $\mathfrak{X}$. *As usual, to fix the* *ideas, we take* $(\mathfrak{X}, \mathcal{S})$ *to be the Borel line and take the limits in* h *as* $0 < h \to 0$.

However, the properties of the tr.pr. to be established and the proofs given are valid for all state spaces $\mathfrak{X}$ such that the diagonal (which consists of all points (x, x)) in $\mathfrak{X} \times \mathfrak{X}$ belong to the product σ-field and, consequently, all singletons $\{x\}$ belong to $\mathcal{S}$ as sections of the diagonal. Also they remain valid when the tr.pr. $P_t(x, S)$ is only a measure in S bounded by 1 in lieu of being a pr. in S. We leave the search for corresponding extensions of properties in 39.2 to the reader.

Denote by $\dot{P}_s(x, S)$ the derivative of $P_t(x, S)$ with respect to t at $t = s$, provided it exists. Note that the derivative $\dot{P}_0(x, S)$ at $t = 0$ is necessarily a right derivative. Set

$$q(x) = \lim_{h \to 0} \frac{1 - P_h(x, \{x\})}{h}, \quad q(x, S) = \lim_{h \to 0} \frac{P_h(x, S)}{h}, \quad x \notin S,$$

whenever the limits exist, and make the convention that

$$q(x, S) = q(x, S\{x\}^c);$$

then

$$\dot{P}_0(x, S) = \lim_{h \to 0} \frac{P_h(x, S) - I(x, S)}{h} = q(x, S) - q(x)I(x, S).$$

Formal differentiation of Kolmogorov's equation with respect to s at $s = 0$ and with respect to t at $t = 0$ (followed by the change of s into t) yields the *backward* and the *forward* equations

$$\text{(B)} \qquad \dot{P}_t(x, S) = \int_{\{x\}^c} q(x, dy) P_t(y, S) - q(x) P_t(x, S)$$

$$\text{(F)} \qquad \dot{P}_t(x, S) = \int P_t(x, dy) q(y, S) - \int_S P_t(x, dy) q(y)$$

Formal solutions (by formal substitution) of these equations are given by

$$\bar{P}_t(x, S) = \sum_{n=0}^{\infty} P_t^{(n)}(x, S)$$

where

$$P_t^0(x, S) = e^{-q(x)t} I(x, S)$$

and

$$P_t^{(n+1)}(x, S) = \int_0^t ds \int_{\{x\}^c} e^{-q(x)s} q(x, dy) P_{t-s}^{(n)}(y, S)$$

or, alternatively,

$$P_t^{(n+1)}(x, S) = \int_0^t ds \int P_s^{(n)}(x, dy) \int_{S\{x\}^c} q(y, dz) e^{-q(z)(t-s)}.$$

In probabilistic language, $\overline{P}_t(x, S)$ is the pr. of transition from x into S in time t in finitely many steps (see the probabilistic interpretation of $q(x)$ and $q(x, S)$ at the end of the section). Thus, whenever there is a possibility of such a transition but not in finitely many steps, it may be expected that $\overline{P}_t(x, S)$ will be a measure in S smaller than 1 in lieu of a pr. in S.

The foregoing formal discussion brings into light the problems to be considered: to begin with, the problem of existence and properties of tr.pr. derivatives, hence of $q(x)$ and $q(x, S)$, and of the corresponding sample properties. Since existence of $\dot{P}_0(x, S)$ implies the *continuity condition*

$$\text{(C)} \qquad P_h(x, S) \to I(x, S),$$

this condition will be assumed at the start. This is the Doblin-Doob approach to the analysis of Markov evolution. The analytical problem of existence, unicity, and tr.pr. properties of solutions of the backward and forward equations—the Kolmogorov-Feller approach—will be left out for it fits within the wider and more powerful Hille-Yosida-Feller-Dynkin semi-group approach.

39.1 Differentiation of tr.pr.'s. The Doblin-Doob results under condition (C) were improved by Kolmogorov who established the existence and finiteness of the function $q(x, S)$ for countable state spaces and his result was extended by Kendall to noncountable state spaces of the general type described above, under a parallel restriction of "σ-uniformity."

Note that condition (C) is equivalent to $P_h(x, \{x\}) \to 1$or, on account of lemma **c** below, to the continuity in t of the tr.pr.

a. *For every section D_x of $D \in \mathcal{S} \times \mathcal{S}$ and in particular for sections $\{x\}$ of the diagonal, the function $P_t(x, D_x)$ is Borelian in x.*

Proof. The class of sets D for which the assertion holds is closed under finite summations and monotone passages to the limit by sequences. It contains all rectangles $S \times S' \in \mathcal{S} \times \mathcal{S}$ since the function $P_t(x, (S \times S')_x) = I(x, S)P_t(x, S')$ has the asserted property. Therefore, it contains the minimal field $\mathcal{S} \times \mathcal{S}$ generated by the rectangles.

b. *The function $P_t(x, \{x\})$ is supermultiplicative in t.*

For, by Kolmogorov's equation,

$$P_{s+t}(x, \{x\}) \geqq \int_{\{x\}} P_s(x, dy)P_t(y, \{x\}) = P_s(x, \{x\})P_t(x, \{x\}).$$

c. *Under* (C), *the function* $g(t) = -\log P_t(x, \{x\})$ *exists and is finite and subadditive in* t, *and the function* $P_t(x, S)$ *is uniformly continuous in* t *uniformly in* S.

Proof. Let $P_h(x, \{x\}) \to 1$. Then, by **b**, $P_t(x, \{x\}) \geqq P_{t/n}{}^n(x, \{x\}) > 0$ for n sufficiently large. Therefore, $g(t)$ exists and is finite, and supermultiplicativity in **b** becomes subadditivity of $g(t)$. The first assertion is proved. By Kolmogorov's equation

$$\Delta = P_{t+h}(x, S) - P_t(x, S)$$

$$= \int_{\{x\}^c} P_h(x, dy)P_t(y, S) - (1 - P_h(x, \{x\})P_t(x, S)$$

so that

$$-(1 - P_h(x, \{x\}) \leqq \Delta \leqq P_h(x, \{x\}^c) = 1 - P_h(x, \{x\}).$$

Therefore,

$$| P_s(x, S) - P_t(x, S) | \leqq 1 - P_{|s-t|}(x, \{x\}) \to 0$$

uniformly in t and in S as $s - t \to 0$, and the second assertion is proved.

d. Point differentiation lemma. *Under* (C),

$$\frac{1 - P_h(x, \{x\})}{h} \to q(x) \leqq \infty$$

where the function $q(x)$ *is Borelian, and* $P_t(x, \{x\}) \geqq e^{-q(x)t}$.

Proof. If $q(x)$ exists, then it is Borelian as limit of sequences of Borel functions corresponding to $h = h_n \to 0$.
Fix x, set $g(t) = -\log P_t(x, \{x\})$ and note that, by **c**,

$$0 \leqq g(t) < \infty, \quad g(s + t) \leqq g(s) + g(t), \quad g(+0) = g(0) = 0.$$

Given $t > 0$ and $h > 0$, take $n = [t/h]$ so that $t = nh + \theta$, $0 \leqq \theta < h$ and, by subadditivity,

$$\frac{g(t)}{t} \leqq \frac{ng(h)}{t} + \frac{g(\theta)}{t} = \frac{g(h)}{h}\frac{nh}{t} + \frac{g(\theta)}{t}.$$

Therefore, $g(t)/t \leqq \liminf_{h\to 0} g(h)/h$ and

$$\limsup_{h\to 0} \frac{g(h)}{h} \leqq \sup_{t>0} \frac{g(t)}{t} \leqq \liminf_{h\to 0} \frac{g(h)}{h}.$$

Thus, $q(x) = \lim_{h\to 0} \frac{g(h)}{h}$ exists with

$$q(x) = \sup_{t>0} \frac{g(t)}{t}.$$

The second assertion follows by

$$P_t(x, \{x\}) = e^{-\frac{g(t)}{t}t} \geqq e^{-q(x)t},$$

and the first one follows by

$$\frac{1 - P_h(x, \{x\})}{h} = \frac{1 - e^{-g(h)}}{h} = (1 + o(1)) \frac{g(h)}{h} \to q(x).$$

The proof is terminated.

We say that a state x is *absorbing*, *instantaneous*, or *steady* according as $q(x) = 0$, $q(x) = \infty$, or $0 < q(x) < \infty$. We say that a set $U = [x: q(x) \leqq c]$ with c finite is *q-bounded*. Since the function $q(x)$ is Borelian, q-bounded sets are state sets and, since $1 - P_h(x, \{x\}) \leqq 1 - e^{-ch} \to 0$ uniformly in $x \in U$, the continuity condition $P_h(x, \{x\}) \to 1$ holds uniformly on every q-bounded set. The same is true on every finite state set even if it has instantaneous states. In general, let $\mathfrak{U}$ be the class of all *uniform continuity state sets*—on each of which (C) holds uniformly. Clearly $\mathfrak{U}$ is closed under taking finite unions and state subsets of its sets and we shall use these closure properties without further comment. Unless otherwise stated, *we denote the sets of $\mathfrak{U}$ by U, with or without affixes.*

e. *Under* (C), *for every x and every uniform continuity state set $U \not\ni x$,*

$$\frac{P_h(x, U)}{h} \to q(x, U)$$

finite bounded by $q(x)$, Borelian in x and a measure in U.

Proof. If the function $q(x, U)$ exists, then it is Borelian in x as limit of sequences of Borel functions corresponding to $h = h_n \to 0$ and it is bounded by $q(x)$ since

$$0 \leqq \frac{1 - P_h(x, \{x\})}{h} - \frac{P_h(x, U)}{h} \to q(x) - q(x, U).$$

Let $V \in \mathfrak{U}$ be such that $U + \{x\} \subset V$, (say, $V = U + \{x\}$). Given a positive $\epsilon < \frac{1}{3}$ there exists a positive $t_0 = t_0(V, \epsilon)$ such that, for $t \leqq t_0$, $P_t(y, \{y\}) > 1 - \epsilon$ for all $y \in V$ hence $P_t(x, U) < \epsilon$. Set

$$P'_h(x, S) = P_h(x, S), \quad P'_{(j+1)h}(x, S) = \int_{U^c} P'_{jh}(x, dy) P_h(y, S).$$

In probabilistic language, $P'_{(j+1)h}(x, S)$ is the pr. of transition from x into S in time $(j + 1)h$ avoiding U at times $h, \cdots, jh$. From this probabilistic interpretation, or directly by induction, it follows that

$$\text{(1)} \qquad P_{kh}(x, S) = \sum_{j=1}^{k-1} \int_U P'_{jh}(x, dy) P_{(k-j)h}(y, S) + P'_{kh}(x, S).$$

Given h and $t \leqq t_0$, let $n = [t/h]$ so that $nh = t - \theta \leqq t_0$, $0 \leqq \theta < h$. Since $P'_{kh}(x, S) \geqq P'_{(k-1)h}(x, \{x\}) P_h(x, S)$ and, by (1) with $S = U$,

$$P_{nh}(x, U) = \sum_{k=1}^{n} \int_U P'_{kh}(x, dy) P_{(n-k)h}(y, U),$$

it follows that

$$P_{nh}(x, U) \geqq \sum_{k=1}^{n} P'_{(k-1)h}(x, \{x\}) \int_U P_h(x, dy) P_{(n-k)h}(y, U)$$

hence

$$\text{(2)} \qquad P_{nh}(x, U) \geqq \sum_{k=1}^{n} P'_{(k-1)h}(x, \{x\})(1 - \epsilon) P_h(x, U).$$

It also follows that

$$\epsilon > P_{nh}(x, U) \geqq (1 - \epsilon) \sum_{k=1}^{n} P'_{kh}(x, U)$$

hence

$$\text{(3)} \qquad \sum_{k=1}^{n} P'_{kh}(x, U) \leqq \frac{\epsilon}{1 - \epsilon}.$$

On account of (3) and (1) with $S = \{x\}$ and $k \leqq n$,

$$1 - \epsilon < P_{kh}(x, \{x\}) \leqq \sum_{j=1}^{k-1} P'_{jh}(x, U) + P'_{kh}(x, \{x\})$$

$$\leqq \frac{\epsilon}{1-\epsilon} + P'_{kh}(x, \{x\})$$

hence

$$P'_{kh}(x, \{x\}) \geqq \frac{1-3\epsilon}{1-\epsilon}. \tag{4}$$

Similarly, on account of (2) and (4),

$$P_{nh}(x, U) \geqq n\left(\frac{1-3\epsilon}{1-\epsilon}\right)(1-\epsilon)P_h(x, U)$$

or

$$\frac{P_h(x, U)}{h} \leqq \frac{1}{1-3\epsilon}\,\frac{P_{nh}(x, U)}{nh}.$$

Therefore, on account of **c**, letting $h \to 0$ then $t \to 0$,

$$\limsup_{h\to 0} \frac{P_h(x, U)}{h} \leqq \frac{1}{1-3\epsilon} \liminf_{t\to 0} \frac{P_t(x, U)}{t}$$

and letting $\epsilon \to 0$,

$$q(x, U) = \lim_{h\to 0} \frac{P_h(x, U)}{h}$$

exists with

$$q(x, U) \leqq \frac{1}{1-3\epsilon}\,\frac{P_t(x, U)}{t}, \quad t \leqq t_0(V, \epsilon), \quad U + \{x\} \subset V. \tag{5}$$

It follows that $q(x, U) < \infty$ is a measure in U since it is clearly finitely additive and, as $U_n \downarrow \emptyset$, for $t_0 = t_0(U_1 + \{x\}, \epsilon)$,

$$q(x, U_n) \leqq \frac{1}{1-3\epsilon}\,\frac{P_{t_0}(x, U_n)}{t_0} \to 0.$$

The proof is terminated.

Note that given x and $V \in \mathfrak{U}$, for $x \notin U \subset V$, $\dfrac{P_h(x, U)}{h} \to q(x, U)$ uniformly in $U \subset V$. For, if $h \leqq t_0(V \cup \{x\}, \epsilon)$ then, by (5),

$$\frac{P_h(x, U')}{h} - q(x, U') \geqq -3\epsilon q(x, U') \geqq -3\epsilon q(x, V), \quad U' \subset V,$$

and

$$\frac{P_h(x, U)}{h} - q(x, U)$$

$$= \frac{P_h(x, V)}{h} - q(x, V) + q(x, V - U) - \frac{P_h(x, V - U)}{h}$$

$$\leqq \frac{P_h(x, V)}{h} - q(x, V) + 3\epsilon q(x, V),$$

so that

$$\sup_{U \subset V} \left| \frac{P_h(x, U)}{h} - q(x, U) \right| \leqq 3\epsilon q(x, V) + \left| \frac{P_h(x, V)}{h} - q(x, V) \right| \to 0.$$

What precedes, together with our convention $q(x, S) = q(x, S\{x\}^c)$, yields

A. TR.PR.'S DIFFERENTIATION THEOREM. *Under the continuity condition, the derivative $\dot{P}_0(x, U)$ at $t = 0$ of the tr.pr. $P_t(x, U)$ exists for every state and every uniform continuity state set U. In fact, for every state x and every uniform continuity state set V,*

$$\frac{P_h(x, U) - I(x, U)}{h} \to q(x, U) - q(x)I(x, U) = \dot{P}_0(x, U)$$

uniformly in $U \subset V$; the nonnegative function $q(x) \leqq \infty$ is Borelian; the function $q(x, U) = q(x, U\{x\}^c) < \infty$ bounded by $q(x)$ is Borelian in x and a finite measure in U, with

$$q(x, U) \leqq \frac{1}{1 - 3\epsilon} \frac{P_t(x, U\{x\}^c)}{t}$$

for all $U + \{x\} \subset V$ and $t \leqq t_0(V, \epsilon)$, $0 < \epsilon < \frac{1}{3}$.

If the function $q(x)$ is bounded then $\mathfrak{X}$ is q-bounded and the continuity condition is *uniform* (for all $x \in \mathfrak{X}$). Conversely, if the continuity condition is uniform, that is, $\mathfrak{X} \in \mathfrak{U}$ then, by **A**, $q(x) = q(x, \mathfrak{X}) \leqq 1/(1 - 3\epsilon)t_0$ for $t_0 = t_0(\mathfrak{X}, \epsilon)$ and the function $q(x)$ is bounded. Thus

COROLLARY. *The function $q(x)$ is bounded if and only if the continuity condition is uniform. Then the tr.pr.'s differentiation theorem is valid with $U \in \mathfrak{U}$ replaced by $S \in \mathcal{S}$, and $q(x, \mathfrak{X}) = q(x)$.*

If the function $q(x)$ is finite, hence $\mathfrak{X} = [x: q(x) < \infty]$, then there exists a countable partition of $\mathfrak{X}$ into uniform continuity state sets U_j

(say, $U_j = [x: j - 1 \leqq q(x) < j]$). In general, whenever there exists such a countable partition $\mathfrak{X} = \sum U_j$, we say that the continuity condition is *σ-uniform.* For example, if the set of instantaneous states is countable $[x: q(x) = \infty] = (x_1, x_2, \cdots)$, then the continuity condition is σ-uniform; it suffices to take $U_j = [x: j - 1 \leqq q(x) < j] + \{x_j\}$. In particular, if the state space is countable then the continuity condition is σ-uniform. Thus, we may consider σ-uniformity as a "natural" transposition of this property of countable state spaces to general state spaces.

Let $\bar{q}(x) = \sup_{U \in \mathfrak{U}} q(x, U)$ and note that there is always a sequence $V_n \uparrow V = \bigcup V_n$ and contained in $\mathfrak{U}$ such that $q(x, V_n) \uparrow \bar{q}(x)$.

f. Extension lemma. *If $\bar{q}(x) < \infty$ then, for this x, the measure $q(x, U)$ in U extends to a measure $q(x, S)$ in S, and the extension is finite with $q(x, \mathfrak{X}) = \bar{q}(x)$.*

If the continuity condition is σ-uniform then, for every x, the measure $q(x, U)$ in U extends to a measure $q(x, S)$ in S, and the extension is unique.

Proof. We use, without further comment the closure properties of $\mathfrak{U}$ and the fact that a nondecreasing sequence of measures $\mu_n \uparrow \mu$ on $\mathcal{S}$ converges to a measure μ on $\mathcal{S}$; it suffices to note that, as $n \to \infty$ then $m \to \infty$,

$$\sum \mu(S_j) \geqq \mu_n(\sum S_j) \to \mu(\sum S_j),$$

while

$$\mu(\sum S_j) \geqq \mu_n(\sum S_j) \geqq \sum_{j=1}^{m} \mu_n(S_j) \to \sum \mu(S_j)$$

Since

$$q(x, V_n) \uparrow \bar{q}(x) < \infty, \quad V_n \uparrow V, \quad V_n \in \mathfrak{U},$$

it follows, by **A**, that

$$q(x, UV^c) = q(x, UV^c + V_n) - q(x, V_n) \leqq \bar{q}(x) - q(x, V_n) \to 0$$

and

$$q(x, U) = q(x, UV) + q(x, UV^c) = q(x, UV) = \lim_n q(x, UV_n).$$

Therefore, the measure $q(x, S)$ in S defined by

$$q(x, SV_n) \uparrow q(x, S) = q(x, SV)$$

is an extension of the measure $q(x, U)$ in U, with

$$q(x, \mathfrak{X}) = q(x, V) = \bar{q}(x) < \infty.$$

The first assertion is proved.

If the continuity condition is σ-uniform, then $\mathfrak{X} = \sum U_n$, or $V_n = \sum_{k=1}^{n} U_k \uparrow \mathfrak{X}$, $V_n \in \mathfrak{U}$. Therefore, $q(x, UV_n) \uparrow q(x, U)$ and the extension of the measure $q(x, U)$ in U to a measure $q(x, S)$ in S is determined by the necessary condition $q(x, SV_n) \uparrow q(x, S)$. The second assertion is proved, and the proof is terminated.

B. EXTENDED TR.PR.'S DIFFERENTIATION THEOREM. *Under the continuity condition, if $\bar{q}(x) = q(x) < \infty$ (or under the σ-uniform continuity condition, if $q(x, \mathfrak{X}) = q(x) < \infty$) then, for this x, the derivative $\dot{P}_0(x, S)$ exists for every S. In fact,*

$$\frac{P_h(x, S) - I(x, S)}{h} \to q(x, S) - q(x)I(x, S) = \dot{P}_0(x, S)$$

uniformly in S and $q(x, S)$ is a finite measure in S.

Proof. It suffices to prove the assertion for $x \notin S$ so that the term in $q(x)$ disappears. According to the hypothesis and the extension lemma, the measure $q(x, U)$ in U extends to a measure $q(x, S)$ in S with

$$q(x, \mathfrak{X}) = \bar{q}(x) = q(x) < \infty$$

and

$$q(x, SV_n) \uparrow q(x, S) = q(x, SV), \quad V_n \uparrow V, \quad V_n \in \mathfrak{U}.$$

It follows that

$$\left|\frac{P_h(x, S)}{h} - q(x, S)\right| \leq \left|\frac{P_h(x, SV_n)}{h} - q(x, SV_n)\right| + \left|\frac{P_h(x, SV_n^c)}{h} - q(x, SV_n^c)\right| \to 0$$

uniformly in S. For, by the tr.pr.'s differentiation theorem, for n fixed, the first term on the right converges to zero uniformly in SV_n hence in S as $h \to 0$, and the upper bound of the second term below

$$\frac{P_h(x, V_n^c\{x\}^c)}{h} + q(x, V_n^c) = \frac{1 - P_h(x, \{x\})}{h} - \frac{P_h(x, V_n\{x\}^c)}{h} + q(x, V_n^c)$$

contains no S and converges to $q(x) - q(x, V_n)$ then to zero as $h \to 0$ then $n \to \infty$.

What precedes is valid with $V = \mathfrak{X}$ under the σ-uniform continuity condition and $q(x, \mathfrak{X}) = q(x)$. The proof is terminated.

COROLLARY. *If the two functions $\bar{q}(\cdot)$ and $q(\cdot)$ coincide and there are no instantaneous states, then the derivative $\dot{P}_t(x, S)$ exists and is finite and continuous in $t \geqq 0$ for every x and every S, and the backward equation holds:*

$$\dot{P}_t(x, S) = \int_{\{x\}^c} q(x, dy) P_t(y, S) - q(x) P_t(x, S).$$

For, by Kolmogorov's equation,

$$\frac{P_{t+h}(x, S) - P_t(x, S)}{h} = \int_{\{x\}^c} \frac{1}{h} P_h(x, dy) P_t(y, S)$$

$$- \frac{1 - P_h(x, \{x\})}{h} P_t(x, S) \to \int_{\{x\}^c} q(x, dy) P_t(y, S) - q(x) P_t(x, S)$$

upon using the following propositions:

g. *If finite measures μ_n on $\mathcal{S}$ converge to a finite measure μ on $\mathcal{S}$ and g on $\mathfrak{X}$ is a bounded Borel function, then $\int g\, d\mu_n \to \int g\, d\mu$.*

It suffices to note that g can be approximated up to any given $\epsilon > 0$ by simple functions $g' = \sum_{j=1}^{m} x_j I_{S_j}$ uniformly bounded by some finite constant c so that, as $n \to \infty$ then $\epsilon \to 0$,

$$\left| \int g\, d\mu - \int g\, d\mu_n \right| \leqq \int | g - g' |\, d\mu + \left| \int g'\, d\mu - \int g'\, d\mu_n \right|$$

$$+ \int | g' - g |\, d\mu_n \leqq \epsilon\mu(\mathfrak{X}) + c \sum_{j=1}^{m} | \mu S_j - \mu_n S_j | + \epsilon\mu_n(\mathfrak{X}) \to 0.$$

According to this proposition, the foregoing passage to the limit as $h \to 0$ is valid. Furthermore

If for $t \geqq 0$, the function $g(t)$ is continuous and its right derivative $\dot{g}^+(t)$ exists and is continuous, then the derivative exists (and coincides with the right derivative).

For, setting $h(t) = \int_0^t \dot{g}^+(s)\, ds$ so that $\dot{h}(t) = \dot{g}^+(t)$, the assertion reduces to the classical one that a continuous function $g(t) - h(t)$ whose right derivative vanishes is a constant, hence has a vanishing derivative.

Since after the foregoing passage to the limit as $h \to 0$, the existence of a right derivative $\dot{P}_t^+(x, S)$ equal to a continuous function in t is established, the term "right" may be omitted on account of the proposition just established. The proof of the corollary is completed.

39.2 Sample functions behavior. Let $X_T = (X_t, t \geqq 0)$ be a separable r.f. and let $\mathcal{B}(X_T)$ be the σ-field of events defined on X_T. We denote by $X_{s+T} = (X_{s+t}, t \geqq 0)$ the translate by s of X_T. Let g be a numerical Borel function on the sample space of X_T such that the exp. of $g(X_T)$ exists. The translate by s of $g(X_T)$ is $g(X_{s+T})$ and B_s—the translate by s of an event $B \in \mathcal{B}(X_T)$—is defined by $I_{B_s} = (I_B)_s$. We denote by P_t the distribution of X_t: $P_t(S) = P[X_t \in S]$.

Throughout this subsection, we assume that X_T is a stationary regular Markov r.f. with tr.pr. $P_t(x, S)$, unless otherwise stated. To be precise: there exists a family $(P^x, x \in \mathfrak{X})$ of pr.'s on $\mathcal{B}(X_T)$ and regular versions of c.pr.'s and c.exp.'s below—the only ones we shall use—such that, for every $x \in \mathfrak{X}$, $S \in \mathcal{S}$, $B \in \mathcal{B}(X_T)$, $0 \leqq s \leqq t < \infty$, the Markov property holds:

$$\text{(M)} \qquad P(B_s \mid X_r, r \leqq s) = P(B_s \mid X_s)$$

and is stationary:

$$\text{(S)} \qquad P(B_s \mid X_s = x) = P(B \mid X_0 = x) = P^x B$$

with tr.pr. $P_t(x, S)$:

$$\text{(Tr.)} \qquad P(X_t \in S \mid X_0 = x) = P^x[X_t \in S] = P_t(x, S).$$

Upon denoting by E^x the exp. which corresponds to P^x and approximating Borel functions by simple Borel ones, it follows that

$$E(g(X_{s+T}) \mid X_r, r \leqq s) = E(g(X_{s+T}) \mid X_s),$$
$$E(g(X_{s+T}) \mid X_s = x) = E^x g(X_T).$$

Note that upon conditioning by $X_{s_1}, \cdots, X_s$, $s_1 \leqq \cdots \leqq s$, we may replace in what precedes X_r, $r \leqq s$, by $X_{s_1}, \cdots, X_s$.

We also assume that, unless otherwise stated, the continuity condition holds:

$$\text{(C)} \qquad P_h(x, S) \to I(x, S),$$

equivalently, $P_h(x, \{x\}) \to 1$ or, by 39.1**c**, $P_t(x, S)$ is uniformly continuous in t uniformly in S.

a. *X_T is continuous in pr., in fact, for every x, $P^x[X_{t+h} \neq X_t] \to 0$ for every $t \geqq 0$ and $P^x[X_{t-h} \neq X_t] \to 0$ for every $t > 0$.*

Proof. The first assertion follows from the second one since, by the dominated convergence theorem,

$$P[|X_{t\pm h} - X_t| \geqq \epsilon]$$
$$= EP(|X_{t\pm h} - X_t| \geqq \epsilon \mid X_0) \leqq EP(X_{t\pm h} \neq X_t \mid X_0) \to 0.$$

Since, by Markov property,

$$P(X_{t+h} \neq X_t \mid X_0) = E\{P(X_{t+h} \neq X_t \mid X_t, X_0) \mid X_0\}$$
$$= E\{P(X_{t+h} \neq X_t \mid X_t) \mid X_0\},$$

it follows that, for $X_0 = x$,

$$P^x[X_{t+h} \neq X_t] = E^x P(X_{t+h} \neq X_t \mid X_t),$$

and, by stationarity,

$$P(X_{t+h} \neq X_t \mid X_t = y) = P(X_h \neq X_0 \mid X_0 = y) = P_h(y, \{y\}^c).$$

Therefore, by the continuity condition which says that $P_h(y, \{y\}^c) \to 0$ and by the dominated convergence theorem,

$$P^x[X_{t+h} \neq X_t] = \int P_t(x, dy) P_h(y, \{y\}^c) \to 0.$$

Similarly, replacing t by $t - h$, by the continuity condition and its implication $P_{t-h}(x, S) \to P_t(x, S)$,

$$P^x[X_t \neq X_{t-h}] = \int P_{t-h}(x, dy) P_h(y, \{y\}^c) \to 0,$$

and the assertion is proved.

The last passage to the limit is based upon the following proposition.

b. *If on $\mathcal{S}$ finite measures μ_n converge to a finite measure μ and on $\mathfrak{X}$ uniformily bounded Borel functions g_n converge to a Borel function g, then $\int g_n \, d\mu_n \to \int g \, d\mu$.*

For, by Egorov's theorem, given $\epsilon > 0$ there exists S with $\mu S^c < \epsilon$ and $s_n = \sup_{x \in S} | g_n(x) - g(x) | \to 0$ and, by 39.1g, as $n \to \infty$ then $\epsilon \to 0$,

$$\left| \int g_n \, d\mu_n - \int g \, d\mu \right| \leqq \left| \int g(d\mu_n - d\mu) \right| + s_n\mu_n S + 2c\mu_n S^c \to 0.$$

Note that the assumptions "measure" μ and "Borel function" g may be omitted, for they follow from the convergence assumptions.

c. Duration of stay lemma. *The pr. that starting from x at time s, X_T stays in x during time t is given by*

$$P(X_{s+r} = x, 0 \leqq r \leqq t \mid X_s = x) = e^{-q(x)t}.$$

Proof. Since X_T is separable and continuous in pr., we can replace $[0, \infty)$ by a countable set S dense in $[0, \infty)$, say, the set of dyadic numbers jh_n, $h_n = (\frac{1}{2})^n$. Thus, the sets

$$[X_{s+r} = x, 0 < r < t] = [X_{s+r} = x, 0 < r < t, r \in S]$$

are events. Because of stationarity, it suffices to prove the asserted relation for $s = 0$.

Let $k_n = [t/h_n]$ so that $k_n h_n \to t$ and, to simplify the writing, drop the subscripts n so that $h = h_n \to 0$ and $k = k_n \to \infty$ as $n \to \infty$. By Markov property and continuity condition $(P(X_t = x \mid X_{kh} = x) \to 1)$.

$$p_h = \prod_{j \leqq k} P(X_{jh} = x \mid X_{(j-1)h} = x) \to P(X_s = x, 0 \leqq s \leqq t \mid X_0 = x)$$

and, by stationarity and 39.1 **d** (see its proof),

$$p_h = P_h^k(x, \{x\}) = \exp\left\{\frac{\log P_h(x, \{x\})}{h} kh\right\} \to \exp\{-q(x)t\}.$$

The proof is terminated.

Upon introducing the *duration of stay* $\tau(\omega)$ of $X_T(\omega)$ in some state and considering the three cases $q(x) = 0$, $q(x) = \infty$, and $0 < q(x) < \infty$, **c** yields

A. Duration of stay theorem. *The duration of stay τ is a random time, not necessarily finite, with $P^x[\tau > t] = e^{-q(x)t}$.*

In particular, outside P^x-null events, when at some fixed time X_T takes the value x, then it stays in x forever when x is absorbing, or leaves it at once

when x is instantaneous, or stays in it for some (random) time and then leaves it when x is steady.

The last statement explains the classification of states x into absorbing, instantaneous, or steady, according as $q(x) = 0$, $q(x) = \infty$, or $0 < q(x) < \infty$.

The sets $B_n = [X_{s+t} = X_s, 0 \leqq t < 1/n] \uparrow B = \bigcup B_n$ are events since X_T is separable, and so is their limit B—the set to which correspond the sample functions remaining constant for some positive time after s. By **A** and the dominated convergence theorem,

$$PB \leftarrow PB_n = \int P_s(dx)e^{-q(x)/n}$$

$$= \int_{[q(\cdot)<\infty]} P_s(dx)e^{-q(x)/n} \rightarrow P_s[q(\cdot) < \infty].$$

Similarly, $C_n = [X_{s+t} \neq X_s$ for some $t \leqq n] \uparrow C$—the set to which correspond sample functions having a discontinuity some time after s, and

$$PC \leftarrow PC_n = \int P_s(dx)(1 - e^{-q(x)n})$$

$$= \int_{[q(\cdot)>0]} P_s(dx)(1 - e^{-q(x)n}) \rightarrow P_s[q(\cdot) > 0].$$

In particular

COROLLARY. *If there are no instantaneous states then, after any given time s, almost all sample functions are constant for some positive times, finite or infinite.*

If there are no absorbing states then, after any given time s almost all sample functions have a discontinuity at some finite time, positive or not.

If there are only steady states then, after any given time s, almost all sample functions are constant for some finite positive time and then have a discontinuity.

At first sight, if there are no instantaneous states, then we expect almost all sample functions to stay constant for some positive times, then jump and remain constant for some positive times, and so on, unless they get into some absorbing state and then stay there forever. Thus, we expect that almost all the sample functions will have a finite or infinite sequence of *isolated jumps*, that is, preceded and followed by time intervals of constancy. Yet, more complicated discontinuities may occur unless some restrictions are imposed. To begin, we shall study those sample functions $X_T(\omega)$ whose first discontinuity time τ after any given s is of the type: $X_T(\omega)$ is constant up to a finite positive time $\tau(\omega)$, is discontinuous at $\tau(\omega)$, and is constant with a different value after $\tau(\omega)$ for some time $h(\omega) > 0$. Because of the separability implications 35.2(3°), these simple discontinuities are isolated jumps, provided we neglect a null set of sample functions. From now on, we assume $g(\cdot)$ finite and we complete our measures so that subsets of null sets are null.

Fix $t > 0$, a steady state x, and a uniform continuity state set $U \not\ni x$. Let $D_{n,h}$ be the set of ω's such that $X_r(\omega) = x$ for $0 \leqq r \leqq \frac{kt}{n}$ and $X_r(\omega) = y$ for $\frac{(k+1)t}{n} \leqq r \leqq \frac{(k+1)t}{n} + h$, for some $k < n$ and some $y \in U$. According to **A**,

$$P^x D_{n,h} = \sum_{k=1}^{n-1} e^{-q(x)kt/n} \int_U P_{t/n}(x, dy) e^{-q(y)h}$$

$$= \frac{e^{-q(x)t/n} - e^{-q(x)t}}{(1 - e^{-q(x)t/n})/(t/n)} \int_U \frac{1}{t/n} P_{t/n}(x, dy) e^{-q(y)h}.$$

Therefore, by 39.1**A** and 39.1**g**, as $n \to \infty$,

$$P^x D_{n,h} \to p_h = \frac{1 - e^{-q(x)t}}{q(x)} \int_U q(x, dy) e^{-q(y)h},$$

then, as $h \to 0$,

$$p_h \to p = (1 - e^{-q(x)t}) \frac{q(x, U)}{q(x)}.$$

Let $\underline{D}_h = \liminf_n D_{n,h}$, $\overline{D}_h = \limsup_n D_{n,h}$ and note that $\underline{D}_h \uparrow \underline{D} = \bigcup_n D_{1/n}$, $\overline{D}_h \uparrow \overline{D} = \bigcup_n \overline{D}_{1/n}$ as $h \to 0$. Since, by the Fatou-Lebesgue theorem, $P^x \underline{D}_h \leqq p_h \leqq P^x \overline{D}_h$, it follows that

$$P^x \underline{D} \leqq p \leqq P^x \overline{D}.$$

Let D_h be the set of ω's such that $X_r(\omega) = x$ for $0 \leqq r < \tau(\omega) < t$ and $X_r(\omega) = y$ for $\tau(\omega) < r < \tau(\omega) + h$ for some $y \in U$. Then $D_h \uparrow D$ as $h \to 0$ and D is the set of all such ω with some $h = h(\omega) > 0$. Thus, D corresponds to the set of those sample functions which have an isolated jump from x into U at some time less than t. According to the above definitions, if $\omega \notin D_h$ then $\omega \in D_{n,h}$ for finitely many values of n only,

hence $\omega \notin \bar{D}_h$. Thus, $\bar{D}_h \subset D_h$ and $\bar{D} \subset D$. Similarly, if $\omega \in D_h$ then, for $n > 2t/h$ sufficiently large, there exists a $k < n$ such that $\frac{kt}{n} \leqq \tau(\omega) < \frac{(k+1)t}{n}$ and $X_r(\omega) = y$ for $\frac{(k+1)t}{n} \leqq r \leqq \frac{(k+1)t}{n} + h - \frac{t}{n}$ for some $y \in U$, hence $\omega \in D_{n,h/2}$ for n sufficiently large. Thus, $D_h \subset D_{n,h/2}$ and $D \subset \underline{D}$. It follows from $\underline{D} \subset \bar{D}$ that $\bar{D} = D = \underline{D}$. Therefore, $P^xD = p$ since $P^x\underline{D} \leqq p \leqq P^x\bar{D}$.
The preceding discussion remains valid when U is replaced by S and $q(x, S) = \lim_{h \to 0} P_h(x, S)/h$ exists for all $S \not\ni x$; by 39.1**B**, it is so when $\bar{q}(x) = \sup_{U \in \mathfrak{U}} q(x, U) = q(x)$ since $q(x) < \infty$. Thus, *when* $q(\cdot)$ *is finite*

d. *The sample functions starting from a steady state x at time s, which remain constant for some positive times less than t then jump into a uniform continuity state set $U \not\ni x$ and remain constant for some times, correspond to a set D with $P^xD = (1 - e^{-q(x)t})q(x, U)/q(x)$. If, moreover, $\bar{q}(x) = q(x)$ then what precedes is valid with S in lieu of U.*

We recall that S denotes any state set while U denotes only the uniform continuity ones.
We make the following convention: $q(x, S)/q(x) = 0$ when $q(x) = 0$.

B. ISOLATED JUMPS THEOREM. *Let X_T be in a state x at time s and $q(\cdot)$ be finite. Then*
The pr. that there be a sample discontinuity in the finite or infinite interval $(s, s + u)$ and the first one be an isolated jump into $U \not\ni x$ is given by $(1 - e^{-q(x)u})q(x, U)/q(x)$. If there is a sample discontinuity in $(s, s + u)$, (and when x is steady, a.s. there is at least one after s), then the pr. that the first one be an isolated jump into U is given by $q(x, U)/q(x)$.
If, moreover, $\bar{q}(x) = q(x)$ then what precedes holds with S in lieu of U, and, when x is steady, a.s. there is a first discontinuity which is an isolated jump.

Proof. The first assertion and the one with S in lieu of U replaces **d** together with the convention about absorbing states, and the second assertion follows by **A**. The assertion about an isolated jump without specifying into which set means that the pr. of an isolated jump from x into $\{x\}^c$ is one and results from $q(x, \{x\}^c) = q(x)$. The proof is terminated.

At first sight, once isolated jumps occur, the same stationary Markov evolution starts anew. However, this means that we can use the random time τ, to be precise $\tau + 0$, as a present or past moment in the relations (M), (S), and (Tr.) at the beginning of this subsection. Since these relations pertain to all states and all state sets, we shall have to assume not only that there are no instantaneous states but also that the two functions $q(x)$ and $\bar{q}(x)$ coincide. It will suffice to prove that $\tau + 0$ is a stationary Markov time of X_T, that is, the relation

$$(\mathrm{SM_{st}}) \quad P(X_{\tau+t} \in S \mid X_r, r \leqq \tau + s) = P_{t-s}(X_{\tau+s}, S), \quad 0 \leqq s < t < \infty$$

has meaning and is valid. Thus, we shall have first to show that all $X_{\tau+t}$, $t \geqq 0$ are r.v.'s for τ finite, that is, on $\Omega^\tau = [\tau < \infty]$. The conditioning by $(X_r, r \leqq \tau + s)$ means then conditioning by the σ-field in Ω^τ of all events $A \subset \Omega^\tau$ such that $A[\tau \leqq t] \in \mathfrak{B}(X_r, r \leqq t + s)$. The proof will be based upon 38.4**d**—the only result we require in Section 38.

e. Isolated jump time lemma. *Let the functions $q(x)$ and $\bar{q}(x)$ coincide and be finite. Then the first isolated jump time τ ($\tau + 0$ to be precise) is a stationary Markov time of X_T.*

Proof. Assume that there are only steady states so that, by **A**, τ is a r.v. with $P^x[\tau > t] = e^{-q(x)t}$. By its definition, τ is a time of X_T, that is, $[\tau \leqq t] \in \mathfrak{B}(X_r, r \leqq t)$. For, if we know a sample function $(X_r(\omega), r \leqq t)$ up to time t inclusive, then we know whether it left the state $X_0(\omega)$ or not during this time interval.

To prove the assertion, we subdivide $[0, \infty)$ into intervals of length $h_n = (\frac{1}{2})^n$, denote by $\tau_n(\omega)$ the first of the subdivision points which follows $\tau(\omega)$, approximate functions of τ by functions of τ_n and let $n \to \infty$. The following immediate properties will be used without further comment: $\tau < \tau_n \leqq \tau + h_n$, $\tau_n \to \tau + 0$, $[\tau_n + t - h_n, \tau_n + t] \downarrow \{\tau + t\}$, and, knowledge of $\tau(\omega)$ implying that of $\tau_n(\omega)$, τ_n is an elementary time of X_T hence, by 38.4**d**, is a stationary Markov time of X_T. The property to be established and to play a central role is that, for every $\tau + t$, $t \geqq 0$, almost all sample functions $X_T(\omega)$ have a time interval of constancy at $\tau(\omega) + t$ that is, on $[\tau(\omega) + t, \tau(\omega) + t + h(\omega)]$, $h(\omega) > 0$.

The constancy property at $\tau + 0$ is immediate. For, by definition of an isolated jump almost all sample functions have a time interval of constancy from $\tau + 0$. Therefore, the limit $X_{\tau(\omega)+0}(\omega) = X_{\tau_n(\omega)}(\omega)$ from some $n = n(\omega)$ on exists, and $X_{\tau+0}$ is a r.v.

Given $t > 0$, we take n sufficiently large so that $t - h_n > 0$. The event $B_n = [X_{\tau_n+r} = z, r \in [t - h_n, t], z \in S]$ corresponds to the set of those sample functions $X_T(\omega)$ which are constant in S during the time interval $[\tau_n + t - h_n, \tau_n + t]$. Thus,

$$P(B_n \mid X_{\tau_n} = y) = \int_S P_{t-h_n}(y, dz)e^{-q(z)h_n} \to P_t(y, S)$$

and, moreover, $P(B_n \mid X_{\tau_n}(\omega)) = P(B_n \mid X_{\tau(\omega)+0}(\omega))$ from some $n = n(\omega)$ on, because of the time interval of constancy at $\tau + 0$. Since $B_n \downarrow B = \bigcap B_n$, it follows that

$$PB \leftarrow PB_n = \int P(B_n \mid X_{\tau(\omega)+0}(\omega))P(d\omega) + o(1) \to \int P_{\tau+0}(dy)P_t(y, S).$$

Similarly, if the event C_n corresponds to the set of all sample functions $X_T(\omega)$ which at time $\tau_n(\omega) + t - h_n$ are either in S or in some state $z \in S^c$ and leave z within time h_n, then

$$P(C_n \mid X_{\tau_n} = y)$$
$$= P_{t-h_n}(y, S) + \int_{S^c} P_{t-h_n}(y, dz)(1 - e^{-q(z)h_n}) \to P_t(y, S)$$

and, setting $C = \liminf C_n$,

$$PC \leqq \liminf PC_n = \lim PC_n = PB.$$

Since $B_n \subset [X_{\tau+t} \in S] \subset C_n$ so that $B \subset [X_{\tau+t} \in S] \subset C$, it follows that $[X_{\tau+t} \in S]$ differs from B by a null event; furthermore, $PB = 1$ when $S = \mathfrak{X}$. Thus, $X_{\tau+t}$ is a r.v., almost all sample functions have a time interval of constancy at $\tau + t$, and $P(B_n \mid X_{\tau_n(\omega)}(\omega)) \to P(X_{\tau+t} \in S \mid X_{\tau(\omega)+0}(\omega))$. Therefore, there is a regular version of c.pr. $P(X_{\tau+t} \in S \mid X_{\tau+0}) = P_t(X_{\tau+0}, S)$. This result is valid for every $X_{\tau+s}$, $s < t$, and what precedes applies with $X_{\tau+s}$ in lieu of $X_{\tau+0}$; in particular, we can take $P(X_{\tau+t} \in S \mid X_{\tau+s}) = P_{t-s}(X_{\tau+s}, S)$.

Since τ_n is a stationary Markov time of X_T and $A \in \mathfrak{B}(X_r, r \leqq \tau + s) \subset \mathfrak{B}(X_r, r \leqq \tau_n + s)$, it follows that

$$PA[X_{\tau+t} \in S] \leftarrow PAB_n$$
$$= \int_A P(B_n \mid X_{\tau_n+s})\, dP \to \int_A P(X_{\tau+t} \in S \mid X_{\tau+s})\, dP.$$

Thus,

$$PA[X_{\tau+t} \in S] = \int_A P_{t-s}(X_{\tau+s}, S)\, dP, \quad A \in \mathfrak{B}(X_r, r \leqq \tau + s),$$

that is, the integral form of (SM_{st}), hence (SM_{st}), are valid. So far, we assumed that there were only steady states so that $\Omega^\tau = [\tau < \infty]$ was an a.s. event. If there are also absorbing states and $P\Omega^\tau > 0$, then what precedes applies upon replacing all events by their intersections with Ω^τ. If $P\Omega^\tau = 0$, then what precedes has no content but almost all sample functions remain constant forever from time $\tau + 0$ on, and we may consider this property as the trivial degenerate form of the proposition. The proof is terminated.

A *q-pair* of functions $q(x)$, $q(x, S)$ will be called *regular* if they are finite, nonnegative, Borelian in x, and $q(x, S)$ is a measure in S with $q(x, \{x\}) = 0$ and $q(x, \mathfrak{X}) = q(x)$. We say that such a pair is *bounded* if the functions are bounded. We say that such a pair *derives* from a tr.pr. $P_t(x, S)$ if the function $\dot{P}_0(x, S) = q(x, S) - q(x)I(x, S)$. In fact, then, by the corollary of 39.1**B**, the derivative $\dot{P}_t(x, S)$ exists and is continuous in t, the backward equation holds, and the tr.pr. obeys the σ-uniform continuity condition; if, moreover, the q-pair is bounded then, by the corollary of 39.1**A**, the tr.pr. obeys the uniform continuity condition.

C. SAMPLE STEP FUNCTIONS THEOREM. *Let a q-pair of functions $q(x)$, $q(x, S)$ be regular.*

If the q-pair derives from the tr.pr. $P_t(x, S)$ of a separable stationary Markov r.f. $X_T = (X_t, t \geqq 0)$, then there exists a random time τ_θ of accumulation of isolated jumps, not necessarily infinite, and almost all sample functions $X_T(\omega)$ are step functions in $[0, \tau_\theta(\omega))$. If, moreover, the q-pair is bounded, then almost all sample functions are step functions.

Conversely, the q-pair derives from at least one tr.pr. $P_t(x, S)$ of a separable stationary Markov r.f. $X_T = (X_t, t \geqq 0)$ with a corresponding random time τ_θ. If, moreover, τ_θ is a.s. infinite, in particular, if the q-pair is bounded, then the tr.pr. is unique.

Proof. We use without further comment the isolated jumps theorem and the strong Markov jump time lemma.

1° If there are no absorbing states (that is, if the q-functions are positive), then there is a sequence of finite positive random times $\tau_1, \tau_2, \cdots$ such that almost all sample functions $X_T(\omega)$ are constant on $[0, \tau_1(\omega))$, $(\tau_1(\omega), \tau_1(\omega) + \tau_2(\omega))$, $\cdots$, with different values in any two consecutive intervals; we set $\tau_0(\omega) + 0 = 0$. If there are absorbing states, then, whenever $X_T(\omega)$ is in such a state for the first time—at some $\tau_{n-1}(\omega) + 0$, then it stays there forever so that $\tau_n(\omega) = \infty$ and we set $\tau_{n+1}(\omega) = \tau_{n+2}(\omega) = \cdots = \infty$.

In either case, the sum of the series $\tau_\theta(\omega) = \sum \tau_n(\omega)$ of positive terms exists and is finite or infinite. If $\tau_\theta(\omega) = \infty$, then the sample function is a step function. If $\tau_\theta(\omega) < \infty$, then we know only that $X_T(\omega)$ is a step function on $[0, \tau_\theta(\omega))$. Thus, the sample functions which correspond to the set $[\tau_\theta = \infty]$ are step functions.

In particular, if the q-pair is bounded by $c < \infty$, then $P[\tau_\theta = \infty] = 1$ so that almost all sample functions are step functions. For, then

$$P[\tau_n \geqq t] = \int P_{\tau_{n-1}+0}(dx)P^x[\tau_n \geqq t] \geqq e^{-ct}$$

for every n and every $t > 0$, so that $P(\limsup [\tau_n \geqq t]) \geqq e^{-ct}$, hence infinitely many τ_n are $\geqq t$ with pr. $\geqq e^{-ct}$ and, thus, $P[\tau_\theta = \infty] \geqq e^{-ct} \to 1$ as $t \to 0$. The direct assertions are proved.

2° Conversely, given a regular q-pair, we construct a separable stationary Markov r.f. X_T with a tr.pr. from which the q-pair derives, upon following the pattern set by what precedes.
We select $\tau_0 = 0$, $\xi_0 = X_0$, τ_1, $\xi_1 = X_{\tau_1}, \cdots$, as follows: the r.v. ξ_0 is chosen arbitrarily and for $n > 0$, given the preceding choices, we choose τ_{n+1} and ξ_{n+1} so that

$$P(\tau_{n+1} \geqq t \mid \tau_0, \xi_0, \cdots, \tau_n, \xi_n) = e^{-q(\xi_n)t}$$

$$P(\xi_{n+1} \in S \mid \tau_0, \xi_0, \cdots, \tau_n, \xi_n, \tau_{n+1}) = q(\xi_n, S)/q(\xi_n)$$

and, whenever $q(\xi_n(\omega)) = 0$, we take $\tau_{n+1}(\omega) = \tau_{n+2}(\omega) = \cdots = \infty$, $\xi_{n+1}(\omega) = \xi_{n+2}(\omega) = \cdots = \xi_n(\omega)$; we set $X_t(\omega) = \xi_n(\omega)$ for $t \in \left[\sum_{k=1}^{n} \tau_k(\omega), \sum_{k=1}^{n+1} \tau_k(\omega)\right)$ and $\tau_\theta(\omega) = \sum \tau_n(\omega)$. Thus X_t is defined for all $t \in [0, \tau_\theta)$.

If $P[\tau_\theta = \infty] = 1$, then $X_T = (X_t, t \geqq 0)$ is so defined. If $P[\tau_\theta = \infty] < 1$, then we continue the construction with τ_θ as with τ_0: we choose an arbitrary r.v. ξ_θ independent of the ξ_n, τ_n, choose $\tau_{\theta+1}$ with $P(\tau_{\theta+1} \geqq t \mid \xi_0, \tau_0, \cdots, \tau_\theta) = e^{-q(\xi_\theta)}$, set $X_t(\omega) = \xi_\theta(\omega)$ for $t \in [\tau_\theta(\omega), \tau_{\theta+1}(\omega))$, and so on, starting over, if necessary, at the new accumulation points of jump times with r.v.'s with distribution P_θ of ξ_θ. It is intuitive that this defines $X_T = (X_t, t \geqq 0)$, but we shall not prove it, for the proof requires the use of ordinals.

X_T is a stationary Markov r.f. and the q-pair derives from its tr.pr., as follows: Note that $P[\tau > t] = e^{-qt}$ implies that $P(\tau > s + t \mid \tau > s) = e^{-qt}$ for any $s > 0$. This means that if we stop the construction when we reach a $\tau > s$ and start it anew at s in lieu of at 0 but use X_s in lieu of ξ_0, then the τ and ξ which follow have the same distribution as when the construction was not stopped. Thus, $P(X_{s+t} \in S \mid X_r, r \leqq s) =$

$P(X_{s+t} \in S \mid X_s)$ and is independent of s, and the above assertion is true.

If $P[\tau_\theta = \infty] = 1$ then, up to the choice of $\xi_0 = X_0$, the r.f. X_T is the only one which conforms to what precedes 2°. Therefore, the tr.pr. is unique and the q-pair derives from it. If $P[\tau_\theta < \infty] < 1$, then the constructed X_T and its tr.pr. depend upon the choice of P_θ.

The converse assertions are proved, and the proof is terminated.

§ 40. MARKOV SEMI-GROUPS

40.1 Generalities. Markov semi-groups on G characterize stationary Markov laws of evolution. Their analysis requires introduction of analytical concepts (limits, continuity, integration, differentiation) in G.

We recall the notation to be used throughout. Unless otherwise stated, times r, s, t, with or without affixes, are points of $T = [0, \infty)$, states x, y, z, with or without affixes, are points of a locally compact separable metric state space $\mathfrak{X}$, sets S, with or without affixes, are topological Borel sets—sets of the σ-field $\mathcal{S}$ generated by the class of open sets in $\mathfrak{X}$, and $V_x(\epsilon)$ are open spheres of radius ϵ centered at x.

To fix the ideas, *we take the state space $\mathfrak{X}$ to be a Borel set in R with the usual topology in it.* What follows extends at once to the general case.

The space G is the Banach space of all bounded Borel f.'s g on $\mathfrak{X}$ with the uniform norm $\| g \| = \sup_x | g(x) |$ for every $g \in G$. The space Φ is the Banach space of all bounded signed measures φ on $\mathcal{S}$ with the variation norm $\| \varphi \| = \operatorname{Var} \varphi = \varphi^+(\mathfrak{X}) + \varphi^-(\mathfrak{X})$; in particular, all pr. measures δ_x which degenerate at x belong to Φ. The elements φ of Φ may be considered as linear functionals on G:

$$\varphi(g) = (\varphi, g) = \int \varphi(dx) g(x), \quad \varphi \in \Phi.$$

However, Φ is not the adjoint space of G; it is only a "reciprocal" subspace of it, that is, such that $\| g \| = \sup_{\|\varphi\| \leq 1} (\varphi, g)$ for every $g \in G$ (follows upon using the δ_x). It ought to be noted that Φ is the adjoint of the subspace $C_0(\subset G)$ of bounded continuous functions on $\mathfrak{X}$ vanishing at infinity.

We introduce two concepts of limit or types of convergence in G. Let $t \to t_0$.

Strong convergence means uniform pointwise convergence: If $g_t(x) \to g(x)$ uniformly in $x \in \mathfrak{X}$, we say that g_t *converges strongly* to g or that g is *strong limit* of g_t, and we write $g_t \xrightarrow{s} g$. Thus, $g_t \xrightarrow{s} g$ is equivalent to $\| g_t - g \| \to 0$, that is, to convergence in norm.

Φ-weak convergence means bounded pointwise convergence: If $g_t(x) \to g(x)$ for every $x \in \mathfrak{X}$ and the g_t are uniformly bounded, we say that g_t *converges Φ-weakly* to g or that g is *Φ-weak limit* of g_t, and we write $g_t \xrightarrow{w} g$. Clearly, strong convergence implies weak convergence, and it is easily seen that $g_t \xrightarrow{w} g$ is equivalent to $(\varphi, g_t) \to (\varphi, g)$ for every $\varphi \in \Phi$ (use the Banach-Steinhaus uniform boundedness theorem). Thus, if we limit ourselves to the subspace C_0 (so that Φ is its adjoint space), then Φ-weak convergence becomes the usual "weak" convergence in C_0. This explains the "Φ-weak" terminology.

To each of the foregoing concepts of limit correspond concepts of continuity, differentiation, and integration. Let g_t be a function in $t \in [a, b] R$ with values in G. Let $t' \to t$ and $0 < h \to 0$. The function g_t is strongly continuous at t if $g_{t'} \xrightarrow{s} g_t$, and it is *strongly differentiable* at t if $(g_{t'} - g_t)/(t' - t)$ converges strongly, necessarily to an element of G—to be called *strong derivative* of g_t at t and to be denoted by Dg_t. If $t' = t + h$ (or $t' = t - h$), then the derivative is from the right (or left) and denoted by D^+g_t (or D^-g_t); the derivative at a (or b) is necessarily from the right (or left). We drop "at t" when the foregoing properties hold for every t. The same definitions apply upon replacing "strong" by "Φ-weak," "s" by "w" and "D" by "D'." Clearly, strong (Φ-weak) differentiability implies strong (Φ-weak) continuity. In fact

a. *Φ-weak differentiability implies strong continuity.*

For, if $(g_{t'} - g_t)/(t' - t)$ converges weakly hence boundedly, then $\| g_{t'} - g_t \| \leqq c \, | t' - t | \to 0$.

The function g_t is *strongly integrable* on a bounded interval $[c, d)$ if its Riemann sums converge strongly in the usual way. The limit is then necessarily an element of G to be called the *strong integral* of g_t on $[c, d)$ and denoted by $\int_c^d g_t \, dt$

Strong integrals on unbounded intervals are defined by strong passages to the limit exactly as for the improper Riemann integrals. The usual properties of Riemann integrals remain valid: change of variables, additivity, integrability of strongly continuous functions on bounded intervals and also on unbounded intervals when these functions are bounded in norm by numerical functions integrable on these intervals. Similarly, the inequality $\left\| \int_c^d g_t \, dt \right\| \leqq \int_c^d \| g_t \| \, dt$, remains valid, and the convergence property: as $h \to 0$,

$$\frac{1}{h}\int_c^{c+h} g_t\,dt \xrightarrow{s} g_c \quad \text{when} \quad g_t \xrightarrow{s} g_c \quad \text{as} \quad t \to c + 0.$$

The *Φ-weak integral* $g_{[c,d)} \in G$ is defined by $g_{[c,d)}(x) = \int_{[c,d)} g_t(x)\,dt$ for functions g_t measurable in (x, t) and bounded in norm by numerical Lebesgue integrable functions. The convergence property holds: as $h \to 0$, $g_{[c,c+h)} \xrightarrow{w} g_c$ when $g_t \xrightarrow{w} g_c$ as $t \to c + 0$, and the Fubini theorem with finite measures μ on $\mathfrak{X}$ applies:

$$\int_{\mathfrak{X}} \mu(dx) \left\{\int_{[c,d)} g_t(x)\,dt\right\} = \int_{\mathfrak{X}\times[c,d)} \mu(dx) g_t(x)\,dt$$
$$= \int_{[c,d)} \left\{\int_{\mathfrak{X}} \mu(dx) g_t(x)\right\} dt.$$

Let T, with or without affixes, denote an *endomorphism on G*— a linear bounded mapping on G to G:

$$T(ag + a'g') = aTg + a'Tg', \quad \|Tg\| \leqq c\|g\|, \quad c < \infty.$$

The smallest possible value of c as g varies, is $\|T\| = \sup_{\|g\| \leqq 1} \|Tg\|$ or the *norm* of T. If $g \geqq 0 \Rightarrow Tg \geqq 0$ we say that T is nonnegative, and if $\|T\| \leqq 1$ we say that T is a *contraction.* Multiplication by scalars, addition, and multiplication of endomorphisms, defined by

$$(aT)g = a(Tg), \quad (T + T')g = Tg + T'g, \quad (TT')g = T(T'g)$$

yield endomorphisms. It follows that the space $\mathcal{E}$ of our endomorphisms is linear and with the foregoing norm becomes a Banach space. Furthermore, multiplication of endomorphisms commutes with their multiplication by scalars, is distributive with respect to addition, and $TI = IT$ where I is the identity mapping. Thus $\mathcal{E}$ is an "algebra with unit I" and since $\|TT'\| \leqq \|T\| \cdot \|T'\|$ it is a "Banach algebra:"

b. *The Banach space of endomorphisms on a Banach space is a Banach algebra.*

Let $t \to t_0$. In $\mathcal{E}$ on our space G we have at our disposal the usual convergence in norm $\|T_t - T\| \to 0$ or *uniform convergence* and the types of convergence induced by those in G: *strong convergence* $T_t \xrightarrow{s} T$ meaning $T_tg \xrightarrow{s} Tg$ for every $g \in G$ and *Φ-weak convergence* $T_t \xrightarrow{w} T$

meaning $T_t g \xrightarrow{w} Tg$ for every $g \in G$. To each of these types of convergence correspond types of limits hence of continuity and of integrals. For example: If T_t is uniformly continuous in $t \in [c, d]$, then the uniform integral $\int_c^d T_t\, dt$ defined as uniform limit of corresponding Riemann sums exists and is an endomorphism. The strong and Φ-weak integrals are induced by those for $g \in G$: the strong integral is defined by $\left(\int_c^d T_t\, dt\right) g = \int_c^d T_t g\, dt$ for all $g \in G$, and the Φ-weak integral is defined by

$$\left(\int_{[c,d)} T_t\, dt\right) g = \int_{[c,d)} T_t g\, dt$$

for all $g \in G$.

If an endomorphism T on G and an endomorphism U on Φ are such that, for all $g \in G$, $\varphi \in \Phi$,

$$(\varphi, Tg) = \int \varphi(dx) Tg(x) = \int U\varphi(dx) g(x) = (U\varphi, g),$$

then we shall set $U\varphi = \varphi T$, write the above relation $(\varphi, Tg) = (\varphi T, g)$, and say that T is Φ-*adjoint*—its adjoint on the adjoint space of G leaves Φ invariant. Clearly

c. *Endomorphisms T and strong passages to the limit commute.*

But generally this is not true of Φ-weak passages to the limit. However

c'. *Φ-adjoint endomorphisms T commute with Φ-weak passages to the limit.*

For, $g_t \xrightarrow{w} g \Rightarrow (\varphi, Tg_t) = (\varphi T, g_t) \to (\varphi T, g) = (\varphi, Tg) \Rightarrow Tg_t \xrightarrow{w} Tg$.

We are now ready for the introduction of Markov endomorphisms. Let $P(x, S)$ denote Borel functions in $x \in \mathfrak{X}$ and pr.'s in $S \in \mathcal{S}$ or, more generally, measures in S bounded by 1. To every $P(x, S)$ there corresponds an endomorphism T on G and an endomorphism U on Φ, to be called *Markov endomorphisms*, defined by

$$Tg(x) = \int P(x, dy) g(y), \quad U\varphi(S) = \int \varphi(dx) P(x, S).$$

Clearly, T and U are nonnegative contractions, and when $P(x, S)$ is a pr. in S so that $P(x, \mathfrak{X}) = 1$ for all x then $\| T \| = \| U \| = 1$. Either T or U determines $P(x, S)$; it suffices to take $g(\cdot) = I(\cdot, S)$ or $\varphi(\cdot) =$

$I(x, \cdot)$. In fact, T is Φ-adjoint (to U) since

$$(\varphi, Tg) = \int \varphi(dx) P(x, dy) g(y) = (U\varphi, g).$$

We shall concentrate on Markov endomorphisms T on G and have

A. Markov endomorphisms criterion. *An endomorphism T on G is Markovian if and only if it is a nonnegative Φ-adjoint contraction.*

Proof. The "only if" assertion is contained in what precedes. As for the "if" assertion, let T be a nonnegative contraction Φ-adjoint (to U). Set $P(x, S) = \Delta_x(S) = I(x, S)T$ (where the last term stands for $U\varphi(S)$ with $\varphi(\cdot) = I(x, \cdot)$). Since every φT is a measure so is Δ_x and, from $(\varphi, Tg) = (\varphi T, g)$ it follows that

$$Tg(x) = \int I(x, dy) Tg(y) = \int P(x, dy) g(y).$$

In particular, $TI(x, S) = P(x, S) \in G$ and, since our Tg is a nonnegative contraction and $I(x, S)$ is bounded by 1, it follows that $P(x, S)$ is a nonnegative Borel function in x bounded by 1. The proof is terminated.

Let $P_t(x, S)$ be a stationary tr.pr. except that in lieu of $P_t(x, \mathfrak{X}) = 1$ we assume only that $P_t(x, \mathfrak{X}) \leqq 1$, unless otherwise stated. In terms of Markov r.f.'s this assumption may mean that its r.v.'s when numerical may take infinite values with positive pr.'s. According to what precedes, $P_t(x, S)$ as a Borel function in x and a measure in S bounded by 1 determines and is determined by a Markov endomorphism T_t on G defined by

$$T_t g(x) = \int P_t(x, dy) g(y), \quad g \in G.$$

There remains the stationary Kolmogorov equation, which links the values of the tr.pr. for different values of t and which, by

$$T_{s+t} g(x) = \int P_{s+t}(x, dz) g(z) = \int P_s(x, dy) P_t(y, dz) g(z) = T_s T_t g(x),$$

is equivalent to the *semi-group property* $T_{s+t} = T_s T_t$.

We say that this family of Markov endormorphisms, which is in a one-to-one correspondence with a stationary tr.pr. hence with the corresponding law of evolution of a stationary Markov process, is a *Markov semi-group*. Unless otherwise stated, semi-groups are semi-groups of endomorphisms on G and all endomorphisms are on G. If a semi-group consists of Φ-adjoint (nonnegative) (contraction) endomorphisms we say that it is a *Φ-adjoint* (*nonnegative*) (*contraction*) semi-group. If, moreover, all functions $T_t g(x)$ are Borelian in (x, t) we say that the semi-group is *Borelian*.

B. Markov semi-groups criterion. *A semi-group is a Markov semi-group if and only if it is a nonnegative Φ-adjoint contraction semi-group.*

A Markov semi-group is Borelian if and only if the corresponding stationary tr.pr. is Borelian.

Proof. The first assertion follows from **A**. The second assertion follows from the integral form of $T_t g(x)$, the "only if" part upon taking $g(\cdot) = I(\cdot, S)$ and the "if" part upon approximating g by simple functions.

40.2 Analysis of semi-groups. While our concern is with Markov semi-groups, the concepts and general properties below are valid for more general contraction semi-groups of endomorphisms on Banach spaces G, and they are stated accordingly. As usual, the limits in h are taken as $0 < h \to 0$, and if a property holds at all t, *we drop* "at t".

Let $(T_t, t \geqq 0)$ be a contraction semi-group:

$$T_{s+t} = T_s T_t, \quad T_0 = I, \quad \| T_t \| \leqq 1, \quad s, t \geqq 0.$$

A set G_0 in G is *invariant* (by the semi-group) if all $T_t G_0 \subset G_0$; in other words there is a restriction of the semi-group on G_0 to G_0. The whole space G and the singleton which consists of the origin of G are trivially invariant.

We denote by G_c the set on which our semi-group is strongly continuous at $t = 0$: $g \in G_c \Leftrightarrow T_h g \xrightarrow{s} g$.

a. Strong continuity lemma. *G_c is the set on which the contraction semi-group is strongly continuous, and it is an invariant Banach subspace.*

Proof. Clearly G_c contains the strong continuity set of the semi-group. But G_c is also contained in this set and is invariant. For, $(T_h - I)T_t = T_t(T_h - I)$ and, for every $g \in G_c$,

$$\| T_{t+h}g - T_t g \| = \| T_t(T_h - I)g \| \leqq \| (T_h - I)g \| \to 0,$$

$$\| T_t g - T_{t-h} g \| = \| T_{t-h}(T_h - I)g \| \leqq \| (T_h - I)g \| \to 0.$$

G_c is obviously closed under linear combinations. It is also closed under strong passages to the limit by sequences $g_n \xrightarrow{s} g$, $g_n \in G_c$, hence is a Banach space. For, as $h \to 0$ then $n \to \infty$,

$$\| T_h g - g \| \leqq \| T_h(g - g_n) \| + \| (T_h - I)g_n \| + \| g_n - g \|$$

$$\leqq 2\| g - g_n \| + \| (T_h - I)g_n \| \to 0.$$

The proof is terminated.

We denote by G_d the set on which the semi-group is strongly differentiable at $t = 0$: $g \in G_d \Leftrightarrow D_h g \xrightarrow{s} Dg$; $D_h = (T_h - I)/h$. Clearly $G_d \subset G_c$. The (*strong*) *differentiation operator* D on G_d to G is also called the (*strong*) *infinitesimal operator* or the *generator* of the semi-group. For, it generates the semi-group at least on G_c, as will be seen later. Clearly, D is a linear operator on the obviously linear space G_d. But in general, D is not bounded and G_d is not a Banach subspace. Note that if $g \in G_c$ and a, b are finite then the strong integral $g_a{}^b = \int_a^b T_t g\, dt$ exists. For, the function $T_t g$ is then strongly continuous in t. Furthermore $g_a{}^b \in G_d$ since

$$(T_h - I)g_a{}^b = \int_a^b T_{t+h} g\, dt - \int_a^b T_t g\, dt = \int_b^{b+h} T_t g\, dt - \int_a^{a+h} T_t g\, dt$$

implies that

$$D_h g_a{}^b \xrightarrow{s} (T_b - T_a)g = Dg_a{}^b;$$

in particular

$$G_d \ni \frac{1}{h}\int_0^h T_t g\, dt \xrightarrow{s} g \in G_c.$$

b. STRONG DIFFERENTIATION LEMMA. *G_d is the set on which the contraction semi-group is strongly differentiable and it is an invariant set dense in G_c with $DT_t = T_t D$ on it.*

Proof. Clearly, G_d contains the strong differentiability set of the semi-group. But G_d is also contained in this set and is invariant with $DT_t = T_t D$ on it. For, if $g \in G_d$, then

$$D_h(T_t g) = (T_{t+h} - T_t)g/h = T_t(D_h g) \xrightarrow{s} T_t(Dg) = D(T_t g),$$

hence $T_t g \in G_d$ and $D(T_t g) = T_t(Dg)$, and

$$(T_t - T_{t-h})g/h = T_{t-h}(D_h g) = T_{t-h}(D_h g - Dg) + T_{t-h}Dg \xrightarrow{s} T_t(Dg).$$

Finally, G_d is dense in G_c since every $g \in G_c$ is strong limit of elements $\frac{1}{h}\int_0^h T_t g\, dt$ of G_d. The proof is terminated.

We replace now strong limits by Φ-weak limits. If we try to proceed as for the strong continuity, we find that left limits have to be left out and we need the functional form $(\varphi, g_n) \to (\varphi, g)$ of $g_n \overset{w}{\to} g$ hence, to use 40.1**c**′, we assume that the semi-group is Φ-adjoint. If we try to proceed as for the strong differentiability, we find that in order to introduce Φ-weak integrals to establish the closure property, we have to assume that the semi-group is Borelian. We are then led without any difficulty to the following definitions and propositions. Denote by G'_c the set on which the semi-group is Φ-weakly continuous at $t = 0$: $g \in G'_c \Leftrightarrow T_h g \overset{w}{\to} g$.

a′. Φ-WEAK CONTINUITY LEMMA. *G'_c is the set on which the Φ-adjoint contraction semi-group is Φ-weakly rightcontinuous, and it is an invariant Banach subspace.*

Denote by G'_d the set on which the semi-group is Φ-weakly differentiable at $t = 0$: $g \in G'_d \Leftrightarrow D_h g \overset{w}{\to} D'g$. The *$\Phi$-weak differentiation operator* D' is also called the *Φ-weak infinitesimal operator* or the *Φ-weak generator* of the semi-group. Then

b′. Φ-WEAK DIFFERENTIATION LEMMA. *G'_d is the set on which the Φ-adjoint contraction semi-group is Φ-weakly right differentiable, and it is an invariant set with $D'T_t = T_t D'$ on it.*

If, moreover, the semi-group is Borelian, then the Φ-weak closure of G'_d contains G'_c.

The four spaces so distinguished are related by

c. INCLUSION AND CLOSURE LEMMA. *The invariant continuity and differentiation sets of a Φ-adjoint contraction semi-group are ordered by the inclusions $G_d \subset G'_d \subset G_c \subset G'_c$, and G_c is the strong closure of G_d.*

If, moreover, the semi-group is Borelian, then these four sets have common Φ-weak closure G'_c.

Proof. The first part follows from the preceding lemmas except for $G'_d \subset G_c$ which follows from 40.1**a**. The second part follows from the asserted ordering and the last assertion in **b′**.

Instead of replacing "strong" by "Φ-weak" we may replace it by "uniform." Then much more is true as follows. Let E be an endomorphism on a Banach space and let E^n be its n-th iterate ($E^0 = I$).

Clearly, the exponentials e^{tE} defined by $e^{tE} = \sum_{n=0}^{\infty} \frac{t^n}{n!} E^n$ are endomorphisms bounded by $e^{t\|E\|}$ and form a semi-group—*the exponential semi-group generated by E*. This semi-group is uniformly continuous and uniformly differentiable with "uniform differential operator" E, since

$$e^{(t+h)E} - e^{tE} = e^{tE}(e^{hE} - I), \quad e^{tE} - e^{(t-h)E} = e^{(t-h)E}(e^{hE} - I)$$

and

$$\left\| \frac{e^{hE} - I}{h} - E \right\| \leqq (e^{h\|E\|} - 1 - h\| E \|)/h \to 0.$$

d. Uniform continuity and differentiation lemma. *If the contraction semi-group is uniformly continuous at $t = 0$ on an invariant Banach subspace G_u, then the semi-group is exponential on it.*

Proof. G_u is invariant; and the semi-group is uniformly continuous on it since

$$\| T_{t+h} - T_t \| \leqq \| T_h - I \| \to 0 \text{ and } \| T_t - T_{t-h} \| \leqq \| T_h - I \| \to 0$$

on G_u. From here on we consider the semi-group on the invariant Banach space G_u only. Clearly, the uniform integral $I_h = \int_0^h T_s\, ds$ exists and $I_h/h \xrightarrow{u} I$. Therefore, for $h = h_0$ sufficiently small, I_{h_0} has an inverse $I_{h_0}^{-1}$. Let $E = (T_{h_0} - I)I_{h_0}^{-1}$. It follows from

$$(T_t - I)\int_0^{h_0} T_s\, ds = \int_t^{t+h_0} T_s\, ds - \int_0^{h_0} T_s\, ds = (T_{h_0} - I)\int_0^t T_s\, ds$$

that

$$T_t - I = E\int_0^t T_s\, ds,$$

hence $D_h = EI_h/h \xrightarrow{u} E$. Also, proceeding by induction,

$$T_t - I = \sum_{k=1}^{n} \frac{t^k}{k!} E^k + E^{n+1} \int_0^t \frac{(t-s)^n}{n!} T_s\, ds$$

where the norm of the right side summation is bounded by $e^{t\|E\|}$ and that of the remaining term is bounded by $t^{n+1}\| E \|^{n+1}/(n+1)! \to 0$. Thus $T_t = \sum_{n=0}^{\infty} \frac{t^n}{n!} E^n = e^{tE}$, and the proof is terminated.

Local characterization. The purely analytical problem relative to Markov processes (regular and stationary) corresponding to stationary tr.pr.'s, hence to Markov semi-groups, can now be stated as follows: Characterize Markov semi-groups in terms of their local properties. It is a specialization of the same problem for general contraction semi-groups to nonnegative Φ-adjoint endomorphisms on our function space G. Thus, we first treat the general case (then the results extend at once to general Banach spaces G), and specialize it later. The problem may be decomposed as follows: The *existence problem* is that of characterizing infinitesimal operators of contraction semi-groups, the *unicity problem* is that of characterizing those infinitesimal operators which determine their semi-group, and the *generation problem* is that of constructing the corresponding semi-groups. However, at best we may hope for answers on the strong or Φ-weak closures of the domains of the infinitesimal operator. Thus appears the *extension problem:* find conditions under which extensions of semi-groups exist and are unique on domains containing the space G with which we are concerned.

In the numerical case, where the contractions are endomorphisms on R, the semi-group property reduces to the classical functional equation $f(s+t) = f(s)f(t)$ with $f(0) = 1, |f(t)| \leqq 1$. The only continuous and, in fact, the only measurable solutions are exponential: $f(t) = e^{td}$ where $d \leqq 0$. In the general case, the *formal* solution of the equation $\frac{d}{dt} T_t = DT_t$ with $T_0 = I$ is similarly exponential $T_t = e^{tD}$ which would require that the infinitesimal operator D be an endomorphism. Yet, an infinitesimal operator D, while not necessarily an endomorphism, is always the limit of endomorphisms $D_h = (T_h - I)/h$. This fact, as well as the numerical case and the formal approach, lead us to expect that, at least on the strong closure G_c of the domain G_d of D, the corresponding semi-group would be determined by D and could be represented as limit of exponential semi-groups. We shall show that these expectations are justified. First, we have to introduce the necessary tool—the Laplace transform or "resolvent" of a contraction semi-group.

The (*strong*) *resolvent* R_λ, on the (strong) continuity subspace G_c of the contraction semi-group $(T_t, t \geqq 0)$ is defined by the strong integrals

$$R_\lambda g = \int_0^\infty e^{-\lambda t} T_t g \, dt, \quad g \in G_c, \quad \lambda \in (0, \infty).$$

$R_\lambda g$ exists and belongs to the Banach subspace G_c, since the integrand is strongly continuous and is bounded in norm by $e^{-\lambda t}\|g\|$ whose integral on $[0, \infty)$ is $\|g\|/\lambda$. Clearly R_λ is linear. Thus R_λ is an endomorphism on G_c with $\|\lambda R_\lambda\| \leqq 1$. The basic role of the resolvent is due to

e. Resolvent lemma. *The resolvent R_λ of the contraction semi-group with infinitesimal operator D on G_d is the inverse of the one-to-one mapping $\lambda I - D$ of G_d onto G_c.*

Proof. If $g \in G_c$ then, from

$$T_h R_\lambda g = \int_0^\infty e^{-\lambda t} T_{t+h} g\, dt$$

$$= e^{\lambda h} \int_h^\infty e^{-\lambda t} T_t g\, dt = e^{\lambda h}\left(R_\lambda g - \int_0^h e^{-\lambda t} T_t g\, dt\right),$$

it follows that

$$D_h R_\lambda g = \frac{e^{\lambda h} - 1}{h} R_\lambda g - \frac{e^{\lambda h}}{h} \int_0^h e^{-\lambda t} T_t g\, dt \xrightarrow{s} (\lambda R_\lambda - I)g,$$

hence $R_\lambda g \in G_d$. If $g \in G_d (\subset G_c)$ then, moreover,

$$D_h R_\lambda g = R_\lambda D_h g \xrightarrow{s} R_\lambda D g.$$

Therefore,

$$DR_\lambda = \lambda R_\lambda - I \text{ on } G_c, \quad R_\lambda D = \lambda R_\lambda - I \text{ on } G_d,$$

equivalently,

$$(\lambda I - D) R_\lambda = I \text{ on } G_c, \quad R_\lambda(\lambda I - D) = I \text{ on } G_d.$$

The proof is terminated.

The theorem which follows will lead us to the "natural" appearance of spaces of continuous functions

A. Unicity theorem. *Let D with its domain G_d be the infinitesimal operator of a contraction semi-group.*

(i) *D determines the semi-group on its strong continuity space G_c and determines the semi-group on the Φ-weak closure G' of G_c when the contractions are Φ-adjoint.*

(ii) *The unique solution of the equation* $\frac{df_t}{dt} = Df_t, f_t \in G_d$, *which is bounded, reduces to* $g \in G_d$ *for* $t = 0$ *and has a strongly continuous strong derivative, is* $f_t = T_t g$, $g \in G_d$.

Proof 1° Given D on the strong differentiability set G_d, the strong continuity space G_c is determined as strong closure of G_d and then, by **e**, R_λ on G_c is determined. Therefore, by the classical unicity theorem for numerical Laplace transforms $R_\lambda g(x) = \int_0^\infty e^{-\lambda t} T_t g(x)\, dt$ of continuous functions $T_t g(x)$ in t, these functions are determined. The G_c-assertion is proved, and the G'-assertion follows since Φ-adjoint endomorphisms commute with Φ-weak passages to the limit.

2° According to **b**, $f_t = T_t g$, $g \in G_d$ is a solution of the stated equation with the asserted properties. The "unicity" assertion will follow if we show that such a solution, which vanishes for $t = 0$ (in lieu of reducing to g), vanishes for all t. Let $g_t = e^{-\lambda t} f_t$ and note that, according to the hypotheses made, $g_0 = 0, \frac{dg_t}{dt} \in G_c$ and $\| g_s \| \to 0$ as $s \to \infty$. On account of the stated equation, we have $\frac{dg_t}{dt} = (D - \lambda I) g_t$ hence, by **e**, $g_t = -R_\lambda \frac{dg_t}{dt}$. It follows that for all $\lambda (> 0)$, as $s \to \infty$,

$$\int_0^\infty e^{-\lambda t} f_t\, dt \overset{s}{\leftarrow} \int_0^s g_t\, dt = -R_\lambda g_s \overset{s}{\to} 0.$$

Therefore, by the unicity theorem recalled above, $f_t = 0$. The proof is terminated.

Upon replacing "strong" by "Φ-weak," parallel definitions and slightly more involved but similar arguments yield without difficulty

e′. Φ-WEAK RESOLVENT LEMMA. *The Φ-weak resolvent R'_λ of a Φ-adjoint contraction Borel semi-group is the inverse of the one-to-one mapping $\lambda I - D'$ of G'_d onto G'_c.*

A′. Φ-WEAK UNICITY THEOREM. *Let D' with its domain G'_d be the Φ-weak infinitesimal operator of a Φ-adjoint contraction Borel semi-group.*

(i) *D' determines the semi-group on the common Φ-weak closure G' of its continuity and differentiation spaces.*

(ii) *The unique solution of the equation* $\frac{d^+ f_t}{dt} = D' f_t$, $f_t \in G'_d$, *which is bounded and reduces to* $g \in G'_d$ *for* $t = 0$ *is measurable in* (x, t) *and continuous in* t *and has a right continuous right derivative bounded on every finite interval of values of* t, *is* $f_t = T_t g$, $g \in G'_d$.

The foregoing unicity theorems lead at once to the following question: when is the closure of a subset of G the whole of G? For then, the semi-group is completely determined—on G—and the extension problem is solved in the affirmative. An answer is immediately available if we recall the Baire definition of Borel functions: The class of Borel functions on a Euclidian space $\mathfrak{X}$ is the closure of its subclass of continuous functions under pointwise passages to the limit of sequences. The space C of bounded continuous functions on $\mathfrak{X}$ is a Banach subspace of the Banach space G of bounded Borel functions on $\mathfrak{X}$. Thus, G is the Φ-weak closure of its Banach subspace C.

B. C-EXTENSION THEOREM. *If the* Φ*-weak closure of the strong* (Φ*-weak*) *continuity subspace of a* Φ*-adjoint* (*Borel*) *contraction semi-group contains the subspace* C *then the semi-group is completely determined by its strong* (Φ*-weak*) *infinitesimal operator. In particular, the semi-group is completely determined when* C *is invariant and the semi-group is strongly* (*weakly*) *continuous on* C.
Moreover, if the contraction Borel semi-group leaves C_0 *invariant and is* Φ*-weakly continuous on* C_0, *then it is* Φ *adjoint, strongly continuous on* C_0, *and, in fact, "weak" and "strong" concepts coincide in* C_0.

Proof. The determination assertion results from the unicity theorems and the Φ-weak closure property of C. The infinitesimal operators assertion results from the continuity assertion since then, by the resolvent lemma, R_λ is a one-to-one mapping of $C \cap G_c$ onto $C \cap G_d$ and of $C \cap G'_c$ onto $C \cap G'_d$.
Since Φ is the adjoint of C_0, the "Φ-adjoint" assertion is immediate. The "weak"-"strong" assertion follows from 40.3**d**. The proof is terminated.

The stated problems are answered in terms of (strong) infinitesimal operators as follows.

Let $(T_{\lambda,t}, t \geqq 0)$ be a family in $\lambda > 0$ of strongly continuous contraction semi-groups on a common invariant Banach space H. Take all limits as $\lambda, \mu \to \infty$, unless otherwise stated. We say that these *semi-groups converge strongly* if on H there exist endomorphisms T_t, necessarily contractions, such that $T_{\lambda,t} \xrightarrow{s} T_t$ uniformly in $t \leqq a$ for any finite a: $\| T_{\lambda,t}\, g - T_t g \| \to 0$ uniformly in $t \leqq a$, for every $g \in H$. It is easily seen that such convergence implies and is implied by the corresponding mutual convergence $T_{\lambda,t} - T_{\mu,t} \xrightarrow{s} 0$ uniformly in $t \leqq a$, and that if either convergence holds on a set dense in H then it holds on H. Furthermore, the limits T_t form a contraction semi-group. Finally, because of the uniformity condition, it is strongly continuous on H and, for $\nu > 0$, $g \in H$, as $\lambda \to \infty$,

$$R_{\nu,\lambda} g = \int_0^\infty e^{-\nu t} T_{\lambda,t} g\, dt \xrightarrow{s} \int_0^\infty e^{-\nu t} T_t g\, dt = R_\nu g.$$

f. Semi-groups convergence lemma. *On a Banach space H, if endomorphisms D_λ commute, $\| e^{tD_\lambda} \| \leqq 1$, and $D_\lambda \xrightarrow{s} D$ on a set H' dense in H, then the strongly (in fact, uniformly) continuous exponential contraction semi-groups $(e^{tD_\lambda}, t \geqq 0)$ converge to a strongly continuous contraction semi-group $(T_t, t \geqq 0)$ whose infinitesimal operator is D on $H_d \supset H'$.*

Proof. We use two elementary relations applicable to commuting endomorphisms:

$$| \alpha^n - \beta^n | \leqq n | \alpha - \beta |, \quad | \alpha |, \quad | \beta | \leqq 1,$$

$$\frac{1}{h}(e^{h\alpha} - 1)x \to \alpha x, \; h \to 0.$$

Let $g \in H'$, $t \leqq a$, and exclude the trivial case $D_\lambda g = D_\mu g$. As $n \to \infty$,

$$\| (e^{tD_\lambda} - e^{tD_\mu}) g \| \leqq a \left\| \frac{e^{(t/n)(D_\lambda - D_\mu)} - I}{t/n} g \right\| \to a \| (D_\lambda - D_\mu) g \|.$$

As λ, $\mu \to \infty$, the last expression converges to 0, and the convergence assertion follows.

For any λ, let R_λ^ν be the resolvent of the semi-group of endomorphisms e^{tD_ν} and let R_λ be that of the limit semi-group, so that $R_\lambda^\nu \xrightarrow{s} R_\lambda$ as $\nu \to \infty$. Since

$$R_\lambda^\nu(\lambda I - D)g = R_\lambda^\nu(\lambda I - D_\nu)g + R_\lambda^\nu(D_\nu - D)g,$$

where on the right side the first term is g and the norm of the second term is bounded by $\| (D_\nu - D)g \| / \lambda \to 0$ as $\nu \to \infty$, it follows that $R_\lambda(\lambda I - D)g = g$ and the infinitesimal operator assertion follows by the resolvent lemma. The proof is terminated.

We are now ready for the basic Hille-Yosida

C. Infinitesimal operators criterion. *A linear transformation D on a linear subset H_d of a Banach space H is the infinitesimal operator of one and only one strongly continuous contraction semi-group $(T_t, t \geqq 0)$ on H, if and only if*

(i) *The inverse R_λ of the transformation $\lambda I - D$ exists and is an endomorphism on H with $\| \lambda R_\lambda \| \leqq 1$ for every $\lambda > 0$.*

(ii) $\lambda R_\lambda \xrightarrow{s} I$ *on H as $\lambda \to \infty$ or* (ii′) *H_d is dense in H.*
Then the exponential semi-groups $(e^{tD_\lambda}, t \geqq 0)$, where $D_\lambda = \lambda(\lambda R_\lambda - I)$, converge strongly to the semi-group $(T_t, t \geqq 0)$ as $\lambda \to \infty$.

Proof. Let $\lambda \to \infty$. Properties (ii) and (ii′) are equivalent under (i): Let $\lambda R_\lambda g \xrightarrow{s} g$ for every $g \in H$. Then, from $\lambda R_\lambda g \in H_d$ it follows that H_d is dense in H. Conversely, let H_d be dense in H. Since $\| \lambda R_\lambda \| \leqq 1$ implies that for every $g \in H_d$

$$\| (\lambda R_\lambda - I)g \| = \| R_\lambda Dg \| \leqq \| Dg \| / \lambda \to 0,$$

hence the contractions $\lambda R_\lambda \xrightarrow{s} I$ on H_d, it follows that $\lambda R_\lambda \xrightarrow{s} I$ on H.
The "only if" and the "unicity" assertions result from **a** and **A**. The "if" assertion then results from the convergence one which we prove by showing that, because of (i), (ii), and (ii′), the semi-groups convergence lemma applies: Since, by (i), the R_λ hence the D_λ commute and $\| \lambda R_\lambda \| \leqq 1$ hence $\| e^{tD_\lambda} \| \leqq e^{-\lambda t} e^{\lambda t \| \lambda R_\lambda \|} \leqq 1$, on account of (ii′) it suffices to show that $D_\lambda \xrightarrow{s} D$ on H_d. But, by (i), $R_\lambda(\lambda I - D) = I$ on H_d hence, by (ii), $D_\lambda = \lambda R_\lambda D \xrightarrow{s} D$ on H. The proof is terminated.

Corollary. *Let H be an invariant subspace of G. Under the conditions of the Hille-Yosida theorem, $T_t \geqq 0$ for all $t > 0 \Rightarrow \lambda R_\lambda \geqq 0$ for all $\lambda > 0 \Rightarrow e^{tD_\lambda} \geqq 0$ for all $t, \lambda > 0 \Rightarrow T_t \geqq 0$ for all $t > 0$.*

For, the first implication results from the definition of R_λ, the second one follows from $e^{tD_\lambda} = e^{-\lambda t} \sum_{n=0}^{\infty} \frac{(\lambda t)^n}{n!} (\lambda R_\lambda)^n$, and the third one is obtained by letting $\lambda \to \infty$.

The definitions and properties in this subsection apply to semi-groups restricted to any given *invariant subspace* $H \subset G$, provided they are relativized accordingly, that is, $G, G_c, G'_c, G'_d, G', \cdots$ are replaced by their intersections with H: $H, H \cap G_c$, and so on. However, there is a difficulty in the Φ-weak case: Φ-weak integrals, supposed to exist, of functions $g_t \in H$ may not belong to H, and some restriction is needed to eliminate this difficulty. For example, it suffices that H be the space

C. This brings out once more the advantage of invariant C. In fact, our study became fast restricted to invariant subspaces selected according to the semi-group continuity requirements. Thus we are led to classify our semi-groups according to their invariant subspaces selected according to suitable requirements. What precedes, and the C-unicity theorem show the convenience of the class for which C is invariant.

40.3 Markov processes and semi-groups. We apply what precedes to the Markov case, searching for a probabilistic interpretation of the concepts and properties.

Let X_T, $T = [0, \infty)$, be a stationary regular Markov process. There is a one-to-one correspondence between the stationary law of evolution $P^{X_0} = (P^x, x \in \mathfrak{X})$ of X_T, the tr.pr. $P_t(x, S) = P(X_{s+t} \in S \mid X_s = x)$, the tr.d.f. $F_t^x(y) = P(X_{s+t} < y \mid X_s = x)$, the tr.ch.f. $f_t^x(u) = E(e^{iuX_{s+t}} \mid X_s = x)$, and the Markov semi-group $(T_t, t \geqq 0)$ with

$$T_tg(x) = \int P_t(x, dy)g(y) = E^xg(X_t), \quad g \in G,$$

or, to emphasize stationarity,

$$T_tg(X_s) = E\{g(X_{s+t}) \mid X_s\}.$$

The Markov endomorphisms are adjoint to endomorphisms on the space Φ, defined by

$$\varphi T_t(S) = \int \varphi(dx)P_t(x, S) = \int \varphi(dx)P^x[X_t \in S]$$

hence translating distributions $\varphi = P_s$ of X_s according to

$$P_sT_t = P_{s+t}.$$

Let the tr.pr., equivalently the semi-group, be Borelian. Form the function

$$\lambda R_\lambda(x, S) = \lambda \int_0^\infty e^{-\lambda t}P_t(x, S)\, dt, \quad \lambda > 0.$$

It is a Borel function in x and a pr. in S. Therefore, the endomorphism λR_λ defined by

$$\lambda R_\lambda g(x) = \int \lambda R_\lambda(x, dy)g(y), \quad g \in G,$$

is Markovian. Since by interchanging the integrations

$$\lambda R_\lambda g(x) = \lambda \int_0^\infty e^{-\lambda t}T_tg(x)\, dt,$$

endomorphisms R_λ appear as an extension on G of the resolvents of $(T_t, t \geqq 0)$. Furthermore, the iterates $(\lambda R_\lambda)^n$ are Markov endomorphisms defined by

$$(\lambda R_\lambda)^n g(x) = \int (\lambda R_\lambda)^{(n)}(x, dy) g(y), \quad g \in G,$$

with

$$(\lambda R_\lambda)^{(n)}(x, S) = \int (\lambda R_\lambda)^{(n-1)}(x, dy) \lambda R_\lambda(y, S).$$

It follows that every e^{tD_λ} with $D_\lambda = \lambda(\lambda R_\lambda - I)$ is a Markov endomorphism defined by

$$e^{tD_\lambda} g(x) = \int (e^{tD_\lambda})(x, dy) g(y), \quad g \in G,$$

with

$$(e^{tD_\lambda})(x, S) = e^{-\lambda t}\left(1 + \frac{\lambda t}{1!} \lambda R_\lambda(x, S) + \frac{\lambda^2 t^2}{2!} (\lambda R_\lambda)^{(2)}(x, S) + \cdots\right).$$

Thus, the exponential Markov semi-groups $(e^{tD_\lambda}, t \geqq 0)$ appear as extensions of the exponential semi-groups of the infinitesimal operators criterion. Clearly, what precedes applies to the relativization on an invariant subspace H.

Finally, $D_h = (T_h - I)/h$ given by

$$D_h g(x) = \int (P_h(x, dy) - I(x, dy)) g(y)/h = E^x(g(X_h) - g(x))/h, \quad h > 0,$$

or, to emphasize stationarity,

$$D_h g(X_s) = E\{g(X_{s+h}) - g(X_s) \mid X_s\}/h$$

may be thought of as a "mean speed" operator. If

$$E\{g(X_{s+h}) - g(X_s) \mid X_s\}/h \to Dg(X_s), \quad g \in G',$$

the "speed" operator D on G' appears as an extension of the infinitesimal operators of $(T_t, t \geqq 0)$.

We pursue our specialization to Markov semi-groups. We recall that C denotes the subspace of G formed by bounded continuous functions on the state space $\mathfrak{X}$. *We denote by C_u the subspace of uniformly continuous functions on $\mathfrak{X}$ and by C_0 that of continuous functions on $\mathfrak{X}$ vanishing at infinity*, and have

$$C \supset C_u \supset C_0.$$

For any locally compact space $\mathfrak{X}$, $g \in C_0$ means that for every $\epsilon > 0$ there is a compact $K_\epsilon \subset \mathfrak{X}$ such that $|g(x)| < \epsilon$ for $x \notin K_\epsilon$; we write $g(x) \to 0$ as $x \to \infty$. (Note that if $\mathfrak{X}$ is compact there is no "point at infinity.") If $\mathfrak{X}$ is compact, the above condition is void and, also, continuity becomes uniform. Thus

If $\mathfrak{X}$ is compact, then $C = C_u = C_0$.

The restriction of a Markov semi-group on $H \subset G$ is defined by

$$T_t g(x) = \int P_t(x, dy) g(y), \quad g \in H$$

and will be called *Markovian on H*.

a. *The restriction of a Markov semi-group on $H \supset C_0$ determines the semi-group.*

Proof. Upon letting g vary over G in the above integral representation of T_t on H, the semi-group so extended is Markovian. This Markovian extension is unique because T on C_0 determines the tr.pr. $P_t(x, S)$ as follows. Let $[a, b) \subset \mathfrak{X}$ be a bounded interval and take $g_n(x) = 1$ for $a \leqq x < b - 1/n$, $= 0$ for $x \leqq a - 1/n$ and $x \geqq b$, and linear for $a - 1/n \leqq x \leqq a$ and for $b - 1/n \leqq x \leqq b$. Thus, $g_n \in C_0$, $g_n \to I_{[a,b)}$ and, by the dominated convergence theorem,

$$T_t g_n(x) = \int P_t(x, dy) g_n(y) \to P_t(x, [a, b)).$$

Therefore, T_t on C_0 determines $P_t(x, S)$ on the class of all bounded intervals $S = [a, b)$ hence on the Borel field $\mathcal{S}$. The lemma is proved.

A. Markov infinitesimal operators criterion. *A linear transformation D on a subset H_d of a Banach subspace $H \supset C_0$ to H is the infinitesimal operator of one and only one Markov semi-group $(T_t, t \geqq 0)$ on G strongly continuous on invariant H, if and only if on H*

(i) *The inverse R_λ of the transformation $\lambda I - D$ exists and λR_λ is a Markov endomorphism for every $\lambda > 0$*

(ii) $\lambda R_\lambda \xrightarrow{s} I$ *on H as $\lambda \to \infty$ or* (ii′) *H_d is dense in H.*

Then, on G, the Markov extension λR_λ is determined and the exponential Markov semi-groups $(e^{tD_\lambda}, t \geqq 0)$, where $D_\lambda = \lambda(\lambda R_\lambda - I)$, converge strongly to the semi-group $(T_t, t \geqq 0)$ as $\lambda \to \infty$.

Proof. According to the infinitesimal operators criterion, the proposition holds on H when we drop "Markov" and "$\supset C_0$" therein. Since the semi-group $(T_t, t \geqq 0)$ on H is strongly continuous, hence Borelian, by the foregoing discussion and lemma, from $H \supset C_0$ it follows that we have:

Markov $(T_t, t \geqq 0)$ on $G \Rightarrow$ Markov λR_λ on $H \Rightarrow$ Markov $(e^{tD_\lambda}, t \geqq 0)$ on $H \Rightarrow$ Markov $(e^{tD_\lambda}, t \geqq 0)$ on $G \Rightarrow$ Markov $(T_t, t \geqq 0)$ on G.

The proof is terminated.

The spaces C, C_0, C_u of continuous functions appear in the convergence of laws criteria:

Let $F, f, \hat{f}$, with same affixes if any, denote corresponding d.f.'s, ch.f.'s, and integral ch.f.'s. According to the complete (weak) convergence criterion and the Helly-Bray theorem (extended lemma), we have the equivalences (convergence of d.f.'s is, as usual, up to additive constants);

$$F_n \xrightarrow{c} F \Leftrightarrow f_n \to f \Leftrightarrow \int g\, dF_n \to \int g\, dF, \quad g \in C$$

$$F_n \xrightarrow{w} F \Leftrightarrow \hat{f}_n \to \hat{f} \Leftrightarrow \int g\, dF_n \to \int g\, dF, \quad g \in C_0.$$

Note that the integral ch.f.'s correspond to the subset of functions $g \in C_0$ defined by $g(x) = \dfrac{e^{iux} - 1}{ix}$, $u \in R$, and the ch.f.'s correspond to the subset of functions $g \in C$ defined by $g(x) = e^{iux}$, $u \in R$. Since the last subset is also in C_u, we may replace C by C_u in what precedes.

The last equivalent forms of convergence of laws leads to the general concept of *H-convergence* of laws: $F_n \xrightarrow{H} F$ meaning that $\int g\, dF_n \to \int g\, dF$ for every $g \in H$, and $F_n^x \xrightarrow{H} F^x$ *uniformly in* x meaning that $\int g\, dF_n^x \to \int g\, dF^x$ uniformly in x for every $g \in H$. But for a stationary regular Markov process X_T with c.pr. $(P^x, x \in \mathfrak{X})$, hence c.d.f.'s $F_t^x(y) = P^x[X_t < y]$, we have

$$\text{(G)} \qquad T_t g(x) = E^x g(X_t) = \int g\, dF_t^x, \quad g \in G.$$

Therefore, to H-convergence of laws correspond continuity properties of its semi-group on H, as follows. Let the limits in h be taken as $0 < h \to 0$ and note that $F_0^x(y) = 0$ or 1 according as $y \leqq x$ or $y > x$. $F_h^x \xrightarrow{H} F_0^x$ corresponds to $T_h g \xrightarrow{w} g$ for every $g \in H$, that is, $H \subset G'_c$—

the weak continuity space of our semi-group. Similarly, $F_h^x \xrightarrow{\text{H}} F_0^x$ uniformly in x corresponds to $T_h g \xrightarrow{\text{s}} g$ for every $g \in H$, that is, $H \subset G_c$ —the strong continuity space of our semi-group. This emphasizes the distinguished role of spaces C_0, C_u, C, when the usual weak or complete convergence of laws are considered.

The spaces of continuous functions also appear under "natural" requirements for semi-groups: stability and continuity of evolution. According to the celebrated Hadamard principle, a "well set" evolution problem is stable, in the sense that small variations of initial data lead to small variations in the evolution. The weakest probabilistic interpretation would be in terms of individual laws given the initial state. To be precise, let X_T, $T = [0, \infty)$, be a regular process. We say that the process is *stable* if the laws $\mathfrak{L}(X_t \mid X_0 = x)$ are continuous in x for every t, equivalently, if for every x and every t, as $x' \to x$

$$F_t^{x'} \xrightarrow{\text{c}} F_t^x \quad \text{or} \quad f_t^{x'} \to f_t^x \quad \text{or} \quad \int g\,dF_t^{x'} \to \int g\,dF_t^x, \quad g \in C.$$

For a stationary regular Markov process X_T and its semi-group $(T_t, t \geqq 0)$, it suffices to use (G) to obtain

b. *A Markov semi-group leaves C invariant if, and only if, the corresponding Markov process is stable.*

For, if $x' \to x$ and $g \in C$, then stability implies that

$$T_t g(x') = \int g\,dF_t^{x'} \to \int g\,dF_t^x = T_t g(x),$$

and invariance of C implies that

$$\int g\,dF_t^{x'} = T_t g(x') \to T_t g(x) = \int g\,dF_t^x.$$

What precedes remains valid with "w" and "C_0" in lieu of "c" and "C" except that we obtain $T_t g \in C$ in lieu of $T_t g \in C_0$. Thus, in order that C_0 be invariant, we have to add the requirement $T_t g(x') = \int g\,dF_t^{x'} \to 0$ as $x' \to \pm\infty$ for $g \in C_0$, equivalently $F_t^{x'}[a, b) \to 0$ as $x' \to \pm\infty$ for every bounded $[a, b)$. We may interpret the conditions so obtained as *weak stability including infinity* of the process, in the following sense: $F_t^{x'} \xrightarrow{\text{w}} F_t^x$ for every t and x including $x = \pm\infty$, upon setting $\text{Var}\, F_t^{\pm\infty} = P^{\pm\infty}[|X_t| < \infty] = 0$. Thus

b′. *A Markov semi-group leaves C_0 invariant if, and only if, the corresponding process is weakly stable including infinity.*

The other "natural" requirement is that of continuous evolution. The weakest probabilistic interpretation would be that of continuity of laws as $h \to 0$: $F_h^x \to F_0^x$, ($F_0^x(y) = 0$ or 1 according as $y < x$ or $y > x$), that is, $P_h(x, (x - \epsilon, x + \epsilon)^c) \to 0$ or $P_h(x, (x - \epsilon, x + \epsilon)) \to 1$ for every $\epsilon > 0$; equivalently $T_h g(x) = \int g\, dF_h^x \to g(x)$, $g \in C$, that is, $C \subset G'_c$. Thus

c. *A Markov semi-group is Φ-weakly continuous on C at $t = 0$ if, and only if, the corresponding Markov process is continuous in law at $t = 0$: $P_h(x, (x - \epsilon, x + \epsilon)) \to 1$ for every x and every $\epsilon > 0$.*

If we consider C_0 and C_u, then uniform conditions appear. We shall denote by K closed bounded intervals in the state space and set $V_x(\epsilon) = (x - \epsilon, x + \epsilon)$, whenever convenient; in fact, what follows is valid with K interpreted as a compact set.

c′. *A Markov semi-group is strongly continuous on C_0 at $t = 0$ only if $P_h(x, (x - \epsilon, x + \epsilon)) \to 1$ uniformly in $x \in K$ for every K and every $\epsilon > 0$.*

Proof. Since every K is compact, it can be covered by a finite number of $V_{x_k}(\epsilon/4)$. Thus every $x \in K$ belongs to some $V_{x_k}(\epsilon/4)$, so that $V_x(\epsilon) \supset V_{x_k}(\epsilon/2)$ and the "only if" assertion reduces to $P_h(x, V_{x_k}(\epsilon/2)) \to 1$ uniformly in $x \in V_{x_k}(\epsilon/4)$. By hypothesis, $T_h g(x) = \int P_h(x, dy) g(y) \to g(x)$ uniformly in x for every $g \in C_0$. Since we can select $g \in C_0$ such that $0 \leqq g \leqq 1$, $g = 1$ on the closure $\overline{V}_{x_k}(\epsilon/4)$ and $g = 0$ on $V_{x_k}^c(\epsilon/2)$, it follows then that, uniformly in $x \in V_{x_k}(\epsilon/4)$,

$$P_h(x, V_{x_k}(\epsilon/2)) \geqq \int P_h(x, dy) g(y) \to g(x) = 1.$$

The assertion is proved.

c″. *A Markov semi-group is strongly continuous on C_u at $t = 0$ if $P_h(x, (x - \epsilon, x + \epsilon)) \to 1$ uniformly in x for every $\epsilon > 0$.*

Proof. Let $g \in C_u$. If $P_h(x, (x - \epsilon, x + \epsilon)) \to 1$ uniformly in x then, from $\| g \| \leqq c < \infty$ and

$$T_h g(x) - g(x)$$
$$= \int_{|y-x|<\epsilon} (g(y) - g(x))\, dF_h^x(y) + \int_{|y-x|\geqq\epsilon} (g(y) - g(x))\, dF_h^x(y) + (\text{Var}\, F_h^x - 1)g(x),$$

it follows that as $h \to 0$ then $\epsilon \to 0$

$$\sup |\, T_h g(x) - g(x) \,| \leqq \sup_{|y-x|<\epsilon} |\, g(y) - g(x) \,| + 3c \sup_x P_h(x, (x - \epsilon, x + \epsilon)^c) \to 0,$$

hence $T_h g \xrightarrow{s} g$.

We combine now the preceding lemmas **b** and **c** into

B. C-INVARIANCE AND CONTINUITY CRITERION. *A Markov semi-group leaves the space C invariant and is Φ-weakly rightcontinuous on it if, and only if, the corresponding process is stable and continuous in law at $t = 0$: $F_t^{x'} \to F_t^x$ as $x' \to x$ and $P_h(x, (x - \epsilon, x + \epsilon)) \to 1$ as $h \to 0$, for every x and every $\epsilon > 0$.*

If the state space $\mathfrak{X}$ is compact, then $C = C_u = C_0$ and **b**′, **c**′ and **c**″ apply. In fact,

d. INVARIANCE AND CONTINUITY LEMMA. *Let C be invariant and $\mathfrak{X}$ be compact or let C_0 be invariant and $\mathfrak{X}$ be only locally compact. Then on the invariant subspace, Φ-weak rightcontinuity implies strong continuity.*

Proof. 1° Let C be invariant and $\mathfrak{X}$ be compact. For every $g \in C$, form

$$g_\lambda(x) = \lambda R_\lambda g(x) = \int_0^\infty \lambda e^{-\lambda s} T_s g(x)\, ds, \quad \lambda > 0,$$

and note that $g_\lambda \in C$; for, by C invariance and the dominated convergence theorem, $g_\lambda(x_n) \to g_\lambda(x)$ as $x_n \to x$. In fact, $g_\lambda \in C_c$: By a straightforward computation,

$$T_h g_\lambda(x) = \lambda e^{\lambda h} \int_h^\infty e^{-\lambda s} T_s g(x)\, ds$$

so that, for $g \geqq 0$, $e^{-\lambda h}\, T_h g_\lambda(x) \uparrow g_\lambda(x)$ as $h \downarrow 0$. However, on a compact space monotone pointwise convergence implies uniform convergence (Dini's lemma). Therefore, $T_h g_\lambda \xrightarrow{s} g_\lambda$ for $g \geqq 0$, hence for any $g = g^+ - g^-$.

Since $\mathfrak{X}$ is compact, Φ is adjoint to C. Therefore, by Riesz's representation theorem, any bounded linear functional is of the form

$$\varphi(g) = \int \varphi(dx)g(x), \quad \varphi \in \Phi, \quad g \in C.$$

In particular,

$$\varphi(g_\lambda) = \int_0^\infty \lambda e^{-\lambda s} f(s)\, ds = \int_0^\infty e^{-u} f(u/\lambda)\, du$$

where

$$f(s) = \int \varphi(dx) T_s g(x).$$

When the semi-group is Φ-weakly rightcontinuous, $f(s)$ is right continuous in s so that, as $\lambda \to \infty$,

$$f(s/\lambda) \to f(0) = \varphi(g), \quad \text{hence} \quad \varphi(g_\lambda) \to \varphi(g).$$

Thus, if there exists a $\varphi(\cdot)$ which vanishes on C_c, hence on all g_λ, then it vanishes on all $g \in C$. Therefore, by the Hahn-Banach theorem, $C_c = C$.

2° Let C_0 be invariant and $\mathfrak{X}$ only locally compact. The preceding argument applies upon noting that $g_\lambda(x) \to 0$ as $|x| \to \infty$. Or, to compactify $\mathfrak{X}$ into $\overline{\mathfrak{X}} = \mathfrak{X} + \{\infty\}$, set

$$\overline{P}_t(\infty, \{\infty\}) = 1, \quad \overline{P}_t(\infty, \mathfrak{X}) = 0$$

and, for $x \in \mathfrak{X}$, $S \in \mathfrak{S}$, set

$$\overline{P}_t(x, S) = P_t(x, S), \quad \overline{P}_t(x, \{\infty\}) = 1 - P_t(x, \mathfrak{X}).$$

This function is a tr.pr. and the corresponding Markov semi-group $\overline{T}_t$ leaves $\overline{C} = C(\overline{\mathfrak{X}})$ invariant: For, every function $\overline{g} \in \overline{C}$ is of the form $\overline{g} = g + c$ with $g \in C_0$, c constant, and

$$\overline{T}_t g = T_t g, \quad \overline{T}_t c = c.$$

Furthermore, Φ-weak rightcontinuity of T_t on C_0 implies that of $\overline{T}_t$ on $\overline{C}$ and 1° applies. The proof is terminated.

We say that weak stability including infinity is weak stability *uniform at infinity* if at infinity $P_t(x', K) \to 0$ as $x' \to \pm\infty$ for every K uniformly in $t \leqq t_0$ arbitrary but finite.

B′. C_0-INVARIANCE AND CONTINUITY CRITERION. *A Markov semi-group leaves the space C_0 invariant and is strongly continuous on it if, and only if, the corresponding process is weakly stable uniformly at infinity and*

$P_h(x, (x - \epsilon, x + \epsilon)) \to 1$ as $h \to 0$ uniformly in $x \in K$ for every K and every $\epsilon > 0$.

Proof. By **b′** and **c′**, C_0-invariance and weak stability including infinity are equivalent and strong continuity implies the uniform in $x \in K$ condition. Therefore, denoting the strong continuity at $t = 0$ condition by (i), the uniform at infinity condition by (ii), and the uniform in $x \in K$ condition by (iii), it remains to prove that under C_0-invariance, (i) implies (ii) and (ii) and (iii) imply (i). Thus, let C_0 be invariant.

1° Given K, form

$$g_\lambda(x) = \lambda R_\lambda g(x) = \int_0^\infty \lambda e^{-\lambda s} T_s g(x)\, ds, \quad \lambda > 0,$$

where $g \in C_0$ is positive and exceeds 1 on K. If (i) holds, then $g_\lambda \in C_0$ (g_λ is a strong integral),

$$T_t g_\lambda(x') = e^{\lambda t} \int_t^\infty \lambda e^{-\lambda r} T_r g(x')\, dr \leqq e^{\lambda t} g_\lambda(x'),$$

and $g_\lambda(x)$ converges to $g(x)$ uniformly in x as $\lambda \to \infty$ (apply **A** or, directly, set $\lambda s = u$ and note that $T_{u/\lambda} g(x) \to g(x)$ uniformly in x). Therefore, $g_{\lambda_0} \geqq 1$ on K for λ_0 sufficiently large, so that for $t \leqq t_0 < \infty$, as $x' \to \pm\infty$

$$P_t(x', K) \leqq T_t g_{\lambda_0}(x') \leqq e^{\lambda_0 t_0} g_{\lambda_0}(x') \to 0,$$

and (ii) holds.

2° Let (ii) and (iii) hold. To establish strong continuity on C_0 at $t = 0$, it suffices to prove it on the subspace $C_{00} \subset C_0$ of those continuous functions which vanish outside K's. For, given $\delta > 0$, every $g \in C_0$ can be decomposed into $g' + g''$ with $g' \in C_{00}$ and $|g''| < \delta$, so that as $h \to 0$ then $\delta \to 0$,

$$\| T_h g - g \| \leqq \| T_h g' - g' \| + 2\delta \to 0$$

provided $\| T_h g' - g' \| \to 0$. Similarly, because of (ii), $P_h(x, K) \leqq \delta$ for sufficiently small h and for x outside a sufficiently large interval, and we can assume C_{00} invariant under T_h. Thus, we can take every g and $T_h g$ arbitrarily small outside some corresponding K sufficiently large.

Let $x, y \in K$, and let c be a bound of g. Since K is compact, given $\delta > 0$ we can select $\epsilon > 0$ sufficiently small so that $|g(y) - g(x)| < \delta$ for $|y - x| < \epsilon$. It follows, by (iii), that uniformly in x, as $h \to 0$

then $\delta \to 0$,

$$| T_h g(x) - g(x) | \leqq \left| \int_{|y-x|<\epsilon} P_h(x, dy)(g(y) - g(x)) \right|$$
$$+ | g(x) | \{1 - P_h(x, (x - \epsilon, x + \epsilon))\}$$
$$+ \left| \int_{|y-x|\geqq \epsilon} P_h(x, dy) g(y) \right|$$
$$\leqq \delta + 2c\{1 - P_h(x, (x - \epsilon, x + \epsilon))\} \to 0.$$

Thus (i) holds, and the proof is terminated.

We illustrate the foregoing concepts and properties in characterizing the infinitesimal operators of Markov semi-groups which correspond to stationary continuous decomposable processes.

Let $(Z_t, t \geqq 0)$ be a stationary continuous decomposable process. Let F_t and $e^{t\psi}$, $\psi = (\alpha, \Psi)$ be the i.d. d.f.'s and ch.f.'s of the $Y_t = Z_t - Z_0$. Our process is a stationary regular Markov process with $Z_t - Z_0$ independent of Z_0, $P^x = P$ and

$$F_t^x(z) = P(Z_t < z \mid Z_0 = x) = P(Y_t < z - x) = F_t(z - x).$$

It follows that $F_t^{x'} \xrightarrow{c} F_t^x$ as $x' \to x$ and $F_t^x \to 0$ or 1 according as $x \to +\infty$ or $-\infty$, that is, $F_t^x[a, b) \to 0$ for every bounded $[a, b)$; thus, the process is stable and weakly stable at infinity. Also $P_h(x, (x - \epsilon, x + \epsilon)^c) = P[| Y_h | \geqq \epsilon] \to 0$ uniformly in x for every $\epsilon > 0$, for the process is continuous in law. Therefore, according to the preceding propositions, the corresponding semi-group defined by

$$T_t g(x) = \int g(z)\, dF_t(z - x) = \int g(x + y)\, dF_t(y), \quad g \in G$$

leaves the space C_0 (also C) invariant and is strongly continuous on it. Thus, the Markov infinitesimal operators criterion applies, and we can and do limit ourselves to C_0. Let g vary over C_0. The resolvent R_λ is defined by

$$\lambda R_\lambda g(x) = \int_0^\infty \left\{ \int g(x + y)\, dF_t(y) \right\} \lambda e^{-\lambda t}\, dt = \int g(x + y)\, dG_\lambda(y),$$

where the function $G_\lambda = \int_0^\infty \lambda e^{-\lambda t} F_t\, dt$ is a d.f. (weighted by the exponential d.f. family of d.f.'s F_t) with ch.f.

$$g_\lambda(u) = \int_0^\infty \lambda e^{-\lambda t} e^{t\psi(u)}\, dt = \lambda/(\lambda - \psi(u)).$$

The operator D_λ of the infinitesimal operators criterion is given by

$$D_\lambda g(x) = \lambda(\lambda R_\lambda - I)g(x) = \int (g(x+y) - g(x))\lambda\, dG_\lambda(y),$$

and the infinitesimal operator D (in C_0) is the strong limit as $\lambda \to \infty$ of D_λ on its domain of existence. Since, as $\lambda \to \infty$,

$$(\alpha, \Psi) = \psi \leftarrow \lambda\left(\frac{\lambda}{\lambda - \psi} - 1\right) = \int (e^{iuy} - 1)\lambda\, dG_\lambda(y) = \psi_\lambda = (\alpha_\lambda, \Psi_\lambda)$$

where

$$\alpha_\lambda = \int \frac{y}{1+y^2}\lambda\, dG_\lambda(y), \quad d\Psi_\lambda(y) = \frac{y^2}{1+y^2}\lambda\, dG_\lambda(y),$$

the convergence theorem 22.1 **D** applies and

$$\alpha_\lambda \to \alpha, \quad \Psi_\lambda \xrightarrow{c} \Psi.$$

In order to use this convergence property, we make appear α_λ and Ψ_λ in the integral representation of $D_\lambda g(x)$. Proceeding formally, it takes the form

$$D_\lambda g(x) = g'(x)\alpha_\lambda + \int h_g(x, y)\, d\Psi_\lambda(y)$$

where

$$h_g(x, y) = \left(g(x+y) - g(x) - \frac{y}{1+y^2} g'(x)\right)\frac{1+y^2}{y^2}$$

is defined by continuity at $y = 0$ to be $h_g(x, 0) = \frac{1}{2}g''(x)$. Always formally, the convergence property then yields

$$D_\lambda g(x) \to Dg(x) = \alpha g'(x) + \int h_g(x, y)\, d\Psi(y).$$

Thus, we are led to consider the operator D on the set $C''_0 \subset C_0$ of all twice differentiable $g \in C_0$ with $g', g'' \in C_0$. Note that C''_0 is dense in C_0 so that the infinitesimal operator on C''_0 determines the Markov semi-group. We use the preceding notation for the Ito-Neveu theorem

C. *The infinitesimal operators D in C_0 of Markov semi-groups corresponding to stationary continuous decomposable processes are of the form*

$$Dg(x) = \alpha g'(x) + \int h_g(x, y)d\Psi(y), \quad g \in C''_0.$$

Proof. If $g \in C''_0$, then the function $h_g(x, y)$ is defined and is bounded and continuous in x and y with $h_g(x, 0) = \frac{1}{2}g''(x)$ and, by elementary computations,

(i) $h_g(x, y) \to h_g(x, 0)$ uniformly in x, as $y \to 0$
(ii) $h_g(x', y) \to h_g(x, y)$ uniformly in y, as $x' \to x$
(iii) $h_g(x, y) \to 0$ uniformly in $|y| \leqq a$ (arbitrary but finite), as $x \to \pm\infty$.

Using these properties, it suffices to go over the foregoing formal argument to verify that it is valid for $g \in C''_0$.

§ 41. SAMPLE CONTINUITY AND DIFFUSION OPERATORS

41.1 Strong Markov property and sample rightcontinuity. The basic results to be established here permit us to recognize the strong Markov property and lead to Dynkin's blending of the semi-group and sample analysis, to be performed in the next subsection.

The first basic theorem is a generalized formulation (by Yushkevitch) of the corollary to 38.4**A**.

A. STRONG MARKOV PROPERTY THEOREM. *Let $X_T = (X_t, t \geqq 0)$ be stationary Markovian with measurable state space $(\mathfrak{X}, \mathcal{S})$, tr.pr. $P_t(x, S)$, and corresponding semi-group $(T_t, t \geqq 0)$ on the space G of bounded $\mathcal{S}$-measurable functions g on $\mathfrak{X}$.*

If X_T and its tr.pr. are Borelian, then X_T is strongly Markovian whenever there exists a topology in $\mathfrak{X}$ such that the subspace C of functions g continuous in this topology is Φ-weakly dense in G and

(i) *the semi-group $(T_t, t \geqq 0)$ leaves C invariant,*

(ii) *the sample functions $X_T(\omega)$, $\omega \in \Omega$, are rightcontinuous in this topology.*

Proof. Let τ be an arbitrary time of X_T and let $s, t \geqq 0$ be arbitrary degenerate times. Since X_T and its tr.pr. are Borelian, because of the strong Markov equivalence theorem 38.4**A**(ii), it suffices to prove the validity for $\tau < \infty$ of the strong semi-group property

$$T_{\tau+t}g = T_\tau T_t g, \quad g \in G.$$

Thus, in what follows, we restrict ourselves to $\Omega^\tau = [\tau < \infty]$ without further comment.

Let $g \in C$. The elementary times $\tau_n = \sum_{k=1}^{\infty} \frac{k}{2^n} I_{\left[\frac{k-1}{2^n} \leq \tau < \frac{k}{2^n}\right]}$ converge to τ from the right. Since X_T is sample rightcontinuous (in the selected topology), $X_{\tau_n+s} \to X_{\tau+s}$, so that $g(X_{\tau_n+s}) \to g(X_{\tau+s})$ and, consequently,

$$T_{\tau_n+s}g(x) = E^x g(X_{\tau_n+s}) \to E^x g(X_{\tau+s}) = g(x).$$

The elementary times τ_n are times of X_T, hence, by 38.4**d**, are its Markov times and $T_{\tau_n+t}g = T_{\tau_n}T_t g$. But $T_t g \in C$ hence

$$T_{\tau+t}g \leftarrow T_{\tau_n+t}g = T_{\tau_n}(T_t g) \to T_\tau T_t g.$$

Since C is Φ-weakly dense in G and Markov endomorphisms commute with Φ-weak passages to the limit, it follows that $T_{\tau+t}g = T_\tau T_t g$ for all $g \in G$. The theorem is proved.

Particular cases. The advantage of the foregoing general formulation lies in the freedom of choice of the topology, whether or not there is already one in $\mathfrak{X}$; the state sets $S \in \mathcal{S}$ are not necessarily topological Borel sets in the topology to be selected. The price of freedom is the requirement that C be dense in G and X_T and its tr.pr. be Borelian. This price is reduced when the freedom of choice is restricted, as follows.

1° *If the tr.pr. is Borelian and the state sets are topological Borel sets in a given metric topology in* $\mathfrak{X}$, *then* X_T *is strongly Markovian whenever* (i) *and* (ii) *hold in a topology at least as fine as the given metric one.*

Note that the recalled corollary enters into this particular case upon selecting the given topology.

To pass from the general formulation **A** to the particular case, let C and C' be the subspaces of those functions belonging to G which are continuous in the new topology $\mathcal{O}$ and in the given metric topology $\mathcal{O}'$, respectively. By hypothesis, $\mathcal{O} \supset \mathcal{O}'$ hence $C \supset C'$. But the state sets $S \in \mathcal{S}$ are now topological Borel sets in the metric topology $\mathcal{O}'$ and the functions $g \in G$ are $\mathcal{S}$-measurable. Therefore C' and *a fortiori* C is Φ-weakly dense in G. Also, the sample functions being rightcontinuous in $\mathfrak{O}$ are rightcontinuous in $\mathfrak{O}'$. Right continuity in t and measurability in ω of $X_t(\omega)$ imply that X_T is Borelian

The finest topology in a space is the trivial discrete topology $\mathfrak{D}$—in which all singletons hence all sets are open. With this topology, only condition (ii) remains:

2° *If the sample functions of X_T are rightcontinuous in the discrete topology in the state space, then X_T is strongly Markovian.*

For, in the discrete topology all functions on $\mathfrak{X}$ are continuous, so that the density and invariance conditions are trivially true. Sample right continuity in $\mathfrak{D}$ implies that, for every $x \in \mathfrak{X}$, $P_h(x, \{x\}) \to 1$ as $h \to 0$, hence $P_h(x, S) \to I(x, S)$ and, by the Kolmogorov equation, $P_{t+h}(x, S) = \int P_t(x, dy)P_h(y, S) \to P_t(x, S)$. Rightcontinuity in t and measurability in ω or x imply that X_T and its tr.pr. are Borelian.

The second basic theorem, essentially due to Dynkin, is as follows.

B. MARKOV TIME THEOREM. *Let stationary Markovian X_T and its semi-group $(T_t, t \geqq 0)$ be Borelian, with extended resolvent R_λ on G, and infinitesimal operators D on G_d and D' on G'_d. If τ is a Markov time of X_T then, whatever be $g \in G$,*

$$E^x\{e^{-\lambda\tau}R_\lambda g(X_\tau)\} = E^x\int_\tau^\infty e^{-\lambda u}g(X_u)\,du$$

and, for $g \geqq 0$, $(-e^{-\lambda(\tau+t)}R_\lambda g(X_{\tau+t}), t \geqq 0)$ is a submartingale on $(\Omega, \mathfrak{A}, P^x)$.

If, moreover, $E^x\tau < \infty$ then, for $g \in G_d$,

$$E^x g(X_\tau) = g(x) + E^x\int_0^\tau Dg(X_t)\,dt$$

and the same is true with D' and G'_d in lieu of D and G_d.

Proof. We have $T_u g(x) = E^x g(X_u)$ and, upon interchanging the integrations,

$$R_\lambda g(x) = \int_0^\infty e^{-\lambda u}T_u g(x)\,du = E^x\int_0^\infty e^{-\lambda u}g(X_u)\,du.$$

Since a Markov time τ of a stationary Markovian X_T is stationary and

$e^{-\lambda\tau} = 0$ on $[\tau = \infty]$, it follows that

$$E^x\{e^{-\lambda\tau}R_\lambda g(X_\tau)\} = E^x\left\{e^{-\lambda\tau}E\left(\int_0^\infty e^{-\lambda u}g(X_{\tau+u})\,du \mid X_\tau\right)\right\}$$

$$= E^x\int_\tau^\infty e^{-\lambda u}g(X_u)\,du,$$

and the first asserted equality is proved. Similarly, for $s \leqq t$,

$$E^x\{e^{-\lambda(\tau+t)}R_\lambda g(X_{\tau+t}) \mid X_{\tau+r}, r \leqq s\}$$
$$= E^x\left\{\int_{\tau+t}^\infty e^{-\lambda u}g(X_u)\,du \mid X_{\tau+r}, r \leqq s\right\}$$

so that, for $g \geqq 0$, the left side is no larger than

$$E^x\left\{\int_{\tau+s}^\infty e^{-\lambda u}g(X_u)\,du \mid X_{\tau+r}, r \leqq s\right\} = e^{-\lambda(\tau+s)}R_\lambda g(X_{\tau+s}),$$

and the submartingale assertion follows.

Finally, by the first asserted equality, for any $g' \in G$,

$$E^x\{e^{-\lambda\tau}R_\lambda g'(X_\tau)\} = R_\lambda g'(x) - E^x\int_0^\tau e^{-\lambda u}g'(X_u)\,du.$$

Therefore, if $g' = (\lambda I - D)g$ with $g \in G_d$ hence $R_\lambda g' = g$, then

$$E^x\{e^{-\lambda\tau}g(X_\tau)\} = g(x) + E^x\int_0^\tau e^{-\lambda u}Dg(X_u)\,du - E^x\int_0^\tau \lambda e^{-\lambda u}g(X_u)\,du.$$

Since $|g| \leqq c < \infty$, the last term is bounded by $cE^x(1 - e^{-\lambda\tau}) \leqq c\lambda E^x\tau$ so that, letting $\lambda \to 0$, if $E^x\tau < \infty$ then

$$E^x g(X_\tau) = g(x) + E^x\int_0^\tau Dg(X_u)\,du.$$

Similarly for D' and G'_d in lieu of D and G_d, and the last assertion is proved. The proof is terminated.

The Markov time theorem applies to all degenerate times $t \geqq 0$ for our stationary Markovian X_T. It applies to all times of X_T, provided X_T satisfies the requirements of theorem **A**. These requirements are of two different kinds: **A**(i) is relative to the corresponding Markov semigroup and, thus, is in terms of the tr.pr. **A**(ii) requires rightcontinuity of sample functions and, thus, is not in terms of the tr.pr. Yet, the primary datum in the investigation and use of Markov property is the tr.pr. This leads us to a search for tr.pr.'s, equivalently, Markov semi-groups, to which correspond sample rightcontinuous X_T.

Given a tr.pr. on a measurable state space $(\mathfrak{X}, \mathcal{S})$, the preliminary question is whether there exists a corresponding X_T. The answer is as follows: According to the existence theorem 38.2**A**, X_T exists when $\mathfrak{X}$ is the Euclidean line and $\mathcal{S}$ is the σ-field of its topological Borel sets. The proof based upon the Tulcea theorem 8.3**A** extends trivially to any finite-dimensional Borel space or Borel subset thereof and, in fact, is valid in the abstract case below.

Let $\mathfrak{X}$ be a separable locally compact metric space with metric "d," and let $\mathcal{S}$ be its σ-field of topological Borel sets; we shall denote by $V_x(\epsilon)$ the sphere $[y: d(x, y) < \epsilon]$. The proof of the separability existence theorem 35.2**B** remains valid and, thus, there exists a separable for closed sets $\tilde{X}_T$ equivalent to X_T, hence Markovian with same tr.pr.

We are now ready for our problem. What follows applies to any state space of the above nature and is couched in corresponding terms. In particular, if x_n goes out of any compact $K \subset \mathfrak{X}$, we write $x_n \to \infty$ and denote by $\mathfrak{X} + \{\infty\}$ the one point compactification of $\mathfrak{X}$ (in the case of a Borel line, "∞" denotes "$\pm\infty$" lumped together). However, as usual, to fix the ideas, we take $(\mathfrak{X}, \mathcal{S})$ to be the Borel line, without further comment. The results are essentially those of Kinney (with modifications due to Blumenthal and Maruyama). We emphasize the methods: the direct method and the martingales method.

From now on, X_T is Borel stationary Markovian and separated for closed sets, with tr.pr. $P_t(x, S)$ and semi-group $(T_t, t \geqq 0)$. As usual, the limits in h are taken as $h \to 0$.

a. SAMPLE LIMITS LEMMA. *Let $T_h g(x) \to g(x)$ for every $x \in \mathfrak{X}$ and every $g \in C_0$.*

If the $T_h g \in C$ or the convergence is uniform in $x \in K$ for every compact K, then almost all sample functions of X_T have at any time at most one left and one right limit value belonging to $\mathfrak{X}$.

If the convergence is uniform in $x \in \mathfrak{X}$, then almost all sample functions of X_T have at any time left and right limits belonging to $\bar{\mathfrak{X}}(= \mathfrak{X}$ or $\mathfrak{X} + \{\infty\}$ according as $\mathfrak{X}$ is compact or not).

Proof. It suffices to give the proof for sample functions which start at any given $x \in \mathfrak{X}$, that is, on the pr. space $(\Omega, \mathcal{A}, P^x)$. Because of separability of X_T, all limits along $[0, \infty)$ may and will be taken along some fixed countable separating set, without further comment. Take $g \geqq 0$.

1° Let $g \in C_0$. By theorem **B**, $(-e^{-\lambda t}R_\lambda g(X_t), t \geqq 0)$ is a semi-martingale. Since it is bounded, the semi-martingales sample continuity theorem applies. Thus, neglecting a null event throughout the rest of the proof, at any time all sample functions $R_\lambda g(X_T(\omega))$ have left and right limits for any g and any rational $\lambda > 0$. Since

$$\lambda R_\lambda g(x) = \int_0^\infty \lambda e^{-\lambda t} T_t g(x)\, dt = \int_0^\infty e^{-u} T_{u/\lambda} g(x)\, du,$$

it follows that, as $\lambda \to \infty$, $\lambda R_\lambda g(x) \to g(x)$ for every x or uniformly in $x \in K$, according as $T_h g(x) \to g(x)$ for every x or uniformly in $x \in K$. Let, say, $t' \uparrow t$; what follows applies as well to $t' \downarrow t$.

2° Suppose that $X_{t'}(\omega)$ has distinct limit values $x' \neq x''$ belonging to $\mathfrak{X}$.

Let the $T_h g \in C$. There is a $g \in C_0$ such that $g(x') \neq g(x'')$, hence $\lambda R_\lambda g(x') \neq \lambda R_\lambda g(x'')$ for a sufficiently large rational λ. This contradicts the existence of a unique limit value of $\lambda R_\lambda g(X_{t'}(\omega))$.

Let the convergence be uniform in $x \in K$ for every compact K. There is a $g \in C_0$ with $g = 1$ on K' and $g = 0$ on K'', where K' and K'' are disjoint compact neighborhoods of x' and x'', respectively. Thus, given $\epsilon > 0$, for a sufficient large rational λ independent of $x \in K' + K''$, $|\lambda R_\lambda g(x) - g(x)| < \epsilon$, while there are two sequences $t'_n, t''_n \uparrow t$ such that $g(x'_n) \to 1$, $g(x''_n) \to 0$, where $x'_n = X_{t'_n}(\omega)$, $x''_n = X_{t''_n}(\omega)$. Therefore, as $n \to \infty$, then $\epsilon \to 0$,

$$1 \leftarrow |g(x'_n) - g(x''_n)| \leqq |\lambda R_\lambda g(x'_n) - \lambda R_\lambda g(x''_n)| + 2\epsilon \to 0.$$

and we reach a contradiction.

Finally, let the convergence be uniform in $x \in \mathfrak{X}$. Upon compactifying $\mathfrak{X}$ as in 40.3**d**, the preceding case applies (a fortiori, if $\mathfrak{X}$ is already compact). The proof is terminated.

The first part of **a** yields (use 40.3**c** for the first hypothesis)

a′. *For every $\epsilon > 0$, let $P_h(x, V_x(\epsilon)) \to 1$ for every $x \in \mathfrak{X}$ and C be invariant, or let $P_h(x, V_x(\epsilon)) \to 1$ uniformly in $x \in K$ for every compact K.*

Then, for almost all $\omega \in \Omega$ and all $t > 0$, as $t' \uparrow t$ and as $t' \downarrow t$, $X_{t'}(\omega)$ converges to some $x \in \mathfrak{X} + \{\infty\}$ or $X_{t'}(\omega)$ has two limit values: $x \in \mathfrak{X}$ and ∞.

When $\mathfrak{X}$ is compact, the point at infinity disappears. Otherwise to eliminate the point at infinity and, thus, have at any time left and right limits belonging to $\mathfrak{X}$, it suffices that the sample functions be bounded. In fact, uniform stability at infinity suffices:

b. *If $P_t(x, K) \to 0$ as $x \to \infty$ uniformly in t on every finite time interval for every compact K, then almost all sample functions of X_T are bounded on every finite time interval.*

If $\mathfrak{X}$ is compact, then the sample functions are necessarily bounded, and we interpret the hypothesis as trivially true.

Proof. Let $t > 0$ and compact $K_n \uparrow \mathfrak{X}$. Suppose there are on $[0, t]$ unbounded sample functions corresponding to an event of pr. δ. It suffices to prove that $P[X_t \in K] \leqq 1 - \delta$ for any given compact K, for then

$$1 - \delta \geqq P[X_t \in K_n] \to P[X_t \in \mathfrak{X}] = 1$$

and $\delta = 0$. Let q_n be the supremum of $P_s(x, K)$ over all $s \leqq t$ and over all $x \notin K_n$; by hypothesis, $q_n \to 0$.

Let $t_0 < \cdots < t_m$ be points of an arbitrary finite set T' in $[0, t]$. Given K_n, let $\tau(\omega) = t_j$ where t_j is the first of these points for which $[X_{t_j}(\omega) \in K_n]$ or $\tau(\omega) = \infty$ if there are no such points. Thus

$$[\tau < \infty] = [X_{t_j} \notin K_n \text{ for some } t_j \in T'], \quad [\tau = \infty] = [X_{t_j} \in K_n \text{ for all } t_j \in T'].$$

The simple time τ is a time of X_T, hence is its Markov time. Therefore,

$$P[\tau < \infty, X_t \in K] = E\{I_{[\tau<\infty]} P_{t-\tau}(X_\tau, K)\} \leqq q_n P[\tau < \infty]$$

and

$$P[X_t \in K] \leqq P[\tau = \infty] + q_n.$$

Apply the standard separability procedure: take a sequence of finite subsets T' of a separating set of $X_{[0,t]}$ converging increasingly to this set. It follows that

$$P[X_t \in K] \leqq P[X_s \in K_n, 0 \leqq s \leqq t] + q_n \leqq 1 - \delta + q_n \to 1 - \delta,$$

and the assertion is proved. The proof is terminated.

Remark. The above method of proof is direct. But we may also use the martingales method, as follows: For positive $g \in C_0$,

$$T_t g(x) = \int_{K+K^c} P_t(x, dy) g(y) \leqq c P_t(x, K) + \sup_{y \notin K} g(y)$$

so that, by hypothesis, $T_t g(x) \to 0$ as $x \to \infty$. It follows that the unbounded sample functions on $[0, a]$ correspond to the event $A = [\inf_{t \leqq a} R_\lambda g(X_t) = 0]$ and, by 36.1**b**(ii) applied to the submartingale formed by $-e^{-\lambda t}R_\lambda g(X_t)$, $\int_A e^{-\lambda a}R_\lambda g(X_a)dP = 0$. The integrand is positive so that $PA = 0$.

Under the condition of **b** and either one of the conditions of **a′**, almost all sample functions of X_T have at any time left and right limits belonging to $\mathfrak{X}$. Since X_T is separable for closed sets, almost all sample functions are continuous except for countable sets of strict jumps. Furthermore, the weakest condition: $P_h(x, V_x(\epsilon)) \to 1$ for every $x \in \mathfrak{X}$ and every $\epsilon > 0$ implies that X_T (that is, every r.f. belonging to it) is right continuous in pr., since for every x

$$P^x[d(X_t, X_{t+h}) \geqq \epsilon] = \int P_t(x, dy)P_h(y, V_y^c(\epsilon)) \to 0,$$

hence for every initial distribution P_0

$$P[d(X_t, X_{t+h}) \geqq \epsilon] = \int P_0(dx)P^x[d(X_t, X_{t+h}) \geqq \epsilon] \to 0.$$

Thus, if S separates X_T and we set $\tilde{X}_t = X_t$ for $t \in S$ and $\tilde{X}_t = \lim_{t' \downarrow t} X_{t'} (t' \in S)$ for $t \notin S$, ($\tilde{X}_t = \lim_{t' \uparrow t} X_{t'}$ for $t \notin S$), then $\tilde{X}_T$ separated for closed sets by S is equivalent to X_T. Hence, $\tilde{X}_T$ has same tr.pr. and almost all its sample functions are right (left) continuous except perhaps at those points of S which are fixed discontinuity points of X_T (where $X_t(\omega)$ may coincide with $X_{t+0}(\omega)$ or with $X_{t-0}(\omega)$ according to the choice of ω); to eliminate this last obstacle for sample right (left) continuity, we may replace the X_t by the $X_{t+0}(X_{t-0})$ and, in fact, a.s. sample continuity may also be imposed upon X_T using 35.3**A** and the particular cases which follow it. However, it will be more instructive to give a direct answer (which overlaps the preceding lemmas) to the problem of sample right (left) continuity.

C. MARKOV SAMPLE RIGHTCONTINUITY THEOREM. *Let a separated for closed sets X_T be stationary Markovian with tr.pr. $P_t(x, S)$. Let either of the two following conditions hold: as $h \to 0$*

(i) $P_h(x, V_x(\epsilon)) \to 1$ *uniformly in $x \in \mathfrak{X}$ for every $\epsilon > 0$*

(ii) $P_h(x, V_x(\epsilon)) \to 1$ *uniformly in $x \in K$ for every compact K and every $\epsilon > 0$, and $P_t(x, K) \to 0$ as $x \to \infty$ uniformly in t on every finite time interval for every compact K.*

Then, X_T is a.s. continuous, almost all its sample functions are continuous except for countable sets of strict jumps, and there exists a separated for closed sets equivalent X_T with almost all sample functions right (left) continuous and left (right) limits for every t.

Proof. Let T' be a time subset with first element t_0. Let $\epsilon > 0$ and $k = 1, \cdots, n$. Let $\omega \in A_n(T') \Leftrightarrow X_T(\omega)$ has n oscillations greater than ϵ on T', that is, there exist n pairs (t'_k, t''_k) of points $t'_k < t''_k \leqq t'_{k+1}$ of T' such that $d(X_{t'_k}(\omega), X_{t''_k}(\omega)) > \epsilon$. For every $\omega \in \Omega$, let $\tau_0(\omega) = t_0$ and let $\tau_k(\omega)$ be the first point t of T' following $\tau_{k-1}(\omega)$ for which $d(X_t(\omega), X_{\tau_{k-1}(\omega)}(\omega)) > \epsilon/2$ or $\tau_k(\omega) = \infty$ if there is no such point. We have

$$A_n(T') \subset B_n(T') = [\tau_0 < \cdots < \tau_n < \infty]$$

and if T' is a finite set, then the τ_k are simple times of X_T, hence are its Markov times.

1° Suppose (i) holds: Given $\epsilon, \delta > 0$, there exists an $h = h(\epsilon, \delta) > 0$ such that $P_{h'}(x, V_x^c(\epsilon/4)) < \delta$ for all $h' \leqq h$ and all $x \in \mathfrak{X}$. Let $I = [a, b]$ be an interval of length h, set $P_\tau^x C = P(C \mid X_\tau = x)$, and let p_n be the supremum of $P_{\tau_0}^x[\tau_0 < \cdots < \tau_n < \infty]$ over all finite sets $T' \subset I$ and over all $x \in \mathfrak{X}$. For any such T'

$$P_{\tau_0}^x[\tau_0 < \cdots < \tau_n < \infty] = E_{\tau_0}^x\{I_{[\tau_0 < \tau_1 < \infty]} P(\tau_1 < \cdots < \tau_n < \infty \mid X_{\tau_1})\}$$

so that $p_n \leqq p_{n-1}p_1$ and, by induction, $p_n \leqq p_1^n$. Furthermore,

$$P_{\tau_0}^x[\tau_0 < \tau_1 < \infty, X_b \in V_x(\epsilon/4)]$$
$$\leqq E_{\tau_0}^x\{I_{[\tau_0 < \tau_1 < \infty]} P(X_b \notin V_{X_{\tau_1}}(\epsilon/4) \mid X_{\tau_1})\} < \delta,$$

so that

$$P_{\tau_0}^x[\tau_0 < \tau_1 < \infty] < \delta + P_{\tau_0}^x[X_b \notin V_x(\epsilon/4)] < 2\delta,$$

hence $p_1 < 2\delta$ and $p_n < (2\delta)^n$. Therefore, upon applying the standard separability procedure,

$$PA_n(I) \leqq PB_n(I) < (2\delta)^n.$$

Thus, taking $n = 1$,

$$P\,[\sup_{t', t'' \in I} d(X_{t'}, X_{t''}) > \epsilon] < 2\delta,$$

and $a, \epsilon, \delta > 0$ being arbitrary, it follows that X_T is a.s. continuous. On the other hand, taking $\delta < \frac{1}{2}$,

$$\sum_{n=1}^{\infty} PA_n(I) \leqq \sum_{n=1}^{\infty} PB_n(I) < \sum (2\delta)^n < \infty$$

so that, by the Borel-Cantelli lemma, on I almost all sample functions of X_T have only a finite number of oscillations greater than ϵ and of consecutive oscillations greater than $\epsilon/2$. Since every finite time interval $[0, c]$ is covered by a finite number of intervals of positive length h, it follows that on $[0, c]$ almost all sample functions of X_T are bounded and have left and right limits. The equivalence assertion results then from the discussion which precedes the theorem.

2° Suppose (ii) holds. In fact, in lieu of its second part, we suppose only the boundedness conclusion of **b**. The preceding argument applies but with restrictions to suitable compacts. First, given ϵ, $\delta > 0$ and $[0, c]$, take $c' > c$ and select a compact K so that the set of sample functions which do not stay in K on $[0, c']$ corresponds to an event of pr. less than δ (thus $P[X_t \in K] > 1 - \delta$ for all $t \leqq c'$); then take a compact K' containing all $V_x(\epsilon/2)$, $x \in K$ (decreasing ϵ if necessary). There exists an $h = h(\epsilon, \delta, K') > 0$ such that $P_{h'}(x, V_x^c(\epsilon/4)) < \delta$ for all $h' \leqq h$ and all $x \in K'$; decrease h if necessary so that $h < c' - c$ and a finite number of intervals $I \subset [0, c']$ covers $[0, c]$.

Let p_n be the supremum of $P_{\tau_0}{}^x[\tau_0 < \cdots < \tau_n < \infty, X_{\tau_0} \in K, \cdots, X_{\tau_{n-1}} \in K]$ over all finite sets $T' \subset I$ and over all $x \in K$. Upon proceeding as in (i), the relation $p_n \leqq p_1{}^n$ is still valid. Similarly, for $x \in K$,

$$P_{\tau_0}{}^x[\tau_0 < \tau_1 < \infty, X_{\tau_0} \in K, X_{\tau_1} \in K', X_b \in V_x(\epsilon/4)] < \delta$$

so that

$$P_{\tau_0}{}^x[\tau_0 < \tau_1 < \infty, X_{\tau_0} \in K]$$
$$< \delta + P_{\tau_0}{}^x[X_{\tau_1} \notin K'] + P_{\tau_0}{}^x[X_b \notin V_x(\epsilon/4)] < 3\delta,$$

hence $p_1 < 3\delta$ and $p_n < (3\delta)^n$.

Upon proceeding as in (i) and using $P[\tau_0 < \tau_1 < \infty] \leqq P[\tau_0 < \tau_1 < \infty, X_{\tau_0} \in K] + P[X_{\tau_0} \notin K]$ it follows that

$$P[\sup_{t',t'' \in I} d(X_{t'}, X_{t''}) > \epsilon] \leqq p_1 P[X_a \in K] + P[X_a \notin K] < 4\delta,$$

and X_T is a.s. continuous on $[0, c]$; since c is arbitrarily large, X_T is a.s. continuous. Similarly, on $[0, c]$, almost all of those sample functions which stay in K have left and right limits; since the others correspond to an event of pr. less than δ and δ is arbitrarily small, the restriction to K can be removed. The equivalence assertion follows. Quasileftcontinuity, page 656, completes this subsection.

41.2 Extended infinitesimal operator. Let $X_T = (X_t, t \geqq 0)$ be stationary Markovian with Borel tr.pr. $P_t(x, S)$ and corresponding Borel semi-group $(T_t, t \geqq 0)$ on the space G of bounded measurable functions on a measurable state space with metric "d." Let D on G_d and D' on G'_d be the strong and the weak infinitesimal operators of the semi-group (we also say "of X_T"). We intend to solve in $D'g$ the integral relation (41.1**B**)

$$E^x g(X_\tau) - g(x) = E^x \int_0^\tau D'g(X_t)\, dt$$

where τ is a Markov time of X_T with $E^x\tau < \infty$, and D' can be replaced by D.

If τ is an ordinary time h, then, dividing both sides by h and letting $h \to 0$, we fall back upon the definition of D'. Yet, this is a "ready-to-wear" approach, in the sense that sample properties of X_T are not taken into account. On the other hand, if we use a random time τ of X_T defined in terms of its sample properties, the approach is fitted to the process—it is "made to measure." The analogy with Riemann versus Lebesgue integration is visible but not adequate. For, τ must be a Markov time of X_T and the approach becomes restricted to strong Markov processes. Furthermore, loosely speaking, the tailored τ would be the time X_T spends in smaller and smaller neighborhoods of x. To be precise, we take for τ the time τ_U that X_T takes to hit the complement of an open neighborhood U of x and let its diameter $|U| \to 0$. According to 38.4**c**, τ_U is a time of X_T when X_T is sample rightcontinuous. Thus from now on, *X_T is strongly Markovian and sample right continuous.*

The integral relation also requires that $E^x\tau_U$ be finite. The simple conditions below will suffice.

Let U, with or without subscripts, denote open sets and, given U_0, set $m(y) = E^y\tau_{U_0}$, $y \in \mathfrak{X}$.

a. *If there exist $s > 0$, $\delta > 0$ such that $P^x[\tau_{U_0} \leqq s] > \delta$ for all $x \in U_0$, then $m(x) = E^x\tau_{U_0} < s/\delta < \infty$ for all $x \in U_0$. If $x \in U \subset U_0$ and $m(x) < \infty$, then $E^x\tau_U < \infty$ and $m(x) = E^x\tau_U + E^x m(X_{\tau_U})$.*

Proof. Let $A_t = [\tau_{U_0} > t]$. Since $A_{t+s} = A_t(A_s)_t$, where $(A_s)_t$ is the translate by t of A_s, and $X_t(\omega) \in U$ for $\omega \in A_t$, the first hypothesis yields

$$P^x A_{t+s} = E^x(I_{A_t} P^{X_t} A_s) < (1 - \delta) P^x A_t.$$

It follows, by induction, that $P^x A_{ns} < (1 - \delta)^n$ and the first assertion is proved by

$$E^x\tau_{U_0} \leqq \sum_{n=0}^{\infty} \int_{ns}^{(n+1)s} P^x[\tau_{U_0} > t]\, dt < s \sum_{n=0}^{\infty} P^x A_{ns} < s/\delta.$$

Since $U \subset U_0$, it follows that $\tau_U \leqq \tau_{U_0}$ and the translate by τ_U of τ_{U_0} is $\tau_{U_0} - \tau_U$. Therefore, by the second hypothesis, $E^x\tau_U \leqq E^x\tau_{U_0} = m(x) < \infty$ and

$$E^x(\tau_{U_0} - \tau_U) = E^x(E^{X_{\tau_U}}\tau_{U_0}) = E^x m(X_{\tau_U}),$$

hence

$$m(x) = E^x\tau_U + E^x(\tau_{U_0} - \tau_U) = E^x\tau_U + E^x m(X_{\tau_U}),$$

The second assertion is proved, and the proof is terminated.

Let U denote open neighborhoods of x, and set

$$\tilde{D}g(x) = \lim_{|U|\to 0} \frac{E^x g(X_{\tau_U}) - g(x)}{E^x\tau_U} \text{ if } E^x\tau_U < \infty \text{ for some } U,$$

$$\tilde{D}g(x) = 0 \text{ if } E^x\tau_U = \infty \text{ for all } U;$$

it suffices to use the first form with the convention that the ratio therein is 0 when $E^x\tau_U = \infty$.

Denote by $\tilde{G}_d$ the set of those functions $g \in G$ for which $\tilde{D}g$ exists and belongs to G. We say that $\tilde{D}$ on $\tilde{G}_d$ is the *extended infinitesimal operator* of X_T or of its semi-group. Note that by **a**, setting $m(z) = E^z\tau_{U_0}$, $U \subset U_0$, and $P_{\tau_U}(x, S) = P^x[X_{\tau_U} \in S]$ hence $T_{\tau_U}g(x) = \int P_{\tau_U}(x, dy)g(y)$, we have

$$\frac{E^x g(X_{\tau_U}) - g(x)}{E^x\tau_U} = \frac{T_{\tau_U}g(x) - g(x)}{E^x\tau_U} = -\frac{\int P_{\tau_U}(x, dy)\{g(y) - g(x)\}}{\int P_{\tau_U}(x, dy)\{m(y) - m(x)\}}$$

with the same convention as above.

According to the integral relation, when some $E^x\tau_U < \infty$ then

$$\text{(C)} \qquad \tilde{D}g(x) = \lim_{|U|\to 0} \frac{1}{E^x\tau_U} E^x\int_0^{\tau_U} D'g(X_t)\, dt, \quad g \in G'_d,$$

in the sense that if either of the sides exists so does the other and both are equal. Thus, if the right side is $D'g(x)$ then $\tilde{D}g(x) = D'g(x)$ and our problem may be enlarged into a search for conditions under which the last equality holds, whether or not $E^x\tau_U$ are finite.

The equality is trivially true when x is an *absorbing* state, that is, almost all sample functions which start at x stay there forever. For then, $E^x\tau_U = \infty$ for all U and $\tilde{D}g(x) = 0$ by definition, while $P^x[X_t = x] = 1$ for all t implies that $T_t g(x) = g(x)$ for all t and $D'g(x) = 0$ by definition.

In fact, we may expect the equality to hold when almost all sample functions stay at x for some positive time. Similarly when $D'g$ is continuous at x provided $E^x\tau_U < \infty$ for some U. For then, given $\epsilon > 0$, $|D'g(y) - D'g(x)| < \epsilon$ for all $y \in U$ with $|U|$ sufficiently small, hence $|D'g(X_t) - D'g(x)| <$ for $t < \tau_U$ and

$$\left| E^x \int_0^{\tau_U} (D'g(X_t) - D'g(x))\, dt \right| < \epsilon E^x \tau_U.$$

However, we are concerned with the process and not with some of its r.f.'s such as those which start at some specified state. Then the required continuity of $D'g$ leads to considering "stable processes" and the required time interval of constancy leads to considering "jump processes." We intend to show that in either case, the term "extended infinitesimal operator" is justified, in the sense that $\tilde{D} \supset D'$, that is, the domain of $\tilde{D}$ under consideration contains that of D'; since always $D' \supset D$, we shall have $\tilde{D} \supset D' \supset D$.

We say that a stationary Markovian process X_T is a *jump process* if, for every $\omega \in \Omega$ and $t \geqq 0$, there exists an $h_0 > 0$ such that $X_t(\omega) = X_{t+h}(\omega)$ for $0 \leqq h < h_0$, equivalently, if X_T is sample rightcontinuous in the discrete topology (in the state space). Since our concern here is with stationary Markov processes, we reserve the term "jump process" for such processes. It follows at once from the definition that for a jump process X_T

$$P_h(x, \{x\}) \to 1, \quad P_h(x, S) \to I(x, S),$$

$$P_{t+h}(x, S) \to P_t(x, S), \quad T_{t+h}g(x) \to T_t g(x),$$

and X_T and its tr.pr. are Borelian. Furthermore, according to 41.1**A**(2°), X_T is strongly Markovian.

Let τ_1 be the first *jump time* of X_T: $\tau_1(\omega)$ is the time $X_t(\omega)$ first hits the complement of the singleton $\{X_0(\omega)\}$—the smallest open neighborhood of the state $X_0(\omega)$ in the discrete topology. The time τ_1 of the stationary strongly Markovian X_T is its stationary Markov time and $A_{t+s} = A_s(A_t)_s$ where $A_t = [\tau_1 > t]$. Since $X_s(\omega) = X_0(\omega)$ for $\omega \in A_s$, it follows that

$$P^x A_{t+s} = E^x(I_{A_s} P^{X_s} A_t) = P^x A_t P^x A_s,$$

and the nonnegative bounded nonincreasing function $p^x(t) = P^x A_t$ obeys the classical functional equation $p^x(t + s) = p^x(t)p^x(s)$, $t, s \geqq 0$. Therefore, $P^x[\tau_1 > t] = e^{-q(x)t}$, $q(x) \geqq 0$, and $q(x) < \infty$ since $q(x) = \infty \Leftrightarrow P^x[\tau_1 > 0] = 0$ (x is *instantaneous*)—contrary to the definition of a jump process. Thus $E^x\tau_1 = 1/q(x) > 0$ and $E^x\tau_1 = \infty \Leftrightarrow q(x) = 0 \Leftrightarrow x$ is absorbing. To summarize

b. *If X_T is a jump process and τ_1 is its first jump time, then*

$$P^x[\tau_1 > t] = e^{-q(x)t}, \quad 0 \leqq q(x) < \infty, \quad x \in \mathfrak{X}, \quad t \geqq 0,$$

and $E^x\tau_1 = 1/q(x) > 0$ is infinite or finite according as x is absorbing or is not.

In the discrete topology, with metric d defined by $d(x, y) = 0$ or 1 according as $x = y$ or $x \neq y$, every set $U \ni x$ is an open neighborhood of x and its diameter $|U| = 0$ or 1 according as U reduces to the singleton $\{x\}$ or has other points besides x. Therefore, we have to set $\tau_U = \tau_1$ in the definition of $\tilde{D}$ and, using **b**, it becomes

$$\tilde{D}g(x) = \frac{E^x g(X_{\tau_1}) - g(x)}{E^x\tau_1} = q(x)\int P_{\tau_1}(x, dy)\{g(y) - g(x)\}.$$

On the other hand, since $X_t = X_0$ for $t < \tau_1$,

$$q(x)E^x\int_0^{\tau_1} D'g(X_t)\,dt = D'g(x), \quad g \in G'_d.$$

Thus

$$\tilde{D}g(x) = D'g(x), \quad g \in G'_d$$

(whether x is absorbing or not), and

A. **Jump processes extension theorem.** *If X_T is a jump process then $\tilde{D} \supset D' \supset D$ and, for every $x \in \mathfrak{X}$,*

$$\tilde{D}g(x) = q(x)\int P_{\tau_1}(x, dy)\{g(y) - g(x)\}, \quad g \in \tilde{G}_d \supset G'_d \supset G_d.$$

Note that for jump processes $\tilde{D}$ is an integral operator, and if the function $q(\cdot)$ is bounded then $\tilde{G}_d = G$.

Stable stationary Markov processes have been defined and investigated in 40.3: Their semi-groups $(T_t, t \geqq 0)$ leave the space C invariant, that is, transform bounded continuous functions g into bounded continuous functions T_tg, for every t. It suffices to consider these semi-groups on the invariant subspace C. Thus, the domains of D, D', $\tilde{D}$ are to be replaced by their intersections C_d, C'_d, $\tilde{C}_d$ with C. In defining $\tilde{D}$, we restricted ourselves to stationary Markov processes with Borel tr.pr. and sample rightcontinuity, for short to *rightcontinuous* processes. We also required strong Markov property, but (as in the case of jump processes) this requirement is superfluous in the case of stable processes;

according to 41.1**A**(1°), rightcontinuous stable processes are strongly Markovian. Since the domains of the corresponding semi-groups are restricted to C, hence all $D'g$ are continuous, the discussion which follows the definition of $\tilde{D}$ shows that then $\tilde{D} \supset D' \supset D$, provided $E^x\tau_U < \infty$ for nonabsorbing states $x \in U$. In fact

c. *Let X_T be stable. If $P^x[X_s \in U] = \delta > 0$ for some $s > 0$ and open U, then $P^y[X_s \in U] > \delta/2$ for some open $U_0 \ni x$ and all $y \in U_0$.*

Let X_T be stable rightcontinuous: If x is nonabsorbing, then $m(y) = E^y\tau_{U_0} < \infty$ for some open $U_0 \ni x$ and all $y \in U_0$.

Proof. Set $V_n = [z: d(z, U^c) \geqq 1/n]$, so that $V_n \uparrow U$ and $P^x[X_s \in V_n] \to P^x[X_s \in U] = \delta$. Thus, for $\delta > 0$ there is an n such that $P^x[X_s \in V_n] > 3\delta/4$. For this n, define the bounded continuous function g by $g(y) = nd(y, U^c)$ or 1 according as $y \notin V_n$ or $y \in V_n$. If X_T is stable, then $g' = T_s g$ is also bounded and continuous and there exists $U_0 \ni x$ such that $g'(y) > g'(x) - \delta/4$ for $y \in U_0$. Then, from

$$g'(x) = \int P_s(x, dy)g(y)$$

and the definition of g, it follows that for $y \in U_0$.

$$P^y[X_s \in U] > g'(y) > g'(x) - \delta/4 > P^x[X_s \in V_n] - \delta/4 > \delta/2.$$

The first assertion is proved.

Let X_T be stable right continuous. If x is nonabsorbing, then there exists an open U with $d(x, U) > 0$ and $P^x[X_s \in U] = \delta > 0$ for some $s > 0$ and, by the first assertion, there exists an open $U_0 \ni x$ that we can take disjoint from U, such that $P^y[X_s \in U] > \delta/2$ for all $y \in U_0$. Therefore, for all $y \in U_0$,

$$P^y[\tau_{U_0} > s] < P^y[X_s \in U_0] < 1 - \delta/2$$

and, by **a**, $m(y) = E^y\tau_{U_0} < \infty$. The second assertion is proved, and the proof is terminated.

Let $m(z) = E^z\tau_{U_0}$, $U \subset U_0$, and recall that $P_{\tau_U}(x, S) = P^x[X_{\tau_U} \in S]$.

B. RIGHTCONTINUOUS STABLE PROCESSES EXTENSION THEOREM. *If X_T is stable right continuous, then $\tilde{D} \supset D' \supset D$ and*

$$\tilde{D}g(x) = -\lim_{|U| \to 0} \frac{\int P_{\tau_U}(x, dy)\{g(y) - g(x)\}}{\int P_{\tau_U}(x, dy)\{m(y) - m(x)\}}, \quad g \in \tilde{C}_d,$$

or $\tilde{D}g(x) = 0$ *according as* x *is nonabsorbing or absorbing; if the state space is compact, then* $\tilde{D} = D' = D$.

Proof. The inclusion and limit assertions result from the discussion which follows the definition of $\tilde{D}$ and the restriction of the semi-group on C.

To prove the last relation, note that, by 40.3**B** and **d**, $D' = D$ and $C'_c = C_c = C$. Since $\tilde{D} \supset D'$, it suffices to prove that $\tilde{D} \subset D'$, that is, $\tilde{C}_d \subset C'_d$. Let $g \in \tilde{C}_d$ and set

$$\tilde{g} = (I - \tilde{D})g, \quad g' = R_1\tilde{g} = \int_0^\infty e^{-t}T_t\tilde{g}\,dt.$$

Since $\tilde{g} \in C = C'_c$, it follows that $g' \in C'_d$ and $\tilde{g} = (I - D')g'$. Thus, $\tilde{g} = (I - \tilde{D})g'$ because of $\tilde{D} \supset D'$. Therefore,

$$(I - \tilde{D})f = 0, \quad f = g - g'.$$

$\mathfrak{X}$ being compact, f attains its supremum on $\mathfrak{X}$ for some x_0 and the defining relation for $\tilde{D}$ implies that $\tilde{D}f(x_0) \leqq 0$, hence $f \leqq 0$; similarly, $-f \leqq 0$. Thus, $f = 0$, that is, $g = g' \in C'_d$. The proof is terminated.

We say that X_T is a *continuous stable* process if it is stable rightcontinuous and also sample leftcontinuous. The sample functions of a continuous stable process being continuous, the time τ_U is the time X_T first reaches the closed set U^c. Therefore, X_{τ_U} belongs to the boundary U' of U and the integrals in **B** can be taken over U' only:

$$\tilde{D}g(x) = -\lim_{|U| \to 0} \frac{\int_{U'} P_{\tau_U}(x, dy)\{g(y) - g(x)\}}{\int_{U'} P_{\tau_U}(x, dy)\{m(y) - m(x)\}}.$$

This expression is similar to the one which gives the ordinary Laplacian operator in terms of averages on spherical surfaces, except that there the averaging is with respect to a uniformly distributed measure. This leads to considering the foregoing operator as a generalized elliptic differential operator of second order or, in physical terms, a (*general*) *diffusion operator*, to be denoted by $\mathfrak{D}$. According to the convention made, we have, for every $x \in \mathfrak{X}$ and $g \in \tilde{C}_d$,

$$\mathfrak{D}g(x) = \lim_{|U| \to 0} \frac{1}{E^x\tau_U} \int_{U'} P_{\tau_U}(x, dy)\{g(y) - g(x)\}.$$

It follows that if $g = g'$ in a neighborhood of x then $\mathfrak{D}g(x) = \mathfrak{D}g'(x)$, and if g attains a relative minimum at x, then $\mathfrak{D}g(x) \geqq 0$. The foregoing terminology is further justified because, on sufficiently smooth functions, $\mathfrak{D}$ can be written as an ordinary elliptic differential operator, as follows.

C. DIFFUSION OPERATORS THEOREM. *Let $\mathfrak{D}$ be a diffusion operator and let $g_k, g_jg_k, j = 1, \cdots, n$, belong to the domain $\tilde{C}_d$ of $\mathfrak{D}$, in some neighborhood of a state x.*

If $f(y_1, \cdots, y_n)$ is twice continuously differentiable in a neighborhood of $(g_1(x), \cdots, g_n(x))$, then $\mathfrak{D}f(g_1(x), \cdots, g_n(x))$ exists and equals

$$\sum_{k=1}^{n} a_k \frac{\partial f}{\partial g_k} + \sum_{j,k=1}^{n} b_{jk} \frac{\partial^2 f}{\partial g_j \partial g_k}$$

where the derivatives are taken at $(g_1(x), \cdots, g_n(x))$,

$$a_k = \mathfrak{D}(g_k - g_k(x))(x), \quad b_{jk} = \mathfrak{D}(g_j - g_j(x))(g_k - g_k(x))(x)$$

and the b_{jk} form a nonnegative type matrix.

Proof. We have

$$f(g_1, \cdots, g_n) - f(g_1(x)), \cdots, g_n(x)) = \sum_k \frac{\partial f}{\partial g_k} h_k + \sum_{j,k} \frac{\partial^2 f}{\partial g_j \partial g_k} (1 + \delta_{jk}) h_j h_k$$

where the derivatives are taken at $(g_1(x), \cdots, g_n(x))$, $h_k = g_k - g_k(x)$ and $\delta_{jk}(x') = \delta_{jk}(g_1(x'), \cdots, g_n(x')) \to 0$ as the $h_k \to 0$, hence as $x' \to x$. Thus, the ratio in the defining expression of $\mathfrak{D}$ can be written as a sum of three terms: As $|U| \to 0$, the first term

$$\sum_k \frac{\partial f}{\partial g_k} \frac{1}{E^x \tau_U} \int_{U'} P_{\tau_U}(x, dy) h_k(y) \to \sum_k \frac{\partial f}{\partial g_k} a_k,$$

the second term

$$\sum_{j,k} \frac{\partial^2 f}{\partial g_j \partial g_k} \frac{1}{E^x \tau_U} \int_{U'} P_{\tau_U}(x, dy) h_j(y) h_k(y) \to \sum_{j,k} \frac{\partial^2 f}{\partial g_j \partial g_k} b_{jk},$$

and the squares of the summands of the third term

$$\left| \frac{1}{E^x \tau_U} \int_{U'} P_{\tau_U}(x, dy) h_j(y) h_k(y) \delta_{jk}(y) \right|^2$$

$$\leqq \max_{y \in U'} |\delta_{jk}(y)|^2 \left| \frac{1}{E^x \tau_U} \int_{U'} P_{\tau_U}(x, dy) h^2{}_j(y) \right| \left| \frac{1}{E^x \tau_U} \int_{U'} P_{\tau_U}(x, dy) h^2{}_k(y) \right|$$

converge to 0, since the first factor converges to zero and the two others are bounded. Finally

$$\sum_{j,k} b_{jk}\lambda_j\lambda_k = \mathfrak{D}(\sum_k \lambda_k h_k(x))^2 \geqq 0,$$

since the function on which $\mathfrak{D}$ operates attains a relative minimum at x. The proof is terminated.

To conclude this general discussion of the extended infinitesimal operator, let us mention that whenever it is a true extension of the weak infinitesimal operator, the last one is obtained by supplying further information. This information usually takes the form of boundary conditions. Thus, loosely speaking, the extended infinitesimal operator describes the behavior of the process *before* it reaches the boundary of the domain on which it is considered.

41.3 One-dimensional diffusion operator. In the one-dimensional case, the diffusion operator takes a specific differential form, and we proceed to establish this fundamental result of Feller, following Dynkin.

Let X_T be continuous stable with a one-dimensional interval state space $\mathfrak{X} = [\alpha, \beta]$ and the σ-field $\mathcal{S}$ of topological Borel sets in it. We take $U = (x_1, x_2) \ni x$; its boundary consists of the two endpoints to which correspond the pr.'s $p_{x_1} = P^x[X_{\tau_{(x_1,x_2)}} = x_1]$ and $p_{x_2} = P^x[X_{\tau_{(x_1,x_2)}} = x_2]$. Let $g \in \tilde{C}_d$. We know that

I. *If x is absorbing then* $\mathfrak{D}\, g(x) = 0$.

If x_0 is not absorbing, we can select $(x_1', x_2') \ni x_0$ so that, for all $x \in (x_1', x_2')$, $m(x) = E^x\tau_{(x_1',x_2')}$ is finite and

$$(\mathfrak{D}_1) \qquad \mathfrak{D}g(x) = -\lim_{\substack{x_1 \to x-0 \\ x_2 \to x+0}} \frac{p_{x_1}(g(x_1) - g(x)) + p_{x_2}(g(x_2) - g(x))}{p_{x_1}(m(x_1) - m(x)) + p_{x_2}(m(x_2) - m(x))}.$$

Formally, upon applying L'Hospital's rule, we obtain

$$\mathfrak{D}g(x) = -\frac{p(x)}{p(x)m''(x) + 2p'(x)m'(x)}g''(x) - \frac{2p'(x)}{p(x)m''(x) + 2p'(x)m'(x)}g'(x)$$

so that $\mathfrak{D}$ appears as a differential operator of second order (which may degenerate). In fact, we shall establish that $\mathfrak{D}$ is a generalized differential operator. For this purpose, we classify the states as follows:

x is a *right passage point* or a *left passage point* if $P^x[X_t > x] > 0$ or

$P^x[X_t < x] > 0$ for some t, and x is a *passage point* if it is a right and a left passage point. Note that if x is neither a right nor a left passage point, then it is absorbing.

II. *At one-sided passage points, $\mathfrak{D}$ is a generalized differential operator of first order.*

If x is a right but not a left passage point then $p_{x_1} = 0$ and

$$\mathfrak{D}g(x) = -\lim_{x_2 \to x+0} \frac{g(x_2) - g(x)}{m(x_2) - m(x)} = -D_m{}^+g(x).$$

If x is a left but not a right passage point then $p_{x_2} = 0$ and

$$\mathfrak{D}g(x) = -\lim_{x_1 \to x-0} \frac{g(x_1) - g(x)}{m(x_1) - m(x)} = -D_m{}^-g(x).$$

To find the differential form in the case of (two-sided) passage points, we require more information about the properties of such points. The arguments to be used are similar to those which preceded the one-dimensional case and, therefore, will be shortened.

Let $\tau_y(\omega) = \inf [t: X_t(\omega) = y]$ if this set is not empty and $\tau_y(\omega) = \infty$ otherwise. Thus, $\tau_y(\omega) = \tau_{[\alpha,y)}$ or $\tau_{(y,\beta]}$ according as $X_0(\omega) < y$ or $X_0(\omega) > y$; note that the subscript sets are open in $\mathfrak{X} = [\alpha, \beta]$. Denote by $p(x, y) = P^x[\tau_y < \infty]$ the pr. starting at x to reach y in a finite time, and denote by $p(x, y, z) = P^x[\tau_y < \tau_z]$ the pr. starting at x to reach y before reaching z. Note that $p(x, y, z) + p(x, z, y) \leqq 1$ and that x is not a right passage point if and only if $p(x, y) = 0$ for all $y > x$.

a. (i) *If $a < x < y < b$ or $a > x > y > b$, then*

$$p(y, a) = p(y, x)p(x, a), \quad p(y, a, b) = p(y, x, b)p(x, a, b),$$
$$p(x, a, b) = 0 \Rightarrow p(x, a) = 0.$$

(ii) *If x is a right passage point, then there exists $(x', x'') \ni x$ such that $p(x', x'') > 0$, and if all $x \in [a, b]$ are right passage points then $p(a, b) > 0$.*

Proof. 1° Let, say, $a < x < y < b$. Since

$$[X_0 = y, \tau_a < \infty] = [X_0 = y, \tau_x < \infty] \cap [\tau_a < \infty]_{\tau_x}$$

and $X_{\tau_x} = x$, it follows that

$$P^y[\tau_a < \infty] = E^y(I_{[\tau_x < \infty]}P^{X_{\tau_x}}[\tau_a < \infty]) = P^x[\tau_a < \infty]P^y[\tau_x < \infty].$$

This proves the first equality in (i), and similarly for the second equality.

Let $a < x' < x < x'' < b$. Let S' be the set of states $z' < x'$ and let

S'' be the set of states $z'' > x''$. We say that $X_T(\omega)$ crosses exactly n times from S'' into S' in time t if there exist n and only n pairs (t''_k, t'_k) with $0 \leqq t''_1 < t'_1 < t''_2 < \cdots < t'_n \leqq t$ such that $X_{t'_k} \in S'$ and $X_{t''_k} \in S''$. Let $\omega \in A \Leftrightarrow X_0(\omega) = x$ and $\tau_a(\omega) < \infty$, and let $\omega \in A_n \Leftrightarrow \omega \in A$ and $X_T(\omega)$ crosses exactly n times from S'' into S' in time $\tau_a(\omega)$, so that $A = \bigcup_{n=0}^{\infty} A_n$. Let $\tau_n(\omega) = \inf[t: X_t(\omega) = x$ and $X_T(\omega)$ crosses exactly n times from S'' into S' in time $t]$ if this set is not empty and $\tau_n(\omega) = \infty$ otherwise, so that $A_n \subset (A_0)_{\tau_n}$. If $p(x, a, b) = P^x A_0 = 0$, then, from $X_{\tau_n} = x$, it follows that $P^x(A_0)_{\tau_n} = 0$, hence $P^x A_n = 0$ and $P^x A = 0$. This proves the last assertion in (i).

2° If x is a right passage point, that is, $P^x[X_t > x] > 0$ for some t, then there exists an $x'' > x$ such that $P^x[X_t > x''] > 0$, hence there exists a neighborhood $(x - \epsilon, x + \epsilon)$ such that $P^y[X_t > x''] > 0$ for all $y \in (x - \epsilon, x + \epsilon)$. Therefore, $p(x', x'') > 0$ for any $x' \in (x - \epsilon, x)$. This proves the first assertion in (ii). If all $x \in [a, b]$ are right passage points then, by what precedes, there exist intervals $(x', x'') \ni x$ such that $p(x', x'') > 0$. These open intervals cover $[a, b]$, hence a finite number of them (x'_k, x''_k), $k \leqq n$, covers $[a, b]$. It follows from the first equality in (i) that $p(a, b) > 0$ and the second assertion in (ii) is proved. The proof is terminated.

b. *Let $p(a, b) > 0$. If $a < b$ then all $x \in [a, b)$ are right passage points, all $E^x\tau_{(a,b)}$ are finite, and $p(x, b, a) \to 1$ as $x \to b - 0$; if $a > b$ then $p(x, b, a) \to 1$ as $x \to b + 0$.*

Proof. Let $x \in [a, b)$. If x is not a right passage point then, by **a**, $p(a, b) = p(a, x)p(x, b) = 0$ contrary to the hypothesis. The first assertion is proved.

Since for $X_0 = y$ with $y \in [a, x]$

$$[\tau_b > t]_{\tau_x} \subset [\tau_b > t], \quad X_{\tau_x} = x, \quad \tau_{(a,b)} = \min(\tau_a, \tau_b) \leqq \tau_b,$$

it follows that

$$P^x[\tau_{(a,b)} > t] \leqq P^x[\tau_b > t] = E^a(P^{X_{\tau_x}}[\tau_b > t]) \leqq P^a[\tau_b > t],$$

where, letting $t \to \infty$,

$$P^a[\tau_b > t] \to P^a[\tau_b = \infty] = 1 - p(a, b) < 1.$$

Therefore, for some t,

$$P^x[\tau_{(a,b)} > t] \leqq P^a[\tau_b > t] = 1 - \delta < 1$$

and, by 41.2**a**, $E^x\tau_{(a,b)} < \infty$. This proves the second assertion.

Let $a < x < x_1 < \cdots < x_n \uparrow b$ and $X_0 = x$, hence $\tau_{x_1} < \cdots < \tau_{x_n} \uparrow \tau \leqq \tau_b$. Set $A = [\tau_b < \tau_a]$ and $A_n = [\tau_{x_n} < \tau_a]$ so that $A \subset \bigcap A_n$. Assume there is an $\omega \in \bigcap A_n - A$. Either $\tau(\omega) < \infty$ or $\tau(\omega) = \infty$. But, by sample continuity, on $[\tau < \infty]$, $X_{\tau_b} = b \leftarrow x_n = X_{\tau_n} \to X_\tau$, hence $\tau = \tau_b$. Therefore $\tau(\omega) < \infty$ implies $\omega \in A$ which contradicts the above assumption. Thus $\tau(\omega) = \infty$ so that $\tau_a(\omega) \geqq \tau_b(\omega) \geqq \tau(\omega) = \infty$, since $\omega \notin A$. It follows that $\tau_{(a,b)} = \infty$ while, by **a**, $p(a, b) > 0$ implies $E^x\tau_{(a,b)} < \infty$. Therefore $P^x(\bigcap A_n - A) = 0$,

$$p(x, x_n, a) = P^x A_n \to P^x A = p(x, b, a) = p(x, x_n, a)p(x_n, b, a)$$

and $p(x, b, a) > 0$, since otherwise $p(x, b) = 0$ and $p(a, b) = p(a, x)p(x, b) = 0$. It follows that $p(x_n, b, a) \to 1$, and similarly when $b < a$. This proves the last assertion. The proof is terminated.

c. *Let $p(a, b)p(b, a) > 0$ and let x vary over $[a, b]$.*

(i) *The function $p(x) = p(x, b, a)$ is continuous and increasing with $p(x) \to 0$ as $x \to a + 0$ and $p(x) \to 1$ as $x \to b - 0$, and*

$$p(x, x_1, x_2) = (p(x_2) - p(x))/(p(x_2) - p(x_1)), \quad a < x_1 < x < x_2 < b.$$

(ii) *The function $-m(x) = -E^x\tau_{(a,b)}$ is continuous and, with respect to the function $p(x)$, it is convex and has an increasing left continuous left derivative $s^-(x) = -D_p^-m(x)$ and an increasing right continuous right derivative $s^+(x) = -D_p^+m(x)$ which coincide on their continuity set $C(s)$.*

Note that, if the hypothesis holds then all points of (a, b) are passage points and if all points of $[a, b]$ are passage points then the hypothesis holds.

Proof. We use **a** without further comment.

1° Let $a < x < y < b$, so that

$$p(x) = p(x, b, a) = p(x, y, a)p(y, b, a) = p(x, y, a)p(y),$$

hence $p(x) \leqq p(y)$. If $p(x) = p(y)$ then either $p(x) = 0$ or $p(x, a, y) = 1 - p(x, y, a) = 0$. In the last case, $p(x, a) = 0$ and $p(b, a) = p(b, x)p(x, a) = 0$ contrary to the hypothesis. In the first case, $p(x, b) = 0$ and $p(a, b) = p(a, x)p(x, b) = 0$ contrary to the hypothesis. Thus $p(x) < p(y)$, and the first assertion in (i) is proved.

The limit assertions in (i) result from the corresponding ones in **b**, upon using for the first one the relation $p(x, a, b) = 1 - p(x)$. Similarly, as $x \to y - 0$, $p(x, y, a) \to 1$, hence

$$p(x) = p(x, y, a)p(y) \to p(y),$$

the function $p(x)$ is left continuous and, analogously, the function $1 - p(x) = p(x, a, b)$ is rightcontinuous. This proves the continuity assertion in (i). Since

$$p(x, x_1, x_2) = p(x, a, x_2)/p(x_1, a, x_2) = (1 - p(x, x_2, a))/(1 - p(x_1, x_2, a))$$

and

$$p(x, x_2, a) = p(x)/p(x_2), \quad p(x_1, x_2, a) = p(x_1)/p(x_2),$$

the last assertion in (i) follows.

2° If $a \leqq x_1 < x < x_2 \leqq b$ then

$$m(x) - p(x, x_1, x_2)m(x_1) - p(x, x_2, x_1)m(x_2) = E^x\tau_{(x_1, x_2)} > 0$$

so that, by (i),

$$m(x) > \frac{p(x_2) - p(x)}{p(x_2) - p(x_1)} m(x_1) + \frac{p(x) - p(x_1)}{p(x_2) - p(x_1)} m(x_2)$$

and the function $-m(x)$ is convex with respect to the function $p(x)$. The remaining assertions of (ii) result from this convexity, upon transforming it into ordinary convexity as follows. By (i), the function $p(x)$ has an inverse function $q(y)$. Set $n(y) = -m(q(y))$ and note that if $0 \leqq y_1 < y < y_2 \leqq 1$ then $a \leqq x_1 = q(y_1) < x < x_2 = q(y_2) \leqq b$. Thus

$$n(y) < \frac{y_2 - y}{y_2 - y_1} n(y_1) + \frac{y - y_1}{y_2 - y_1} n(y_2)$$

and the function $n(y)$ bounded from above (by 0) is convex. Since

$$s^-(x) = -D_p{}^-m(x) = D_y{}^-n(y), \quad s^+(x) = -D_p{}^+m(x) = D_y{}^+n(y),$$

the assertions follow from the corresponding properties of the function $n(x)$. The proof is terminated.

A. One-dimensional diffusion operator theorem. *Let X_T be continuous stable and let all $x \in [a, b]$ be passage points. Let $g \in \tilde{C}_d$ and*

$$m(x) = E^x\tau_{(a,b)}, \quad s^-(x) = -D_p{}^-m(x), \quad s^+(x) = -D_p{}^+m(x).$$

On (a, b), the function $D_p{}^-g(x)$ exists and is left continuous, the function $D_p{}^+g(x)$ exists and is right continuous, they coincide on $C(s)$, and

$$\mathfrak{D}g(x) = D_{s^-}{}^+D_p{}^-g(x) = D_{s^+}{}^-D_p{}^+g(x).$$

Proof. We apply **c** without further comment. Set

$$g(x, y) = \frac{g(y) - g(x)}{p(y) - p(x)}, \quad m(x, y) = \frac{m(y) - m(x)}{p(y) - p(x)}.$$

Relation ($\mathfrak{D}_1$) becomes

$$\text{(1)} \qquad \mathfrak{D}g(x) = -\lim_{\substack{x_1 \to x-0 \\ x_2 \to x+0}} \frac{g(x, x_2) - g(x, x_1)}{m(x, x_2) - m(x, x_1)}.$$

Since

$$\lim_{\substack{x_1 \to x-0 \\ x_2 \to x+0}} (m(x, x_2) - m(x, x_1)) = \lim_{x_2 \to x+0} m(x, x_2) - \lim_{x_1 \to x-0} m(x, x_1) = s^-(x) - s^+(x)$$

it follows that

$$\lim_{x_1 \to x-0} g(x, x_1) - \lim_{x_2 \to x+0} g(x, x_2) = D_p{}^-g(x) - D_p{}^+g(x)$$

exists,

$$\text{(2)} \qquad (s^+(x) - s^-(x))\mathfrak{D}g(x) = D_p{}^+g(x) - D_p{}^-g(x)$$

and the last difference reduces to zero on $C(s)$. Set

$$\begin{aligned} G(y) &= g(y) - g(x) - D_p{}^-g(x)(p(y) - p(x)) \\ &= (p(y) - p(x))(g(x, y) - D_p{}^-g(x)), \\ M(y) &= -m(y) + m(x) - s^-(x)(p(y) - p(x)) \\ &= (p(y) - p(x))(-m(x, y) - s^-(x)). \end{aligned}$$

Note that $M(x) = 0$, for $y > x$

$$D_p{}^-M(y) = s^-(y) - s^-(x) > 0, \quad D_p{}^+M(y) = s^+(y) - s^-(x) > 0,$$

and, upon letting $x_1 \to x - 0$ in (1),

$$\mathfrak{D}g(x) = \lim_{y \to x+0} \frac{G(y)}{M(y)}.$$

To obtain the asserted form of $\mathfrak{D}g(x)$, it will suffice to prove that L'Hospital's rule applies, as follows. First, we show that the limits of $h^+(y)$ and $h^-(y)$ as $y \to x + 0$ exist and are the same, upon setting

$$h^+(y) = D_p{}^+G(y)/D_p{}^+M(y), \quad h^-(y) = D_p{}^-G(y)/D_p{}^-M(y).$$

Suppose that $\liminf_{y \to x+0} h^+(y) < \limsup_{y \to x+0} h^+(y)$ so that there exist distinct c', c'' which lie between these limits. If c is either c' or c'' and $f(y) = G(y) - cM(y)$, then $D_p{}^+f = D_p{}^+G - cD_p{}^+M$ changes signs in (x, y) for any $y > x$. Thus, there exist sequences y_n, $y'_n \to x + 0$ such that f attains relative maxima at the y_n and relative minima at the y'_n.

Therefore, $\mathfrak{D}f(y_n) \leqq 0$, $\mathfrak{D}f(y'_n) \geqq 0$ and, from $\mathfrak{D}f = \mathfrak{D}g - c$ it follows that $\mathfrak{D}g(y_n) \leqq +c$, $\mathfrak{D}g(y'_n) \geqq +c$, hence $\mathfrak{D}g(x) = +c$. Since c is either of two distinct numbers c', c'', this is impossible. Thus,

$$\lim_{y \to x+0} h^+(y) = \lim_{y \to x+0} \frac{D_p{}^+g(y) - D_p{}^-g(x)}{s^+(y) - s^-(x)}$$

exists, and similarly

$$\lim_{y \to x+0} h^-(y) = \lim_{y \to x+0} \frac{D_p{}^-g(y) - D_p{}^-g(x)}{s^-(y) - s^-(x)}$$

exists. Since $D_p{}^-g(y) = D_p{}^+g(y)$ on the everywhere dense set $C(s)$, the two limits coincide. Set now

$$F(z) = G(z)M(y) - G(y)M(z), \quad z \in [x, y]$$

and note that $F(x) = F(y) = 0$ so that F attains either its maximum or its minimum at some $z' \in (x, y)$. In the first case, $D_p{}^-F(z') \geqq 0 \geqq D_p{}^+F(z')$, hence

$$h^+(z') \leqq \frac{G(y)}{M(y)} \leqq h^-(z')$$

and in the second case, these inequalities are reversed. As $y \to x + 0$, hence $z' \to x + 0$, the extreme terms converge to the same limit while the middle term converges to $\mathfrak{D}g(x)$. Therefore,

$$\begin{aligned}\mathfrak{D}g(x) &= \lim_{y \to x+0} \frac{D_p{}^+g(y) - D_p{}^-g(x)}{s^+(y) - s^-(x)}\\ &= \lim_{y \to x+0} \frac{D_p{}^-g(y) - D_p{}^-g(x)}{s^-(y) - s^-(x)} = D_{s^-}{}^+D_p{}^-g(x).\end{aligned}$$

The function $s^+(x)$ being right continuous, it follows that

$$D_p{}^+g(x + 0) - D_p{}^-g(x) = (s^+(x) - s^-(x))\mathfrak{D}g(x)$$

and, taking into account (2), $D_p{}^+g(x + 0) = D_p{}^+g(x)$, that is, $D_p{}^+g$ is right continuous. Similarly for the remaining assertions. The proof is terminated.

Note that the passage points form an open set $\sum_j (a_j, b_j)$ and what precedes applies to any $[a, b] \subset (a_j, b_j)$. Furthermore, the foregoing form of $\mathfrak{D}g$ can be rewritten as follows:

III. *If the $x \in [a, b]$ are passage points, then*

$$\mathfrak{D}g(x) = D_s D_p g(x), \quad x \in C(s)$$

$$\mathfrak{D}g(x) = \frac{D_p g(x+0) - D_p g(x-0)}{s(x+0) - s(x-0)}, \quad x \notin C(s).$$

Together, I, II, and III determine the one-dimensional diffusion operator.

COMPLEMENTS AND DETAILS

Unless otherwise stated, the state space $(\mathfrak{X}, \mathfrak{S})$ is a separable locally compact metric space, $\mathfrak{S}$ is its σ-field of topological Borel sets, and $P_t(x, S)$, $x \in \mathfrak{X}$, $S \in \mathfrak{S}$, $t > 0$, is a stationary tr.pr. with $P_t(x, \mathfrak{X}) \leqq 1$.

1. $P_t(x, \mathfrak{X})$ is nonincreasing as $t > 0$ increases.

2. Let $\tau \geqq 0$ be a r.v. not necessarily a.s. finite. Let X_t be a r.v. defined on $[t < \tau]$. $X_T = (X_t, t \geqq 0)$ is a r.f. of "lifetime τ."

If $P_t(x, \mathfrak{X}) \equiv 1$, then there exists a Markov r.f. of infinite lifetime with the given tr.pr. and an arbitrary initial distribution on $\mathfrak{S}$. If $P_t(x, \mathfrak{X}) \leqq 1$ and $P_h(x, \mathfrak{X}) \to 1$ as $h \to 0$ for every x, then there exists a Markov r.f. X_T of possibly finite lifetime with positive pr., with the given tr.pr. and arbitrary initial distribution on $\mathfrak{S}$:

Add an isolated point at infinite "∞" and determine a tr.pr. $P'_t(x', S')$, where the sets S' are sets S and sets $S + \{\infty\}$, so that it coincides with $P_t(x, S)$ for $x' = x$ and $S' = S$ and $P'_t(x', \{\infty\}) = 1$ or $1 - P_t(x, \mathfrak{X})$ according as $x' = \infty$ or $x' \neq \infty$. Construct X_T as above. Complete $\mathfrak{B}(X_s, s \leqq t)$. Define $\tau(\omega)$ as the supremum of all rational r for which $X_r(\omega) \neq \infty$. "Curtail" X_T to have lifetime τ upon replacing the domain of X_t by $[t < \tau]$ and then replace $X_t(\omega) = \infty$ by an arbitrary $x_0 \in \mathfrak{X}$.

3. For all (x, S), if $P_h(x, S) \to I(x, S)$ as $h \to 0$, then $\{1 - P_h(x, \mathfrak{X})\}/h$ converges to a finite limit: Introduce the point at infinity and $P'_t(x', S')$ and apply section 39.1.

4. Apply section 39 to Markov processes with a finite number of states under the continuity condition and to Markov processes with a countable number of states under the uniform continuity condition.

5. Let $P_t(x, S)$, $t > 0$, be $\mathfrak{S} \times \mathfrak{I}$-measurable ($\mathfrak{I}$ is the σ-field of Lebesgue sets in $(0, \infty)$).

a) $P_t(x, S)$ is continuous in t if and only if there exists a finite measure μ on $\mathfrak{S}$ such that $P_t(x, S)$ is μ-continuous in S:

If $P_t(x, S)$ is continuous in t, take $\mu S = \int_0^\infty e^{-t} P_t(x, S)\, dt$. If μ exists, given t and S note that for $0 < \epsilon < t < t'$ and $|h_n| < \epsilon$

$$\iint_\epsilon^t | P_{s+h_n}(x, S) - P_s(x, S) |\, ds\mu(dx) = \int (dx) \int_\epsilon^t | P_{s+h_n}(x, S) - P_s(x, S) |\, ds,$$

where the inner integral converges to zero as $h_n \to \infty$ and, for some subsequence h'_n, $P_{s+h'_n}(x, S) \to P_s(x, S)$ for some $s \in (\epsilon, t)$ a.e. in μ, hence a.e. in $P_{t-s}(x, \cdot)$.

Thus,

$$P_{t+h'_n}(x, S) = \int P_{t-s}(x, dy) P_{s+h'_n}(y, S) \rightarrow P_t(x, S).$$

What if the state space is the Borel line and $P_t(x, S) = 1$ or 0 according as $x + t$ belongs or not to S?

b) Let $P_t(x, S)$ be right continuous in t. Then it is continuous in t: Note that to find μ finite such that $P_t(x, S)$ is μ-continuous in S, it suffices to know that $P_t(x, S)$ is t-measurable and that if $P_t(x, S) = 0$ for a.e. t, then $P_t(x, S) \equiv 0$.

If $P_t(x, S) = \int_S P_t(x, y)\mu(dy)$ where μ is σ-finite and $p_t(x, y)$ is (x, y)-measurable, then $P_t(x, S)$ is continuous in t.

c) If $\lim_{h \downarrow 0} P_h(x, S)$ exists for all (x, S), then $P_t^+(x, S) = \lim_{h \downarrow 0} P_{t+h}(x, S)$ exists, is a tr.pr. continuous in t, and coincides with $P_t(x, S)$ except for countably many values of t: Note that $P_t^+(x, S)$ is right continuous in t and that $P_t(x, S)$ has at most countably many discontinuity points in t.

6. Let X_T be a Poisson r.f. with the sample functions selected to be left continuous, and let $\tau(\omega)$ be the infimum of those t for which $X_t(\omega) = 1$. Then the Markov r.f. X_T is stable and τ is a time of X_T but is not its Markov time.

7. Let $\alpha_n, \beta_n > 0$, $p_n = \alpha_n/(\alpha_n + \beta_n)$, $q_n = \beta_n/(\alpha_n + \beta_n)$. Let $X_n = (X_n(t), t \geqq 0)$, $n = 1, 2, \cdots$, be independent Markov r.f.'s with two states 0 and 1, $X_n(0) = 0$, and

$$P(X_n(t + h) = 1 \mid X_n(t) = 0) = \alpha_n h + o(h),$$

$$P(X_n(t + h) = 0 \mid X_n(t) = 1) = \beta_n h + o(h).$$

a) $P(X_n(t) = 0 \mid X_n(0) = 0) \geqq q_n$, $P(X_n(s) = 0, t \leqq s \leqq t + h \mid X_n(t) = 0) = e^{-\alpha_n h}$. If $\prod q_n > 0$, that is, $\sum p_n < \infty$ then, at every time t, a.s. $X_n(t) = 0$ for almost all n.

b) If $\sum \lambda_n = \infty$, then $P(X_n(s) = 0, t \leqq s \leqq t + h$, for all $n \geqq m \mid X_n(t) = 0) = 0$ for every m.

Let $X = (X(t), t \geqq 0)$ be the joint r.f. $(X_1(t), X_2(t), \cdots, t \geqq 0)$. X is a Markov r.f. If $\sum p_n < \infty$ then X has only a countable number of states. If, moreover, $\sum \alpha_n = \infty$ then all these states are instantaneous.

c) Analytically, let the state space consist of sequences $x = (x_1, x_2, \cdots)$, $y = (y_1, y_2, \cdots)$, $\cdots$, of 0's and 1's with finitely many 1's. Set $p_t^{(n)}(0, 0) = q_n + p_n e^{-(\alpha_n+\beta_n)t}$, $p_t^{(n)}(1, 1) = p_n + q_n e^{-(\alpha_n+\beta_n)t}$, $p_t^{(n)}(0, 1) = 1 - p_t^{(n)}(0, 0)$, $p_t^{(n)}(1, 0) = 1 - p_t^{(n)}(1, 1)$. The function $P_t(x, y) = \prod_n p_t^{(n)}(x_n, y_n)$ is a tr.pr. which obeys the continuity condition, and ${}_0\dot{P}_0(x, x) = -\infty$.

8. Let P, Q, with or without affixes, be (pr.) distributions on $\mathfrak{X}$ and set

$$d(P, P') = \operatorname{Var}(P - P') = \int | P(dx) - P'(dx) |.$$

a) d is a metric and the space of distributions is complete in this metric.

Let X_T, $T = [0, \infty)$, be stationary Markovian with tr.pr. $P_t(x, S)$, $P_t(x, \mathfrak{X}) \equiv 1$, and distributions P_t, P'_t (of X_t) corresponding to initial distributions P_0, P'_0 (of X_0).

b) $d(P_t, P'_t)$ does not increase as t increases, whatever be P_0 and P'_0.

c) If $d(P_t, \bar{P}) \to 0$ as $t \to \infty$ for some P_0, then $\bar{P}$ is invariant (that is, $\bar{P}_t = \bar{P}$ for all t).

Suppose that (H): for every $\epsilon > 0$ there exist $C \in \mathfrak{J}$, a distribution Q, positive numbers a, b, t_1, and there exists t_0 for every P_0 such that

(i) $aQ(S) \leqq P_{t_1}(x, S)$ for all $x \in C$, $S \subset C$, (ii) $P_tC \geqq 1 - \epsilon$ for all $t \geqq t_0$, (iii) $P_tS \leqq bQ(S) + \epsilon$ for all $S \subset C$ and $t \geqq t_0$.

Then (C): there exists a unique invariant distribution $\bar{P}$ which is "ergodic" (that is, such that $d(P_t, \bar{P}) \to 0$ whatever be P_0):

d) Under (H), $d(P_t, P'_t) \to 0$ as $t \to \infty$ whatever be P_0, P'_0. Thus, there exists at most one invariant distribution and when it exists, it is ergodic.

e) Under (H), $d(P_{t_m}, P_{t_n}) \to 0$ as $t_m, t_n \to \infty$ whatever be P_0 and $t_n \to \infty$.

f) Are conditions (H) necessary for (C) to hold?

9. *Quasileftcontinuity* (Blumenthal, Hunt, Meyer). Let $T = [0, \infty)$, X_t be $\mathfrak{B}_t$-measurable, $\mathfrak{B}_t \uparrow$, τ_n, τ be times of $\mathfrak{B}_T$ and g be nonnegative measurable on $\mathfrak{X}$.

a) *$(X_T, \mathfrak{B}_T)$ is Markovian with tr. pr. $P_t(x, S)$ $\Leftrightarrow$ $(P_{t-s}(X_s, S), \mathfrak{B}_t, 0 \leqq s \leq t)$ are martingales.* What about adding "strongly"?

Let $(X_T, \mathfrak{B}_T)$ be Markovian with semi-group $(T_t, t \geqq 0)$. Then $(g(X_t), t \geqq 0)$ is a supermartingale $\Leftrightarrow$ g is supermedian, i.e., $g \geqq T_tg$.

We say that g is *excessive* (uniformly) if it is supermedian and $T_hg \to g$ ($T_hg \to g$ uniformly and g is bounded); g (or X_T) is *quasileftcontinuous* (*qlc*) if $\tau_n \uparrow \tau$ a.s. $\Rightarrow g(X_{\tau_n}) \to g(X_\tau)$ (or $X_{\tau_n} \to X_\tau$) a.e. on $[\tau < \infty]$.

b) Let $(X_T, \mathfrak{B}_T)$ be strongly Markovian with almost all sample functions rightcontinuous with left limits; we can take $\mathfrak{B}_t = \mathfrak{B}_{t+}$.

Lemma. Let g be bounded with $g(X_{\tau_n}) \to Y$ a.e. on $[\tau < \infty]$ as $\tau_n \uparrow \tau$ a.s. If $T_hg \to g$ uniformly, then $Y = E(X_\tau \mid \mathfrak{B}_\tau^-)$ where $\mathfrak{B}_\tau^-$ is the σ-field over the $\mathfrak{B}_{\tau_n}$:

Reduce to bounded τ (replacing τ_n by $\inf(\tau_n, t)$ with arbitrary $t \in T$). Let $\sigma_n = \sup(\tau_n + h, \tau)$ so that, as $n \to \infty$, $P[\sigma_n \neq \tau_n + h] \to 0$ and $\Delta_2 = g(X_{\tau_n+h}) - g(X_{\sigma_n}) \neq 0$ with pr. $\to 0$. $\Delta_1 = g(X_{\tau_n}) - g(X_{\tau_n+h}) = T_hg(X_{\tau_n})$ and $\Delta_3 = g(X_{\sigma_n}) - g(X_\tau) = T_{\sigma_n - \tau}g(X_\tau)$ with $\sigma_n - \tau \leqq h$ converge vers zero uniformly in n, as $h \to 0$. Thus, given $\epsilon > 0$, $B \in \mathfrak{B}_{\tau_k}$, then fixing h sufficiently small, for all n sufficiently large, $\left|\int_B g(X_{\tau_n}) - \int_B g(X_\tau)\right| < \epsilon$. Letting $n \to \infty$ then $\epsilon \to 0$, $\int_B Y = \int_B g(X_{\tau_n})$; and this equality extends on $\mathfrak{B}_\tau^-$.

Uniformly excessive g on qlc X_T are qlc: For, lemma applies to bounded supermartingale $g(X_{\tau_n})$ and X_τ is $\mathfrak{B}_\tau^-$-measurable by qlc of X_T.

If the semi-group $(T_t, t \geqq 0)$ is strongly continuous on invariant C_0, then X_T is *qlc:*

Reduce to compact $\mathfrak{X}$ (proceeding as in 40.3**d**). For $g \in C$, Y exists by existence of left limits and lemma applies. If $P[Y \neq X_\tau] > 0$, there is a sphere $V_X(r)$ with $PA = P[Y \in V_X(r), X_\tau \neq V_X(2r)] > 0$. With $g \in C$, $g \leqq 1$, $g = 1$ on $V_X(r)$, $g = 0$ on $V_X{}^c(2r)$, integrating $Y = E(g(X_\tau) \mid \mathfrak{B}_\tau^-)$ on $[Y \in V_X(r)]$ yields $PA \leqq 0$-contradiction.

c) Qlc strongly Markovian processes with almost all sample functions rightcontinuous with left limits are *standard* (*Markovian*) in applications to potential theory. Combine with 40.3 and 41.1 to find conditions for existence of equivalent standard Markov processes.

BIBLIOGRAPHY

Titles of articles the results of which are included in cited books by the same author are omitted from the bibliography. Roman numerals designate books, and arabic numbers designate articles.

The references below may pertain to more than one part or chapter of this book.

INTRODUCTORY PART: ELEMENTARY PROBABILITY THEORY

I. Bernoulli, J. *Ars conjectandi* (1713).

II. Bernstein, S. *Theory of probabilities*—in Russian (1946).

III. Borel, E. *Principes et formules classiques* (1925).

IV. Chung, K. L. *Notes on Markov chains*—mimeographed, Columbia University (1951).

V. Delteil, R. *Probabilités géometriques* (1926).

VI. Feller, W. *An Introduction to probability theory and its applications* (1950).

VII. Fréchet, M. *Méthode des fonctions arbitraires. Théorie des événements en chaine dans le cas d'un nombre fini d'états possibles* (1938).

VIII. Fréchet, M. *Les probabilités associees à un systeme d'événements compatibles et dependants* **I** (1940), **II** (1943).

IX. Gnedenko, B. *Course in theory of probabilities*—in Russian (1950).

X. Laplace. *Traité des probabilités* (1801).

XI. Loève, M. *Probability methods in physics*—mimeographed, University of California (1948).

XII. Markov. *Wahrscheinlichkeitsrechnung*—translated from Russian (1912).

XIII. Parzen, E., *Modern probability theory and its applications* (1960).

XIV. Perrin, F. *Mécanique statistique quantique* (1939).

XV. Neyman, J. *First course on probability and statistics* (1950).

XVI. Uspensky, J. *Introduction to mathematical probability*.

1. Kac, M. Random walk in the presence of absorbing barriers. *Ann. Math. Stat.* **16** (1945).
2. Kolmogorov, A. Markov chains with a countable number of possible states—in Russian. *Bull. Math. Univ. Moscow* **1** (1937).
3. Loève, M. Sur les systèmes d'événements. *Ann. Univ. Lyon* (1942).
4. Polyà, G. Sur quelques points de la théorie des probabilités. *Ann. Inst. H. Poincaré* **1** (1931).

PART ONE: NOTIONS OF MEASURE THEORY

I. Banach, S. *Théorie des opérations lineaires* (1932).

II. Bourbaki, N. *Eléments de mathématique* (1939–).

III. Glivenko, J. *Stieltjes integral*—in Russian (1936).
IV. Fréchet, M. *Espaces abstraits* (1928).
V. Hahn and Rosenthal. *Set functions* (1948).
VI. Halmos, P. *Measure theory* (1950).
VII. Hausdorff, F. *Mengenlehre* (1914).
VIII. Lebesgue, H. *Leçons sur l'intégration et la recherche des fonctions primitives* (1928).
IX. Riesz, F. and Sz-Nagy, B. *Leçons d'analyse fonctionnelle* (1952).
X. Saks, S. *Theory of the integral* (1937).

CHAPTER I. SETS, SPACES, AND MEASURES

1. Andersen, E. S. and Jessen, B. On the introduction of measures in infinite product sets. *Danske Vid. Selsk, Mat-Fys. Medd.* **22** (1946).
2. Daniell, P. J. Functions of limited variation in an infinite number of dimensions. *Ann. Math.* **21** (1919–1920).
3. Doubrosky, V. On some properties of completely additive set-functions and passage to the limit under the integral sign. *Izv. Ak. Nauk SSSR* **9** (1945).
4. Kelley, J. L. Convergence in topology. *Duke Math. J.* (1952).

CHAPTER II. MEASURABLE FUNCTIONS AND INTEGRATION

5. Bochner, S. Integration von Funktionen, deren Werte die Elemente eines Vektorraumes sind. *Fund. Math.* **20** (1933).
6. Dunford, N. Integration in general analysis. *Trans. Amer. Math. Soc.* **37** (1942).
7. Kolmogorov, A. Untersuchungen über den Integralbegriff. *Math. Ann.* **103** (1930).
8. Nikodym, O. Sur une généralisation des integrales de Radon. *Fund. Math.* **15** (1930).
9. Robbins, H. Convergence of distributions. *Ann. Math. Stat.* **19** (1948).
10. Tulcea, I. Mesures dans les espaces produits. *Atti Accad. Naz. Lincei Rend.* **7** (1949, 1950).
11. Scheffé, H. A useful convergence theorem for probability distributions. *Ann. Math. Stat.* **18** (1947).

PART TWO: GENERAL CONCEPTS AND TOOLS OF PROBABILITY THEORY

I. Bochner, S. *Fourierische integrale* (1932).
II. Cramér, H. *Mathematical methods of statistics* (1946).
III. Fréchet, M. *Généralités sur les probabilités. Eléments aléatoires* (1937).
IV. Kolmogorov, A. *Grundbegriffe der Wahrscheinlichkeitsrechnung* (1933).
V. Lévy, P. *Calcul des probabilités* (1925).
VI. Lukacz, E. *Characteristic functions* (1960).

CHAPTER III. PROBABILITY CONCEPTS

1. Bienaymé. Considérations à l'appui de la découverte de Laplace sur la loi des probabilités dans la méthode des moindres carrés. *C. R. Acad. Sci. Paris* **37** (1853).

2. Cantelli, F. P. Una teoria astratta del calcolo delle probabilitá. *Ist. Ital. Attuari* **3** (1932).
3. Lomnicki, A. Nouveaux fondements du calcul des probabilités. *Fund. Math.* **4** (1923).
4. Steinhaus, H. Les probabilités denombrables et leur rapport à la théorie de la mesure. *Fund. Math.* **4** (1923).
5. Tchebichev. Des valeurs moyennes. *J. de Math.* **12** (1867).

CHAPTER IV. DISTRIBUTION FUNCTIONS AND CHARACTERISTIC FUNCTIONS

6. Bochner, S. Monotone Funktionen, Stieltjes Integrale, und harmonische Analyse. *Math. Ann.* **108** (1933).
7. Bray, H. E. Elementary properties of the Stieltjes Integral. *Ann. Math.* **20** (1919).
8. Dugué, D. Analyticité et convexité des fonctions caractéristiques. *Ann. Inst. H. Poincaré* **XXII** (1952).
9. Fortet, R. Calcul des moments d'une fonction de repartition à partir de sa characteristic. *Bull. Sc. Math.* **68** (1944).
10. Fréchet, M. and Shohat, J. A proof of the generalized second limit theorem in the theory of probability. *Trans. Am. Math. Soc.* **33** (1931).
11. Helly, E. Ueber lineare Funktionaloperationen. *Sitz. Nat. Kais. Akad. Wiss.* **121** (1912).
12. Kawata, T. and Udagawa, M. On infinite convolutions. *Kadai Math. Sem.* **3** (1949).
13. Marcienkiewicz. Sur les fonctions indépendantes. *Fund. Math* **31** (1939).
14. Marcienkiewicz. Sur une propriété de la loi de Gauss. *Math. Zeit.* **44** (1939).
15. Parzen, E. On uniform convergence of families of sequences of random variables. *Univ. Calif. Publ. Stat.* **2** (1954).
16. Polya, G. Remark on characteristic functions. *Proc. First Berkeley Symp. on Stat. and Prob.* (1949).
17. Zygmund, A. A remark on characteristic functions. *Ann. Math. Stat.* **18** (1947).

PART THREE: INDEPENDENCE

I. Cramér, H. *Random variables and probability distributions* (1937).
II. Gnedenko, B. and Kolmogorov, A. *Limit distributions for sums of independent random variables*—in Russian (1950).
III. Kac, M. *Statistical independence in probability, analysis and number theory* (1959).
IV. Khintchine, A. *Asymptotische Gesetze der Wahrscheinlichkeitsrechnung* (1933).
V. Khintchine, A. *Limit laws for sums of independent random variables*—in Russian (1938).
VI. Lévy, P. *Théorie de l'addition des variables aléatoires* (1937).

CHAPTER V. SUMS OF INDEPENDENT RANDOM VARIABLES

1. Birnbaum, Z. and Zuckerman, H. An inequality due to Hornich. *Ann. Math. Stat.* **15** (1944).
2. Borel, E. Les probabilités dénombrables et leurs applications arithmétiques. *Rend. Circ. Mat. Palermo* **27** (1909).

3. Brunk, H. D. The strong law of large numbers. *Duke Math. J.* **15** (1948).
4. Cantelli, F. P. Su due applicazioni di un teorema di G. Boole. *Rend. Accad. Naz. Lincei* **26** (1917).
5. Chung, K. L. The strong law of large numbers. *Proc. Second Berkeley Symp. on Stat. and Prob.* (1951).
6. Daly, J. On the use of the sample range in an analogue of Student's test. *Ann. Math. Stat.* **17** (1946).
7. Feller, W. The general form of the so-called law of the iterated logarithm. *Trans. Am. Math. Soc.* **54** (1943).
8. Kolmogorov, A. Ueber die Summen durch den Zufall bestimmter unabhängiger Grössen. *Math. Ann.* **99** (1928), **102** (1929).
9. Prokhorov, U. V. On the strong law of large numbers—in Russian. *Dokl. Akad. Nauk USSR* **69** (1949).
10. Robbins, H. On the equidistribution of sums of independent random variables. *Proc. Am. Math. Soc.* **4** (1953).
11. Tukey, J. An inequality for deviations from medians. *Ann. Math. Stat.* **16** (1945).

CHAPTER VI. CENTRAL LIMIT PROBLEM

12. Bawly, G. Über einige Verallgemeinerungen der Grenzwertsätze der Wahrscheinlichkeitsrechnung. *Mat. Sbornik* **43** (1936).
13. Berry, A. The accuracy of the Gaussian approximation to the sum of independent variates. *Trans. Am. Math. Soc.* **49** (1941).
14. Doblin, W. Sur les sommes d'un grand nombre de variables aléatoires indépendants. *Bull. Sc. Math.* **63** (1939).
15. Doblin, W. Sur l'ensemble de puissances d'une loi de probabilité. *Ann. Ec. Norm. Sup.* (1947).
16. Esseen, G. Fourier Analysis of distribution functions. A mathematical study of the Laplace-Gaussian law. *Acta Math.* (1944).
17. Feller, W. Ueber den Zentralen Grenzwertsatz der Wahrscheinlichkeitsrechnung. *Math. Zeit.* **40** (1935), **42** (1937).
18. Feller, W. Ueber das Gesetz der grossen Zahlen. *Acta Univ. Szeged* **8** (1937).
19. Feller, W. On the Kolmogoroff-P. Lévy formula for infinitely divisible distribution functions. *Proc. Yougoslav Acad. Sc.* **82** (1937).
20. de Finetti, B. Sulle funzioni a incremento aleatorio. *Rend. Accad. Naz. Lincei* **10** (1929).
21. Liapounov, A. Nouvelle forme du theorème sur la limite de la probabilité. *Mem. Acad. St-Petersbourg* **11** (1900), **12** (1901).
22. Lindeberg, J. W. Ueber das Exponentialgesetz in der Wahrscheinlichkeitsrechnung. *Math. Zeit.* **15** (1922).
23. Loève, M. Nouvelles classes de lois limites. *C. R. Ac. Sc.* (1941), *Bull. Soc. Math.* (1945).
24. Loève, M. Ranking limit problem. *Second Berkeley Symp.* (1956).
25. Loève, M. A l'intérieur du problème limite central. *Ann. Inst. Stat. Paris* (1958).
26. Raikov, D. A. On the decomposition of Gauss and Poisson laws—in Russian. *Izv. Akad. Nauk USSR* **2** (1938).

PART FOUR: DEPENDENCE

I. Blanc-Lapierre & Fortet, *Theorie des fonctions aléatoires* (1953).
II. Kaczmarz and Steinhaus. *Orthogonalreihen* (1935).
III. Hopf, E. *Ergodentheorie* (1937).
IV. Ville, J. *Etude critique de la notion du collectif* (1939).
V. Wold, H. *A study in the analysis of stationary time series* (1938).

CHAPTER VII. CONDITIONING

1. Blackwell, D. Idempotent Markov chains. *Ann. of Math.* **43** (1942).
2. Cogburn, R. Asymptotic properties of stationary sequences. *Univ. of Calif. Publ. Stat.* **3** (1960).
3. Doblin, W. Sur les propriétés asymptotiques de mouvement régi par certains types de chaines simples. *Bull. Math. Soc. Roum. Sc.* **39** (1937).
4. Doblin, W. Eléments d'une théorie générale des chaines simples constantes de Markov. *Ann. Sc. Ecole Norm. Sup.* **57** (1940).
5. Halmos, P. and Savage, L. J. Application of the Radon-Nikodym theorem to the theory of sufficient statistics. *Ann. Math. Stat.* **20** (1949).
6. Nagaev, C. B. Some limit theorems for stationary Markov chains. *Teorya Veroyatn.* **2** (1957).
7. Neyman, J. Su un teorema concernente le cosiddette statistiche sufficienti. *Ist. Ital. Attuari.* **6** (1935).
8. Ueno, T. Some limit theorems for temporally discrete Markov processes. *J. Univ. Tokyo* **7** (1957).

CHAPTER VIII. FROM INDEPENDENCE TO DEPENDENCE

9. Andersen, E. S. and Jessen, B. Some limit theorems on set functions. *Danske Vid. Selsk. Mat-Fys. Medd.* **25** (1948).
10. Blackwell, D. and Dubins, L. Merging opinions with increasing information. *Ann. Math. Stat.* **33** (1962).
11. Loève, M. Etude asymptotique des sommes de variables aléatoires liées. *J. de Math.* **24** (1945).
12. Loève, M. On sets of probability laws and their limit elements. *Univ. of Calif. Publ. Stat.* **1** (1950).
13. Robbins, H. The asymptotic distribution of the sum of a random number of random variables. *Bull. Am. Math. Soc.* **54** (1948).
14. Snell, J. L. Applications of martingale system theorems. *Trans. Am. Math. Soc.* **73** (1952).

CHAPTER IX. ERGODIC THEOREMS

15. Birkhoff, G. D. Proof of the ergodic theorem. *Proc. Nat. Ac. Sc. U.S.A.* **17** (1931).
16. Blum, J. R. and Hanson, D. L. On invariant probability measures. *Pacific J. of Math.* **10** (1960).
17. Doob, J. L. Asymptotic properties of Markov transition probabilities. *Trans. Am. Math. Soc.* **63** (1948).
18. Dunford, N. Spectral theory. *Trans. Am. Math. Soc.* **54** (1943).
19. Dunford, N. and Miller, D. S. On the ergodic theorem. *Trans. Am. Math. Soc.* **60** (1946).

20. Dowker, Y. Invariant measure and the ergodic theorem. *Duke Math. J.* **14** (1947).
21. Halmos, P. Approximation theories for measure preserving transformations. *Trans. Am. Math. Soc.* **55** (1944).
22. Halmos, P. An ergodic theorem. *Proc. Nat. Ac. Sc. U.S.A.* **32** (1946).
23. Hartman, S., Marczewski, E. and Ryll-Nardzewski, C. Théorèmes ergodiques et leurs applications. *Coll. Math.* **2** (1951).
24. Hurewicz, N. Ergodic theorem without invariant measure. *Ann. of Math.* **44** (1944).
25. Khintchine, A. Zu Birkhoffs Lösung des Ergodenproblems. *Math. Ann.* **107** (1933).
26. Loève, M. On almost sure convergence. *Proc. Sec. Berkeley Symp. on Stat. and Prob.* (1951).
27. von Neumann, J. Proof of the quasi-ergodic hypothesis *Proc. Nat. Ac. Sc. U.S.A.* **18** (1932).
28. Oxtoby, J. C. On the ergodic theorem of Hurewicz. *Ann. of Math.* **49** (1948).
29. Ryll-Nardzewski, C. On the ergodic theorems. *Studia Math.* **12** (1951, 1952).
30. Riesz, F. Sur la théorie ergodique. *Comm. Math. Helv.* **17** (1945), **19** (1947).
31. Yosida, K. and Kakutani, S. Operator-theoretical treatment of Markov's process and mean ergodic theorem. *Ann. of Math.* **42** (1941).

CHAPTER X. SECOND ORDER RANDOM PROPERTIES

32. Cramér, H. On the theory of stationary random processes. *Ann. Math.* **41** (1940).
33. Cramér, H. On harmonic analysis in certain functional spaces. *Ark. Mat. Astr. Fys.* **28** (1942).
34. Cramér, H. A contribution to the theory of stochastic processes. *Second Berkeley Symp.* (1951).
35. Karhunen, K. Über lineare Methoden in der Wahrscheinlichkeitsrechnung. *Ann. Acad. Sci. Fenn.* **37** (1947).
36. Khintchine, A. Korrelationstheorie der stationäre stochastischen Prozesse. *Math. Ann.* **109** (1934).
37. Kolmogorov, A. Stationary sequences in Hilbert space—in Russian. *Bull. Math. Univ. Moscow* **2** (1941).
38. Loève, M. Fonctions aléatoires de second ordre. *C. R. Acad. Sci.* **220** (1945), **222** (1946); *Rev. Sci.* **83** (1945), **84** (1946).
39. Maruyama, G. The harmonic analysis of stationary stochastic processes. *Mem. Fac. Sci. Kyusyu Univ.* **4** (1949).
40. Slutsky, E. Sur les fonctions aléatoires presque periodiques et sur la decomposition des fonctions aléatoires stationaires en composantes. *Act. Sci. Ind.* **738** (1938).

PART FIVE: ELEMENTS OF RANDOM ANALYSIS

I. Chung, K. L. *Markov chains with stationary transition probabilities* (1960).
II. Doob, J. L. *Stochastic processes* (1953).
III. Dynkin, E. B. *Foundations of Markov processes*—in Russian (1959).

IV. Hille and Phillips. *Functional Analysis and Semi-groups* (1957).
V. Lévy, P. *Processus stochastiques et mouvement brownien* (1948).

CHAPTER XI. FOUNDATIONS; MARTINGALES AND DECOMPOSABILITY

1. Dobrushin, R. L. On the Poisson law for distributions of particles in space—in Russian, *Ukrain. Math. J.* **8** (1956).
2. Dobrushin, R. L. The continuity condition for sample martingale functions—in Russian, *Teorya Veroyatn.* **3** (1958).
3. Dynkin, E. B. Criteria of continuity and of absence of discontinuities of the second kind for trajectories of a Markov random process—in Russian, *Izv. Ak. Nauk SSSR* **16** (1952).
4. Ito, K. On stochastic processes I (Infinitely divisible laws of probability), *Jap. J. Math.* **18** (1942).
5. Meyer, P. A. On decomposition of martingales, *Ill. J. of Math.* **6** (1962).
6. Slutsky, E. Alcune proposizioni sulla teoria delle funzioni aleatorie, *Giorn. Ist. Ital. Attuari* **8** (1937).
7. Wiener, N. Differential space, *J. Math. Phys. M.I.T.* **2** (1923).

CHAPTER XII. MARKOV PROCESSES

7. Austin, Blumental, and Chacon. On continuity of transition functions, *Duke Math. J.* **25** (1958).
8. Blackwell, D. Another countable Markov process with only instantaneous states, *Ann. Math. Stat.* **29** (1958).
9. Blumental, R. An extended Markov property, *Trans. Am. Math. Soc.* **85** (1957).
10. Doblin, V. Sur certains mouvements aléatoires discontinus, *Skand. Akt.* **22** (1939).
11. Doob, J. L. Topics in the theory of Markov chains, *Trans. Am. Math. Soc.* **52** (1942).
12. Doob, J. L. Markov chains—denumerable case, *Trans. Am. Math. Soc.* **58** (1945).
13. Doob, J. L. Brownian motion on a Green space, *Teorya Veroyatn.* **2** (1957).
14. Dynkin, E. B. Markov processes and semi-groups of operators—in Russian, *Teorya Veroyatn.* **1** (1956).
15. Dynkin, E. B. Infinitesimal operators of Markov processes—in Russian, *Teorya Veroyatn.* **1** (1956).
16. Dynkin, E. B. One-dimensional continuous strong Markov processes—in Russian, *Teorya Veroyatn.* **4** (1959).
17. Dynkin and Yushkevitch. Strong Markov processes—in Russian, *Teorya Veroyatn.* **1** (1956).
18. Feller, W. Zur Theorie der Stochastischen Prozesse (Existenz und Eindeutigkeitssätze), *Math. Ann.* **113** (1936).
19. Feller, W. On the integro-differential equations of purely discontinuous Markov processes, *Trans. Am. Math. Soc.* **48** (1940); *Errata Ibid.* **58** (1945).
20. Feller, W. Semi-groups of transformations in general weak topologies, *Ann. of Math.* **57** (1953).
21. Feller, W. The general diffusion operator and positivity preserving semi-groups in one dimension, *Ann. of Math.* **60** (1954).

22. Feller, W. On second order differential operators, *Ann. of Math.* **61** (1955).
23. Fortet, R. Les fonctions aléatoires du type de Markov associées à certaines equations lineaires aux dérivées partielles du type parabolique, *J. Math. Pures Appl.* **22** (1943).
24. Hunt, G. A. Markov processes and potentials, *Ill. J. of Math.* **1** (1957), **2** (1958).
25. Hunt, G. A. Markov chains and Martin boundaries, *Ill. J. of Math.* **4** (1960).
26. Ito, K. On stochastic differential equations, *Mem. Am. Math. Soc.* **4** (1951).
27. Kendall, D. G. Some analytical properties of continuous stationary Markov transition functions, *Trans. Am. Math. Soc.* **78** (1955).
28. Kinney, J. R. Continuity properties of sample functions of Markov processes, *Trans. Am. Math. Soc.* **74** (1953).
29. Kolmogorov, A. N. Über die analytischen Methoden in der Wahrscheinlichkeitsrechnung, *Math. Ann.* **104** (1931).
30. Kolmogorov, A. N. On some problems concerning the differentiability of the transition probabilities in temporally homogeneous Markov processes having a denumerable set of states—in Russian, *Uch. zapiski MΓY*, **148** (1951).
31. Lévy, P. Systèmes markoviens et stationaires: cas dénombrable, *Ann. Ec. Norm.* **68** (1951), 69 (1952).
32. Maruyama, G. Continuous Markov processes and stochastic equations, *Rend. Circ. Mat. Palermo* **4** (1955).
33. Maruyama, G. On the strong Markov property, *Mem. Kyushu Univ.* **13** (1959).
34. Maruyama and Tanaka. Some properties of one-dimensional diffusion processes, *Mem. Kyushu Univ.* **11** (1957).
35. Meyer, P. A. Fonctionnelles multiplicatives et additives de Markov, *Ann. Inst. Fourier* **12** (1962).
36. Neveu, J. Théorie des semi-groupes de Markov, *Univ. of Calif. Publ. Stat.* **2** (1958).
37. Ray, D. Resolvents, transition functions and strongly Markovian processes, *Ann. of Math.* **70** (1959).
38. Sevastianov, B. A. An ergodic theorem for Markov processes and its application to telephone systems with refusals—in Russian, *Teorya Veroyatn.* **2** (1957).
39. Yushkevitch, A. A. On strong Markov processes, *Teorya Veroyatn.* **2** (1957).

INDEX